PHASED ARRAY ANTENNAS

PROCEEDINGS OF THE 1970 PHASED ARRAY ANTENNA SYMPOSIUM

Sponsored by:

U.S. Army Advanced Ballistic Missile Defense Agency
M.I.T. Lincoln Laboratory
P.I.B. Electrophysics Department

June 2, 3, 4, 5, 1970

Polytechnic Institute of Brooklyn
Long Island Graduate Center
Farmingdale, New York

MODERN FRONTIERS IN APPLIED SCIENCE

Gallium Arsenide Microwave Bulk and
Transit-Time Devices
by Lester F. Eastman

Laser Applications
by W. V. Smith

Microwave Engineers Handbook
by Ted Saad

Logarithmic Video Amplifiers
by R. S. Hughes

Parallel Coupled Lines and Directional Couplers
by Dr. Leo Young

Microwave Filters Using Parallel Coupled Lines
by Dr. Leo Young

Data Modem Selection and Evaluation Guide
by V. V. Vilips

Design, Performance and Application of Microwave
Semiconductor Control Components
by Kenneth E. Mortenson and Jose M. Borrego

Adaptive Electronics
by Dr. Wolfgang Gaertner

Gallium Arsenide Microwave Bulk and
Transit-Time Devices
by Lester F. Eastman

Avalanche Transit-Time Devices
by Dr. George I. Haddad

Infrared to Millimeter Wavelength Detectors
by Frank R. Arams

ARTECH HOUSE, INC.
610 Washington Street
Dedham, Massachusetts 02026
affiliated with Horizon House-Microwave, Inc.

PHASED ARRAY ANTENNAS

Edited by:

Dr. Arthur A. Oliner

Dr. George H. Knittel

Polytechnic Institute of Brooklyn

ARTECH HOUSE, INC.
Dedham, Massachusetts 02026
affiliated with Horizon House-Microwave, Inc.

Printed and bound in the United States of America.

Library of Congress catalog card number: 73-189392

PREFACE

Phased-array antennas represent perhaps the most active and important topic in the antenna field today. These antennas are the most complex and versatile of all antennas, and are called upon to perform tasks such as the detection and tracking of many objects present in the sky at the same time. Their most obvious use is in sophisticated ABM systems, where no other antenna type can replace them, but they are also finding increasing use in shipboard systems, airborne systems, and satellite communication systems, and they are being designed for other applications such as airport traffic control.

In order to determine the state of the art and to provide a comprehensive up-to-date source of information on phased-array antennas, a Phased-Array Antenna Symposium was held on June 2-5, 1970, at the Polytechnic Institute of Brooklyn Graduate Center in Farmingdale, New York. The symposium was sponsored by the U.S. Army Advanced Ballistic Missile Defense Agency, the M.I.T. Lincoln Laboratory, and the P.I.B. Electrophysics Department. There were 325 registrants from 120 different organizations.

A phased-array symposium was previously held in 1964 at Rome Air Development Center, Rome, New York. However, since the most important developments in the field were made after 1964, it was decided to hold another symposium in 1970. Many of the papers presented at this symposium consisted of invited surveys by authorities on the various topics. The newer work was represented by both invited and contributed papers. In addition, panel discussions were held on topics involving controversial aspects. The organizers of the technical program and the authors of the papers included the most active and well known contributors to the field.

The symposium covered all aspects of phased-array antennas, with the topics ranging from the theory and design of array elements to problems involving manufacturing techniques. Included were comparisons of requirements for various applications (land-based, ship, airborne, and satellite), basic theoretical interpretations of certain performance difficulties, measurement methods for array elements, new types of antenna elements, feed networks and beam-steering methods, phase-shifter technology, comparisons between planar and nonplanar arrays, and the impact of solid state technology on array design.

At the time of the symposium a Digest, containing short versions of the papers, was published and distributed to all registrants. This Proceedings of the Phased-Array Antenna Symposium contains the full-length versions of the papers; it is the only publication in which these papers will appear. All of the papers presented at the symposium are published in this volume and they appear as full-length versions except for those few which were never written full-length for one reason or another. In those cases, the short Digest-version of the paper has been included. This Proceedings also includes one paper by Hanfling and Giarrizzo, which was presented at the symposium but which did not appear in the Digest.

Three panel discussions were held at the symposium, on the topics "Manufacturing Problems and Techniques", "Why Conformal Arrays?", and "Impact of Solid State Devices on Array Design". It was not possible to include a written record of these discussions either in the Digest or in the Proceedings.

All of the papers published in this Proceedings have been carefully reviewed, primarily by the members of the Technical Program Committee. The reviewers comments were considered by us and then mailed to the authors with instructions for modifications of their papers, to ensure that the published papers would be of good quality and free of serious error.

The papers published on the following pages represent a tremendous effort on the part of many competent, knowledgeable people whose goal was to assemble in one volume a comprehensive source of information on phased-array antennas. We sincerely appreciate the time and effort devoted by the authors to their work; it has been our privilege and benefit to be associated with such a fine, dedicated group. We know that this volume will be a great help to those who work with phased-array antennas in the future. Because there have been few basic new developments since the 1970 symposium, the contents of this volume are still highly valuable and remarkably up-to-date.

We want to express our gratitude to Miss Jean Beirne, who very efficiently handled the many organizational and administrative details both for the symposium and during the preparation of these Proceedings. In addition, thanks are due to Mr. Jerome Fox and Prof. Saul W. Rosenthal for their advice and assistance with the many considerations associated with the publication of this volume.

Arthur A. Oliner
George H. Knittel

April 1972

SYMPOSIUM COMMITTEES

General Chairman: Arthur A. Oliner
Secretary: Bliss L. Diamond
Treasurer: Carl Blake

STEERING

U. S. ARMY ADVANCED BALLISTIC MISSILE DEFENSE AGENCY

Lindsey B. Anderson   Frank Rouffy

M. I. T. LINCOLN LABORATORY

Carl Blake   Bliss L. Diamond

P. I. B. ELECTROPHYSICS DEPARTMENT

George H. Knittel   Arthur A. Oliner

PUBLICITY AND INVITATIONS

Chairman: Leon Schwartzman

Robert J. Adams   George H. Knittel
Moses Diab   Saul W. Rosenthal
Jerome Fox   Frank Rouffy

DIGEST

Chairman: Henry W. Redlien

D. Cermignani   C. Schlotterhausen

PROCEEDINGS

Jerome Fox

LOCAL ARRANGEMENTS

Saul W. Rosenthal

TECHNICAL PROGRAM

Chairman: Louis Stark
Session Organizer: Theo C. Cheston
Session Organizer: Peter W. Hannan
Session Organizer: Willard T. Patton
Session Organizer: Leon J. Ricardi

Noach Amitay   Alexander Hessel
Carl Blake   George H. Knittel
Bliss L. Diamond   Arthur A. Oliner
Robert C. Hansen   George D.M. Peeler
Leon Schwartzman

CONTENTS

SECTION I ▪ INTRODUCTION

CHARACTERISTICS OF ACOUSTIC ARRAYS AND DEEP-OCEAN ACOUSTIC PROPAGATION 1
J. A. McCarthy, Bell Telephone Laboratories, Whippany, N.J.

BMD REQUIREMENTS FOR PHASED ARRAY RADARS 6
R. S. McCarter, Bell Telephone Laboratories, Whippany, N.J.

SECTION II ▪ TYPES AND APPLICATIONS OF PHASED-ARRAY ANTENNAS

SURVEY OF GROUND BASED PHASED ARRAY ANTENNAS 9
A. C. Schell, Air Force Cambridge Research Laboratories, Bedford, Mass.

SURVEY OF PHASED ARRAY ACCOMPLISHMENTS AND REQUIREMENTS FOR NAVY SHIPS 15
M. I. Skolnik, Naval Research Laboratory, Washington, D.C.

SURVEY OF AIRBORNE PHASED ARRAY ANTENNAS 21
J. F. Rippin, Jr., Air Force Avionics Laboratory, (AFSC), Wright Patterson Air Force Base, Oh.

PHASED ARRAY ANTENNAS FOR APPLICATIONS ON SPACECRAFT 30
W. H. Kummer, Hughes Aircraft Company, Culver City, Cal., and A. Kampinsky, NASA/Goddard Space Flight Center, Greenbelt, Md.

DESIGN, PERFORMANCE AND COST CONSIDERATIONS FOR SOLID-STATE ARRAYS 43
C. Blake, Lincoln Laboratory, Massachusetts Insiitute of Technology, Lexington, Mass.

SECTION III ▪ INTRODUCTION AND ANALYSIS METHODS FOR RADIATING ELEMENTS

COMPARISON OF ARRAY ELEMENT TYPES 51
L. Stark, Hughes Aircraft Company, Fullerton, Cal.

METHODS OF PHASED ARRAY ANALYSIS 68
N. Amitay and C. P. Wu, Bell Telephone Laboratories, Inc., Whippany, N.J. and V. Galindo, TRW Systems, Redondo Beach, Cal.

INFINITE ARRAYS OF SUB-ARRAY ANTENNAS 83
W. Wasylkiwskyj, Institute for Defense Analyses, Arlington, Va., and W. K. Kahn, The George Washington University, Washington, D.C.

ANALYSIS OF INFINITE PLANAR ARRAY OF RECTANGULAR WAVEGUIDES BY GENERALIZED SCATTERING MATRIX APPROACH 91
S. W. Lee, N. S. Wong, and R. Tang, Hughes Aircraft Company, Fullerton, CA.

SECTION IV ▪ SPECIAL PROBLEMS OF RADIATING ELEMENTS

SURFACE-WAVE EFFECTS AND BLINDNESS IN PHASED-ARRAY ANTENNAS 107
A. A. Oliner, Polytechnic Institute of Brooklyn Graduate Center, Farmingdale, N.Y.

EDGE EFFECTS IN DIELECTRIC COVERED PHASED-ARRAYS 113
J. B. Andersen and F. Hass, Technical University of Denmark, Lyngby, Denmark.

ON THE EFFECTS OF ELIMINATING THE PASSIVE ELEMENTS FROM A THINNED ARRAY 121
G. A. Arredondo, Bell Telephone Laboratories, Inc., Whippany, NJ. and R. L. Knuckey,
Western Electric Company, Inc., on loan to Bell Telephone Laboratories, Inc., Whippany, N.J.

SECTION V ▪ DESIGN PROCEDURES FOR RADIATING ELEMENTS

SMALL ARRAYS—THEIR ANALYSIS AND THEIR USE FOR THE DESIGN OF ARRAY ELEMENTS 127
B. L. Diamond, Lincoln Laboratory, Massachusetts Institute of Technology, Lexington, Mass.

A SURVEY OF THE SIMULATOR TECHNIQUE FOR DESIGNING A RADIATING ELEMENT 132
H. A. Wheeler, Wheeler Laboratories, Smithtown, N.Y.

A NEW PROCEDURE FOR THE DESIGN OF A WAVEGUIDE ELEMENT FOR A PHASED-ARRAY ANTENNA 149
B. L. Diamond, Lincoln Laboratory, Massachusetts Institute of Technology, Lexington, Mass. and
G. H. Knittel, Polytechnic Institute of Brooklyn Graduate Center, Farmingdale, N.Y.

SECTION VI ▪ IMPROVEMENT TECHNIQUES FOR RADIATING ELEMENTS

WIDE-ANGLE IMPEDANCE MATCHING OF PHASED-ARRAY ANTENNAS—
A SURVEY OF THEORY AND PRACTICE 157
G. H. Knittel, Polytechnic Institute of Brooklyn, Graduate Center, Farmingdale, N.Y.

THE DESIGN OF A WIDE BAND WIDE SCAN—ANGLE WAVEGUIDE RADIATING ELEMENT 173
G. A. Arredondo, P. E. Butzien, and R. W. Humes, Bell Telephone Laboratories, Inc., Whippany, N.J.

MULTIMODE PHASED ARRAY ELEMENT FOR WIDE SCAN ANGLE IMPEDANCE MATCHING 178
N. S. Wong, R. Tang, S. W. Lee and W. R. Jones, Hughes Aircraft Company, Fullerton, Cal.

A NEW FLUSH MOUNTED ANTENNA ELEMENT FOR PHASED ARRAY APPLICATION 187
E. V. Byron, The Johns Hopkins University, Applied Physics Laboratory, Silver Spring, Md.

DESIGN OF A PHASED ARRAY RADIATING FACE FOR PREVENTION OF PERFORMANCE
DEGRADATION IN THE PRESENCE OF RAIN 193
J. D. Hanfling and V. A. Giarrizzo, Raytheon Co., Bedford, Mass.

SECTION VII ▪ FEEDING AND PHASING SYSTEMS

PHASED ARRAY FEED SYSTEMS, A SURVEY 197
R. T. Hill, NAVSEC, Hyattsville, Md.

DIODE AND FERRITE PHASER TECHNOLOGY 212
D. H. Temme, Lincoln Laboratory, Masaachusetts Institute of Technology, Lexington, Mass.

BEAM STEERING OF PLANAR PHASED ARRAYS 219
T. C. Cheston, The Johns Hopkins University, Applied Physics Laboratory, Silver Spring, Md.

QUANTIZATION AND REFLECTION LOBE DISPERSION 222
G. J. Buck, Hughes Aircraft Company, Culver City, Cal.

THE STEP RECOVERY DIODE: AN ANALOG PHASE SHIFTER 230
J. P. Weir Jr., and R. C. Kent, Teledyne Rayan Aeronautical Co., San Diego, Cal.

TIME SCANNED ARRAY RADAR 235
G. J. Vogel, B. J. Lyons, Rome Air Development Center, Rome, N. Y.,
K. Parsons and L. W. Beachler, Bendix Communications Division, Townson, Md.

THE DESIGN OF HYBRID MULTIPLE BEAM FORMING NETWORKS 240
K. H. Hering, TRW Systems Group, Los Angeles, Cal.

SECTION VIII ▪ ARRAY SIGNAL BANDWIDTH

BANDWIDTH CRITERIA FOR PHASED ARRAY ANTENNAS 243
J. Frank, Technology Service Corporation, Silver Spring, Maryland

SURVEY OF TIME-DELAY BEAM STEERING TECHNIQUES 254
R. Tang, Hughes Aircraft Co., Fullerton, Cal.

TRANSIENT RESPONSE OF CENTER-SERIES FED ARRAY 261
R. R. Kinsey and A. L. Horvath, General Electric Co., Syracuse, N.Y.

SECTION IX ▪ NON-PLANAR ARRAYS

MUTUAL COUPLING EFFECTS IN CIRCULAR ARRAYS ON CYLINDRICAL SURFACES – APERTURE DESIGN IMPLICATIONS AND ANALYSIS 273
A. Hessel, Polytechnic Institute of Brooklyn Graduate Center, Farmingdale, N.Y.

A SURVEY OF CIRCULAR SYMMETRIC ARRAYS 292
J. H. Provencher, Naval Electronics Laboratory Center, San Diego, Cal.

CONFORMAL ARRAYS ON SURFACES WITH ROTATIONAL SYMMETRY 301
G. V. Borgiotti and Q. Balzano, Raytheon Co., Bedford, Mass.

REALIZED GAIN FUNCTION FOR A CYLINDRICAL ARRAY OF OPEN-ENDED WAVEGUIDES 315
J. C. Sureau, Grumman Aerospace Corp., Bethpage, N.Y. and A. Hessel, Polytechnic Institute of Brooklyn Graduate Center, Farmingdale, N.Y.

BASIC THEORETICAL ASPECTS OF SPHERICAL PHASED ARRAYS 323
H. E. Schrank, Westinghouse Electric Corp., Baltimore, Md.

SECTION X ▪ CONSIDERATIONS IN DESIGN AND EVALUATION

EVALUATION OF LARGE PHASED-ARRAY ANTENNAS 329
L. Schwartzman and F. J. Esposito, Sperry Gyroscope Div., Great Neck, N.Y., and
C. Blake, Lincoln Laboratory, Massachusetts Institute of Technology, Lexington, Mass.

LIMITED SCAN ARRAYS 330
W. T. Patton, Radio Corporation of America, Moorestown, N.J.

PATTERN MEASUREMENTS OF PHASED-ARRAYED ANTENNAS BY FOCUSING INTO THE NEAR ZONE 344
W. E. Scharfman and G. August, Stanford Research Institute, Menlo Park, Cal.

A METHOD OF LOCATING DEFECTIVE ELEMENTS IN LARGE PHASED ARRAYS 351
P. L. Ransom and R. Mittra, University of Illinois, Urbana, Ill.

DUAL BEAM ANTENNA A UNIQUE WAVEGUIDE PHASED ARRAY WITH INDEPENDENTLY STEERED BEAMS 358
D. K. Alexander and B. A. Sichelstiel, Westinghouse Electric Corporation, Baltimore, Md.

A SINGLE-PLANE ELECTRONICALLY SCANNED ANTENNA FOR AIRBORNE RADAR APPLICATIONS 366
D. J. Lewis, J. R. Lee, and D. K. McCarty, Hughes Aircraft Company, Culver City, Cal.

PHYSICAL DESIGN CONSIDERATIONS FOR AIRBORNE ELECTRONIC-SCANNING ANTENNAS 372
M. H. Rosengard, Hughes Aircraft Company, Culver City, Cal.

MECHANICAL DESIGN CONSIDERATIONS OF A PHASED-ARRAY ANTENNA 379
L. T. Embrekts, Sperry Gyroscope Div., Great Neck, N.Y.

AUTHOR INDEX *381*

Characteristics of Acoustic Arrays and Deep-Ocean Acoustic Propagation

JOHN A. McCARTHY

Bell Telephone Laboratories, Inc.
Whippany, New Jersey

Acoustic arrays in sonar systems are directly analogous to electromagnetic arrays in radar. However, environmental factors introduce significant differences; these can be appreciated from consideration of the fundamentals of deep-ocean acoustic propagation, ambient noise and array performance factors.

The principles of sonar and of acoustic array design are very similar to the principles of radar and electromagnetic array antennas. This is not at all surprising since both involve wave phenomena and have almost the same order of magnitude of wavelength. (Radars tend to have somewhat shorter wavelengths). Thus, there is a sonar equation which directly parallels the radar equation. There is a similar behavior with respect to the laws of diffraction and refraction, and commonality is achieved down to such fundamental rules as "the gain of the array is less than or equal to the number of elements multiplied by the gain of an individual element."

However, when the principles of sonar are examined in detail, numerous differences from the radar case become apparent and become very important. Most of these differences are associated directly with the ocean environment. It is interesting to consider some of the specifics of acoustic arrays in the deep ocean and some of the ways in which they differ from electromagnetic arrays. There are a few obvious examples such as the difference in velocity and time scale of the process. Acoustic energy propagates nearly six orders of magnitude more slowly than electromagnetic energy, and events take place in the acoustic environment on a time scale which can be characterized by tens of minutes rather than by microseconds. However, the particular factors which are extremely important to sonar design and which represent major differences from radar are the transmission characteristics of the deep ocean and the nature of the acoustic noise background.

In the electromagnetic case energy propagates roughly in a straight line path through a nearly homogeneous medium. However, in the case of acoustic transmission in the deep ocean, the situation becomes much more complex because of refraction. The velocity of sound is a function of water depth in the ocean. Typically, the velocity is high both near the surface and very deep in the ocean with a minimum at some intermediate ("axis") depth, as shown in Figure 1. Below a few thousand feet, the velocity increases with increasing pressure (i.e., depth).

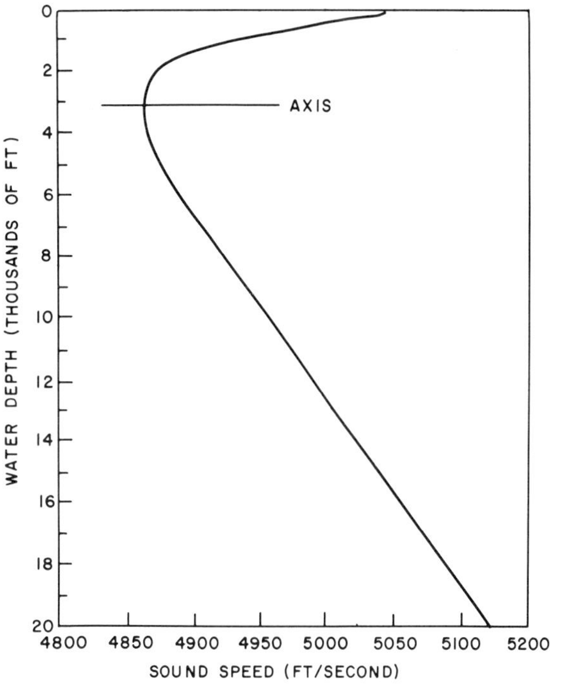

Fig. 1 Typical deep-ocean velocity profile.

Nearing the surface, the temperature increase causes a velocity increase which typically overrides the pressure effect. The effect of the velocity minimum is to create a main sound channel which is most significant in deep ocean acoustic propagation. Rays (acoustic paths) which penetrate deep are refracted upward, while acoustic rays at less than axis depth are refracted downward. Thus, there is what might be considered a very large waveguide tending to trap and hold acoustic energy.

Figure 2 shows the effect of the sound channel on the propagation paths. We should note that the vertical scale is exaggerated by a factor of 15 with respect to horizontal scale. The figure shows the rays

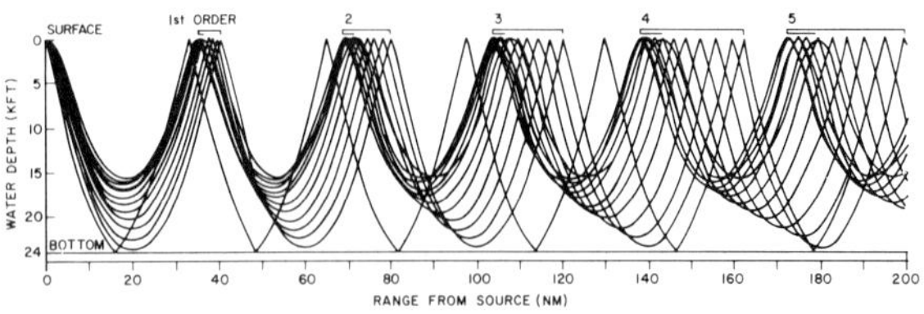

Fig. 2 Deep-ocean acoustic propagation.

from a surface or near-surface acoustic source as they propagate out to 200 miles. Initially, all rays are refracted or reflected downward. Some hit the bottom -- we haven't shown all of the bottom paths because those hitting the bottom steeply incur a high loss and dissipate energy very rapidly. A few which hit the bottom at near-grazing angles are reflected without major loss and can contribute significant energy. The remainder of the rays are refracted or reflected at the surface, go deep in the ocean, reaching the deepest point at a distance of about 15 to 20 miles, then are refracted upward and toward the surface, once again refracted (and, in some cases, reflected) downward and continue to cycle out, sometimes to very long ranges. This presents the peculiar problem illustrated in Figure 3, which is a representation of the computed propagation loss from a source at zero on the range axis and 60-foot depth to any point on the figure. There are acoustic paths which result in only moderate loss at points as far as 40-75 miles from a source, while there are poorly coupled intermediate points, perhaps within 15 or 20 miles of the source, which show very high transmission loss.

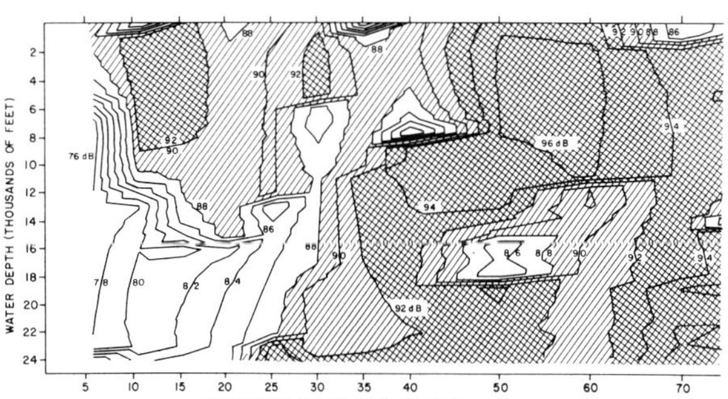

Fig. 3 Computed loss in propagation from a near-surface source. The lines are contours of equal loss in 2 dB intervals as labelled. The single-hatched area has a path loss above 88 dB, and the double-hatched above 92 dB.

In addition to variations in received energy, there is a serious multipath problem. If one refers back to Figure 2 and looks at the second or third cycle, for example, he will note that the rays tend to stay concentrated in bundles, and that there is a strong tendency for rays to cross each other. The resulting multipath problem is nearly always present and nearly always causes difficulty. The problem is further complicated by the fact that these paths change with time, from minute to minute, from day to day, and tremendously from season to season. Thus acoustic energy is transmitted through a variable-refraction ocean environment which causes continuously changing multipath propagation.

One other important feature of this type of transmission at longer ranges is that most of the rays tend to come from a generally horizontal direction, since the steeper-angled rays are lost by multiple reflections from the bottom. Figure 4 is a representation of the primary directions of arrival of acoustic energy from ranges of the order of 100 miles, shown as a function of receiver depth. For very shallow and very deep detectors, the energy from distant near-surface sources is confined rather closely around the horizontal. At intermediate depths, it is divided into two bundles of rays, corresponding to those which are going downward and those which are coming upward. In no case, however, are there predominant paths which arrive at angles larger than about 20 degrees from the horizontal.

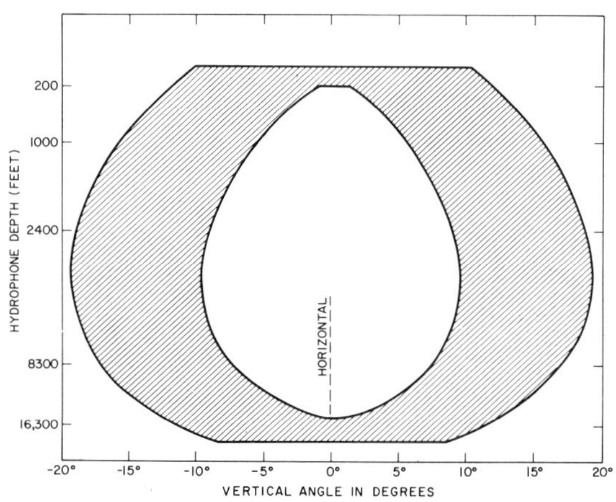

Fig. 4 Vertical angle of arrival of acoustic energy from a distance (~100 nm) source. Energy arrives almost entirely in the angular range indicated by the hatching.

Obviously, this description of the arrival angles of energy at long ranges applies not only to signals, but to noise generated at or near the surface. This fact is illustrated by Figure 5 which displays measured noise intensity as a function of arrival angle in the vertical plane. Thus another departure from the radar case has been introduced in that the array does not operate upon signal energy having a unique arrival angle, superimposed upon a generally isotropic noise field. The usual concept of array gain as signal advantage over isotropic noise is no longer completely valid.

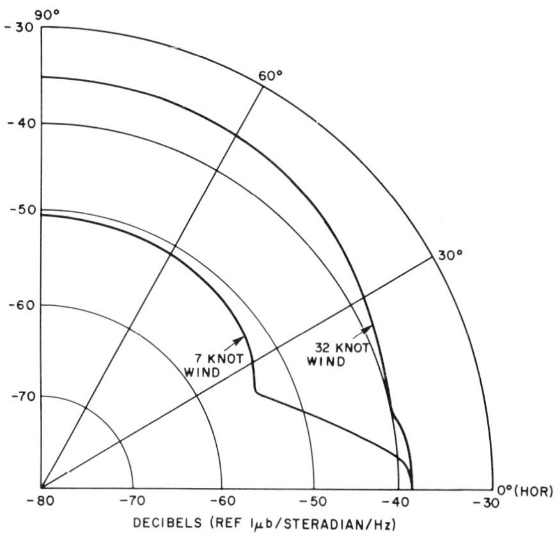

Fig. 5 Angular distribution in the vertical plane of ambient noise measured in the deep ocean under conditions of light-to-moderate shipping. The distribution is seen to approximate isotropy only when there are high surface winds. The frequency is 200 Hz.

A modest vertical aperture ($3\lambda - 6\lambda$) can provide some S/N improvement by reduction of the isotropic component of noise. A larger vertical aperture might be useful in improving performance in the multipath environment. However, this step is fundamentally quantized, i.e., one must go to sufficiently large vertical aperture so that individual paths can be selected. From detailed ray tracing, it can be calculated that typical multipath resolution would require a beamwidth of a fraction of a degree. Therefore, the systems engineer is faced with a crude quantization of vertical aperture: 0λ (linear horizontal array), $\sim 6\lambda$ (modest vertical aperture), or $\geqslant 60\lambda$ (single path resolution). However, the construction of near-vertical structures in the deep ocean presents extremely difficult engineering and economic problems. The comparison of potential S/N performance to the cost and risk usually leads to choice of the horizontal array for long-range systems.

These, then, are some of the first order considerations which are relevant to transmission and structure problems. The other significant factor in the environment is noise. There are many sources of environmental noise, as indicated in Figure 6. At low frequencies, seismic noise tends to dominate. In the frequency regime of the order of 100 cycles, the major noise source is shipping. Above a few hundred cycles, the dominant effect becomes the noise contributed by waves and weather. There may also be flow noise if there is a strong current past the array or if the array is mounted on a moving ship. In some cases biological noise, which can appear throughout the band, is also significant.

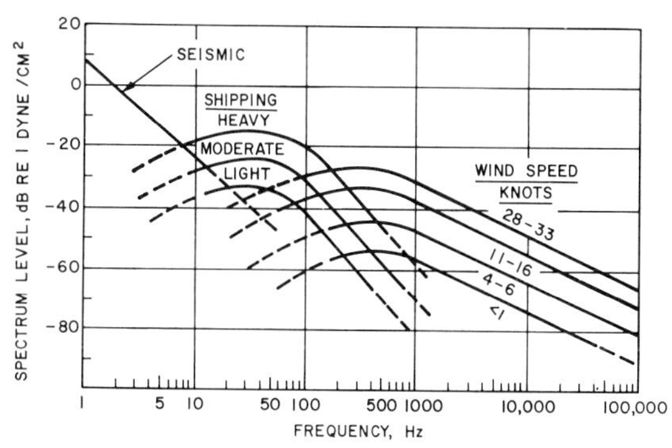

Fig. 6 Average deep-water ambient noise.

Here, again, there is a large difference from the radar case. In most radar applications involving phased arrays, the dominant noise is circuit noise introduced at the individual antenna elements rather than external ambient noise. In sonar, however, the noise levels resulting from the ocean environment are far above the noise which will be introduced by properly designed detectors. Therefore, sonar differs from radar in that one does not attempt to collect the maximum possible energy; it suffices to sample the pressure.

Since the noise power level decreased with increasing frequency, the use of the highest reasonable frequency appears to be indicated. However, high frequency performance is limited by frequency dependent attenuation of the ocean. The attenuation varies from a value of the order of .005 dB per kiloyard (path length) at 200 cycles to more than ten times that value at 1000 cycles. At a one-way range of 100 miles the corresponding losses are about 1 and 10^+ dB. For typical conditions the difference in noise at these two frequencies is only about 5 dB. Thus, there is a net <u>disadvantage</u> of the order of 4 dB in going from 200 Hz to 1000 Hz. Further increase in frequency simply increases the loss penalty. Therefore, the region from 200 to 1000 Hz is generally good for long-range systems with a slight preference for the lower part of the band.

The next problem for the sonar designer is the selection of bandwidth. In the case of submarine detection, one would like at least sufficient resolution to distinguish submarines from other large objects, particularly those which are mobile, e.g. whales or schools of porpoise. Resolution on the order of 25 to 50 feet is appropriate. Since the velocity of sound in water is of the order of 5,000 feet per second, this implies a bandwidth of approximately 100–200 Hz for the pulse. In addition, one may require frequency diversity for pulse-to-pulse isolation or for assistance in studying the shape of returns, so that two bandwidth increments are indicated. This, too, is an extraordinary difference from the radar case in that we are considering bandwidths which are of the order of 50% or more of the center design frequency. For these and other reasons it is not unusual for the acoustic array designer to be faced with the problem of designing arrays which can operate satisfactorily over nearly a decade in frequency.

A typical system might, for example, be designed to operate over a frequency range of 4 to 1. At the lower frequencies, the major performance factors influencing array length are "gain" and directivity; at the high end the problem is minimization of grating lobe effects. A compromise solution is an array length which provides an (average) inter-element spacing of $\frac{\lambda}{4}$ at the low end with a resulting loss of 3 dB in "gain". At the high end the spacing is λ, so that only one grating lobe would be in real space for a uniform array not steered to endfire. (At endfire, a second grating lobe would appear). However, a space tapered design is usually preferable to a uniform array. The taper is designed to serve two purposes. First is the usual reduction of lobes close to the main beam; the second purpose is to depress (and therefore spread) the grating lobe to minimize ambiguity in target direction. A space taper which accomplishes both purposes in a systematic (and calculable) way is a geometric progression in which the inter-element spacing increases by a constant factor for each element from the center to the extremes of the array. For example, the pattern of a 50-element linear array with a ratio of maximum to minimum spacing of 2 (spacing increase of about 3% per element) is illustrated in Figure 7. The near-in sidelobe level is reduced (from a uniform array performance) by about 8 dB for the odd-numbered lobes and 1-2 dB for the even-numbered lobes. The grating lobe, which would be at sin θ = 1 in the figure, is reduced by about 12 dB but is spread out in angle and appears as a "plateau."

Further increase in the degree of the taper does not provide additional advantage. For example, the pattern for a spacing ratio of 4 is illustrated in Figure 8. Some of the near-in lobes are higher than those of a uniform array; the "grating lobe plateau" is depressed an additional 3 dB but occupies twice the angle space.

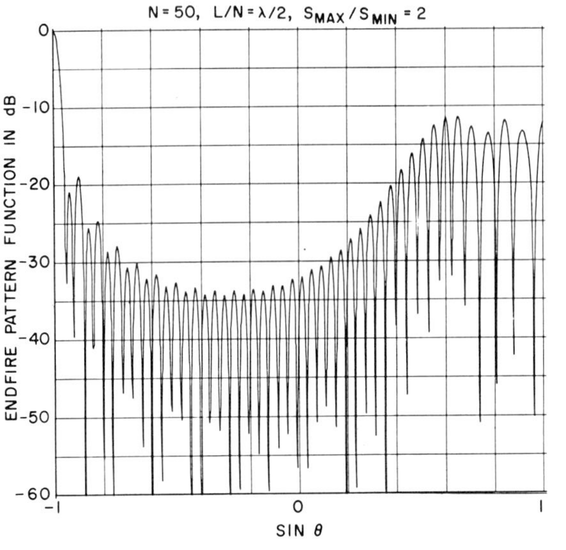

Fig. 7. Linear 50-element array pattern for geometric space taper with a ratio of maximum-to-minimum spacing of 2, and average inter-element spacing of λ/2.

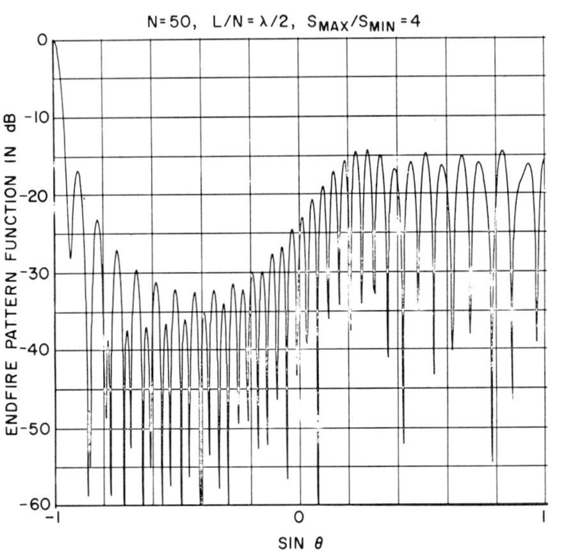

Fig. 8 Linear 50-element array pattern for geometric space taper with spacing ratio of 4 and average inter-element spacing of λ/2.

In summary, the general laws of physics and engineering which apply to radar antennas also apply to acoustic antennas in the ocean. However, the nature of long-distance deep-ocean transmission results in a non-isotropic medium, and therefore directive noise patterns and signal transmission paths. In addition, it produces a complicated time-variable multipath environment. Inaccessibility and the hostile nature of the physical environment make it expensive and difficult to use a planar array which could solve some of the problems. The noise in the system is dominated by environmental noise, with a significant contribution (in the deep ocean) from shipping. This provides a non-white noise that is not stationary in time. Considerations of the noise spectrum and of energy loss as a function of frequency, lead the designer to a frequency of only a few hundred cycles for a long-range deep-ocean system, while considerations of target resolution may lead him to require bandwidths which are of the same order of magnitude as the center frequency of operation. The net outcome is that the designer is likely to come up with a solution which is a linear array (or a planar array with modest vertical aperture) and a design utilizing a space taper of sampling-type elements.

BMD Requirements for Phased Array Radars

R. S. McCARTER
Bell Telephone Laboratories, Inc.
Whippany, New Jersey

Introduction

Phased arrays have been proposed for use in ballistic missile defense for about a decade. Their putative advantages include multifunction capability, fast beam steering, and passive hardness. Several BMD systems using phased array radars have been designed and a few have been built. The purpose of this paper is to describe the considerations that determine requirements placed on phased arrays for this application.

There are many factors that affect the requirements an array radar must meet to be a part of the solution to the BMD problem. Unfortunately, we will not be able to be specific about these requirements in this paper due to its lack of classification. However, it may be of some interest to examine the total system requirements for a BMD system and explore how these requirements are translated into requirements for the radar.

The factors affecting BMD system requirements include: (1) characteristics of the defended sites, (2) the defense objectives to be obtained, (3) characteristics of the threat, (4) the system configurations being considered, (5) the expected battle environment, and (6) availability of installed system components. All of these system requirements affect the requirements for the radar in some way, some being more influential than others. For example, the characteristics of the threat are very strong contributors to the selection of the radar parameters. On the other hand, the influence of defense objectives is felt most strongly in the level of deployment that is required and have more subtle influences on the radar requirements.

Discussion

Characteristics of the Defended Sites

Those characteristics of the defended sites which most strongly influence radar requirements include: (1) size, (2) relative value, (3) hardness, and (4) geography.

Defense Objectives

The defense is usually configured to either prevent damage or to assure survival of a portion of the population, ICBM force, etc. when attacked by the design threat.

Characteristics of the Threat

Properties of the design threat are usually the most important factors in determining the nature of the defense system and the resulting requirements on the radar. Important among the threat characteristics are attack direction, radar cross-section, number and character of objects as a function of time to detonation, number of warheads to be intercepted per interceptor flight time, and penetration aids.

System Configurations

Obviously, a radar configured for an area defense system will be quite different from one designed for the defense of one Minuteman silo. Even within the available options for a single type of defense, such as active defense of Minuteman, there are substantial differences in individual radar requirements.

Battle Environment

There is a dichotomy in the defense of cities and hardened sites. In both city defense and Minuteman defense it is desirable to have the defenses be less vulnerable than the protected sites to make the defense an unattractive target. This objective is relatively easy to meet in city defense; however, it is probably impossible to meet for the defense of Minuteman. The most difficult battle environment occurs when relatively large nuclear weapons burst within the defended site. Such bursts occur when leakage is not prevented, when preferential defense doctrines are employed, or when the attacking weapon bursts just before intercept.

Existing System Components

The existence of deployed system components also interacts with new system requirements: (l) there is a strong tendency to proliferate that which exists because of the cost and time required to establish production capability, training of installers and operators, and separate administrative operation and (2) the new system will probably have to interface with the operation of the existing system.

Synthesis

The requirements on the array radar are derived from the above system considerations, the particular system solution chosen, and economic trades among array antenna parameters, other radar parameters, and additional weapon system resources such as data processing and interceptors. Most array radars in BMD perform multiple functions, usually for both point targets (tracking) and volume coverage (searching), causing many of the antenna parameters to be compromise choices.

Survey of Ground Based Phased Array Antennas

A. C. SCHELL
Microwave Physics Laboratory
Air Force Cambridge Research Labs.
Bedford, Massachusetts

ABSTRACT

This survey of phased array antennas in ground based sensor systems infers three main development directions. These are embodied in the antennas of the AN/FPS-85, HAPDAR, and the AP/TPN-19. The significance of each of these as a type is not the number of elements or the various radar parameters, but the design approach, with its attendant advantages and limitations. The resultant antenna configuration qualifies for a limited range of performance and the design becomes a candidate for those applications that match best to the antenna advantages.

I. INTRODUCTION

The application of phased array technology to ground based sensor systems has led to the creation of several types of antennas to satisfy differing requirements. Over twenty years ago the MUSA system was constructed by the Bell Telephone Laboratories to provide an electronically adjustable beam for short wave reception. After World War II, many concepts for electronic beam steering were investigated, and several electromechanical methods such as the use of circular commutators were developed for beam scanning at moderately high speeds. The impetus behind most of these developments was a series of military requirements for sensor systems that could provide substantially greater search and track capabilities than existing equipment.

The goal of an effective defense against ballistic missiles has brought the greatest challenge to phased array design, and has been the motive behind most of the developments of the past decade. The problems associated with ballistic missile defense have imposed constraints and criteria that have steadily pushed the limits of antenna technology.

However, the utility of phased array antennas for other purposes is emerging as an important consideration. Although for many applications the cost of a phased array is presently too great for the benefit derived from its use, the increase in complexity and data rate of many sensing functions that previously were simply treated has necessitated the introduction of sophisticated equipment, and phased arrays are becoming competitive in these areas. As an example, air traffic control radars have used simple mechanical antennas for many years. During this time there has been an enormous increase in the number of takeoffs and landings, and the types of aircraft range from large commercial transports to STOL aircraft and helicopters. Aircraft speeds have increased, and operations in inclement weather are commonplace. The need for airport equipment that effectively measures and displays flight information is clear and present.

Another area that has presented challenges of an order high enough to favor phased arrays is the surveillance and tracking of space objects. The ranges, rates of travel, and number of objects have imposed requirements for electronic performance that are well matched to phased arrays.

An aspect of phased array operation that is particularly in consonance with present day system design is the inherent capability for computer controlled operation. Not only is the beam of a phased array steered by computer, but the properties of the array permit very efficient use of the available transmitter power. During the time that a transmitted pulse is enroute to a distant target in a given direction, the antenna beam can be steered to cover other shorter range targets, returning to receive the reflected pulse from the distant target at the appropriate interval of time. By careful planning of the radar operation sequence, the interpulse period can be much more efficiently used by a phased array than by a slow mechanical antenna, particularly for very long range operation. The capability of phased arrays to permit computer assignment and control of virtually all aspects of the sensing conducted within the antenna coverage will lead to the incorporation of phased arrays in new equipments that are designed to take advantage of electronic scanning modes to yield higher system performance. Special purpose arrays that provide unique performance roles are inevitable outgrowths of the trend towards integration of the antenna with the system computer.

There are many different types of phased arrays that have been designed and constructed over the years, and there are various ways to catalogue them, depending on the emphasis attached to certain performance aspects. One method of classification is based on the type of beam steering. The categories are time delay steering, phase shifter scanning, frequency scanning, and amplitude scanning. Other categorizations of phased arrays are planar vs. curved surfaces, dipole vs. waveguide, digital (stepped) vs. analog (continuous), multibeam vs. scanning and properties based on polarization, geometry, or phasing. The choice of a particular type of array for a specific application usually represents an engineering judgment of the merits of several closely competing designs. The cost of a given choice changes many of the apparent subtleties into major decisions. For example, the decision to place the power amplifiers before the phase shifters in a transmitting array requires the phase shifter to handle high power, whereas amplification after phase shifting necessitates the use of many transmitting modules. These design choices are continually revised as the current technology alters and improves the availability and performance of components. To a great extent the design of phased arrays is an amalgam of current design practice, available componentry, and an approach towards solving the overall system goal, biased by the designer's experience. As a result, there do not tend to be direct quantitative formulations for each aspect of design, and at present the topic is approached as an art.

Each phased array consists of a large number of radiating elements, a feed network, and a set of components or a sequence of operations by which the beam is steered or a set of beams is created in the desired coverage. The application for which the array is intended has the dominant effect on the type of array and the method of operation. In this paper an attempt is made to discern three directions in which ground-based phased array design has proceeded. These represent three stages of development of phased arrays, but each has its area of applicability and undoubtedly will continue to be used.

The first design is the "wired" array, in which the feed network consists of a set of waveguides or transmission lines with appropriate interconnections. A variant is the use of a large number of modular transmitters, with transmission line interconnections for coherent r.f. excitation and phasing. An example of this type of antenna is the pair of arrays of the AN/FPS-85 Space Track Radar.

The second design is the "optical" array, in which the feed network consists of free space paths between a single horn (or set of horns) and the array face. Phase shifters are introduced at the array face to collimate the aperture field with the appropriate phase front for the desired beam direction. This type of antenna can be built as an active lens or as an active reflector, and it tends to be preferred when the number of square wavelengths in the array face is large, and the wavelength is short. An example of this approach is the HAPDAR array. This type of phased array is being considered for several applications, and it is being used for the missile site radar of the present ABM system.

The third design involves the combination of phased array technology with other types of antennas to produce a hybrid design. System applications that do not require wide angle coverage and that mitigate against conventional phased array approaches can dictate the use of a hybrid antenna. As an example, the Precision Approach Radar of the AN/TPN-19 GCA system uses the combination of a phased array and a reflector to produce a beam that is electronically steered. As specialized applications appear, phased array technology can meet the needs by utilizing basic elements and arrays in new ways. The system limits such as coverage sector, bandwidth, or number of beams can be used to reduce the complexity of the phased array. In so doing, other antenna techniques will be combined with the phased array design to produce the desired results. In the case of a limited sector coverage such as that needed for a PAR, the number of phase shifters and radiators of a hybrid array can be considerably less than the number dictated by gain considerations for a conventional planar array.

II. THE AN/FPS–85 SPACETRACK RADAR ANTENNA

The AN/FPS-85 is a large UHF phased array antenna used for the detection, identification and tracking of space objects. The radar was constructed at Eglin Air Force Base in Northwest Florida by the Bendix Corp., and is operated by the Air Defense Command of the USAF. Operation of the radar is by computer control, and a complex series of search and track functions are carried out by the computer program in accordance with periodically updated instructions from the computer at the ADC Space Defense Center.

The phased array antennas of the FPS-85 occupy the inclined southern face of the radar building. The transmit array consists of 5184 pear-shaped dipoles with directors, arranged on a 72x72 square grid with 0.55λ interelement spacing. The beamwidth of the transmit array on boresight is 1.4^{o} at the center frequency of 442MHZ. Each element is fed by a transmitter module that converts and amplifies the phased UHF signal to a peak power of 10 kw. The transmit array is uniformly illuminated and has a 10MHZ bandwidth. Two views of the FPS-85 radar are shown in Fig. 1 and Fig. 2.

Fig. 1 The front face of the AN/FPS-85 radar, showing the two arrays. (Courtesy of Rome Air Development Center)

Fig. 2 A side view of the FPS-85 (Courtesy of Rome Air Development Center)

The receive array consists of 19,500 crossed dipoles in a square grid within an octagonal perimeter approximately 190 ft. in diameter. There are 4660 identical receiver modules connected to elements in a pattern that provides space tapering. Inactive elements are terminated. The beam forming network generates a cluster of nine beams on reception. The IF bandwidth is 1MHZ.

A common beam steering network is shared by the transmit and receive arrays. Operating at a frequency of 100 MHZ, the 7-bit diode phase shifters generate row and column phase gradients that are multiplied with the receive signals and the low power transmit signal.

The AN/FPS-85 radar contains an extensive amount of automatic test and checkout equipment. System status is constantly assessed by a series of hardware checks and operation tests, and there are monitor panels, such as the transmitter calibration board that displays the transmitter modules that are out of tolerance at anytime.

This radar is a very successful realization of a planar "wired" phased array radar. It is in operation today and its performance exceeds the design specifications (1). In size and capability it represents a landmark of phased array technology.

III. THE HAPDAR ARRAY

The use of r.f. transmission line circuits for the interconnection of phased array radiators results in a complex and expensive feed network when a large number of array elements are employed. The number of connections per unit area of aperture rises with the square of frequency, and at microwave frequencies the feed network tends to be lossy. In order to realize a phased array with a large number of elements more economically, free space paths can be used within the antenna as part of the field collimation. A radar that has successfully demonstrated this approach is the HAPDAR, or Hard Point Demonstration Array Radar, built at the White Sands Missile Range by the Sperry Rand Corp. for the Advanced Research Projects Agency. The antenna consists of an active lens and a five-horn feed. The side of the lens facing the feed horn, termed the collector face, is a triangular array of printed dipoles shown in Fig. 3. The diameter of the lens face is 41λ, as is the distance from the face to the feed horn. The outside or radiating face is a triangular array of horns with 0.676λ interelement spacing.

Between the faces are printed circuit combiners that couple together the outputs of pairs of over half of all collectors, and three - bit digital diode phase shifters that collimate the field and provide the wavefront phase tilt necessary for the desired beam direction. The distribution of the 2:1 combiners is chosen to provide a space tapering of the aperture in addi-

Fig. 3 The collector face of the HAPDAR array (Courtesy of Sperry Gyroscope Co.)

tion to the 9 dB edge taper of the feed horn. Most of the remaining collector elements are connected directly to a radiating horn through a phase shifter, and some are terminated. Fig. 4 shows the components of the active lens. Approximately one-half of

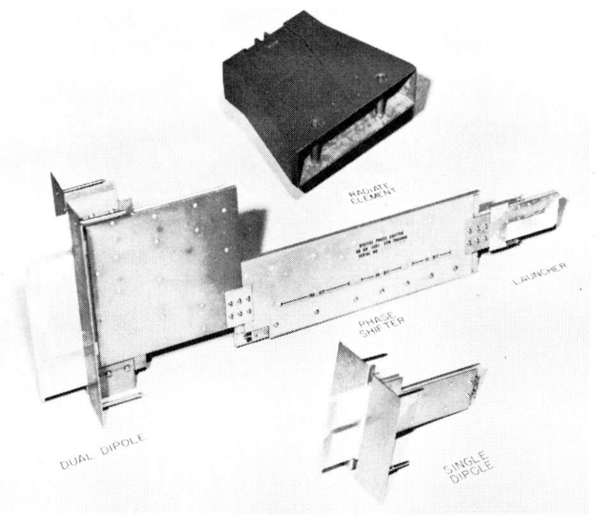

Fig. 4 The components of the HAPDAR lens (Courtesy of Sperry Gyroscope Co.)

the elements of the radiating face are connected to the collector face; the remainder are terminated. Construction of the radiating face is shown in Fig. 5. The phase shifters perform the functions of changing the spherical wavefront of the horn to a planar front, and introducing the desired tilt to the wave-

Fig. 5 The assembly of the radiating face of the HAPDAR lens. (Courtesy of Sperry Gyroscope Co.)

front. As a result, the regular pattern of error that can occur with a planar array is distorted into a random-appearing phase function that does not produce high grating lobes. The power gain of the antenna is 35.9 dB, including 1.2 dB phase shifter loss and 2.4 dB loss due to spillover, mismatch, and other causes. The calculated peak sidelobe level is 28 dB. The horn feed provides monopulse in both azimuth and elevation in addition to the sum pattern. The HAPDAR array has been found to provide stable and accurate operation for radar usage. A thorough series of tests have validated this design (2). The completed antenna is shown in Fig. 6.

Fig. 6 The completed HAPDAR array. (Courtesy of Sperry Gyroscope Co.)

The principles of quasi-optical feed design can be applied to a reflector as well as a lens configuration. A space-fed reflective array antenna consists of a feed horn and an array of elements, each of which is connected to a phase shifter followed by a short circuit. The horn illuminates the array, and the phase of each reflected signal is adjusted by the phase shifters to provide the desired pattern (3).

The quasi-optical feed design technique is increasingly being used for radar application, particularly as the technology is developed for producing low cost modular array elements and phase shifters. A prominent example of a space-fed array is the antenna for the missile site radar of the present anti-ballistic missile system. This radar employs a space-fed lens antenna that is similar to the HAPDAR Array (4).

IV. THE AN/TPN–19 PRECISION APPROACH RADAR

The antennas discussed in the previous sections are designed to cover a wide angular sector. Not all applications require wide sector coverage, and this can be reflected in the antenna design. By careful matching to the criteria dictated by the system use, the antenna design can yield performance gains or cost and complexity reductions. As an example, the stated requirements for the scan sector coverage of the precision approach radar of a new USAF tactical GCA system are 14° x 20°. However, gain considerations for the x-band antenna would dictate the use of approximately 3000 elements in a conventional planar waveguide phased array. In order to substantially reduce the number of array elements, the Raytheon Company has developed a reflector - array combination for the AN/TPN-19 PAR. The design approach is based on the idea of an active subreflector, in which the conventional subdish of a reflector antenna is replaced by a reflectarray that simulates a movable feed (5). This has been modified by the incorporation of a space-fed lens in place of the reflectarray. The array lens consists of 824 elements arranged in staggered columns along a cylindrical surface. A multimode feed horn illuminates the back face of the lens with linear polarization, and the front face radiates circular polarized fields towards a hyperboloidal reflector, approximately 9' x 12' in size. The array illuminates an area on the reflector that in turn produces a beam in the desired direction. The reflector and array geometry were chosen to minimize the performance degradation as the beam is steered away from boresight. The operation of this antenna can be viewed alternatively on reception from a geometrical optics standpoint. Rays striking the reflector from a given direction are reflected towards a focal region. The phased array is situated to intercept these rays and to combine the received fields with the appropriate phases.

The design of this type of antenna involves compromises in terms of reflector and array size, number of elements, and performance limits. A fundamental limit appears to be that the minimum number of elements for full pattern control is given by the coverage sector solid angle divided by the solid angle beamwidth of the antenna. In practice, the number of elements used is considerably greater. A photograph of the breadboard model of the AN/TPN-19 PAR antenna is shown in Fig. 7.

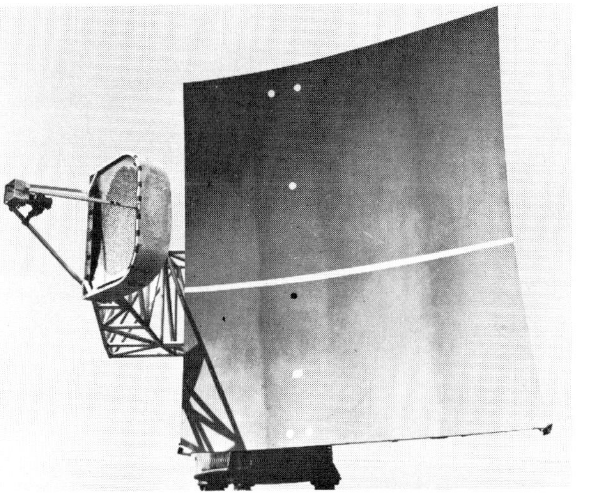

Fig. 7 The breadboard test model of the AN/TPN-19 PAR antenna. (Courtesy of Raytheon Co.)

The design of limited sector coverage antennas is an area where the employment of hybrid antenna techniques can be used to advantage. A hybrid design combines two or more antenna techniques to obtain the desired result. It is not necessary that the combination involve a reflector; a lens or possibly a second array could be used with a phased array to electronically scan a beam over a limited sector. For the AN/TPN-19 PAR, a dominant consideration is system weight; however, other systems may have bandwidth, scan rate or sequence, coverage, or siting restrictions that permit the use of hybrid arrays.

V. CONCLUSIONS

Phased array antennas are incorporated into ground based sensor systems in operational use, and they are planned for other systems under development or being deployed. Each phased array design differs from another for reasons associated with the applications requirements, the state of technology, and the designer's approach to the problem. For a large, very high power UHF array the designers of the FPS-85 elected to use modular transmitters and receivers in a planar "wired" array. For a lower cost, higher frequency radar the HAPDAR designers chose to use an "optical" approach for an active lens. For a limited scan X-band approach radar the TPN-19 designers are using a "hybrid" array consisting of an active lens and a reflector. Each of these design approaches represents a trend that favors a range of application, frequency, coverage, bandwidth, and operational sequence. As the uses of phased arrays are increased, the specialized designs will increase, and there will appear many more design types that will fill in the spectrum of approaches to the electronic scanning of the beam of a ground based antenna.

REFERENCES

(1) J. E. Reed, "The AN/FPS-85 Radar System", Proc. IEEE, 57, 1 pp 324-335, March 1969

(2) P. J. Kahrilas, "HAPDAR - An Operational Phased Array Radar", Proc. IEEE, 56, 11, pp. 1967-1975, November 1968

(3) D. G. Berry et al., "The Reflectarray Antenna," IEEE Trans. Ant & Prop., AP-11, pp 645-652, November 1963

(4) C. M. Johnson, "Ballistic - Missile Defense Radars", IEEE Spectrum, 7, 3, pp 32-41, March 1970

(5) C. F. Winter, "Phase Scanning Experiments with Two-Reflector Antenna Systems", Proc. IEEE, 56, 11, pp 1984-1999, November 1968

Survey of Phased Array Accomplishments and Requirements for Navy Ships

MERRILL I. SKOLNIK
Radar Division
Naval Research Laboratory
Washington, D. C.

Synopsis

Shipboard phased array antenna requirements and applications are reviewed and a listing of the Navy's accomplishments with operational phased arrays is given. Laboratory developments are also briefly mentioned. Success in the past has been mainly the result of the judicious use of frequency scanning by itself for one-coordinate beam positioning or in conjunction with phase scanning for two-coordinate systems.

Introduction

A modern Naval vessel carries an impressive assortment of electromagnetic antennas for such purposes as radar surveillance, radar weapon control, navigation, air traffic control, communications, IFF, direction-finding, and electronic warfare. The phased array antenna must be considered as a potential candidate for those systems that require directive antenna beams for transmission or reception of electromagnetic energy. The Navy was the first service to deploy the electronically scanned phased array antenna in operational systems and was for a long time the only military service with operational phased array antennas. It was more than a decade ago that the Navy introduced electronic scanning into the Fleet as a vital part of shipboard radar designed for high data-rate 3-D air-search application.

It is possible to apply the phased array to any kind of ship, but the two major classes of ships that have been generally considered for phased arrays are the guided-missile destroyer and the aircraft carrier. These two types of ships are also more likely to be nuclear powered than smaller ships. On nuclear powered ships the physical placement of antennas generally is more convenient than on ships with stacks. With both the guided-missile destroyer and the aircraft carrier it is primarily the radar application for which the phased array has received the most attention. The radars of a guided missile destroyer must provide surveillance and tracking of air targets at long range at all altitudes within the radar horizon. In addition, provision must be made for IFF and guidance must be provided the defensive missiles. On an aircraft carrier, the phased array might have to perform the same functions as on a guided missile destroyer as well as the functions normally required for the control and the landing of the carrier's aircraft.

The purpose of this paper is to review the past and present applications of phased arrays on board Navy ships. The research and development being conducted in this area of technology by the Navy laboratories is also reviewed. The major effort in Navy phased arrays has been for radar and this paper will therefore concentrate for the most part on this important application.

Rationale for the Phased Array

Two characteristics of the phased array that are unique when compared to other antenna techniques are its ability to transmit high power and its ability to perform rapid, agile beam positioning. On board ship the ability to radiate the extremely large powers that might be possible with a phased array is seldom utilized. The transmitter average powers that have been or can be employed with conventional shipboard long-range radar systems are not that much different from the transmitter average powers that have been employed or proposed with phased array radars. Although there can be a large amount of prime power installed on a modern ship there are limits to the total that can be allocated to the radar. Furthermore, restrictions on space and weight would make extremely large powers undesirable since other needs might have to be sacrificed. Another consideration that mitigates against large power is that the transmitter powers of modern Navy radars are probably adequate to do their intended job. Thus, the ability of a phased array to radiate extremely large powers generally is not needed for most of the Navy applications of radar and would be difficult to accommodate on board ship if it were.

Although the potential advantage of high power does not hold much attraction, the ability of a phased array antenna to generate beams that can be positioned rapidly so as to perform near-simultaneous tracking of many targets with high data rate, as well as perform surveillance in a weapon-control radar system, is probably the chief characteristic of a phased array of interest for shipboard radar. A corollary advantage of beam flexibility is that the stabilization of the antenna beam required of many shipboard radars can be performed electronically with the phased array beam-steering rather than with the mechanical movement of large masses as is done with conventional radar.

Three other advantages sometimes claimed for shipboard phased arrays are that their aperture distributions can be controlled better than with reflector antennas, they can be sited lower on the ship so that the topside weight and moment can be reduced, and their shape makes it easier to harden against blast than other antenna shapes. These advantages may be utilized in particular applications but generally they offer little in themselves to be the dominant reason for choosing a phased array. It is seldom that the improved sidelobes achieved by better control of the aperture is worthwhile since sidelobes on a ship-mounted radar are often determined more by the surrounding structure than the aperture illumination; the siting of the array lower on the ship can be harmful if it reduces its low-altitude coverage and the hardness of an antenna against blast or other hostile action need be no better than any other vital part of the ship's system.

Thus it is primarily the unique beam agility characteristics of a phased array that makes it worthwhile considering for shipboard application.

In addition to its advantages, the phased array comes with some liabilities. Its cost and complexity are usually significantly greater than other antenna types and the software production for the array control system can be costly and time consuming. These liabilities are not always insignificant and they are the reason the phased array antenna has not seen greater application.

Phased Arrays for Shipboard Radar

One of the first, if not the first, application of phased array technology to operational radar was the FH MUSA, or Mk 8, a 42 element S-band array of polyrod radiators, Fig. 1 (1). The 42 elements were arranged in 3 rows of 14 radiators each, with the 3 vertical elements tied together so that the antenna was basically a 14 element linear array. The phase shifters were of the Fox rotary waveguide type (2) that were mechanically driven so as to scan ±9 degrees at a rate of 10 per sec. At this scan rate the radar provided an almost continuous map-like display and was used for main-battery fire control on capital ships. The original need that the Mk 8 was supposed to satisfy disappeared with the disappearance from the Navy of the heavy gun-equipped capital ship. Although the Mk 8 utilized mechanical phase shifters rather than the electronic phase shifters that form the heart of today's interest in array technology, it is included in this review of the Navy's work in phased arrays since it was the first operational application of a phased array and saw wide use.

Fig. 1 FH MUSA, or Mk 8, polyrod phased array radar with mechanical phase shifters. Used for main-battery fire control on capital ships.

As the demand increased for the detection and high-data-rate tracking of aircraft targets in three coordinates (range, azimuth and elevation) the Navy turned to electronic scanning methods for achieving a 3-D capability with a single antenna. Since frequency scanning is one of the most simple, reliable, and relatively inexpensive forms of electronic beam steering it was natural that it was the first electronic method to be employed in operational radars. The AN/SPS-39, 48, and 52 radars for designation and aircraft-control aboard guided missile destroyers were the result. These radars rapidly position a narrow pencil beam in elevation by means of frequency scanning and rotate mechanically to scan in azimuth. Such radars are employed on anti-aircraft guided-missile ships which carry Talos, Terrior, or Tartar missiles.

The AN/SPS-39 (as well as the related AN/SPS-42) utilizes a parabolic cylinder reflector antenna fed by a 40-element linear array with a frequency-sensitive serpentine delay line, Fig. 2. The pencil beam radiation pattern can be varied through an angle of more than 100 degrees with a frequency excursion of less than 10 percent. The vertical scanning period is of the order of milliseconds. The power output is

Fig. 2 AN/SPS-39 frequency scanned antenna.

not linear with frequency, but is deliberately shaped so that power increases with frequencies corresponding to a lower beam angle and hence long range. The pulse repetition rate also varies as a function of the elevation angle to compensate for the longer ranges at lower angles. The variation in pulse rate is approximately 5 to 1. The flexibility of electronic beam steering is taken advantage of to stabilize the antenna without the need for heavy mechanical gyro-stabilization equipment above deck. Roll and pitch signals from a ship's main gyro are supplied to a computer which automatically modifies the elevation beam scanning program and presents the true position of targets on the radar scopes. The above-deck equipment weight is about 2800 pounds. The first prototype equipment manufactured by the Hughes Aircraft Company was delivered to the Navy in August, 1957. Since that time more than 100 such radars have been placed aboard ships of the U. S. and other nations.

The AN/SPS-39 will be replaced in the future by two other radars. The AN/SPS-48 (3 & 4) (Fig. 3) will be used on large ships and the AN/SPS-52 (Fig. 4) on small ships. Both radars employ planar array antennas that scan in elevation by means of frequency and in azimuth by mechanical rotation. The first shipboard model of the AN/SPS-48 developed by ITT Gilfillan, Inc. was delivered to the Navy for evaluation in July 1962. It is designed to provide the necessary target data to support the Navy ship missile weapon control systems as well as aircraft-intercept control. This radar transmits multiple beams to achieve long range and a high data rate and employs flexible power management. The antenna weighs 4500

Fig. 3 AN/SPS-48 frequency scanned antenna.

Fig. 4 AN/SPS-52 frequency scanned antenna mounted on an Australian ship.

pounds and is fed by a serpentine delay line.

The AN/SPS-52 built by the Hughes Aircraft Company utilizes a planar-array antenna that generates a narrow pencil beam that is stepped in elevation by frequency. Its antenna and pedestal weigh 3240 pounds. A digital computer replaced the previous analog computer for radar beam control. The first antenna system was delivered in February 1966. This radar has been used on the ships of several other countries as well as those of the U. S.

The first operational "billboard" phased arrays

were the AN/SPS-32 and AN/SPS-33 radars on the nuclear carrier USS ENTERPRISE and the nuclear cruiser USS LONG BEACH (Fig. 5). The planar antennas are mounted in pairs on each of the four sides of the ship's square superstructure and provide 360 degree azimuth coverage and elevation coverage to nearly the zenith. Figure 6 is a closer view of the two antennas. The antenna at the left is for the AN/SPS-32 radar. It generates a fan-shaped beam that is frequency scanned in azimuth to provide surveillance of targets several hundred miles away. The antenna on the right is for the AN/SPS-33 radar. Its pencil-shaped beam is steered in elevation by frequency scan but utilizes electronic phase shifters for steering the beam in azimuth. Hundreds of targets can be tracked automatically by the computer-controlled AN/SPS-33.

Fig. 5 USS LONG BEACH, nuclear-powered missile cruiser with the "billboard" AN/SPS-32 and AN/SPS-33 antennas mounted on the superstructure.

Fig. 6 AN/SPS-32 antenna (left) under construction and (right) AN/SPS-33 antenna.

The development of the AN/SPS-32 by the Hughes Aircraft Co. was started in 1956 and the AN/SPS-33 in 1958. These radars have been well received by the operating forces. At the present time the LONG BEACH's radar is being upgraded and modernized.

The AN/SPS-33 was a bold venture at the time it was conceived and devloped. The combination of frequency and phase scanning in a 3-D radar was chosen because it was more economical and practical than phase scanning in both planes. It can be said that frequency scanning has proven itself a successful method of electronic scanning in operational radar systems.

In the late fifties, the Navy started the development of a sophisticated 3-D radar known as the AN/SPG-59, or Typhon radar. This computer-controlled radar was supposed to perform the functions of surveillance, tracking and missile guidance. The Typhon project development was cancelled, however, as being too large and complex for full reliability. Its complexity resulted in large part from trying to perform several major radar functions from a single aperture at a single frequency.

The Navy is now embarked on the development of the Aegis (formerly called ASMS) surface-to-air

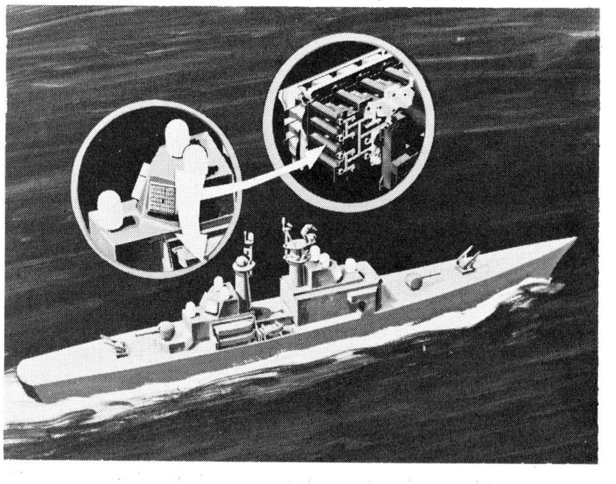

Fig. 7 Artist's conception of a phased array radar system for the control of a surface-to-air missile system.

missile defense system for the new guided missile ships scheduled to join the Fleet in the mid-1970's. This system employs an electronically scanned phased array radar called MFAR consisting of four separate faces, configured as shown in the artist's conception of Fig. 7. MFAR detects and tracks multiple targets at long ranges and transmits guidance commands to interceptor missiles. The development of this radar is being undertaken by RCA.

Other Phased Array Navy Applications

Although most of the work in phased arrays has been for radar, there has been interest in applying this technology to shipboard IFF, Electronic Warfare systems, and microwave communications including communications via satellites. It has also been proposed to integrate several electronic systems functions with the use of a single multifunction array antenna aperture performing the functions normally accomplished with separate antenna systems.

Phased Array Development in Navy Laboratories

The amount of phased array development conducted in Navy laboratories has not been large when compared with that sponsored by other services, but its accomplishments have not been insignificant. The Naval Electronics Laboratory (NELC) and the Naval Research Laboratory (NRL) have achieved considerable success in the development of circularly symmetric array antennas (5 & 6). Circularly symmetric arrays provide pattern symmetry in azimuth and the potential for physically locating the antennas around the ship's mast.

The 2000 element S-band reflectarray radar shown in Fig. 8 has been used at NRL as a test radar to explore problems associated with the applicability of phased arrays for Navy shipboard application. NRL has also been engaged in exploring the applicability of solid state techniques to phased arrays. Both NRL and NELC have been engaged in the development of new phased array techniques including arrays capable of radiating at more than one frequency.

Fig. 8 2000-element S-band reflectarray radar located at the Naval Research Laboratory Chesapeake Bay field site.

The Applied Physics Laboratory of The Johns Hopkins University although not a government laboratory, is supported in large part by the Navy and has done outstanding pioneering development in phased array technology including wideband arrays, subarrays and component technology (7 & 8).

Systems Applications Considerations

The Navy's past success in employing electronic scanning can be attributed to the relative simplicity of frequency scanning for one-coordinate beam steering and phase-frequency for two-coordinate steering. Phase scanning is a strong competitor to frequency scanning and will be used in applications where it has merit over frequency scan. The radar applications of electronic beam steering that appear most attractive in the Navy are those where high data-rate 3-D target information is needed. Long-range surveillance, however, often is better accomplished by more conventional radar methods.

The flexibility of phased arrays has led to their consideration for use as multifunction radars. The inclusion of functions other than radar has also been proposed for integrated electronic systems. In general, restraint must be exercised when considering the use of phased arrays for operational multifunction radars. Consider, for example, the application where it is necessary to detect and track aircraft as in a missile defense system. Normally the detection and tracking functions are accomplished with two separate radars. Conventional aircraft-surveillance radars are generally at the lower microwave frequencies, they have large antennas, and they survey a large volume with a low data rate but must be capable of handling a large number of targets both real and clutter. Conventional aircraft-tracking radars are generally at the higher microwave frequencies, they have small antennas, and they obtain precision position information on a relatively few targets, each in a restricted volume, at a high data rate. The distinctive radar equations (9) for the surveillance and tracking situations lend theoretical support to these differences that have been practiced ever since the early days of radar development. Thus it is for good reasons that surveillance radars look different than tracking radars. When a single phased array radar that operates at a single frequency is required to perform both the surveillance and the tracking functions a compromise must be made between the needs of surveillance and tracking. Such a compromise may result in a more costly system than if the two functions were performed by separate radars.

Discussion

Phased arrays have been employed in the Navy to perform some unique capability such as rapid, flexible beam steering for radar. They have been used in spite of the added cost compared to more conventional systems. Disproportionate costs may be acceptable when attempting to exploit a new technique but as knowledge increases and the art matures, phased arrays when applied to operational systems must be competitive. When deploying operational systems, cost effectiveness should be foremost and this generally leads to the conservative approach. On the other hand it would be hoped that expenditures for the research and development of phased arrays utilize a different criterion. In research and development, imagination and boldness should be the driving forces and the only important bounds should be those provided by Nature.

The Navy's accomplishments in the application of phased arrays have been of significance, especially considering the relatively low level of research and development funds that have been available for this purpose. It is hoped that the Navy will be able to continue to contribute significantly to the advancement of this important area of electronic technology.

REFERENCES

(1) G. G. Harvey, ed., "Report of Conference on Rapid Scanning," MIT Radiation Laboratory Report 54-27, June 15, 1943.

(2) A. G. Fox, "An Adjustable Waveguide Phase Changer," Proc. IRE, Vol.35, pp 1489-1498, December, 1947.

(3) W. J. Evanzia, "Faster, Lighter 3-D Radar in Sight for Tactical Warfare," Electronics, Vol.39, pp 80-88, June 27, 1966.

(4) D. C. Bailey, "AN/SPS-48 Radar Development," NAVSHIPS Technical News, pp 16-18, September, 1966.

(5) J. H. Provencher, "A Survey of Circular Symmetric Arrays," Polytechnic Institute of Brooklyn Phased-Array Antenna Symposium, 1970.

(6) B. Sheleg, "A Matrix-Fed Circular Array for Continuous Scanning," Proc. IEEE, Vol.56, pp 2016-2027, November, 1968.

(8) J. Frank, H. M. Grady, J. H. Kuck, and C. A. Shipley, "A 256-Element Phased Array Development," in preparation (Applied Physics Laboratory).

(7) J. Frank, "Phased Array Antenna Development," Applied Physics Laboratory Technical Memorandum TG-882, March, 1967.

(9) D. K. Barton, "Radar System Performance Charts," IEEE Trans., Vol. MIL-9, pp 255-263, July-October, 1965.

Survey of Airborne Phased Array Antennas

JAMES F. RIPPIN, JR.
Air Force Avionics Laboratory (AFSC)
Wright Patterson Air Force Base, OH

INTRODUCTION

This paper provides a brief status report of exploratory and advanced development programs that include airborne phased array antennas designed for use in radar and communication systems. The programs reviewed are shown in Table I and include those supported by the Air Force Avionics Laboratory (AFAL) and the Naval Air Systems Command (NASC). This paper will discuss planar phased arrays, conformal arrays, and distributed apertures. A planar phased array is one where the array elements, which usually consist of a radiator/phase shifter/driver assembly, are mounted to form a planar radiating surface. An airborne conformal array has a radiating surface that conforms to (or follows) the contoured aircraft skin; this contour may be in one or two dimensions. A distributed aperture is an array of elements, each of which contain a source of radiated RF energy, receiver RF circuits, and appropriate phase shifters and filters. Many applications require combinations of the general types of phased array antennas enumerated above; distributed arrays can be arranged in a planar or conformal configuration, and a planar array can utilize an active module for each element.

STATUS OF RADAR ARRAY PROGRAMS

The stimulus for the airborne distributed array concept was provided by the Molecular Electronics for Radar Applications (MERA) program (1), which has demonstrated the practicality of volume fabrication processes and assembly techniques for seven different types of microwave integrated circuits required to provide active transmitter-receiver modules for an all solid state, X-band, electronically

TABLE I

PROGRAM	CONTRACTOR	CONTRACT NUMBER
Molecular Electronics for Radar Applications (MERA)	Texas Instruments Inc.	AF33(615)2525
Radome, Antenna and RF Circuitry (RARF)	Raytheon Company	AF33(615)-3291 and F33615-67-C-1108
Radome, Antenna and RF Circuitry (RARF)	Emerson Electric Company	AF33(615)-3289 and F33615-67-C-1017
Radome Effects on a Radar Phased Array Antenna	Autonetics Division of North American Rockwell Corporation	F33615-68-C-1627 and F33615-70-C-1493
Single Axis Scanned Array	Hughes Aircraft Company	NOw66-0736
Electronically Scanned Airborne Intercept Radar Antenna (ESAIRA)	Hughes Aircraft Company	N00019-67-C-0653
SHF Airborne Distributed Phased Array Antenna	Raytheon Company	F33615-68-C-1592
Airborne UHF Receiving Antenna System	Texas Instruments Inc.	F33615-69-C-1706

scanned phased-array radar. Each of the initial 604 array modules provides approximately 600 milliwatts peak power for 20 microseconds at 9 GHz using a times-four multiplier. Each module contains separate 4-bit, S-band diode phase shifters for the transmit and receive modes. An S-band stripline RF manifold provides uniform amplitude distribution during the transmit cycle; the 500 MHz receive manifold provides a modified Taylor amplitude distribution designed for -25 dB peak sidelobes. Two axis monopulse is provided in the receive mode. Expected gain values are 31 dB on transmit and 28 dB on receive. Preliminary pattern testing of a 4 x 4 and a 10 x 28 MERA subarray has been completed with predicted results being obtained (2). Antenna pattern range testing of the complete MERA array is in progress. Of particular interest will be any data obtained concerning effects of phase coherency on the antenna patterns.

Present Weapon System requirements show need for a smaller, lighter, and more reliable radar package compatible with a digitally controlled total avionics system. The key to such a radar package is an antenna subsystem capable of interlaced multimode operation at high data rates; in most cases the answer is an electronically scanned phased array capable of rapid inertialess beam positioning, electronic beam shaping, and probably operation at multiple polarizations to provide rain clutter rejection and/or target enhancement. Under the 698DF Advanced Development Program of AFAL, two such Ku-band brassboard antenna subsystems with 28 inch diameter apertures are being prepared for flight testing; requirement for an electronically scanned IFF array is also included in the program. These complementary RARF (radome, Antenna and RF Circuitry) programs are being conducted by Raytheon Company (3) and Emerson Electric Company (4). Raytheon selected an optically fed reflect-array design with a four-horn feed, while Emerson selected an optically fed transmission type array with a 12-element feed. Both contractors utilize ferrite phase shifters with integrated circuit drivers which are digitally controlled. The Raytheon units are analog phase shifters, while the Emerson units are reciprocal, latching phase shifters which utilize a flux transfer type driver circuit. Each contractor's array elements have passed MIL-E-5400 environmental tests; element design and qualification of engineering models in accordance with Table II required fabrication of hundreds of elements and approximately one year. Figures 1 and 2 are photographs showing the Raytheon and Emerson RARF elements; each contractor has manufactured at least 4,000 elements. Table III contains the attained characteristics of the Raytheon RARF array element, and Table IV lists the specification goals for the Emerson RARF array element. Raytheon's approach was to procure, in finished form, the element parts and then match, assemble, and test the complete element. Emerson selected the approach of procuring mostly bulk raw materials and fabricating (sawing, grinding, etc.) the complete phase shifter/radiator assembly at their Rantec division; the necessary integrated circuit driver assembly was commercially procured and matched to the Rantec microwave assembly at the Emerson plant.

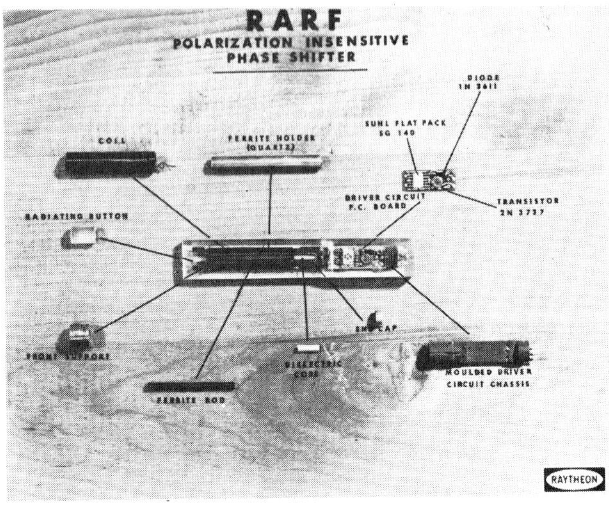

Fig. 1 Raytheon RARF Element.

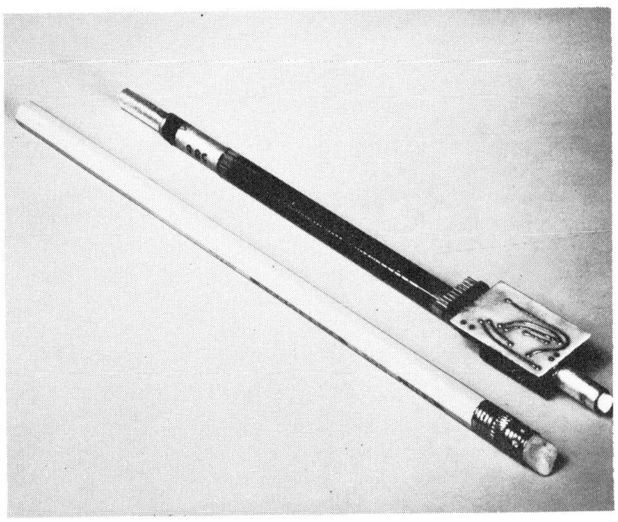

Fig. 2 Emerson RARF Element.

Emerson fabricated and tested a 6-inch diameter, 163 element array which demonstrated their design and array assembly techniques (5). This array was tested as a fully filled vertically polarized array and also as a 50% filled array by replacing half of the elements on a random basis with horizontally

TABLE II

QUALIFICATION TESTS

A. Test Group I (24 Units)

(1) Visual and Mechanical Inspection
(2) Electrical Performance Check
Insertion Phase
Insertion Loss
Switching Characteristics
VSWR

B. Test Group II (10 Units)

(1) Vibration
(2) Electrical Performance Check
(3) Shock
(4) Electrical Performance Check
(5) Moisture Resistance (Humidity)
(6) Electrical Performance Check

C. Test Group III (10 Units)

(1) Temperature Cycling
(2) Electrical Performance Check
(3) Altitude
(4) Electrical Performance Check
(5) Moisture Resistance (Humidity)
(6) Electrical Performance Check

D. Test Group IV (6 Units)*

(1) Temperature Tracking

*4 Units previously used only for Test Group I plus 1 Unit each from Test Groups II and III.

polarized elements. Patterns and gain of pencil, ground mapping, and fan beams were measured at three polarizations and three frequencies; VSWR, beam pointing repeatability and axial ratio were also measured. Patterns were recorded that demonstrated electronic roll stabilization of the fan and ground mapping beams. As of January 1970, the 28-inch diameter Emerson RARF is in final check-out on the antenna pattern range in preparation for a 4-5 month testing program. Figure 3, a block diagram of the Emerson RARF, also applies for the Raytheon design.

Extensive pattern range testing of the Raytheon RARF shown in Figure 4, was completed in early November 1969. Approximately 1000 patterns were recorded using three test frequencies (±5%), three polarizations, and three beam shapes. Over 250 beam pointing tests were performed using optical standards; in addition, 200 gain and axial ratio measurements were recorded. About one-third of all tests or measurements were performed

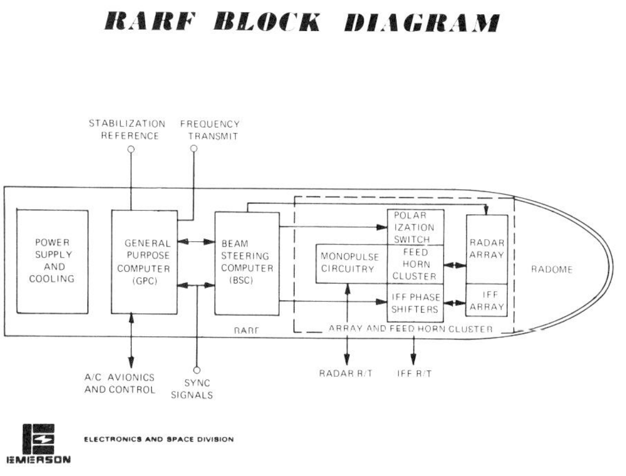

Fig. 3 RARF Block Diagram.

with a radome mounted in front of the aperture. This radome was a standard F-111A production model except that the pitot boom mounting base was removed and replaced with a spherical fiberglass cap. Pencil beam sum pattern peak sidelobes were at least 19 dB below the beam peak with far-out RMS sidelobes approximately 37 dB below the beam peak. The RARF measured pointing errors ranged between 1.0 and 1.5 milliradians. Broadside pencil beam gain was 34 dB. Figure 5 shows panoramic patterns at 10 degree increments within the azimuth scan limits of ±70 degrees and the elevation limits of ±60 degrees. Scanning effects on relative gain and near-in sidelobes can be evaluated from these radome-off patterns. The only significant effect of the radome noted during preparation of the six-volume Raytheon RARF final report (6) was a loss in gain that ranged from 0.5 to 2.0 dB.

Fig. 4 Raytheon RARF Phased Array.

TABLE III

CHARACTERISTICS OF THE RAYTHEON RARF ARRAY ELEMENT

Phase Shift Range	360 Degrees
Insertion Loss	Less than 1.0 db
Frequency	Ku-band, 10% Bandwidth
Phase Error (Reproducibility)	Less than 14 degrees RMS
Phase/Duty Cycle Nonlinearity	30 Degrees Max.
Switching Speed	130 Microseconds, Max.
Element Size (Including Driver)	4.07 inches long/0.43 inch OD
Element Weight (Including Driver)	0.044 Pound
VSWR	Less than 1.5 at any duty cycle
Polarization	LHCP, RHCP, Linear Vertical and Linear Horizontal
Type	Analog Ferrite for a Reflect-Array; Reciprocal Action
Power, Peak	250 Watts
Power, Average	1.3 Watts

TABLE IV

CHARACTERISTICS OF THE EMERSON RARF ARRAY ELEMENT

Phase Range	360 Degrees
Insertion Loss	1.5 db, Max.
Frequency	Ku-Band, 10% Bandwidth
Phase Error	30 Degrees, Max.
Switching Speed	125 Microseconds, Max.
VSWR	Less than 1.5 at Any Duty Cycle
Element Weight (Including Driver)	0.055 Pound
Element Size (Without Driver)	7.4 inches long/0.30 inch OD, Max.
Element Size (Including Driver)	7.4 inches long/0.75 inch OD, Max.
Polarization	Linear, Vertical or Horizontal
Type	Digital Latching, Reciprocal
Power, Peak	1000 Watts
Power, Average	7.0 Watts

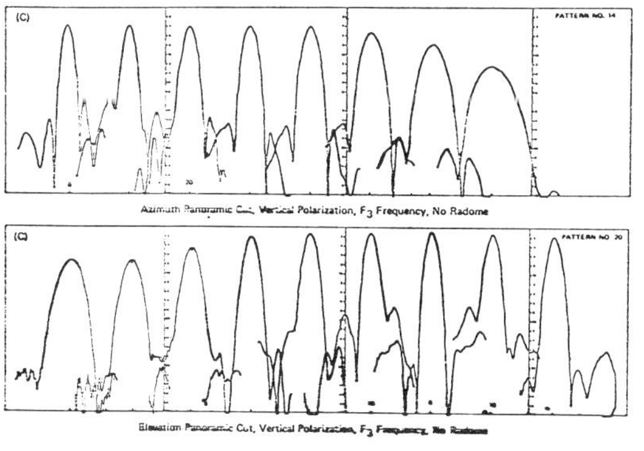

Fig. 5 Raytheon RARF Panoramic Patterns.

Flight testing of the Raytheon RARF as part of a multimode radar should begin in the fall of CY1970; the radar will be mounted in an Air Force C-135 aircraft. The Emerson RARF will be flight-tested as part of a 698DF funded forward-looking advanced multimode radar (FLAMR) beginning in the first quarter of CY1971 using a modified B-47 aircraft.

In order to establish the extent to which the characteristics of an electronically scanned radar array are affected by a radome designed for a high performance aircraft, Autonetics was funded by AFAL to measure the effects of two radomes on their Company developed, 1600 element, VOLPHASE (Volumetric Phase Scanning) X-band array. Data recorded (7) at 8.4, 8.8, and 9.2 GHz at both linear horizontal and circular polarization included the following: antenna patterns, both sum and difference, in the principal planes; radiation distribution (contour) plots; beam pointing accuracy; repeatability of beam shape and beam pointing; ellipticity; depth of null of difference patterns; and VSWR. The two radomes utilized were a solid wall fiberglass F-104 radome from Air Force stock and a fiberglass reinforced ceramic honeycomb sandwich experimental radome. These basic data were recorded typically on antenna patterns; the data were then reduced into various forms of graphs and figures to show the variation of the parameter (such as peak sidelobe level) typically versus scan angle. For comparison a limited number of measurements were taken utilizing mechanically scanned conventional parabolic reflector. Autonetics has been given a follow-on contract to demonstrate minimization of radome effects on various radar antennas when the radome is "tuned" using techniques developed by General Dynamics under an AFAL radome program (8); particular emphasis will be given to minimizing the radome effects on antenna sidelobes. In addition to the VOLPHASE array (peak sidelobes of -15 dB) and parabolic reflector previously used, measurements will include a low sidelobe (peak sidelobes of approximately -30 dB) constrained lens radar array being developed by Autonetics under Air Force Contract F33615-70-C-1596. These tests are scheduled for completion in July 1971.

Under a Navy program, Hughes Aircraft Company has built a one dimensional (elevation) electronically scanned X-band array which will be mounted on an available gimbal and be scanned mechanically in azimuth. Array is built in four quadrants so two axis monopulse beams and a sum beam are available; beam will also be spoiled using phase control for ground-mapping use. Each quadrant is about one foot square and uses vertical slots in the narrow wall of waveguide; 4-bit digital, non-reciprocal ferrite phase shifter is used for each waveguide row. The SCR phase shifter drivers will not be mounted on the gimbal. About 100 phase shifters were built, and 72 are used in the array. This array is now being flight tested.

Hughes also fabricated and pattern-range-tested an X-band reflect-array 36 inches in diameter that utilizes 2-bit (3 diodes) diode phase shifter (9). Array contains 2400 elements and was designed for air-to-air and ground-mapping modes. This array was combined with components of an AN/APQ-72 Radar for roofhouse tracking and ground mapping tests.

STATUS OF COMMUNICATION ARRAY PROGRAMS

Present microwave satellite communication systems for use between aircraft and for links between aircraft and ground terminals have created technology limitations in the airborne terminal subsystem; these limitations are translated into aerodynamic problems which degrade aircraft performance and/or reduce mission capabilities. The hemispherical coverage desired by the aircraft antenna in consideration of freedom of aircraft attitude and aircraft to satellite geometry creates most of these aerodynamic problems. Present SHF airborne terminal antennas consist of a protruding, mechanically scanned reflector or some other form of solid aperture mounted beneath a radome. As a potential solution to this external protrusion and to internal protrusion and weight of more conventional phased arrays, research on a distributed solid-state phased array communication terminal is being performed by Raytheon (10). Figure 6 is a block diagram of a proposed communication terminal. A single aperture laboratory breadboard antenna subsystem designed for mounting on a C-135 aircraft was planned. The development and demonstration of components and subassemblies is now in progress (11). Part of this effort is to fabricate and evaluate four array modules, which contain separate transmit and receive ferrite latching phase shifters, necessary microstrip filters, receiver mixer, IF amplifiers, and a multi-stage, locked oscillator in the transmit section. Goal of this transmit section is to provide 1-watt CW over a 50 MHz operating bandwidth and approximately 8 GHz. Figure 7 shows two views of the first module; a heat pipe is used to cool the transmit section. A second module using a different high power RF source is now being evaluated. The present required array module length, overall array depth, and cooling problems encountered to date appear to prohibit fabrication of a practical, complete breadboard laboratory model conformal distributed array airborne terminal at this time. Present plans for the last ten months of research are to compare a presently attainable distributed array terminal design for a C-135 aircraft with a more conventional (passive) array design which can be built using available technology.

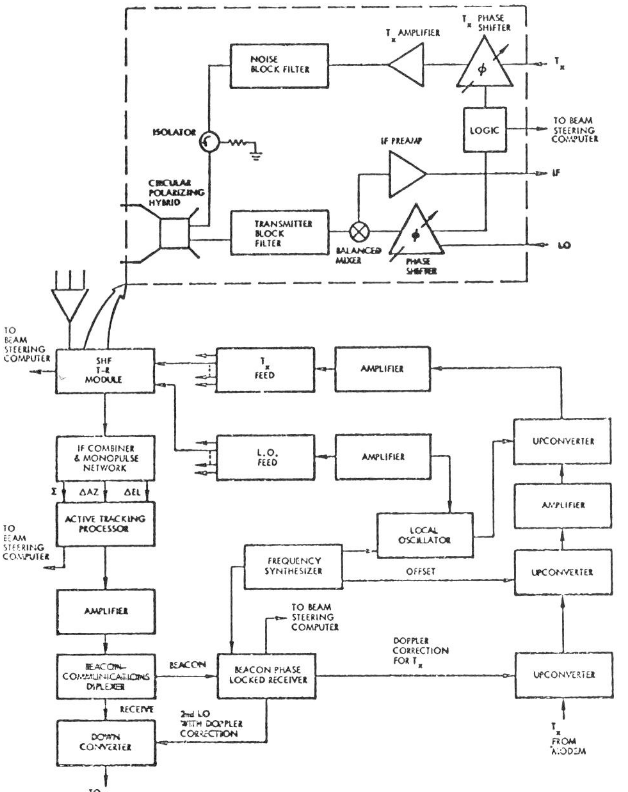

Fig. 6 Proposed SHF Airborne Distributed Array Communication Terminal Block Diagram.

Since a significant portion of avionics equipment serves communication, navigation, and identification (CNI), my discussion would not be complete without a few comments on the still partially undefined airborne phased array antenna problems introduced by this pro-

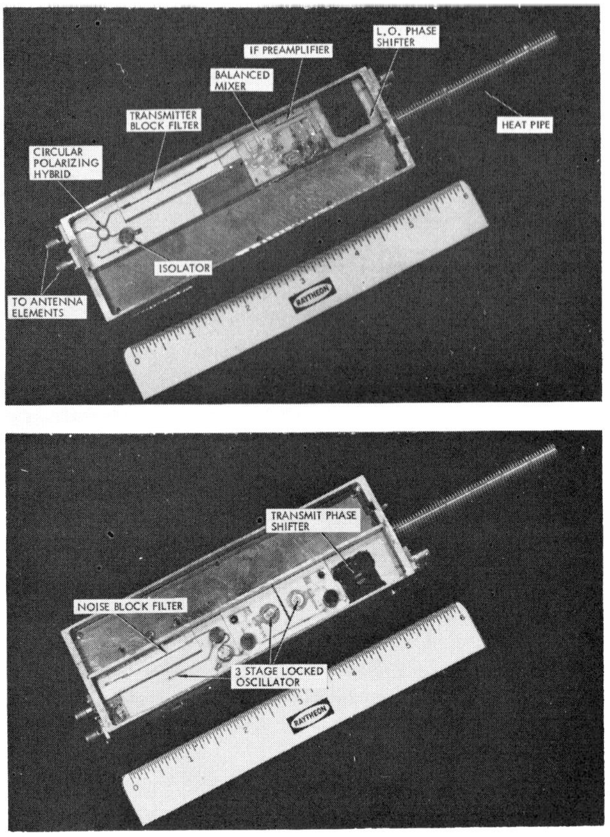

Fig. 7 SHF Distributed Array Module.

posed information transfer system (12). A CNI antenna subsystem must have all the capabilities of the airborne SHF Satellite Communications Terminal previously described and also the following additional capabilities; (1) greater bandwidth because of the navigation and identification functions, (2) greater coverage because of the direct air-to-air identification mode and because of the direct air-to-ground communication mode, and (3) probably multiple receive beams because of the navigation mode. Programs directed toward solving at least some of the receiving function CNI system problems will be initiated with FY71 AFAL exploratory development funds.

The feasibility of an L-band receive only antenna system capable of moderate (10 dB to 15 dB) gain and hemispherical coverage above an aircraft has been demonstrated by Texas Instruments Inc. (13). Antenna subsystem requirements dictated automatic spatial search, lock-on, and tracking of a desired signal; elliptical polarization with a maximum ellipticity ratio of 4 dB was specified. The basic elements of the selected subsystem are the apertures, a monopulse receiver, a phase-shift computer, an aperture selection receiver, and a communication receiver. The antenna apertures consist of two 16-element linear phased arrays and supplementary helix antennas for forward and rear coverage; array designs were developed for complete external or internal mounting of the linear arrays, or a combination of the two. The phased array elements (planar spirals) are followed by modules containing filters, amplifiers, and phase shifters. These are connected to an RF distribution manifold, which has a built-in 25 dB Taylor amplitude taper, and a 180 degree hybrid, giving a sum and difference output signal. The sum and difference signals at the beacon frequency are processed in the array monopulse receiver, which gives an output error signal, and a search or lock signal to the phase-shift computer. The computer transforms the error signal into an updated pointing direction. The aperture selection receiver compares the outputs of all apertures, selects the strongest signal and routes the communication signal from that aperture to the communication receiver. In addition, the aperture selection receiver chooses the phased array with the largest signal, and uses its output to determine the pointing direction for both arrays. This permits use of one computer and monopulse receiver to drive both arrays. Design study included the determination of mounting locations and configurations for C-135 and F-4 aircraft. Choice of location on the F-4 was made primarily from structural installation considerations, with secondary considerations given to effects of the aircraft. The location choice on the C-135 was made from both structural and coverage considerations. Since location of the arrays on the F-4 (one on each side) is much more subject to wing reflections, evaluation was made using a scale model of the aircraft and array. Figure 8 shows the preferred location of the linear array on the F-4; Figure 9 is a sketch showing the external mounting of one L-band linear array. A laboratory breadboard 8-element linear array of square cavity-backed spiral elements mounted on a cylindrical ground plane, followed by eight electronic phase shifters, an RF manifold, hybrid, beacon monopulse receiver and phase shift computer were fabricated and tested by Texas Instruments Inc.

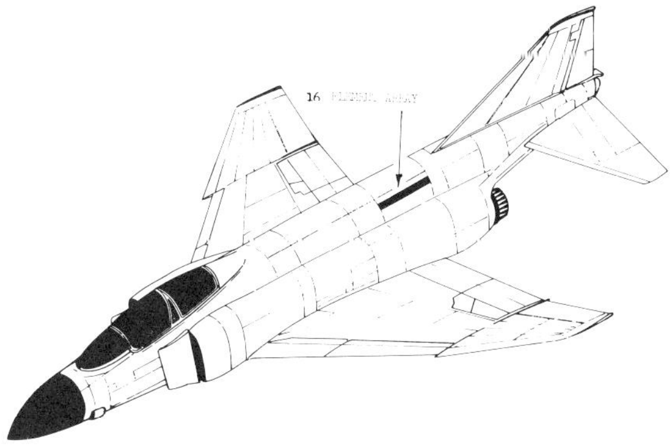

Fig. 8 Proposed Location of L-Band Array on an F-4.

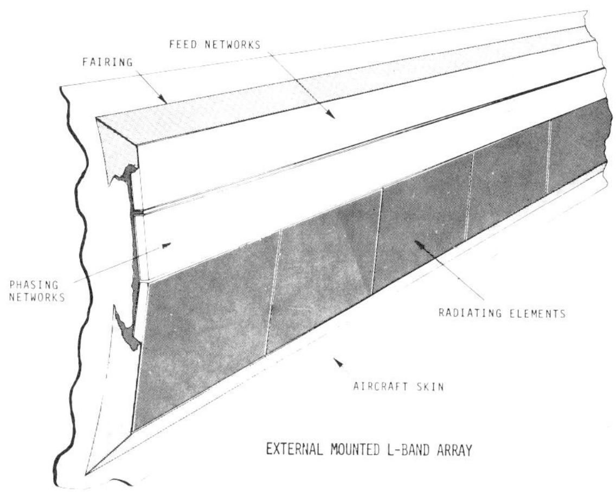

Fig. 9 Sketch of Externally Mounted L-Band Array.

CONCLUSIONS AND RECOMMENDATIONS

Four years of intensive research, design and development by Raytheon and Emerson has produced two engineering model optically-fed Ku-band phased array systems designed for rooftop and subsequent flight evaluation. These arrays will be integrated into brassboard advanced multimode radars that utilize digital computers for mode selection and control and also for processing the return signals. This program included the first consideration of the radome as an integral microwave component of a forward-looking radar antenna subsystem during the complete development cycle. Establishment of the array element (radiator/phase shifter/driver assembly) as the critical component which had to be environmentally qualified to MIL-E-5400 before array fabrication was initiated is considered a necessary and significant decision. Navy-sponsored X-band airborne radar array developments have also provided significant technological progress by demonstrating their compatibility with existing transmitter-receiver components.

A distributed array satellite communication terminal feasibility demonstration program has been initiated, but final results will not be available for evaluation until the middle of CY1971.

Most Air Force planners who discuss Conformal Phased Array Radars or Conformal Phased Array Airborne Satellite Terminals envision an active transmitter at each phased array element. A few of the more obvious problems that must be solved using exploratory and advanced development funds include the following:

a) Improved efficiency in the high power solid state devices.

b) Cooling techniques for the solid state devices, array modules, and the complete array.

c) Determine the phase coherency quality of transmitted signals from a distributed array.

d) Develop array elements with selectable multiple patterns and/or polarization that could help improve effectiveness of a conformal aperture.

e) Study beam steering commands required to steer and shape a beam from an irregular conformal array.

f) Study effects of a typical aircraft skin flexing on the conformal array radiation characteristics and the required beam steering commands.

Because of the potential need for large quantities of phased array elements (radiator/phase shifter/driver assemblies) for Avionics Systems during the next ten years, financial support to develop manufacturing methods for this class of phased array element is needed. However, before a program for such semi-automatic production techniques is finalized, a series of basic questions remain to be answered:

a) Which type of Avionics applications will require large numbers of array elements during the next ten years? Can more than one application be served by a single type of phase shifter?

b) What type of phase shifter characteristics are most needed? Reciprocal vs. Non-reciprocal; Latching vs. Analog; Ferrite vs. Diode; etc.

c) What is the required operating frequency range (or ranges) and bandwidth?

d) Could interchangeable heads and measurement fixtures be designed to make a facility suitable for producing elements for two or more frequency bands?

e) What are acceptable cost ranges for array elements based on various size production runs?

REFERENCES

1. "MERA Modules - How Good in an Array?" by G. C. Bandy, L. J. Handeman and W. F. Hayes; Microwaves, dated August 1969.

2. "Molecular Electronics for Radar Applications," Eighth Interim Report of Contract AF33(615)-2525, dated January 1968 (AD828740) "MERA Final Antenna Study Report," Texas Instruments Inc. Report No. 03-66-123 dated December 1966

3. "RARF Preliminary Design Review Report," Raytheon Report BR-3655, February 1966 (AD-376609); Contract AF33(615)-3291.

4. "Study and Design of an Integrated Radome, Antenna and RF Circuitry (RARF), Vol I Technical," Emerson Report No. 2041, February 1966 (AD-376614L); Contract AF33(615)-3289.

5. "6-Inch RARF Array (Mini-RARF) Test Report," Emerson Report No. 2270 dated July 1969; Contract F33615-67-C-1017.

6. "RARF Final Technical Report, Vol. V System Test Results," AFAL-TR-69-323 dated December 1969; Contract F33615-67-C-1108.

7. "Radome Effects on a Radar Phased Array Antenna," AFAL-TR-69-258 dated October 1969 (AD861829); Contract F33615-68-C-1627.

8. "Circular Polarization Radome Techniques," AFAL-TR-69-39 dated February 1969 (AD-850097) and AFAL-TR-69-164 dated August 1969; Contract F33615-67-C-1492.

9. "Electronically Scanning Airborne Intercept Radar Antenna (ESAIRA)," Hughes Report No. P69-409 dated September 1969, Contract N00019-67-C-0653.

10. "Aircraft to Satellite SHF Communications" by Harold E. Weber; Microwave Journal dated October 1969.

11. "SHF Airborne Distributed Phased Array Antenna," Semiannual Report dated January 1969 (AD395495L), Contract F33615-68-C-1592; Semiannual Report dated July 1969 with same title (AD503456L).

12. "Improvements in CNI - Will They Help Create a New Air System?" by Irving J. Gableman; Astronautics and Aeronautics dated February 1969.

13. "Airborne L-Band Receive Array", AFAL-TR-70-136 dated June 1970; Contract F33615-69-C-1706.

Phased Array Antennas for Applications on Spacecraft

W. H. KUMMER
Aerospace Group
Hughes Aircraft Company
Culver City, Calif.

A. KAMPINSKY
NASA/Goddard Space Flight Center
Greenbelt, Maryland

ABSTRACT

This is a survey paper wherein phased array antenna systems, now operating on spacecraft, are reviewed so as to reveal their basic techniques of beam formation, steering and receive/transmit (transponder) functions. Laboratory developmental antennas, designed for ultimate spacecraft flight qualification, are assessed in terms of the needs of communications, earth resources instrumentation, and area surveillance.

I. INTRODUCTION

The antenna designer has a large choice in the type of antenna that will satisfy the required radiation characteristics from a spacecraft. Many parameters constrain the design of an overall satellite system; thus any antenna system chosen will incorporate many compromises. Prime parameters include type of spacecraft stabilization, type of orbit, antenna coverage, volume available, weight, power consumption, reliability, gain, and number of independent links. Factors such as sidelobe levels, beam agility, wide scan angle in two planes, and tracking accuracy play secondary roles in most applications for communications, but may be emphasized for applications to surveillance and instrumentation.

Improvement in system performance could be achieved if high-gain antennas were used. High gain, however, implies a beam from the satellite antenna which may be narrower than the total required coverage region. Therefore, increasing the antenna gain introduces a new requirement, that of steering the narrow beam to a desired portion of the coverage region. The degree of pointing accuracy required increases as the gain is increased. Two limitations are encountered in steering the beam. First, there is a limitation in the capability of controlling the attitude of a satellite, and second, there is some uncertainty in measuring its attitude at any given time. The first limitation means that very narrow beams would have to be steered continuously to maintain the desired pointing direction as the satellite attitude changes. The second limitation introduces some uncertainty as to exactly where to point the beam with respect to the satellite. In addition, a question also exists of the extent to which the satellite attitude should be controlled in any particular mission. From these remarks, it is evident that antenna gain, beam steering requirements, and satellite orientation control requirements are closely interrelated. Thus, the implications of increased gain are so important that a consideration of antenna techniques must be a part of any overall system design for a mission. The characteristics of these techniques, both electrical and mechanical form essential parts of tradeoffs in system weight, accuracy, and reliability. For a particular mission, practical compromises must be made. These will depend on the techniques available for directing or steering the antenna beam as compared with those for controlling the vehicle attitude.

Mechanically steered antennas have had a long history of application to space vehicles thus this paper will be restricted to a review of electronically steered arrays. Several of the arrays to be discussed are complete transponder systems and their merit has to be judged as complete satellite transponders not as antennas alone when weighed against discrete systems in terms of their total power, weight, volume, and reliability. In the next several sections applications of phased arrays to communications, despun beams, earth observation and instrumentation will be discussed.

II. ARRAYS FOR MULTI-CHANNEL COMMUNICATIONS

Particularly for communications, the need is for

high gain antennas that allow simultaneous access to multiple small users that are spatially dispersed. Applications include data relay satellites, navigation and traffic control satellites and broadcast satellites. Such accommodations are incompatible with spacecraft attitude steering and preclude mechancially actuated antennas. Laboratory demonstration antennas, fulfilling such requirements, now operate on two beam formation principles - the all electronic, self steering, pilot-beam controlled array, and the discrete beam (matrix or lens and feed) array. Both of these systems are applicable to communications and earth resources survey instrumentation which can be carried on NASA ATS F&G, Data Relay Satellite, and Earth Resources Spacecraft, as examples, or on any three-axis-stabilized spacecraft.

It was pointed out in the Introduction that the formation of high-gain beams requires the simultaneous consideration of steering them. There are two generic types: those that require external controls to phase the elements properly and those that are self-steering. The externally controlled systems, such as the conventional phased arrays need external commands to program the scan controller and a phasing network to form and point the beam in the appropriate direction toward the user. The attitude of the satellite array with respect to the location of the user is a parameter needed by the command system. In the self-steering or adaptive system, electronic circuitry senses the phase of incoming signals to phase the array elements.

IIA. SELF–STEERING SYSTEMS

Self-phasing arrays use phase inversion by mixing in order to obtain the required aperture phase progression. The systems used to accomplish the phase inversion may have a variety of forms, each with particular advantages in a given situation. One of the simplest circuits (1), (2) is shown in Figure 1. The

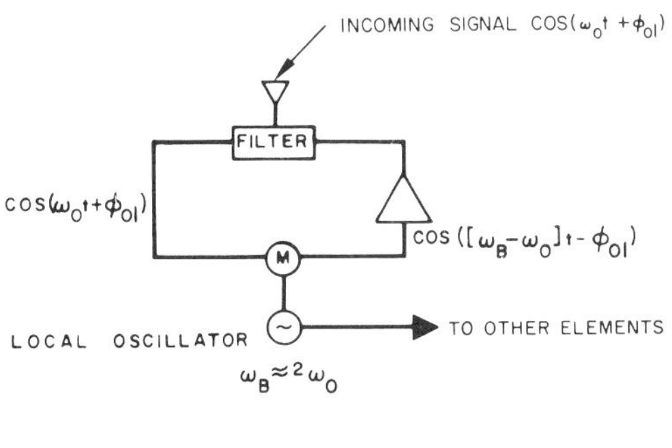

Fig. 1 Elementary Array Module.

basic operating principle may be seen from the figure: the phase of the incoming signal is inverted at each element by the mixer (M) to create the condition necessary for retransmission of a signal back in the direction from which the signal came. Information may be modulated on the local oscillator signal to be sent back in the desired direction. In actual practice it is simpler to obtain large amplifier gains and to perform some of the filtering functions at intermediate frequencies rather than at the radio frequencies.

X-Band Repeater

An engineering model of a self-steering repeater for satellite-to-earth communications has been designed, fabricated, and tested. (3) The system is designed as a relay station between any two pairs of ground stations that lie within a 30-degree cone of coverage. The satellite vehicle is assumed to be in a synchronous orbit and to be gravity gradient stabilized with a ± 5 degree uncertainty in attitude.

The system has two independent FM/FM channels, each receiving in the 8-GHz band and transmitting at 7.3 GHz. Each channel has an RF bandwidth of 125 MHz. The repeater will steer four independent high-gain beams (two for receiving and two for transmitting) along arbitrary directions within the $\pm$15-degree coverage angle. The positions of the beams are controlled by the phase information obtained from CW pilot signals generated by the communicating ground stations. The function of the pilot signals is to supply the phase information at each receiving antenna element. In systems where a carrier is always present it can be used as the pilot signal. The power and bandwidth in the pilot signals is made small compared to that in the information band so as to conserve overall transmitter power from the station generating the pilot. In all the discussions CW pilots will be used. In an actual system these pilots will be appropriately coded so as to reduce RFI and the redirection of the beams in unwanted directions. Beam designations and the frequency bands utilized are illustrated in Figure 2.

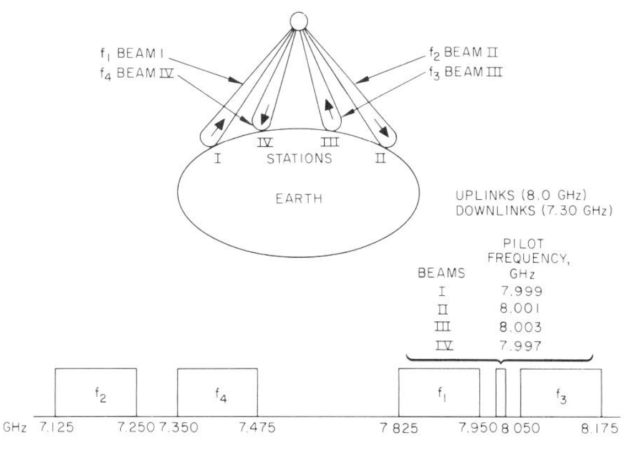

Fig. 2 Beam Designations and Frequency Bands Utilized by the Self-steering Repeater.

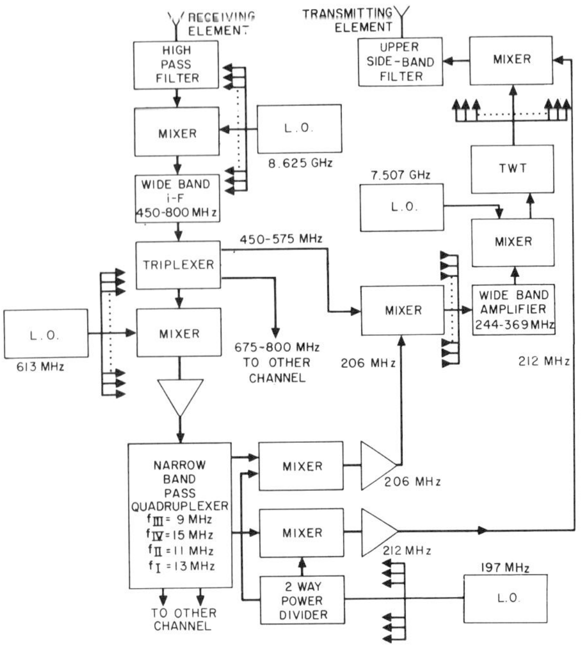

Fig. 3 High-gain Self-steering Engineering Model Schematic (The subscripts of the pilot frequencies correspond to the beams shown in Figure 2.).

For the system shown in Figure 3, receiving pilots, transmitting pilots, and modulated signals are received by the receiving element, passed through a high-pass filter, down-converted to an intermediate frequency, and amplified by a wideband IF preamplifier. After preamplification, the two information bands and the pilots are separated by means of a triplexer filter. The pilots then are down-converted to a second intermediate frequency to allow utilization of very-narrow-band bandpass filters to establish a good signal-to-noise ratio for the pilots. These bandpass filters comprise the quadruplexer which, in addition to limiting the noise bandwidth of the pilot channels, serves to separate the pilot signals. After the pilot signals pass through the quadruplexer, they are up-converted to about 200 MHz to enable them to be mixed with the wideband modulated signals without overlap of the power spectra.

With reference to Figure 3, the information band, 450 to 575 MHz, passes from the triplexer to a wideband mixer.* The receiving pilot signal, 206 MHz also passes to this mixer. If these two signals are mixed and the lower sideband retained, the phase of the resultant IF signal is independent of the relative phase angle of the signals at the elements. The signals from the output of these mixers (one for each element) are in phase and are summed. At the point of summation, the receiving array gain is realized for the information signals.

*The other information band, 675-800 MHz is processed similarly.

The signal is then amplified at the intermediate frequency, upconverted to RF and amplified, and then distributed to the final transmitting mixers. At these mixers the transmitting pilot is mixed with the modulated signal and the upper sideband is selected by the bandpass filter that follows. A modulated signal is produced at a transmitting element; this signal has a phase angle that has the opposite sense from the phase angle of the transmitting pilot at the corresponding receiving element. That is the condition necessary to transmit the information from the antenna system in the direction of the transmitting pilot.[+] In this particular design two arrays were used; one for transmitting and one for receiving (Figure 4); they were scaled in wavelength to avoid beam

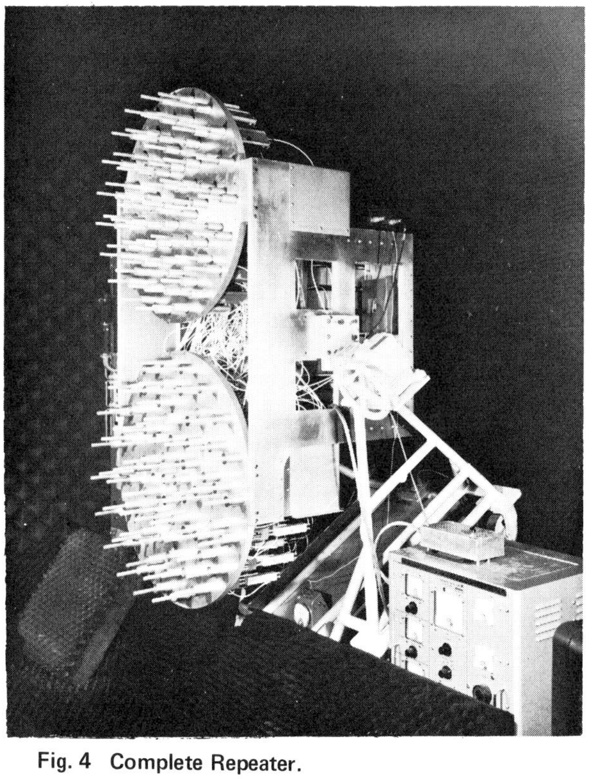

Fig. 4 Complete Repeater.

squint. For systems operating at lower frequencies with requirements for similar gain only one array could be used due to limitations on size and weight and problems of stowage in the launch vehicle. The beam squint needs to be removed for widely spaced transmitting and receiving frequencies (4). Several schemes using electronic circuitry, array scaling and combinations of both have been proposed for obtaining the proper phase scaling (3) (5) (6).

The system test results were generally satisfac-

[+]The other information band is also fed to the same transmitting element through a diplexer which is not shown in Fig. (3).

tory and are indicative of the performance of a high quality communication channel (7). Table I summarizes the characteristics of the engineering model.

Table I. Characteristics of the Engineering Model

Parameter	Design Goals for Engineering Model	Measured Performance
Number of elements, each for two arrays	64	
Number of channels	2	
RF bandwidth, each channel	125 MHz	125 MHz
Guard band, between channels	100 MHz	100 MHz
Total cone angle of coverage	30 degrees	30 degrees
Element gain, minimum	11.6 db	12.4 db
Array gain, minimum	29.8 db	30.4 db
Polarization	Circular	Circular axial ratio = 0.8 db
Average mixer-filter preamplifier noise figure	15.2 db (max)	14.4 db
Effective radiated power per channel	28.0 dbw	27.8 dbw
Ratio of pilot to modulated signal power when 125-MHz bandwidth is utilized	-10 db	*
Attitude readout accuracy (min)	±0.5 degree	±0.5 degree
Power consumption: receiver	32.0 watts excluding LO	–
Power consumption: pilot processor	108.7 watts	182.6 watts
Power consumption: attitude readout	0.9 watt	–
Power consumption: total transmitter	201.1 watts excluding LO	87.9 watts** excluding LO
Prime power (exclusive of power supplies)	324.7 watts	270.5 watts
Total weight		180.0 kg

*Pilot signal ratios much lower than the design values were measured in the TV tests (see [3])

**Power consumption applies to one channel of transmitter. For two-channel operation less than twice this value would be needed.

S-Band Repeater

At the present time a small portion of a two-beam, two-channel S-band phased array based on the principles of the pilot-controlled multibeam array (3) is being developed using all solid state "micromin" circuitry to obtain an estimate of prime power consumption, weight and size in terms of electrical performance to be extrapolated for a full array with a gain of 40 dB (6). Each receiving beam and the resultant transmitting beam point toward the user generating the particular pilot. Data are relayed through the self-steering array toward one earth station. Maximum r-f bandwidth is 10 MHz.

This system includes circuits for acquisition and tracking of the frequencies from the moving user satellites. The rate of change of the doppler is preserved so that coherent range rate measurements can be obtained between the user and the earth station. The program is still in progress, the electrical performance is being evaluated at the present time (6). The characteristics of the channel modules are listed in Tables II and III.

Table II. I-F Module
(After Schwartz, et al.)

Frequency	100 MHz
Number of beams	2
Data bandwidth	10 MHz
Pilot bandwidth	20 Hz
Weight	250 grams
DC power in	533 milliwatts

Table III. S-Band Module
(After Schwartz, et al.)

RF Module	Receive	Transmit
Frequency	2200-2300 MHz	1760-1840 MHz
Number of beams	unlimited	2
Noise figure	8.0 db	
RF power		60 milliwatts/beam
DC power in	2 watts	
Weight	210 grams	

IIB. MULTIBEAM SYSTEMS

Self-Steering Multibeam Repeater

A special configuration of a discretely-controlled multiple-beam array utilized Bulter matrices (8) to form a high-gain antenna that received incident signals from arbitrary directions and processed them with amplification and frequency translation before re-radiating the signals toward selected directions (4). The system had two 4 x 4 arrays of helical elements, two beam-forming matrices (9), and two separate r-f sections, one for transmitting at 4 GHz and one for receiving at 6 GHz. The cone of coverage for this system was 50 degrees. The principal feature of a 4 x 4 beam-forming matrix is availability of 16 distinct beams or beam-pointing directions, any one of which can be selected by feeding energy into the appropriate beam port. The beams overlap to form in space the 4 x 4 cluster that is symmetric about broadside. The operation of the multiple-beam antenna system is described with reference to the schematic diagram of Figure 5.

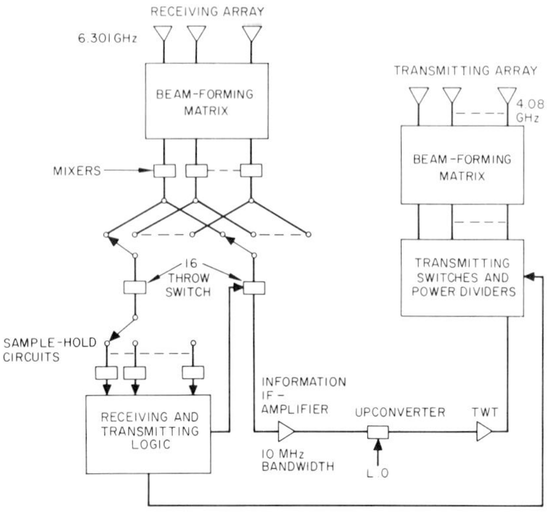

Fig. 5 Multiple-beam Antenna System Schematic.

The 6-GHz receiving system processed both the receiving and transmitting logic. Proper operation of the transmitting system required that the receiving and transmitting antenna beams be identical; the arrays and elements were scaled appropriately. The pilot signals were received by the 6-GHz array and its beam-forming matrix. All the signals from each beam port of the matrix were down-converted to i-f. The pilot signals were passed to a 16 throw sampling switch whose output was directed to the input of one i-f amplifier. Each pilot i-f signal, after proper filtering at the output of the i-f amplifier, was video detected and stored in sample-hold circuits, one for each beam port. These sample-hold circuits were synchronized with the 16 throw switch so that a particular sample-hold circuit then held the voltage of that particular channel until it was sampled again. The output of the sample-hold circuits are proportional to the power received by each antenna beam.

The function of the receiving logic was to determine the antenna beam with the largest receiving pilot signal and to switch the information channel into this beam by the application of a voltage to another 16 throw i-f switch connected to the 10 MHz information channel. The amplified signal was then up-converted; the upper sideband was then passed to the input of the traveling-wave tube (TWT) amplifier. The TWT amplifier raised the signal level up to that required for transmitting from the antenna system.

Because the constant-amplitude plane for each beam forms a circle, the gain will not be constant within the region of coverage. At points at which two adjacent overlapping beams are of equal gain, the gain is down ~4 dB from the peak. At points at which four adjacent overlapping beams are of equal gain, the antenna gain is down about 8 dB from the peak gain. The use of distinct beams also gives rise to noticeable switching transients as the antenna system switches from one beam to another. To eliminate the effect of amplitude variations between adjacent beams and switching transients, the transmitting beam was continuously scanned. The distinct but overlapping beams were utilized to form one continuously moving beam. This improvement was accomplished by using four adjacent overlapping beams with appropriate weightings to form a new beam that continuously scanned to track the transmitting pilot. Although this scanning beam will have minor amplitude variations with scan, it will have no noticeable holes and also no noticeable switching transients.

The beam direction was controlled by digital ferrite switches and ferrite variable power dividers. With proper design only three power dividers and 12 SPDT digital switches were required for the scanning of one beam synthesized from 16 beams (10).

The self-steering multibeam arrays utilizing Bulter matrices require logic circuitry (10) about as complex as that used for conventional phased arrays. However with "micromin circuitry," a significant reduction in weight and power consumption may be expected. Then the required logic for beam selection will find application in the future.

The concept can be extended to multiple independent channels with the use of additional electronics. Larger multibeam arrays with high-gain beams can be envisioned. Bulter matrices, however, become more lossy as the number of beams increases. R-f preamplification can be used on reception and post amplification on transmission to reduce r-f losses. Alternate multiple beam systems can be considered such as multiple-beam feeds and lenses which will be discussed next.

IIC. VARIABLE–COVERAGE COMMUNICATION ANTENNA

This antenna system is designed to provide a variable coverage on earth from synchronous altitude. Coverage can be varied from a 3-degree pencil beam to full-earth coverage. The satellite for this array will be stationary with three axis stabilization.

The system consists of a waveguide lens, a cluster of nineteen feeds and a switching matrix. A waveguide zoned lens of double-concave design with an F/D ratio of 1.0 was chosen (11). The feed cluster is a hexagonal array of conical horns each having a 2-inch aperture and spaced 2 inches apart. Figure 6

Fig. 6 Experimental Lens Antenna (After Dion).

shows the lens antenna. The system operates at X-band with a total bandwidth of about 15 percent. Both senses of polarization are used, one for transmitting and one for receiving.

The directivity at the design frequency for the 19 3-degree beamwidths is 31.5 dB $\pm$1/2 dB. The cross-over level is 4.5 dB between adjacent beams and 8 dB at the point where three overlapping beams have equal gain. Earth coverage directivity with equal energy fed into all beams was 21.5 dB with a peak to peak ripple of 2 dB. All measurements are referred to the beam ports and do not include losses in the switching matrix or connecting networks. At the edge of the frequency band the gain is reduced by about 1 dB for a scan angle of 7 beamwidths from broadside.

A corporate feed with variable power dividers is being designed (Figure 7). It will allow power to be

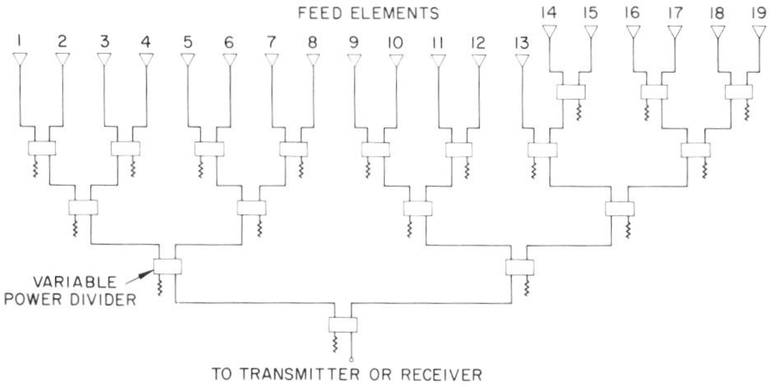

Fig. 7 Proposed Combiner Switch (After Dion).

divided among the 19 feeds in any desired ratio in order to provide for flexible beam shaping and transitions. Nonreciprocal ferrite phase shifters using a single ferrite slug will be employed (Figure 8). A

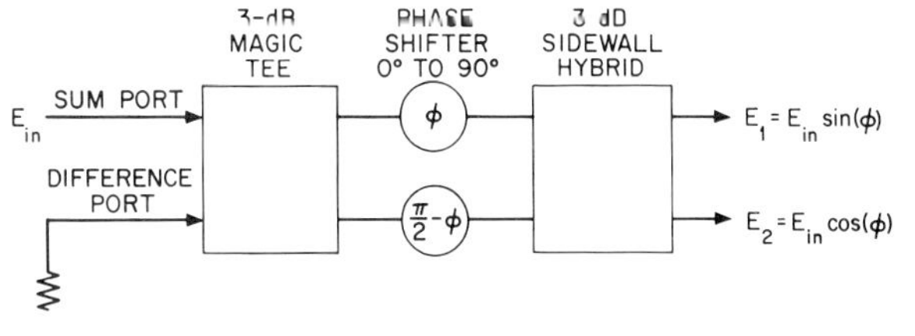

Fig. 8 Power Divider (After Dion).

separate network will be used for transmitting and receiving. The weight of the lens and feed system was 27 kg for the laboratory model. The flight model is expected to weigh 2-1/2 kg. These weights do not include the feed network and power divider.

IID. ELECTRONICALLY CONTROLLED PRIMARY FEED FOR A 30 FOOT PARABOLOID

A multipurpose feed system is being developed for a 30 foot paraboloid which is part of the ATS-F spacecraft. An S-band electronically scanned feed will be used for communications, and monopulse signal tracking. Cylindrical horns, a comparator and a power divider comprise the elements of the system. Single-pole, seven-throw diode switches and circulators comprise the components of the S-band switching system. One set of four switches couple a signal being transmitted from the power divider through four circulators into the four selected feed horns. When a signal is received from a desired direction, the four horns forming a beam closest to that direction are coupled through four circulators to the comparator by a second set of four switches. A logic and switch driver circuit provides the proper switch diode bias states for independent receiving and transmitting beam positions. Beam position commands are received by the logic circuit from a remote command decoder.

One transmitting and one receiving beam are formed and can independently be scanned through a limited coverage area. Beamwidths are about 1 degree.

III. SPIN STABILIZED SPACECRAFT ANTENNAS

Spin stabilized spacecraft antenna requirements, with another major antenna technique category, are exemplified in the flight experience of the NASA Applications Technology Satellites (ATS) and the Air Force Lincoln Laboratory Experimental Satellites

(LES) and the Intermin Defense Communication Satellite Program (IDCSP). These spinning spacecraft carry antennas that are electronically steered 360 degrees in one plane; the beam is counter-spun and remains pointed towards the earth. These antenna systems have performed under operational flight conditions and the design techniques have thus been validated.

IIIA. ATS–1 4 GHz Array

Several despun arrays have been flown on the NASA ATS Satellites. The first one, a cylindrical transmitting array (ATS-1) operates at 4.1 to 4.2 GHz and consists of 16 elements spaced on a circle of one-wavelength radius. (12) Each element of the circular array consists of a collinear array of half-wave dipoles. The power divider that feeds the phase shifters is a stripline assembly and the coupling from the phase shifters to the linear arrays is also a stripline assembly. Figure 9 shows the completed array. The array is fed by

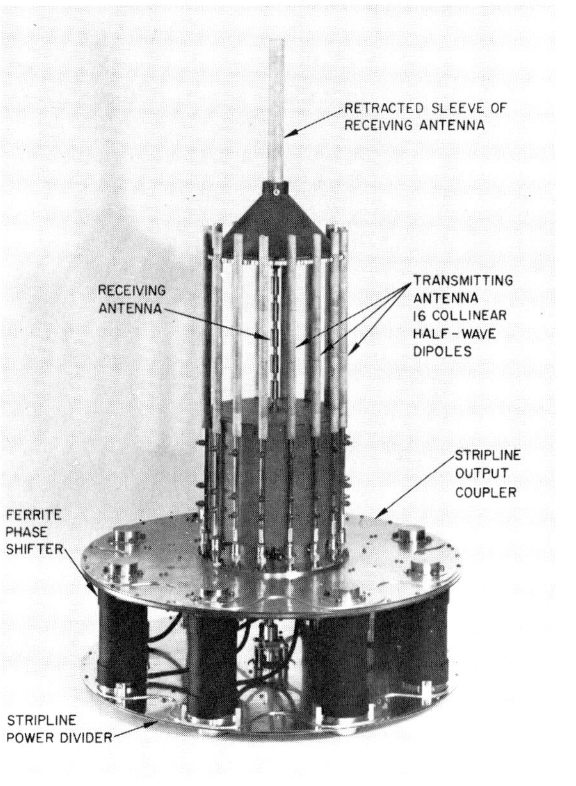

Fig. 9 Applications Technology Satellite Circular Phased Array (After Erhardt, et al.).

a power divider leading to eight ferrite phase shifters. Each phase shifter has two conjugate outputs to feed two diametrically opposite array elements. The phase shifter combines the properties of both the Fox and the synchronous ferrite phase shifter. To scan the array, varying currents are applied to each phase shifter.

The currents are generated to cause a progressive phase shift around the array. This progressive phase shift causes the field to rotate under the control of a clock pulse that keeps the rotation of the r-f field in synchronism with the spacecraft rotation. A sun sensor on the vehicle provides the reference for the control. Ground command can also be used.

The control electronics generates a set of 16 control voltages for the ferrite phase shifters. A system diagram is shown in Figure 10. The system uses

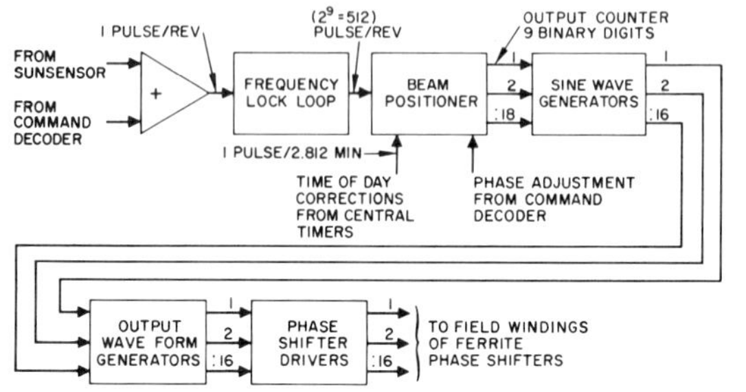

Fig. 10 System Diagram of Control Electronics for Circular Phased Array (After Erhardt, et al.).

a sun sensor as an input reference. A frequency-lock loop generates $2^9 = 512$ pulses for every input pulse. The output of the frequency-lock loop drives the beam positioner, which, in turn drives 16 sine wave generators in the proper phase to position the beam. The sine-wave generators drive output waveform generators and phase-shifter drivers which provide the required waveforms for the coils of the phase shifters.

Table IV

Characteristics of the ATS-I Despun Array and Control Electronis

Weight of complete array as shown in Figure (9) (includes mounting structure to satellite that is not shown)	9.1 Kg
D.C. power into phase shifters (1/4 watt each)	2 watts
Phase shifter drivers (Fig. 10)	
Weight	2 Kg
Power consumption (coil power of phase shifters excluded)	6.8 watts
Number of electronic parts	828
Control Electronics (Fig. 10)	
Weight	2 Kg
Power consumption	4 watts
Number of electronic parts	2064

The minimum array gain as measured on the spacecraft over the band is 13.5 dB (13). This difference, compared with the directivity of 16.4 dB, is due to mismatches, shadowing of elements in the array, stripline losses, element matching network losses, and phase-shifter losses. Combined stripline and phase shifter losses are between 1.5 and 2.0 dB. Loss through a typical phase-shifter is about 1 dB. This figure will vary over the range of operating temperatures by about $\pm 1/4$ to $\pm 1/2$ dB, depending on the operating frequency. The weight, and power consumption of the complete system, are shown in Table IV.

IIIB. ATS–III VHF ARRAY

A VHF transmitting and receiving array was used as a repeater on the ATS-III (14). Eight whips deployed around the body of the spacecraft are used as the array. While the 4 GHz array on ATS-I utilizes a central (TWT) r-f source and power divider, the VHF array is of the distributed type with every radiating element having its own receiver, transmitter, and phase shifter. The phase shifting is accomplished before final amplification. The repeater operates at 149 MHz for receiving and 136 MHz for transmitting. The antenna gain is about 7.5 dB. Each phase shifter consists of an L-C delay line composed of seven cascaded L-C sections of a high-pass filter structure; variable delay is obtained by using varactors as the capacitive elements. The phase shifter is shown in Figure 11. The transmitter power is 7 dBw per element or 16 dBw into the eight element array.

Fig. 11 Phase Shifter Using Helical Printed Circuit Conductors (Courtesy Hughes Aircraft Company).

IIIC. LES–6 DESPUN UHF ARRAYS

Another technique uses a switched multiple beam array. It is being flown on one of the Lincoln Experimental Satellite (LES 6). It operates at UHF, provides circular polarization and consists of 8 slot pairs with concentrically mounted dipoles to obtain circular polarization(15). Upper and lower slots are fed in phase as collinear pairs.

The antenna system for LES-6 consists of two linearly polarized antenna arrays that are orthogonal to each other and fed appropriately to obtain circular polarization. The spacecraft is in the shape of a right circular cylinder one wavelength (λ) in diameter and 1.3λ high. The cylindrical portion of the spacecraft is covered with the usual solar cells. Advantage is taken of the gaps in the solar cells to place the circumferentially polarized slots which are about 0.5λ long. Each slot is backed by a cavity 0.02λ deep which consists of an inner-concentric cylinder divided into 16 cavities, one behind each slot. Dipoles of 0.4λ for LES-6 were spaced about $\lambda/8$ from the cylindrical ground plane. The dipoles were located over the corresponding slots (Figure 12).

Fig. 12 Details of Slot-Dipole Array for LES-6 Spacecraft (Courtesy MIT Lincoln Laboratories).

As the spacecraft rotates, the beam, generated by two pairs of adjacent elements, is scanned in 22-1/2 degree steps, thus creating 16 beam positions. Two beam positions are obtained from each set at adjacent slot-dipole pairs by exciting them in two different phase relationships.

The feed network combines the transmitter, beacon, receiver and radio frequency interference signals as shown in Figure 13. A detailed diagram of the switch matrix is shown in Figure 14. It is composed of two SP4T sections and a DPDT section. P-I-N diodes are used as the switches. The scan controller biases these diodes in the proper fashion so that the beam is scanned sequentially. The phasing and line lengths were optimized for the transmitting frequency. Table V summarizes the performance of the antenna. The gain is measured at the reference input as shown in Figure 13.

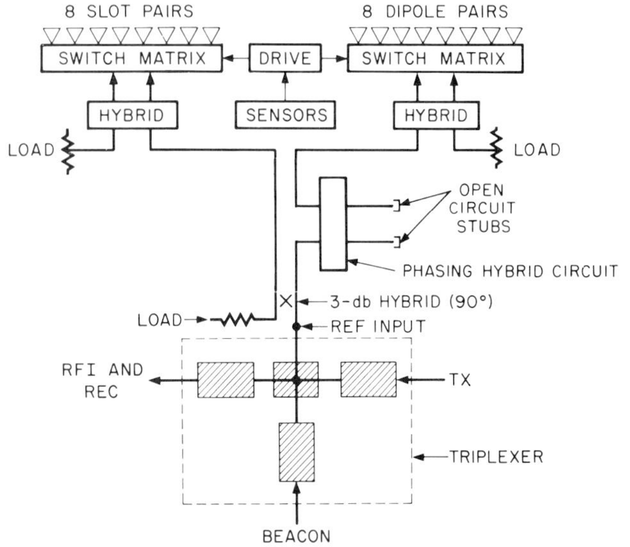

Fig. 13 Despun-antenna System (After Rosenthal, et al.).

Table V

(After Rosenthal, et al.)

Despun-Antenna Gain (db) at Reference Input Shown in Figure 13 Matched Right-Hand Circular Polarization Gain, Average of 16 Beams with Variation Indicated

Frequency	East Edge of Earth	Center of Earth	West Edge of Earth
Transmit	8.6 +0.4 -0.3	9.8 +0.3 -0.4	8.0 +0.3 -0.4
Beacon	+7.1 +0.3 -0.2	+8.4 +0.2 -0.3	+6.9 +0.5 -0.4
Receive	+8.5 +0.8 -1.0	+10.2 +0.3 -0.3	+8.3 +1.0 -1.3

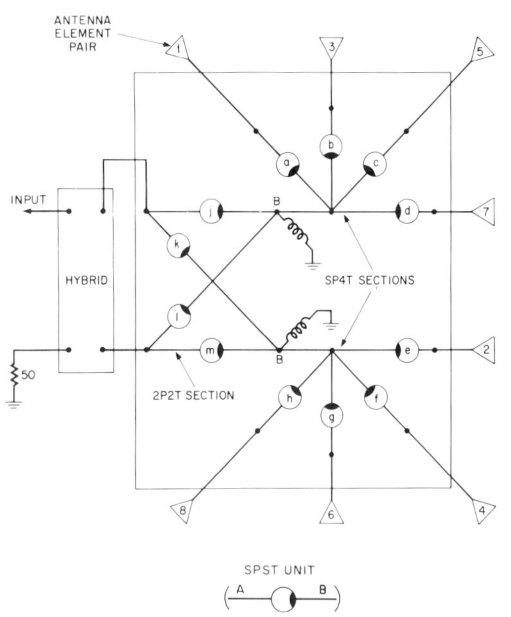

Fig. 14 Switch Matrix (After Rosenthal, et al.).

IV. MILLIMETER WAVE COMMUNICATION ARRAYS

Future satellite requirements indicate that frequencies beyond 12 GHz will be employed in operating systems. Studies have been conducted to survey the state of the art and to come up with possible implementations (16).

The major components that need development include low noise front ends, efficient transmitters and phase shifting devices. In conjunction with the aforementioned study, a 4-bit digital ferrite phase shifter was developed at 35 GHz with an insertion loss of 1.5 dB for 360-degree phase shift (Figure 15): a pulse of less than 20 milliseconds with a drive power of 5 amps at 2 volts is required; the device weighed 59 grams with a size of 3/4 by 3/4 by 2-1/2 inches and a bandwidth of 3 percent. (Westinghouse model WAX-35, April, 1967.) Since then major efforts have been underway to improve the other components required for satellite systems operating at millimeter waves. Extensive work has been performed on multibeam electronically scanned lens systems at 35 GHz.

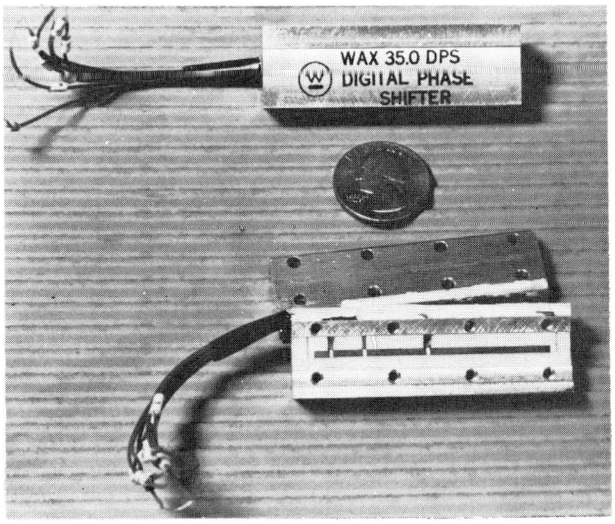

Fig. 15 Westinghouse 4-bit 35 GHz Phase Shifter (Courtesy NASA/Goddard Space Flight Center).

V. EARTH OBSERVATION INSTRUMENTATION

The application of electronic beam formation and steering to earth observation is illustrated by a planar array design that was qualified as a radiometric instrument at 19 GHz on jet aircraft and is now being qualified for spacecraft operation for the Nimbus D (17).

The antenna consists of a group of 49 parallel edge-slotted traveling arrays each with 36 radiating slots. The aperture is 14 by 16 inches. The arrays and the feeds are of the nonresonant traveling-wave type. For scanning in one plane a set of analog ferrite phase shifters of the Spencer-Reggia type are used. Scan of ±50 degrees from broadside has been achieved; the beamwidth was 3 degrees, gains of 32 to 36 dB were achieved. Beam efficiencies of better than 90 percent were achieved;* the antenna losses were 1.0 dB and the aperture efficiency about 79 percent. The phase shifters had an insertion loss of about 0.6 dB maximum. Input d-c power was 62 watts for the total instrument with about 20 watts for the antenna system (3/4 watt per phase shifter), with a 1/2 millisecond time between beam position. Overall system noise figure was 6.0 to 7.0 dB; accuracy was 2.0° K with a 0.7 degree sensitivity for the radiometer. The bandwidth was 300 MHz; a photograph of the array is shown in Figure 16. A second planar array with twice the physical aperture is in development for NASA Nimbus E. Figure 17 shows a sketch of both arrays (18). A third design allows a radiometer receiver and a short-pulse radar to share the same antenna system as an Earth Resources Spacecraft Sensor at 19 GHz.

*Beam efficiency is defined as the ratio of the power in the main lobe to the total received power.

Fig. 16 Photograph of Electronically Scanned Planar Phased Array (After Louapre and Fujioka).

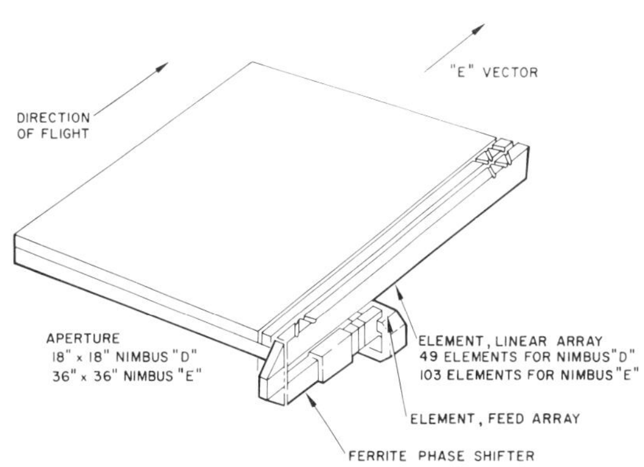

Fig. 17 Electronic-scanning Radiometer 19.35 GHz Antenna System (Courtesy Aerojet General Corp.).

The aperture control technique is extendable for operation over the entire 0.1 to 35-GHz microwave region from a one dimensional scan to two dimensional scanning.

VI. RADAR

A laboratory breadboard array of 16 elements has been fabricated and tested for application to a NASA Manned Spacecraft rendezvous radar. The 6 GHz array was designed to scan ±45 degrees in two dimensions (19).

The novel feature was the all solid state C-band transmit/receive module. It used as a phase shifter a step recovery diode (SRD). The bias of the diode is adjusted to obtain a limited amount of phase shift. Since the output of the diode is rich in harmonics, the output is tuned to a harmonic of the input frequency giving an automatic multiplication of the phase shift. In the receiving portion a multiplication of eight times was used. Linearity between phase shift and applied voltage was good. The output signal (4 milliwatts) from the phase shifter serves as the local oscillator for the mixer. In the transmitter, the multiplication in the SRD phase shifter is two. Further multiplication by a factor of five is achieved by a fixed-bias SRD. Pulsed output power was 300 mw/module with a duty cycle of 0.005. A photograph is shown in Figure 18.

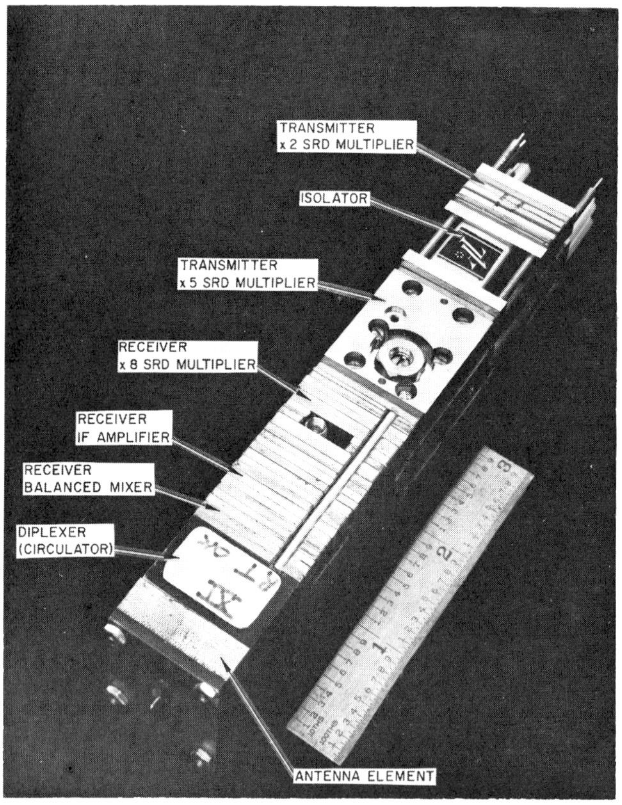

Fig. 18 Assembled C-band Transmitting/Receiving Module (Courtesy Teledyne Ryan Aeronautical).

VII. THE FUTURE OF SPACECRAFT PHASED ARRAYS

It becomes fairly evident that there are many applications for which electronically steered arrays are the best solution for present and future missions. The technology for these arrays is evolving rapidly. The answers to several critical questions must be obtained before their full potential can be realized:

a) Electronically steerable antenna systems must be considered not only on the basis of an antenna aperture but on the basis of a communications systems that embraces a complete receiving system as well as transmitter, modulator, frequency converter and power amplifiers. These systems have already been shown to be competitive with the noise figures, stability, bandwidth, dispersion qualities and reliability of the discrete subsystems, receivers, modulators and transmitters.

b) The continuous upgrading and application of microcircuits to the fabrication of the multiple channels of the electronically steerable antenna designs must produce systems of weight and power consumptions competitive with the discrete subsystems - provided that comparisons are made on a <u>total</u> performance level.

c) The continual frequency increases for operation at millimeter wavelengths is dependent on reduction of mixer noise figures, transmitter efficiency upgrading, frequency stabilities and low insertion losses for all subsystem components. A better understanding of the propagation losses due to rain must also be obtained for operational systems.

d) The demonstration of a total reliability figure for operation in a spacecraft environment is a critical development task.

e) Other aspects in the overall system design must include the consideration of the forward scatter or multipath problems that will be encountered (20). Solutions have been proposed using antenna techniques, electronic processing techniques and modulation techniques such as spread-spectrum. (20) (21) The problem of the capture of the channel by the multipath signal in self-steering systems has been investigated and several solutions have been proposed. However, the quantitative nature of the forward scatter for surfaces such as the ocean, needs further investigation before the techniques can be applied with certainty. (22)

VIII. CONCLUSIONS

What is evident from this review is that the basic principles and techniques of electronic steering have been demonstrated to conclusive levels, in both the laboratory developmental models and in spacecraft flights, for communications and survey instrumentation applications.

Spacecraft carrying electronically despun beam steering systems under operational flight conditions have validated the design techniques and have been flight qualified for as long as 3 years. What needs to be done now is to apply the best of microminiaturization technology to circuitry, components, and solid state devices so as to provide spacecraft antenna systems which are of operational quality and which can perform tasks which no mechanically-actuated antenna system can accomplish.

REFERENCES

(1) B. A. Sichelstiel, W. M. Waters, and J. A. Wild, "Self-Focusing Array Research Model," IEEE Trans. Antennas and Propagation, Vol. AP-12, pp. 150-154, March 1964.

(2) C. Y. Pon, "Retrodirective Array Using the Heterodyne Technique," IEEE Trans. Antennas and Propagation, Vol. AP-12, pp. 176-180, March 1964.

(3) W. H. Kummer and R. A. Biagenheier, "A High-Gain Self-Steering Microwave Array," Proc. IEEE, Vol. 56, pp. 2028-2038, November 1968.

(4) W. H. Kummer and A. T. Villeneuve, "Spacecraft Antenna Systems, Phase I," Final Report, P65-35, NASA/Goddard Space Flight Center Contract, NAS 5-3545, January 1965, pp. 145-166.

(5) A. F. Seaton and R. A. Birgenheier, Unpublished Work, Hughes Aircraft Company, Culver City, May, 1968

(6) L. Schwartz, J. Smith, B. Popovich, L. R. Dod, "Data Relay Satellite Phased Array," AIAA Paper 70-424, April 1970.

(7) R. A. Birgenheier and A. R. Sisson, "A High-Gain Self-Steering Microwave Repeater," Final Engineering Report, Vol. II, NASA/GSFC Contract NAS 5-10101, Report No. P69-403, September 1969.

(8) J. Butler and R. Lowe, "Beam Forming Matrix Simplifies Design of Electronically Scanned Antennas," Electronic Design, Vol. 9, pp. 170-173, 1961.

(9) W. H. Kummer, "Self-Steering Arrays for Satellite Applications," in Communication Satellite Systems Technology, Progress in Astronautics and Aeronautics, No. 19, R. B. Marsten, Ed. New York: Academic Press, 1966, pp. 573-592.

(10) W. H. Kummer and R. A. Birgenheier, "Spacecraft Antenna Systems," Final Engineering Report, NASA/GSFC Contract NAS 5-3 545, Report P66-66, March 1966.

(11) A. R. Dion, "Variable Coverage Communication Antenna for LES-7," AIAA Paper 70-423, April 1970.

(12) H. R. Erhardt, A. Gerson, and D. P. Mead, "The Advanced Syncom Communication Antenna System: A Directive Array for Spin Stabilized Satellites," IRE National Symposium on Space Electronics and Telemetry Record, 1963.

(13) Phazed Array Transmitting Antenna and PACE, NASA/GSFC Contract NAS 5-3823, October 1966, HAC Document SSD 60196R.

(14) VHF Repeater Experiment for ATS/C, NASA/GSFC Contract NAS-5-10290, HAC Report P67-56, February 1967.

(15) M. L. Rosenthal, M. E. Devane and B. F. La Page, "VHF Antenna Systems for Spin-Stabilized Satellites, Trans. IEEE, AP-17, pp. 433-451, July 1969.

(16) Study Program on 30-100 GHz Electronically Steerable Antenna Systems, ADTEC, NASA/GSFC Contract NAS 5-10256.

(17) M. E. Louapre and J. K. Fujioka, "A High-Efficiency Electronically Scanned K-Band Phased Array for Spaceborne Radiometric Applications," Proc. IEEE, Vol. 56, pp. 2010-2016, November 1968.

(18) NASA/GSFC Contract 5-2115 with Microwave Division of Aerojet General, El Monte, California.

(19) Engineering Model of an Electronically Scanned Phased Array Antenna System, Final Report, Ryan Company, Report No. 59867-1, NASA/MSC Contract NAS 9-7626, May 1969.

(20) S. H. Durrani and H. Staras, "Multipath Problems in Communications Between Low-Altitude Spacecraft and Stationary Satellites, RCA Re-

view, pp. 77–105, March 1968.

(21) LAS/DRSS Multipath Transmission Study, NASA/GSFC Contract NAS 5-11602, Hughes Aircraft Company, 1970 (to be issued).

(22) W. H. Kummer, A. T. Villeneuve, and K. R. Weiner, "Literature Review for the Multipath Study of a Low-Altitude Satellite Utilizing a Data Relay Satellite System, NASA/GSFC Contract NAS 5-11602, June 1969.

Design, Performance, and Cost Considerations for Solid-State Arrays*

CARL BLAKE
Lincoln Laboratory
Massachusetts Institute of Technology
Lexington, Massachusetts

Introduction

The solid-state, phased-array radar (SSPAR) has been studied for about five years. Since all modern phased-array radars make extensive use of solid-state devices throughout the system except for the transmitter, the SSPAR differs from the conventional phased-array radar in that the microwave transmitter is also entirely solid state. This paper will be concerned primarily with this difference.

Following a review of past and current SSPAR programs, solid-state device development will be scrutinized for hints of limitations and potential in SSPAR applications. This exercise provides the guidelines for the predictions of viable SSPAR design configurations. Simple technical feasibility arguments are presented that indicate a SSPAR could have sensitivity comparable to many existing mechanically and electronically steered radars. A more detailed, but still much simplified analysis, is described that begins to quantify the parameters of a viable SSPAR. The role of the antenna designer in the development of the SSPAR antenna element is discussed. The paper concludes with a status report and a prediction of the probable developments in the critical SSPAR technology over the next several years.

Background

SOLID STATE PHASED ARRAY RADAR (SSPAR) PROGRAMS

Program Name	Start Date	Sponsor	$ × 10^6
MERA	1964	Air Force	3.0
MAIR	1966	Navy	0.5
PAR Option	1967	Army	—
CAMEL	1967	ARPA-Army	1.0
Tri Service			
RASSR	1970	Air Force	—

Fig. 1 Solid State Phased Array Radar (SSPAR) Programs.

Figure 1 indicates the major SSPAR programs either completed or now underway. The start date of each program, the sponsor, and the level of support are also given. The number of programs and variety of sponsors reflects the intense interest in this subject. This interest is partially rooted in the belief, bordering on faith, that solid-state is the wave of the future and that microwave vacuum tubes will eventually succumb to solid state devices as have their lower frequency cousins. To better evaluate the validity of this faith one is wise to review the history of transistors and integrated circuits.

SOLID STATE ASCENSION

Digital Circuits

Low-Power Audio Amplifiers

Servo and Control – Opamps

Consumer Portable Radios

Power Audio Amplifiers – 20 dBW

TV Tuners

Portable and Mobile Communications – 20 dBW

Telemetry Transmitters

Satellite Transmitters

TV Receivers and Video Equipment

Microwave Communications – 0 dBW

Sonar Transmitters – 40 dBW

Power Conversion

Fig. 2 Solid State Ascension.

*This work was sponsored by the Department of the Army

Figure 2 traces the encroachment of the transister into the electronics field. It began with the computer industry. Since this industry was itself an infant, it had not adopted rigid standards nor had it made heavy investments in factories and equipment that would become obsolete with the introduction of the transistor. Furthermore, it was in dire need of a cheap, reliable, long-life, active device. The new solid-state technology and the dynamic glamorous computer industry constituted an unprecedented example of temporal serendipity. Transistor technology and the transistor industry entered a period of frenzied growth unique in the annals of engineering and business. Soon transistors were replacing vacuum tubes in consumer products. As power handling capability increased and the frequency range was extended, transistors took over high-fidelity audio systems, servomechanism and control systems, mobile radio and television receivers. In reviewing this history, it appears that the growth of the power handling capability of a single amplifying transistor chip has leveled off at about 20 dBw while the high-frequency capability pushes ever upward. The ceiling on power per chip is due to a combination of problems that occur as chips become very large including low yields, awkward impedance levels, but primarily troublesome thermal designs. A few solid-state devices such as the silicon-control rectifiers (SCRs) and high-power diodes break this ceiling by virtue of their high efficiency. The high power level of the sonar transmitter indicated in Figure 2 is in part due to SCR efficiency and in part because the system is a phased array. The 40 dBw figure given is that for the entire system. This sonar array is probably the first operational SSPAR. The observed 20 dBw limit on single transistor chips tends to carry over to all semiconductors of comparable efficiency when operated with pulse durations that are long relative to the thermal time constants of the device in question. In all cases considered herein, it is assumed that this situation prevails. The situation is quite different with vacuum tubes. The vacuum tube has a decided advantage in a radar application in terms of its thermal time constant in that a vacuum tube design can be optimized for almost any duty cycle.

The reasons for the spectacular success of the transistor are outlined in Figure 3. Once the techniques for automated transistor manufacturing were mastered, the transistor became a cheaper, more reliable, and often more efficient replacement for the vacuum tube. The transistor had unique features such as no warm-up time and complimentary symmetry, but these were secondary. Reduced cost was the driving force behind its encroachment in all fields of electronics and will assuredly be the basis for whatever success it achieves in the SSPAR.

WHY SS ASCENSION?

LOWER COST!

Initial Device and Fabrication Cost

Efficiency

Reliability

Maintainability

UNIQUE PERFORMANCE

No Warm-Up

Complementary Symmetry

WEAKNESSES

Thermal Limitations

Nuclear Vulnerability

Fig. 3 Why SS Ascension.

There remain a few bastions indicated in Figure 4 where vacuum tubes continue to flourish. Three factors that limit the effectiveness of the solid-state devices are in evidence--high voltage, high power, and high frequency. The high voltage has little direct relevance to the SSPAR except perhaps in the distribution of dc power. The remaining two factors are decidedly relevant.

VACUUM-TUBE BASTIONS

HIGH VOLTAGE SWITCHING

HIGH SPEED, HIGH POWER SWITCHING

HIGH FREQUENCY, HIGH POWER AMPLIFICATION

Fig. 4 Vacuum-Tube Bastions.

Solid-State Designs

Two design approaches have been advanced for replacing vacuum tubes with solid-state devices as primary sources of microwave power in phased-array radars and in microwave systems in general. In the most straightforward of these, the solid-state devices are distributed appropriately throughout a network such that the network has one low-level input and one high-level output. (1, 2, 3) Such a configuration is frequently referred to as a "solid-state bottle." Proponents of this approach argue that it is the natural way to bring solid-state microwave power devices into existing systems, since it involves little more than plugging the solid-state bottle into the vacuum-tube socket and modifying the power supply appropriately to accommodate the solid-state devices. The implementation of such a design is indicated in Fig. 5. (4) Clearly, a great deal of expense is involved in the power-dividing and power-combining networks. Most such networks are sensitive to the phase and amplitude balance of the signals being combined. Losses due to input-signal imbalance and to the hybrid combining circuits decrease the efficiency of the combination of devices below that ach-

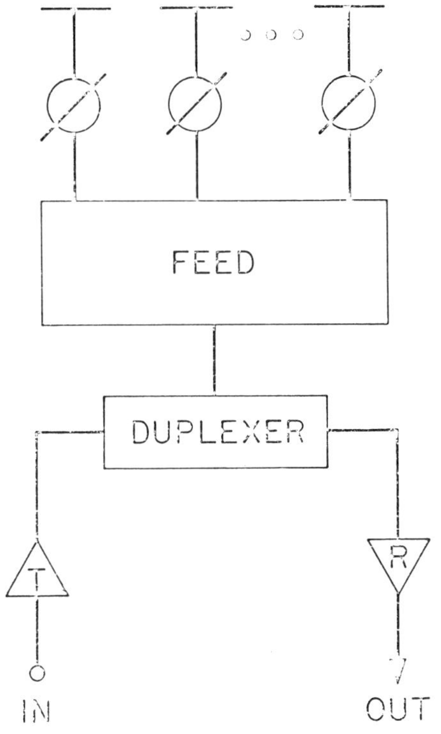

Fig. 5 Array or Subarray Feed Approach.

ievable with a single device. Figure 6 indicates a possible feed approach for applying the "solid-state bottle" to a phased-array radar. One is struck by the double penalty that has been levied upon the power-generating devices by first combining power into a single feed point, and then power dividing so as to feed the individual elements.

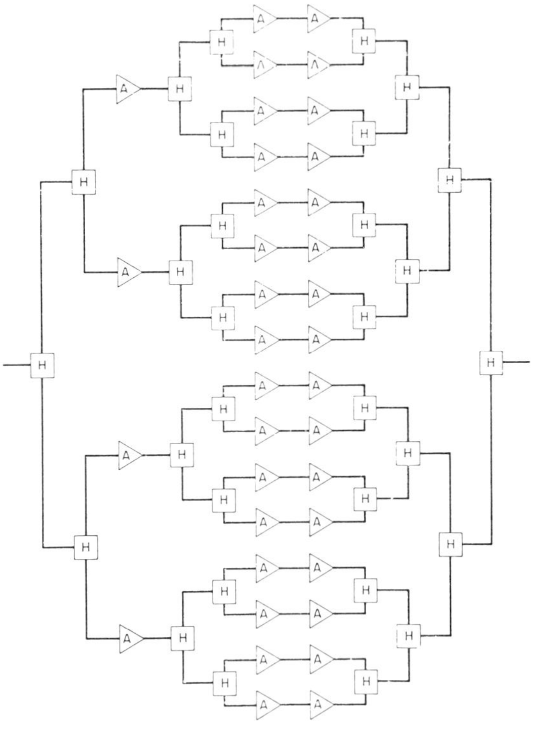

Fig. 6 A 10-Watt, 2-GHz Amplifier System Using Hybrids.

The alternative approach is to avoid the costly and cumbersome problem of combining the power of a myriad of devices, by capitalizing upon the distribution of power over a large area that invariably takes place in directive, high-gain radiating systems. The electronically scanned phased-array transmitter is especially well suited to this approach. (5, 6) In this configuration, solid-state devices compete with vacuum tubes on grounds that are economically most attractive to the solid-state approach. The higher costs of solid-state devices in terms of dollars per watt are at least partially offset by the microwave power losses incurred in distributing the microwave power from the "big bottle" over the radiating aperture. Figure 7 indicates the system configuration typical of this approach, and Figure 8 depicts the principal constituents of a transceiver module that might be employed in an all-solid-state phased-array radar. The peak power level encountered at the radiating element in a typical vacuum-

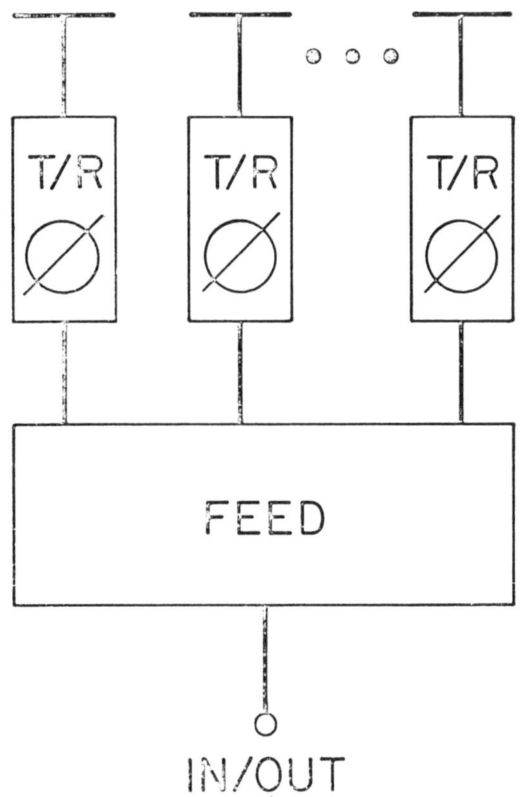

Fig. 7 Elemental Electronics Approaches.

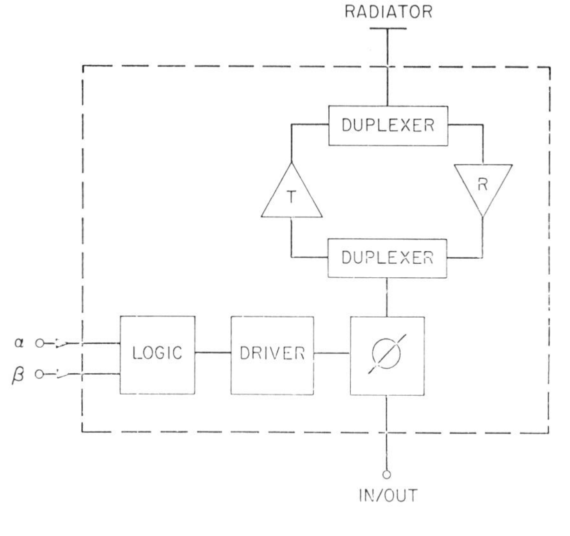

Fig. 8 All Solid State Module.

tube-powered phased array radar is 30 dBw. As we shall indicate later in this paper, a transistor-powered phased array to be technically and economically feasible would operate with elemental peak power levels of about 20 dBw. (7) A module capable of supplying this level of power can be packaged and cooled in a relatively compact fashion. Since many such modules are required even for a single modest-sized phased-array radar, production runs would be large enough to justify a substantial amount of automation and productizing to minimize module costs. Figure 9 is a photograph of a breadboard module that typifies the technology that could be invoked to provide these low-cost modules.

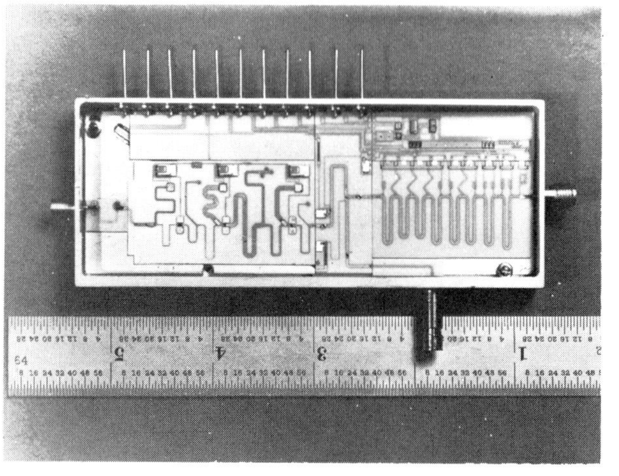

Fig. 9 Solid State Phased Array Radar Receiver Module.

Solid-state microwave power devices impose constraints upon their system parameters that may limit performance and increase costs. The thermal time constant of solid-state devices is typically a few tens of microseconds or less. If the duration of the generated pulse exceeds this thermal time constant, the device is essentially operating in a CW mode, and it is economically advantageous to use as high a duty cycle as possible. A concomitant of this constraint is the requirement for long pulses and high pulse compression ratios, both of which tend to increase the cost of the associated signal and data processors and to aggravate minimum range and clutter problems.

Another nontrival difference between solid-state power devices and vacuum tubes is the magnitude of the power supply voltages required. Solid-state devices operate from low-voltage sources. In small systems, this may be an advantage, e.g. satellite transmitters. For earth-based systems requiring many kilowatts of average power, the low voltage entails current levels and energy storage problems that are decidedly awkward. One solution to these problems is to proliferate many small power supplies. This solution is straightforward, though somewhat expensive.

Plausibility Considerations

If one assumes a reasonable number (<10) of devices in parallel driving each radiating element, power levels of a hundred watts peak appear feasible at microwave frequencies with higher power levels possible as better power combining circuits evolve. Figure 10 reviews the technical and economic factors. A typical element occupies an area of $\lambda^2/3$. At 100 w per element, the power per λ^2 is 300 w. A 1° beam corresponds to about 3000 λ^2 or 1 Mw of peak power. A survey of fielded radars reveals many dish radars operate at or below this level, but most phased-array radars exceed it. Peak power is not the best criteria of radar performance. It is chosen here because transistors are peak power limited. If it is assumed that a high duty ratio (20–30%) is possible, the average power levels of the SSPAR are more reasonable even when compared with phased-array radars.

SSPAR PLAUSIBILITY CONSIDERATIONS

TECHNICAL

SSPAR Power	100 W per Element
Area per Element	$\lambda^2/3$
Power Density	300 W per λ^2
Total Power, 1° Beam	1 MW

ECONOMIC (ANTENNA)

Vacuum Tube Cost (TWT 5-Percent Duty)	600 + 0.5 p
SSPAR	250 + 2.0 p

Fig. 10 SSPAR Plausibility Considerations.

Turning to the economic factors, consider an L-band, TWT powered (big-bottle) radar driving an antenna steered by high-power diode phase shifters. These phase shifters and their associated drivers cost \$600 each ready to install. The cost of the transmitter power normalized to the phase shifter output is \$0.30 per w. The per element cost is the sum \$600 + 0.3 p where p is the elemental peak power. Several cost factors should also be included but will be ignored such as the power supplies, prime power and cooling costs associated with the drivers, and the transmitter costs which are fixed; i.e. independent of power level, which should be prorated over the total number of elements. The L-band solid-state modules cost about \$250 + 2p. The peak power per element is much lower than that encountered with the vacuum-tube phase shifters, but with a large duty cycle, the spread in average power per element is reduced to roughly one order of magnitude. Even this difference of power does not give a fair comparison of the two systems since each element contributes to the aperture and the measure of performance must adequately reflect both power and aperture. Figure 11 sketches how one can affect this measure.

PARAMETRIC RELATIONS

SYSTEM REQUIREMENTS

$$S/N_o = \frac{EG^2\lambda^2\sigma}{(4\pi)^3 R^4 L\, KT_{sys}}$$

$$G = Ng$$

Search, Measured σ, Track

$$P = \frac{K_s}{N} + \frac{K_x}{N^2} + \frac{K_T}{N^3}$$

$$p = \frac{K_s}{N^2} + \frac{K_x}{N^3} + \frac{K_T}{N^4}$$

COSTS

$$S_{\mu s} = A + BN + CP = A + (B + Cp)\, N$$

$$S_M = B + Cp$$

For Minimum Cost Systems

$$1 \leqslant \frac{B}{Cp} \leqslant 3$$

$$\frac{B}{3C} \leqslant p \leqslant \frac{B}{C}$$

Fig. 11 Parametric Relations.

Two parametric relations are needed: one to describe the system requirements; the other to depict the costs. The former is the radar equation and relations derived therefrom. The functions of the radar in a particular application are broken down into search, cross-section measurements, and tracking. In so doing, the radar equation takes the form indicated for p, the elemental peak power, in Figure 11. The constants are determined by the respective search, cross-section measuring, and tracking functions.

Cost models for the microwave subsystem of most phased-array radar systems can be approximated quite accurately by the simple linear equation of Figure 11. The two variables N and p can be coupled using the system requirements equation. From this a cost minimization can be deduced. The constants B and C are derived from a great many items in the cost model and should not be construed directly to be measures of module cost. However, the ratio of B to C will be quite closely the ratio of fixed model costs to the cost of power per peak watt. One can therefore deduce using conventional minimization techniques that the peak power per module for minimum microwave hardware cost will be $\frac{B}{3C} \le p \le \frac{B}{C}$. At the going price of \$250 + 2p, the optimum power range is from 40 to 125 w which, fortuitously, is within the state of transistor technology. The same relationship applied to the vacuum tube system indicates optimum powers in the range of 750 to 2000 w and is in reasonable agreement with recently fielded systems.

Figure 12 depicts another L-band comparison; the SSPAR and a module configuration using a gridded tube for the output power amplifier. The figure also indicates the more precise approach to the optimization which is achieved using the detailed radar cost model and calculating the curve of cost vs number of elements. Note that the optimum is relatively broad and the cost penalty for using less than optimum power is not very severe. Also note that the gridded vacuum tube appears to be slightly less expensive. This small difference is meaningless because of the uncertainties in the numbers used in the cost model.

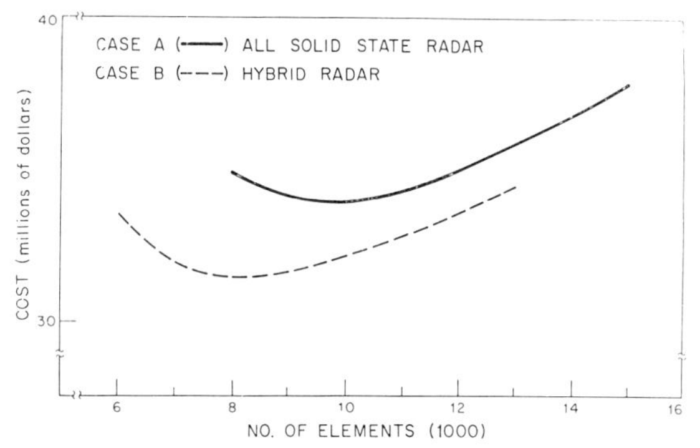

Fig. 12 Solid State Radar Costs.

From the start, solid-state microwave power amplifying devices have been and are today the critical components of the SSPAR. Figure 13 shows current costs and projected costs for 1975 of solid-state devices and gridded tubes such as might be used in transceiver modules. The indicated price floor is a consequence of the combination of a maximum practical size per chip and the cost of packaging, testing, distributing and all overhead charges associated with manufacturing and distributing the devices. The prices for the diodes are relevant only if circuits can be developed harnessing a high efficiency mode in a reproducible and practical avalanche-diode amplifier. At the pricing levels indicated, the cost of the microwave power devices comes into a reasonable balance with other transmitter components. Under these conditions, systems which can employ the long pulses and high duty ratio of the SSPAR look quite competitive.

MICROWAVE POWER DEVICES

	Cost (\$/W)		Efficiency (Percent)	
	Current	Projected	Current	Projected
Transistors				
UHF	1.00	0.20	60	70
L-Band	2.00	0.35	50	60
S-Band	—	3.00	—	30
Diodes				
UHF		0.30		60
L-Band		0.30		60
S-Band		0.30		60
C-Band		0.35		50
X-Band		0.50		40
Gridded Tubes (5% Duty)				
UHF	0.03		50	
L-Band	0.06		40	
S-Band	0.10		30	

Fig. 13 Radar Initial Investment Costs.

Antenna Design Considerations

RELEVANT ANTENNA CONSIDERATIONS

1. LOW COST
2. LOW PEAK AND AVERAGE POWER
3. ANTENNA ELEMENT INTEGRAL WITH MODULE
4. GOOD MATCH WITH SCAN

Fig. 14 Relevant Antenna Considerations.

The use of solid-state devices as microwave

power sources presents the antenna element designer with an outstanding opportunity to apply his talents and ingenuity. The relevant characteristics are indicated by Figure 14. The most economically attractive form of the solid-state transmitter puts the device in immediate proximity to the radiating element. The overriding consideration becomes that of minimizing the loss between the element and the solid-state device. Ideally, the element becomes an integral part of the circuit. The MERA module shown in Figure 15, is an example of this technique. One of the current pressing problems is that of finding a way to handle the mutual coupling among elements such that an antenna can be steered electronically while maintaining an element VSWR that does not present an intolerable mismatch to the solid-state device at some scan angles. Figure 16 indicates the variation of reflection coefficient with scan angle for one type of element. (8) With variations as large as those indicated, the solid-state devices must be isolated from the elements, as indicated in Figure 17. The added cost, weight, and loss of this isolation constitutes a serious penalty. Its elimination through improved element design would substantially enhance the viability of the solid-state concept.

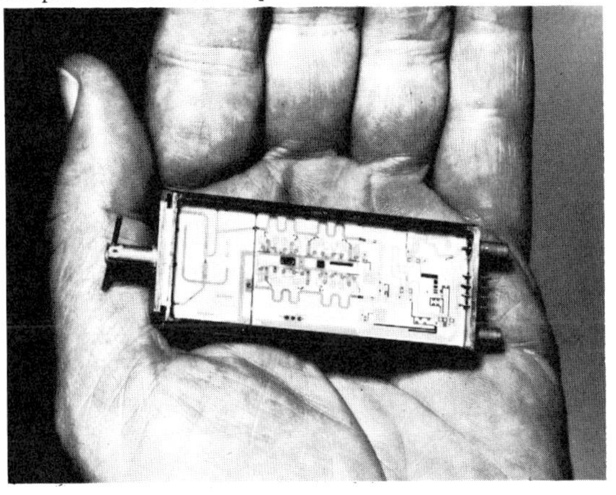

Fig. 15 Microwave Power Devices.

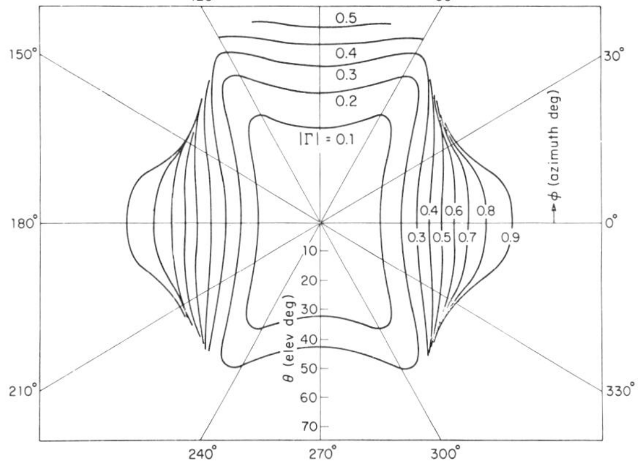

Fig. 16 Typical Reflection Coefficient vs. Scan Angle (Antenna Coordinates).

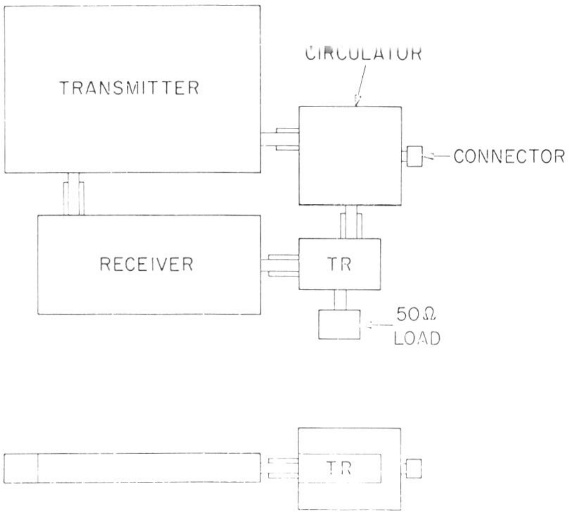

Fig. 17 Typical Solid State Module Layout.

Conclusions

The all-solid-state microwave system will slowly find its way into radar and communications systems. The crucial component remains the solid-state microwave power source. The microwave power transistor leaves little to be desired in reliability, operating efficiency and life expectancy. Unfortunately, it is currently limited to frequencies below 2 GHz and is incapable of handling enough power for many applications. The ideal solid-state power source would retain all the virtues of the present day transistor but would operate throughout the microwave spectrum and accommodate peak powers of several hundred watts with pulse lengths of up to several hundred microseconds. Most important of all, it would sell for a small fraction of the cost of current microwave power transistors. In any case, the success of the solid-state module will be dependent in large part upon the ingenuity with which it can be coupled to the radiating element. The design of this element is critical both in terms of interfacing with the solid-state device and in satisfying the antenna system requirements. Herein lies one of the more important challenges to antenna technology in the years ahead.

REFERENCES

1. Hines, M. E., "Network Integration Approaches for Multiple-Diode High Power Microwave Generation," Digest of 1968 G-MTT IEEE International Microwave Symposium, May 1968, pp. 46-53.

2. Bailey, R. L., W. P. Bennett, L. F. Heckman, I. E. Martin, "An All-Transistor, 1 Kilowatt, High-Gain UHF Power Amplifier," 1969 International Microwave Symposium, Digest of Technical Papers, May 1969, pp. 149-153.

3. Staiman, D., and M. Breese, "200W Solid-State UHF Amplifier," 1969 International Microwave Symposium, Digest of Technical Papers May 1969, pp. 445-449.

4. Blum, S. C., "A Low S-Band Solid-State Amplifier," Solid-State Circuits Conference, February 1968.

5. Robinson, A.S., "Future Trends in Phased-Array Radars," 1969 NEREM Record, Session 22, November 1969.

6. Johnson, G. D. and Hyltin, T. M., "Application of Microelectronic Technology to Radar and Communications Systems," 1969 NEREM Record; Session A-3, November 1969.

7. Blake, C., "Transistors versus Vacuum Tubes in UHF Phased-Array Radars, "Microwave Journal, 11, September 1968, pp. 61-65.

8. Diamond, B. L., private communication.

Comparison of Array Element Types

L. STARK

Hughes Aircraft Company
Ground Systems Group
Fullerton, Calif.

A phased array antenna may be thought of as consisting of essentially three basic parts, which are (1) the array feeding system, (2) the phase shifters, and (3) the radiating elements. When the array is subdivided in this way the design problems may be readily formulated for each part and the overall problem is reduced to a set of more basic and elementary problems which can then be attacked. The radiating part of a phased array antenna is one of these basic parts and it constitutes an interesting problem for study because the elements not only radiate into the space in front of the array to form a beam as intended, but they also radiate signals into each other. This coupling action produces results which affect element driving point impedance, and hence, power transfer to the antenna beam, and it also affects the angular distribution of energy in the radiation pattern.

Basic studies and hardware developments in array element design have been pursued since about 1950 and since 1960 the scope and interest has increased markedly. The present paper will review some of the phased array element designs which have been developed over the total period with an emphasis on those designs which have been reduced to practice.

Array element types which have been employed are dipoles and open-ended waveguide radiators. Typical examples of these cases are shown in Figures 1 and 2.

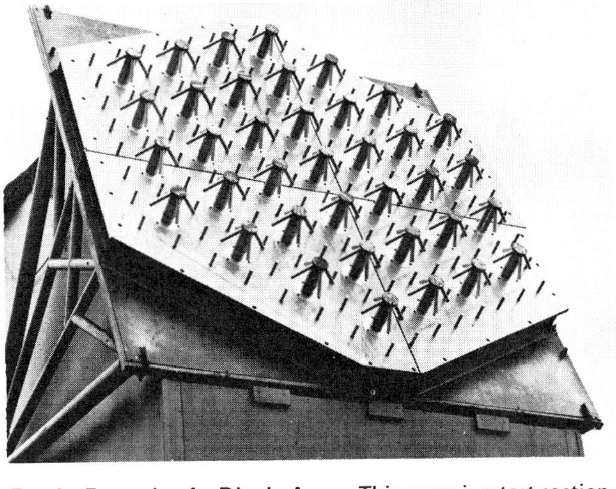

Fig. 1 Example of a Dipole Array. This array is a test section of an array containing a much larger number of elements. (Photo courtesy of Bell Telephone Laboratories.).

Fig. 2 Example of an array of open-ended waveguide radiators. The total array is divided into six subarrays, each of which is driven by an amplifier.

Waveguide slot elements milled into the broad wall or narrow wall of the waveguide, as shown in Figure 3, have also been used. In the case just cited, phased array operation is obtained by imparting relative phase shift between linear waveguide arrays, producing beam scanning along only one coordinate.

The primary basis for the progress that has been achieved in element design has been the steadily growing ability to calculate with increasing accuracy the driving point impedance of the elements, not only at broadside, but for any scan angle of the

Fig. 3 Example of a waveguide slot array. The main feed linear array will be connected at the lower left.

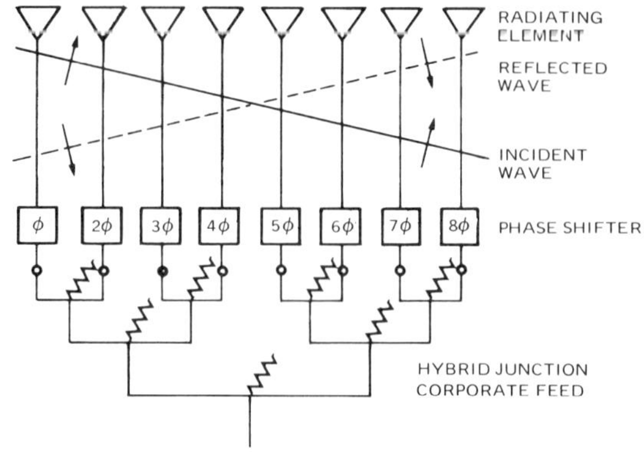

Fig. 4 This figure shows the reflected wavefront from the radiating elements by joining the points of constant phase in the feed system. A corporate feed system employing hybrid junctions is shown.

array. With this knowledge and a knowledge of the controlling factors in impedance variation, we have been able to minimize this variation. A second basis for progress, important for empirically developed radiating elements, is the concise development of the concept of the active element pattern (1). Employing this concept provides a powerful tool in minimizing the impedance variation in empirically developed elements. Reducing the impedance variation has been important because the element driving point impedances are the terminating impedances for the array feeding system and therefore interact with the array feeding system in determining the aperture illumination of the phased array. The errors produced by an incorrect set of terminating impedances will, as a minimum, reduce the efficiency of power transfer to the elements due to a reflection loss and may further have the more serious effect of producing amplitude and phase errors, which will in turn modify the angular distribution in the radiation patterns. Different types of feed systems have varying degrees of sensitivity to the effect of mismatched elements. One way of studying this interaction problem is to consider the effect of the reflected wavefront from the mismatched array as it impinges on the array phasing and feeding system from the mirror angle, as shown in Figure 4. This approach sets up the initial scattering problem and the reflected plane wave approximation for the reflected wavefront is valid so long as the reflection coefficient of each array element is the same, which is a good large-array approximation. The scattering of the reflected wavefront from the various junctions in the feed may then be systematically analyzed. If the feed system is an optical device the scattering may be similarly analyzed and it may be possible to develop schemes for absorbing the reflections. This subject will not be pursued here but it is mentioned that the best that can be done to minimize scattering effects is to employ feed systems in which the elements are fed from independent generators, or from a branching network of hybrid junctions (2), as shown in Figure 4, or from a network of directional couplers as shown in Figure 5. These isolating feed systems prevent the reflected wave from any element

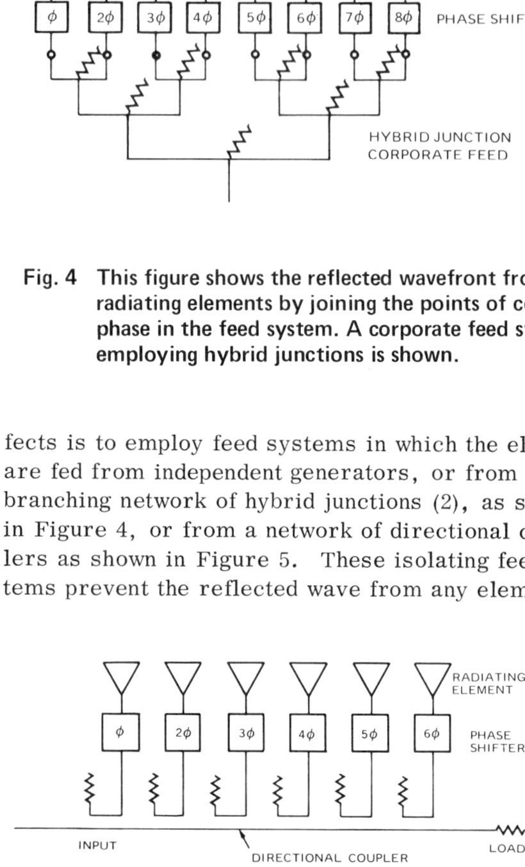

Fig. 5 The figure shows an array feed system employing directional couplers.

employing directional couplers.

from being re-radiated from any other element. The angular distribution of energy in the radiation pattern is thus not modified by errors in the feed system. Some reference listings on other types of feed systems are given here (3, 4) and these show the interesting effects on the radiation pattern when this type of interaction is not prevented. When isolating feed systems are used the change in element impedance for different radiation angles produces only a reflection loss. The reflection loss is, however, angularly dependent so that there is still an effect on the angular distribution of radiation in the pattern. The energy transfer to the main beam is a function of radiation angle, as mentioned above, and the energy transfer in the sidelobe directions for any given main beam scan angle is also weighted by the angularly dependent reflection coefficient. This situation may be viewed on receive to readily describe the overall effect on the pattern. By doing this, one develops the concept of the active element pattern, as was done in Reference 1. Let us briefly re-state the basic idea of the active element pattern here. With a plane wave incident on the array at an angle θ from the array normal the power received at the element terminal input is proportional to the projected area per element of the array and the transmission coefficient, $1 - |\Gamma|^2$, of the array aperture. These multiplying factors on the energy received occur before the feed sums the energy from the individual elements and thus have no regard for the main beam or sidelobe directions. The incident wave, at this point, sees only an infinite regular array and has no knowledge of the feed interconnections. The feed system then further determines the array response pattern by summing the signals after introducing differential phase shift. In an isolating feed system there are no further interacting factors in determining the angular response pattern. Viewed in this manner, the feed summing pattern may be called the array factor and the remaining factors are grouped together and called the active element factor. The active element factor is thus

$$G = \frac{4 \pi A \cos \theta}{\lambda^2} \left(1 - |\Gamma_\theta|^2 \right),$$

where A is the area per element of the array, and
λ is the wavelength

The active element pattern is a fixed, angularly dependent weighting function on the response of the array. It is a fixed function in the sense that it is not dependent on the main beam scan angle. It is evident that since the transmission factor, $1 - |\Gamma|^2$, may be greater in the direction of a sidelobe than in the main beam direction even when main beam is on broadside, the element impedance variation can have the effect of increasing the sidelobe level. This pattern effect, obtained when employing an isolating feed system, was very surprising when it was first noted (5) but is now clearly understood as an effect which can be present.

In writing the above expression for active element pattern and assuming that the transmission coefficient, $1 - |\Gamma|^2$, is the same for each element, it is assumed that all elements are identical and the edge effect in the array can be ignored. The array size required in order to ignore the edge effect is discussed by Diamond in another paper in this Volume (6) and is found to depend upon the rate of coupling decay between elements. The array size is of the order of 10 x 10 elements for elements consisting of dipoles over a ground plane. It is assumed for most of the discussions to follow that the arrays we are dealing with are always large enough to ignore the edge effect so that essentially all elements in the array may be designed alike. However, even in the case of small arrays, one can make a good case for using an element designed on the infinite array basis. If one examines the impedance variation with position in moving in from the edge of a large array one finds that the driving point impedance traverses a circle of ever decreasing radius, which must converge to the impedance of the infinite array value as shown in Figure 6. Thus, the infinite array element design provides the best single

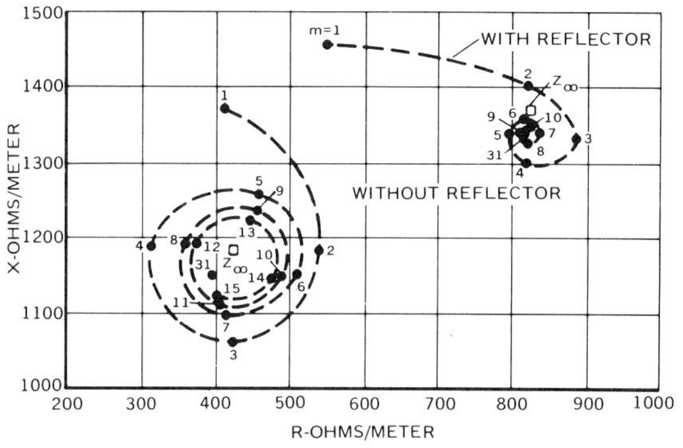

Fig. 6 Driving point impedances of an infinite wire array scanned to 30° off normal in the H plane. No. 1 wire is at array edge opposite beam scan direction.

compromise element design for a small array and is much more practical than designing each element individually.

The Element Lattice

The element lattice characterizes an infinite array as well as a finite array. In the infinite array the lattice dimensions determine the wavelengths of the space harmonic plane waves relative to the fundamental wave (7). In the finite array the lattice dimensions determine the direction cosines of the (infinite set of) grating lobes relative to the direction of the main beam. The space harmonic waves of the infinite array propagate in the same directions as the pointing directions of the grating lobes of the finite array. (In the infinite array the plane waves never diffract into spherical waves forming antenna beams; thus, there is the minor distinction between grating lobes and space harmonic waves.)

The grating lobe positions are significant in array design since, in all cases of practical importance, only a single beam is desired in real space. The lattice should be designed to permit only a single beam in real space over the scan angle range of the main beam. The relationship between the array lattice geometry and that of the peak directions of the grating lobes is a very simple one and easy to show in a picture. Figures 7a and 7b show the relationship. The grating lobe lattice, when plotted in direction cosine space, is obtained by rotating the array element lattice by 90° and applying a scale factor. The mathematical relationship between the array lattice and grating lobe lattice vectors is (8),

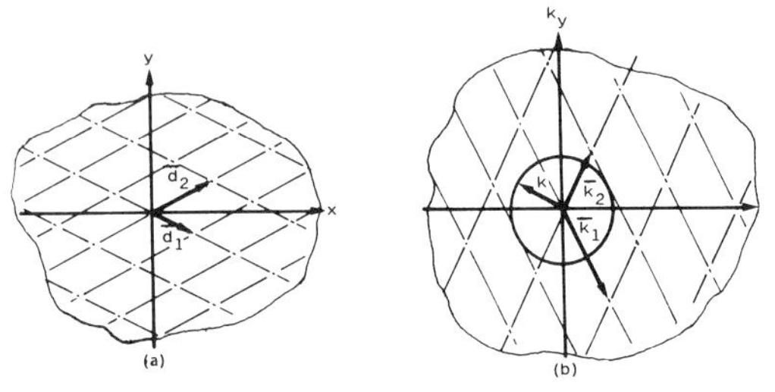

Figs. 7a, b (a) is element lattice and (b) is grating lobe lattice.

$$\left(\frac{k_x}{k}\right)^2 + \left(\frac{k_y}{k}\right)^2 \lesseqgtr 1 \text{ is visible space.}$$

$$\overline{k_1} = -\frac{2\pi}{A}(\overline{1_z} \times \overline{d_2})$$

$$\overline{k_2} = \frac{2\pi}{A}(\overline{1_z} \times \overline{d_1})$$

where

$A = 1_z \cdot (d_1 \times d_2)$ = area per element, and

$\overline{k_1}$ and $\overline{k_2}$ are the grating lobe lattice vectors, as shown in Figure 7b.

These formulas readily and concisely determine the grating lobe lattice. As the main beam of the array is scanned, the assembly of grating lobes scans as in a rigidly connected assembly (9). In order that no higher order beams enter real space, they must be excluded from crossing the boundary circle defining real space, which is the unit circle around the origin. It is evident from Figure 7b that the shape of the contour of maximum scan angle of the main beam which just excludes any grating lobe from touching the unit circle of real space, can be determined for any given grating lobe lattice shape. This property may be used in array lattice design to tailor the lattice shape to the maximum scan angle requirements in a few principle directions. By so determining the array lattice, the maximum element spacing is obtained, thus reducing the number of elements required in the array.

In arrays consisting of elements which have apertures that are larger than a half wavelength, one must provide a margin of safety in spacing the unwanted grating lobes from the unit circle. Whereas, the original reason for introducing the array and grating lobe lattices here was to fix conditions which prevented the formation of more than one antenna beam, the position of the grating lobe lattice is also related to resonances in element driving point impedance. Analyses have shown that element driving point impedance singularities may occur for scan angles producing grating lobes near the unit circle and the larger the element the greater the spacing margin that must be applied. Some practical guidelines resulting from analysis of this effect will be presented later.

Examples of Element Types

Some of the results that have been achieved in practical element design over the last few years will now be discussed. The basic array element types which have been used in antenna designs are dipoles and open-ended waveguides. Waveguide slot radiators have also been used as mentioned previously. The emphasis in the following discussions will be on dipole and open-ended waveguide elements, which can be independently phased and provide phased array beam scanning in two angular coordinates. Dipoles are ordinarily used at frequencies of 1.0 GHz and below where tolerances permit and also power handling permits. Waveguide radiators are most often used above these frequencies.

Dipole Radiators

A great deal of sustained analysis of dipole arrays has been carried out by the array group at M.I.T., Lincoln Laboratory (10), (11), (12), (13). These investigations used P.S. Carter's mutual coupling formulas for the analyses and were undertaken for the purpose of establishing trends in performance and determining key parameters in controlling design. The grating lobe theory for infinite arrays (7) was also used in their analysis and found to agree with results using Carter's formulas. An experimental investigation was also carried out employing a 10 x 10 slot-fed dipole array in a test program. The array is shown in Figure 8. The agreement between theory and experiment was good, providing some significant guidelines for dipole array design. The results showed two key parameters in controlling impedance in phased dipole arrays. The element lattice dimension and the dipole-to-ground plane spacing were found to be highly controlling parameters. It was found that

Fig. 8 10 X 10 slot-fed dipole array used in Lincoln Laboratory array investigations.

the lattice dimension should not only be chosen to exclude grating lobes but, by making it smaller still and moving the grating lobes further into imaginary space, one can obtain improved wide angle impedance match. Shrinking the lattice size drives the cutoff space harmonic waves (7) further below cutoff so that they store less energy. This produces less reactance change and better wide angle impedance match. This effect is shown by the results plotted in Figure 9. Each curve is for a different lattice spacing and the dashed curve is for the limiting case of an infinitesimal spacing. As the lattice spacing becomes smaller and smaller, one notes from the curves that the mismatch at any scan angle inside the grating lobe angle is reduced. The limiting case of infinitesimal elements having infinitesimal spacing produces no reactance change with scan angle and the scan angle mismatch is caused by a resistance change only. The study of the infinitesimally spaced array of dipoles (current sheet array) and the method of relating this array to a periodic array was first carried out in a paper by Wheeler (9) in which he developed the concept of the grating lobe series.

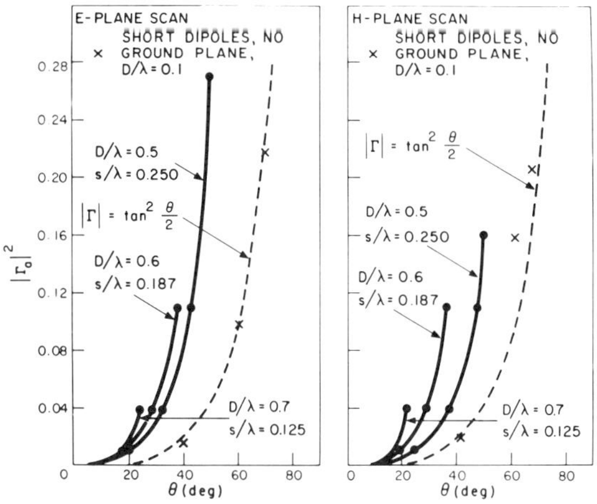

Fig. 9 Plot of percent reflected power vs. scan angle for various array lattice dimensions. Dipole to ground plane spacing was adjusted to equalize the VSWR in the two scan planes at maximum scan angle.

The results of Fig. 9 are very interesting and, among other things, they disprove the notion that increased mutual coupling, which comes from closer element spacing, produces more scan angle mismatch. This notion was prevalent in the first days of array analysis. A cost question that these results must raise for every array design is to what extent is an increased number of array elements in a design worthwhile in reducing scan angle mismatch.

The other influential parameter found from the Lincoln Laboratory studies is the dipole-to-ground plane spacing. It was determined that the dipole-to-ground plane spacing should be adjusted to equalize the mismatches in the E and H scan planes at maximum scan angle. This has the effect of minimizing the maximum VSWR and producing the best result. A parameter which does not have much effect on scan angle mismatch is the dipole length, if it is assumed that the initial condition of element match at broadside is inserted.

A reasonable expectation of what practical performance results are possible can be extracted from the curves of Figure 9. For the halfwave spaced array having a ground plane spacing optimized for a maximum scan angle of 45 degrees, we note that the VSWR is less than 2.0 over the 45-degree scan range. Therefore, 2:1 VSWR at 45-degree scan should be an achievable result for dipole arrays. It is probable that this result may be improved upon still further

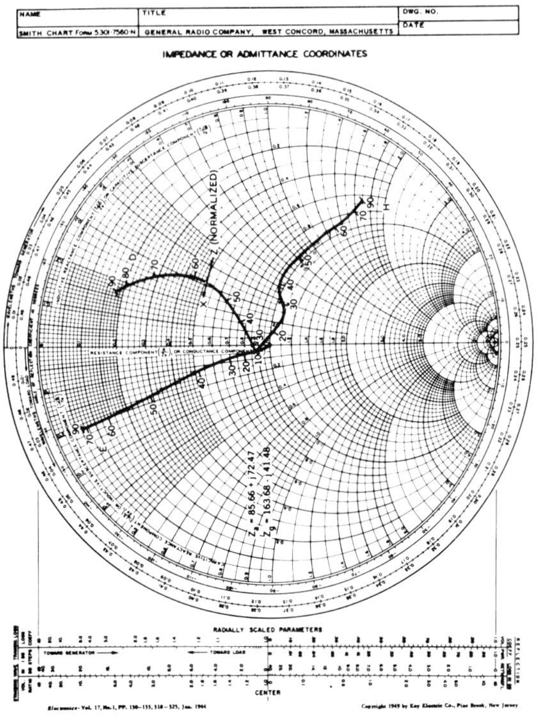

Fig. 10 Plot of impedance vs. scan angle for a dipole array. Element spacing is $\lambda/2$ in both directions and element to reflector spacing is $\lambda/4$.

by matching the elements at a scan angle other than broadside. The actual impedance variation on which the VSWR curves of Figure 9 were based is shown in Figure 10. Since the E and H plane scan curves are not diametrically opposite and the diagonal plane scan curve lies in between, a more optimum normalizing impedance located at the centroid of the locus of impedance points would produce still better results than shown in Figure 9. The centroid would be determined for the impedance curves truncated at 45° scan angle. Also shown in Figure 10 is the point, Z_a, which is the impedance of a single isolated element over an unobstructed ground plane. It can be proved that this point is at the centroid of the loci of the impedance curves when the maximum scan angle represents the full range of element phasing angles, ±180°. This fact has been recognized by the workers at Sylvania Electric Corporation (14) and forms a basis for their design procedure for dipole type elements or others which are essentially transparent when open-circuited. Based on this procedure, a matching network is designed which matches an isolated element to the generator impedance or the transmission line. This same network then provides the best average match for an array of elements of this type when phase scanned over ±180°. These results should then be verified in a small array.

An example of a dipole type of element which has achieved very good results for low reflected power in a phased array is shown in Figure 11. This element was developed at the General Electric Company. The dipole wings are bent back at a 45-degree angle in this design. The balun and feeding arrangement for the crossed dipoles are shown in the figure. The predicted performance of this element in a phased array was obtained by first measuring the

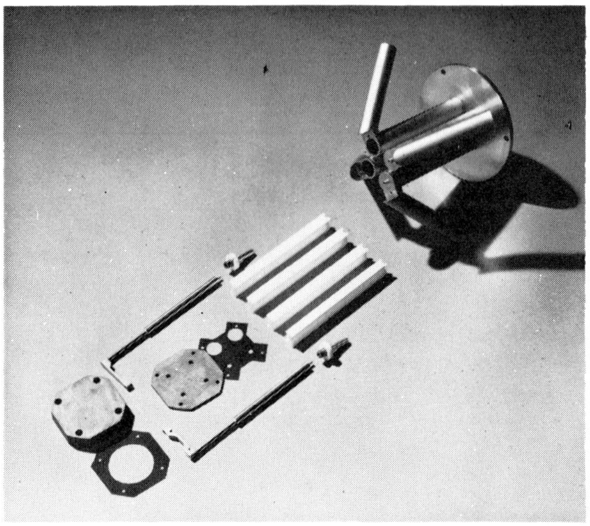

Fig. 11 Photo of a dual-polarized element for a dipole array.

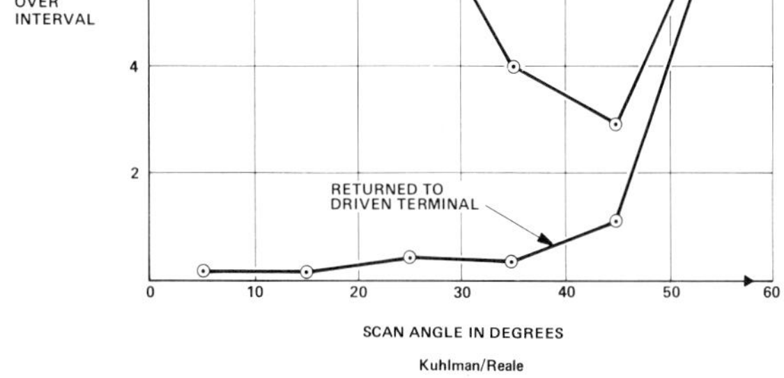

Fig. 12 Plot of reflected power vs. scan angle for the dipole element shown in Figure 11.

mutual scattering coefficients in a 91-element array and then computing the driving point impedance when all elements are energized with a progressive phase shift. The results obtained are shown in Figure 12, which is a plot of reflected power vs. scan angle when the element is driven as a circularly polarized radiator. The lower curve in this figure is for the reflected power returned to the driven terminal, which is one sense of circular polarization, and the upper curve is for the sum of the powers returned to both circularly polarized terminals. Each point in the plot denotes the average reflected power over the 10-degree sector it is plotted in. The maximum reflected power in each interval is seldom greater than the average value plotted. The plot shows that less than about ten percent of the power is reflected for scan angles out to 55 degrees. This percentage reflection is equivalent to a VSWR of 2.0 in a linearly polarized antenna and represents an excellent result. By using this dipole structure, which has bent-back wings, and optimizing performance at a scan angle off broadside (approximately 45 degrees), the workers at GE were able to obtain results which are somewhat better than the results for the simple straight dipole quoted earlier. The grating lobe angle for this array lattice is at 64 degrees. In the early design stages of this array, parallel dipole wings of conventional design were used but a large resonant mismatch was produced at 58 degrees scan angle. After using modified dipoles with bent-back wings the mismatch was removed and it was also noted that the mutual coupling was reduced by a few dB throughout the array. The posts which can be seen in Figure 11 placed normal to the ground plane between the elements are for the purpose of improving the circularity of polarization at scan angles beyond 40 degrees or so and are empirically designed. Using these posts, axial ratio is less than 4.0 dB over the entire scan range and is less than 3.0 dB over most of the range.

Another example of a dipole element design for a phased array is shown in Figure 13. This design utilizes a thick dipole in an array application in the 400 MHz band. This design was carried out at the Bendix Radio Division. A single linearly polarized element is used for the transmit element and a dual linear polarized element for receive. The guiding philosophy used in this element design was to reduce the coupling measured between neighboring elements (15). The dipole shape was patterned after Schelkunoff's result for the dipole shape having constant normal electric field distribution over the dipole surface (16). The intent here was to achieve constant distributed capacity and thus decrease the large end-capacity which would couple adjacent dipoles more strongly. The effectiveness of this approach could be demonstrated by comparing the performance of the shaped dipole with other dipole designs but there is not sufficient data to do so. A director rod was also employed to reduce coupling to the nearest neighbor (which rules out the comparison noted above). Since the bandwidth required for the element was not large, the director rod could be tuned to reduce coupling to a very small value, approximately -25 dB. After coupling was thus reduced, the mutual impedances were measured using a microwave bridge in a 7 x 7 element test array.

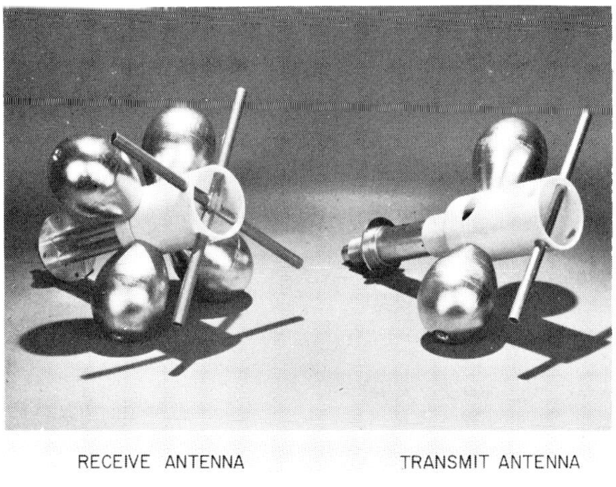

Fig. 13 Thick dipole element employing a parasitic element to reduce mutual coupling.

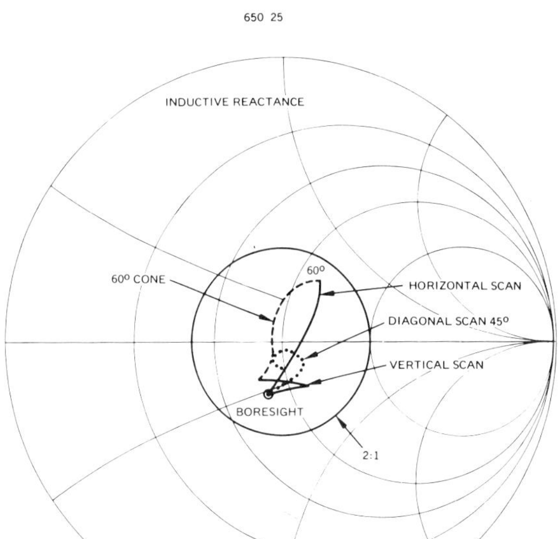

Fig. 14 Plot of driving point impedance vs. scan angle for the dual polarized dipole shown in Figure 13.

This data formed the basis for the large operational phased array element performance and is plotted in Figure 14. It is seen that VSWR is less than 2.0 at 60 degrees scan angle, which is an excellent result.

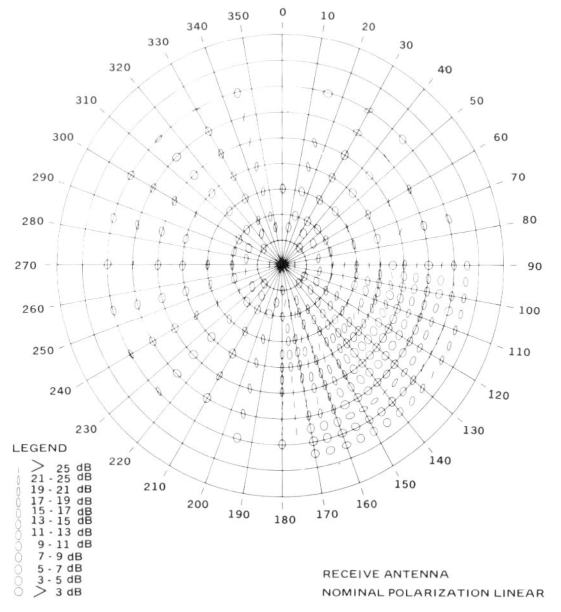

Fig. 15 Polarization of the array vs. scan angle using the dual polarized dipole element shown in Figure 13.

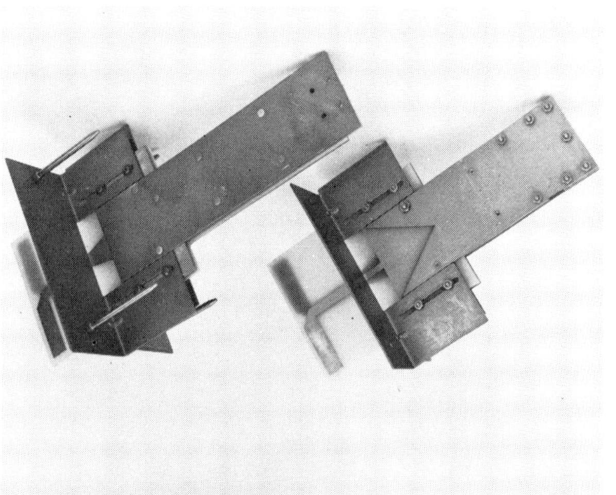

Fig. 16 Photo of printed circuit dipoles developed at Sperry Rand Corporation.

In measuring the mutual impedances in the 7 x 7 element test array it was found that the energy which was coupled to the cross-polarized dipoles produced a negligible effect and was therefore ignored. The polarization vs. scan angle performance of the test array is shown in Figure 15.

Another type of construction for dipole elements is shown in Figure 16. This shows an etched circuit dipole design for an L-band array application carried out at Sperry Gyroscope Division. A simple method of fabricating the balun was used in this design. By tapering away one of the ground planes, the transition is gradually made to a balanced two-wire strip line which then feeds the dipole. This printed circuit design also has the advantage that it will connect directly to a printed circuit phase shifter with no intermediate co-axial connector. This element was used on the spherical-wave side of a feed-through phased array lens. The maximum angle of incidence of the wavefront from the array normal was 30 degrees, at the edge of the lens. The phased array element was matched to VSWR less than 1.65 over a 12 percent frequency band and the scan angle was within 30 degrees from broadside.

Another example of printed circuit dipole array element design is shown in Figure 17. Shown here

Fig. 17 Section of a printed circuit array developed for the AN/SPS-32 antenna.

are three columns of radiators, each containing nine elements, which were studied for the AN/SPS-32 antenna. Strip dipoles and feed lines are shown bonded to one side of a plastic panel. The conjugate dipole wing and the other half of the feed line is on the other side of the plastic panel. The entire panel assembly is mounted about $\lambda/8$ above a ground plane and a baffle is placed between columns. Each of the three columns shown is actually a linear array with all elements fed in phase and having a prescribed amplitude distribution along the column which is obtained by design of the transformer sections. Each column is actually divided into two parts and fed through a coaxial T section. A phase reversal of 180 degrees of the dipoles, permitting $\lambda/2$ spacing, is obtained by attaching the dipole wings to the feed lines in an alternating fashion. Differential phase shift is imparted between the columns of radiators to provide electronic scanning in one dimension. This type of element construction is especially appropriate when the "elements" of the array are linear arrays, as is the case here, since then the elements and the feed lines can be combined in a simple, easy to produce, package. It was found in this application that feeding the balanced strip line from a coaxial line did not require a balun, rendering the construction quite simple.

Waveguide Radiators

Open-ended waveguides have evolved as a very practical element configuration for phased arrays for a number of reasons. Some of the more important ones are listed below.

(1) Waveguide radiators form a natural extension of the waveguide sections in which the phase shifters are placed.

(2) The performance of waveguide elements can be accurately predicted by analysis and calculation can play a dominant role in the design (17).

(3) The calculated performance of the element in a large array can be verified by measurement in a simple phased array simulator. This technique is discussed in detail elsewhere in this volume (18).

(4) When the radiator is designed using full benefit of current knowledge, the performance is good. This, of course, provides the strongest basis for the use of waveguide radiators.

Knowledge of waveguide radiators has gained tremendously over the last five years. One is now able to calculate performance in advance of construction for practical waveguide element designs. Experimental verification can, in addition, be simply executed by employing the concept of the active element pattern (1) and the use of a phased array simulator (18). Thus, by a combination of theory and experiment, both based on well-founded array concepts, the array element design task has been substantially reduced to a science.

Recent progress of a substantial nature has been made over the past five years in the analysis of array elements by extending the simple modal analysis of (7 and 21) to carry out an essentially exact solution of the boundary value problem for waveguide arrays. These new analyses show results for element impedance behavior which are not shown by the simple modal theory. The newer results recognize the contribution of higher modes in the element aperture which become important when larger elements are used. Thus, although they are not significant for the case of half wavelength elements (such as dipoles) they are significant for larger waveguide elements. The exact analysis shows that whereas an array may use small elements and scan over an angle which goes out to the grating lobe angle (the first grating lobe touches the unit circle in Figure 7b) before the reflection coefficient becomes very large, an array using larger elements will experience a large reflection at scan angles less than this value. In general, the larger the element, the further the scan limit is moved towards broadside. An important consequence of this phenomenon is that the array lattice dimensions may not be chosen from a grating lobe consideration alone, as was the accepted design procedure earlier. The design of a matched element must include a reduced lattice dimension as a parameter right from the start of the design and probably the best and most economical way to carry this out is by computer analysis (17). Some of this analysis has already been carried out, furnishing general guidelines for element design. Rectangular waveguide arrays have been analyzed to determine the element size and lattice spacing limitations that will exclude an impedance resonance inside the grating lobe angle (22). The results of this analysis are summarized in Table I. This table shows the largest element size that can be permitted for various lattice dimensions. The part of the table on the left deals with a rectangular lattice and on the right deals with a triangular or brick lattice. The triangular lattice results are further subdivided into two columns, one for a large Y dimension, as defined in Figure 18, and another for small Y dimensions. It is very interesting to note from Table I that the element sizes permitted are such that conventional waveguide sizes can be used for any rectangular lattice and for triangular lattices having small E plane dimensions. We usually must rule out the rectangular lattice, however, because it does not permit a wide enough scan angle without a grating lobe. By using the guidelines in this table an element can be selected for further study and development by computer or experimental techniques.

Examples of Waveguide Array Element Designs

An example of array design using open-ended waveguide radiators is shown in the Figure 19. This array uses a triangular lattice of relatively small elements. The element dimensions are 0.30λ by 0.58λ and the lattice dimensions are 0.425λ

Fig. 19 Array of open-ended waveguide radiators. The matching sheet-radome in front of the array has been removed.

Rectangular Lattice		Triangular Lattice $D_y > 0.45$	Triangular Lattice $D_y < 0.45$
$B < 0.5$ and $A < 0.75$	for	$D_y \geq 0.75$ $A < 0.7$ $B < 0.4$;	 $A < 0.75$ $B \leq 0.4$
	for	$0.75 \geq D_y \geq 0.6$ $A < 0.65$ $B < 0.40$;	
	for	$0.6 \geq D_y \geq 0.45$ $A < 0.6$ $B < 0.4$	

Table I.

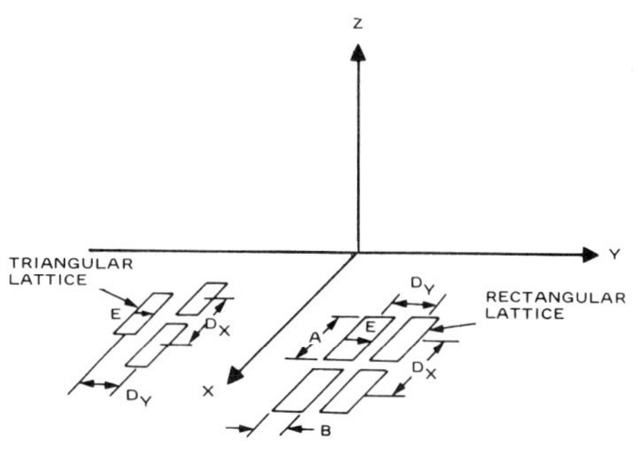

$A = \frac{a}{\lambda}$, $B = \frac{b}{\lambda}$ are waveguide dimensions normalized to the wavelength

D_x and D_y are lattice dimensions normalized to the wavelength

Fig. 18 Sketch of rectangular and triangular array lattices showing notation for Table I.

by 0.64λ . This array is steered by ferrite phase shifters which are located just behind the radiating elements. The corporate feed network occupies most of the rear part of the array and the driver circuits for the phase shifters are located below the cover at the right rear side of the array. This array configuration is typical of a constrained fed array for medium to high power sea or land-based operation. The antenna was developed at Hughes Aircraft Company. Impedance matching of the elements in this array was accomplished by using a flat sheet radome in front of the array. A thick sandwich configuration was used. Measured performance using a waveguide simulator showed that the VSWR was less than 2.5 for scanning to 50 degrees in both the E and H planes and over a 10 percent frequency band. The use of relatively small radiating elements prevented an impedance resonance from occurring.

Essentially the same scan angle mismatch performance has been obtained by other workers in the field for this type of rectangular waveguide array configuration. An array built at the Applied Physics Laboratory of Johns Hopkins University provides another example of this type of performance. The array developed at Johns Hopkins was comprised of six subsections, or subarrays, one of which is shown in Figure 20. The elements were matched with a thin, high di-electric constant radome spaced in front of the array (23). The subarrays could be arranged side by side in a line or in a 3 x 2 arrangement to form a nearly square array. Measurements of element mismatch were obtained approximately by measuring the mismatch at the

Fig. 20 Photo of a subarray developed at Applied Physics Laboratory of Johns Hopkins University. A matching sheet radome is removed for this photo.

input to a subarray. The basis for this technique rests primarily on the fact that when non-reciprocal phase shifters are used the sum of the forward and return phase shifts through each phase shifter is a constant independent of phase shifter setting. Since each element reflection coefficient is (ideally) the same, it can be shown that all the reflected energy is returned to the main input of the corporate feed and the VSWR in this line is the same as the VSWR at the element, after correcting for the effect on VSWR of the loss in the corporate feed. To the extent that the ideal performance required of the phase shifters and elements is realized, this method provides a simple method of measuring element VSWR. This technique can provide the measuring tool for proceeding with a purely experimental element development in those cases where the element is not amenable to calculation. The Johns Hopkins array was matched in this manner using a thin high dielectric constant radome in front of the array. This method yielded a VSWR less than 3.0 for a scan angle out to 60 degrees and a frequency band of 10 percent.

Another example of a waveguide array design is shown in Figure 21. This photo shows a test array used during the development of the AN/SPS-33 antenna at Hughes Aircraft Company. The element for this array was matched by an experimental procedure since the techniques for calculating performance did not exist at the time of development. The lattice is triangular and the lattice dimension was chosen to provide a good margin of safety in excluding grating lobes, the margin amounting to a 10 percent reduction in element spacing

Fig. 21 Test array studied during development of AN/SPS-33 antenna at Hughes Aircraft Company.

from that which produced a grating lobe at the maximum scan angle. The element size in the array is 0.184λ by 0.78λ at the highest frequency of operation. The narrow E plane dimension permitted a sinuous feed to be placed between columns of radiators which then provided for frequency scanning in elevation. The array elements were matched by means of a plane radome of appropriate thickness spaced correctly in front of the array. An inductive post was used inside of the element in order to center the impedance curve. Impedance data was obtained experimentally for various radomes by means of a slotted line placed just behind the aperture of a central element. The final results for element VSWR are shown in Figure 23.

A unique type of element must be designed when there is a requirement for dual linear polarization or circular polarization. When dual linear polarization is required in an element, the element must be reasonably large electrically in both dimensions in order to support propagation. Various techniques have been employed to reduce the element physical size in order to reduce the lattice dimension and maintain the grating lobe angle at some reasonably large value. One size reduction technique employs the use of a ridge loaded waveguide as shown in Fig-

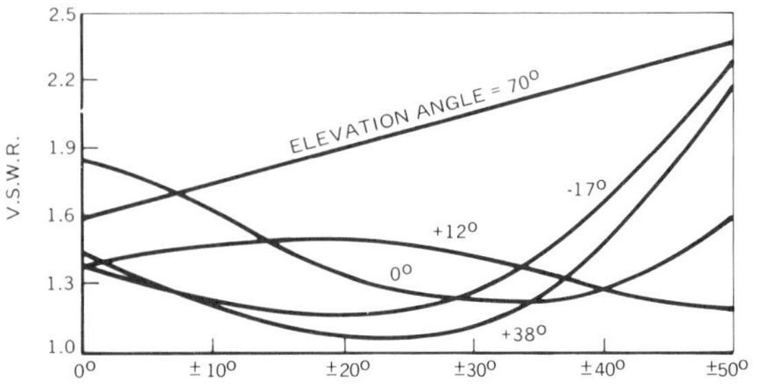

Fig. 22 Plot of VSWR vs. azimuth scan angle for an element of the AN/SPS-33 test array. Elevation angle is plotted as a parameter.

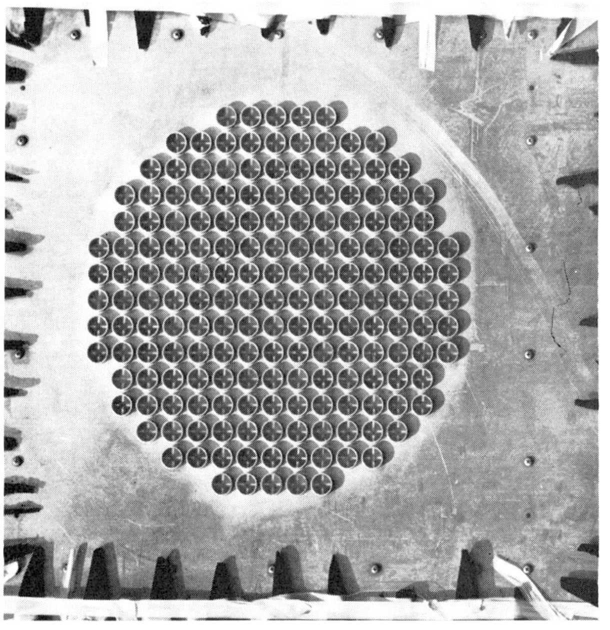

Fig. 23 Photo of an array of quadruple ridge-loaded circular waveguide elements.

ure 24. This photo shows an S-band test array developed at Hughes Aircraft Company. Care must be exercised in such designs to exclude the TE_{21} mode by using symmetrical excitation.

Another interesting size reduction technique, which was employed by the workers at Sylvania and Bell Telephone Laboratories, is to use an open-ended section of coaxial line having a large inner conductor. Since the desired $TE_{1,1}$ mode begins to propagate when the average circumference is one wavelength, the average diameter of the annular region is then approximately one third of a wavelength at cutoff. This readily permits fitting the element in a half wavelength lattice. A sketch of an early configuration of this element is shown in Figure 25. Mode filtering tech-

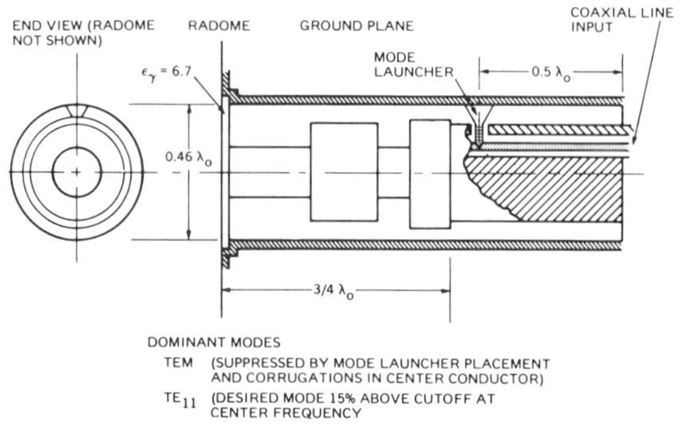

Fig. 24 Sketch of a dual polarized element using large sized coaxial line.

niques were used to suppress the excitation and radiation of the TEM mode. This was accomplished by the half wave spacing of the probes from the end plate and by the use of a filter section which is selective in guide wavelength. A modified form of the element is shown in Figure 26. The performance

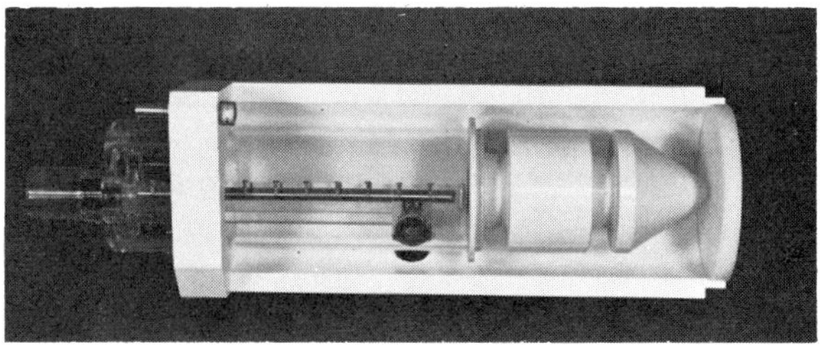

Fig. 25 Model of modified version of element shown in Figure 25.

of this element is discussed by Amitay, et al in (24) and is generally such that less than 15 percent of the power is returned for scan angles to 50 degrees over the lower part of the percent frequency band.

A third design method for a dual polarized element has found an advantage in not loading the waveguide cross section but in actually employing a waveguide section which is below cutoff. This element design is shown in Figure 27. This element was developed at Wheeler Laboratories. The waveguide below cutoff section is between the crossed-dipole launcher and the aperture. Of course, the length of the below cutoff section is not large. By employing a small element such as this, the impedance resonance effect was avoided. In addition, the design procedure was simplified since the elementary grating lobe series could be used for design calcu-

Fig. 26 Photo of a dual polarized element employing a short length of waveguide below cutoff.

lations (25). This element is matched to VSWR less than 2.0 over about a 50-degree scan range and a 13 percent bandwidth.

Another method of providing a small cross section dual polarized element was developed at Bendix Radio Division. The element consists of a strip dipole superimposed on an open-ended waveguide radiator and is shown in Figure 28. The tracking of the shapes of

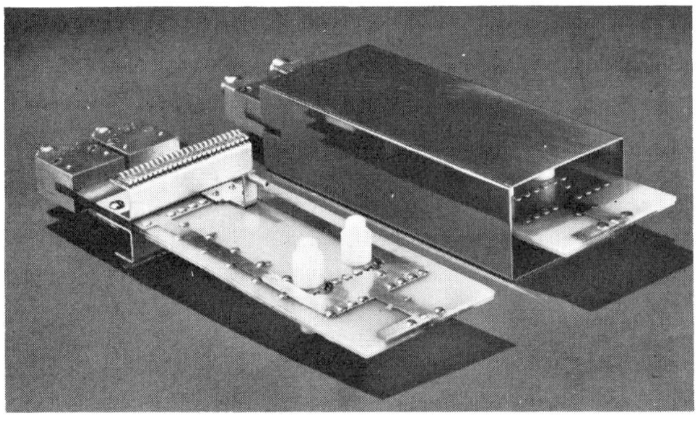

Fig. 27 Photo of dual polarized radiator employing a dipole superimposed on a waveguide radiator.

the active element patterns of the separate radiators was quite good out to about 50 degrees, as demonstrated in a test array using 136 radiators. When excited for circular polarization the element axial ratio vs. scan angle is as shown in Figure 29. Plots are shown for the two principle planes of the array.

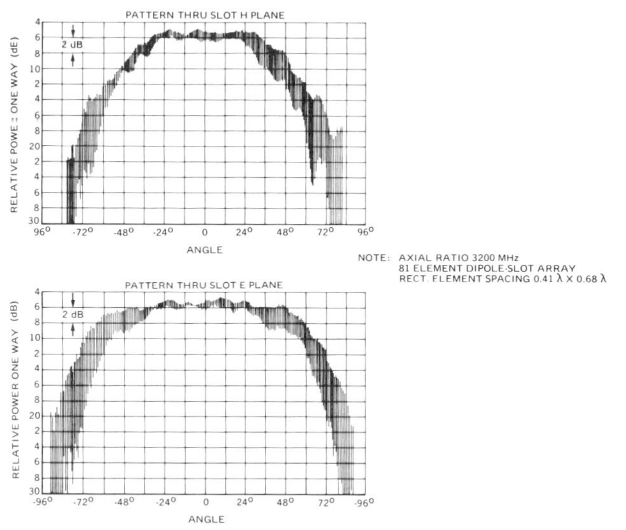

Fig. 28 Plot of axial ratio vs. scan angle for element of Figure 28. Axial ratio indication obtained by rotating a linearly polarized transmitter.

Waveguide Slot Array Element Design

Waveguide slot array elements were among the earliest used for microwave antenna arrays. Their use was pioneered at the National Research Council of Canada (26) and at Hughes Aircraft Company. Their early use was in non (electronic) scanning arrays and they were employed to achieve a high degree of aperture distribution control. By the mid-1950's these arrays were used in frequency scanned electronic scanning systems and they were later used in phase scanned systems. The effects of mutual coupling caused by beam scanning were found not to be serious for these arrays since the radiating elements are essentially half wave radiators. In addition, the feed systems employed are not sensitive to changes in element impedance.

A typical waveguide slot array antenna configuration capable of one-dimensional electronic scanning is shown in Figure 30. The linear waveguide slot arrays employ edge slots so that the waveguides may be stacked sufficiently closely to obtain wide angle scanning without a grating lobe entering real space. Usually enough bandwidth is required in the array and the arrays are sufficiently long that it is necessary to employ a traveling wave linear array design for each of the waveguide arrays (27). The space between waveguide arrays is blocked by a metal wall to prevent radiation in the backward direction. The metal wall must be placed some distance back from the front surface of the waveguides because the edge slots wrap around the waveguides into the broad walls to some extent.

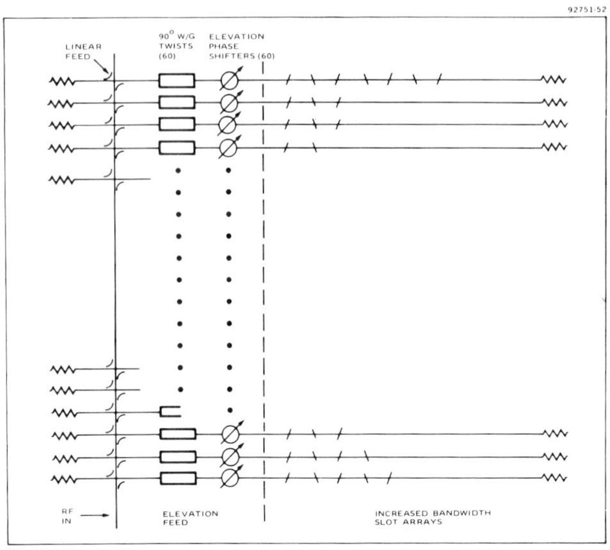

Fig. 29 Sketch of a waveguide slot array capable of electronic scanning in one dimension.

The design of the waveguide slot elements for an electronic scanning array actually proceeds along the same lines as for the design of a fixed beam array (27). The additional decision to be made for a scanning array is to select the scan angle at which the slot data is experimentally obtained. To briefly state the slot design method, a small test assembly is built which contains five to ten identical arrays of identical slots which are resonantly spaced along the waveguide. The slots are alternated in inclination angle to obtain a beam at broadside. The arrays may be of reduced length and contain of the order of ten slots. All slots are together varied in length until an input admittance measurement to an array shows that the slots are resonant, i.e., that the admittance at the plane of a slot is pure real. Because the slots are resonantly spaced, this admittance measurement yields the sum of the admittances of all slots in the array. This resonant length is the slot length employed in the array design. Now to obtain aperture distribution control along the arrays in the ultimate antenna design, the slot conductance must be varied. This is accomplished by varying the slot inclination angle. It has been well established that the conductance obeys a $\sin^2 \theta$ law,

$$G = G_o \sin^2 \theta$$

and the constant G_c may be obtained from the array input impedance measurement at resonance. The small test array thus supplies the essential data for the array design. There is a slight variation in slot length required to achieve resonance as the slot inclination angle changes. Usually three or four separate slot angles in different test arrays will provide enough data for accurate interpolation to obtain resonance at any slot angle. The choice of phasing between linear arrays for the obtaining of the slot data should be that inter-array phasing resulting in the scan angle at which best performance is desired. It is mentioned, however, that array performance does not deteriorate rapidly with scan angle.

With the array designed for the proper aperture distribution in this manner, at some preferred inter-array phasing, the effect of beam scanning is of interest. The effect of a variation in mutual coupling in the scan plane will be to alter the slot admittances presented along each waveguide linear array. To some extent, the slots will become non-resonant and also the conductance will change, which will alter the energy tap-off rate. This problem has not been analyzed theoretically in any substantial depth. However, actual measurements show the effects to be small insofar as modifying the radiation patterns in the plane along the linear waveguide arrays. This is demonstrated by the data in Figure 31, showing radiation patterns along the arrays for scan angles of 1.7°, 30.3°, and 48° in the orthogonal plane. Apparently all slot conductances are changed by approximately the same factor, producing essentially the same aperture distribution. The change in phase angle of radiation from each slot due to a departure from resonance is either nearly the same for all slots or is small.

The effect on radiation patterns in the scan plane may be indicated by the active element pattern of a waveguide linear array in the scan plane. Two such patterns are shown at two frequencies for an S-band array in Figures 32 and 33. It is noted that the active element pattern is less than 1.0 dB below the ideal pattern at scan angles of 50 degrees or less. When all arrays are excited together as a phased array the gain loss indicated by the active element pattern is accounted for by an increase in the power into the loads terminating the waveguide linear arrays. There is essentially no reflection loss at the input to the traveling wave arrays since these lightly loaded waveguides remain essentially always matched. It would be interesting to correlate a measurement of power into the load with the measured active element patterns, however, the measured data is not available.

In general, the performance of these waveguide slot arrays in one-dimensional electronic scanning systems can be seen to be quite favorable.

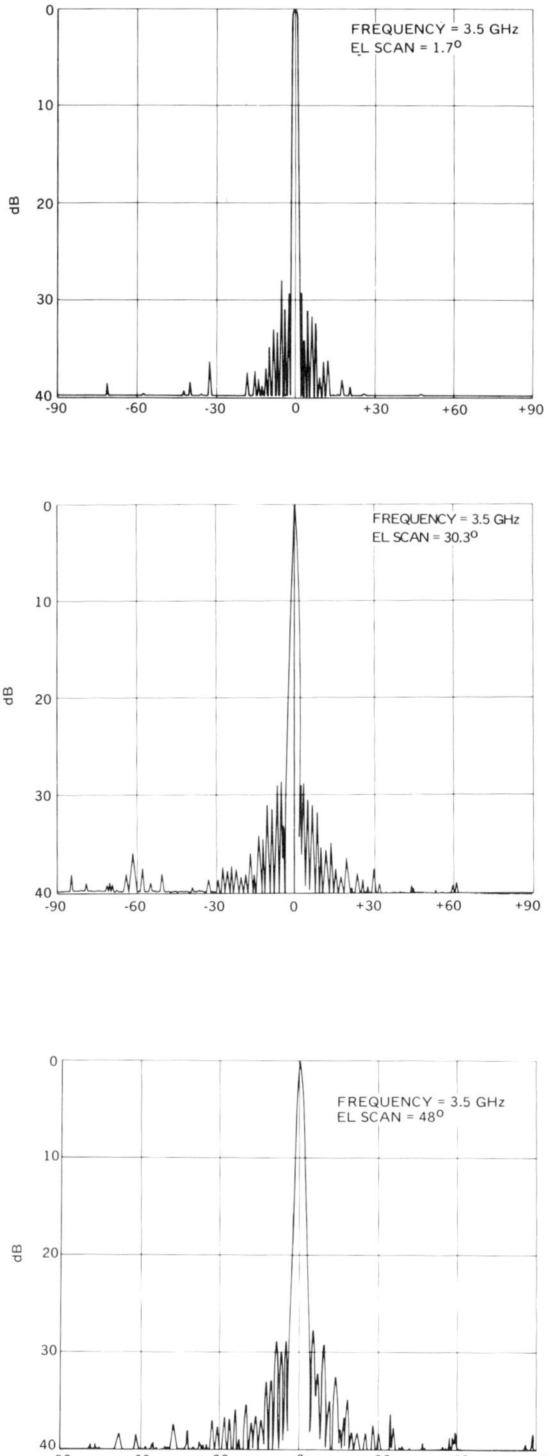

Fig. 30 Radiation pattern along the waveguide array when arrays are scanned in orthogonal plane by(a) 1.7°, (b) 30.3° and (c) 48°.

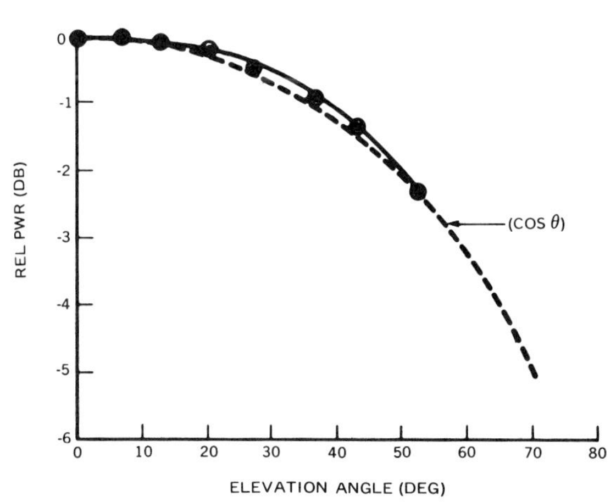

Fig. 31 Active element pattern of a waveguide linear array imbedded in other waveguide arrays, frequency = 3.6 GHz.

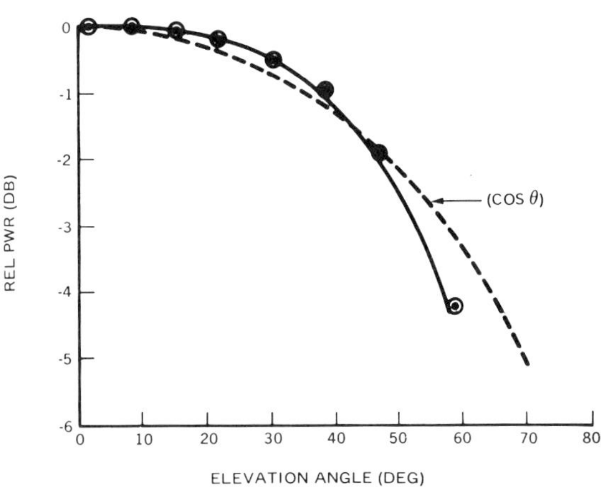

Fig. 32 Active element pattern of a waveguide linear array imbedded in other waveguide arrays, frequency = 3.5 GHz.

REFERENCES

(1) P. W. Hannan, "The element-gain paradox for a phased-array antenna, "IEEE Trans. Antennas and Propagation, vol. AP-12, pp. 423-433, July 1964.

(2) L. Stark, "A helical line scanner for beam steering a linear array, " IRE Trans. Antennas and Propagation, vol. AP-5, pp. 211-216, April 1957.

(3) L. A. Kurtz and R.S. Elliot, "Systematic errors caused by the scanning of antenna arrays: Phase shifters in the branch lines," Technical Memo 359, Hughes Aircraft Company, Culver City, California, 1 May 1953.

(4) J. L. Allen and B. L. Diamond, "Mutual coupling in array antennas," MIT Lincoln Lab Tech. Report 424, p. 49, 4 October 1966.

(5) L. Parad, "Mutual coupling effects on array radiation patterns," M. I. T. Research Lab. of Electronics, Cambridge, Mass. Rept. 5, May 31, 1957.

(6) B. L. Diamond, "Small arrays-their analysis and their use for the design of array elements," pp. 127 this volume

(7) L. Stark, "Radiation impedance of a dipole in an infinite planar phased array," Radio Sci., vol. 1 (new series), pp. 361-377, March 1966.

(8) B. L. Diamond, "A generalized approach to the analysis of infinite planar array antennas," Proc. IEEE, vol. 56, No. 11, pp. 1837-1851, November 1968.

(9) H.A. Wheeler, "The grating-lobe series for the impedance variation in a planar phased-array antenna," IEEE Trans. Antennas and Propagation (Communications), vol. AP-13, pp. 825-827, September 1965.

(10) J. L. Allen, et al, "Phased array radar studies, 1 July 1959 to 1 July 1960," Technical Report 228, Lincoln Laboratory, M. I. T., DDC 249470 H 335, 12 August 1960.

(11) J. L. Allen, et al, "Phased array radar studies, 1 July 1960 to 1 July 1961," Technical Report 236, Lincoln Laboratory, M. I. T., DDC 271724 H 474, 13 November 1961.

(12) J. L. Allen, et al, "Phased array radar studies, 1 July 1961 to 1 January 1963," Technical Report 299, Lincoln Laboratory, M. I. T., DDC 417572, 20 February 1963.

(13) J. L. Allen, et al, "Phased array radar studies, 1 January 1963 to 1 July 1964," Technical Report 381, Lincoln Laboratory, M. I. T., DDC 629363, 31 March 1965.

(14) L. Parad, Private Communication, 24 November 1969.

(15) W. E. Rupp, "Antenna mutual impedance bridge," Microwave J. 5, 95, August 1962.

(16) Schelkunoff, S. A. and Friis, H. T., "Antennas Theory and Practice," John Wiley & Sons, Inc. New York, 1952 p. 318.

(17) B. L. Diamond and G. H. Knittel, "A new procedure for the design of a waveguide element for a phased-array antenna," pp. 149 this volume.

(18) H. A. Wheeler, "A survey of the simulator technique for designing a radiating element, pp. 132 this volume .

(19) P. S. Carter, "Circuit relations in radiating systems and application to antenna problems," Proc. IRE, vol. 20, pp. 1004-1041, June 1932.

(20) "ICBM Early Warning Antenna Studies, Final Report," Astia Doc. AD 474 670, pp. 15-19, 31 October 1957.

(21) S. Edelberg and A. A. Oliner, "Mutual coupling effects in large antenna arrays: pt. I - Slot arrays," IRE Trans. Antennas and Propagation, vol. AP-8, pp. 286-297, May 1960.

(22) A. G. Ehlenberger, L. Schwartzman and L. Topper, "Design criteria for linearly polarized waveguide arrays," Proc. IEEE, vol. 56, No. 11, pp. 1861 - 1872, November 1968.

(23) E. G. Magill and H. A. Wheeler, "Wide-angle impedance matching of a planar array antenna by a dielectric sheet," IEEE Trans. Antennas and Propagation, vol. AP - 14, pp. 49-53, January 1966. (Comments, vol. AP-14, pp. 636-637, September 1966.)

(24) N. Amitay, P. E. Butzien, R. C. Heidt, "Match optimization of a two-port phased array element," IEEE Trans. GAP, Vol. AP-16, No. 1, p. 47, January 1968.

(25) H. A. Wheeler, "A systematic approach to the design of a radiator element for a phased-array antenna," Proc. IEEE, vol. 56, No. 11, pp. 1940-1951, November 1968.

(26) W. H. Watson, "The physical principles of waveguide transmission and antenna systems," Clarendon Press, Oxford, 1947.

(27) I. P. Kaminow, R. J. Stegen, "Waveguide slot array design," Hughes Aircraft Company Tech. Memorandum 348, 1 July 1954.

Methods of Phased Array Analysis

N. AMITAY and C. P. WU

Bell Telephone Laboratories, Inc.
Whippany, New Jersey

V. GALINDO

TRW Systems
Redondo Beach, Calif.

ABSTRACT

Various approaches to phased array analysis are reviewed and compared. These include both the circuit (element by element) method and the modal (field theory) approach. Rigorous formulation of the boundary value problems is discussed for finite and infinite arrays of current-carrying wire elements and aperture type elements. Various methods of solution are reviewed and ways of ascertaining their validity are outlined.

I. Introduction

Experimental and theoretical investigations (1-53) have clearly indicated that mutual coupling cannot be ignored in phased arrays, especially in the case of closely spaced array elements, since it strongly affects the radiation and reflection characteristics. In a finite sized array, the radiation patterns and the reflection characteristics of the various elements also depend on the elements' position because of the proximity of some of the elements to the edge of the array. In large arrays, however, the bulk of the inner core elements behave almost uniformly and the important aspects of this behavior may be well approximated by the identical behavior of elements in a uniformly illuminated infinite array. Thus, infinite arrays can serve as useful models for the analysis of large planar arrays.

The infinite array model provides a basis for various experimental design methods. (8-16) It also lends itself to a rigorous formulation and simplification in the analysis of the boundary value problem associated with a broad class of phased array elements that have meaningful and practical applications. (8,17-27) The case of nonuniform illumination and certain finite array edge effects can be derived from the solution of the uniformly illuminated infinite array problem. (20, 28-31)

In this work we shall briefly discuss various general methods for the analysis of phased array antennas. These methods can be broadly classified into two categories, namely, the element by element (circuit) approach, and the field theory (modal) approach. The application of the above methods as well as more specific methods will be illustrated in more detail for two broad classes of practical array elements. These are the linear antenna (current-carrying wire) elements and the aperture (waveguide type) antenna elements.

II. General Methods of Phased Array Analysis

As we shall later see, only a very limited class of arrays has an exact analytical solution of the associated boundary value problem. In most practical cases, a numerical or a costly experimental method is required in order to determine the reflection and radiation characteristics of phased arrays.

In the early stages of phased array analysis, numerical solutions were hampered by the low storage capacity and speed of the existing digital computers. Consequently, simplifying assumptions were incorporated in earlier analysis that produced a reasonably qualitative understanding of certain phenomena. With the rapid advance and growth of digital computers, the situation has changed. More sophisticated and accurate methods have and are being utilized for the solutions of rigorously formulated boundary value problems associated with phased array antennas.

II. 1 The Element by Element (Circuit) Approach

Consider the uniformly illuminated planar array of elements of Figure 1. The elements are arranged in a periodic rectangular grid.* The array elements may consist of radiating apertures imbedded in a

*A rectangular grid is illustrated for simplicity. Any doubly periodic grid can be treated similarly.

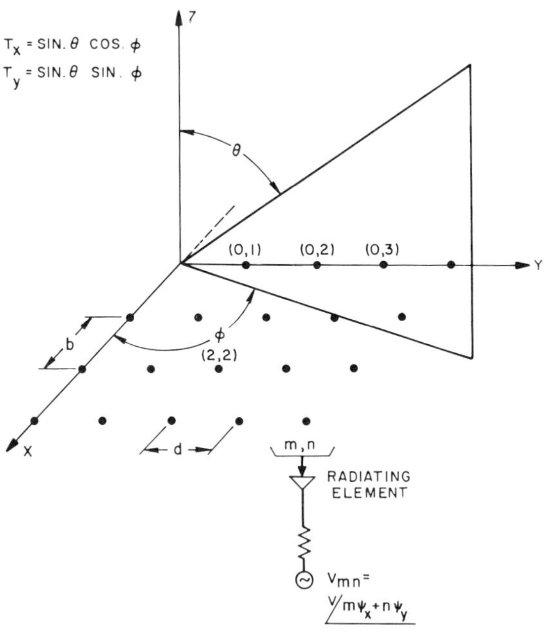

Fig. 1 Planar array model.

perfectly conducting ground plane or current-carrying wire elements (e.g. monopoles, dipoles, etc.) in free space or above a ground plane. The voltages of the exciting generators are given by

$$V_{mn} = V\, e^{j\left(m\psi_x + n\psi_y\right)}; \quad \begin{cases} \psi_x = kb \sin\theta \cos\phi = kbT_x \\ \psi_y = kd \sin\theta \sin\phi = kdT_y \end{cases} \tag{1}$$

where ψ_x and ψ_y are the phase shifts between adjacent elements required to scan a beam to the (θ, ϕ) direction, or (T_x, T_y) in directional cosine space. The free space propagation constant is $k = 2\pi/\lambda$ where λ is the wavelength. The terminal relationship of an active array element (i.e. the variation of the active impedance Z_a, admittance Y_a or reflection coefficient R_a) is a linear superposition of the contributions from all the array elements.

$$Z_a = \sum_{(m)} \sum_{(n)} z_{mn}\, e^{j\left(m\psi_x + n\psi_y\right)} \tag{2a}$$

$$Y_a = \frac{1}{Z_a} = \sum_{(m)} \sum_{(n)} y_{mn}\, e^{j\left(m\psi_x + n\psi_y\right)} \tag{2b}$$

$$R_a = \frac{1 - Y_a}{1 + Y_a} = \sum_{(m)} \sum_{(n)} C_{mn}\, e^{j\left(m\psi_x + n\psi_y\right)} . \tag{2c}$$

The knowledge of either the mutual impedances $\{z_{mn}\}$, or mutual admittances $\{y_{mn}\}$ or mutual coupling (scattering) coefficients $\{c_{mn}\}$ determines the active element terminal relationship as a function of scan. For finite arrays, Z_a, Y_a and R_a in Equations (2a)-(2c) vary from element to element while for infinite arrays these relationships become Fourier series transforms and are independent of the location of the element in the array. Relationships of Equations (2a)-(2c) must be augmented by some methods, either analytical or experimental, to determine z_{mn}, y_{mn} or C_{mn}. Analytical methods must, of course, involve certain field theoretical analysis which accounts for the particular element geometry and mode of excitation. Basically, the array problem is reduced to the auxiliary problems of determining z_{mn}, y_{mn} or C_{mn} between various pairs of array elements.

II. 2 The Modal (Field Theory) Approach

This approach treats the whole array as a boundary value problem. The effects of mutual coupling are intrinsically accounted for in this approach.

Infinite arrays are viewed as infinite periodic structures. Due to the periodicity of the grid, the active element aperture field (or current distribution) has to be evaluated only in a single periodic cell, Figure 2. Utilizing Floquet's theorem for periodic structures, we derive a set of orthogonal free space modes

$$\left\{ \Psi_{pmn}(x,y)\, e^{-j\Gamma_{mn} z} \right\}$$

for the representation of the electromagnetic field at the aperture and in free space $(z \geqslant 0)$

$$\vec{\Psi}_{pmn}(x,y) = \vec{\Omega}_{pmn}\left(\psi_x,\psi_y,b,d\right)\exp\left[-j\left(k_{xm}x + k_{yn}y\right)\right] \quad (3)$$

where p = 1 denotes a TE mode and p = 2 denotes a TM mode

and

$$k_{xm} = \frac{2\pi m-\psi_x}{b}; \quad k_{yn} = \frac{2\pi n-\psi_y}{d}; \quad k_{xm}^2 + k_{yn}^2 + \Gamma_{mn}^2 = k^2$$

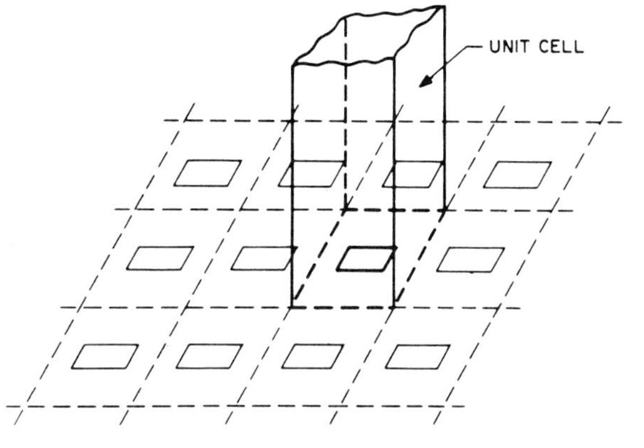

Fig. 2 Unit cell in a two-dimensional periodic array.

$m,n = -\infty, \ldots, -1,0,1, \ldots, \infty$. Exp $(j\omega t)$ time dependence is assumed. Depending on whether Γ_{mn} is real or imaginary, $\vec{\psi}_{mn} e^{j\Gamma_{mn}z}$ corresponds to a propagating or an evanescent plane wave. For normal array operation, the element spacing is adjusted such that Γ_{oo} is real, corresponding to the main beam. The rest of Γ_{mn} corresponding to the grating lobes are purely imaginary. The free space fields are directly related to the aperture field (or current distribution). The active reflection (or impedance) and radiation characteristics (including polarization characteristics) can be readily derived from the aperture field (or current distribution).

In earlier approaches, a scan invariant distribution was preassigned as an approximation to the aperture field (or current). Although the analysis is greatly simplified, the validity and applicability of this method is limited.

A rigorous approach, which is widely used at present, consists of the formulation of the problem in terms of an operator equation (e.g., integral, matrix, etc.) for the unknown field or current distribution at the aperture. Numerical (and, in special cases, analytical) solutions of this equation yield all the pertinent information regarding the array characteristics.

In the subsequent paragraphs we shall illustrate the application of these and other methods to broad classes of array elements.

III. Phased Arrays of Linear Elements

This class includes the current-carrying elements such as monopoles, dipoles, and other "current sheet" type array elements.

Early analysis of finite or infinite dipole arrays (32, 33) relied on the element by element approach. The array terminal relationships were characterized by the mutual impedances, Equation (2a). All open-circuited dipoles in the array were assumed to be "invisible."* Mutual impedances between all possible pairs of elements were, in effect, evaluated outside the array environment from Carter's formulas. (3) In addition, in this first order interaction calculation, simple current distributions (e. g. constant or sinusoidal) across the dipoles were assumed. The array radiation characteristics were related to an isolated element radiation pattern. (6, 40) Numerical results will be shown later.

An alternate approach (18) to infinite dipole arrays (or their duals-thin slots in a ground plane (3)) is based on the modal approach. The elements consist of prescribed distribution of unidirectional current sheets periodically placed on an aperture plane (with or without a ground plane below them), Figure 3. The current sheet distribution, $K_x(x,y)$ is assumed to be invariant with scan. Expressing $K_x(x,y)$ in terms of Equation (3), we write the various field quantities at the apertures as

$$K_x(x,y) = \sum_{m,n=-\infty}^{\infty}\sum K_{mn}\, e^{-j[k_{xm}x + k_{yn}y]} \quad (4)$$

*This "invisibility" is a property of a minimum scattering antenna. (39, 40)

$$E_x(x,y) = \sum_{m,n=-\infty}^{\infty}\sum E_{mn}\, e^{-j[k_{xm}x + k_{yn}y]} \qquad (5)$$

$$H_y(x,y) = \left(\frac{\varepsilon}{\mu}\right)^{\frac{1}{2}} \sum_{m,n=-\infty}^{\infty}\sum \frac{k\Gamma_{mn}}{k_{xm}^2 - k^2} E_{mn}\, e^{-j[k_{xm}x + k_{yn}y]} \qquad (6)$$

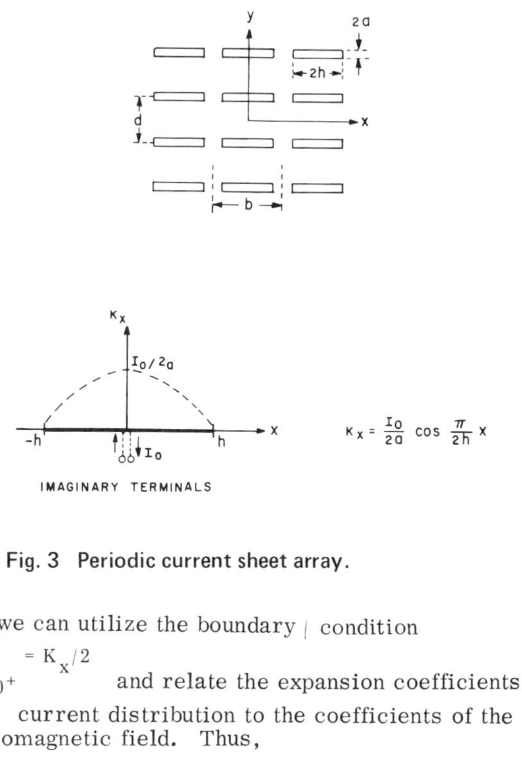

Fig. 3 Periodic current sheet array.

Here we can utilize the boundary condition $H_y\Big|_{z=0^+} = K_x/2$ and relate the expansion coefficients of the current distribution to the coefficients of the electromagnetic field. Thus,

$$E_{mn} = \frac{K_{mn}}{2}\left(\frac{\mu}{\varepsilon}\right)^{\frac{1}{2}} \frac{k_{xm}^2 - k^2}{k\Gamma_{mn}} . \qquad (7)$$

The coefficients K_{mn} can be readily obtained from the Fourier analysis of the assumed K_x, Equation (4). The far field (for a single beam in real space) is proportional to E_{00}. The active impedance characteristic at a pair of imaginary terminals, Figure 3, is obtained from the ratio of the complex power flow per unit cell and the mean square value of the current at the terminals.

$$Z_a = R_a + jX_a = \frac{-\frac{1}{2}\int_{(dipole)} \vec{E}\cdot\vec{K}^* dxdy}{\frac{1}{2}\left|I_o\right|^2} . \qquad (8)$$

Following the above mentioned procedure, one obtains a single expression for R_a and a slowly convergent infinite series for X_a. (3, 18) This method was applied to infinitesimal dipole and half-wave dipole arrays. In the case of half-wave dipole arrays, excellent agreement was found between the modal approach and the element by element approach for a 65x149-element array. For all practical purposes, the numerical results for R_a obtained by these two methods, coincide. (3) Many terms, corresponding to higher order Floquet's modes, have to be taken into account in the modal approach. In Fig-

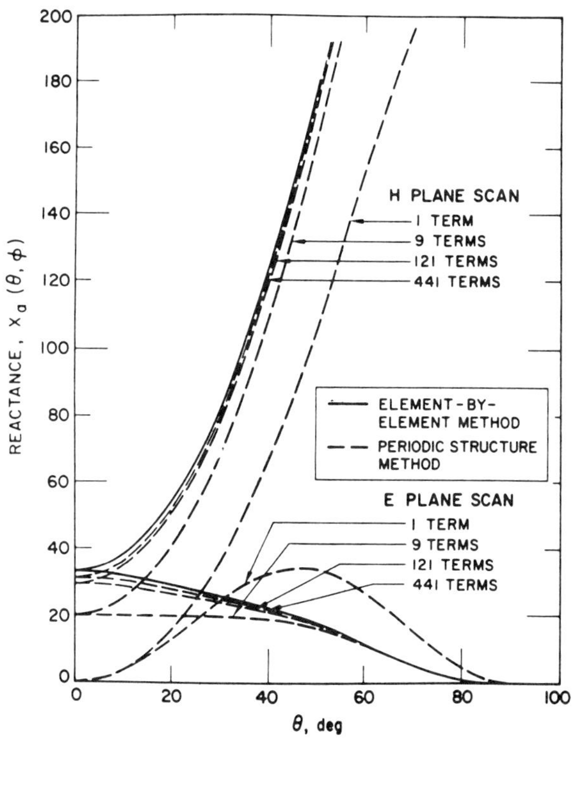

Fig. 4 Active reactance versus scan in E- and H-planes for an array of λ/2 dipoles located λ/4 above a ground plane. Square grid with λ/2 spacing. Prescribed sinusoidal current distribution. (From Oliner and Malech[3]).

ure 4 we compare X_a as computed by the two methods in two planes of scan. As can be seen, the agreement is very good provided that a sufficient number of terms is used in the modal approach. A faster convergence can be obtained when we evaluate the variation of X_a about its value at broadside. (35) The similarity between the current distributions and their invariance with scan in these two approaches suggest that these two approaches may not be completely independent. Further investigation into the interdependence of these approaches may prove useful. The validity of the assumption of prescribed sinusoidal current distribution will be commented upon later.

A more rigorous approach to the analysis of finite and infinite arrays of dipoles is the integral equation formulation. (25, 26, 27) We shall present here the more general approach (25, 26) based upon Hallen's method. Figure 5a shows a planar array of dipoles arranged

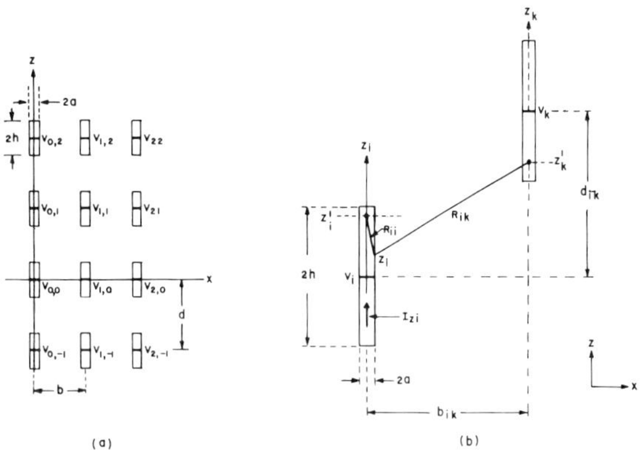

Fig. 5 (a) Planar array of dipoles.
(b) Definition of coordinates.

(for simplicity) in a rectangular grid. The length of each dipole is 2h with a radius a. The dipoles are driven by voltage sources V_{mn} across an infinitesimal gap at their centers. With $ka << 1$ the currents can be assumed as unidirectional (in the z-direction). Therefore, the vector potential has a z component, A_z, only. In Figure 5b we show certain dipole coordinates where, for simplicity, a single subscript replaces the double subscript of Figure 5a. Along the surface of the ith dipole, A_{zi} satisfies the following differential equation

$$\left(\frac{\partial^2}{\partial z_i^2} + k^2\right) A_{zi}(z_i) = -j \frac{k^2}{\omega} V_i \delta(z_i) \;. \tag{9}$$

The solution of (9) is

$$A_{zi}(z_i) = -j\sqrt{\mu\varepsilon} \left\{ C_{iA} \cos kz_i + C_{iB} \sin kz_i + \frac{V_i}{2} \sin k|z_i| \right\} \tag{10}$$

where C_{iA} and C_{iB} are constants, unknown yet. The vector potential can also be expressed in terms of all the currents in the array,

$$A_{zi}(z_i) = \frac{\mu}{4\pi} \sum_{k=1}^{N} \int_{-h}^{h} I_{zk}(z_k') \frac{e^{-jkR_{ik}}}{R_{ik}} dz_k' \tag{11}$$

where $R_{ik} = \sqrt{(d_{ik} + z_k' - z_i)^2 + b_{ik}^2}$ and N is the total number of elements in the array.

In equating Equations (10) and (11), we obtain a set of N-coupled integral equations for the element currents I_{zi},

$$\sum_{k=1}^{N} \int_{-h}^{h} I_{zk}(z_k') \frac{e^{-jkR_{ik}}}{R_{ik}} dz_k'$$

$$= -\frac{4\pi j}{\sqrt{\mu/\varepsilon}} \left\{ C_{iA} \cos kz_i + C_{iB} \sin kz_i + \frac{V_i}{2} \sin k|z_i| \right\} ;$$

$$i = 1, 2, \ldots, N \;. \tag{12}$$

For a uniformly illuminated infinite array with a linear phase taper, Equation (1), Equation (12) reduces to a single integral equation (26) of the form

$$\int_{-h}^{h} K(z, z', \psi_x, \psi_y, b, d) I_z(z') dz'$$

$$= -\frac{4\pi j}{\sqrt{\mu/\varepsilon}} \left\{ C_1 \cos kz + C_2 \sin kz + \frac{V}{2} \sin k|z| \right\} , \tag{13}$$

where $K(z,z',\psi_x,\psi_y,b,d)$ is the kernel of the integral equation. The vanishing of the currents at the ends of the dipoles

$$I_z(\mp h) = 0 \tag{14}$$

will be imposed later in the solution of Equations (12) and (13).

It is important to note a basic difference between Equations (12) and (13). The kernel of Equation (12) is a function of the array geometry only and the excitation (and phasing) appear in the free term on the right-hand side. In contradistinction, the kernel of Equation (13) is a function of the scan angle and geometry.

Equations (12) and (13) cannot be solved analytically, and one has to resort to numerical methods for a solution. One of the most widely used techniques is the method of moments. (34) In this method we expand the unknown current (or field) in terms of a linearly independent set of expansion (basis) function $\{f_p(z);\ p = 1, 2, \ldots, M\}$,

$$I_z(z') = \sum_{p=1}^{M} a_p f_p(z') \ . \tag{15}$$

After the substitution of Equation (15) into Equation (13), we take (M+2) moments* of the resulting functional equation with a linearly independent set of testing (or weighting) functions $\{g_q(z)\ ;\ q = 1, 2, \ldots, M+2\}$.† The integral equation, (13), is reduced to a set of (M+2) linear algebraic equations of the following form

$$\sum_{p=1}^{M+2} K_{qp} a_p = b_q \ ; \qquad q = 1,2,\ldots,M+2 \ , \tag{16}$$

for the M unknown expansion coefficients a_p and the two constants C_1 and C_2. The various unknowns may be obtained from Equation (16) through matrix inversion or, alternatively, by iterative procedures. Note that $\|K_{pq}\|$, which corresponds to Equation (13), is scan dependent. Therefore, Equation (16) has to be evaluated and solved over and over again for each beam pointing direction. This can be quite costly. On the other hand, the matrix $\|K_{qp}\|$ corresponding to the finite array case, Equation (12), is scan independent. Therefore, once this matrix and its inverse have been evaluated, the currents in the array elements can be determined for various illuminations $V_{m,n}$. However, the size of the array to be analyzed by this method is restricted due to the fact that the current of each dipole in the array has to be expanded separately in the form of Equation (15), resulting in a matrix of order N(M+2) in Equation (16). Similar observations will be made later for waveguide arrays.

*With a properly defined scalar product.

†When $f_p(z) = g_p(z)$, this method is known as the Ritz-Galerkin method.

Although the sets of expansion and weighting functions in the method of moments seem arbitrary, their choice is very important from the standpoint of accuracy and efficiency. Economically, the feasibility of solving a problem may sometimes depend upon the ease and accuracy with which $\|K_{qp}\|$ can be evaluated and the conditioning of this matrix. Generally speaking, in choosing the sets of expansion and weighting functions, it is desirable to incorporate as much a-priori knowledge about the problem as possible. For example, the condition of Equation (14) can be incorporated by utilizing sets of functions which vanish at $z = \mp h$. Sets of functions which satisfy Maxwell's equations and the pertinent boundary conditions including edge conditions when applicable should be employed whenever possible.

In any numerical method, it is imperative to ascertain the validity of the results and their relevance to the problem that we are trying to solve. (53) General quantitative measures and criteria for this purpose have not yet been developed. We shall briefly indicate various alternatives for checking numerical solutions which are commonly used.

It has been found that in the Ritz-Galerkin and related methods (36, 37) energy conservation and reciprocity relations are necessary checks, but far from sufficient.

Convergence tests consist of increasing the number of expansion and weighting functions, M in Equation (15), and observing the change in $I_z(z)$. This test may provide a partial and necessary check. It should be applied with care in the vicinity of field singularities where convergence difficulties may be encountered.

The rate of convergence (with M) of the solution usually depends on the particular choice of expansion and weighting functions. Therefore, if repeated solutions of the problem with two independent sets of expansion and weighting functions yield close results, it is reasonable to assume that the solutions are accurate.

One can test how well the boundary conditions are satisfied at certain interfaces, (54) although general quantitative criteria have yet to be developed.

In some cases, by adjusting certain parameters, the boundary value problem can be reduced to a simplified problem which has a rigorous analytical solution. This rigorous solution can be used as a partial standard of comparison to test the numerical solution.

A comparison of the solution with other approximate analytical or numerical solutions obtained by completely independent methods is a very important check.

Finally, experimental verification, whenever possible, is an excellent and independent check.

For the infinite (or finite) dipole array, we can satisfy Equation (14) by the following expansion

$$I_z(z) = \sum_{m=1}^{M} a_m \sin\left[\frac{m\pi(z-h)}{2h}\right]; \quad M \to \infty . \tag{17}$$

After substituting Equation (17) into Equation (13), we weight the resulting functional equation with the set $e^{j\ell\pi z/h}$. This is accomplished by multiplying both sides of the equation by $e^{j\ell\pi z/h}$ and integrating over the interval $-h \leqslant z \leqslant h$. An alternate method known as the five-terms method, (26) can be used to solve Equation (13) when $h/\lambda \lesssim 5/8$. The current is assumed as

$$I_z(z) \approx \sum_{m=1}^{5} Q_m J_m(z) \tag{18}$$

with

$$J_1(z) = \sin k|z| - \sin kh \; ; \quad J_2(z) = \cos kz - \cos kh \; ;$$

$$J_3(z) = \cos \frac{kz}{2} - \cos \frac{kh}{2} \; ; \quad J_4(z) = \sin kz - \frac{z}{h} \sin kh \; ;$$

$$J_5(z) = \sin \frac{kz}{2} - \frac{z}{h} \sin \frac{kh}{2} .$$

Equation (18) satisfied Equation (14) as well. Upon substitution of Equation (18) into Equation (13) and point matching the resulting equation at the seven points, $z = 0, \mp h/3, \mp 2h/3, \mp h$,* we obtain a 7x7 set of linear equations which can be easily solved for Q_m and C_1, C_2.

*This is equivalent to weighting the functional equation over the interval $-h \leqslant z \leqslant h$ with the set of impulse functions $\delta(z)$, $\delta(z-h/3)$, $\delta(z+h/3)$, $\delta(z+2h/3)$, $\delta(z-2h/3)$, $\delta(z-h)$, $\delta(z+h)$.

It was found (26) that covergence was achieved for $M \geqslant 25$ in Equation (17) even for full-wave dipoles. The solution does not converge near the driving point due to the assumed nature of the excitation. A comparison of this method with the five-term method has shown a close agreement of the calculated current distributions. This demonstrates the importance of the choice of expansion and weighting functions as well as the dependence of the rate of convergence upon this choice.

We shall now comment on the adequacy of the sinusoidal current approximation used earlier. Figure 6 shows the active admittance, $Y_a = G + jB$, in the H-plane of scan of an infinite planar array of half-wave dipoles. As can be seen, there is a difference in the admittance calculated by the two methods. A different integral equation formulation of the infinite array of half-wave dipoles and its solution by the Fourier transform method (27) has shown a similar discrepancy with respect to the sinusoidal current approximation. These results suggest that the sinusoidal current approximation may not yield sufficiently accurate results.

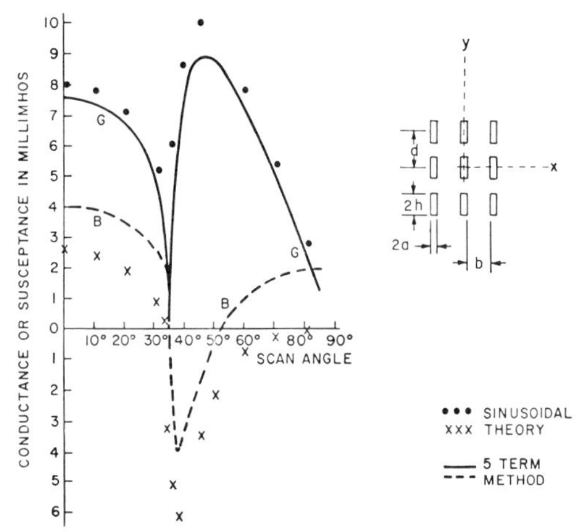

Fig. 6 Active admittance versus scan in H-plane of a half-wave dipole in an infinite planar phased array. h/λ = 0.25, a/λ = 0.007022, d/λ = 0.6, b/λ = 0.25. (From Chang, 26).

It should be pointed out that, in all the methods of analysis of dipole arrays, the effects of a practical dipole feeding network were neglected. Further work in this area may prove fruitful.

IV Phased Array of Aperture–Type Elements

Aperture type arrays refer to waveguide radiators imbedded in a ground plane. The waveguide apertures may be partially covered by thin metallic irises. The use of either dielectric sheaths to cover the array and/or dielectric plugs to load the waveguides can be incorporated in the analysis. Figure 7 depicts an array of circular waveguides arranged in a periodic grid along skewed coordinates s_1 and s_2.

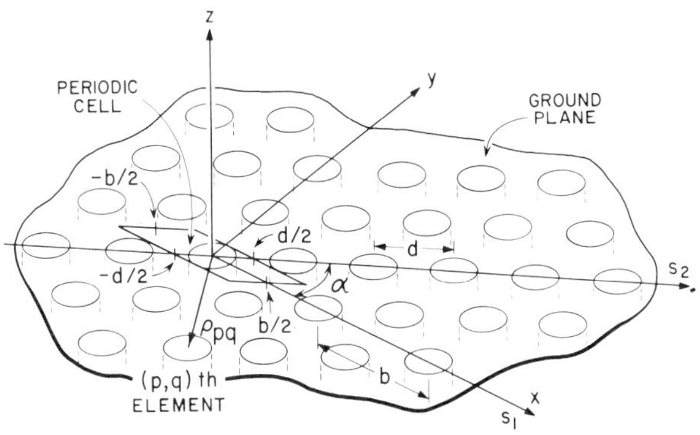

Fig. 7 Circular waveguide array geometry.

For convenience, we shall divide the following discussion into finite and infinite arrays. The latter category will be considered first. Many methods have been developed for the analysis of infinite arrays through a rigorous formulation of the boundary value problem. They may be classified as

1) Residue calculus (or Wiener-Hopf) technique (20)
2) Modified residue calculus (41)
3) Generalized scattering approach (42, 43)
4) Mode matching method (21, 23, 48, 49)
5) Complex power (22)
6) Integral equation (12, 19, 24, 38, 50)
7) Grating lobe series (17, 35)

The boundary value problems associated with phased arrays are in general rather complex. Rigorous analytical solutions to these problems exist only for a rather limited number of cases. Such a special case is a rectangular waveguide array under certain idealizing assumptions. In the array shown in Figure 8, if we assume that the waveguide walls are infinitisimally thin, and the voltage applied to the (m, n)th element is given by

$$V_{m,n} = V e^{j(m\psi_x)} \quad \text{for H plane scan}$$

or $\qquad\qquad\qquad\qquad\qquad\qquad\qquad$ (19)

$$V_{m,n} = V e^{j(\pm m\pi + n\psi_y)} \quad \text{for quasi-E plane scan}$$

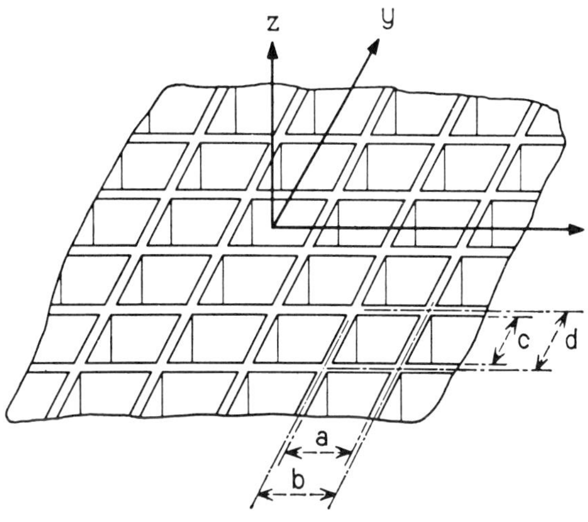

Fig. 8 An infinite array of rectangular waveguides.

The problem becomes equivalent to a linear array of parallel-plate waveguides. Exact solutions for the parallel-plate array can be obtained by a residue calculus (or Wiener-Hopf) technique. (20) An infinite set of linear equations is derived by the mode matching method or by the application of the moments method to the pertinent integral equation. It turns out that this infinite set of equations has a special form and thus can be solved exactly by applyi ng the Cauchy residue theorem. The exact analytical solution has been utilized to calculate the reflection and transmission coefficients as functions of scan, determine the mutual coupling between the elements, and examine the asymptotic behavior of the coupling coefficients.

The results for the parallel plate arrays are obviously very limited in scope. In order to extend the usefulness of the method, a modified residue calculus method (41)has been developed. This approach relaxes the restriction of thin waveguide wall assumption, and also allows a solution when the waveguides are completely loaded with dielectrics. Although the method has extended the scope of the residue calculus technique, it has been applied only to scalar one-dimensional problems. Furthermore, while this method converges very rapidly, it is not an exact solution in the classical sense of the word.

An approximate method which has been devised to solve two-dimensional array problems is to apply the so-called generalized scattering matrix approach. (42, 43) To do this, with reference to Figure 8, a set of waveguide walls, say those parallel to the xz plane, is extended by a certain amount to create a parallel-plate region. The outcome of this arrangement, as shown in Figure 9, is that we now have an array in which

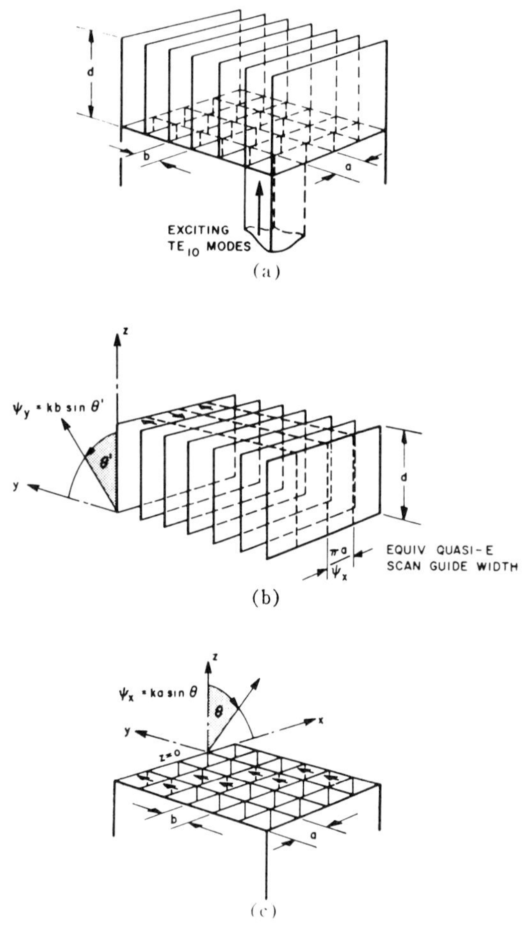

Fig. 9 (a) Composite planar array.
(b) Quasi-E plane scanning.
(c) H-plane scanning.

rows of rectangular waveguides radiate into parallel plate regions, and then the latter radiates into space. We thus have a cascade of two discontinuities. One of the advantages of adopting such an arrangement is that each discontinuity corresponds to a scalar one-dimensional problem, and is easier to handle mathematically. In particular, if the waveguide walls are assumed to be thin, each discontinuity has an exact analytical solution as discussed. The aggregate effect of the composite array may be determined by using scattering matrix formalism. When the separation between these two discontinuities is large, the higher order modes generated at each discontinuity sufficiently decay before they reach the other one and only interaction between the propagating modes need to be considered. On the other hand, if the separation is small, the interaction between some of the higher order modes must be taken into account. In the latter situation the generalized scattering approach has to be used. (42, 43) This method has been successfully used in a number of investigations. In fact, a recent study (42) seems to indicate that the process coupled with a limiting procedure (letting the height of the parallel plate go to zero) may be used to analyze the original array problem.

The formulation of methods 4) through 7) has a common starting point. First, suitable representations of the tangential fields for the region both inside and outside the waveguides are introduced. Inside the waveguides, the fields are best expressed in terms of the waveguide modes, $\vec{\Phi}_{pmn}(x,y)$. In general, these are orthonormal vector mode functions. The first subscript denotes the mode type; p = 1 for the TE modes and p = 2 for TM modes, while the other two subscripts are the usual modal indices. One can always rearrange this set of modes in a consistent way utilizing a single subscript, $\vec{\Phi}_i(x,y,)$ Under the assumption that each waveguide is excited by the first I modes, the tangential electric and magnetic fields may be described as

$$\vec{E}_t(x,y;z\leq 0) = \sum_{i=1}^{I}\left(A_i e^{-j\gamma_i z} + R_i e^{j\gamma_i z}\right)\vec{\Phi}_i(x,y) + \sum_{i=I+1}^{\infty} D_i e^{j\gamma_i z}\vec{\Phi}_i(x,y) \qquad (20)$$

$$-\hat{z}\times\vec{H}_t(x,y;z\leq 0) = \sum_{i}^{I} y_i\left(A_i e^{-j\gamma_i z} - R_i e^{j\gamma_i z}\right)\vec{\Phi}_i(x,y) - \sum_{i=I+1}^{\infty} y_i D_i e^{j\gamma_i z}\vec{\Phi}_i(x,y)$$

where R_i and D_i are the unknown modal coefficients to be determined, and y_i and Υ_i are the modal admittances and propagation constants, respectively. Because of the periodicity of the array geometry and excitation, the fields outside the waveguides must satisfy the Floquet theorem and thus may be expanded in terms of Floquet modes, Equation (3).

$$\vec{E}_t(x,y,z\geq 0) = \sum_{p=1}^{2} \sum_{m,n=-\infty}^{+\infty} B_{pmn}\, e^{-j\Gamma_{mn}z}\, \vec{\Psi}_{pmn}(x,y)$$

$$-\hat{z}\times\vec{H}_t(x,y,z\geq 0) = \sum_{p=1}^{2} \sum_{m,n=-\infty}^{+\infty} Y_{pmn} B_{pmn}\, e^{-j\Gamma_{mn}z}\, \vec{\Psi}_{pmn}(x,y) \qquad (21)$$

Here, B_{pmn} are the unknown expansion coefficients, Y_{pmn} are modal admittances, and Γ_{mn} are the propagation constants.

The fields in the two respective regions must be continuous at the common aperture (z = 0). When this requirement is imposed, we obtain two equations relating the unknown modal coefficients of the two different regions. One can eliminate one set of unknowns by utilizing the orthonormality of each set of modal functions. The result is an infinite linear system of equations with the modal coefficients as unknowns. In practice, this infinite system of equations has to be truncated, so that approximate solutions can be obtained. The solutions thus obtained can be very accurate, however. The process described above is called the mode matching technique, or modal analysis method. (21, 23)

Instead of matching the tangential electric and magnetic field across the aperture separately, one can equate the complex power flow across the aperture. (22) This step yields a relation equating a quadratic form of the modal coefficients in one region to that of the other. A condition for the existence of solution is that the modal coefficients be the solution of an infinite and linear system of equations, which turns out to be identical to that obtainable by the mode matching method. Therefore, the above two methods are actually equivalent as might be anticipated.

A more general method for formulating phase array problems is the integral equation technique. This method differs from the mode matching approach in that the tangential electric field is expressed in terms of the tangential magnetic field or vice versa before the continuity conditions are imposed. This step can be accomplished by utilizing the orthonormality of the modes. When the boundary conditions are finally enforced, one obtains the desired integral equation. There are different forms into which the integral equations may be cast. A pair of frequently used Fredholm integral equations of the first kind are shown below.

$$2\sum_{i=1}^{I} A_i y_i \vec{\Phi}_i(x,y)$$

$$\iint_{(\text{aperture})} \left\{ \sum_{i=1}^{\infty} y_i \vec{\Phi}_i(x,y)\vec{\Phi}_i(x',y') + \sum_{p=1}^{2} \sum_{p,m,n=-\infty}^{\infty} Y_{pmn}\vec{\Psi}_{pmn}(x,y)\vec{\Psi}^{*}_{pmn}(x',y') \right\} \cdot \vec{E}_t(x',y')\,dt \qquad (22)$$

$$2\sum_{i=1}^{I} A_i z_i \vec{\Phi}_i(x,y)$$

$$= \iint_{(\text{unit cell})} \left\{ \sum_{i=1}^{\infty} z_i \vec{\Phi}_i(x,y)\vec{\Phi}_i(x',y') + \sum_{p=1}^{2} \sum_{p,m,n=-\infty}^{\infty} Z_{pmn}\vec{\Psi}_{pmn}(x,y)\vec{\Psi}^{*}_{pmn}(x',y') \right\} \cdot \vec{H}_t(x',y')\,dt \qquad (23)$$

where

$$Y_{1mn} = 1/Z_{1mn} = \frac{\Gamma_{mn}}{\omega\mu},$$

$$Y_{2mn} = 1/Z_{2mn} = \frac{\omega\epsilon}{\Gamma_{mn}}$$

and the definition of $\{\vec{\Phi}_i\}$ is extended to the entire periodic cell, i.e.,

$$\vec{\Phi}_i = \begin{cases} \vec{\Phi}_i & \text{inside the waveguide} \\ 0 & \text{elsewhere} \end{cases}$$

The tangential fields at the aperture plane are

$$\vec{E}_t(x,y) = \vec{E}_t(x,y,z)\Big|_{z=0}, \qquad \vec{H}_t(x,y) = \vec{H}_t(x,y,z)\Big|_{z=0}.$$

Note that Equation (22) is defined over the waveguide aperture while Equation (23) is defined over the entire periodic cell. As was remarked earlier, the kernel of the above integral equations are scan dependent because Z_{pmn} and $\vec{\psi}_{pmn}(x,y)$ are functions of scan.

The reflection coefficients R_i and far field radiation characteristics E_θ and E_ϕ may be obtained (23, 23) from the solution of either Equation (22) or Equation (23). Thus,

$$R_i = \iint_{\text{aperture}} \vec{E}_t(x,y)\cdot\vec{\Phi}_i(x,y)\,da - A_i$$

$$\left.\begin{aligned} E_\theta &= \left(1 - T_x^2 - T_y^2\right)^{-1/2} \iint_{\text{aperture}} \vec{E}_t(x,y)\cdot\vec{\Psi}^{*}_{100}(x,y)\,da \\ E_\phi &= -\iint_{\text{aperture}} \vec{E}_t(x,y)\cdot\vec{\Psi}^{*}_{200}(x,y)\,da \end{aligned}\right\} \qquad (24)$$

The functional form of Equations (22) and (23) remains unchanged when the array is covered by dielectric sheaths or loaded with dielectric plugs. (44, 45) However, when thin metallic irises are present at the waveguide apertures, Equation (22) is preferrable, since only the domain of integration has to be modified accordingly. Equation (23) is no longer valid and a new formulation for the magnetic field is necessary.

The integral equations may be solved by the method of moments, which has already been discussed. The set $\vec{\Phi}_i(x,y)$ seems a natural choice for expansion and testing in the solution of Equation (22), while the set $\vec{\psi}_{pmn}(x,y)$ for Equation (23). In addition to satisfying Maxwell's equations and the boundary conditions, the orthonormality of these sets can be utilized. Following this procedure, one obtains a set of equations identical to those derived by the mode matching or complex power methods. The latter methods may therefore be regarded as special cases of the integral equation technique.

Variational expressions for the input impedance or admittance can be derived from the integral equations provided that their kernels are hermitian or complex symmetric. The kernels arising from phased array problems may be shown to be complex symmetric when the geometry of the waveguide and the array grid exhibit point symmetry. (46, 47) Under such conditions and a single mode, $\vec{\Phi}_1$, excitation the variational expression for the array input admittance is given by

$$y_1\left(\frac{1-R}{1+R}\right)\left[\iint_A \vec{\Phi}_1(x,y)\cdot\vec{E}_t(x,y)\,dxdy\right]^2$$

$$= \iint_{\text{aperture}} dxdy\,\vec{E}_t(x,y)\cdot\iint_{\text{aperture}} dx'dy'\left\{\sum_{i=2}^{\infty} y_i\vec{\Phi}_i(x,y)\vec{\Phi}_i(x',y') + \sum_{p=1}^{2}\sum_{m,n=-\infty}^{\infty} Y_{pmn}\vec{\Psi}_{pmn}(x,y)\vec{\Psi}^*_{pmn}(x',y')\right\}\cdot\vec{E}_t(x',y') \quad (25)$$

When an assumed aperture field distribution, called the trial function, is substituted into the above equation, the input admittance is readily evaluated.† The incident field is quite often used to approximate the true aperture field. Thus, let

$$\vec{E}_t(x,y) \approx \alpha\vec{\Phi}_1(x,y)$$

where α is an unknown constant being equal to $1 + R$. The input admittance is therefore given by

$$y_1\left(\frac{1-R}{1+R}\right) \approx \sum_{p=1}^{2}\sum_{m,n=-\infty}^{\infty} Y_{pmn}\left|C^1_{pmn}\right|^2 \quad (26)$$

with

$$C^1_{pmn} = \iint_{\text{aperture}} \vec{\Phi}_1(x,y)\cdot\vec{\Psi}_{pmn}(x,y)\,dxdy \quad (26A)$$

This result is sometimes referred to as the grating lobe series. (35, 21) While the method has been used on a number of occasions to yield satisfactory results, it has been found to fail in various situations. There is no test which can be performed a priori to determine when the grating lobe series is applicable. In order to avoid any unforeseen difficulty, one may use the Rayleigh-Ritz procedure to determine a better approximation of the unknown aperture field. This process leads to a system of linear equations for determining the un known expansion coefficients of the aperture field. This system of equations is identical to that obtained by the Ritz-Galerkin method of solution of the integral equation, provided that the same set of expansion functions is used in both methods. This fact demonstrates that the stationarity property is inherent in the various formulations.

Examples showing that the incident field approximation is not adequate are illustrated in Figure 10. Figure 10a shows the convergence of solution with increasing number of modes used to approximate the aperture field. Figure 10b shows the reflection characteristics of a circular waveguide array in the vicinity of grating lobe incipience along the circular arc ABC. While Equation (26) would predict a total reflection because the admittance of the pertinent TM mode becomes infinite, the actual calculation (24) exhibits an entirely different behavior. This result and an examination of Equation (22) indicate that the pertinent TM modal coefficient vanishes at the grazing angles.

The derivation of the integral equation for infinite arrays suggests that in the formulation of finite array problems only modification of the field representations outside the waveguides is needed. This can be accomplished by using a Green's representation. Finite arrays imbedded in infinite ground planes are particularly easy to handle, because the Green's function for radiation into half space is readily available. For example, for a finite array of parallel plate waveguides shown in Figure 11, the tangential magnetic field in free space $z \geqslant 0$ can be expressed in terms of the aperture electric field by using a combination of the source equivalence theorem and imagy theory. (51) For the E plane of

†It should be noted that convergence properties of the series in Equations (2) and (25) are, in general different.

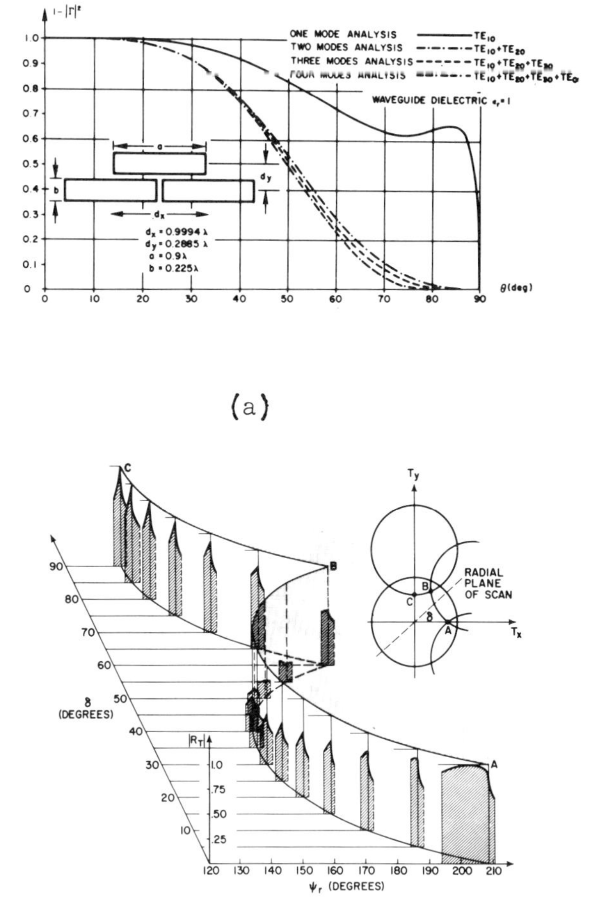

Fig. 10 (a) Normalized transmitted power in the 45° plane of scan of a rectangular waveguide array. (From Borgiotti, 23)
(b) Total reflection, R_T, of a circular waveguide array versus steering phase, ψ_r, in the vicinity of grating lobe incipience. (From Amitay and Galindo 24).

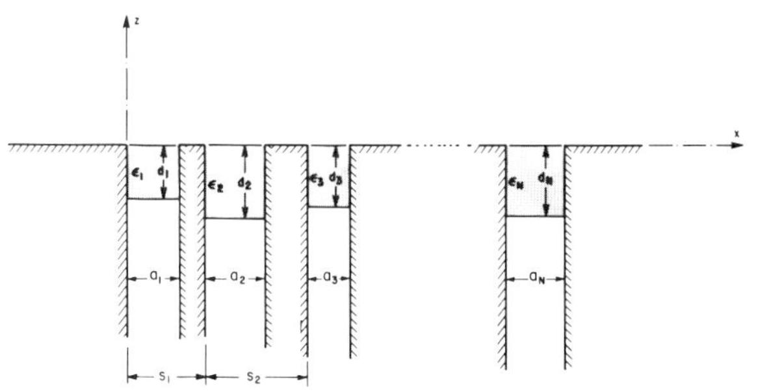

Fig. 11 A finite array of parallel plates in an infinite ground plane.

scan, only the TM modes are excited. The result is

$$H_x(y) = -\frac{\omega\varepsilon_o}{2}\int_{\Sigma A_i} H_o^{(2)}\left(k|y-y'|\right) E_y(y')\, dy' \tag{27}$$

For the H-plane of scan, only TE modes are excited, resulting in

$$H_y(y) = -\frac{1}{2\omega\mu_o}\left(\frac{\partial^2}{\partial y^2}+k^2\right)\int_{\Sigma A_i} H_o^{(2)}\left(k|y-y'|\right) E_x(y')\, dy' \tag{28}$$

The integrals in Equations (27) and (28) are defined over all the waveguide apertures. The field inside the waveguide may be described in a manner similar to the infinite array case. Matching the boundary condition at the waveguide aperture leads to the desired integral equation (38)

$$2\sum_{p=1}^{N}\alpha_p y_{po}\phi_{po}(y) = \int_{\Sigma A_i}\left\{\sum_{p=1}^{N}\sum_{n=0}^{\infty} y_{pn}\phi_{pn}(y)\phi_{pn}(y') + \frac{\omega\varepsilon_o}{2}H_o^{(2)}\left(k|y-y'|\right)\right\}E_y(y')\, dy' \tag{29}$$

for E plane scan and

$$2\sum_{p=1}^{N}\alpha_p y_{p1}(y) = \left(\frac{\partial^2}{\partial x^2}+k^2\right)\int_{\Sigma A_i}\left\{\sum_{p=1}^{N}\sum_{n=1}^{\infty}\frac{y_{pn}}{\gamma_{pn}^2}\phi_{pn}(y)\phi_{pn}(y') + \frac{1}{2\omega\mu_o}H_o^{(2)}\left(k|y-y'|\right)\right\}E_x(y')\, dy' \tag{30}$$

for the H plane scan.

Notice that the first subscript of the waveguide modal functions ϕ_{pn} refers to the relative location of the element in the array. The extension of the formulation to the situation of loading the waveguides with dielectric plugs follows a similar procedure. When a dielectric sheath covers the array the derivation is somewhat less straightforward. Because of the lack of a closed form Green's function for such a case, a Fourier transform type of representation will have to be used. (52)

In addition to imbedding a finite array in an infinite ground plane, finite arrays may be placed either in a passive infinite array environment or in a corrugated structure. (28, 29, 30) It can be shown that solutions to such problems may be derived in terms of the solution of the uniformly illuminated infinite array. Alternatively, a direct integral equation formulation of this problem is also available.

V. Conclusions

Various approximate and rigorous methods for the analysis of phased array antennas have been reviewed and compared. The availability of high speed digital computers has facilitated meaningful quantitative array analysis in terms of rigorously formulated boundary value problems. Infinite and finite arrays of both current carrying wire elements and aperture type elements have been analyzed extensively by this approach. The salient feature and differences between finite and infinite arrays analyses are discussed.

Finite array analysis is useful especially in assessing the edge effects due to the finiteness of the array as well as the absence of elements in thinned arrays. Infinite array analysis is useful in predicting the behavior of the central core elements in medium to large sized arrays.

As was pointed out, the feed structure of dipole arrays has not been adequately taken into account in the various analyses. An examination of this aspect may prove fruitful. It has been suggested that a phased array of below cut-off waveguide elements may possess attractive electrical and mechanical characteristics. This is another area for possible future investigation. Recently, there has been considerable interest in conformal arrays (see this issue). Approximate analyses of cylindrical arrays have been carried out. Extension of some of the techniques developed for planar array analysis to conformal arrays seems worthwhile. Although many methods have been developed to ascertain the validity of computer-generated numerical solutions, no single approach except perhaps experimental verification seems to be absolutely reliable. Techniques for obtaining quantitative measure of ascertaining the validity of numerical solutions are wide open areas in which further research is very much needed.

REFERENCES

1. Amitay, N., Cook, J.S., Pecina, R.G. and Wu, C.P., "On Mutual Coupling and Marching Conditions in Large Planar Arrays", IEEE G-AP International Symposium Digest, pp. 150-156, 1964.

2. Farrell, G. F., Jr., and Kuhn, D. H., "Mutual Coupling Effects of Triangular-Grid Arrays by Modal Analysis," IEEE Trans. Antennas and Propagation; vol. AP-14, pp. 652-654, September 1966.

3. Oliner, A. A. and Malech, R. G., "Microwave Scanning Antennas" (R. C. Hansen, Editor) Chapters 2, 3, 4, Vol. II, Academic Press, Inc., New York, 1966.

4. Diamond, B. L., "Resonance Phenomena in Waveguide Arrays," 1967 IEEE G-AP Inaternational Symposium Digest, pp. 110-115.

5. Hannan, P. W., "The Element-Gain Paradox for a Phased-Array Antenna," IEEE Trans. Antennas and Propagation, vol. AP-12, pp. 423-433, July 1964.

6. Allen, J. L. and Diamond, B. L., "Mutual Coupling in Array Antennas," Lincoln Laboratories, M.I.T. Tech. Report No. 424, October 4, 1966.

7. Lechtreck, L. W., "Effects of Coupling Accumulation in Antenna Arrays", IEEE Trans. Antennas and Propagation, vol. AP-16, pp. 31-37, January 1968.

8. Wheeler, H. A., "The Radiation Resistance of an Antenna in an Infinite Array or Waveguide," Proc. I.R.E., vol. 36, pp. 478-487, April 1948

9. Hannan, P. W., Meier, P. J. and Balfour, M. A., "Simulation of Phased Array Antenna Impedance in Waveguide," IEEE. Trans. Antennas and Propagation, vol. AP-11, pp. 715-716, November 1963.

10. Hannan, P. W. and Balfour, M. A., "Simulation of a Phased Array Antenna in Waveguide," IEEE Trans. Antennas and Propagation, vol. AP-13, pp. 342-353, May 1965.

11. Balfour, M. A., "Phased Array Simulators in Waveguide for a Triangular Arrangement of Elements," IEEE Trans. Antennas and Propagation, vol. AP-13, pp. 475-476, May 1965.

12. Galindo, V. and Wu, C. P., "Integral Equations and Variational Expressions for Arbitrary Scanning of Regular Infinite Arrays", IEEE Trans. Antennas and Propagation, vol. AP-14, pp. 392-394, May 1966.

13. Amitay, N., "Improvement of Planar Array Match by Compensation Through Contiguous Element Coupling," IEEE Trans. Antennas and Propagation, vol. AP-14, pp. 580-586, September 1966

14. Amitay, N., Butzien, P. E. and Heidt, R. C., "Match Optimization of a Two Port Phased Array Antenna Element," IEEE Trans. Antennas and Propagation, vol. AP-16, pp. 47-57, January 1968.

15. DuFort, E. C., "A Design Procedure for Matching Volumetrically Scanned Waveguide Arrays," Proc. IEEE, Vol. 56, pp. 1851-1860, November 1968.

16. Tang, R. and Wong, N. S., "Multimode Phased Array Element for Wide Scan Angle Impedance Matching," Proc. IEEE, vol. 56, pp. 1951-1959, November 1968.

17. Edelberg, S. and Oliner, A. A., "Mutual Coupling Effects in Large Antenna Arrays, part I - Slot Arrays," IEEE Trans. Antennas and Propagation, vol. AP-8, pp. 286-297, May 1960.

18. Stark, L., "Radiation Impedance of a Dipole in Infinite Arrays," Hughes Aircraft Co., Tech. Report No. FL60-230 May 1960. Also Radio Science, vol. 1, pp. 361-379, March 1966.

19. Galindo, V. and Wu, C. P., "Numerical Solutions for an Infinite Phased Array of Rectangular Waveguides with Thick Walls," IEEE Trans. Antennas and Propagation, vol. AP-14, pp. 149-158, March 1966.

20. Wu, C. P. and Galindo, V., "Properties of a Phased Array of Rectangular Waveguide with Thin Walls," IEEE Trans. Antennas and Propagation, vol. AP-14, pp. 163-173, March 1966.

21. Farrell, G. F., Jr., and Kuhn, D. H., "Mutual Coupling in Infinite Planar Arrays of Rectangular Waveguide Horns," IEEE Trans. Antennas and Propagation, vol. AP-16, pp. 405-414, July 1968.

22. Diamond, B. L., "A Generalized Approach to the Analysis of Infinite Planar Array Antennas," Proc. IEEE, vol. 56, pp. 1837-1851, November 1968.

23. Borgiotti, G. V., "Modal Analysis of Periodic Planar Phased Arrays of Apertures," Proc. IEEE, Vol. 56, pp. 1881-1892, November 1968.

24. Amitay, N. and Galindo, V., "The Analysis of Circular Waveguide Phased Arrays," Bell Syst. Tech. Journal, vol. 47, pp. 1903-1931, November 1968.

25. King, R. W. P., Mack, R. B. and Sandler, S. S., "Arrays of Cylindrical Dipoles," Cambridge University Press, New York, 1968.

26. Chang, V. W. H., "Infinite Phased Dipole Arrays," Proc. IEEE, vol. 56, pp. 1892-1900, November 1968.

27. Van Koughnett, A. L. and Yen, J. L., "Properties of a Cylindrical Antenna in an Infinite Planar or Collinear Array," IEEE Trans. Antennas and Propagation, vol. AP-15, pp. 750-757, November 1967.

28. Galindo, V., "A Generalized Approach to a Solution of Aperiodic Arrays and Modulated Surfaces," Trans. IEEE Antennas and Propagation, vol. AP-16, pp. 424-429, July 1968.

29. Karjala, D. S., and Mittra, R., "Radiation from Some Periodic Structures Excited by a Waveguide," IEEE Electronic Letters, vol. 1, pp. 111-112, June 1965.

30. Lee, S. W., "Radiation from an Infinite Aperiodic Array of Parallel Plate Waveguides," IEEE Trans. Antennas and Propagation, vol. AP-15, pp. 598-606, September 1967.

31. Hessel, A. and Hochstadt, H., "Plane Wave Scattering from a Modulated-Corrugated Structure," 1966 URSI Symposium, Palo Alto, California

32. Carter, P. S., Jr., "Mutual Impedance Effects in Large Beam Scanning Arrays," IEEE Trans. Antennas and Propagation, vol. AP-8, pp. 276-285, May 1960.

33. Diamond, B. L., "Phased Array Radar Studies, January 1963 to July 1964," Lincoln Laboratories, M. I. T. Tech. Report no. TR 381, 1965.

34. Harrington, R. F., "Field Computation by Moment Method," MacMillan Co., New York, 1968.

35. Wheeler, H. A., "The Grating-Lobe Series for the Impedance Variation in a Planar Phased Array Antenna," IEEE Trans. Antennas and Propagation, vol. AP-14, pp. 707-714, November 1966. (Correction, vol. AP-15, p. 235, March 1967.)

36. Jones, D. S., "A Critique of the Variational Method in Scattering Problems," IRE Trans. on Antennas and Propagation, vol. AP-4, pp. 297-301, July 1956.

37. Amitay, N. and Galindo, V., "On Energy Conservation and the Method of Moments in Scattering Problems," IEEE Trans. on Antennas and Propagation, vol. AP-17, pp. 722-729, November 1969.

38. Wu, C. P., "Numerical Solutions for the Coupling Between Waveguides in Finite Arrays," Radio Science, vol. 4, pp. 245-253, March 1969.

39. Kahn, W. K. and Kurss, H., "Minimum Scattering Antennas," IEEE Trans. in Antennas and Propagation, vol. AP-13, pp. 671-675, September 1965.

40. Wasylkiwskyj, W. and Kahn, W. K., "Mutual Coupling and Element Efficiency for Infinite Linear Arrays," Proc. IEEE, vol. 56, pp. 1901-1907, November 1968.

41. Lee, S. W. and Mittra, R., "Radiation from Dielectric-Loaded Arrays of Parallel-Plate Waveguides," IEEE Trans. Antennas and Propagation, vol. AP-16, pp. 513-520, September 1968.

42. Lee, S. W., Wong, N., and Tang, R., "Analysis of Infinite Planar Array of Rectangular Waveguides by Generalized Scattering Matrix Approach," This issue.

43. DuFort, E. C., "A Scattering Matrix Method for solving Waveguide Array Impedance Problems," Radio Sci., vol. 3, pp. 459-467, May 1968.

44. Wu, C. P. and Galindo, V., "Surface Wave Effects on Di-Electric Sheathed Phased Arrays of Rectangular Waveguides," Bell Syst. Tech. J. vol. 47, pp. 117-142, January 1968.

45. Galindo, V. and Wu, C. P., "Dielectric Loaded and Covered Rectangular Waveguide Phased Arrays," Bell Syst. Tech. J. vol. 47, pp. 93-116, January 1968.

46. Wu, C. P., "Note on Integral Equations and Variational Expressions for Arbitrary Scanning of Regular Infinite Arrays," IEEE Trans. Antennas and Propagation, vol. 16, pp. 136-138, January 1968.

47. Galindo, V., "On the Stationary Properties of the Integral Equations of Infinite Phased Arrays," IEEE Trans. Antennas and Propagation vol. AP-17, pp. 359-360, May 1969.

48. Schaedla, W. H., "Equipment Circuit Formulation for an Array of Phased Waveguide Apertures," IEEE Trans. Antennas and Propagation, vol. AP-18, pp. 28-33, January 1970.

49. Lee, S. W. and Jones, W. R., "On the Supression of Radiation Nulls in Waveguides Phased Arrays," URSI Digest, pp. 69-70, April 1970.

50. Amitay, N. and Galindo, V., "Characteristics of Dielectric Loaded and Covered Circular Waveguide Phased Arrays," IEEE Trans. Antennas and Propagation, vol. AP-17, pp. 722-729, November 1969.

51. Harrington, R. F., "Time Harmonic Electromagnetic Fields," New York, McGraw-Hill, 1960.

52. Wu, C. P., "Integral Equation Solutions for the Radiation from a Waveguide Through a Dielectric Slab," IEEE, Trans. Antennas and Propagation vol. AP-17, pp. 733-740, November 1969.

53. Amitay, N., Galindo, V., and Wu, C. P., "Phased Array Antennas," to be published.

54. Cole, W. J., Nagelberg, E. R. and Nagel, C. M., "Iterative Solution of Waveguide Discontinuity Problems," Bell System Tech. Journal, vol. 46, pp. 649-672, March 1967.

Infinite Arrays of Sub-Array Antennas

W. WASYLKIWSKYJ

Institute for Defense Analyses
Arlington, Virginia

W. K. KAHN

The George Washington University
School of Engineering and Applied Science
Washington, D. C.

ABSTRACT

A unified approach to the network analysis of infinite regular arrays of antennas is presented. This analysis deals with the properties of the collection of radiating elements (array properties) as distinguished from the evaluation of particular antenna characteristics through the solution of an electromagnetic boundary value problem. It is necessary that the network parameters characterizing the interaction among the elementary antennas be known.

The periodically repeated constituent of the array is itself regarded as a sub-array, the sub-array taking a variety of particular forms. With the appropriate choice of sub-array, the same formulation yields as special cases the uniform linear, planar, and cylindrical arrays as well as such arrays comprising several heterogeneous elements or multimode elementary antennas. Resonance minima which may be arbitrarily placed in angle and varied in depth are shown to occur as circuit properties of an array network. The points are illustrated employing an array of dipoles.

I. INTRODUCTION

A unified approach to the network analysis of infinite regular arrays of antennas is presented. This analysis deals with the properties of the collection of radiating elements (array properties) as distinguished from the evaluation of particular antenna characteristics through the solution of an electromagnetic boundary value problem. It is necessary that the network parameters characterizing the interaction among the elementary antennas be known.

The periodically repeated constituent of the array is itself regarded as a sub-array, the sub-array taking a variety of particular forms. With the appropriate choice of sub-array, the same formulation yields as special cases the uniform linear, planar, and cylindrical arrays as well as such arrays comprising several heterogeneous elements or multimode elementary an-tennas. Three possibilities are illustrated in Figs. 1a through 1c. Fig. 1a shows an in-line arrange-

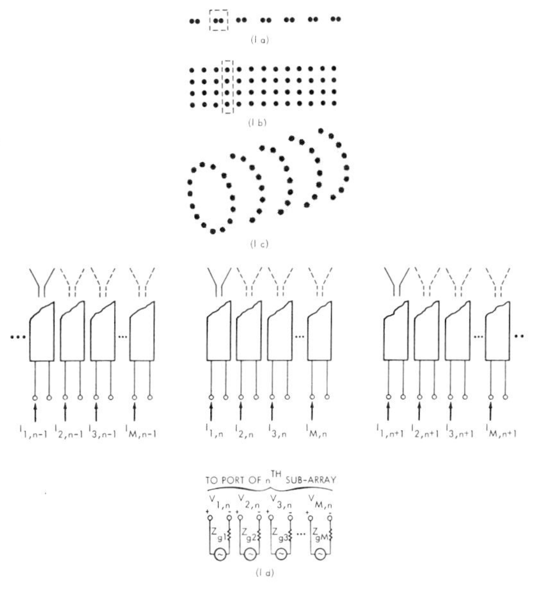

Fig. 1 Infinite Array of Sub-Array Antennas.

ment of 2-element sub-arrays, a typical sub-array is delineated by the dashed line. Similarly, Fig. 1b shows an arrangement of linear sub-arrays, each sub-array extending in a direction orthogonal to the principal array axis. The planar array illustrated is finite in one dimension; however, this restriction to finite sub-arrays is not essential. Fig. 1c shows a cylindrical array. Each circular array may conveniently be regarded as a sub-array, bringing such (glide symmetric) cylindrical arrays within the purview of the general scheme. The network problem for the antenna terminal quantities is formulated as an infinite order finite-difference equation. The solution is obtained as a matrix generalization to arrays of sub-array antennas of the technique presented in Reference 1.

The finite-difference formulation is applicable to arrays of quite general antennas and is not restricted to apertures in a conducting plane or cylinder. It is only necessary that the network parameters characterizing the interaction among the elementary antennas be known. For minimum scattering antennas, these network parameters may be calculated conveniently and exactly from the elementary radiation patterns, without reference to the structural type of the elements. (1, 7) The particular sub-array chosen for purposes of illustration consists of two inclined dipole radiators. Only one dipole of each sub-array is excited; the remaining dipoles act as passive parasitic radiators. Quite deliberately, no physical supporting structure for the dipoles is specified. However, the parasitic dipole radiators only are interconnected in a manner to be described. The array of fed dipoles, when excited with uniform amplitude (linear phase taper), will be found to exhibit peaks in the active reflection coefficient, or, equivalently, resonance nulls or more generally minima in the pattern of a single excited element within the array of sub-arrays environment. (2, 3, 4) It is therefore demonstrated that such resonances, existing purely as circuit properties of the array network, can in principle occur generally in linear, planar or cylindrical arrays as opposed to being necessarily limited to specific kinds of structural configurations. (5, 13) Further, it will be seen that these resonance minima (reflection coefficient maxima) may be arbitrarily placed and varied in depth. Conversely, the ability to control and place rapid variations in reflection coefficient with scan angle holds promise with respect to matching out variations found in any given case. (8, 11)

II. DIFFERENCE EQUATION FORMULATION FOR AN ARRAY OF SUB-ARRAY ANTENNAS

Consider an infinite, uniformly spaced linear array composed of identical M-element sub-arrays as shown in Fig. 1d. The terminal voltages and currents satisfy

$$\underline{V}_m = \sum_n Z_{mn} \underline{I}_n \qquad (1a)$$

where Z_{mn} is an MXM (open-circuit) impedance matrix coupling the mth and the nth sub-arrays; $\underline{V}_m$ and $\underline{I}_n$ are column matrices

$$\underline{V}_m = \begin{bmatrix} V_{1,m} \\ V_{2,m} \\ \cdot \\ \cdot \\ \cdot \\ V_{m,n} \end{bmatrix} \qquad \underline{I}_n = \begin{bmatrix} I_{1,n} \\ I_{1,n} \\ \cdot \\ \cdot \\ \cdot \\ I_{m,n} \end{bmatrix}$$

The matrix Z_{mn} has the form

$$Z_{mn} = \begin{bmatrix} Z_{11,mn} & Z_{12,mn} & \cdots & Z_{1m,mn} \\ Z_{21,mn} & Z_{22,mn} & \cdots & Z_{2m,mn} \\ \cdots & \cdots & \cdots & \cdots \end{bmatrix}$$

It follows from Lorentz reciprocity that

$$Z_{mn} = \widetilde{Z}_{nm}$$

wherein $\sim$ denotes the transposed matrix. The translational symmetry of the infinite linear array implies that Z_{mn} is, in fact, only a function of the difference $\nu = m - n$; thus

$$Z_\nu = \widetilde{Z}_{-\nu}$$

Let Z_g be the diagonal matrix containing internal impedances of the driving generators, i.e.,

$$Z_g = \operatorname{diag}\left[Z_{g1}\ Z_{g2}\ \ldots\ Z_{gm}\right] = R_g + jX_g\ .$$

Then the array may alternatively be regarded as being excited by incident waves $\underline{a}_n$

$$2R_g^{1/2}\ \underline{a}_n = \underline{V}_n + Z_g \underline{I}_n\ . \qquad (2a)$$

Since the internal impedances Z_{gn} are, in general, complex the reflected waves $\underline{b}_n$ are given by

$$2R_g^{1/2}\ \underline{b}_n = \underline{V}_n - Z_g^* \underline{I}_n\ . \qquad (2b)$$

Explicitly then, the currents $\underline{I_n}$ produced by the incident waves $\underline{a_n}$ satisfy

$$\lim_{N \to \underline{\alpha}} \sum_{\nu = -N}^{N} Z_\nu \underline{I_{n+\nu}} + Z_g \underline{I_n} = 2R_g^{1/2} \underline{a_n} . \tag{1b}$$

The solution of Equation (1b) for $\underline{I_n}$ with arbitrary excitations $\underline{a_n}$ may be obtained by recognizing that the $\underline{I_n}$ and $\underline{a_n}$ must be coefficients of (matrix) Fourier series (cf. (1) for the case M = 1). Thus, substituting

$$\underline{I_n} = \frac{1}{2\pi}\int_{-\pi}^{\pi} \underline{I}(\xi) e^{-jn\xi}\, d\xi , \tag{3a}$$

$$\underline{a_n} = \frac{1}{2\pi}\int_{-\pi}^{\pi} \underline{A}(\xi) e^{-jn\xi}\, d\xi , \tag{3b}$$

into (1) yields (4)

$$\underline{I_n} = \frac{1}{2\pi}\int_{-\pi}^{\pi} \left\{ \left[Q(\xi) + Z_g\right]^{-1} 2R_g^{1/2} \left(\sum_{m=-\infty}^{\infty} \underline{a_m}\, e^{jm\xi} \right) \right\} e^{-jn\xi}\, d\xi ,$$

where the matrix $Q(\xi)$ is given by

$$Q(\xi) = \sum_{\nu=-\infty}^{\infty} Z_\nu e^{-j\nu\xi} \tag{5}$$

When only the first element of each sub-array is excited (all others being terminated in $Z_{g\ell}$) one sets in (4)

$\tilde{\underline{a}}_m = [a_{1,m}\ 0\ 0\ .\ .\ .\ 0]$.

This yields

$$I_{\ell,n} = \frac{1}{2\pi}\int_{-\pi}^{\pi}\left\{ \frac{2\sqrt{R_{g1}}\det \hat{q}_{1\ell}(\xi)(-1)^{\ell+1} \sum_{m=-\infty}^{\infty} a_{1,m} e^{jm\xi}}{\det Q(\xi)} \right\} e^{-jn\xi}\, d\xi , \tag{6}$$

$\ell = 1, 2, .\ .\ .\ M$,

where the matrix $\hat{Q}(\xi)$ is given by

$$\hat{Q}(\xi) = \hat{Q}(\xi) + Z_g \tag{7}$$

and $\hat{q}_{1\ell}$ is the minor of $\hat{Q}$ obtained by striking out the 1st row and ℓth column. The sub-array elements $\ell = 2, 3, .\ .\ .$ M can be either physical parasitic radiators (employed, e.g., to effect an improved match of the array with scan (4) or equivalent radiators corresponding to higher order evanescent modes in the input transmission line. (2, 5) In the following, the sub-array element $\ell = 1$ will be referred to as the driven element.

Two types of array excitations are of particular importance: one in which only one element in each sub-array is excited (all others being terminated in Z_g), while another corresponds to the excitation of all $\ell = 1$ elements by equal amplitude waves with a linear phase taper. In the former case, the currents at the driven elements are proportional to the mutual and self-coupling (scattering) coefficients. Thus, denoting these by S_{nm} and employing Equation (2), one obtains

$$S_{nm} = \left.\frac{b_{1,n}}{a_{1,m}}\right|_{a_{1,n}=0,\ n \neq m} = \sqrt{\begin{matrix} -\sqrt{R_{g1}}\dfrac{I_{1,n}}{a_{1,m}} , & n \neq m \\ 1 - \sqrt{R_{g1}}\dfrac{I_{1,m}}{a_{1,m}} . & \end{matrix}} \tag{8}$$

Setting $a_{1,m} = 0$, $m \neq n$ and $a_{1,n} = 1$ in (6) one also has

$$S_{nm} = \delta_{nm} - \frac{\sqrt{R_{g1}}}{\pi}\int_{-\pi}^{\pi} \frac{\det \hat{q}_{1\ell}(\xi)}{\det \hat{Q}(\xi)} e^{-j(n-m)\xi} d\xi . \tag{9}$$

When the array is excited with uniform amplitudes and a linear phase taper one sets $a_{1,m} = e^{-jm\alpha}$

(α the phasing constant) and (6) yields

$$I_{\ell,n} = \frac{2\sqrt{R_{g1}}\det \hat{q}_{1\ell}(\alpha)(-1)^{\ell+1}}{\det \hat{Q}(\alpha)} e^{-jn\alpha} , \quad \ell = 1, 2\ .\ .\ .\ M . \tag{10}$$

Employing (10) with $\ell = 1$ together with (2a) one obtains the input (active) impedance looking into the typical driven element:

$$Z_{1n}(\alpha) = \frac{\det \hat{Q}(\alpha)}{\det \hat{q}_{11}(\alpha)} - Z_{g1} . \tag{11}$$

Expanding the determinent, yields

$$Z_{1n}(\alpha) = Q_{11}(\alpha) + \frac{\sum_{\ell=2}^{M} Q_{1\ell}(\alpha)\det \hat{q}_{1\ell}(\alpha)(-1)^{1+\ell}}{\det \hat{q}_{11}(\alpha)} \tag{12}$$

where Q_{11} , $Q_{1\ell}$ are (scalar) elements of Q. If the array is matched at broadside then the active reflection coefficient $\rho(a)$ is

$$\rho(a) = \frac{Z_{1n}(a) - Z_{1n}(0)}{Z_{1n}(a) + Z_{1n}^{*}(0)} \quad . \qquad (13)$$

In Equation (12), the active impedance is expressed as a sum of two terms: $Q_{11}(a)$ which corresponds to the active impedance of the driven elements in the absence of parasitic radiators and a sum of terms taking account of coupling among the elements of a typical sub-array. Equation (12) holds quite generally and has, in fact, been derived from pure circuit considerations, i.e., without reference to structural properties of the radiators comprising each sub-array. If one is dealing with the special case of single mode elements, then in the range of a for which no grating lobes appear in the visible region, $Q_{11}(a)$ will generally not exhibit sharp resonances and if resonance phenomena in the active reflection coefficient do appear they must necessarily be associated with the second term in Equation (12). In particular, the possibility of such resonances exists whenever det $\hat{q}_{11}(a)$ possesses sharp minima (maxima). This has been noted by Borgiotti (5) in connection with a planar array of apertures where, in the present context, the parasitic radiators correspond to evanescent modes in the feed waveguides. The point being made here is that such resonances can be described adequately in pure network terms. The above considerations apply also to sub-array elements which are minimum-scattering (M.S.) antennas (1, 7) since they are completely defined in terms of their radiation patterns (without references to physical structures). For such antennas, the $Q_{ij}(a)$ can be readily computed since the mutual impedances entering into the definitions of Z_ν, eq. (1), are given explicity in terms of element patterns (7). Moreover, arrays of M.S. antennas with the same element radiation patterns as certain single mode elements (in particular, planar apertures) can be treated rigorously (7). In the following, a linear array of two-element sub-arrays consisting of two M.S. (dipole-type) antennas will be analyzed.

III. A LINEAR ARRAY OF DIPOLE SUB–ARRAYS

Fig. 2a shows a linear array of two-element dipole sub-arrays. The dipoles are inclined at an

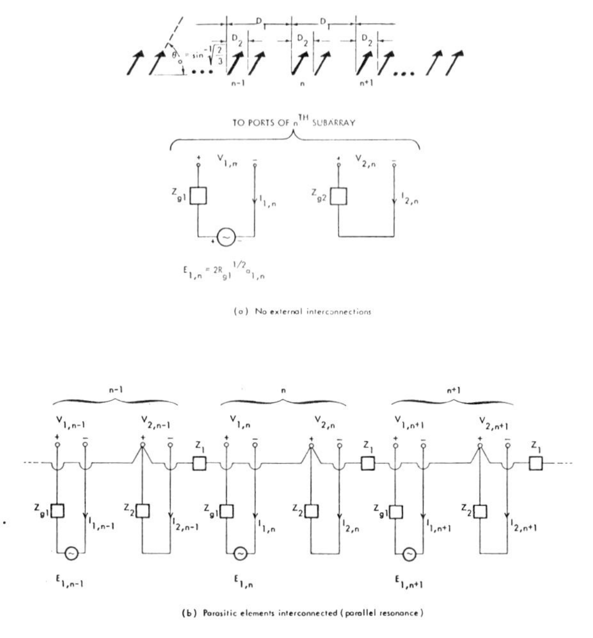

Fig. 2 Linear Dipole Array.

angle $\theta_0 = \sin^{-1}\sqrt{\frac{2}{3}}$ with the array axis. The active impedance for a linear array (M = 1) containing dipoles with this particular orientation has been computed in (1). The active impedance has been shown to exhibit the interesting property of having a constant real part in the visible region if the inter-element spacing does not exceed $\lambda/2$. In addition, the imaginary part remains small throughout most of the visible region so that nearly perfect match is attained for a large range of scan angles. In the following, the effect on the active impedance looking into the port of the driven element is examined.

Employing (12) with M=2, one obtains (14)

$$Z_{1n}(a) = Q_{11}(a) - \frac{Q_{12}(a)\, Q_{21}(a)}{Q_{22}(a) + Z_{g2}} \quad .$$

The coupling impedance matrix Z_ν for a two element sub-array is

$$Z_\nu = \begin{bmatrix} Z_{11}(\nu) \\ Z_{21}(\nu) \end{bmatrix} \begin{bmatrix} Z_{12}(\nu) \\ Z_{22}(\nu) \end{bmatrix} \quad .$$

Employing the results of [1] and noting that the parasitic and driven elements are identical, one obtains

$$Z_{11}(\nu) \equiv Z_{22}(\nu) = \begin{cases} 1 & ;\ \nu = 0 \\ \dfrac{je^{-jkD_1|\nu|}}{kD_1} & ;\ \nu \neq 0\ . \end{cases} \tag{16}$$

Similarly, the cross-coupling terms may be written

$$Z_{12}(\nu) \equiv \zeta(k|\nu D_1 + D_2|) = \frac{je^{-jk|\nu D_1 + D_2|}}{k|\nu D_1 + D_2|}\ , \tag{17a}$$

$$Z_{21}(\nu) \equiv \zeta(k|\nu D_1 - D_2|) = \frac{je^{-jk|\nu D_1 - D_2|}}{k|\nu D_1 - D_2|}\ . \tag{17b}$$

By virtue of (ζ)

$$Q_{11}(a) = Q_{22}(a) = \sum_{\nu=-\infty}^{\infty} Z_{11}(\nu)\, e^{-j\nu a}\ , \tag{18a}$$

$$Q_{12}(a) = \sum_{\nu=-\infty}^{\infty} Z_{12}(\nu)\, e^{-j\nu a}\ ,$$

$$Q_{21}(a) = \sum_{\nu=-\infty}^{\infty} Z_{21}(\nu)\, e^{-j\nu a} = Q_{12}(-a) \tag{18b}$$

Using (16), (18a) is readily summed to yield

$$Q_{11}(a) = Q_{22}(a) =$$

$$\begin{cases} \dfrac{\pi}{kD_1} - \dfrac{j}{kD_1}\,\ell n\left\{4 \sin\dfrac{|kD_1 + a|}{2} \sin\dfrac{|kD_1 - a|}{2}\right\}, & |a| < kD_1 < \pi\ , \\ -\dfrac{j}{kD_1}\,\ell n\left\{4 \sin\dfrac{|kD_1 + a|}{2} \sin\dfrac{|kD_1 - a|}{2}\right\}, & |kD_1| < |a| < \pi\ . \end{cases} \tag{19}$$

Equation (19) is, of course, identical with the active impedance in the absence of a parasitic element as given in (1).

The series for $Q_{12}(a)$, i.e., Equations (17) and (18b), cannot be summed in closed form. By summing all the terms having a $1/\nu$ dependence one obtains a representation which is more suitable for numerical computation than the original series. This is described in the Appendix. The result is

$$Q_{12}(a) = Q_{21}(-a) = \frac{je^{-jkD_2}}{kD_2} - \cos kD_2 + h(D_1;D_2) \tag{20}$$

$$+ \frac{j}{kD_1}\left[a \sin kD_2 - e^{jkD_2}\,\ell n\, 2 \sin\frac{|kD_1 - a|}{2} - e^{-jkD_2}\,\ell n\, 2 \sin\frac{|kD_1 + a|}{2}\right]$$

$$+ \frac{jD_2}{kD_1} \sum_{\nu=1}^{\infty}\left[\frac{e^{-j[\nu(kD_1 - a) - kD_2]}}{\nu(\nu D_1 - D_2)} - \frac{e^{-j[\nu(kD_1 + a) + kD_2]}}{\nu(\nu D_1 - D_2)}\right],$$

where

$$h(D_1, D_2) = \begin{cases} \dfrac{\pi}{kD_1} \cos kD_2 & ;\ |a| < kD_1\ , \\ \dfrac{-j\pi}{kD_1} \sin kD_2 & ;\ kD_1 < a < \pi\ , \\ \dfrac{j\pi}{kD_1} \sin kD_2 & ;\ -\pi < a < -kD_1\ . \end{cases}$$

In the absence of external interconnections among the parasitic elements, Fig. 2a, Z_{g2} in Equation (14) is a fixed (i.e., scan independent) terminating impedance. If the parasitic elements are interconnected through external impedances Z_1 and Z_2 as shown in Fig. 2b, Z_{g2} in Equation (14) is replaced by the equivalent impedance $Z_{go}(a)$ which depends on Z_1 and Z_2 as well as on the phasing constant a (8). Referring to Fig. 2b, the voltage at the port of the parasitic element of the nth sub-array satisfies the following difference equation:

$$\frac{V_{2,n} - V_{2,n-1}}{Z_1} + \frac{V_{2,n}}{Z_2} + I_{2,n} = \frac{V_{2,n+1} - V_{2,n}}{Z_1}\ . \tag{21}$$

For excitations (of the eigen-form*) $e^{-jn\alpha}$ one has

$$V_{2,n} = V_2(\alpha)\, e^{-jn\alpha},\ I_{2,n} = I_2(\alpha)\, e^{-jn\alpha} \quad . \tag{22}$$

The above together with (21) yields

$$Z_{go}(\alpha) = \frac{V_{2,n}}{I_{2,n}} = -\frac{V_2(\alpha)}{I_2(\alpha)} = \frac{Z_1 Z_2}{Z_1 + 4Z_2 \sin^2 \frac{\alpha}{2}} \quad . \tag{23}$$

If one assumes that Z_1 and Z_2 are pure reactances, i.e.,

for scan angles $\alpha_\circ$ satisfying

$$X_1 + 4X_2 \sin^2 \frac{\alpha_\circ}{2} = 0 \quad . \tag{24}$$

The resonance point can be placed at any predetermined scan angle. Fig. 3 shows the variation of the magnitude of the active reflection coefficient $\rho(\alpha)$ for the array in Fig. 2 looking into sub-array element 1 for several values of $\alpha_\circ$ and $X_2=1$. For each $\alpha_\circ$, the array is matched at broadside and the separation of sub-arrays is $D_1 = \lambda/4$ so that the visible region is $|\alpha| < \pi/2$; D_2 is chosen to be $D_1/10$. For comparison a curve of $|\rho(\alpha)|$ in the absence of the parasitic element is included, and is seen to vary monotonically in the visible region. The pronounced peaks in the reflection coefficient occur at the resonant points $\alpha_\circ$.

*As previously pointed out in connection with finite arrays (9), for eigen-excitations the radiation pattern obtained from an array is undisturbed by mutual coupling, i.e., the normalized pattern is precisely the same as would be calculated from a zero order calculation in which mutual coupling is neglected. The point just made is crucial. It extends the correspondence in scan space of visible and invisible regions for a particular phase progression of impressed antenna currents to a rigorous correspondence for a particular phase progression of antenna excitations by generators with finite internal impedance in which normally mutual coupling would be expected to play a role (10).

Even though the reflection at $\alpha_\circ$ is quite high, it is not total, i.e., $|\rho(\alpha_\circ)| \neq 1$. Indeed, the maximum value attained by $|\rho(\alpha)|$ near $\alpha_\circ$ may be adjusted by varying the reactance X_2. This is demonstrated in Fig. 4.

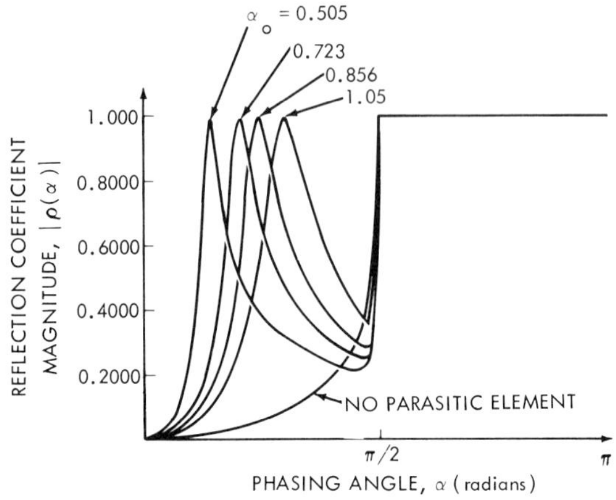

Fig. 3 Variation of Magnitude of Reflection Coefficient with Scan ($X_2 = 1$).

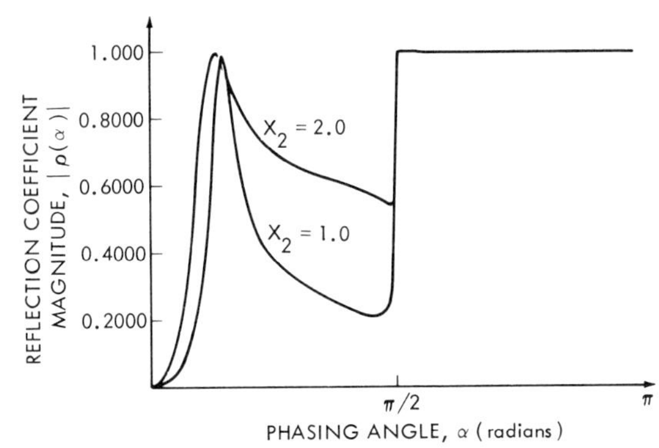

Fig. 4 Variation of Magnitude of Reflection Coefficient with Scan for Different Tuning Conditions ($\alpha_o = .505$).

The lack of total reflection at $\alpha_\circ$ is due to the general character of $Q(\alpha)$, which, for each α, represents the impedance matrix of a dissipative network, the dissipation being, of course, loss of energy due to radiation. Unless the equivalent network for $Q(\alpha)$ degenerates into a pure shunt structure this dissipation precludes total reflection. Although the amount of radiated power may be very small, nevertheless it is not identically zero except in the case cited.

The rapid variations introduced into $\rho(a)$ by the parasitic element in the present example may in other circumstances be utilized for matching purposes. Were excited elements themselves to exhibit resonant behavior, (as is known to occur in the case of multi-mode elements or an array covered by a dielectric sheet) interconnecting networks of the type shown in Figs. 2b and 5 together with a parasitic radiator could be employed to compensate for such resonances. In particular,

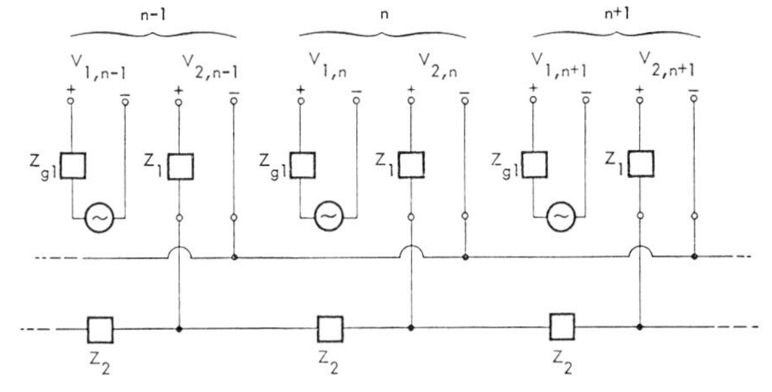

Fig. 5 Parasitic Element Interconnections for Series Resonance.

if $Q_{11}(a)$ exhibits a series resonance at a_o then, by virtue of Equation (14), the network in Fig. 2b is appropriate which itself is parallel resonant at ρ_o , as is evident from Equation (23). On the other hand, were $Q_{11}(a)$ to exhibit a parallel resonance, then the proper interconnecting network is that in Fig. 5. One can readily show that in this case

$$Z_{go}(a) = \frac{Z_2 + 4Z_1 \sin^2 \frac{a}{2}}{4 \sin^2 \frac{a}{2}} , \qquad (25)$$

which is seen to be series resonant whenever the numerator vanishes. For such compensations to be effective, the parasitic element should not introduce additional resonances, i.e., $Q_{22}(a)$ in Equation (14) must be a slowly varying function of a . This will usually be the case for single mode elements, or more generally, for minimum scattering antennas.

REFERENCES

(1) W. Wasylkiwskyj and W. K. Kahn, "Mutual Coupling and Element Efficiency for Infinite Linear Arrays," Proc. IEEE, Vol. 56, No. 11, pp. 1901-1907 (November 1968).

(2) G. F. Farrel, Jr., and D. H. Kuhn, "Mutual Coupling in Infinite Planar Arrays of Rectangular Waveguide Horns," IEEE Trans. on Antennas and Propagation, Vol. AP-16, pp. 405-414, (July 1968).

(3) G. H. Knittel, A. Hessel and A. A. Oliner, "Element Pattern Nulls in Phased Arrays and Their Relation to Guided Waves," Proc. IEEE Vol. 56, No. 11, pp. 1822-1836, November 1968).

(4) A. Hessel and G. H. Knittel, "A Loaded Ground Plane for the Elimination of Blindness In a Phased Array Antenna," 1969 G-AP International Symposium, (Digest pp. 163-169).

(5) G. V. Borgiotti, "Modal Analysis of Periodic Planar Phased Arrays of Apertures," Proc. IEEE, Vol. 56, No. 11, pp. 1881-1900, November 1968).

(6) B. L. Diamond, "A Generalized Approach to the Analysis of Infinite Planar Array Antennas," Proc. IEEE, Vol. 56, No. 11, pp. 1837-1851, (November 1968).

(7) W. Wasylkiwskyj and W. K. Kahn, "Theory of Mutual Coupling Among Minimum Scattering Antennas," IEEE Trans. Antennas and Propagation, Vol. AP-18, pp. 209-216, (March 1970).

(8) P. W. Hannan, P. S. Lerner and G. H. Knittel, "Impedance Matching a Phased-Array Antenna Over Wide Scan Angles by Connecting Circuits," IEEE Transactions on Antennas and Propagation, Vol. AP-13, No. 7, pp. 28-34, (January 1965).

(9) W. K. Kahn and W. Wasylkiwskyj, "Coupling Radiation and Scattering by Antennas," Proc. Symp. on Generalized Networks, Vol. XVI, Polytechnic Institute of Brooklyn, (April 1966).

(10) W. Wasylkiwskyj and W. K. Kahn, "A Formulation of the Interrelations Among Mutual Coupling, Element Efficiency, Active Impedance and Patterns in Linear Arrays," Alta Frequenza, Vol. XXVIII, N. Speciale, pp. 295-298, (May 1969).

(11) P. W. Hannan, "Proof that a Phased-Array Antenna Can be Impedance Matched for all Scan Angles," Radio Science, Vol. 2, (New Series) No. 3, pp. 361-369, (March 1967).

(12) M. Abramowitz and I. A. Stegun, eds., "Handbook of Mathematical Functions," National Bureau of Standards, Applied Mathematics Series 55, pp. 1005-1006, (1964).

(13) V. Galindo, "A Generalized Approach to a Solution of Aperiodic Arrays and Modulated Surfaces," IEEE Transactions, AP-16, pp. 424-429, (July 1968).

APPENDIX

TRANSFORMATION OF THE SERIES FOR $Q_{12}(a)$ INTO A MORE RAPIDLY CONVERGENT REPRESENTATION

Employing eqs. (17) and(18b), the series representation for $Q_{12}(a)$ is

$$Q_{12}(a) = \frac{j}{k} \sum_{\nu=-\infty}^{\infty} \frac{e^{-j(k|D_2+\nu D_1|+\nu a)}}{|D_2+\nu D_1|} \quad . \qquad (A1)$$

The above can be written in the alternate form

$$Q_{12}(a) = \frac{je^{-jkD_2}}{kD_2} + \frac{j}{k}\sum_{\nu=1}^{\infty} \frac{e^{-j\nu(kD_1+a)}}{\nu D_1 + D_2} e^{-jkD_2} + \frac{j}{k}\sum_{\nu=1}^{\infty} \frac{e^{-j\nu(kD_1-a)}}{\nu D_1 - D_2} e^{+jkD_2} \quad . \qquad (A2)$$

Since the terns for large ν decrease as $1/\nu$, the series converges slowly and is not useful for numerical computations. A more rapidly convergent representation is obtained by employing the following identity:

$$\frac{1}{\nu D_1 = D_2} = \sum_{\ell=0}^{p-1} \frac{(\mp D_2)^\ell}{(\nu D_1)^{\ell+1}} + \left(\mp\frac{D_2}{\nu D_1}\right)^p \frac{1}{\nu D_1 \pm D_2}, \qquad (A3)$$

which holds for arbitrary integers p.

Substituting the above into the infinite series in (A2) yields the following:

$$\sum_{\nu=1}^{\infty} \frac{e^{-j\nu(kD_1\pm a)}}{\nu D_1 \pm D_2} = \sum_{\nu=1}^{\infty}\left(\mp\frac{D_2}{\nu D_1}\right)^p \frac{e^{-j\nu(kD_1\pm a)}}{(\nu D_1 \pm D_2)} \qquad (A4)$$

$$+\sum_{\ell=0}^{p-1}\left[\frac{\left(\mp D_2\right)^\ell}{D_1^{\ell+1}} \sum_{\nu=1}^{\infty} \frac{e^{-j\nu(kD_1\pm a)}}{\nu^{\ell+1}}\right] .$$

The magnitudes of the terms in the first infinite series on the right of (A4) are seen to decay with ν as $1/\nu^{p+1}$ and hence improved convergence is obtained by increasing p. The second term on the right of (A4) is a sum of p infinite series which are slowly convergent. However, for p = 1 the series has a known closed form solution which is [9] :

$$f_1(a) \equiv \sum_{\nu=1}^{\infty} {}' \frac{e^{-j\nu(kD_1+a)}}{\nu} = \qquad (A5a)$$

$$= j\left(\frac{kD_1 + a(2n-1)\pi}{2}\right) - \ell n \left[2\sin\frac{kD_1 + a + 2n\pi}{2}\right],$$

$$f_2(a) \equiv \sum_{\nu=1}^{\infty} \frac{e^{-j\nu(kD_1-a)}}{\nu} = \qquad (A5b)$$

$$= j\left(\frac{kD_1 - a + (2m-1)\pi}{2}\right) - \ell n \left[2\sin\frac{kD_1 - a + 2m\pi}{2}\right],$$

where n, m are natural numbers satisfying

$$0 < kD_1 + a + 2m\pi < 2\pi \quad , \qquad (A6)$$

$$0 < kD_1 - a + 2m\pi < 2\pi \quad . \qquad (A7)$$

Employing (A5) and setting p = 1 in (A4), one obtains eq. (20). For $p > 1$ one has series of the form

$$\sum_{\nu=1}^{\infty} \frac{e^{j\nu(kD_1 \pm a)}}{\nu^p} ,$$

which can be generated by repeated integrations (with limits, e.g., between 0 and a) of the left side of (A5) with respect to a. The right side of (A5) then yields definite integrals of the logarithmic function which can be computed numerically. (For p = 2 the resulting integral, known as Clausen's integral, has been tabulated [12] .)

Analysis of Infinite Planar Array of Rectangular Waveguides by Generalized Scattering Matrix Approach

S. W. LEE, N. S. WONG, and R. TANG

Hughes Aircraft Company
Ground Systems Group
Fullerton, Calif.

ABSTRACT

This paper studies the radiation properties of an infinite planar array of rectangular thin-walled waveguide using the generalized scattering matrix technique. By extending the waveguide walls along one direction to an infinitesimally small distance, the two-dimensional lattice at the array aperture may be viewed as two scattering junctions. Each junction is made of a one-dimensional lattice and its scattering properties can be found exactly by the residue calculus method. Tracing the multiple scattering between these two junctions gives rise to a solution in the form of Neumann series, while it can be summed up by inverting a moderate-size matrix. Numerical results are presented to illustrate various transmission and reflection properties of two typical arrays. The theoretical value for the active element pattern is compared with experimental data measured from an array with 284 elements, and they are in reasonably good agreement.

1. Introduction

In the recent years, a considerable amount of theoretical work has been devoted to the analysis of an infinite array of rectangular waveguides arranged in a periodic lattice (1) (Figure 1). Apart from several approximate methods* the first rigorous result was obtained by Wu and Galindo, (5) who considered an array arranged in a rectangular lattice (d=0 in Figure 1) with scan in the H-plane only. Under these conditions, the problem can be reduced to a two-dimensional one (i.e., an array of parallel-plates), and is solved exactly by the residue calculus method. However, for a general waveguide array where the periodic lattice may be triangular and the beam can be pointed at an arbitrary direction in space, the analysis of the array problem becomes much more complicated, mainly due to the following two factors:

*Such as the method based on the grating lobe series, (2, 3) and that based on one-mode element. (4)

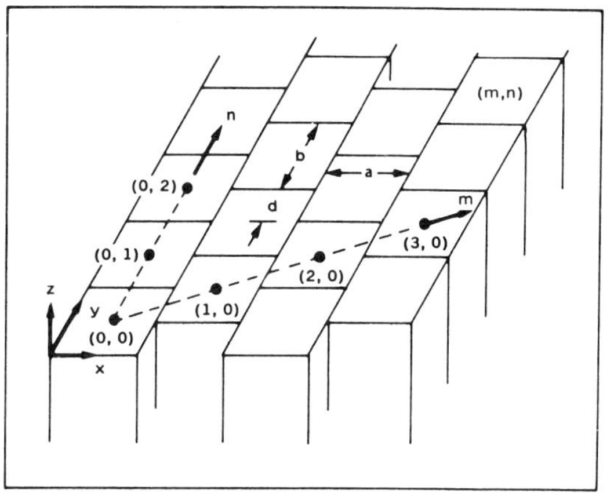

Fig. 1 An infinite array of rectangular waveguides with staggered spacing.

(i) The waveguide discontinuity at the aperture introduces field variations along both x and y directions. Consequently, the number of modes needed for the aperture matching is doubly infinite, as opposing to a simple infinite set of modes encountered in a two-dimensional problem. The doubly infinite series in the modal field presents a quite serious problem for numerical computations.

(ii) The TE and TM modes are coupled at the aperture. Consequently the scattering problem involves vector field, instead of an (equivalent) scalar one which is usually the prerequisite for a problem to be analytically solvable.

To date, there are only numerical solutions available to the general array problem. These include the work of Farrell and Kuhn, (6) Diamond, (7) Campbell and Jones (8) etc. Their procedure involves first a direct match of the four tangential field components at the aperture. Through a judicious application of Fourier analysis and elimination, it leads to a matrix equation where the order of the matrices is doubly infinite and each element in the matrices is a doubly infinite series. Finally, the matrices are truncated at a finite order and are inverted numerically with the help of a computer. The major difficulty with such a numerical approach lies in the complexity of the computer programming, and in the lengthy machine time required to perform the series summations and matrix inversions.

The present paper suggests a systematic scheme to attack the array problem with a view to improve the computation effort. Instead of inverting the matrices resulted from the aperture matching in a brute-force manner, we try to process the problem as far as possible along the analytical route, while an inversion of a moderate-sized matrix is needed only at the very last step as a means to sum up a Neumann series. Referring to Figures 1 and 2, we regard the

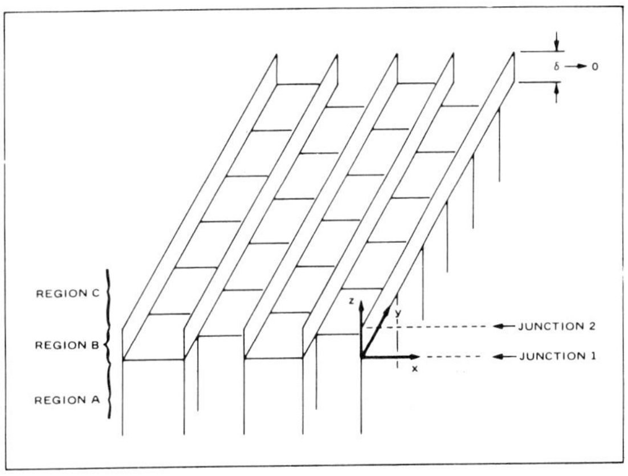

Fig. 2 Scattering junctions.

two-dimensional lattice at the aperture (z=0) is made of two one-dimensional junctions, one on top of the other with "zero" separation. The interaction between the two junctions is accounted for by tracing the multiple scatterings in a way suggested by Pace and Mittra, (9) which results in a Neumann series.* The advantage of such an approach lies in the fact that the three dimensional, vector field array problem is now transformed into two, two-dimensional, scalar field scattering problems at the two junctions. Furthermore, for the case of thin-walled waveguides considered in this paper, the scattering matrices at each junction can be solved analytically by the residue calculus method (or the Wiener-Hopf technique). Therefore, we may derive a formally exact solution to the general array problem in terms of a Neumann series. The rapid convergence of the Neumann series of similar type has been demonstrated in a large number of electromagnetic boundary value problems where the exact solutions are available. (11)

*A similar approach was used by DuFort (10) for studying a waveguide array with corrugated plates.

This paper is divided into seven sections. In Sections 2 and 3, we identify the scattering junctions, define the scattering matrices, and derive the Neumann series from a multiple scattering consideration. Then in the next two sections, the various scattering matrices are found explicitly by using the residue calculus method. In Section 6, we present some typical numerical results and compare them with an experiment. Finally, a conclusion of the present work and suggestion for future extension are included in Section 7.

2. Formulation of the Generalized Scattering Matrix

The geometry of an infinite array of rectangular waveguides with staggered spacing is shown in Figure 1. The excitation of the waveguides in cophasal, i.e., uniform in amplitude and progressive in phase. Explicitly, let us denote the excitation in a reference guide, say the (0, 0)'th element in Fig. 1, by $\psi^{(i)}$, then the excitation in the (m, n)'th waveguide is given by

$$\psi^{(i)} \exp\left[-jm\left(u+\frac{d}{b}v\right)-jnv\right] \tag{2.1}$$

where $u = k_o a \sin\theta_o \cos\varphi_o$, $v = k_o b \sin\theta_o \sin\varphi_o$, and $k_o = \omega\sqrt{\mu_o \epsilon_o}$.

(θ_o, φ_o) is the direction of the main beam, (a x b) is the waveguide dimension, and d the element displacement between two adjacent columns. The time dependence exp $(+j\omega t)$ is suppressed throughout.

In order to identify the scattering junctions in the present problem in a convenient manner, let us tacitly extend the waveguide walls along the y-direction from $z = 0^-$ to $z = \delta$, which δ is an infinitesimally small number, i.e. $\delta \to 0$. Referring to Fig. 2, the array now consists of two junctions. At junction 1, there is an array (along y-direction) of semi-infinite strips of width "a" sandwiched by two parallel-plates at x = pa and x=(p+1) a for p = 0, ±1, ±2, . . . , while junction 2 is an array of parallel-plates along x-direction. For a given incident field, the wave will bounce back and forth between these two junctions to give rise to multiple scatterings, and it is based on this concept that the following formulation is developed.

Let the incident field $\psi^{(i)}$ be arrived at junction 1, corresponding to point 1 in Fig. 3. It is scattered there,

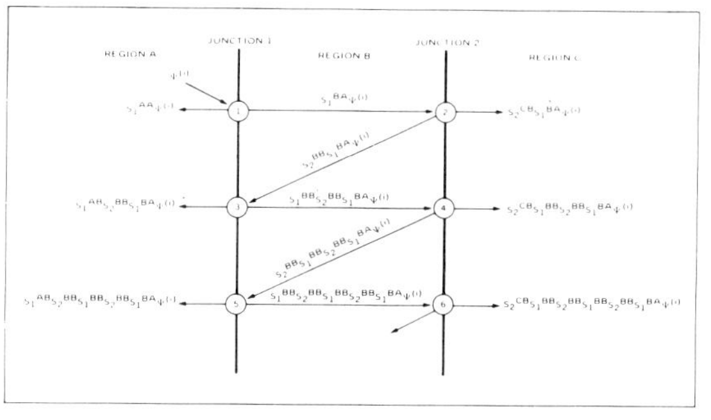

Fig. 3 The multiple scattering scheme at junctions 1 and 2 due to the incident field $\psi^{(i)}$.

which generates a reflected field symbolically denoted by $S_1^{AA}\psi^{(i)}$ in region A. The second A in the superscript of the operator S_1^{AA} signifies the incident field is from region A, and the first A signifies the scattered field is into region A. The subscript 1 in S_1^{AA} implies that the scattering occurs at junction 1. In addition to the reflected field, the scattering at point 1 also generates a transmitted field into region B, denoted by $S_1^{BA}\psi^{(i)}$. This transmitted field travels an infinitesimal distance δ and arrives at junction 2, corresponding to point 2 in Fig. 3. It is scattered again, and gives rise to a transmitted field $S_2^{CB}S_1^{BA}\psi^{(i)}$ into region C and a reflected field $S_2^{BB}S_1^{BA}\psi^{(i)}$ into region B. This process continues an infinite number of times. Then, the solution of the reflected field $\psi^{(r)}$ inside the waveguides is exactly the sum of all the contributions in region A, namely,

$$\Psi^{(r)} = S_1^{AA}\Psi^{(i)} + S_1^{AB}S_2^{BB}S_1^{BA}\Psi^{(i)} + S_1^{AB}S_2^{BB}S_1^{BB}S_2^{BB}S_1^{BA}\Psi^{(i)} + S_1^{AB}S_2^{BB}S_1^{BB}S_2^{BB}S_1^{BB}S_2^{BB}S_1^{BA}\Psi^{(i)} + \ldots \quad (2.2)$$

which is a Neumann series, and can be summed up with the result

$$\Psi^{(r)} = S_1^{AA}\Psi^{(i)} + S_1^{AB}S_2^{BB}[I-S_1^{BB}S_2^{BB}]^{-1} S_1^{BA}\Psi^{(i)} \quad (2.3)$$

where I is an identity operator. Similarly, it can be shown that the transmitted field $\Psi^{(t)}$ in the open half space is the sum of all the contributions in region C and is given by,

$$\Psi^{(t)} = S_2^{CB}[I-S_1^{BB}S_2^{BB}]^{-1} S_1^{BA}\Psi^{(i)} \quad (2.4)$$

So far, we have obtained the symbolical solutions of the problem in Equations (2.3) and (2.4). It remains to find out explicitly the operators S_1^{AA}, S_1^{AB}, ... etc., which will be referred to as (generalized)* scattering matrices in the following discussions.

To determine the various scattering matrixes, one needs to solve the scattering problems at junction 1 and 2. It is important to realize that at each junction, the structure varies only in one direction. For example, the geometry at junction 1 changes only along the y-axis, and therefore, the x-variation of the field remains undisturbed there. Thus, we have reduced our three-dimensional problem into two, two-dimensional problems!

3. Field Representations

Before the scattering matrices can be written down in definite forms, the conventions about the field representations in regions A, B, and C have to be specified. Since the field is periodic, one needs only to consider the field in the reference unit cell, $0 < x < a$ and $0 < y < b$.

First, let us consider the field in region A, the waveguide region $z < 0$. It is convenient to decompose the field into TE and TM with respect to x (not z). This is due to the fact that the TE and TM fields become uncoupled only if the field is decomposed with respect to the direction where the structure is uniform. Then, the complete field is derivable from two vector potentials, i.e.,

$$\text{TE:} \quad \vec{A} = \hat{e}_x \psi \quad (3.1)$$

$$\text{TM:} \quad \vec{F} = \hat{e}_x \overline{\psi} \quad (3.2)$$

through the following formulas

$$\vec{E} = -\nabla \times (\hat{e}_x \overline{\psi}) + \frac{1}{j\omega\epsilon_0} \nabla \times \nabla (\hat{e}_x \psi) \quad (3.3)$$

$$\vec{H} = \nabla \times (\hat{e}_x \psi) + \frac{1}{j\omega\epsilon_0} \nabla \times \nabla \times (\hat{e}_x \psi) \quad (3.4)$$

The conventional scattering matrix includes only the propagating modes. However, the matrix used here includes both propagating and evanescent modes.

Note that we have used the same letters for TE and TM fields except that bars are placed on the letters corresponding to the TM field. Since the field in region A has to satisfy the boundary conditions at the waveguide walls, the vector potentials take the following forms:

$$\psi_A = \sum_{m=1}^{\infty} \sum_{n=o}^{\infty} \psi_{mn}^A \sin\left(\frac{m\pi}{a}x\right) \cos\left(\frac{n\pi}{b}y\right) e^{\pm\Gamma_{mn}z} \tag{3.5}$$

$$\bar{\psi}_A = \sum_{m=o}^{\infty} \sum_{n=1}^{\infty} \bar{\psi}_{mn}^A \cos\left(\frac{m\pi}{a}x\right) \sin\left(\frac{n\pi}{b}y\right) e^{\pm\Gamma_{mn}z} \tag{3.6}$$

where:

$$\Gamma_{mn} = \sqrt{\left(\frac{m\pi}{a}\right)^2 + \left(\frac{n\pi}{b}\right)^2 - k_o^2}$$

Throughout this paper, the square root sign should be interpreted such that $\sqrt{a} = +\sqrt{a}$ if $a > 0$, and $\sqrt{a} = +j\sqrt{a}$ if $a < 0$. Then the field in region A can be represented by a column matrix of the following form,

$$\Psi_A = [(\psi_{mo}^A), (\psi_{m1}^A), (\psi_{m2}^A), \ldots\ldots, (\bar{\psi}_{0n}^A), (\bar{\psi}_{1n}^A), (\bar{\psi}_{2n}^A), \ldots\ldots]^T \tag{3.7}$$

where T is the transpose operator; (ψ_{mn}) and $(\bar{\psi}_{mn})$ are row matrices, and are explicitly given by

$$(\psi_{mn}^A) = [\psi_{1n}^A, \psi_{2n}^A, \psi_{3n}^A, \psi_{4n}^A, \ldots\ldots] , \tag{3.8}$$

for $n = 0, 1, 2, 3 \ldots$

$$(\bar{\psi}_{mn}^A) = [\bar{\psi}_{m1}^A, \bar{\psi}_{m2}^A, \bar{\psi}_{m3}^A, \bar{\psi}_{m4}^A, \ldots\ldots] , \tag{3.9}$$

for $m = 0, 1, 2, 3, \ldots$.

In other words, the mode coefficients of the vector potentials are arranged in the column matrix ψ_A in the order of

$[TE_{10}, TE_{20}, TE_{30}, \ldots TE_{11}, TE_{21}, TE_{31}, \ldots TE_{12}, TE_{22}, TE_{32}, \ldots . TM_{01}, TM_{02}, TM_{03}, \ldots . TM_{11}, TM_{12}, TM_{13}, \ldots TM_{21}, TM_{22}, TM_{23} \ldots]^T$

Clearly, the arrangement in Equation (3.7) is only one of many possible choices. It is a matter of "bookkeeping," and any proper arrangement will lead to the same final results. It should also be mentioned that ψ_A in Equation (3.7) is a (1 x N) matrix, where N is the number of modes, propagating or evanescent, in the waveguide. Since there are infinitely many modes in a waveguide, ψ_A is an infinite column matrix.

In a similar fashion, we can represent the field in region B, the infinitesimal transition region $0 < z < \delta$, by an infinite column matrix ψ_B. Following the Floquet's theorem, the vector potentials in region B take the forms:

$$\psi_B = \sum_{m=1}^{\infty} \sum_{q=-\infty}^{\infty} \psi_{mq}^B \sin\left(\frac{m\pi}{a}x\right) e^{-j\beta_q y} e^{\pm\gamma_{mq}z} \tag{3.10}$$

$$\bar{\psi}_B = \sum_{m=o}^{\infty} \sum_{q=-\infty}^{\infty} \bar{\psi}_{mq}^B \cos\left(\frac{m\pi}{a}x\right) e^{-j\beta_q y} e^{\pm\gamma_{mq}z} \tag{3.11}$$

where:

$$\beta_q = \frac{2q\pi + v}{b} ,$$

$$\gamma_{mq} = \sqrt{\left(\frac{m\pi}{a}\right)^2 + \beta_q^2 - k_o^2}$$

The complete field is again derivable from these two vector potentials from (3.3) and (3.4). Then, we write

$$\Psi_B = [(\psi_{mo}^B), (\psi_{m1}^B), (\psi_{m(-1)}^B), \ldots, (\bar{\psi}_{oq}^B), (\bar{\psi}_{1q}^B), (\bar{\psi}_{2q}^B) \ldots]^T \tag{3.12}$$

which represents the field in region B.

Finally, consider the field representation in region C, the open half-space $\delta < z < \infty$. The forms of the vector potentials can be written down immediately according to Floquet's theorem, namely,

$$\psi^C = \sum_{p=-\infty}^{\infty} \sum_{q=-\infty}^{\infty} \psi_{pq}^C e^{-j\alpha_{pq}x} e^{-j\beta_p y} e^{\pm\Omega_{pq}z} \tag{3.13}$$

$$\bar{\psi}^C = \sum_{p=-\infty}^{\infty} \sum_{q=-\infty}^{\infty} \bar{\varphi}_{pq}^C e^{-j\alpha_{pq}x} e^{-j\beta_p y} e^{\pm\Omega_{pq}z} \tag{3.14}$$

where:

$$\alpha_{pq} = \frac{2p\pi + u}{a} - \frac{2q\pi}{a}\frac{d}{b}$$

$$\Omega_{pq} = \sqrt{\alpha_{pq}^2 + \beta_q^2 - k_o^2}$$

Then, we use

$$\Psi_C = [(\psi_{po}^C), (\psi_{p1}^C), (\psi_{p(-1)}^C), \ldots, (\bar{\psi}_{oq}^C), (\bar{\psi}_{1q}^C), (\bar{\psi}_{(-1)q}^C), \ldots]^T \tag{3.15}$$

for the field representation in region C.

With the field representations in the three regions defined by Equations (3.7), (3.12), and (3.15), the various scattering matrices can now be formed in a unique fashion as will be detailed in the next two sections.

4. Scattering Matrices at Junction 1

There are four scattering matrices at junction 1, namely (S_1^{BB}, S_1^{AB}), and (S_1^{AA}, S_1^{BA}). Let us consider the first two in some detail. Recalling the formulation in Section 2, the scattering matrices S_1^{BB}, S_1^{AB} are defined such that, for an incident field $\psi_B^{(i)}$ from region B, the reflected field $\psi_B^{(r)}$ and the transmitted field $\psi_A^{(t)}$ are given by

$$\Psi_B^{(r)} = S_1^{BB}\Psi_B^{(i)} \tag{4.1}$$

$$\Psi_A^{(t)} = S_1^{AB}\Psi_B^{(i)} \tag{4.2}$$

since $\psi_B^{(i)}, \psi_B^{(r)}$, and $\psi_A^{(t)}$ are infinite column matrices as discussed in Section 3, it is clear that S_1^{BB} and S_1^{AB} are square matrices of infinite order, and their elements are simply the scattering coefficients of various modes at junction 1 due to various incident modes from region B. We shall now proceed to find them explicitly.

First consider the incident TE_{mq} mode at junction 1:

$$\psi_B^{(i)} = \sin\left(\frac{m\pi}{a}x\right) e^{-j\beta_q y} e^{+\gamma_{mq} z}, \quad \text{for } z > 0 \tag{4.3}$$

where m = 1, 2, 3, ..., and q = 0, ±1, ±2, ... The amplitude of the vector potential of the incident mode is always taken to be unity. Since the structure has no discontinuity along x direction, the x-dependence of the incident field will be maintained as the wave propagates across the aperture at junction 1. Therefore, the scattered field should consist of TE alone. Let us denote the reflected field by,

$$\psi_B^{(r)} = \sum_{p=-\infty}^{\infty} e_{mp}^{(q)} \sin\left(\frac{m\pi}{a}x\right) e^{-j\beta_p y} e^{-\gamma_{mp} z}, \quad \text{for } z > 0 \tag{4.4}$$

and the transmitted field by,

$$\psi_A^{(t)} = \sum_{n=o}^{\infty} s_{mn}^{(q)} \sin\left(\frac{m\pi}{a}x\right) \cos\left(\frac{n\pi}{b}y\right) e^{\Gamma_{mn} z}, \quad \text{for } z < 0 \tag{4.5}$$

where $\{e_{mp}^{(q)}\}$ and $\{s_{mn}^{(q)}\}$ are unknowns to be determined. The (q)'s in their superscript signify that the incident field has a y-dependence of exp $(-j\beta_q y)$.

At aperture z = 0, there are three tangential field components to be matched, i.e. E_y, H_x, and H_y, and this procedure would yield three equations. It can be easily shown that only two of them are independent, which are given in the following, for $0 \leqslant y \leqslant b$,

$$e^{-j\beta_q y} + \sum_{p=-\infty}^{\infty} e_{mp}^{(q)} e^{-j\beta_p y} = \sum_{n=o}^{\infty} s_{mn}^{(q)} \cos\left(\frac{n\pi}{b}y\right) \tag{4.6}$$

$$\gamma_{mq} e^{-j\beta_q y} - \sum_{p=-\infty}^{\infty} e_{mp}^{(q)} \gamma_{mp} e^{-j\beta_p y} = \sum_{n=o}^{\infty} s_{mn}^{(q)} \Gamma_{mn} \cos\left(\frac{n\pi}{b}y\right) \tag{4.7}$$

Multiplying (4.6) and (4.7) by cos $(\ell\pi y/b)$, and integrating with respect to y from o to b, there results two infinite sets of linear equations:

$$\sum_{p=-\infty}^{\infty} \frac{\beta_p}{\Gamma_{m\ell} - \gamma_{mp}} e_{mp}^{(q)} + \frac{\beta_q}{\Gamma_{m\ell} + \gamma_{mq}} = 0 \tag{4.8}$$

$$\sum_{p=-\infty}^{\infty} \frac{\beta_p}{\Gamma_{m\ell} + \gamma_{mp}} e_{mp}^{(q)} + \frac{\beta_q}{\Gamma_{m\ell} - \gamma_{mq}} + \frac{j\,\epsilon_\ell^o\, b\Gamma_{m\ell}}{1-(-1)^\ell e^{-jv}} s_{m\ell}^{(q)} = 0$$

$$\text{for } \ell = 0, 1, 2, \ldots \tag{4.9}$$

where $\epsilon_\ell^o = 2$, if $\ell = 0$, and $\epsilon_\ell^o = 1$, if $\ell \neq 0$.

The above two sets of simultaneous equations can be solved exactly by the well-known residue-calculus method. The essential steps for obtaining such a solution will be outlined below.

Consider a contour integral in the complex w-plane,

$$\oint_P \frac{f(w)}{\Gamma_{m\ell} - w}\, dw \quad , \qquad \text{for} \quad \ell = 0, 1, 2, \ldots \tag{4.10}$$

where the function f(w) is required to satisfy the following conditions.

(i) f(w) is an analytical function of w except for the first order poles at $w = -\gamma_{mq}$ and γ_{mp}, for $p = 0, \pm 1, \pm 2, \pm 3, \ldots$

(ii) f(w) has zeros at $w = \Gamma_{m\ell}$, for $\ell = 0, 1, 2, 3, \ldots$

(iii) f(w) has algebraic behavior as $|w| \to \infty$. More precisely, $f(w) \sim |w|^{-1/2}$ as $|w| \to \infty$ in accordance with the edge condition.

(iv) $\lim_{w \to -\gamma_{mq}} (w + \gamma_{mq})\, f(-\gamma_{mq}) = \beta_q$

If P denotes an infinitely large circular closed path in the complex w-plane, the integral in (4.10) yields the following equation:

$$\sum_{p=-\infty}^{\infty} \frac{f'(\gamma_{mp})}{\Gamma_{m\ell} - \gamma_{mp}} + \frac{\beta_q}{\Gamma_{m\ell} + \gamma_{mq}} = 0 \, , \qquad \text{for} \quad \ell = 0, 1, 2, \ldots \tag{4.11}$$

where $f'(\gamma_{mp})$ is the residue of f(w) at $w = \gamma_{mp}$, i.e.,

$$f'(\gamma_{mp}) = \lim_{w \to \gamma_{mp}} (w - \gamma_{mp})\, f(\gamma_{mp}) \tag{4.12}$$

Comparing (4.8) with (4.12), one immediately may identify that

$$e^{(q)}_{mp} = \frac{f'(\gamma_{mp})}{\beta_p} \quad , \qquad \text{for} \quad p = 0, \pm 1, \pm 2, \ldots \tag{4.13}$$

To obtain the unknown $\{ s^{(q)}_{m\ell} \}$ in (4.9), let us consider another contour integral,

$$\oint_P \frac{f(w)}{\Gamma_{m\ell} + w}\, dw \quad , \qquad \text{for} \quad \ell = 0, 1, 2, \ldots \tag{4.14}$$

Because of the conditions (i) through (iv), the above integral yields the following equations:

$$\sum_{p=-\infty}^{\infty} \frac{f'(\gamma_{mp})}{\Gamma_{m\ell} + \gamma_{mp}} + \frac{\beta_q}{\Gamma_{m\ell} - \gamma_{mq}} + f(-\Gamma_{m\ell}) = 0 \quad , \quad \text{for} \quad \ell = 0, 1, 2, \ldots \tag{4.15}$$

Comparing (4.9) with (4.15), it follows,

$$s^{(q)}_{m\ell} = \frac{1 - (-1)^{\ell} e^{-jv}}{j\, \epsilon^{o}_{\ell} b\, \Gamma_{m\ell}}\, f(-\Gamma_{m\ell}) \quad , \qquad \text{for} \quad \ell = 0, 1, 2, \ldots \tag{4.16}$$

Thus, the solutions of the two infinite sets of simultaneous equations in (4.8) and (4.9) have been reduced to the construction of the function f(w). Once f(w) is known, $\{ e^{(q)}_{mp} \}$, and $\{ s^{(q)}_{m\ell} \}$ are obtainable from Equations (4.13), and (4.16), respectively.

The construction of f(w) is quite standard. Omitting the detail, we simply give the final result,

$$f(w) = \beta_q\, e^{-(w + \gamma_{mq})\, b \ln 2/\pi}\, \frac{1}{w + \gamma_{mq}}\, \frac{(\Gamma_{mo} - w)}{(\Gamma_{mo} + \gamma_{mq})}\, \frac{(\gamma_{mo} + \gamma_{mq})}{(\gamma_{mo} - w)} \tag{4.17}$$

$$\prod_{n=1}^{\infty} \frac{(w - \Gamma_{mn})\, e^{wb/n\pi}}{(-\gamma_{mq} - \Gamma_{mn})\, e^{-\gamma_{mq} b/n\pi}} \prod_{n=1}^{\infty} \frac{(-\gamma_{mq} - \gamma_{mn})(-\gamma_{mq} - \gamma_{m(-n)})\, e^{-\gamma_{mq} b/n\pi}}{(w - \gamma_{mn})\,(w - \gamma_{m(-n)})\, e^{wb/n\pi}}$$

From (4.13), the modal coefficients of the reflected field in (4.4) are obtained,

$$e^{(q)}_{mo} = \frac{\beta_q}{\beta_o}\, e^{-(\gamma_{mo} + \gamma_{mq}) \frac{b}{\pi} \ln 2}\, \frac{(\gamma_{mo} - \Gamma_{mo})}{(\gamma_{mq} + \Gamma_{mo})} \tag{4.18a}$$

$$\prod_{n=1}^{\infty} \left[\frac{(\gamma_{mo} - \Gamma_{mn})(-\gamma_{mq} - \gamma_{mn})(-\gamma_{mq} - \gamma_{m(-n)})}{(-\gamma_{mq} - \Gamma_{mn})(\gamma_{mo} - \gamma_{mn})(\gamma_{mo} - \gamma_{m(-n)})} \right]$$

$$e^{(q)}_{mp} = \frac{\beta_q}{\beta_p}\, e^{-(\gamma_{mp} + \gamma_q) \frac{b}{\pi} \ln 2}\, \frac{(\gamma_{mp} - \Gamma_{mo})}{(\gamma_{mq} + \Gamma_{mo})}\, \frac{(-\gamma_{mq} - \gamma_{mo})}{(\gamma_{mp} - \gamma_{mo})}\, \frac{(\gamma_{mp} - \Gamma_{m|p|})}{(-\gamma_{mq} - \Gamma_{m|p|})} \tag{4.18b}$$

$$\frac{(-\gamma_{mq} - \gamma_{m(-p)})}{(\gamma_{mp} - \gamma_{m(-p)})} \prod_{n=1}^{\infty\,(|p|)} \left[\frac{(\gamma_{mp} - \Gamma_{mn})\,(-\gamma_{mq} - \gamma_{mn})\,(-\gamma_{mq} - \gamma_{m(-n)})}{(-\gamma_{mq} - \Gamma_{mn})\,(\gamma_{mp} - \gamma_{mn})\,(\gamma_{mp} - \gamma_{m(-n)})} \right]$$

for $p = \pm 1, \pm 2, \pm 3, \ldots$

where the superscript $(|p|)$ in Equation (4.18b) signifies that the term $n = |p|$ = the absolute value of p, is omitted.

For the transmitted field in Equation (4.5), the modal coefficients are obtained from Equations (4.16) and (4.17), namely,

$$
s_{m\ell}^{(q)} = \frac{1-(-1)^{\ell} e^{-jv}}{j\epsilon_{\ell}^{o}\, b\Gamma_{m\ell}} \beta_q \, e^{(\Gamma_{m\ell}-\gamma_{mq})\frac{b}{\pi}\ln 2} \cdot \frac{(\Gamma_{mo}+\Gamma_{m\ell})(\gamma_{mo}+\gamma_{mq})}{(\Gamma_{mo}+\gamma_{mq})(\gamma_{mo}+\Gamma_{m\ell})(\gamma_{mq}-\Gamma_{m\ell})} \prod_{n=1}^{\infty} \frac{(\Gamma_{mn}+\Gamma_{m\ell})(\gamma_{mn}+\gamma_{mq})(\gamma_{m(-n)}+\gamma_{mq})}{(\Gamma_{mn}+\gamma_{mq})(\gamma_{mn}+\Gamma_{m\ell})(\gamma_{m(-n)}+\Gamma_{m\ell})}, \tag{4.19}
$$

for $\ell = 0, 1, 2, \ldots.$

This completes the solution for the scattering problem at junction 1 due to the incidence of a unit amplitude TE_{mq} mode from region B.

Next we need to consider the incident TM_{mq} mode from region B at junction 1. Omitting the details, the final solutions are

$$
\tilde{e}_{mo}^{(q)} = (-1)\, e^{-(\gamma_{mo}+\gamma_{mq})\frac{b}{\pi}\ln 2} \prod_{n=1}^{\infty} \left[\frac{(\Gamma_{mn}-\gamma_{mo})(\gamma_{mn}+\gamma_{mq})(\gamma_{m(-n)}+\gamma_{mq})}{(\Gamma_{mn}+\gamma_{mq})(\gamma_{mn}-\gamma_{mo})(\gamma_{m(-n)}-\gamma_{mo})}\right] \tag{4.20a}
$$

$$
\tilde{e}_{mp}^{(q)} = e^{-(\gamma_{mp}+\gamma_{mq})\frac{b}{\pi}\ln 2} \, \frac{\gamma_{mo}+\gamma_{mq}}{\gamma_{mo}-\gamma_{mp}} \, \frac{\gamma_{mp}-\Gamma_{m|p|}}{\gamma_{mq}+\Gamma_{m|p|}} \, \frac{\gamma_{m(-p)}+\gamma_{mq}}{\gamma_{m(-p)}-\gamma_{mp}} \tag{4.20b}
$$

$$
\prod_{n=1}^{\infty}(|p| \left[\frac{(\Gamma_{mn}-\gamma_{mp})(\gamma_{mn}+\gamma_{mq})(\gamma_{m(-n)}+\gamma_{mq})}{(\Gamma_{mn}+\gamma_{mq})(\gamma_{mn}-\gamma_{mp})(\gamma_{m(-n)}-\gamma_{mp})}\right]
$$

for $p = \pm 1, \pm 2, \pm 3, \ldots.$

$$
\tilde{s}_{m\ell}^{(q)} = \frac{(\ell\pi)\,[(-1)^{\ell} e^{-jv}-1]}{b^2\Gamma_{m\ell}(\gamma_{mq}-\Gamma_{m\ell})} \, \frac{\gamma_{mo}+\gamma_{mq}}{\gamma_{mo}+\Gamma_{m\ell}} \, e^{(\Gamma_{m\ell}-\gamma_{mq})\frac{b}{\pi}\ln 2} \prod_{n=1}^{\infty} \left[\frac{(\Gamma_{mn}+\Gamma_{m\ell})(\gamma_{mn}+\gamma_{mq})(\gamma_{m(-n)}+\gamma_{mq})}{(\Gamma_{mn}+\gamma_{mq})(\gamma_{mn}+\Gamma_{m\ell})(\gamma_{m(-n)}+\Gamma_{m\ell})}\right]
$$

for $\ell = 1, 2, 3, \ldots$

So far, we have obtained all the elements for the generalized scattering matrix S_1^{BB} and S_1^{AB}. It remains to arrange them in a proper order. Recalling the field representation of region B in equation (3.12), it follows immediately from Equation (4.1) that S_1^{BB} takes the following form:

$$
S_1^{BB} = \left[\begin{array}{cccc|cccc}
[e_{mo}^{(0)}] & [e_{mo}^{(1)}] & [e_{mo}^{(-1)}] & \cdots & & & & \\
[e_{m1}^{(0)}] & [e_{m1}^{(1)}] & [e_{m1}^{(-1)}] & \cdots & & & & \\
[e_{m(-1)}^{(0)}] & [e_{m(-1)}^{(1)}] & [e_{m(-1)}^{(-1)}] & \cdots & & 0 & & \\
\cdot & \cdot & \cdot & \cdots & & & & \\
\cdot & \cdot & \cdot & \cdots & & & & \\
\cdot & \cdot & \cdot & \cdots & & & & \\
\cdot & \cdot & \cdot & \cdots & & & & \\
\hline
 & & & & [\tilde{e}_{op}] & 0 & 0 & \\
 & & & & 0 & [\tilde{e}_{1p}] & 0 & \\
 & 0 & & & 0 & 0 & [\tilde{e}_{2p}] & \cdots \\
 & & & & \cdot & \cdot & \cdot & \cdots \\
 & & & & \cdot & \cdot & \cdot & \cdots \\
 & & & & \cdot & \cdot & \cdot & \cdots
\end{array}\right] \tag{4.23}
$$

where:

$$
[e_{mq}^{(p)}] = \begin{bmatrix} e_{1q}^{(p)} & & & 0 \\ & e_{2q}^{(p)} & & \\ & & e_{3q}^{(p)} & \\ 0 & & & \ddots \end{bmatrix}, \quad \text{for } p, q = 0, \pm 1, \pm 2, \ldots
$$

$$
[\tilde{e}_{mp}] = \begin{bmatrix} \tilde{e}_{mo}^{(0)} & \tilde{e}_{mo}^{(1)} & \tilde{e}_{mo}^{(-1)} & \cdots \\ \tilde{e}_{m1}^{(0)} & \tilde{e}_{m1}^{(1)} & \tilde{e}_{m1}^{(-1)} & \cdots \\ \tilde{e}_{m(-1)}^{(0)} & \tilde{e}_{m(-1)}^{(1)} & \tilde{e}_{m(-1)}^{(-1)} & \cdots \\ \cdot & \cdot & \cdot & \cdots \\ \cdot & \cdot & \cdot & \cdots \end{bmatrix}, \quad \text{for } m = 0, 1, 2, \ldots
$$

The scattering matrix S_1^{AB} has the same form as S_1^{BB} except that the $[e_{mq}^{(p)}]$ is replaced by $[s_{mn}^{(p)}]$, and $[\tilde{e}_{mp}]$ by $[\tilde{s}_{m\ell}]$, where

$$
[s_{mn}^{(p)}] = \begin{bmatrix} s_{1n}^{(p)} & & & 0 \\ & s_{2n}^{(p)} & & \\ & & s_{3n}^{(p)} & \\ 0 & & & \ddots \end{bmatrix}, \quad \begin{array}{l} \text{for } p = 0, \pm 1, \pm 2, \ldots \\ n = 0, 1, 2, \ldots \end{array} \tag{4.24}
$$

$$[\tilde{s}_{m\ell}] = \begin{bmatrix} \tilde{s}^{(0)}_{m1} & \tilde{s}^{(1)}_{m1} & \tilde{s}^{(-1)}_{m1} & \cdots \\ \tilde{s}^{(0)}_{m2} & \tilde{s}^{(1)}_{m2} & \tilde{s}^{(-1)}_{m2} & \cdots \\ \tilde{s}^{(0)}_{m3} & \tilde{s}^{(1)}_{m3} & \tilde{s}^{(-1)}_{m3} & \cdots \\ \cdot & \cdot & \cdot & \cdots \\ \cdot & \cdot & \cdot & \cdots \end{bmatrix}, \quad \text{for } m = 0, 1, 2, \ldots \tag{4.25}$$

This completes the solution of S_1^{BB} and S_1^{AB}.

The other two scattering matrices (S_1^{AA}, S_1^{BA}) can be determined by considering the scattering problem at junction 1 with an incident wave from region A. The final results are summarized below. The form of S_1^{AA} is exactly the same as that of S_1^{BB} in (4.23) except that $[e^{(p)}_{mq}]$ is replaced by $[a^{(n)}_{m\ell}]$; and $[\tilde{e}_{mp}]$ by $[\tilde{a}_{m\ell}]$, where

$$[a^{(n)}_{m\ell}] = \begin{bmatrix} a^{(n)}_{1\ell} & & & O \\ & a^{(n)}_{2\ell} & & \\ & & a^{(n)}_{3\ell} & \\ O & & & \ddots \end{bmatrix}, \quad \text{for } n = 0, 1, 2, \ldots, \; \ell = 0, 1, 2, \ldots \tag{4.26}$$

$$[\tilde{a}_{m\ell}] = \begin{bmatrix} \tilde{a}^{(1)}_{m1} & \tilde{a}^{(2)}_{m1} & \tilde{a}^{(3)}_{m1} & \cdots \\ \tilde{a}^{(1)}_{m2} & \tilde{a}^{(2)}_{m2} & \tilde{a}^{(3)}_{m2} & \cdots \\ \tilde{a}^{(1)}_{m3} & \tilde{a}^{(2)}_{m3} & \tilde{a}^{(3)}_{m3} & \cdots \\ \cdot & \cdot & \cdot & \cdots \\ \cdot & \cdot & \cdot & \cdots \end{bmatrix}, \quad \text{for } m = 0, 1, 2, \ldots \tag{4.27}$$

Also, S_1^{BA} is in the same form as S_1^{BB} in (4.23) except that $[e^{(p)}_{mq}]$ is replaced by $[b^{(n)}_{mq}]$, and $[\tilde{e}_{mp}]$ by $[\tilde{b}_{mp}]$, where

$$[b^{(n)}_{mq}] = \begin{bmatrix} b^{(n)}_{1q} & & & O \\ & b^{(n)}_{2q} & & \\ & & b^{(n)}_{3q} & \\ O & & & \ddots \end{bmatrix}, \quad \text{for } n = 0, 1, 2, \ldots, \; q = 0, \pm 1, \pm 2, \ldots \tag{4.28}$$

$$[\tilde{b}_{mp}] = \begin{bmatrix} \tilde{b}^{(1)}_{mo} & \tilde{b}^{(2)}_{mo} & \tilde{b}^{(3)}_{mo} & \cdots \\ \tilde{b}^{(1)}_{m1} & \tilde{b}^{(2)}_{m1} & \tilde{b}^{(3)}_{m1} & \cdots \\ \tilde{b}^{(1)}_{m(-1)} & \tilde{b}^{(2)}_{m(-1)} & \tilde{b}^{(3)}_{m(-1)} & \cdots \\ \cdot & \cdot & \cdot & \cdots \\ \cdot & \cdot & \cdot & \cdots \end{bmatrix}, \quad \text{for } m = 0, 1, 2, \ldots \tag{4.29}$$

The various scattering coefficients in (4.26) through (4.29) are explicitly given below:

$$a^{(n)}_{mo} = \left[\frac{1-(-1)^n e^{jv}}{1-e^{jv}}\right] e^{(\Gamma_{mo}+\Gamma_{mn})\frac{b}{\pi}\ln 2}\, \frac{\Gamma_{mo}-\gamma_{mo}}{\Gamma_{mn}+\gamma_{mo}} \prod_{\ell=1}^{\infty} \left[\frac{(\gamma_{m\ell}-\Gamma_{mo})(\gamma_{m(-\ell)}-\Gamma_{mo})(\Gamma_{m\ell}+\Gamma_{mn})}{(\gamma_{m\ell}+\Gamma_{mn})(\gamma_{m(-\ell)}+\Gamma_{mn})(\Gamma_{m\ell}-\Gamma_{mo})}\right] \tag{4.30a}$$

$$a^{(n)}_{ms} = \left[\frac{1-(-1)^n e^{jv}}{1-(-1)^s e^{jv}}\right] e^{(\Gamma_{ms}+\Gamma_{mn})\frac{b}{\pi}\ln 2} \cdot \frac{(\Gamma_{mn}+\Gamma_{mo})(\gamma_{mo}-\Gamma_{ms})(\gamma_{ms}-\Gamma_{ms})(\gamma_{m(-s)}-\Gamma_{ms})}{(\Gamma_{ms}-\Gamma_{mo})(\gamma_{mo}+\Gamma_{mn})(\gamma_{ms}+\Gamma_{mn})(\gamma_{m(-s)}+\Gamma_{mn})} \prod_{\ell=1}^{\infty}{}^{(s)} \left[\frac{(\gamma_{m\ell}-\Gamma_{ms})(\gamma_{m(-\ell)}-\Gamma_{ms})(\Gamma_{m\ell}+\Gamma_{mn})}{(\gamma_{m\ell}+\Gamma_{mn})(\gamma_{m(-\ell)}+\Gamma_{mn})(\Gamma_{m\ell}-\Gamma_{ms})}\right] \tag{4.30b}$$

for $s = 1, 2, 3, \ldots\ldots$

$$b^{(n)}_{mq} = \frac{\beta_q}{2bj\,\gamma_{mq}}\, \frac{1-(-1)^n e^{jv}}{\Gamma_{mn}-\gamma_{mq}}\, e^{(\Gamma_{mn}-\gamma_{mq})\frac{b}{\pi}\ln 2} \cdot \frac{(\Gamma_{mn}+\Gamma_{mo})(\gamma_{mq}+\gamma_{mo})}{(\gamma_{mq}+\Gamma_{mo})(\Gamma_{mn}+\gamma_{mo})} \prod_{\ell=1}^{\infty} \left[\frac{(\gamma_{mq}+\gamma_{m\ell})(\gamma_{mq}+\gamma_{m(-\ell)})(\Gamma_{mn}+\Gamma_{m\ell})}{(\Gamma_{mn}+\gamma_{m\ell})(\Gamma_{mn}+\gamma_{m(-\ell)})(\gamma_{mq}+\Gamma_{m\ell})}\right] \tag{4.31}$$

for $q = 0, \pm 1, \pm 2, \ldots$

$$\tilde{a}^{(n)}_{ms} = \frac{n}{s}\, \frac{1-(-1)^n e^{jv}}{1-(-1)^s e^{jv}}\, e^{(\Gamma_{ms}+\Gamma_{mn})\frac{b}{\pi}\ln 2} \cdot \frac{(\Gamma_{ms}-\gamma_{mo})(\gamma_{ms}-\Gamma_{ms})(\gamma_{m(-s)}-\Gamma_{ms})}{(\Gamma_{mn}+\gamma_{mo})(\gamma_{ms}+\Gamma_{mn})(\gamma_{m(-s)}+\Gamma_{mn})} \prod_{\ell=1}^{\infty}{}^{(s)} \left[\frac{(\gamma_{m\ell}-\Gamma_{ms})(\gamma_{m(-\ell)}-\Gamma_{ms})(\Gamma_{m\ell}+\Gamma_{mn})}{(\gamma_{m\ell}+\Gamma_{mn})(\gamma_{m(-\ell)}+\Gamma_{mn})(\Gamma_{m\ell}-\Gamma_{ms})}\right], \tag{4.32}$$

for $s = 1, 2, 3, \ldots$

$$\tilde{b}^{(n)}_{mq} = \frac{1}{2b\gamma_{mq}} \left(\frac{n\pi}{b}\right) [1-(-1)^n e^{jv}]\, e^{(\Gamma_{mn}-\gamma_{mq})\frac{b}{\pi}\ln 2}\, \frac{1}{\Gamma_{mn}-\gamma_{mq}}\, \frac{\gamma_{mo}+\gamma_{mq}}{\gamma_{mo}+\Gamma_{mn}} \cdot \prod_{\ell=1}^{\infty} \left[\frac{(\gamma_{m\ell}+\gamma_{mq})(\gamma_{m(-\ell)}+\gamma_{mq})(\Gamma_{m\ell}+\Gamma_{mn})}{(\gamma_{m\ell}+\Gamma_{mn})(\gamma_{m(-\ell)}+\Gamma_{mn})(\Gamma_{m\ell}+\gamma_{mq})}\right] \tag{4.33}$$

for $q = 0, \pm 1, \pm 2, \ldots$

5. Scattering Matrix at Junction 2

For the purpose of the present paper,* only (S_2^{BB}, S_2^{AB}) are needed at junction 2. First consider the case of the incident TE_{mq} mode from region B,

$$\psi_B^{(i)} = \sin\left(\frac{m\pi}{a}x\right) e^{-j\beta_q y}\, e^{-\gamma_{mq} z}, \quad \text{for } z < \delta$$

for $m = 1, 2, 3, \ldots$, and $q = 0, \pm 1, \pm 2, \ldots$. Since we are interested only in the limiting case $\delta \to 0$, we will set $\delta = 0$ in the following discussion.

*If we consider the problem of scattering by an infinite array, then (S_2^{BC}, S_2^{CC}) are also needed.

The scattered field due to the incidence in Equation (5.1) consists of both TE and TM (with respect to x). Let us denote the potentials of the reflected field by

$$\left.\begin{array}{ll} \text{TE:}\ \psi_B^{(r)} = \sum_{\ell=1}^{\infty} c_{\ell q}^{(m)} \sin\frac{\ell\pi}{a}x\, e^{-j\beta_q y}\, e^{+\gamma_{\ell q} z}, & (5.2a) \\ \text{TM:}\ \bar{\psi}_B^{(r)} = \sum_{\ell=0}^{\infty} d_{\ell q}^{(m)} \cos\frac{\ell\pi}{a}x\, e^{-j\beta_q y}\, e^{+\gamma_{\ell q} z}, & (5.2b) \end{array}\right\}\ \text{for } z<0,$$

and those of the transmitted field by,

$$\left.\begin{array}{ll} \text{TE:}\ \psi_C^{(t)} = \sum_{p=-\infty}^{\infty} g_{pq}^{(m)}\, e^{-j\alpha_{pq} x}\, e^{-j\beta_q y}\, e^{-\Omega_{pq} z}, & (5.3a) \\ \text{TM:}\ \bar{\psi}_C^{(t)} = \sum_{p=-\infty}^{\infty} h_{pq}^{(m)}\, e^{-j\alpha_{pq} x}\, e^{-j\beta_q y}\, e^{-\Omega_{pq} z}, & (5.3b) \end{array}\right\}\ \text{for } z>0$$

Perhaps the most convenient way to solve to $\{c_{\ell q}^{(m)}, d_{\ell q}^{(m)}, g_{pq}^{(m)}, h_{pq}^{(m)}\}$ is to first decompose Equation (5.1) into TE and TM with respect to y, since the structure at junction 1 is uniform along y-direction, the scattering of $TE^{(y)}$ and $TM^{(y)}$ are not coupled. Each of the scattering problems with $TE^{(y)}$ and $TM^{(y)}$ fields can be again exactly solved by the residue calculus method. Finally we superimpose the solutions of ($TE^{(y)}$ and $TM^{(y)}$ fields to give the desired coefficients in Equations (5.2) and (5.3). Omitting the details, we give here the final results:

$$c_{\ell q}^{(m)} = R_{\ell q}\left[\beta_q\left(\frac{m\pi}{a}\right)\left(\frac{\ell\pi}{a}\right) C_{\ell q}^{(m)} + \gamma_{mq}\gamma_{\ell q}\,\tilde{C}_{\ell q}^{(m)}\right] \tag{5.4a}$$

$$d_{\ell q}^{(m)} = R_{\ell q}\frac{1}{120\pi}\beta_q\left[\left(\frac{m\pi}{a}\right)\gamma_{\ell q} C_{\ell q}^{(m)} + \left(\frac{\ell\pi}{a}\right)\gamma_{mq}\tilde{C}_{\ell q}^{(m)}\right] \tag{5.4b}$$

$$g_{pq}^{(m)} = R'_{pq}\left[j\beta_q\left(\frac{m\pi}{a}\right)\alpha_{pq}\beta_q D_{pq}^{(m)} - k_o^2\Omega_{pq}\gamma_{mq}\tilde{D}_{pq}^{(m)}\right] \tag{5.4c}$$

$$h_{pq}^{(m)} = R'_{pq}\frac{1}{120\pi}\beta_q\left[-\Omega_{pq}\left(\frac{m\pi}{a}\right) D_{pq}^{(m)} - j\alpha_{pq}\gamma_{mq}\tilde{D}_{pq}^{(m)}\right] \tag{5.4d}$$

where

$$R_{\ell q} = \{[k_o - (\ell\pi/a)^2]\,[k_o^2-\beta_q^2]\}^{-1}$$

$$R'_{pq} = \{(k_o^2-\alpha_{pq}^2)(k_o^2-\beta_q^2)\}^{-1}$$

The four sets of coefficients $(C_{\ell q}^{(m)}, D_{pq}^{(m)})$, and $(\bar{C}_{\ell q}^{(m)}, \bar{D}_{pq}^{(m)})$ appeared in Equations (5.4) and are the (reflected, transmitted) coefficients due to an incident $TE_{mq}^{(y)}$, and $TM_{mq}^{(y)}$ modes from region B, respectively. Their explicit forms are

$$C_{\ell q}^{(m)} = \left[\frac{1-(-1)^m e^{j\alpha_{pq}a}}{1-(-1)^\ell e^{j\alpha_{pq}a}}\right]\left[\exp\left(\frac{\gamma_{\ell q}+\gamma_{mq}}{\pi}\,a\ln 2\right)\right]\frac{\gamma_{mq}+\gamma_{oq}}{\gamma_{\ell q}-\gamma_{oq}} \cdot \frac{(\Omega_{oq}-\gamma_{\ell q})(\Omega_{\ell q}-\gamma_{\ell q})(\Omega_{-\ell q}-\gamma_{\ell q})}{(\Omega_{oq}+\gamma_{mq})(\Omega_{\ell q}+\gamma_{mq})(\Omega_{-\ell q}+\gamma_{mq})} \prod_{n=1}^{\infty}{}^{(\ell)} \frac{(\gamma_{nq}+\gamma_{mq})(\Omega_{nq}-\gamma_{\ell q})(\Omega_{-nq}-\gamma_{\ell q})}{(\gamma_{nq}-\gamma_{\ell q})(\Omega_{nq}+\gamma_{mq})(\Omega_{-nq}+\gamma_{mq})} \tag{5.5}$$

$$D_{pq}^{(m)} = \frac{-j\alpha_{pq}\left[1-(-1)^m e^{j\alpha_{pq}a}\right]}{2a\,\Omega_{pq}(\gamma_{mq}-\Omega_{pq})}\exp\left[(\gamma_{mq}-\Omega_{pq})\frac{a}{\pi}\ln 2\right] \cdot \frac{(\gamma_{oq}+\gamma_{mq})(\Omega_{oq}+\Omega_{pq})}{(\gamma_{oq}+\Omega_{pq})(\Omega_{oq}+\gamma_{mq})} \prod_{n=1}^{\infty}\frac{(\gamma_{nq}+\gamma_{mq})(\Omega_{nq}+\Omega_{pq})(\Omega_{-nq}+\Omega_{pq})}{(\gamma_{nq}+\Omega_{pq})(\Omega_{nq}+\gamma_{mq})(\Omega_{-nq}+\gamma_{mq})} \tag{5.6}$$

$$\bar{C}_{\ell q}^{(m)} = \left(\frac{m}{\ell}\right)\frac{\gamma_{oq}-\gamma_{\ell q}}{\gamma_{oq}+\gamma_{mq}}\,C_{\ell q}^{(m)} \tag{5.7}$$

$$\bar{B}_{pq}^{(m)} = \frac{(m\pi/a)}{(-j\alpha_{pq})}\,\frac{\gamma_{oq}+\Omega_{pq}}{\gamma_{oq}+\gamma_{mq}} \tag{5.8}$$

Next, consider the incident TM_{mq} mode (with respect to x) from region B, which can be treated similarly as TE_{mq} case. Corresponding to (5.4), the results for TM field are

$$\bar{c}_{\ell q}^{(m)} = 120\pi R_{\ell q}\beta_q\left[\left(\frac{\ell\pi}{a}\right)\gamma_{mq}C_{\ell q}^{(m)} + \left(\frac{m\pi}{a}\right)\gamma_{\ell q}\bar{C}_{\ell q}^{(m)}\right] \tag{5.9a}$$

$$\bar{d}_{\ell q}^{(m)} = R_{\ell q}\left[k_o^2\gamma_{mq}\gamma_{\ell q}C_{\ell q}^{(m)} + \beta_q^2\left(\frac{m\pi}{a}\right)\left(\frac{\ell\pi}{a}\right)\bar{C}_{\ell q}^{(m)}\right] \tag{5.9b}$$

$$\bar{g}_{pq}^{(m)} = 120\pi R'_{pq}\beta_q\left[j\alpha_{pq}\gamma_{mq}D_{pq}^{(m)} - \Omega_{pq}\left(\frac{m\pi}{a}\right)\bar{D}_{pq}^{(m)}\right] \tag{5.9c}$$

$$\bar{h}_{pq}^{(m)} = R'_{pq}\left[-k_o^2\gamma_{mq}\Omega_{pq}D_{pq}^{(m)} - j\alpha_{pq}\beta_q^2\left(\frac{m\pi}{a}\right)\bar{D}_{pq}^{(m)}\right] \tag{5.9d}$$

where the coefficients $[C_{\ell q}^{(m)}, D_{\ell q}^{(m)}]$ and $[\bar{C}_{pq}^{(m)}, \bar{D}_{pq}^{(m)}]$ were given in (5.5) through (5.8).

The explicit form of S_2^{BB} is

$$S_2^{BB} = \left[\begin{array}{cccc|cccc} [c_{mo}] & 0 & 0 & \cdot & [\tilde{c}_{mo}^{(o)}] & [\tilde{c}_{mo}^{(1)}] & [\tilde{c}_{mo}^{(2)}] & \cdot \\ 0 & [c_{m1}] & 0 & \cdot & [\tilde{c}_{m1}^{(o)}] & [\tilde{c}_{m1}^{(1)}] & [\tilde{c}_{m1}^{(2)}] & \cdot \\ 0 & 0 & [\;_{m,-1}] & \cdot & [\tilde{c}_{m(-1)}^{(o)}] & [\tilde{c}_{m(-1)}^{(1)}] & [\tilde{c}_{m(-1)}^{(2)}] & \cdot \\ \cdot & \cdot & \cdot & \cdot & \cdot & \cdot & \cdot & \cdot \\ \hline [d_{oo}] & [d_{01}] & [d_{0,-1}] & \cdot & [\tilde{d}_{oq}^{(o)}] & [\tilde{d}_{oq}^{(1)}] & [\tilde{d}_{oq}^{(2)}] & \cdot \\ [d_{10}] & [d_{11}] & [d_{1,-1}] & \cdot & [\tilde{d}_{1q}^{(o)}] & [\tilde{d}_{1q}^{(1)}] & [\tilde{d}_{1q}^{(2)}] & \cdot \\ [d_{20}] & [d_{21}] & [d_{2,-1}] & \cdot & [\tilde{d}_{2q}^{(o)}] & [\tilde{d}_{2q}^{(1)}] & [\tilde{d}_{2q}^{(2)}] & \cdot \\ \cdot & \cdot & \cdot & \cdot & \cdot & \cdot & \cdot & \cdot \end{array}\right]$$

where

$$[c_{mq}] = \begin{bmatrix} c_{1q}^{(1)} & c_{1q}^{(2)} & c_{1q}^{(3)} & \cdot \\ c_{2q}^{(1)} & c_{2q}^{(2)} & c_{2q}^{(3)} & \cdot \\ c_{3q}^{(1)} & c_{3q}^{(2)} & c_{3q}^{(3)} & \cdot \\ \cdot & \cdot & \cdot & \cdot \end{bmatrix} \qquad [\tilde{d}_{mq}^{(\ell)}] = \begin{bmatrix} d_{mo}^{(\ell)} & & O & \cdot \\ & d_{m1}^{(\ell)} & & \cdot \\ & & d_{m(-1)}^{(\ell)} & \cdot \\ & O & \cdot & \cdot \\ & & & \cdot\ \cdot \\ \cdot & \cdot & \cdot & \cdot \end{bmatrix}$$

$$[\tilde{c}_{mq}^{(\ell)}] = \begin{bmatrix} \cdot & 0 & \tilde{c}_{1q}^{(\ell)} & \cdot \\ \cdot & 0 & \tilde{c}_{2q}^{(\ell)} & \cdot \\ \cdot & 0 & \tilde{c}_{3q}^{(\ell)} & \cdot \\ \cdot & \cdot & \cdot & \cdot \end{bmatrix} \qquad [d_{mq}] = \begin{bmatrix} \cdot & \cdot & \cdot & \cdot \\ 0 & 0 & 0 & \cdot \\ d_{mq}^{(1)} & d_{mq}^{(2)} & d_{mq}^{(3)} & \cdot \\ 0 & 0 & 0 & \cdot \\ \cdot & \cdot & \cdot & \cdot \end{bmatrix} \leftarrow \text{s'th row}$$

($\uparrow$ s'th column)

$$s = 2|q| + \frac{1+\text{sgn}.q}{2}, \qquad \text{sgn}\,q = 1, \text{ if } q \geq 0; \text{ and } -1, \text{ if } q < 0$$

The scattering matrix S_1^{CB} takes the same form as S_2^{BB} in (5.10) except that $[c_{mq}]$ is replaced by $[g_{pq}]$, $[\tilde{c}_{pq}^{(\ell)}]$ by $[\tilde{g}_{pq}^{(\ell)}]$, $[\tilde{d}_{mq}^{(\ell)}]$ by $[\tilde{h}_{pq}^{(\ell)}]$, and $[d_{mq}]$ by $[h_{pq}]$.

6. Numerical Computation and Results

The solutions for the reflected and the transmitted field for the array problem under consideration are given in Equations (2.3) and (2.4), respectively. Since all the scattering matrices are of infinite order, it is necessary to truncate them at a finite order N for practical computations. The problem of convergence of the solution with respect to N has been studied extensively in the literature, (11) and generally N may be taken to be a reasonably small number to obtain good accuracy. In the present paper, we have used N = 18 in all computations and the results are satisfactory. As a numerical example, consider an array of square waveguides with the following parameters:

$$a = b = 0.636\lambda_o, \quad d = 0, \quad \theta_o = 34.5^o, \quad \varphi_o = 45^o \quad .$$

We choose the incident wave to be either TE_{10} (with respect to x) or TM_{01}. Due to symmetry, the scattered field should be identical in these two cases. However, in our analysis, only the waveguide walls along y-direction are extended (Figure 2), and the solution is obtained in such a limit. Thus, the symmetry between TE_{10} and TM_{01} incidence is not "built in" the analysis. Hence, a comparison of the numerical results for these two cases provides us with a good check of the accuracy of the solution. By using a (18 x 18) scattering matrix, the results are tabulated in Table I, which is quite satisfactory.

Table I. Comparison of the Reflected Fields due to TE_{10} and TM_{01} Incidences

	TE_{10} Incidence	TM_{01} Incidence
TE_{10}^*	$(0.164, -12.1^o)$	$(0.334, 155.1^o)$
TM_{01}^*	$(0.338, 155.2^o)$	$(0.163, -12.7^o)$
Power Error	0.7%	0.3%

*The modal amplitudes are normalized so that the magnitude square is proportional to power, and the phase is referred to that of the electric field.

In Figures 4 thru 8, we present the solution for an equilaterally triangular array with the parameters

$$a = 0.539\lambda_o, \; b = 0.619\lambda_o, \; d = b/2.$$

The excitation is the dominant TM_{01} mode with electric field polarized in the x-direction. The grating lobe structure in the (θ_o, φ_o) plane is shown in Figure 4. Note that all the near-in grating lobes are situated symmetrically on a circle, and it can be shown (12) that this array is optimum in the sense of minimum elements for a given aperture and scan range $0 \leq \theta_o < 60°$. As may be measured from Figure 4, in the E-plane scan the grazing angle of the (-1, 0) lobe occurs at $\theta_o = 60°$, in the D-plane (diagonal plane $\varphi = 45°$) the grazing angle of the (-1, -1) lobe occurs at $\theta_o = 68°$, while there is no grating lobe in the H-plane scan. The (self) reflec-

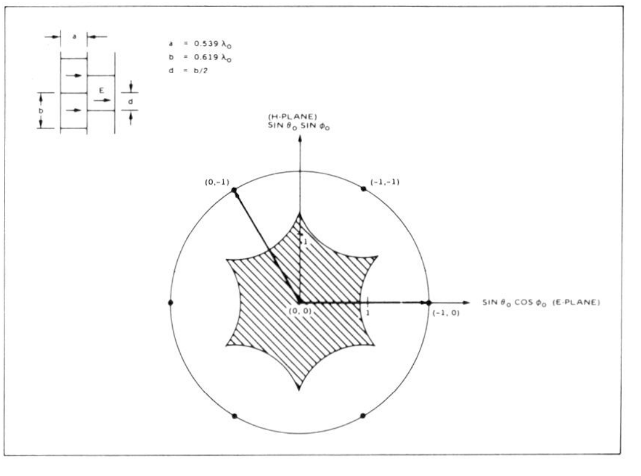

Fig. 4 Grating lobe structure in $(\theta_o \varphi_o)$ plane for an equilaterally triangular array of rectangular waveguides with an incident TM_{01} mode. The shaded region is grating-lobe-free.

tion coefficients for E_x of the TM_{01} mode are presented in Figure 5. Note that a nearly total reflec-

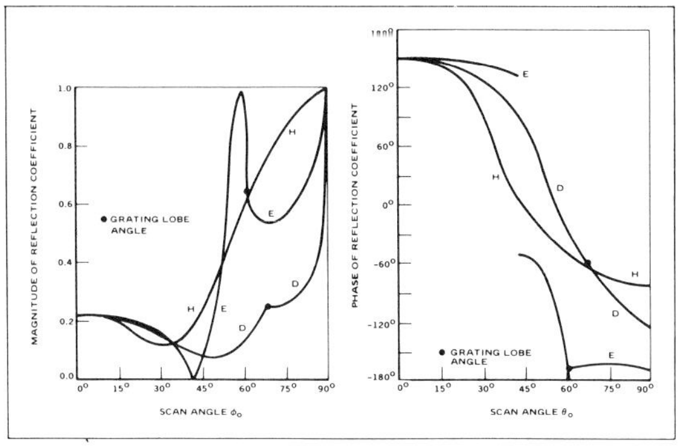

Fig. 5 The self-reflection coefficient of TM_{01} mode for the array in Figure 4. Its magnitude is normalized so that the magnitude square represents power for a unit incident power.

tion occurs at $\theta_o \approx 59^\circ$ in the E-plane scan, which is about one degree before the grazing of the (-1, 0) lobe. The cause for such a total reflection has been discussed extensively in the literature, (13, 14) and will not be investigated further here. In the D-plane scan, this is no total reflection even though the (-1, -1) lobe grazes at $\theta_o = 68^\circ$. In Figure 6, we present the coupling in the waveguide to the other propagating mode TE_{10} whose electric field is polarized in the y-direction. The modal coefficient of TE_{10} in Figure 6 is normalized such that its magnitude square is the power carried by TE_{10} for a unit incident power. In both E and H-plane scan, the coupling is very small (20 dB down from TM_{01} or less). However, TE_{10} mode is quite strongly excited in the D-plane scan, and as a matter of fact, the power in TE_{10} becomes greater than that in the (reflected) TM_{01} for scan beyond $\theta_o = 35^\circ$. Such a strong coupling to the cross-polarized field would generally present a problem for matching in the waveguides. In Figure 7, we present the power transmission coefficient of the main beam, as well as that of the grating lobes. All the power coefficients are normalized with respect to the incident power. Note that, as the azimuthal scan sweeps from the H-plane toward the E-plane the grating lobe angle becomes closer to the broadside (Figure 4), and the power in the grating lobe becomes bigger. As a matter of fact, in the E-plane scan, the power in the grating lobe is nearly equal to that in the main beam for scan beyond the grating lobe angle. It is also interesting to point out that the polarization of the main beam varies somewhat as the beam scans. In the H-plane scan, the polarization of the $\vec{E}$ field is almost entirely in the x-direction, same as the incident (TM_{01}) wave in the waveguide. In the E-plane scan, the $\vec{E}$ field of the main beam is almost entirely in the θ -direction, which is again a continuation of the incident wave. However, in the D-plane scan, the main beam deviates appreciably

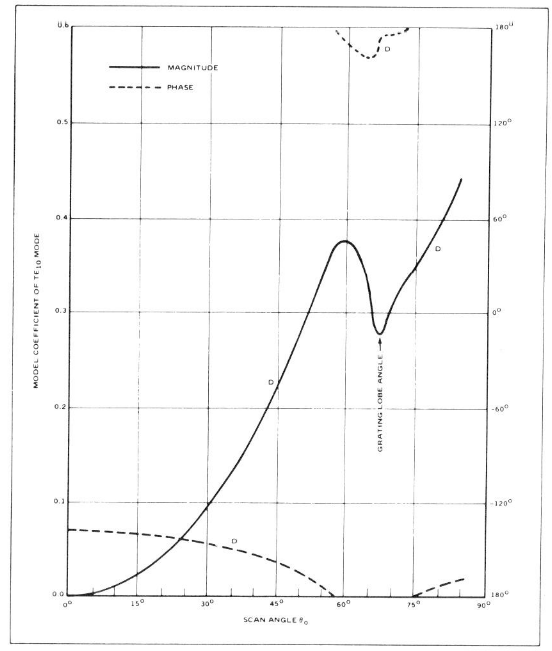

Fig. 6 The modal coefficient of the TE_{10} mode for the array in Figure 4. Its magnitude is normalized so that the magnitude square represents power for a unit incident power. In the E and H-plane scans, the modal coefficient of the TE_{10} mode is negligibly small as compared with that of the TM_{01} mode.

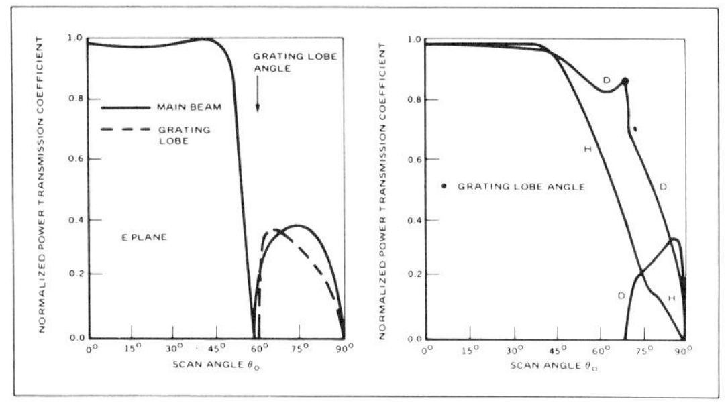

Fig. 7 The power transmission coefficients of the main beam, and the grating lobes for the array in Figure 4.

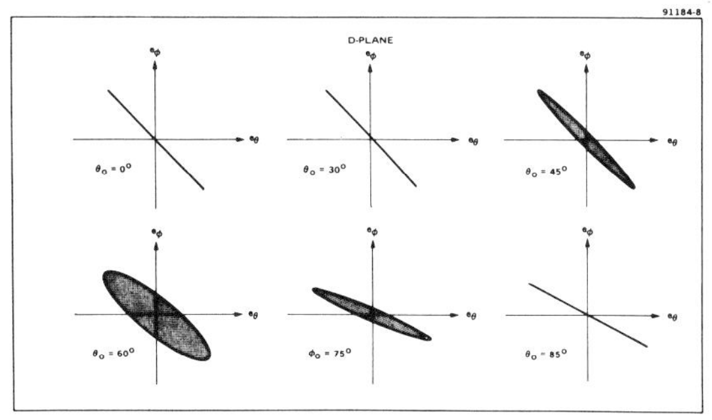

Fig. 8 The polarization ellipse of the main beam for the array in Figure 4. In the E and H-plane scan, the main beam is essentially linearly polarized.

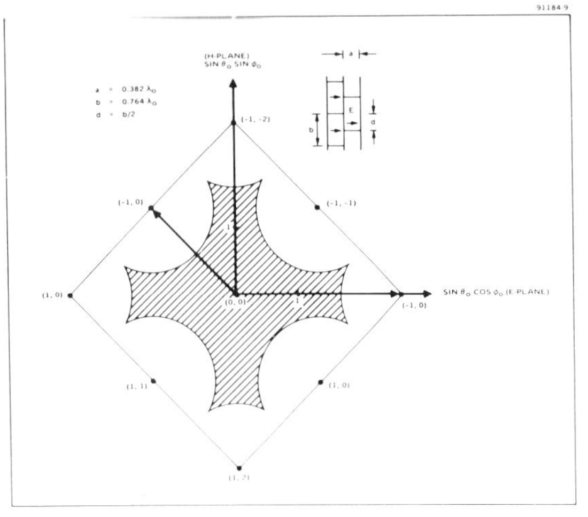

Fig. 9 Grating lobe structure in the $(\theta_0 \varphi_0)$ plane for an array of rectangular (z x 1) waveguides with an incident TM_{01} mode. The shaded region is grating-lobe-free.

Fig. 10 The 64-element experimental array.

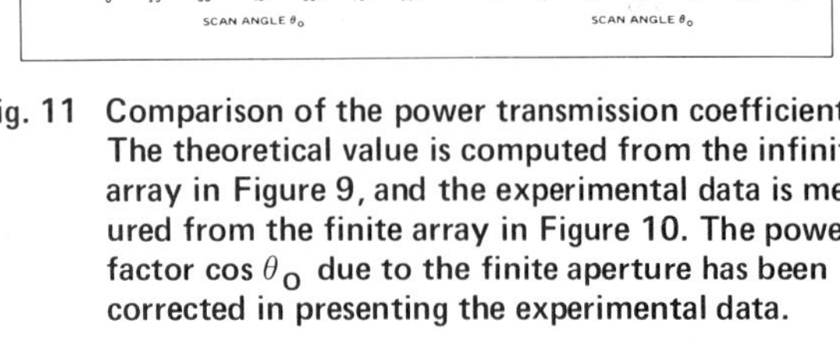

Fig. 11 Comparison of the power transmission coefficient. The theoretical value is computed from the infinite array in Figure 9, and the experimental data is measured from the finite array in Figure 10. The power factor $\cos\theta_0$ due to the finite aperture has been corrected in presenting the experimental data.

from the linear polarization in the direction near the grating lobe angle, as may be seen from various polarization ellipses in Figure 8.

As a final example, we compare the experimental active element pattern measured from a large array with the theoretical value computed in this paper. The array used in the theoretical computation has the following parameter

$$2a = b = 0.764\lambda_0, \quad d = b/2 \quad .$$

The grating lobe structure in the (θ_0, φ_0) plane is shown in Figure 9, which indicates that the array is free of grating lobes in E and H-planes, but the (-1,-1) lobe grazes at $\theta_0 = 57^\circ$ in the D-plane scan. The experimental array consists of (8x8) elements, mounted on a ground plane in the manner shown in Figure 10. The active element pattern was measured at the frequency of 3 GHz. In presenting the experimental result in Figure 11, we have divided the power data by the factor $\cos\theta_0$ to account for the finite size of the experimental array so that they may be compared with theoretical data computed from an infinite array. Generally speaking, the agreement between the theory and the experiment is quite well except for large scan angles when the edge effect of the finite array becomes important. At $\theta_0 = 57^\circ$ in the D-plane scan, the theory predicts a null while the experiment data shows a 5.5 dB dip. It is believed (6, 14) that the dip will become sharp as the number of elements in the experimental array increases.

7. Conclusions

We have presented a scattering matrix approach to attack the problem of infinite planar phased arrays of rectangular waveguides. The main advantage of this method lies in the fact that we are able to reduce the original three-dimensional, vector-field boundary value problem to a couple of two-dimensional, scalar-field problems. Furthermore, for the case of a thin-walled array, the scattering matrices involved can be solved exactly, which not only saves the computation labor but also enhances the accuracy in the subsequent calculations.

The scattering matrix approach is also preferable from a computer programming point of view. By making use of the unitary condition of all the scattering matrices, we may break up the whole program in several parts and check them independently, whereas in the conventional numerical methods such a step-by-step checking scheme is generally

In the present paper, we have treated only the array where the waveguide walls are infinitesimally thin. For a thick-walled array, the scattering matrix approach still applies except that each scattering matrix then has to be solved numerically.

ACKNOWLEDGEMENTS

This program was supported by Air Force Cambridge Research Laboratory under Contract F19628-68-C-0185. The computer program was written by G. Roberts and D. Perry of the Electromagnetics Laboratory, Hughes Ground Systems Group.

REFERENCES

1. See, e.g., The special issue on electronic scanning, Proc. IEEE, V. 56, No. 1, Nov. 1968.

2. H. A. Wheeler, "The grating-lobe series for the impedance variation in a planar phased-array antenna," IEEE Trans., AP-14, pp. 707-714, Nov. 1966.

3. L. Stark, "Radiation impedance of a dipole in an infinite planar phased array," Radio Science, V. 2, pp. 361-377, March 1966.

4. S. Edelberg and A. A. Oliner, "Mutual coupling effects in large antenna arrays: Part I - Slot arrays," IRE Trans., AP-8, pp. 286-297, May 1960.

5. C. P. Wu and V. Galindo, "Properties of a phased array of rectangular waveguides with thin walls," IEEE Trans., AP-14, pp. 163-173, March 1966.

6. G. F. Farrell Jr., and D. H. Kuhn, "Mutual coupling in infinite planar arrays of rectangular waveguide horns," IEEE Trans., AP-16, pp. 405-414, July 1968.

7. B. L. Diamond, "A generalized approach to the analysis of infinite planar array antennas," Proc. IEEE, V. 56, pp. 1837-1851, Nov. 1968.

8. J. J. Campbell and W. R. Jones, private communication.

9. J. Pace and R. Mittra, "Generalized scattering matrix analysis of waveguide discontinuity problems," Quasi-optics, Vol. 24, pp. 177-197, Polytechnic Institute of Brooklyn Press, N. Y. 1964.

10. E. C. DuFort, "A design procedure for matching volumetrically-scanned waveguide arrays," Proc. IEEE, V. 56, pp. 1851-1860, Nov. 1968.

11. R. Mittra and S. W. Lee, Analytical techniques in the theory of guided waves, MacMillan, N. Y. 1970.

12. Y. T. Lo and S. W. Lee, "Affine transformation and its application to antenna arrays," IEEE Trans., AP-13, pp. 890-896, Nov. 1965.

13. C. P. Wu and V. Galindo, "Surface wave effects on dielectric sheathed phased array of rectangular waveguides," Bell Sys. Tech. J., Vol. 47, pp. 117-142, Jan. 1968.

14. G. H. Knittel, A. Hessel, and A. A. Oliner, "Element pattern nulls in phased array and their relation to guided waves," IEEE Proc. Vol. 56, No. 11, pp. 1822-1836, Nov. 1968.

APPENDIX: COMPUTER PROGRAM DESCRIPTION

D. M. PERRY

The program is written for an operational on a G. E. 635 computer. It is written in FORTRAN IV and has numerous calls to Hughes Scientific Subroutines. One nonstandard FORTRAN feature is the use of "namelist" for input. Core requirements are 32K words. A list of routines used by the program follows:

MAIN PROGRAM

CDGHMX

EFMTRX

SWMTRX

SCALE

FUNA

FUNB

FUNAT

FUNBT

DVERS1

DVERS2

PSIAB

PQM

PRAR

PRR

System Routines

CMATS

CMXMP

CSQRT

CABS

CEXP

DLOG10

DEXP

Common Blocks

CONST

DATA

SBBSCB

MATRIX

SGAMMA

NORMAL

INPUTS TO PROGRAM

The namelist feature requires data to be input in cases. Each case is preceeded by a namelist name card.

$INPUT

cc2

The data values, explained before, are entered in cc3–cc72 of cards following the name card. A $ must follow the last data value in the case. Data values are assigned to varables by placing the computer name for the varable on the left side of an equal sign and the numerical value on the right. Data values assignments are separated by commas.

Computer Name	Engineering Name	Nominal Value	Notes
A	$k_o a$	None	
B	$k_o b$	None	
D	$k_o d$	0	Used in cal. of a_{pg}
KPSIA		1	Excitations. For incident TE_{10} mode, KPSIA = KPSIB = 1. For incident TM_{01} mode, KPSIA = KPSIB = 10.
KPSIB		1	
PHIE	φ	None	Azimuthal scan angle
THET1	θ_1	None	First value of polar scan angle
THET2	θ_2	None	Last value of polar scan angle
THETCT		None	Number of equally spaced values of θ to sequence through, including the end points θ_1 to θ_2. If set to 1, only θ_1 is processed.
PRINT		FALSE	Causes printing of the following: initial input, propagation constants, coefficients of polynomial, PSIA and PSIB, B, $\tilde{B}$, C, $\tilde{C}$, D, $\tilde{D}$, G, $\tilde{G}$, H, $\tilde{H}$, S, $\tilde{W}$, E, $\tilde{F}$, AK, $A\tilde{K}$, BK, $B\tilde{K}$.

Surface-Wave Effects and Blindness in Phased-Array Antennas

ARTHUR A. OLINER
Polytechnic Institute of Brooklyn
Graduate Center
Farmingdale, N. Y.

Introduction

Mutual coupling between radiating elements in a phased array can under certain circumstances produce blindness, which is a serious performance problem. A phased-array antenna is "blind" at those scan angles at which no power is radiated from (or received by) the array. The earliest explanations for this effect involved the concept of "surface waves" along the array face. Although these earliest explanations were in fact incorrect, the term "surface wave" stuck and the phenomenon is often still referred to as a "surface-wave effect." To confuse the issue further, it is possible in a highly-limited sense to speak correctly of a surface wave associated with blindness, although this surface wave is different from the originally-postulated one. It is the function of this paper to clarify the confusion surrounding the term "surface wave" in this context, and to present the correct explanations for the blindness phenomenon.

Other names have been used to describe the blindness effect. They include resonances, cancellation effects, element-pattern nulls and leaky-wave effects. All of these terms are correct, but each emphasizes a different aspect of the behavior. One reason for the multiplicity of names is that a phased array can be excited in two entirely different modes of operation. The first is the customary mode in which all of the elements are excited and are appropriately phased so as to produce a radiating beam in a given direction. We then characterize the array performance in terms of the active admittance Y_a or reflection coefficient Γ_a of a typical radiating element. The second occurs when only a single element of the array is excited and all of the other elements are terminated in their generator admittances. The array element pattern, which is characterized by the element gain function $g_e(\theta)$, where θ is the scan angle in a given scan plane, is used to describe this mode of behavior. The characteristic quantities pertinent to these two different methods of exciting the array are related by (1)

$$g_e(\theta) = \cos\theta \, [\, 1 - |\Gamma_a(\theta)|^2 \,] \, . \tag{1}$$

When the reflection coefficient is unity at some scan angle θ, the array radiates (or receives) no power, and it is therefore blind at that angle. From (1) we see that $g_e(\theta) = 0$ at that angle, so that the element pattern has a null at the angle for which the array is blind. The discussion below considers each of these two different modes of operation for the array, and shows that the terms "resonance" and "cancellation effects" properly describe the behavior in terms of the first, or customary, mode mentioned above, while the terms "element-pattern nulls" (this should be clear already) and "leaky-wave effects" are appropriate to the second, or single-element excitation, mode.

Types of Radiating Element for Which Blindness Occurs

The blindness effect has been observed experimentally, both on arrays themselves and on waveguide simulators of arrays, and detected theoretically. Various direct array measurements have been made which exhibit the element-pattern null. Among them are those made by Lechtreck (2), Bates (3), Diamond (4), and Byron and Frank (5). On many of these, the dips obtained were shallow because the arrays were small; Farrell and Kuhn (6) present a pattern measured at the Bendix Corp. which shows a deep null. This pattern is given here as Fig. 1. Waveguide simulator measurements which illustrate the blindness effect were carried out by Hannan (7), Byron and Frank (5), and Gregorwich, Hessel, Knittel and Oliner (8).

Theoretical calculations which exhibit the blindness effect include those by Farrell and Kuhn (6) (9), Diamond (4), Wu and Galindo (10), Amitay and Galindo (11), Borgiotti (12), Amitay and Galindo (13), and Knittel, Hessel and Oliner (14). The different structures on which these various measurements and theoretical analyses were made, and the significance of these investigations, will be included in the full paper.

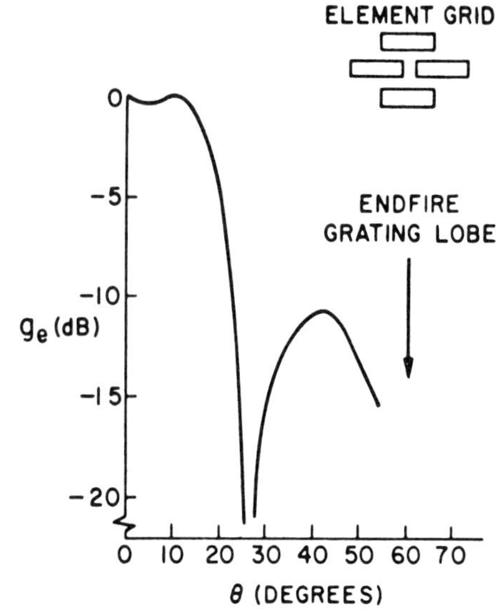

Fig. 1 A measured element pattern which exhibits a null. Endfire grating lobe signifies the main beam angle at which the grating lobe just appears at endfire. (Measured at Bendix Corp. and reported by Farrell and Kuhn[6].)

The large body of work referred to above permits us to recognize that these blindness effects can occur on a variety of structures, when the circumstances are appropriate, and that these structures (when of the aperture type) may be divided into two classes, as shown in Figs. 2 and 3.

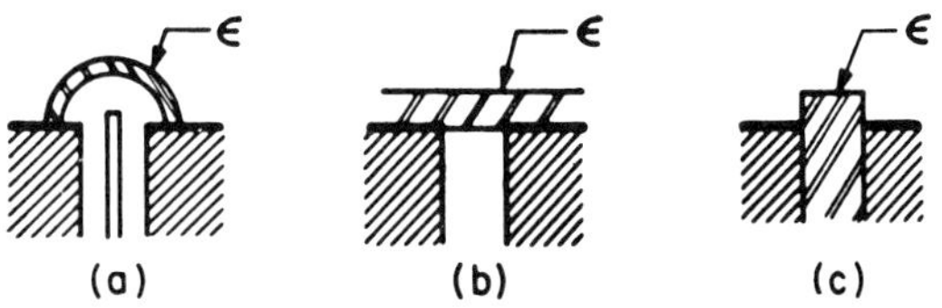

Fig. 2 Examples of radiating elements for which external resonances may occur.

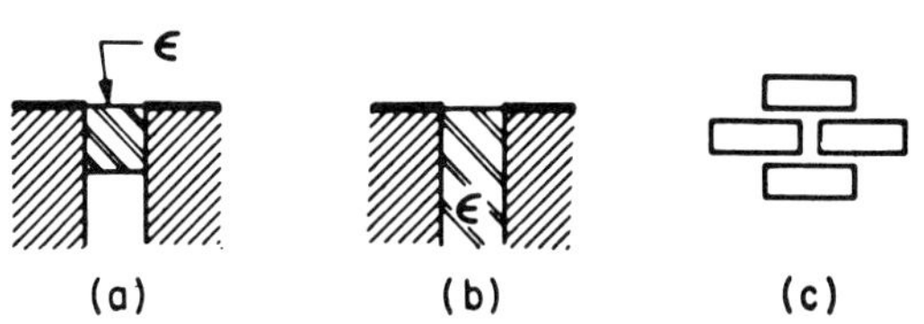

Fig. 3 Examples of radiating elements for which internal resonances may occur.

Corresponding comments for arrays of the wire dipole type will be included in the full paper. As will be seen below, a key role in all of these blindness effects is played by the presence of one (or more) of the higher modes. Depending on the situation, this higher mode will resonate or will partake in a cancellation effect which changes rapidly with change in scan angle, and which thus resembles a resonance. It is therefore appropriate to distinguish, as in Figs. 2 and 3, between two types of such resonance, external resonances and internal resonances, according as the relevant higher mode is in the region outside of the array aperture or within the feed guide region, respectively.

Examples of radiating elements for which external resonances can occur are shown in Fig. 2; (a) was used by Lechtreck, (b) by Bates and by Byron and Frank, and (c) by Hannan. Each of these configurations contains a region which can support the propagation of a higher external mode, even though that mode is below cut-off in the air space beyond. Examples of radiating elements which can support internal resonances are given in Fig. 3. In (a), the higher internal mode would be either propagating in the dielectric and cut off in air (Amitay and Galindo), or just slightly below cut-off in the dielectric. In (b), the higher mode would be just below cut-off in the dielectric region (Diamond). In (c), which is a broadside view of an air-filled "brick" array, the waveguide dimensions are so large that a higher mode is just below cut-off in the waveguide even though there is no dielectric loading (Diamond or Farrell and Kuhn). The examples presented in Figs. 2 and 3 are illustrative; one can easily choose additional examples.

The first structures on which the blindness effect was found were those of Fig. 2. The earliest explanations for the effect (15) (16) involved surface waves which would propagate along the array face, so that cancellation would occur when the phasing was correct. The explanation seemed very reasonable at first, and it produced results which were qualitatively in the right direction, but on closer examination the surface wave as postulated could not exist. Because the spacing between array elements is greater than $\lambda/2$, the surface wave becomes transformed into a leaky wave, and the physical process then becomes quite different. We will return to this point later.

Explanation as a Resonance or a Cancellation Effect

The first correct, but incomplete, explanation (17) relates to the customary mode of operation of the array, in which all elements are excited. The explanation viewed the effect as a resonance and was made in terms of a "ghost mode" hypothesis. The viewpoint was applied to the structure of Fig 2(b), and it postulated that the total reflection of power from the array face corresponded to a resonance of a higher mode in the dielectric region, a mode that could propagate in dielectric but not in air. Since the mode could propagate only in the intermediate region, the dielectric layer, it was termed a "ghost mode," in conformity with an earlier explanation of breakdown in ceramic waveguide windows. This mode capture has later been called a "trapped surface wave" (13). The implications of this nomenclature will also be considered later.

The resonance of the higher mode may also be viewed as a cancellation effect, wherein the coupling between the feed mode and the radiating mode is exactly cancelled by the coupling from the higher mode. This manner of viewing the effect is made much clearer by reference to an equivalent network (17) (14) (18).

The original, partially-correct explanation (17) cannot be applied to the understanding of internal resonances. Furthermore, it is deceptively simple because it assumed a scan-invariant aperture field. Actually the aperture field changes in form as the array is scanned because the total field distribution in space becomes asymmetric. This asymmetry in space reflects itself as an asymmetry in the aperture field which is usually well accounted for by the addition of the first higher order mode in the feed guide, even though this mode may be below cut-off. When conditions are right, the presence of this added mode produces a field cancellation in the aperture at some scan angle, and the active conductance becomes zero, resulting in a complete reflection of the power at that angle. A sketch which illustrates conceptually the cancellation mechanism is given in Fig. 4.

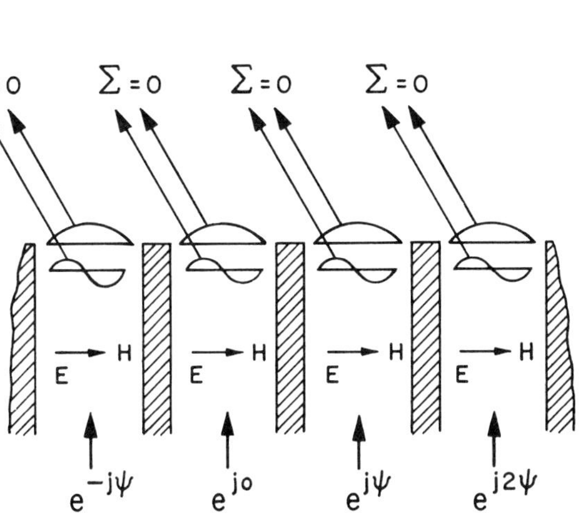

Fig. 4 Sketch showing the cancellations of radiation from the two lowest-order waveguide modes to produce the complete-reflection condition (Courtesy of G. H. Knittel).

Although a fuller discussion of this viewpoint has been presented (18), its implications for both internal and external resonances will be given in the full paper.

Explanation in Terms of Leaky Waves

The resonance, or cancellation, viewpoint given above and illustrated by Fig. 4 applies to the mode of operation in which all elements of the array are excited. When only a single element is excited, and all other elements are terminated in their generator impedances, the blindness effect corresponds to a null in the element pattern, as indicated above. It was proposed originally that this null is due to a cancellation effect produced by a surface wave guided by the passively-terminated elements. Such a surface wave is not possible, but a leaky wave does indeed exist and its presence causes the null.

A leaky wave (19) (20) is a wave which is guided along the array face and leaks energy away from the array face as it travels. Depending upon the array spacing and the frequency, this leakage, or radiation, could occur in either the backward or the forward direction (19). For ordinary phased-array operation, the radiation occurs in the backward direction. This situation is illustrated in Fig. 5.

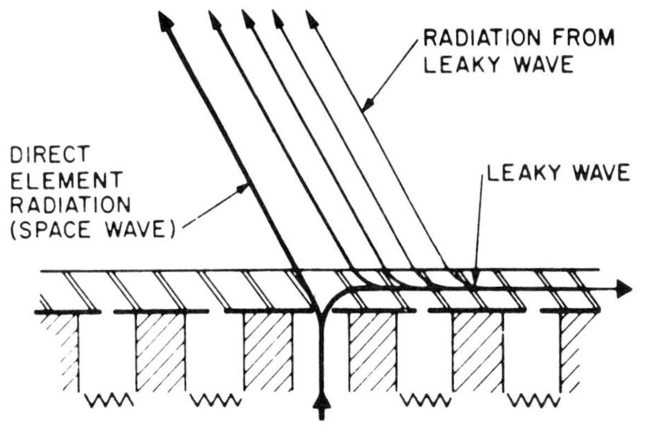

Fig. 5 Sketch showing interference between direct element radiation and that from leaky wave to produce element pattern null on left side of pattern. (From Knittel, Hessel and Oliner[14]).

The single excited element radiates a space wave (in all directions) and it also excites a leaky wave on the neighboring portions of the passively-terminated array. The radiated portion of the leaky wave then interferes with the portion of the space wave radiating in the same direction in such a way as to produce a null in the element pattern. Precisely the same thing occurs on the other side of the array so that a symmetrical pattern is produced. Note that a surface wave along the array face could not serve to produce the cancellation required in achieving the radiation null; the radiating leaky wave offers a simple and physically-satisfying mechanism.

This leaky-wave mechanism was first proposed by Knittel, Hessel and Oliner (14), and they applied it to structures exhibiting external resonances. Knittel (21) has shown that a leaky wave is also present for internal resonances on structures such as those of Fig. 3(c).

Validity of the Term "Surface Wave"

The term "surface wave" has been used loosely in connection with the blindness phenomenon, and a substantial confusion exists in this context. Two recent papers (14) (18) have sought to clarify this confusion by describing the various types of surface wave possible and distinguishing between them. Their conclusions are now summarized in terms of the preceding discussion.

For the mode of operation of the array in which only a single element is excited, and the remaining elements are passively terminated, it was remarked earlier that a surface wave is not possible for customary array spacings; only a leaky wave can be present. Thus, the early surface wave explanations (15) (16) are not valid, and surface waves of that type do not exist.

When all elements are excited, in the customary mode of operation of the array, and the array is considered infinite in extent, no net surface wave can appear along the array face. Even if an edge or a nonperiodic discontinuity were present to excite a guided wave, the array spacing is such that a leaky wave, rather than a surface wave, would be excited. Thus, the only wave type which can be supported by the actual array under either condition of operation is a leaky wave, and surface waves on the actual array are not possible.

However, at exactly the complete-reflection condition, a driven surface-wave-like field is present on the array face. This field, which resembles a surface wave field on an artificial structure, not the actual array, has been referred to loosely as a "surface wave," thus compounding the confusion.

At the precise complete-reflection condition, all power from the source is reflected back from the aperture and, as shown by Wu and Galindo (10), one can place a short-circuiting plate some distance below the aperture plane without disturbing the field. The resulting array structure becomes some form of corrugated surface; the configuration for the case of an array of slots covered by a dielectric layer is shown in Fig. 6.

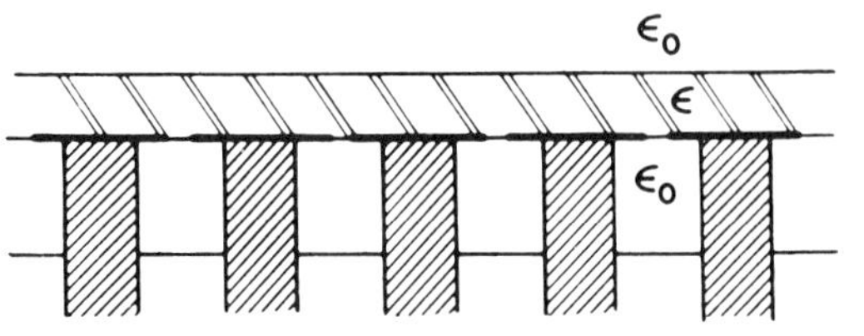

Fig. 6 Modified array structure produced by inserting short-circuiting plates into the phased array at the completion-reflection condition.

Under these conditions, the forced (driven) field existing on the array face is identical with the field of a surface wave mode of the dielectric-loaded corrugated structure shown in Fig. 6. This surface wave is not a mode of the original array, which includes the finite conductance of the generators below the aperture plane. It is a mode of the modified structure which was produced by introducing short-circuiting plates at an appropriate location below the aperture plane. It must also be stressed that this modified structure has meaning only at exactly the scan angle corresponding to the complete reflection. For any deviation from this angle, $|\Gamma_a|$ is no longer unity, and the perfectly-conducting plates can no longer be inserted. The modified structure has then lost its significance.

The forced (or driven) field associated with the condition of complete reflection thus resembles the field of a surface wave on the modified (corrugated) structure. This forced field, however, is not the field of an actual surface wave; it is not an independently-excited wave and it is present only because all the elements are driven from below. It is similar to the evanescent field which is present on the air side of a dielectric-air interface when a wave is incident from the dielectric side under the condition of total reflection. That evanescent wave has sometimes been referred to, inexactly and incorrectly, as a surface wave; in a similar way, the driven field (also of evanescent form above the array plane) occurring at the complete-reflection condition has at times been called a surface wave.

Some recent calculations (13) have shown that the blindness effect need not be total, and that off the cardinal scan planes the value of $|\Gamma_a|$ was found to be less than unity. Under those conditions, the modified structure of Fig. 6 is no longer valid, and hence the "surface wave" referred to above loses its meaning. The leaky wave viewpoint, however, is still applicable.

Conclusions

The blindness effect, a serious performance problem, can occur on a variety of structures, and these structures can be classified according to the resonance which produces the blindness as external or internal, as shown in Figs. 2 and 3.

Depending on the mode of operation of the array, the physical cause of blindness can be expanded in different terms. When all elements of the array are excited, which is the customary mode of operation, the blindness is due to a resonance or cancellation effect, as illustrated by Fig. 4. When only one element of the array is excited and the remaining elements are passively terminated, the blindness effect, which is manifested by a null in the element pattern, is caused by interference between the radiated space wave and a leaky wave which is supported by the passively-terminated elements, as represented in Fig. 5.

Regarding the validity of the term "surface wave" in this connection, it is not possible under the customary conditions of phased array operation for surface waves to be present on the actual array. The early explanations of blindness in terms of surface waves is therefore not valid. A driven surface-wave-like field is present at the precise condition of complete reflection, however, and this field resembles that of a surface wave on a corrugated surface, not on the actual array. Unfortunately, the term "surface-wave effect" has been used very loosely, and the distinctions between the non-existent surface wave on the actual array and the surface-wave-like field which exists only at the complete-reflection condition are often blurred.

The other names for the blindness effect, such as resonance, cancellation effect, element-pattern null or leaky-wave effect, are all meaningful and correct, but they each refer to one or another aspect of the behavior. The term "surface-wave effect" is to be deplored since it is incorrect in one sense and ambiguous in another, and it is indeed unfortunate that its use became widespread before a clear picture was obtained for the cause of the blindness. This writer recommends that the name blindness be used to define the phenomenon since it is the most directly descriptive and unambiguous.

Now that the cause of blindness is understood, the next challenge relates to how array element design should take advantage of this understanding. As a result, we should have elements which not only eliminate blindness but show substantial improvements in wide-angle performance, even to the minimization or elimination of grating lobe effects. Some early progress has already been made along these lines (22) (23), and further comments in this connection will be made by G. H. Knittel in his survey paper in the next session.

Acknowledgment

The research reported in this paper was sponsored by the U. S. Army Advanced Ballistic Missile Defense Agency under Contract DAHC 60-69-C-0110.

References

1. P. W. Hannan, "The element-gain paradox for a phased-array antenna," IEEE Trans., Vol. AP-12, pp. 423-433 (July 1964).

2. I. W. Lechtreck, "Effect of coupling accumulation in antenna arrays," IEEE Trans., Vol. AP-16, pp. 31-37 (January 1968).

3. R. H. T. Bates, "Mode theory approach to arrays," IEEE Trans., Vol. AP-13, pp. 321-322 (March 1965).

4. B. L. Diamond, "Resonance phenomena in waveguide arrays," IEEE G-AP Intl. Symp. Digest, pp. 110-115 (October 1967).

5. E. V. Byron and J. Frank, "'Lost beams' from a dielectric covered phased-array aperture," IEEE Trans., Vol. AP-16, pp. 496-499 (July 1968).

6. G. F. Farrell, Jr. and D. H. Kuhn, "Mutual coupling effects of triangular-grid arrays by modal analysis," IEEE Trans., Vol. AP-14, pp. 652-654 (September 1966).

7. P. W. Hannan, "Discovery of an array surface wave in a simulator," IEEE Trans., Vol. AP-15, pp. 574-576 (July 1967).

8. W. S. Gregorwich, A. Hessel, G. H. Knittel and A. A. Oliner, "A waveguide simulator for the determination of a phased-array resonance," IEEE G-AP Intl. Symp. Digest, pp. 134-141 (September 1968).

9. G. F. Farrell, Jr. and D. H. Kuhn, "Mutual coupling in infinite planar arrays of rectangular waveguide horns," IEEE Trans., Vol. AP-16, pp. 405-414 (July 1968).

10. C. P. Wu and V. Galindo, "Surface wave effects on dielectric sheathed phased arrays of rectangular waveguides," The Bell System Technical Journal, Vol. 47, pp. 117-142 (January 1968).

11. N. Amitay and V. Galindo, "Analysis of circular waveguide phased arrays," URSI Spring Meeting Digest, Washington, DC, p. 103 (April 1968); also, B. S. T. J., Vol. 47, pp. 1903-1931 (November 1968).

12. G. V. Borgiotti, "Modal analysis of infinite periodic phased arrays," URSI Symp. on Electromagnetic Waves, Stresa, Italy, Alta Frequenza, Vol. 38, pp. 255-262 (May 1969).

13. N. Amitay and V. Galindo, "Characteristics of dielectric loaded and covered circular waveguide phased arrays," IEEE Trans., Vol. AP-17, pp. 722-729 (November 1969).

14. G. H. Knittel, A. Hessel and A. A. Oliner, "Element pattern nulls in phased arrays and their relation to guided waves," Proc. IEEE, Vol. 56, pp. 1822-1836 (November 1968).

15. J. L. Allen, "On surface-wave coupling between elements of large arrays," IEEE Trans., Vol. AP-13, pp. 638-639 (July 1965).

16. L. W. Lechtreck, "Cumulative coupling in antenna arrays," IEEE G-AP Intl. Symp. Digest, pp. 144-149 (1965).

17. A. A. Oliner and R. G. Malech, "Speculations on the role of surface waves," in Microwave Scanning Antennas, R. C. Hansen, Ed., (New York: Academic Press, 1966) pp. 308-322.

18. A. A. Oliner, "On blindness in large phased arrays," URSI Symp. on Electromagnetic Waves, Stresa, Italy, Alta Frequenza, Vol. 38, pp. 221-228 (May 1969).

19. A. A. Oliner and A. Hessel, "Guided waves on sinusoidally-modulated reactance surfaces," IEEE Trans., Vol. AP-7, special supplement, pp. S201-S208 (December 1959).

20. T. Tamir and A. A. Oliner, "Guided complex waves, Part I. Fields at an interface," Proc. IEE (London), Vol. 110, pp. 310-324 (February 1963).

21. G. H. Knittel, "Element pattern nulls in phased-array antennas and their relation to guided waves," Ph.D. Thesis, Polytechnic Institute of Brooklyn, New York, 1969.

22. P. W. Hannan and S. P. Litt, "Capacitive ground plane for a phased-array antenna," IEEE G-AP Intl. Symp. Digest, pp. 115-123 (September 1968).

23. A. Hessel and G. H. Knittel, "A loaded ground plane for the elimination of blindness in a phased-array antenna," IEEE G-AP Intl. Symp. Digest, pp. 163-169 (December 1969).

Edge Effects in Dielectric Covered Phased-Arrays

J. BACH ANDERSEN and F. HASS

Technical University of Denmark
Lyngby, Denmark

Abstract

A scattering matrix approach is used for numerical evaluation of the effects of external coupling between the elements in a finite linear array of slots in an infinite ground plane covered by a dielectric slab. Coupling between elements is interpreted as a superposition of two waves, a backward leaky wave and a forward space wave. It is shown how this coupling brings about the well-known element-pattern nulls. In the slab outside the array a surface wave may propagate to either side, the waves being launched by the array. Launching efficiencies have been found as a function of scan angle. When phasing into the element-pattern null to the left of broadside all elements except the edge elements to the left are practically short-circuited. The surface wave to the right, which is launched by the non-short-circuited edge elements, can now propagate without loss over the remaining part of the array and there is a maximum in launching efficiency to the right. Array efficiency, element directivity and main beam direction are found as functions of scan angle and discussed. A substantial beam-pointing error is found for directions about the element pattern nulls and about the endfire directions.

I. Introduction

When analyzing a phased array with a large number of elements spaced equidistantly, the corresponding infinite periodic structure is often chosen as a model for mathematical convenience. No matter how large an array, however, there will always be effects associated with the finite size. These effects are included in the scattering matrix method. Furthermore, with this method we are able to find the effects of increasing the number of elements in the array. With a certain number of elements the performance of the infinite array is approximated.

Anomalous coupling effects in phased arrays are due to the fact that, even though the amplitude of the individual contributions to the coupled field may be small, they may add in phase at specific scan angles and the accumulated effect may be large. The most important phenomenon in this connection is that of the element-pattern nulls. These have been discussed in the literature, (1-5) for the infinite array. When phasing into an element-pattern null the signals coupled back into the feeding lines will cancel the incident signals thereby short-circuiting the antenna gaps. In the finite array, however, the situation is different since the edge elements are not short-circuited. When analyzing a finite array we are able to follow gradually how the null develops and, furthermore, effects peculiar to the finiteness may be studied. Thus this investigation will supplement those dealing with infinite arrays.

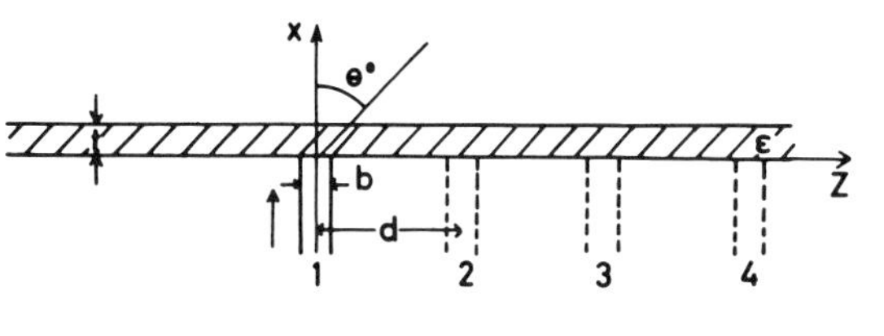

Fig. 1 Linear slot array covered by dielectric.

In order to examine the coupling effects associated with the appearance of element-pattern nulls in finite arrays we have chosen a simple model (Fig. 1). This is a two-dimensional structure consisting of parallel-plate waveguides embedded in an infinite ground plane with equidistant spacing. The ground plane is covered with a dielectric slab bringing about slow wave coupling between the elements, which is essential for the appearance of element-pattern nulls. The waveguides are assumed to be so narrow that the field can be computed from a magnetic line current. Effects associated with apertures comparable in size with the spacing are not considered in this simple model. This model is quite unrealistic, but it is sufficient for investigating some characteristic coupling mechanisms.

The scattering matrix $\bar{\bar{S}}$ is computed from the admittance matrix $\bar{\bar{Y}}$. The elements in the latter have been found and discussed in (8). A review of this paper will be given with some corrections to formulas and figures.

II. The Admittance Matrix

The electromagnetic field problem of a dielectric slab on a ground plane has earlier been dealt with by Tamir and Felsen (6) for a line source with different positions relative to the slab. This was done by using the Fourier-analysis method.

The far field from a single slot in the air region well above the dielectric, that is $\rho >> \lambda$ and t, is given by

$$H_y = \frac{\omega\varepsilon_o}{\sqrt{2\pi}} \frac{e^{i(k_o\rho - \frac{\pi}{4})}}{\sqrt{k_o\rho}} \frac{\cos\theta \cdot e^{-i\psi} \cdot V}{\varepsilon\cos\theta \cdot \left(\cos\psi - i\sqrt{\varepsilon - \sin^2\theta}\,\sin\psi\right)} \tag{1}$$

where

V is the gap voltage,
$\psi = k_o t \sqrt{\epsilon - \sin^2\theta}$,
ρ is the distance to the far field,
t is the thickness of the dielectric,
$k_o = 2\pi/\lambda$ is the wavenumber of free space,
ϵ is the relative permittivity of the dielectric.

The determination of the mutual admittance between two slots is equivalent to finding the current (H_y) at the position of a short-circuited gap when all gaps except the excited one are short-circuited. This means that we have to find the magnetic field at x = 0. This field is a sum of a branch-cut integration corresponding to the continuous spectrum and a number of residue contributions from surface wave poles (Fig. 2), the number being determined by the cutoff frequencies for the surface waves. The cut-off frequencies are given by

$$k_o t \sqrt{\varepsilon - 1} = n \cdot \pi, \quad n = 0,1,2 \ldots \tag{2}$$

In this paper only one surface wave is considered.

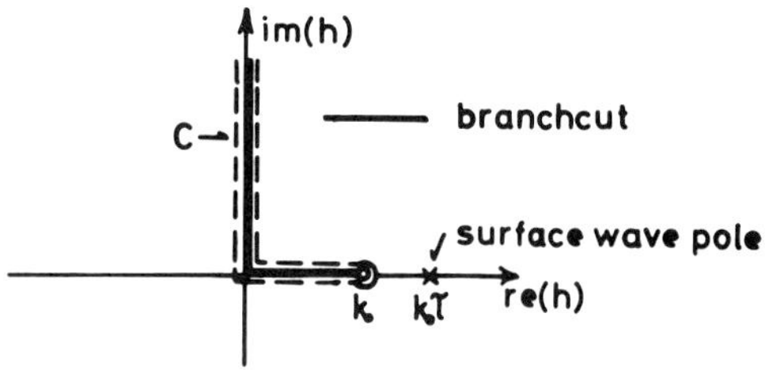

Fig. 2 Complex h-plane with branch-point at h = k_o. Integration path C.

The magnetic field from a surface wave equals

$$H_y(x = 0) = G_{su}\, e^{i\tau k_o z} \cdot V \tag{3}$$

where τ is the relative wavenumber. τ is found from the dispersion relation

$$D = \varepsilon u - v \tan v = 0 \tag{4}$$

where

$$u = k_o t \sqrt{\tau^2 - 1},$$

$$v = k_o t \sqrt{\varepsilon - \tau^2}.$$

G_{su} is the conductance of the surface wave, which may be found as a residue contribution from the corresponding surface wave pole, giving

$$G_{su} = \frac{\varepsilon\omega\varepsilon_o u}{v k_o t} \frac{(\varepsilon^2 u^2 v + v^3)}{(\varepsilon v^2 + \varepsilon u^2 + u v^2 + \varepsilon^2 u^3)} \cdot \tag{5}$$

The branch-cut integration is given by

$$H_y(x=0)/V = Y_{sp}(z) = \frac{1}{2\pi} \int_C (h_y/V)\, e^{ihz}\, dh$$

where

$$h_y/V = \frac{\omega\varepsilon\varepsilon_o}{H_\varepsilon} \frac{H_\varepsilon \cos(H_\varepsilon t) - i\varepsilon H \sin(H_\varepsilon t)}{\varepsilon H \cos(H_\varepsilon t) - i H_\varepsilon \sin(H_\varepsilon t)} \tag{6}$$

with

$$H = \sqrt{k_o^2 - h^2},$$

$$H_\varepsilon = \sqrt{k_o^2 \varepsilon - h^2}.$$

h_y has been found by Tamir and Felsen (6). The above integral is evaluated numerically. It is also called the space wave contribution to the field due to the asymptotic behavior of the integral. Away from any cut-off frequencies we have

$$Y_{sp}(z) = \frac{\omega\epsilon_0}{\sqrt{2\pi}} \frac{\epsilon^2}{(\epsilon-1)} \frac{e^{i(k_oz - 3\pi/4)}}{\sin^2(k_ot\sqrt{\epsilon-1})} \cdot \frac{1}{(k_oz)^{3/2}}$$

valid for $p_o >> 1$, where

$$p_o = k_ot\frac{\epsilon-1}{2\epsilon^2}\tan^2(k_ot\sqrt{\epsilon-1}) \quad . \tag{7}$$

Close to cut-off frequencies (including the one for $\epsilon \to 1$) p_o tends to zero and

$$Y_{sp}(z) = \frac{\omega\epsilon_o}{\sqrt{2\pi}} \frac{e^{i(k_oz - \frac{\pi}{4})}}{\sqrt{k_oz}} \quad . \tag{8}$$

In general the space wave along x = o will decay rather fast depending on the parameters involved, but the phase velocity will always be close to the velocity of light in free space.

The mutual admittance may then be expressed as a sum of two terms, the space wave and the surface wave,

$$Y_{ij} = Y_{sp}(z_j - z_i) + G_{su}e^{i\tau k_o(z_j - z_i)} . \tag{9}$$

For the self-conductance we get

$$G = G_{sp}(0) + G_{su} \tag{10}$$

where $G_{sp}(0)$ is the real part of $Y_{sp}(0)$. We then assume that the self-susceptance is tuned to zero by some external means, hence

$$Y_{ii} = G . \tag{11}$$

The scattering matrix can now be derived ($\bar{\bar{E}}$ is the unit matrix)

$$\bar{\bar{S}} = (Y_o\bar{\bar{E}} - \bar{\bar{Y}})(Y_o\bar{\bar{E}} + \bar{\bar{Y}})^{-1} \tag{12}$$

The matrix $\bar{\bar{S}}$ connects the reflected waves (b_i) with the incident waves (a_i) in the feeding lines of characteristic admittance Y_o

$$\bar{b} = \bar{\bar{S}} \cdot \bar{a} . \tag{13}$$

III. One Element Excited

It is well known that the gain in a specific direction for a large array is N times the element gain in the same direction, N being the number of elements. The element gain is found by exciting a center element, the feed lines being terminated with the actual admittances. The element gain is determined both by the individual element and by the coupling between the elements. The element gain concept versus coupling coefficients has been discussed by Hannan, (7) We shall only consider the case where the feed lines are terminated in the characteristic admittance. When this is the case the coupling coefficients simply equal the scattering matrix elements, which were found above.

We shall not excite a center element, but the left edge element. In this way it is easier to interpret the numerical results and thereby determine the characteristic mechanisms involved in external coupling. As for the coupling coefficients, the difference will be small though.

III a. Coupling Coefficients

The coupling coefficients S_{j1} have been computed for a specific case, where element No. 1 is an edge element. The result is shown in Figs. 3 and 4 (curve (3)) for N = 25, where N is the total number of slots. The plane of reference for the phase is x = 0. It is found that the "reflection" from the end of the array can only be detected at the last two elements (j = 24 and 25), in other words S_{j1} is independent of N as long as $N > j+2$. In Fig. 3 the points have been connected by a continuous curve. Only the points are meaningful; nevertheless the curve is convenient for the interpretation. We shall show how this curve can be interpreted in terms of propagating waves.

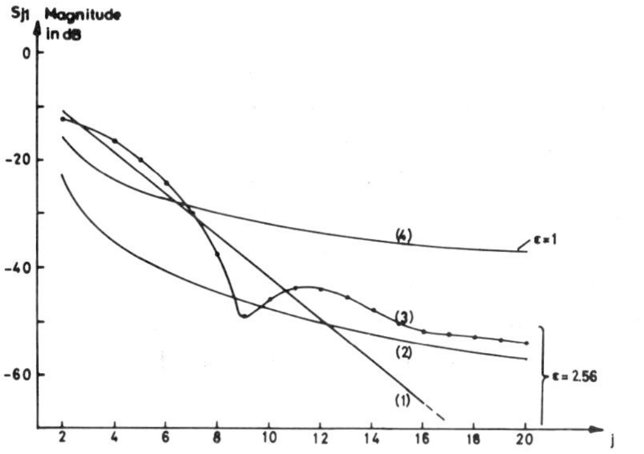

Fig. 3 Magnitude of coupling coefficients. (1) Only surface wave in Y_{ij}. (2) Only space wave in Y_{ij}. (3) Both waves in Y_{ij}. (4) Array with no dielectric.
$k_od = \pi$, $k_ot = 1$, $k_ob = 1$, $\epsilon = 2.56$.

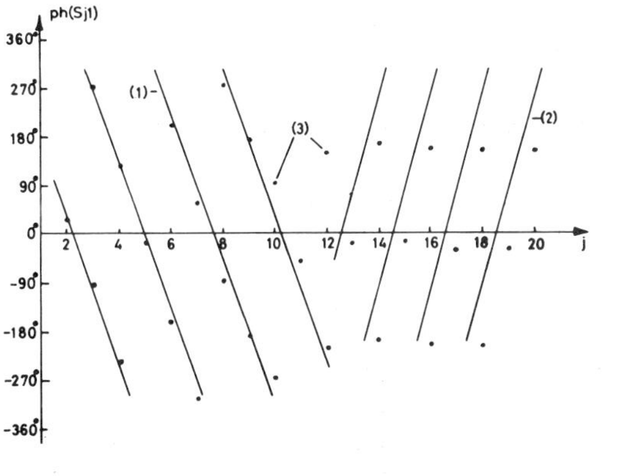

Fig. 4 Phase of coupling coefficients.

First we shall consider the role of the surface wave found in the mutual admittance. When this surface wave, launched from the excited slot, propagates over the array surface it will meet the other slots as a line of scattering discontinuities. As a first approximation the relative phase between the scatterers will equal the phaseshift of the surface wave, hence forming an antenna group with constant spacing d between elements and with a phase progression $\delta = \tau k_o d$ in the propagation direction of the surface wave. This group will radiate a main beam in the direction θ given by

$$k_o d \sin\theta = \delta + 2p\pi \ , \qquad (14)$$

p being all integers. We get

$$\sin\theta = \tau + \frac{2p\pi}{k_o d} \ . \qquad (15)$$

In order to have a solution inside the radiation region, p must be negative. In the usual case of $k_o d = \pi$ (which is treated here) we get for p = -1

$$\sin\theta = \tau - 2 \ . \qquad (16)$$

For $\tau < 2$ ($\tau = 1.216$ for the curves shown) the scattered field radiates in the backward direction, and the surface wave has been turned into a backward leaky wave. The heavy leakage and the negative phase progression are evident from the curves (1) where only the surface wave contributions to the mutual admittance are considered. The leakage is partly due to radiation, partly to losses in the feed line terminations. The leaky wave parameters have also been found by Knittel et al. (5) from the infinite structure. With this method, however, we also get the magnitude of the excitation.

Comparing the curves in Fig. 3 it is found that for j up to 7 curve (1) is close to curve (3), but asymptotically curve (3) is approximated by curve (2). The curves (2) have been computed by neglecting the surface wave in the mutual admittance. Asymptotically the magnitude of this coupling has a decay rate very close to $(k_o z)^{-3/2}$ and the phase progression corresponds to a phase velocity close to that of light in free space, the wave being forward. This indicates that the characteristics of the space wave found in the mutual admittance are only slightly changed by the slot discontinuities. Curve (2) and curve (3) are parallel, not identical, for large j since the exciting voltage is not the same.

There is a transition region where curve (3) in Fig. 3 is approximated neither by curve (1) nor by curve (2). When viewed at discrete points only, the backward leaky wave is completely equivalent with a forward slow wave which is exponentially damped. The element-to-element phaseshift is $\tau k_o d$ and this slow wave will interefere with the forward space wave to produce a slowly varying standing wave pattern. The mentioned transition region constitutes just a pattern of this kind and in fact, by adding curve (1) and curve (2) in Fig. 3 both properly weighted with complex constants (the excitation voltage is not the same) it is possible to construct a good approximation to curve (3).

Curve (4) in Fig. 3 gives the magnitude of the coupling for the case of no dielectric. Close to the excited element the coupling is weaker but asymptotically it is much stronger than in the case of the dielectric slab. The decay rate is $(k_o z)^{-1}$ which disagrees with the conjecture put forward by Galindo and Wu (9) that all arrays have the same asymptotic coupling behaviour, namely $(k_o z)^{-3/2}$ for linear arrays.

III b. Element Patterns (Edge element excited)

From the gap voltages found from the scattering matrix it is easy to find the radiation pattern. In Fig. 5 we have shown the scattered far field, which is defined as the far field with no contribution from the excited element. It is shown how the pattern develops for an increasing number of elements. For $N > 10$ the pattern is not changed significantly. For N = 10 is also shown the case where we only take into consideration the surface wave scattering. There is only a slight difference between the two curves confirming the dominating role of the surface wave for the backward radiated field as discussed in the preceeding section. According to

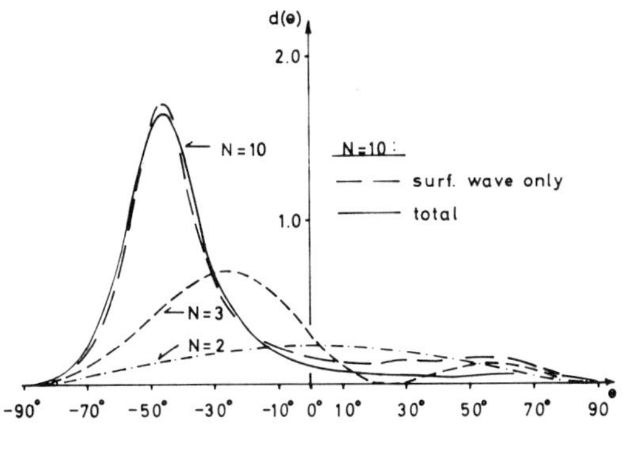

Fig. 5 Scattered far field. Same parameters as in Fig. 3.

equation (16) the main beam direction for the actual parameters ($\tau = 1.216$) is $\theta = -51\overset{\circ}{.}5$, while $\theta_{max} = -46^\circ$ is found from Fig. 5. This difference is due to the attenuation of the wave along x = 0 (Fig. 3 curve (1). The far field corresponding to a leaky wave dominated surface field can be found from a Kirchhoff-Huygens integration. This has been done in (10). The main beam direction found from the decay rate and propagation constant of the leaky wave agrees well with the actual direction.

By adding the field from the excited edge element to the scattered field we get the patterns in Fig. 6. The interference between the two fields produces a pronounced dip in the pattern already for a few elements in the array. The position of this dip is given by equation (16) for a sufficient number of elements. This is a well-known result for an infinite array (5).

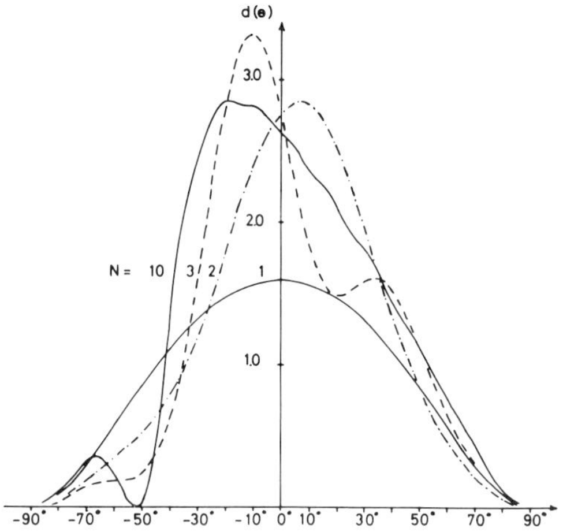

Fig. 6 Element patterns with left edge element excited. Same parameters as in Fig. 3.

IV. All Elements Excited

When phasing into an element pattern null with all elements excited the element gaps will be short-ciruited in the infinite array, while in the finite array the edge elements at one end of the array are not short-circuited. This can be explained by means of the coupling coefficients found in section III. When phasing into the null to the left of broadside we have an element-to-element phase progression δ to the right in the incident signals (equation (14) and (15)),

$$\delta = \tau k_o d + 2p\pi \quad . \tag{17}$$

When this is the case the important part of the coupling to an element in the array is the coupling from the elements to the left. The dominating part of this coupling is due to the forward slow waves. Launched from their respective elements the waves to the right propagate with an element-to-element phaseshift $\beta = \tau k_o d$. Using equation (17) we find

$$\delta = \beta + 2p\pi \quad . \tag{18}$$

Hence all the slow wave contributions to the coupling to one element from elements to the left add in phase, and the coupling will be strong. The only exceptions are the left edge elements which have no coupling from the left. For a large array practically all energy is reflected back into the feed lines, the reflection coefficients being numerically close to one. When this is the case the active gap-admittances must be infinite and the directivity will usually tend to zero for a large array. This is the case here. (If the active gap-conductance remains finite the antenna will be superdirective, when the number of elements are increased).

IVa. Launching Efficiency and Array Efficiency

The short-circuited part of the array surface can now support the surface wave found in section II in the same way as the ground plane outside the array surface. Surface waves are launched by the non-short-circuited edge elements, and there is a maximum in launching efficiency, which is defined as the power in the surface wave relative to the total power leaving the array. The element-to-element phaseshift for a surface wave propagating to the right is $\beta = \tau k_o d$. Hence equation (18) is valid and the waves will add in phase giving a maximum in launching efficiency. In a more realistic model, where only the array would be covered by dielectric, the wave would radiate in the endfire direction. Hence the situation is equivalent to the case of an array with no coating but with a larger distance between the elements such that a normal grating lobe appears. If this array is phased into the element pattern dip a main beam appears at endfire.

In Figure 7 the launching efficiencies are shown for positive scan-angles, δ being the element-to-element phaseshift in the incident signals. At the positive scan angle corresponding to the pattern null there is a maximum in launching efficiency for launching to the left. This is only slightly changed when adding more elements to the array. This is due to the fact that the extra elements are short-circuited too. The array efficiency (percentage of available power transmitted by the gaps), however, will decay in proportion to the number of short-circuited elements. This is seen in Fig. 8. The efficiency is approximately halved when the number of elements is doubled.

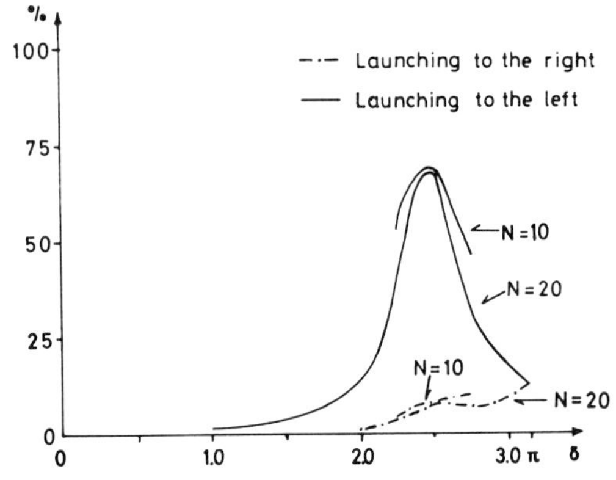

Fig. 7 Launching efficiency. All elements excited with equal amplitudes. Same parameters as in Fig. 3.

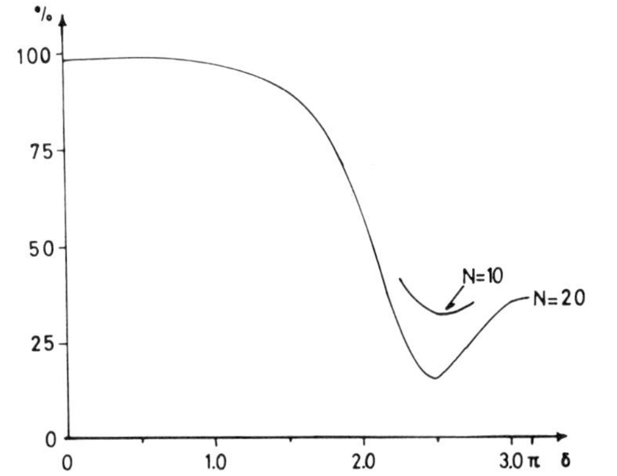

Fig. 8 Array efficiency. All elements excited with equal amplitudes. Same parameters as in Fig. 3.

IV b. Radiation Patterns and Beam Pointing

The directivity patterns for scan angles close to the one corresponding to an element-pattern null are shown in Fig. 9. In a finite array the pattern

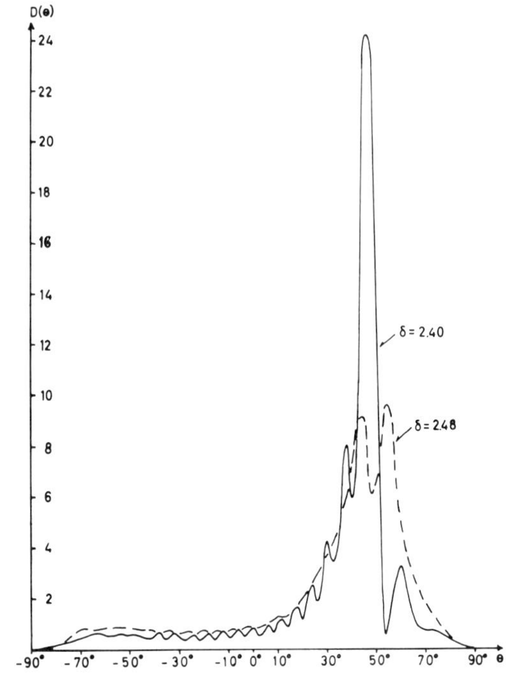

Fig. 9 Array directivities near element pattern null (δ = 2.48). All elements excited with equal amplitudes. Same parameters as in Fig. 3.

"null" is not restricted to one specific scan angle, but the pattern for $\delta = 2.48$ is representative for the patterns in the "null." There is a rapid change in patterns when phasing through the null; this is evident from the patterns in Fig. 9. The pattern for $\delta = 2.40$ is the normal pattern when many elements contribute to the radiation. When phasing into the null only the edge elements contribute thereby producing the broad main beam and the reduced sidelobe-ripple. The dip in the top of the pattern is due to the field produced by the almost short-circuited elements. Though the gap voltages are very small the contributions to the far field add in phase giving the marked dip. Adding extra elements to the array does not change the pattern corresponding to the element pattern null $\delta = 2.48$ significantly since these elements will be practically short-circuited. Hence the element directivity (directivity per element when all elements are excited) in the element pattern null direction will tend to zero for a large array. The element directivity is given in Fig. 10.

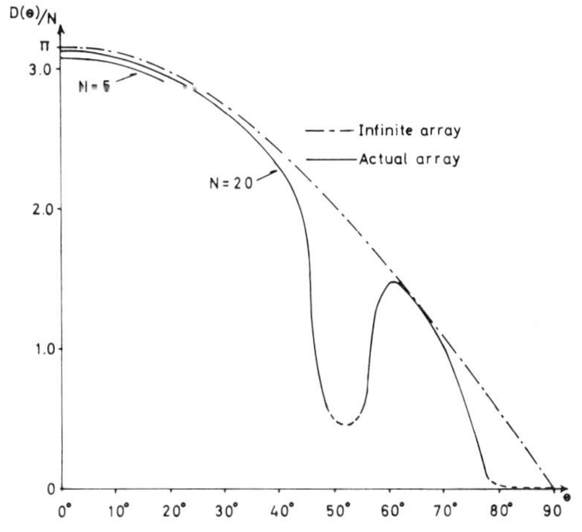

Fig. 10 Element directivities for actual array (same parameters as in Fig. 3) and for infinite array with isotropic elements and no dielectric. All elements excited with equal amplitudes.

The directivity for a large linear group of isotropic radiating antennas, when no anomalous coupling effects are present, is given by

$$D(\theta) = \frac{2\pi \cdot L}{\lambda} \cos\theta \quad . \qquad (19)$$

where

L = N · d is the length of the array.

Equation (19) is also the expression for the directivity of the large uniformly illuminated aperture. For the array in question d = λ/2 and hence

$$D(\theta) = N \cdot \pi \cdot \cos\theta \qquad (20)$$

according to (19). The curve corresponding to (20) is shown in Fig. 10. The directivity given by (19) can be considered an upper limit for non-superdirective antennas. For the array in question the broadside directivity is well approximated even with very few elements in the array; this is seen in Fig. 10.

As already discussed there is a directivity dip due to anomalous coupling effects. However, there is also a dip in directivity near endfire. This is due to the non-isotropic radiation of the individual elements (the N = 1 curve in Fig. 6). Due to the reflections at the dielectric-air interface there is a marked reduction in radiated power near endfire and at endfire itself no power is radiated at all to the air (into the dielectric slab, however, energy is efficiently radiated).

The beam pointing errors are shown in Fig. 11; the heavy curve is the intended direction due to equation (14). The errors are a direct consequence of the directivity decary. For a finite array the beam is not infinitely sharp, especially not at angles close to the element-pattern null and the endfire direction. Hence the decay in radiation towards these directions will shift the position of the main beam maximum as shown in Fig. 11.

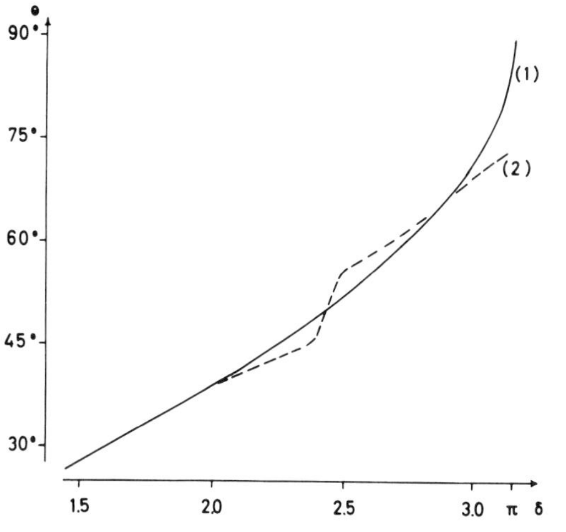

Fig. 11 Main beam direction. All elements excited with equal amplitudes. Same parameters as in Fig. 3. N = 20. (1) Intended direction. (2) Actual direction.

V. Conclusion

It has been shown for a well-known array structure how it is possible in an easy way to find the coupling by using the scattering matrix method. The decay of coupling away from an excited element is shown to be due to an interference between a backward leaky wave and a forward space wave. For a finite array it is shown that when phasing in the element-pattern null, a large part of the transmitted power goes into a surface wave external to the array. A further consequence of the finite size and the anomalous coupling is a certain beam pointing error.

References

(1) R. H. T. Bates, "Mode Theory Approach to Arrays", IEEE Trans. Antennas and Propagation, vol. AP-13, March 1965, pp. 321-322.

(2) J. L. Allen, "On Surface-Wave Coupling between Elements of Large Arrays", IEEE Trans. Antennas and Propagation, vol. AP-13, July 1965, pp. 638-639.

(3) A. A. Oliner and R. G. Malech, "Mutual Coupling in Infinite Scanning Arrays" in "Microwave Scanning Antennas", vol. 2, ed. R. C. Hansen, New York, Academic Press, 1966.

(4) C. P. Wu and V. Galindo, "Surface-Wave effects on dielectric sheathed phased arrays of rectangular waveguide", Bell System Techn. Journ., vol. 47, No. 1, January 1968, pp. 117-142.

(5) G. H. Knittel, A. Hessel and A. A. Oliner, "Element Pattern nulls in phased arrays and their relation to guided waves", Proc. IEEE, vol. 56, No. 11, November 1968, pp. 1822-1836.

(6) T. Tamir and L. B. Felsen, "On lateral waves in slab configurations and their relation to other wave types", IEEE Trans. Antennas and Propagation, vol. AP-13, May 1965, pp. 410-422.

(7) P. W. Hannan, "The element-gain paradox for a phased-array antenna", IEEE Trans. Antennas and Propagation, vol. AP-12, July 1964, pp. 423-433.

(8) J. Bach Andersen and F. Hass, "External Coupling Effects in Phased-Arrays", Proc. of the Fourth-International Colloquim on Microwave Communications, vol. III, pp. ET 2/1 - 2/9, April 1970.

(9) V. Galindo and C. P. Wu, "On the asymptotic decay of coupling for infinite phased arrays", Proc. IEEE, vol. 56, No. 11, 1968, pp. 1872-1880.

(10) T. Tamir and A. A. Oliner, "Complex Guided Waves, Part 2. Relation to Radiation Patterns". Proc. IEEE, vol. 110, No. 2, Feb. 1963.

On the Effects of Eliminating the Passive Elements from a Thinned Array

R. L. KNUCKEY

Western Electric Company, Inc.
On Loan to Bell Telephone Labs., Inc.
Whippany, New Jersey

G. A. ARREDONDO

Bell Telephone Laboratories, Inc.
Whippany, New Jersey

ABSTRACT

A method is presented for the study of the effects on various radar parameters, of replacing the passive elements by ground plane in a thinnned array. The approach consists of treating the amplitude and phase deviations of the element patterns as errors at the element level in a Monte Carlo experiment. An application of this approach to an antenna with 50% thinning is given. The statistics of the errors were estimated experimentally. It is shown that the expected gain degradation is about 0.1 dB and the changes in sidelobe statistics and average element match are negligible.

I. Introduction

The beamwidth and side lobe level of the far-field pattern of an antenna are controlled, for a given aperture size, by the aperture illumination. One way of implementing this illumination in large arrays (a few thousand elements) is to space taper the uniformly-spaced antenna elements, i.e. by not placing excited antenna elements in some of the nodal points of a periodic grid. The density distribution of the excited (active) elements is chosen to be proportional to the illumination function (1-4). The space taper is usually based on some random procedures so as to supress any further increase in sidelobes.

It is common practice to fill the nodal points of the grid which have no active elements with similar but terminated (passive) elements. This is done in order to maintain an identical environment for all the active elements. Implementations of space tapering often lead to a large number of passive elements (about 50% of all elements is quite typical). The potential cost savings due to the elimination of the passive elements is quite apparent, specially in large arrays.

This paper considers the effects of replacing the passive elements by ground plane on the array characteristics. The method used is a Monte Carlo simulation where the changes in the active elements characteristics due to the removal of the passive elements are treated as random variables. These changes are treated, in effect, as statistical element errors.

Allen (6) has shown that mutual coupling affects strongly the patterns of unequally spaced arrays. The passive elements in a thinned array (which has a periodic grid, i.e. the basic array is equally spaced), by providing an identical environment for all active elements, eliminate the need to account for mutual coupling when synthesizing a desired pattern. Removal of the passive elements will disrupt the mutual coupling between active elements. Our object is to estimate the mutual coupling effects on the array performance when the basic array is equally spaced.

The problem considered here was studied experimentally by Esposito and Valentino using the HAPDAR radar (6). Their results and ours are of the same order of magnitude.

The mathematical statement of the problem is given in sections II and III. The approach used to solve this problem is stated in section IV, and an application of the approach to an antenna array with 50% thinning is given in section V.

II. Far-Field Pattern

Consider a planar array with a periodic grid of

N possible element locations, where N is a large number. The far-field radiation pattern, $\vec{F}(\theta,\phi)$ is given by

$$\vec{F}(\theta,\phi) = \sum_{\nu\epsilon Q} a_\nu \vec{P}_\nu(\theta,\phi) e^{j\vec{k}\cdot\vec{x}_\nu} \tag{1}$$

where a_ν is the excitation coefficient, $\vec{P}_\nu(\theta,\phi)$ is the νth active element radiation pattern, $\vec{x}_\nu$ is the position of the νth active element, $\vec{k}$ is the free space propagation vector, and Q is the ordered set of array locations with active elements.

In the absence of passive elements, the element pattern $\vec{P}_\nu(\theta,\phi)$ will be a function of position, and therefore may not be factored out in Equation (1). The principle of pattern multiplication is therefore not applicable. Let $\vec{P}_\nu(\theta,\phi)$ be written in the form

$$\vec{P}_\nu(\theta,\phi) = \vec{P}(\theta,\phi)\left[1 + \alpha_\nu(\theta,\phi)\right]e^{j\psi_\nu(\theta,\phi)} \tag{2}$$

where $\vec{P}(\theta,\phi)$ is the pattern of an element in a large fully-packed regular grid array, and α_ν and ψ_ν are the deviations of the pattern of the νth element from $\vec{P}$ in amplitude and phase respectively. The array far-field pattern can then be written as:

$$\vec{F}(\theta,\phi) = \vec{P}(\theta,\phi) \sum_{\nu\epsilon Q} a_\nu\left[1 + \alpha_\nu(\theta,\phi)\right]e^{j\psi_\nu(\theta,\phi)} e^{j\vec{k}\cdot\vec{x}_\nu} \tag{3}$$

The element pattern deviations can be treated as scan-dependent errors, yet unknown, of the aperture excitation coefficients.

III. Element Match

The active reflection coefficient of the νth active element in a uniformly excited array is given by

$$\Gamma_\nu(\theta,\phi) = \sum_{\mu\epsilon q_\nu} C^\nu_\mu \, e^{j\beta_\mu} \tag{4}$$

where C^ν_μ is the coefficient of coupling between ν^{th} and μ^{th} active elements, β_μ is the steering phase of the μth element, and q_ν is the ordered set of relative active element locations centered at ν.

Since q_ν is in general a function of ν, the active reflection coefficient will be a function of position even if the passive elements are not removed (7). The replacement of passive elements by ground plane results in a change of C^μ_ν and thereby a change in Γ_ν.

IV. Statistical Model

In evaluating the effect of replacing the passive elements by the ground plane, the parameters to consider would typically be: gain degradation, side-lobe level, monopulse slope linearity, antenna match, etc. All these parameters could be evaluated from Equations (3) and (4) if $\{\alpha_\nu(\theta,\phi)\}$, $\{\psi_\nu(\theta,\phi)\}$ and $\{C^\mu_\nu\}$ were known. The periodicity of the grid is destroyed due to the removal of the passive elements. Therefore, analytical solutions even for the simplest antenna element in a large array are practically impossible.

In our approach we consider $\{\alpha_\nu(\theta,\phi)\}$ and $\{\psi_\nu(\theta,\phi)\}$ as independent random variables and thereby treat the effects of removing the passive elements as random errors at the active element level. The following arguments justify treating $\{\alpha_\nu(\theta,\phi)\}$ and $\{\psi_\nu(\theta,\phi)\}$ as random variables. The specific frequency of operation and the beam pointing direction of an operational radar are unknown variables as far as the design of the radar is concerned. The functional dependence of $\{\alpha_\nu\}$ and $\{\psi_\nu\}$ on the frequency and angles, which is also unknown, may be different for every ν, i.e. for every active element. The functions $\{\alpha_\nu\}$ and $\{\psi_\nu\}$ are expected to be small for a well designed element in the scan region of interest. We are dealing then with a large collection of different unknown small functions $\{\alpha_\nu, \psi_\nu\}$ which are sampled in an arbitrary way. The sampled values may hence be treated as discrete random variables in a Monte Carlo simulation (8).

The results obtained using this procedure are not expected to be exact, but rather to yield useful information about the order of magnitude. The statistics of $\{\alpha_\nu\}$ and $\{\psi_\nu\}$ can be determined experimentally in a small test array.

Only the coupling of nearby elements affect the characteristics of an active element (9), hence only a small subarray around each active element needs to be considered. The size of the subarray depends on the type of radiating element. Some elements are coupled stronger than others.

One can then form a sequence of sets $\{q_\nu\}$ where q_ν is the ordered set of active element locations in a small subarray centered about the νth active element. If the size of the sequence $\{q_\nu\}$ is small, one could measure the far-field pattern of the center active element associated with every set q_ν. For most common space tapers and subarray sizes, the size of $\{q_\nu\}$ is likely to be rather large (a few thousand sets). A complete experimental characterization would then not be practical.

Since our purpose is to determine the statistics of $\{\alpha_\nu\}$ and $\{\psi_\nu\}$ rather than their complete description, the set $\{q_\nu\}$ can be sampled at random and thereby determine a subset $\{q_\nu\}_s$ which is used to conduct the experiments. This data can be used to estimate the statistics of $\{\alpha_\nu\}$ and $\{\psi_\nu\}$ which can then be used in a Monte Carlo study of the far-field parameters using Equation (3).

The same approach could be used to study the element match by estimating the statistics of the variations of the coupling coefficients C_ν^μ.

V. An Application of the Method

This study was conducted on an antenna that had a reference aperture of approximately 8000 linearly polarized circular aperture elements packed in an equilateral triangular grid. Space tapering, according to a Taylor weighting of 45 dB with $\bar{n} = 10$ (10), reduced the active element content to approximately 4000. This tapering was constrained to be quadrant-symmetric for reasons not pertinent to this study. Fig.1 shows schematically a quadrant of this aperture. Every area drawn has an equal number of active elements. The ratio of active to passive elements is shown for every area.

We had available a 37 element test array for this study. The determination of the family of sets $\{q_\nu\}$ resulted in a tabulation of 693 sets in each quadrant consisting of active elements having one or more passive elements in the surrounding 36 elements. Using a uniformly distributed random number generator, a random sample of 40 sets was chosen from this family. The average and standard deviation of the active element content of the sample were 58% and 24% respectively. This was the sample $\{q_\nu\}_s$ used in the experiments.

Every set q_ν of the sample $\{q_\nu\}_s$ was obtained in the test array by covering the appropriate active elements with shorting caps. All the measurements were conducted in an anechoic chamber. For every set q_ν in $\{q_\nu\}_s$, the H-plane element pattern and the far field phase at boresight of the center element were measured. We also measured the coefficients of coupling between the center element and every active element in q_ν. All measurements were made at the edges and at the center of a 15% frequency band. Four different configurations plus the reference case of all active elements were repeated several times to establish repeatability. This repeatability was ±1° in phase and ±0.07 dB in amplitude for both the far-field and coupling measurements.

The changes in the element pattern due to the

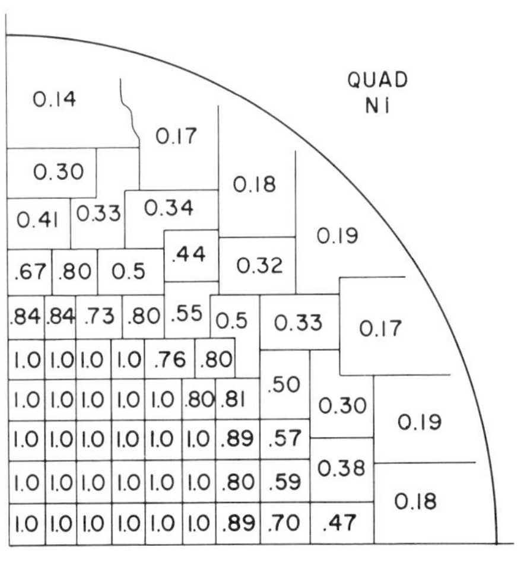

Fig. 1 One Quadrant of a Thinned Aperture.

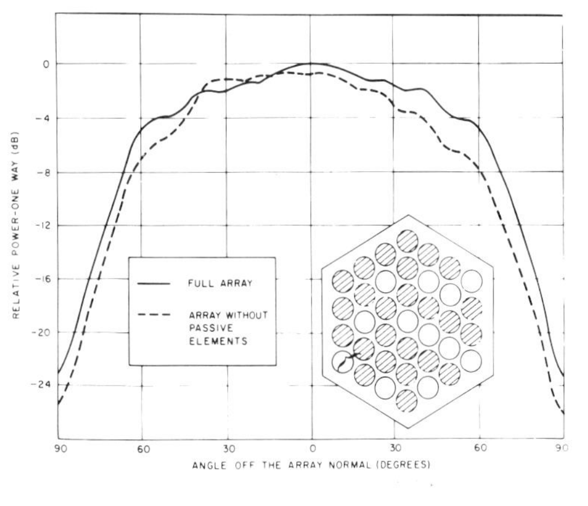

Fig. 2 Typical Element Pattern (Shaded Circles Represent Passive Elements).

covering of passive elements were found to be for most cases of the order of 1 dB. A typical case is shown in Figure 2 where both the element patterns in a subarray with and without passive elements are given for comparison. The case that had the most asymmetry is shown in Figure 3. Here the worst deviation is of the order of 2 dB.

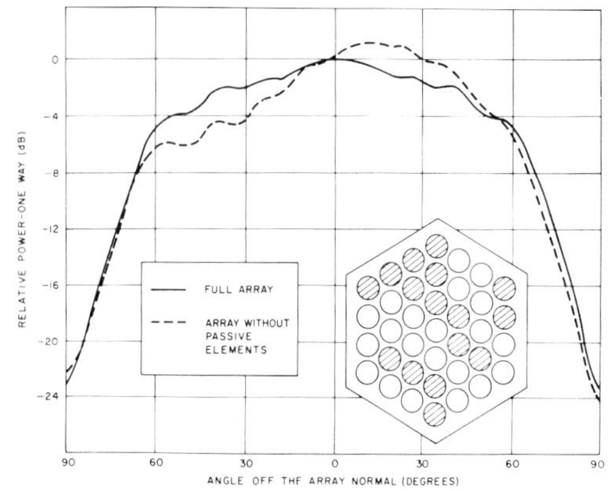

Fig. 3 Asymmetric Element Pattern (Shaded Circles Represent Passive Elements).

The scan and frequency dependence of the amplitude and phase deviations of the element pattern were found

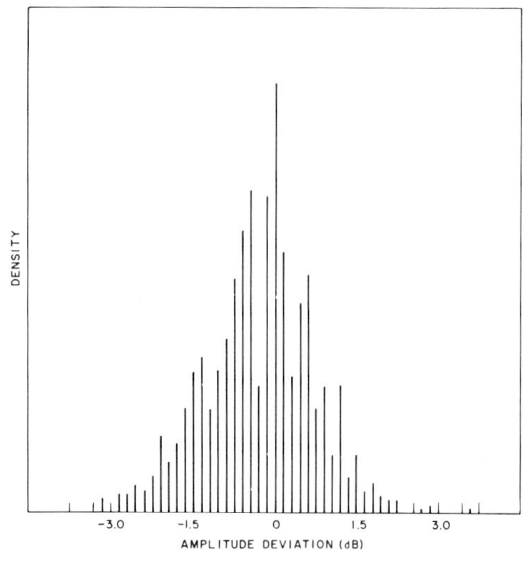

Fig. 4 Amplitude Deviation Histogram for all Frequencies and Angles.

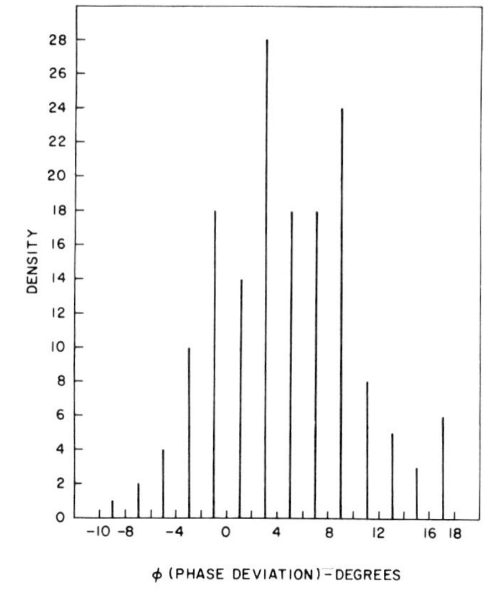

Fig. 5 Phase Deviation Histogram for all Frequencies.

not to be significant in the Monte Carlo runs. The histograms obtained by including the data for all frequencies and all angles within $\pm 60°$ from boresight are given for amplitude in Figure 4 and for phase in Figure 5. The intervals used for the histograms ($2°$ for phase and 0.15 dB for amplitude) were the absolute variations obtained in the repeatability of the data. The amplitude deviations have an rms of 1.3 dB and the phase deviations $7°$. These histograms were used as the probability distribution functions of the amplitude and phase deviations.

The cumulative distribution functions, $P_\psi(\psi)$ and $P_\alpha(\alpha)$, were obtained for the phase and amplitude deviations from their probability density functions (given by the histograms in Figures 4 and 5). These distribution functions were approximated by fifth degree polynomials, using the functions P as the independent variables. The polynomials are given by

$$\psi = -8.94 + 92.19\, P_\psi - 412.65\, P_\psi^2 + 973.97\, P_\psi^3 - 1048.56\, P_\psi^4 + 421.42\, P_\psi^5 \quad (5)$$

where ψ is the far-field phase deviation in degrees, and

$$\alpha = -0.418 + 3.573\, P_\alpha - 19.155\, P_\alpha^2 + 49.919\, P_\alpha^3 - 57.606\, P_\alpha^4 + 24.063\, P_\alpha^5 \quad (6)$$

where α is the far-field amplitude deviation as a voltage ratio.

The Monte Carlo simulation was done using a computer program that calculates the array pattern, gain, and sidelobe statistics of a planar array of point sources with an arbitrary aperture distribution. The program admits complex element excitation coefficients. The unerrored case is computed by placing a unity amplitude, zero phase coefficient on each active element. The effect of removing the passive elements is obtained by injecting amplitude and phase errors in the excitation coefficients. These errors were generated via the inverse method, using Equations (5) and (6), from uniformly distributed random deviates (8).

The results are shown in Table I. Shown are the results of placing errors on all the active elements in the

TABLE I. RESULTS

Remarks	One-Way Gain (dB)	Peak Sidelobe (dB)	Mean Sidelobe (dB)
No errors reference	0.00	0.0	0.0
Errors on all elements	-0.10	+0.2	+0.3
Errors on elements outside 10λ	-0.08	+0.1	+0.2
All other system errors	-0.62		+7.4

array. We also placed errors only on active elements outside a 10λ circle (where λ is the free space wavelength) since inside this circle there were no passive elements. In all cases the variances obtained were very small. For example, for the case of errors outside a 10λ circle the mean one-way gain loss was found to be 0.08 dB with a variance of 0.0008 dB. We hence show only the mean values obtained.

In both cases considered, the gain loss was found to be of the order of 0.1 dB and the change in sidelobe statistics negligible. For the purpose of comparison we include the degradation due to all other element errors typically found in a phased array. In this case the gain loss to be expected is 0.62 dB with an increase of 7.4 dB in the mean sidelobe.

The measured coupling coefficients $\{C_\nu^\mu\}$ had variations that were strong functions of both ν and μ, i.e of the sets q_ν and the relative position of the active elements with respect to the center element. Hence, a meaningful distribution function could not be estimated with our sample size. Our procedure was then to calculate the impedance match of the cases measured using Equation (4) and to consider these results as typical. The worst case is shown in Figure 6 where we plot contours of reflected power as a function of the direction cosines Tx and Ty. As can be seen, there is a marked change in the topography of the reflected power, but there is no

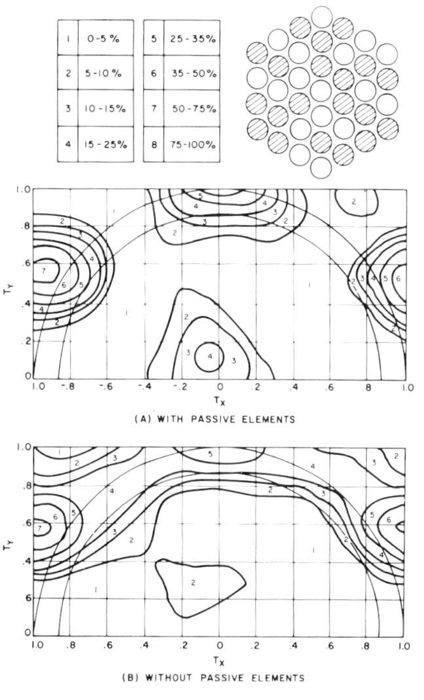

Fig. 6 Returned Power into Center Element. Typical Case.

unacceptable degradation in the match. This general result was obtained in all the cases studied. A case in which an active element is surrounded by many passive elements is shown in Figure 7, where the same observation can be made.

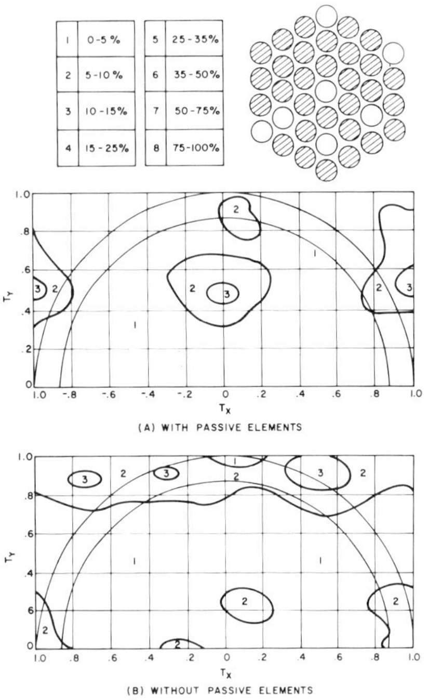

Fig. 7 Returned Power into Center Element. Case with Many Passive Elements.

From our measurements and calculations, we conclude that the system degradation due to the replacement of passive elements by ground plane extensions is small. This general observation is in agreement with the results obtained by Esposito and Valentino (5).

VI. Conclusion

We have presented an approach to the study of the effects on various parameters of removing the passive elements in a thinned array. The approach consists in treating the effects as errors at the element level and using Monte Carlo techniques to study the parameters of interest. The results obtained from the application of this approach to an antenna with 50% thinning indicate that the effects on the array gain, sidelobes and element match are negligible. We conclude that the elimination of the passive elements should be seriously considered in any array radar design.

VII. Acknowledgments

The authors wish to thank J. L. Martin, A. Anastasio and W. G. Splees for their assistance in the Monte-Carlo simulations, and L. H. Hendler and H. A. Oakes for their help in making the experimental measurements, and Dr. N. Amitay for his critical reading of this manuscript.

REFERENCES

(1) Willey, R. E., "Space Tapering of Linear and Planar Arrays", IEEE Trans. on Antennas and Prop., AP-10, 369 (1962).

(2) Skolnik, M.I., Sherman III, J. W. and F.C. Ogg, Jr., "Statistically Designed Density-Tapered Arrays", IEEE Trans. on Antennas and Prop., AP-12, 408 (1964).

(3) Lo, Y.T., "Random Periodic Arrays", Radio Science, 3 (New Series), 425 (1968).

(4) Redlien Jr., H.W., "Density Tapering of Circular Aperture Array Antennas", Radio Science, 3 (New Series), 437 (1968).

(5) Allen, J. L. "A Simple Model for Mutual Coupling Effects on Patterns of Unequally Spaced Arrays", IEEE Trans. on Antennas and Prop., AP-15, 530 (1967).

(6) Esposito, F. and P. Valentino, "HAPDAR Terminated Element Measurement Study", Sperry Gyroscope Division of Sperry Rand Corp., Rep. SGD-31-005, Oct. 1969.

(7) Schwartzman, L., "Element Behavior in a Thinned Array", IEEE Trans. on Antennas and Prop., AP-15, 571 (1967).

(8) Hammersley, J.M. and D. C. Handscomb, Monte Carlo Methods, Wiley: New York (1964).

(9) See for example: Amitay, N., Butzine, P.E. and Heidt, R.C., "Match Optimization of a Two-Port Phased Array Antenna Element", IEEE Trans. on Antennas and Prop., AP-16, 47 (1968)

(10) Taylor, T.T., "Design of Circular Aperture for Narrow Beamwidth and Low Side-lobes", IRE Trans. on Antennas and Prop., AP-8, 17 (1960)

Small Arrays–Their Analysis and Their Use for the Design of Array Elements*

B. L. DIAMOND

Lincoln Laboratory
Massachusetts Institute of Technology
Lexington, Massachusetts

Introduction

Over the years various techniques for the design of phased-array radiating elements have been devised. Three approaches -- the small array, the infinite array simulator in waveguide, and the grating lobe series -- have been outlined by Wheeler (1). A fourth approach, based on accurate infinite array computations of realistic elements, is discussed by Diamond and Knittel (2).

Phased-array simulators (3) and infinite array computations (4) are, of course, very powerful techniques for element design for those elements which possess the symmetry necessary for simulation and/or can be accurately modelled for infinite array computations. Unfortunately, not all element designs satisfy these conditions, so it becomes necessary to do element design by means of small test arrays. In some instances the infinite array modelling is only approximate and it is therefore desirable to test a small array to make certain that there are no unexpected phenomena.

In this paper we will discuss the ways in which small arrays (sizes ranging from 20 to 200 elements) can provide useful design information in addition to that obtained by other means. Attention will first be focused on the analytical procedures and results. Then we will consider experimental techniques together with a possible procedure for element design with a small array.

Analysis of Finite Arrays

Analytical studies of small arrays are presently limited to just a few element types because formulas do not exist for the mutual couplings among most types of elements. Fortunately, analytical mutual impedance (or admittance) formulas do exist, as reasonable first approximations, for some elements of practical importance -- electric or magnetic Hertzian dipoles, small holes in a ground plane, narrow slots in a ground plane, and thin dipoles with or without a ground plane. The slot and thin dipole relations are approximately valid only when the lengths are near or less than the resonant length ($\ell \gtrsim \lambda/2$); (5) however, the examples we consider utilize the simple, sinusoidal current approximation used by Carter (7). Some recent investigations have considered the more complex problem of a finite array of parallel plate waveguides (8) (9) (10), but we will not treat this case.

The analytical procedure involved in the solution of a small array of dipoles (or slots) is very well known (see Refs. 11, 12, 13 for example). Basically, it amounts to computing an impedance, admittance or scattering matrix representing the coupling among all elements in the array and then applying standard techniques of circuit theory to solve for the unknown element excitations (e.g., currents on the dipoles or voltages across the slots), subject to appropriate source constraints. The source constraints range from an ideal current (or voltage) source ("forced" excitation (1), (13)) at each element to a complex feed network with internal cross-coupling (12) (14). For the results presented in this paper, each element is excited by an independent, constant incident power source (a voltage source with a non-zero generator impedance); that is, "free" excitation of the elements is employed (1), (13).

From the computed element excitations with equal amplitude, linear progressively phased incident power at the elements, one can readily obtain the active (or driving point) impedance or admittance of each element. The scan angle dependance of the active impedance, $Z_D(\theta,\phi)$, is obtained by varying the element-to-element phasing. From $Z_D(\theta,\phi)$ and the generator impedance, Z_g, one obtains the scan dependent reflection coefficient

$$\Gamma(\theta,\phi) = \frac{Z_D(\theta,\phi) - Z_g^*}{Z_D(\theta,\phi) + Z_g} . \tag{1}$$

The imaginary part of the generator impedance is assumed to be a tuning reactance which could equally well be added to $Z_D(\theta,\phi)$.

*This work was sponsored by the Department of the Army.

A very valuable experimental tool, the element-gain pattern (11) (12) (13), is related to the dipole self impedance, R_a, the generator resistance R_g, the isolated element pattern $g_{iso}(\theta, \phi)$, and the reflection coefficient as follows:

$$g(\theta, \phi) = \frac{R_a}{R_g} g_{iso}(\theta, \phi) \, | 1 - \Gamma(\theta, \phi) |^2 , \tag{2}$$

for the center element of an array of dipoles. This type of pattern (defined as the pattern of an element in a passive array environment; i.e., all elements, except the one of interest, are passively terminated in their normal generator impedances) is a useful tool because, in a large array, the total array gain varies with scan angle exactly as the element-gain pattern $[G(\theta,\phi) = \eta N_g(\theta, \phi)$, where η is the taper efficiency and N is the number of elements]. Since the array is passively terminated (no feed network), element-gain pattern tests are simple to implement and give a rough estimate of the severity of mutual coupling effects $[\Gamma(\theta, \phi)$ appears in Eq. (2)]. An alternate, well known form for the element-gain pattern, which emphasizes the degradation in gain due to mutual coupling, can be derived for an infinite array (13)

$$g(\theta, \phi) = 4\pi \frac{A_e}{\lambda^2} [1 - | \Gamma(\theta, \phi) |^2] \cos\theta , \tag{3}$$

when only one beam is radiated (all grating lobes outside of visible space). The factor 4π applies for radiation into half-space and is reduced to 2π for radiation into all space. The area per element is denoted A_e.

For the purpose of illustrating the variation of reflection coefficient vs scan angle and element-gain pattern with array size and element type, we present Figs. 1-5.

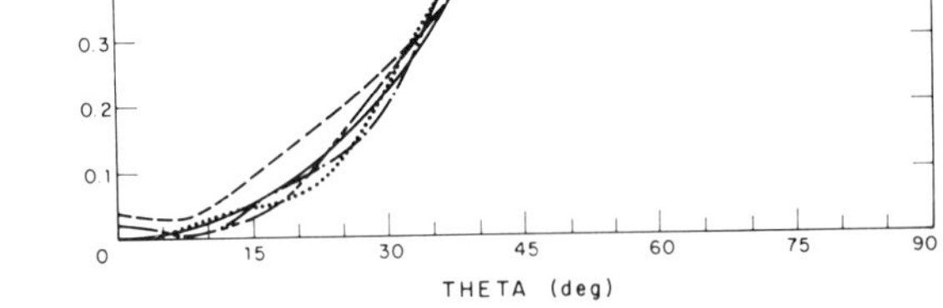

Fig. 1 Dependence of the reflection coefficient variation of the center element on array size; H-plane scan.

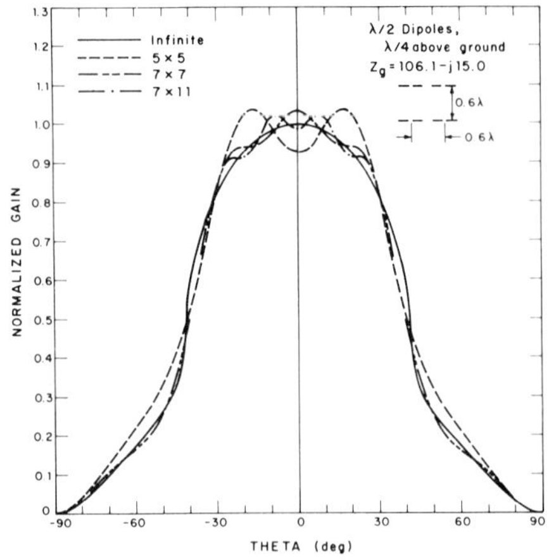

Fig. 2 Dependence of the H-plane element-gain pattern of the center element on array size.

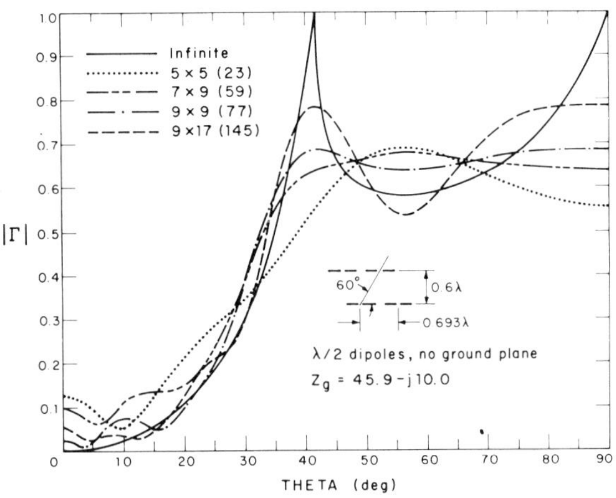

Fig. 3 Dependence of the reflection coefficient variation of the center element on array size; H-plane scan.

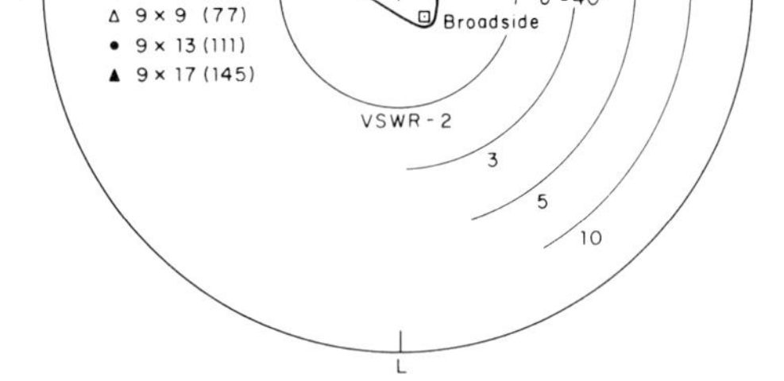

Fig. 4 Comparison of center element mismatch at selected scan angles for various array sizes.

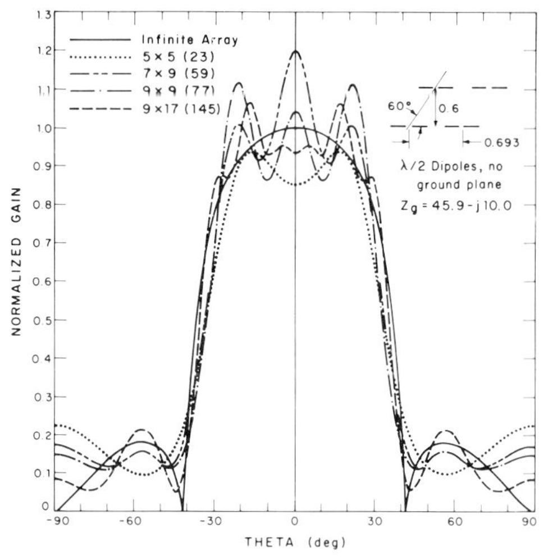

Fig. 5 Dependence of the H-plane element-gain pattern of the center element on array size.

Figs. 1 and 2 apply to a square-grid array of λ/2 dipoles, λ/4 above a ground plane, while Figs. 3 - 5 apply to an equilateral-triangle lattice array of λ/2 dipoles without a ground plane. In both examples the worst case of H-plane scan is emphasized. The grating lobe angle for H-plane scan is 41.8° in each case; it appears as a discontinuity in slope for the infinite array reflection coefficient data in Fig. 1, as a complete reflection for the infinite array data in Fig. 3 and a radiation null in Fig. 5. An inspection of the data contained in these figures shows, as expected, that increasing the array size results in a successively better approximation to a very large or infinite array. However, it is also evident from Figs. 1 and 2 that an array with approximately 25 elements will probably be adequate for the design of a phased-array element which behaves like a dipole above a ground plane. A larger test array would probably be preferable for elements which behave like dipoles without a ground plane (or slots in a ground plane, by duality). It should be noted perhaps that the results in Figs. 3-5 are somewhat pessimistic because most practical elements (e.g., waveguides) do not couple as strongly as do thin dipoles without a ground plane.

Experimental Procedures

In the preceding discussion we have outlined some analytical aspects of finite arrays and we have presented some representative results for small arrays. With this background material we are now prepared to discuss some experimental aspects of small arrays. Basically there are three experimental tests which are commonly employed:

1. Element-gain pattern measurements.

This type of experiment is simple to implement since all elements are passively terminated (no feed network required). Although it gives a quick indication of array gain variation over the scan volume, this test is incomplete for element design because it does not give any direct indication of element mismatch. However, the patterns can be compared with the ideal cos pattern to obtain a crude estimate of mismatch for scan angles less than the grating lobe angle.

2. Coupling coefficient measurements.

In this experiment a source is connected to one element (typically a central element) and the voltage, amplitude and phase relative to the incident voltage, coupled to each element is measured with all other elements passively terminated. These measured coupling coefficients are then used to numerically compute the reflection coefficient as a function of scan angle.

3. Active impedance measurements.

In this test the reflection coefficient of an element is directly measured in the fully excited array environment as the beam is scanned. This approach has the advantage of providing a direct measure of mismatch vs scan angle, but it unfortunately requires construction of a feed network with beam steering capability.

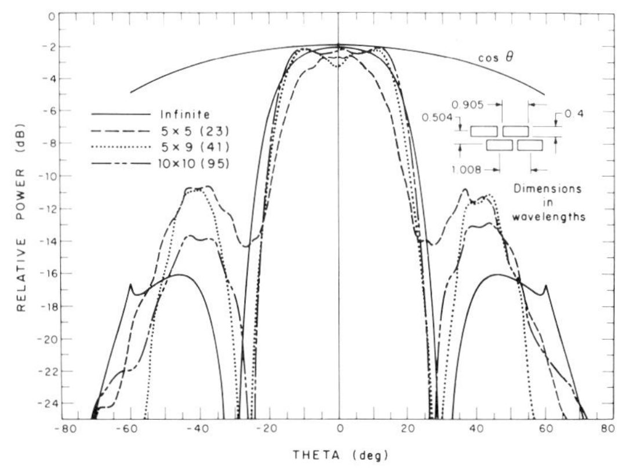

Fig. 6 Experimental H-plane patterns of the center elements of waveguide arrays.

Some measured H-plane element-gain patterns for waveguide arrays of various sizes are shown in Fig. 6. This particular waveguide array exhibits a very strong resonance null well inside the grating lobe angle (60°). Even for the smallest array size, this resonance is evident. Further evidence from Bates (20), Lechtreck (17), Knittel (21), Farrell and Kinsey (22) suggests that element-gain pattern measurements on arrays with 25-50 elements will probably reveal any severe "leaky" wave or resonance effects (23). However, since arrays for element-gain pattern tests are relatively simple and inexpensive to construct, it may be desirable to use a much larger array to be certain that important gain degradation effects will be observed.

Element Design with Small Arrays

The information presented above suggests the following procedure for phased-array element design with small arrays.

1. Choose an element lattice (triangular or rectangular) and a preliminary element design.

2. Approximately match a single element (with a large ground plane for most element types).

3. Construct a small test array containing at least 25 elements and measure element-gain patterns of the center element. If these patterns deviate significantly from the $\cos\theta$ ideal pattern (e.g., Fig. 6), rework the element design and remeasure until the gain loss with scanning is reduced to a tolerable bound (within 5 db of $\cos\theta$, for example).

4. Measure the coupling coefficients among the elements of the test array at three (or more) frequencies in the frequency band of interest.

5. From the measured coupling coefficients compute the mismatch variation with scan angle at each of the measured frequencies.

6. Analytically or experimentally design an appropriate matching network for the elements.

7. Appropriately modify each element in the test array and repeat the above steps. Usually one iteration of the design is sufficient for matching.

There are obviously other schemes for element design with small arrays which can be devised. However, the above procedure is quite efficient in the sense that it utilizes just a simple passive array and the design usually converges very rapidly to a well matched element for a phased array.

Conclusions

From the above theoretical and experimental results, it is evident that an element in even a very small array will exhibit a gross behavior similar to that of an element in a large array. The disagreement between the small array case and the infinite array case is worst in the vicinity of a grating lobe angle and in the vicinity of endfire. (This is to be expected of course due to the non-zero beamwidths of the main beam and grating lobe.) In addition, it is to be noted that isolated elements having element pattern nulls at endfire (dipoles above a ground plane) exhibit less edge effect at the center of a small array than do elements having non-zero patterns at endfire (slots, waveguides); that is, a larger test array should be used with the aperture type elements. Although it does not appear possible to firmly state that a specific array size (for example, a 37 element hexagonal array) will always be adequate, the data presented does indicate that a modest sized array should result in a satisfactory element design.

References

1. H. A. Wheeler, Proc. IEEE 56, 1940-1951 (1968).
2. B. L. Diamond and G. H. Knittel, this book.
3. P. W. Hannan and M. A. Balfour, IEEE Trans. Ant. Prop. 13, 342-353 (May 1965).
4. N. Amitay, V. Galindo and C. P. Wu, this book.
5. R. W. P. King, Theory of Linear Antennas (1956).
6. R. W. P. King, R. B. Mack and S. S. Sandler, Arrays of Cylindrical Dipoles (1968).
7. P. S. Carter, Proc. IRE 20, 1004-1041 (1932).
8. S. W. Lee, IEEE Trans. Ant. Prop. 15, 598-605 (September 1967).
9. C. P. Wu, 1968 IEEE Ant. Prop. Symp. Digest.
10. J. P. Quine, Proc. IEEE 56, 1959-1962 (1968).
11. J. L. Allen, et al, TR-381, M.I.T. Lincoln Laboratory (1965).
12. J. L. Allen and B. L. Diamond, TR-424, M.I.T. Lincoln Laboratory (1966).
13. R. C. Hansen, Microwave Scanning Antennas: vol. II, (1966).
14. L. Schwarzman and L. Topper, IEEE Trans. Ant. Prop. 16, 628-632 (November 1968).
15. T. R. Debski and P. W. Hannan, Microwave Journal 8, 93-96 (June 1965).

16. N. Amitay, P. E. Butzien and R. C. Heidt, IEEE Trans. Ant. Prop. 16, 47-57 (January 1968).

17. L. W. Lechtreck, IEEE Trans. Ant. Prop. 16, 31-36 (January 1968).

18. B. L. Diamond and T. B. Lewis, IEEE Trans. Ant. Prop. 13, 806-7 (September 1965).

19. B. L. Diamond and D. M. Bernella, 1966 IEEE Ant. Prop. Symp. Digest.

20. R. H. T. Bates, IEEE Trans. Ant. Prop. 13, 321-2 (March 1965).

21. G. H. Knittel, private communication (March 1970).

22. G. F. Farrell, Jr. and R. R. Kinsey, private communications (March 1970).

23. A. A. Oliner, this book.

A Survey of the Simulator Technique for Designing a Radiating Element in a Phased-Array Antenna

HAROLD A. WHEELER

Wheeler Laboratories, Inc.
Smithtown, New York

Abstract

The infinite-array environment is taken as a basis for designing a radiating element for a planar periodic array of many elements, to be phased for beam steering. For measurements, this environment is simulated by imaging a small number of elements in the walls of a waveguide. Each simulator represents one sample of scan conditions in the array. The measurements are supplemented with computations of the variations over all scan conditions. While severely restricted in application, this technique still offers great versatility. Various simulators are described, with emphasis on the choice of a configuration for any particular purpose.

I. Introduction

The simulator technique is directed to the practical design of a radiating element for use in a planar periodic array of many elements to be phased for beam steering. It is based on the concept of an infinite array (1). This technique relies on measurements in a simulator including one or a few elements, supplemented by computations before and after the measurements (10). The simulator has reflecting walls in certain symmetry planes of the array, which form a waveguide, usually of rectangular cross-section (2). The computations are based on some modal series such as the "grating-lobe series" (5).

This article will emphasize the opportunities and limitations peculiar to the simulator, and will outline some ways in which it may be supplemented by computations. The result is a technique for evaluating the performance of a radiating element, which forms the basis for arriving at a practical design. The purpose of this article is a survey of the knowledge and experience to date.

One of the principal problems in the design of an efficient element is the impedance matching between radiation space in front of the array and an internal circuit behind the array. The objective is to reduce the reflection at the interface. The reflection coefficient varies with frequency and also varies with phasing for beam steering. Both variations are dependent on the intercoupling between elements in the array. The variation with phasing is peculiar to a phased array and increases with the amount of beam deflection.

There are available various modal series for the computation of reflection at the face of an infinite periodic planar array. However, such theoretical analysis must be supported by experimental techniques for confirmation and also for the measurement of structures too complicated for computation. The simulator is intended to enable such measurements under a few discrete sets of scan conditions. These can be chosen to be typical of the entire range of operation, or at least to provide sampling points from which variations can be computed.

The evaluation of an array element involves three steps to be performed in any of several alternative sequences and iterations (10): (1) The measurement of the element reflection coefficient in one or more simulators (2) representing an equal number of sets of scan conditions: (2) the computation of the variation of reflection coefficient over a specified range of scan conditions at each of several frequencies (5): and (3) the correlation of the measured values for the simulated sample conditions with the computed variation over the range of scan conditions (9). It is possible to measure in each simulator the reflection at the array face, converting this to element impedance or admittance (2), and then to make the computations by means of the "grating-lobe series" (5). Based on this technique of evaluation, there can be formulated a procedure for arriving at a practical design to meet given specifications.

After identifying some terminology to be employed herein, the presentation will continue with the theory and practice of the simulator technique.

II. Terminology.

A. Definitions and Abbreviations.

(1) Relating to the array grid.

Cardinal (C) plane: a plane of symmetry in which there is least spacing of element centers ($\phi = 0°$).

Intercardinal (IC) plane: a plane of symmetry which bisects the least angle between two cardinal planes.

Regular grids: square (S) and equilateral-triangular (T) grids.

(2) Relating to the array element.

Circular element: one having circular symmetry, which requires a pair of like crossed modes terminated alike.

Linear element: one intended for linear polarization and realizing linear polarization at broadside (because of shape of termination).

Active element (in a simulator): one having coupling between the simulator waveguide propagation mode and the element port.
Inactive element (in a simulator): one having no coupling between the simulator waveguide propagation mode and the element port (because of element location or orientation relative to the field of this mode in the waveguide).

(3) Relating to the array operation:

Main beam (MB): the scan direction for which the array is phased.

Grating lobe (GL): any other direction for which the array is (incidentally) phased.

Scan conditions: polarization in the scan plane (E, H, or Eψ), angle of the scan plane from the C plane (C, IC or ϕ), and scan angle from broadside (θ). (See Fig. 1 below.)

Coefficient to be measured: transmission (T) or reflection (R).

B. Symbols.

λ = free-space wavelength at operating frequency
λ_c = cutoff wavelength in waveguide
E, H = electric or magnetic polarization in scan plane ($\psi = 0, 90°$)
TM, TE = waveguide modes for E or H polarization in scan plane
C, IC = cardinal and intercardinal planes of symmetry (see Fig. 1)
a, b = rectangular-waveguide dimensions (see Fig. 1)
a, b = equilateral-triangular-waveguide height in C or IC direction
c = equilateral-triangular-waveguide height (in general)
m, n = waveguide mode numbers
2g = element spacing center-to-center in square or triangular grid
g, h = step between consecutive planes of symmetry (see Fig. 1)
p(= a/g) q(= b/h) = step numbers determining size of waveguide
ψ = angle of E polarization plane from scan plane, in the sense of ϕ ($\psi = \pm 90°$ for H polarization)
ϕ = angle of scan plane from (nearest) C plane
θ = scan angle from broadside (in scan plane)

III. The Infinite Array as a Basis for Designing a Radiating Element.

The simulator technique is based on the behavior of a radiating element in the environment of an infinite array (1) (2). This concept is an accepted basis for the design of an element for a finite array of many like elements. The measurement and computation of the performance of an element in an infinite array gives an approximate evaluation of some of its characteristics in the finite array. This applies to the characteristics which are affected mainly by the internal design and immediate environment, such as transmission and reflection coefficients and power capacity. Not evaluated by this technique, are those characteristics depending mainly on the array size and shape and the number of elements, such as the directive gain and the radiation pattern.

Table I outlines the significance of the infinite-array concept in the simulator technique. By iterative imaging of a few elements in the plane walls of a waveguide, the environment of an infinite array is simulated. Measurements in this environment, and the opportunities and limitations of simulators, are the principal subject of this article

Table I. Features of the Simulator Technique.

A. The design of a radiating element on the basis of an infinite-array environment.

1. Practically a close approximation to the average environment of many elements in a large array.

2. Historically the first technique to make a design on this basis.

B. Evaluation of reflection coefficient, etc., by measurement in a simulator providing an infinite-array environment.

1. Limited to array grid and element configuration that can be imaged in simulator walls.

2. Limited to samples of scan conditions that can be realized in practical simulators.

C. Supplemental evaluation by computation of the variation of reflection coefficient over a range of scan conditions.

1. Used for interpolation and extrapolation from measurements.

2. Practically limited to simple configurations at the array face.

D. The one-port simulator enables experimental design of element internal configuration, without restriction to a computable form, for one or more scan conditions.

IV. The Principles of Simulating an Infinite Array in a Waveguide.

After a decade of experience, the "infinite-array simulator in a waveguide" is now a recognized tool in the design of a radiating element for a phased array (10). If an element meets certain conditions, such a simulator enables its measurement in the environment of an infinite array. Table II outlines some principles underlying this technique.

Table II. Some Principles Underlying the Simulator

A. For a planar, periodic array of many like elements, the element design is based on an infinite array as an approximation to the average environment of the individual elements.

B. If the array grid and the element (or an equivalent element) have in common at least two planes of symmetry, the infinite-array environment can be simulated (for some scan conditions) by image walls forming a rectangular waveguide.

C. Each propagating mode (TM or TE) in the simulator waveguide represents a beam with linear polarization (E or H) in its scan plane.

Table III. Some Configurations Having at Least Two Planes of Symmetry.

A. Array Grid.

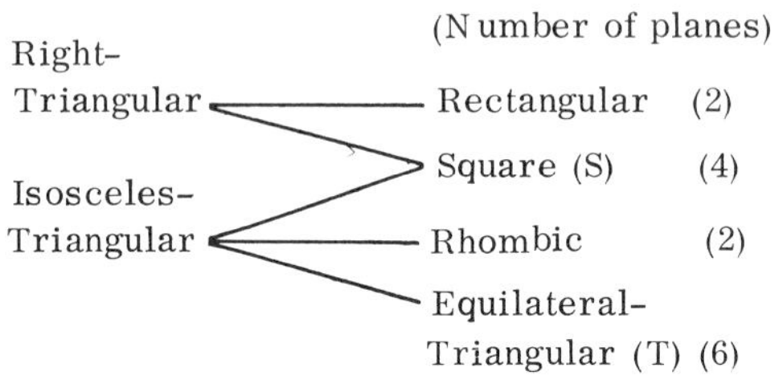

B. Element.

Circular (circular symmetry or an equivalent, with a pair of crossed linear modes terminated alike).

Linear (crossed-axis symmetry with only one linear mode utilized and terminated).

The simulator is made of a waveguide (here called the simulator waveguide) terminated in a section of the array containing a small number of elements (2). The plane walls of the waveguide image these elements to simulate the environment of an infinite array. Each propagating mode in the waveguide may be analyzed into a set of one or more pairs of plane waves, every wave of the set being representative of the same discrete set of scan conditions.

The application of simulators is severely restricted by the requirement of planes of symmetry. Table III outlines the array grids and the element configurations for which valid simulators can be made. The equilateral-triangular grid has the greatest number of planes of symmetry (6) so it offers the greatest versatility in this respect. However, only a circular element can take advantage of all symmetry planes. Otherwise, the number of planes is usually reduced to one pair (2). See Rule 9 below.)

For any one propagating mode in the simulator waveguide, the reflection at the face is measured "looking in" or "looking out". This measurement is valid for one set of scan conditions with linear polarization. The scan conditions are: (1) the scan angle (θ) from broadside; (2) the scan-plane angle (ϕ) from the cardinal (C) plane of symmetry in the array; and (3) the polarization (E or H or mixed) in the scan plane. Since the spacing in terms of wavelength varies with frequency, a single simulator can be used to cover some range of conditions by frequency variation.

The square grid and the equilateral-triangular grid are the most versatile in the application of simulators, so they are taken as the basis for this discussion. The former is the simplest but the latter enables greater area per element while avoiding grating lobes. Fig. 1 shows, for each of these grids, the planes of symmetry and the essentials of simulator waveguides offering unrestricted imaging of a circular element (2) (3).

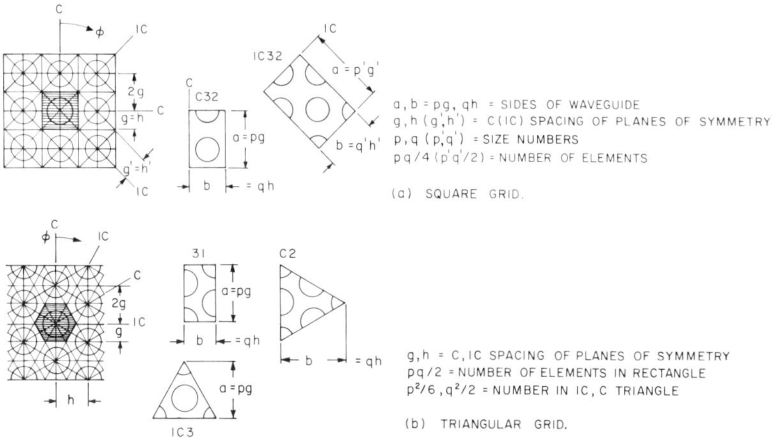

Fig. 1 The array grid and simulator waveguides.

For each grid, the planes of symmetry are identified (C, IC), also their spacing. The element cell is shaded. The scan-plane angle (ϕ) is measured from any C plane in either sense. In each case, the usually designated "element spacing" is 2g in the C direction. It is seen that some elements may be split by the image planes to leave fractional elements in the simulator waveguide.

In Fig. 1(a) for the square grid, there are two significant orientations of a rectangular-waveguide simulator, each identified by C or IC and a pair of size numbers (p, q). There are also two significant orientations of an isosceles-right-triangular waveguide, but they seldom offer an advantage so they are not shown. A C simulator may contain only whole elements but any IC simulator contains some fractional elements.

In Fig. 1(b) for the traingular grid, there is only one distinctive orientation of a rectangular-waveguide simulator, because the sides are parallel respectively to both C and IC planes. It is identified by a pair of size numbers. There are shown also the two possible orientations of an equilateral-triangular-waveguide simulator, each identified by C or IC for its walls and one size number for its height. Any simulator contains some fractional elements.

In a simulator waveguide, as has been mentioned, each propagating mode may be analyzed into one or more pairs of plane waves, every wave representing the same set of scan conditions. Fig. 2 shows the scan planes of each pair of waves in typical waveguides.

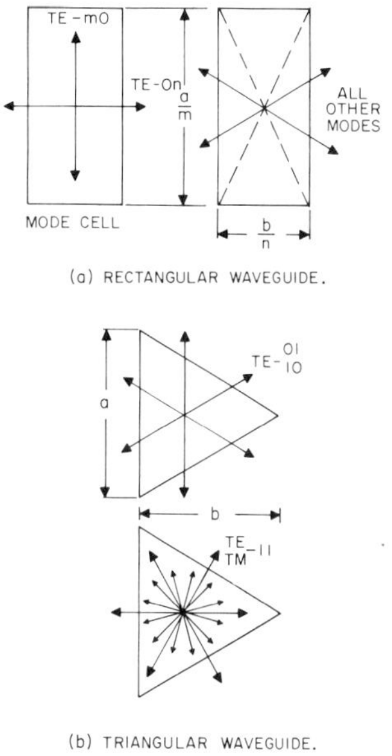

Fig. 2 The scan planes of the component waves in a waveguide.

A rectangular waveguide can be analyzed in terms of an elementary "mode cell" as shown in Fig. 2(a). Each scan plane is identified with a pair of waves (one double arrow) (2). Each non-zero mode has two pairs in mutually symmetrical scan planes.

An equilateral-triangular waveguide, shown in Fig. 2(b), has a more complicated behavior (4). In any mode having one of the lowest two cutoff frequencies there are three pairs of waves (the longer arrows) in scan planes parallel to the three sides. In any one of the higher modes there are six pairs (the shorter arrows) oblique to all sides. The lower modes have been found useful, as will be seen.

Since one simulator operating in one mode can represent only one set of scan conditions, the application of simulators involves a choice of the fewest and most economical simulators to meet some requirements. It becomes a rather complicated game, more like chess than checkers. In general, any one set of scan conditions (with linear polarization) can be simulated to any required degree of approximation, but it may involve a compromise between the degree of approximation and the economy in number of elements. Therefore some formulas will be given as a basis for such a compromise.

Fig. 3 shows a mode cell of a rectangular waveguide with formulas for the relations that are essential to a simulator function (2). Formulas (1), (2) and (3) are in a form for analysis, giving the angles (ϕ, θ) in a specified simulator. If the array is specified, but not the simulator, formulas (4) and (5) are in a form for synthesis of a simulator. In terms of the array spacing and the angles to be simulated, each of these formulas gives a quantity that is to be approximated by the ratio of two integers. The objective is to obtain a close approximation by two integers that are small enough to describe a practical simulator (a few elements and a low mode).

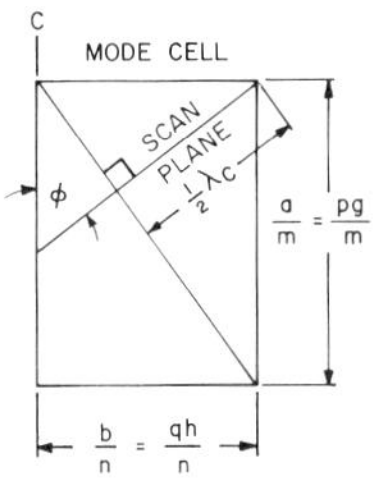

$$\tan\phi = \frac{npg}{mgh} \qquad (1)$$

$$\sin\theta = \lambda/\lambda_c \qquad (2)$$

$$\frac{1}{2}\lambda_c = \frac{(pg)(qh)}{\sqrt{(npg)^2 + (mqh)^2}} \qquad (3)$$

$$\frac{m}{p} = \frac{2g}{\lambda}\cos\phi\,\sin\theta \qquad (4)$$

$$\frac{n}{q} = \frac{2h}{\lambda}\sin\phi\,\sin\theta \qquad (5)$$

CHOOSE SMALLEST INTEGERS TO APPROXIMATE THESE TWO RATIOS.

Fig. 3 Formulas for synthesis of the smallest rectangular waveguide simulator for some specified conditions.

Here are a few rules relevant to this section.

Rule 1 - At any operating frequency, a simulator is assumed to have infinite length between the array face and any probe or junction for coupling with the operating mode. This condition is approximately satisfied if the length is sufficient for great attenuation of all nonpropagating modes that are excited. This requirement is most severe if operating near cutoff of the latter.

Rule 2 - A simulator waveguide for unrestricted imaging in its walls is restricted to a polygon cross-section with internal angles which are integral submultiples of 180° (namely 90°, 60°, 45°, 30°). Such a polygon has 4 or 3 sides. Also, restricted imaging is realized in some 4-sided shapes formed by two triangles in symmetrical relation with a common wall (diamond shapes).

Rule 3 - The relation between the simulated linear polarization and the waveguide modes:

Polarization	ψ	Circular Element	Linear Element
E	0°	TM	--
E/H	Oblique	TM/TE	TM/TE
H	±90°	TE	TE

Note that E polarization from a linear element cannot be simulated exactly in a waveguide with conducting walls. However, there are some simulators offering a close approximation, denoted "near-E" polarization.

A special case should be noted (though it is usually avoided in practice) because it imposes further restrictions on the use of a simulator. This is the case of main beam and grating lobe at equal and opposite scan angles ("symmetrical") (2). One mode in the simulator waveguide then represents two radiating beams with some relationship in time and space. The implications of this condition will not be discussed further here, and such simulators will be omitted because they are not regarded as "useful".

The next section will go into some detail relative to simulators available for various purposes.

V. The Available Configurations and Scan Conditions.

Any one simulator may be described and its operation specified by a table of symbols and numbers. Fig. 4 shows such a table for a rectangular waveguide. The first part (a) describes the simulator and some of its dimensional properties. The second part (b) gives its operational parameters. The relevant characteristics may be abstracted for distinguishing one simulator from another.

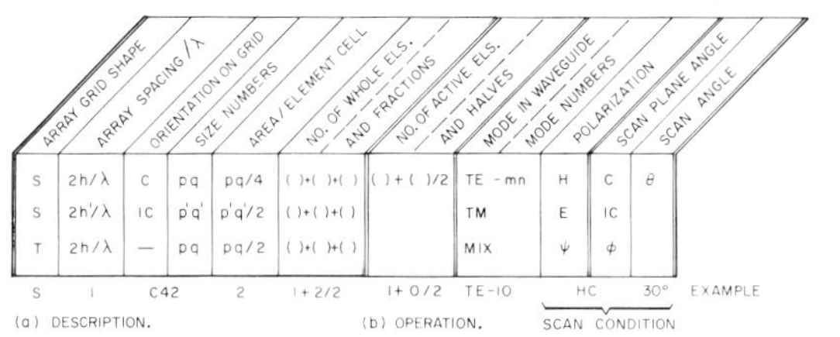

Fig. 4 Description and operation of a simulator in rectangular waveguide.

The scan direction (ϕ, θ) simulated in a rectangular waveguide can be mapped on the sin θ plane. Its polar coordinates are angle ϕ and radius sin θ. Its rectangular coordinates are proportional to the array phasing. The "sector chart" shown in Fig. 5 is based on these relations (2) (10) (11). The scan-plane angle is imaged in a number of sectors of a circle, in such a manner that all the information appears on any one quadrant and even that may be folded into a smaller sector.

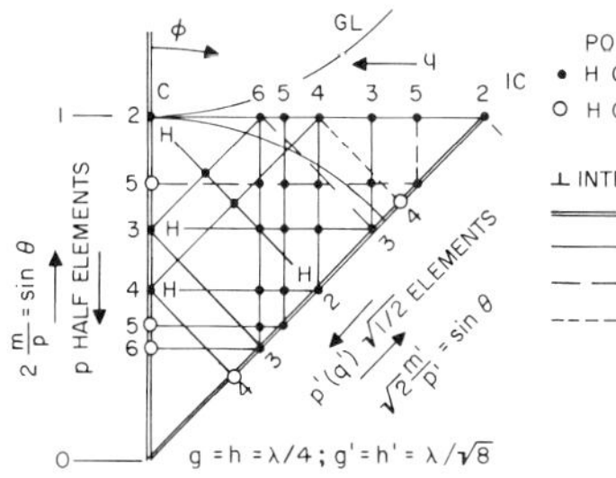

(a) SQUARE GRID (WAVEGUIDES WITH EITHER SIDE NOT TO EXCEED 3/2 WAVELENGTHS).

Fig. 5a Sector charts showing angle coverage of simulators in rectangular waveguide (either mode number not to exceed 3).

On these coordinates, the central unit circle includes the hemisphere of "real space". Nearby other circles correspond to grating lobes (GL) in real space, which usually are to be avoided. These charts are drawn for an array spacing indicated on the figure, which is just small enough to exclude grating lobes in real space for all scan angles ($\theta < 90°$).

Each sector chart has one or more sets of orthogonal straight lines, corresponding to rectangular coordinates as follows, from (4) and (5) in Fig. 3:

$$\text{Vertical coordinate} = \sin\theta\cos\phi = \frac{m}{p}\frac{\lambda}{2g} \quad (6)$$

$$\text{Horizontal coordinate} = \sin\theta\sin\phi = \frac{n}{q}\frac{\lambda}{2h} \quad (7)$$

These two coordinates are seen to be determined separately by different array dimensions, size numbers and mode numbers. Each intersection of two perpendicular lines corresponds to one scan direction. Each line is marked with a size number and coded to a mode number.

Each non-zero mode, indicated by a black dot, is available for E or H polarization. Otherwise a white dot indicates that only H polarization is available. (A dot marked "H" indicates the availability of a "one-port" simulator, yet to be described.)

Each of the sector charts shows all available scan directions in simulators subject to the stated limitations. Usually only those points inside of the unit circle are of practical interest.

In Fig. 5(a) for a square grid, two charts are superposed for C and IC orientation of the waveguide. The latter is a variation of the former. The entire chart is one-half quadrant.

In Fig. 5(b) for a triangular grid, one quadrant is completed with orthogonal sets of straight lines to show the intersections. By folding at C and IC angles, the entire pattern is compressed to one-third quadrant, on which all intersections are mapped.

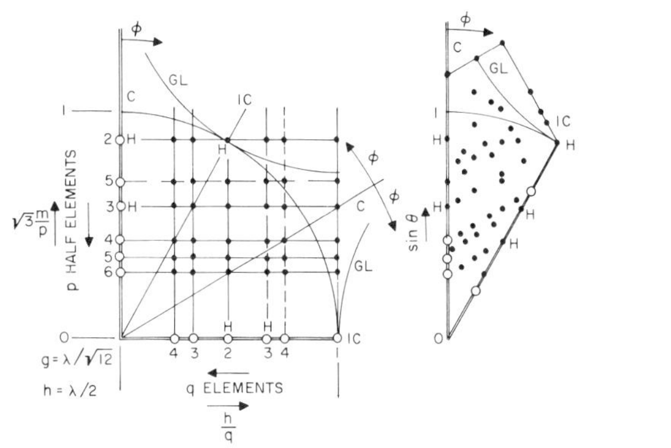

(b) TRIANGULAR GRID (WAVEGUIDES WITH EITHER SIDE NOT TO EXCEED 2 WAVELENGTHS).

Fig. 5b Sector charts showing angle coverage of simulators in rectangular waveguide (either mode number not to exceed 3).

While a rather large number of points appear on each chart, it is noted that one point may correspond to a simulator containing a moderately large number of elements and operated in a difficult mode. The charts may be summarized as follows, as to the number of points within the unit circle:

	(a) S grid	(b) T grid
No. of elements not to exceed:	9	12
No. of points, polarization H only:	5	11
E or H:	16	26

Increasing the element spacing, relative to the values indicated in Fig. 5(a) and (b), has the effect of proportionately decreasing the size of the grids on the chart. Alternatively, the same grids may be used and the scale changed by increasing the size of the unit circle.

The following rules are here stated for their relevance to further discussion of the availability of simulators for some purposes.

Rule 4 - A whole or fractional element in a simulator may be "inactive" in the sense that there is no coupling between the waveguide operating mode and the element port. This is true of a whole or half element in special cases, and is usually true of a lesser fraction of an element (1/3, 1/4 ,1/6, 1/12). In such a case, it is adequate to include in the simulator only the element structure so near the face that it affects the "stored energy" at the interface. In a typical case, just an empty hole may suffice, and even that may have little effect. A whole element is inactive if located in the center of a 11-mode cell. In some of the diagrams, each inactive element may be shaded.

Rule 5 - For a scan angle "near broadside", the simulator requires many elements in one dimension (for H polarization) or in both dimensions (for E polarization). The former is therefore preferred. There is usually little difference in behavior between E and H polarization and between C and IC scan planes, because these conditions merge at broadside. Near-broadside is of particular interest because it may be close to one extreme of the locus on the reflection chart. However, it is furthest removed from the peculiarities of an incipient grating lobe.

Rule 6 - A simulator waveguide so small that it includes only one element is capable of simulating only the condition of symmetrical main beam and grating lobe, and perhaps no beam in the hemisphere of real space, even in the lowest mode (TE-01 or TE-10). This "symmetrical" rule may also be applicable to a waveguide including more than one element if operated in a higher mode (such as TM-11 or TE-11). Here, such examples are regarded as not "useful" and therefore receive little attention.

Rule 7 - An element with a circular hole for radiating crossed-linear polarization may not have circular symmetry in its internal configuration, in which case it may be designed in two steps:

(a) Simulator tests looking in, computations, etc. on a strictly circular element.
(b) Internal design by experiment (in a one-port simulator, looking in) to approximate the same behavior with a crossed-linear structure.

Each simulator test is valid for one linear polarization of radiation.

Rule 8 - In each example (to be shown) including a linear element, a slot radiator is shown in the C or IC plane. There may be used instead a wire radiator perpendicular to the slot (and spaced from a plane reflector) as follows: (There is space for a relatively "short" slot or wire in the C plane, or relatively "long" in the IC plane, as indicated.)

Slot Radiator	Wire Radiator: Square grid	Wire Radiator: Triangular grid
C-plane "short" slot	C-plane "short" wire	IC-plane "long" wire
IC-plane "long" slot	IC-plane "long" wire	C-plane "short" wire

Rule 9 - For a circular element, a valid simulator may be made of any set of image planes, subject to Rule 2 above. The same is true for a square element with crossed polarization in a square grid ("egg crate"). For a linear element, the simulator must be a rectangular waveguide with walls parallel and perpendicular to the element. The same applies strictly to a crossed-linear element, but this restriction may be relaxed for "looking in" if its reaction is substantially independent of its orientation in the array grid. In such a case, the crossed-linear element is said to be "equivalent" to a circular element.

Rule 10 - For a linear element, a single mode in a rectangular waveguide can simulate only conditions having H polarization in a principal scan plane (C or IC); it is a TE mode with one number zero. For any other scan plane (at an angle ϕ) the corresponding polarization (E at an angle ψ) can be simulated by mixing a pair of modes (TM and TE of the same pair of non-zero numbers). In some cases, it may be practical to approximate E polarization by realizing scan conditions of "near-E" polarization (see Rule 3).

Rule 11 - For one set of scan conditions, there may be available more than one form of simulator. It may be more economical to choose a form having fewer active elements, especially fewer active fractional elements. The extreme case (for a single mode in the waveguide) is the "one-port" simulator, having only one active (whole) element. The extreme case for a pair of modes is the "two-half-port" simulator, having only two active half-elements.

Rule 12 - If a set of scan conditions can be simulated in a waveguide of any shape, the same set can be simulated in some rectangular waveguide operated in some mode. In some cases (for circular elements) there is an alternative nonrectangular shape of waveguide for the same conditions, usually a triangular waveguide. The latter may contain fewer active elements and may be operated in a lower mode, in which case it may offer a practical advantage.

Rule 13 - For simulating any one set of scan conditions, the required rectangular waveguide (size and mode) may leave some options such as the following:

(a) A zero mode leaves indeterminate the waveguide size in the E direction, so one may use any size number. Usually just one step, but perhaps two steps for some reason such as (1) symmetry or (2) in a square grid, the use of whole elements only.

(b) An even number of steps may leave a choice in the relative numbers of whole and fractional elements and of active and inactive elements.

(c) If a simulator lacks rotational symmetry, there may be "right-hand" and "left-hand" variations.

Rule 14 - For a square grid, the same square waveguide and included elements can simulate (for a circular element) both principal scan plane (different scan angle) by changing between a TE-01 or TE-10 mode and a TM-11 or TE-11 mode.

Rule 15 - For a triangular grid, certain rectangular waveguides are available for the principal scan planes, as follows:

	C	IC
Zero modes:	TE-m0	Te-0n
Non-zero modes:	p/m = 3q/n	p/m = q/n

For a circular element, the same waveguide may simulate both planes (different scan angles) by interchanging the zero modes, or by changing between mode numbers 1 and 3.

Rule 16 - Increasing the operating frequency in a simulator causes some effects in the same degree as proportionately increasing the ratio of array spacing/wavelength, such as: (a) decreasing the scan angle (θ); (b) increasing the grating-lobe and surface-wave effects.

Rule 17 - The choice of a simulator may be influenced by the availability of one simulator which will provide two or more sets of scan conditions by operating in different waveguide modes.

Rule 18 - In each example to be shown herein, for which the scan angle is stated, it is computed for the frequency and element spacing which places the nearest grating lobes at the edge of the hemisphere of real space (see Fig. 5). (Exception: In each example of the use of simulators, the scan angle may be stated for the midfrequency of use and the actual spacing.)

For simulation of "near-broadside" (small θ), Fig. 6 shows typical simulators for about 14° with a variety of other conditions. (See Rule 5.) Linear elements are shown, keyed to the H polarization for which each simulator is valid. A particular grid and linear element orientation restricts the selection to one of these four configurations. (See Rule 10.)

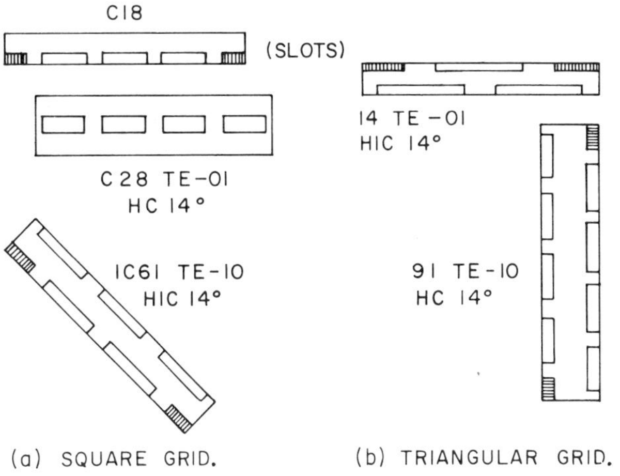

Fig. 6 Simulators for near-broadside (H-polarization).

For simulation of "far-from-broadside" (large θ) without grating lobe, Fig. 7 shows that fewer elements may be used. Three of the four sets of scan conditions can be obtained in a "one-port" simulator. The other is a "two-half-port" operating in the next-to-lowest (TE-02) mode. For linear elements as shown, a particular grid and element orientation restricts the selection to one of these four.

Fig. 8 shows how E polarization "far-from-broadside", with a circular element, may be simulated in different shapes and sizes of waveguides. The two forms (a) and (b) contain the same number of active elements (1 + 3/2) but have different shapes and scan conditions. The two forms (b) and (c) show the same scan conditions simulated in triangular waveguide (lowest TM mode) and rectangular waveguide (third

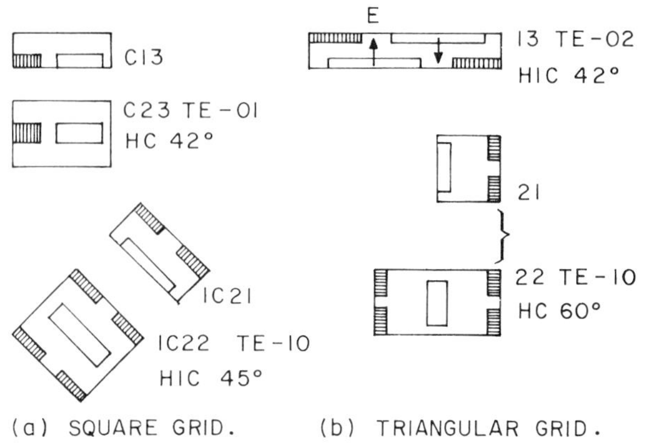

Fig. 7 Simulators for far-from-broadside (H-polarization).

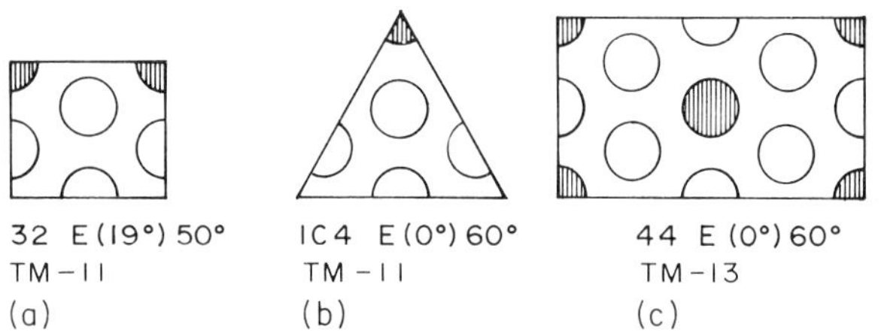

Fig. 8 Comparison of rectangular and triangular waveguides for E polarization (TM modes).

TM mode). While (a) is the simplest, (b) simulates a greater angle with comparable simplicity, and (c) requires a much greater number of active elements (4 + 4/2). If the circular element is actually a crossed-linear element, (a) and (c) may be preferable to (b).

Fig. 9 shows a large variety of simulators for E polarization with circular elements. They include the smallest (fewest elements) available for "far-from-broadside" without grating lobe. Each operates in the lowest TM mode of the waveguide. There are some pairs that simulate in different waveguides the same scan conditions or the same scan angle in different scan planes. This is one situation where a triangular waveguide is particularly interesting and therefore a few are shown, both right-triangular and equilateral-triangular.

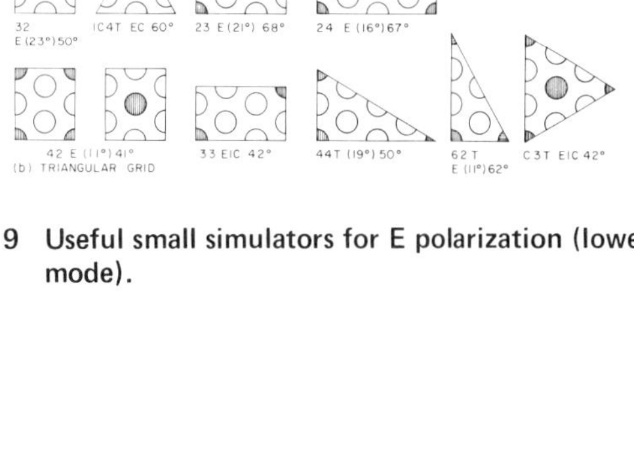

Fig. 9 Useful small simulators for E polarization (lowest TM mode).

Fig. 10 shows a few "one-port" simulators (7) (10) that are remarkable in that each one contains more than one element but only one (whole) element is active in the operating mode. Each one has the property that it can simulate a main beam far-from-broadside with H polarization in a principal plane, and with exclusion of grating lobes. Furthermore, there is only one element port, so measurements can be made "looking out" with no divider (see Fig. 13 below). If a "one-port" simulator can be chosen for experimental design of the interior of an element, it is ideal for minimizing the hardware effort. The configurations shown are all that are known to offer all of these properties in a square or triangular grid. There are a few other configurations that are strictly one-port simulators but lack one or two of the properties outlined for a useful one-port. In Fig. 10, any example in a rectangular waveguide is valid for a linear element in one orientation.

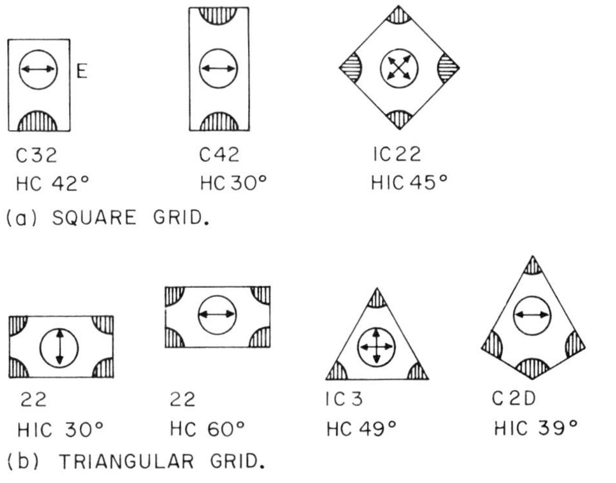

Fig. 10 Useful one-port simulators (H-polarization).

One example in Fig. 10(b) is unusual in that a one-port simulator of diamond cross-section (C2D), containing one active whole element, is formed of two adjoining right triangles, each containing one active half element. The wall angles do not all satisfy the conditions of Rule 2, so its validity is restricted to the indicated E-field direction found in the component right triangles. It offers the same scan conditions as a certain equilateral triangle that contains two active half-elements and therefore is not a "one-port".

Fig. 11 shows the simplest kind of simulator that can offer, for a linear element, any polarization

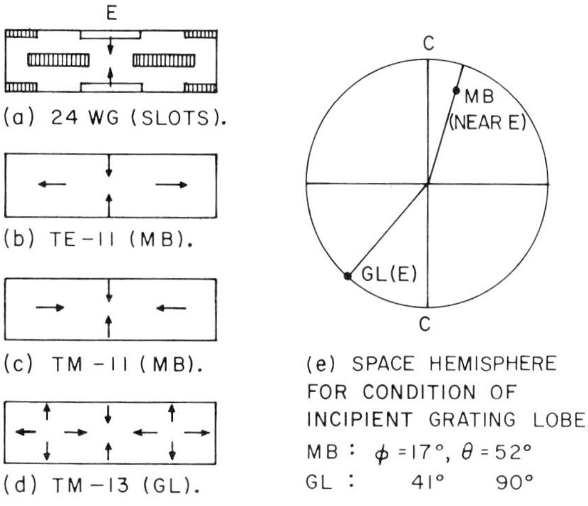

Fig. 11 A multimode waveguide for simulating the polarization from a linear element.

other than H, namely, mixed E and H (coded Eψ). The example shown has been devised and used by R. F. Frazita to evaluate an element with linear polarization (E in the C plane of a triangular grid) but in an oblique scan plane. The objective was to observe the condition of an incipient grating lobe, by operating at or near the cutoff frequency of a higher mode. The simulator configuration (a) has only two active half-elements, having opposite E. The field configuration is diagramed in (b) and (c) for the lowest-order (11) pair of TE and TM modes required to represent the main beam (Rule 10), which has near-E polarization far-from-broadside. It is diagramed in (d) for the higher mode (TM-13) that completely represents the condition of incipient grating lobe (because only E polarization propagates at grazing incidence). In (e) are shown the scan conditions for this frequency of operation, corresponding to the lower and higher modes.

This simulator has come to be appreciated as one of a family of "two-half-port" configurations which are particularly useful for simulating oblique polarization from a linear element. Fig. 12 shows all the known configurations in this family (see Rule 11).

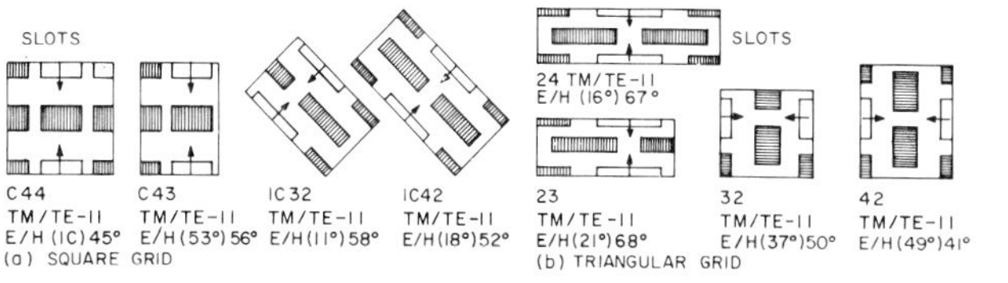

Fig. 12 Useful two-half-port simulators for linear elements.

Their common attraction is the inherent simplicity of exciting both modes (TE and TM) of the same order (11) in the proper ratio for a linear element. Looking out, this ratio is achieved by simply coupling to the two active half-elements in opposition, as indicated. This is the simplest way to simulate other than H polarization for a linear element. For each of the four combinations of array grid and element orientation, there are two configurations offering slightly different scan conditions. A more important difference in the two is the next higher order of coupled modes, depending on odd or even size number. Such modes may be used for observing an incipient grating lobe, as in Fig. 11.

This concludes the discussion of available configurations for simulation of a large variety of scan conditions. All are valid for a circular element. A linear element is restricted to a rectangular cross-section consistent with its orientation relative to the array grid, which leaves a much smaller number of choices with few elements and low-order modes. For a linear element, single-mode simulation offers only H polarization, and dual-mode simulation offers mixed E and H; pure E cannot be simulated (see Rules 3, 9 and 10). Of greatest practical utility are the "one-port" simulators (Rule 11).

VI. Measurements in General.

A simulator is most commonly used to measure the reflection coefficient at the array face "looking in" or "looking out". From such a measurement under several conditions, one may evaluate an equivalent network. The simulator may be used also to measure the transmission coefficient and polarization, although these may be computed from reflection measurements (if grating lobes are excluded from real space). These measurements and computations are straight-forward but may require advanced or unusual techniques. They are here outlined briefly.

Fig. 13 is an outline of the reflection measure-

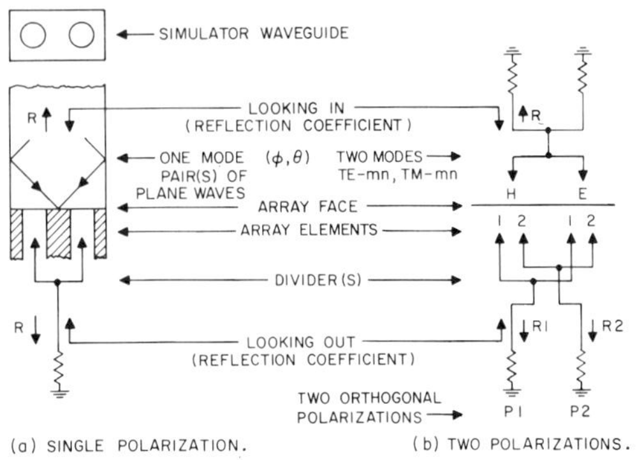

Fig. 13 The measurement of reflection.

ments that may be made in a simulator (2). Each measurement may be made "looking in" or "looking out". One mode in a waveguide (a) is sufficient for the simplest polarization (H or E in the scan plane). For some conditions, two modes (b) with the same mode numbers (m, n) may be mixed to simulate two orthogonal linear polarizations of oblique angles from the scan plane. For measurements looking in, all elements are terminated alike and no divider is needed. For measurements looking out, if there are more than one element ports, a divider is needed. It must provide some ratios of amplitude and one or both polarities, but no phase shifter is needed.

Transmission measurements may be made in either direction between simulator waveguide (operating mode) and element port (one polarization). In general, if the element operates in two crossed modes, elliptic polarization may be evaluated by transmission measurements. A special case is oblique linear polarization from a linear element which has only one polarization but requires a mixture of two modes (of the same order) in the simulator waveguide (Rules 3 and 10).

The same measurements (reflection in one direction only, if desired) may be used to evaluate an equivalent circuit (at any frequency). Fig. 14 shows one such equivalent circuit that is easily evaluated by reflection measurements in only one direction (looking in or looking out) (2). The terminal planes (T_1 and T_2) may be located at any convenient place.

For an array of flush elements, with only free space in front, the array face is a convenient and significant place for T_1.

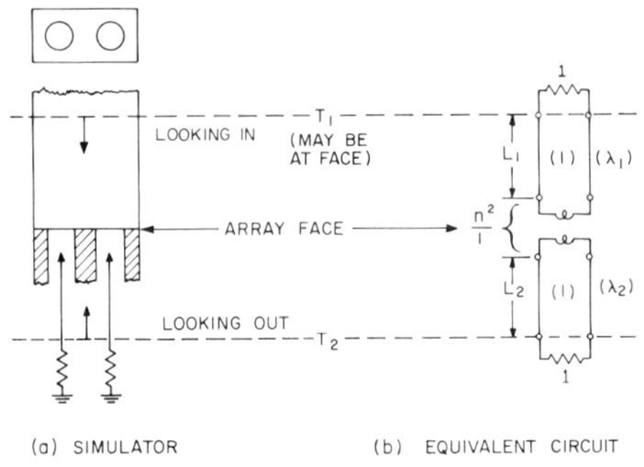

Fig. 14 The evaluation of an equivalent circuit.

L_1, L_2 = length of each line
λ_1, λ_2 = wavelength in each line
n = voltage ratio of equivalent ideal transformer between the lines

By means of the equivalent circuit, the reflection coefficient looking-out may be computed from measurements looking-in, or vice versa. Or the transmission coefficient may be computed.

There will be shown here a few examples of the various practical applications of reflection measurements graphed on the reflection chart. Fig. 15(a) shows, for a triangular grid of flush circular elements, the reflection coefficient at the array face, looking in.

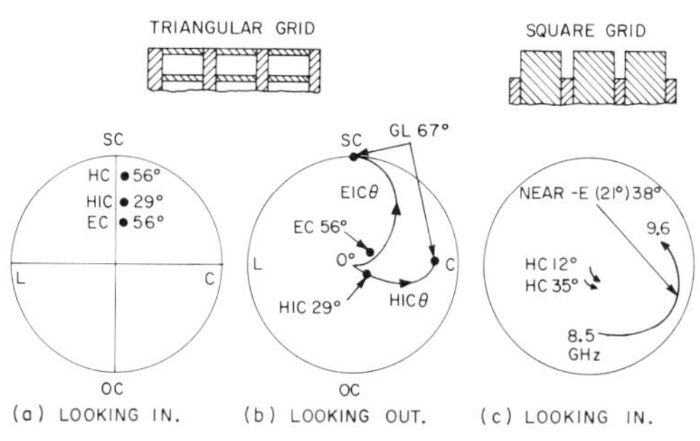

Fig. 15 Examples of measurements shown on the reflection chart.

The element is an undersize circular waveguide loaded with dielectric discs. This is one of the simplest examples of the great difference between E and H polarization in wide-angle scanning (far-from-broadside). In this case, the 56° points are in the C plane, which is far from GL so there is little quadrature effect from the higher terms in the GL series. The three simulators are the 22 (two modes) and IC4 configurations (see Fig. 17, also 8, 9, 10, and 18).

For the same array and simulators, Fig. 15(b) shows the relation between some measured points (29°, 56°) and some curves computed by the GL series (9) (10). The curves show the computed variation of reflection coefficient with scanning from broadside (0°) as a reference at the center of the chart. The measured points are located relative to the same reference by placing the group of points for best fit to the corresponding computed points. Each curve is carried out to the angle of incipient GL (67°). The two curves shown are selected because they show (for one frequency) the points farthest apart for all scan conditions free of GL.

Fig. 15(c) will be discussed in the next section.

In addition to the low-power measurement of impedance properties in laboratory conditions, the simulator has been found useful in testing for high-power breakdown in a closed system. Usually a one-port simulator is selected, and it can be inserted in the resonant ring commonly used for extra-high-power testing. It is convenient for special conditions such as (1) nuclear irradiation to expedite pulse breakdown tests and (2) high or low-pressure air or other gas. Environmental effects, such as a sheet of water or ice on the array face, can be readily evaluated by low-power or high-power tests.

VII. The Measurement of Grating Lobes and Surface-Wave Effects.

In the limit of an infinite array, an incipient grating lobe or surface wave, or scanning 90° from broadside, may cause complete reflection for a discrete set of scan conditions (5) (8) (10). This effect would be a serious deficiency in the performance of a large array but would appear only as partial reflection in an array having a small number of elements. The simulator is suited for critical measurement of such an effect while still requiring only a few elements.

Fig. 11 has been described as a simulator enabling the observation of the effect of an incipient grating lobe. In this case, the effect is complete reflection. (In the case of Fig. 15(b), one of the two computed curves is seen to approach complete reflection at the angle of incipient GL.) If there is also a "slow-wave" or "surface-wave" phenomenon on the array face, this may cause complete or nearly complete reflection at a scan angle somewhat less than the angle of incipient GL. This phenomenon has been observed in this simulator, even in a case where its effect is small.

Fig. 15(c) shows the early (1961) measurements by which the author's group discovered the array "surface-wave" phenomenon, causing complete or nearly complete reflection for some scan conditions (8). In this case, a square grid of circular elements was matched near broadside by means of projecting dielectric plugs, but exhibited large reflection for E polarization at an angle much less than the grating-lobe angle. In subsequent designs, this phenomenon has been avoided by using nonprojecting dielectric, such as a flush window, in conjunction with internal matching (10).

These are examples of the discovery and observation of phenomena that become extreme in the infinite-array environment. It is this environment whose effects can be measured only in the simulator.

VIII. A Design Procedure Based on Computations and Measurements.

A practical design is one that meets some specifications, either definite or objective in nature. For a radiating element, there are some specifications peculiar to a phased array, in that they involve the tolerance of variation of some properties with phasing. For example, the achievement of various mechanical and electrical objectives may have to be compromised with the need for impedance matching or reduction of reflection at the array face. The reflection variation with phasing makes it difficult to realize close matching for wide-angle scanning, and more so over a wide band of frequencies. Beyond the inherent limitations, there are the practical difficulties of testing an element configuration as to its behavior in the environment of a large array.

A design procedure is inherently a sequence of operations by which one may arrive at a practical configuration that will meet the test of evaluation by some techniques, to comply with some specifications. For the design of an array element, the simulator technique includes measurements in one or more simulators, supplemented by some computations. Both are based on the infinite array for approximating some properties of an element in a finite array of many elements.

The element property which is always relevant and which may present the most severe problem is the reflection coefficient. In addition to the usual frequency variation, it is subject to a great variation with phasing for wide-angle scanning. Table IV outlines some operations in which both theoretical computations and experimental measurements may be used together in evaluation of array reflection. The former are relied on for continuous coverage over a range of scan conditions; the latter for shapes that cannot be computed. The first example (A) is ideal for a configuration so simple that it can be completely computed (19), which is seldom the case. It then requires only experimental verification in sample conditions (as in a simulator).

Table IV. Some Operations Available for the Evaluation of Reflection at an Element in an Infinite Array.

	Theory (Modal Series)		Measurement (Simulator)
	Range of scan angle; simple shape can be described and computed.		Samples of scan angle; any shape that can be simulated (actual or equivalent).
A.	Initiate (absolute & variation)	→	Verify samples (absolute & variation)
B.	Initiate (variation)	→	Supplement (abs.) & verify samples (var.)
C.	Supplement (var.) by interpolation & extrapolation	←	Initiate (abs.) one or more samples

The second example (B) requires computations of only the variation with phasing. This depends mainly on the fields in front of the array, and its simple computation is the objective of the GL series (5). Simulator measurements are then used for what cannot be computed, typically the internal design for matching the external radiation space with the impedance at the internal port.

The third example (C) is the opposite extreme, requiring measurements first and computations to fill in afterward.

There will be described here a design procedure based on Table IV-B. It is the procedure which the author now prefers, as offering the greatest opportunity for practical applications. Table V outlines in six steps (A to F) the essentials of the procedure. Fig. 16 shows on the reflection chart the significance of the steps involving computation and measurement.

Table V. A Design Procedure Utilizing the Simulator Technique.

A. Select an array grid and an element external configuration for which the reflection variation with scan conditions can be computed.

B. Compute and graph the reflection variation with scan conditions at each design frequency (looking out, reference at array face).

C. Select a scan condition for which a one-port simulator is available and make this simulator.

D. At each design frequency, compute for this simulator, the reflection corresponding to the center of scan variation (same basis).

E. Design the element internal configuration for closest approximation to matching the center reflection at all design frequencies.

F. Compute and graph the resulting residual reflection over the range of scan conditions at all design frequencies.

Table V-A and B lead to the computation of reflection variation with scan angle from broadside (0°) as a reference. This is graphed (looking out) in Fig. 16(a), which shows simplified triangles intended to include the entire range of scan conditions specified: Scan plane (not indicated, but usually C to IC), polarization (E, ψ, H), scan angle (θ), and frequency (f_1 to f_3). As directed in Table V-C, a scan condition is marked, for which a one-port simulator is available.

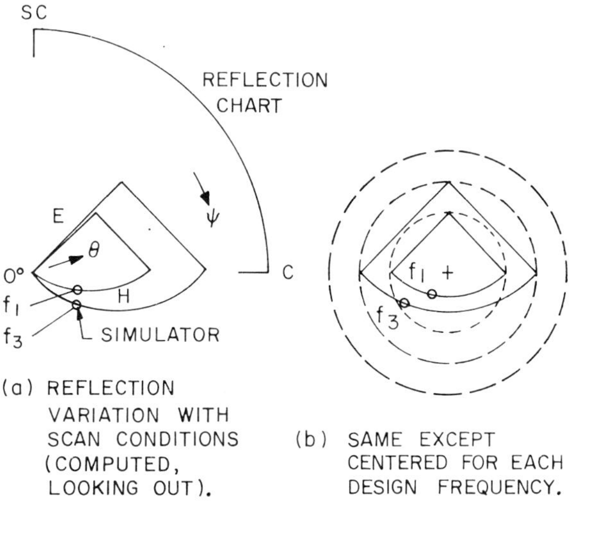

Fig. 16 A design procedure utilizing the simulator technique.

In accordance with Table V-D, the variation at each design frequency is centered as shown in Fig. 16(b). This determines the marked points indicating what reflection coefficient is desired in the simulator. The largest circle indicates a tolerance of reflection, either specified or taken as an objective. Fig. 16(c) shows the same simulator points with corresponding circles representing the design tolerances consistent with holding all variations within the outside tolerance circle in (b). The intermediate point (f_2) is shown because it is needed here. Then, as directed in Table V-E, the element internal configuration is designed by experiment, in view of the objective points and the tolerance circles. The curve is intended to represent a realizable performance that would just fall within the tolerance circles.

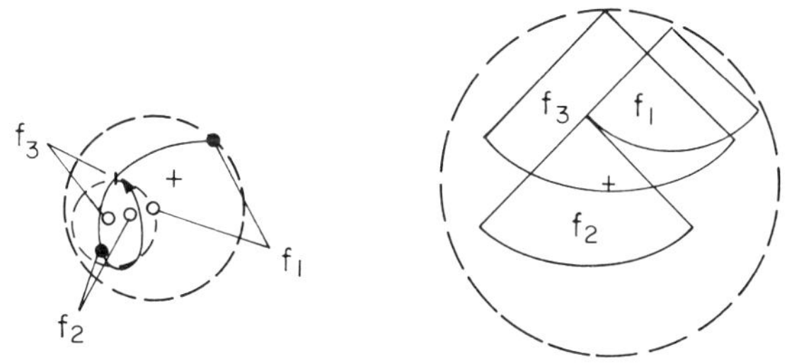

(c) ELEMENT DESIGN REFLECTION IN THE SIMULATOR (MEASURED, LOOKING OUT).

(d) RESULTING RESIDUAL REFLECTION FOR ALL CONDITIONS.

Fig. 16 c & d A design procedure utilizing the simulator technique

Following Table V-F, the resulting spread of reflection coefficient is computed on the basis of Fig. 16(c) and graphed on (d). In this case, the outer limits of reflection are determined by different scan conditions at the three frequencies. A more sophisticated approach might have used tolerance regions other than circles on (c). The result of this procedure is an element configuration with complete evaluation of its reflection variation with wide-angle scanning.

In general, a design procedure is a trial and may be recycled for improvement. Either the same objectives may be achieved in greater degree, or a a different relative weight may be placed on conflicting objectives.

IX. Simulators Used in Practice.

There will be described briefly some of the more significant uses of simulators in various laboratories.

The author's laboratory has been active in the pioneering and development of the simulator technique (2) (10). Fig. 17 shows photographs of the two simulators most used for a triangular grid of circular elements. The one-port simulator in rectangular waveguide (a) is probably the most used of all simulators, for circular or linear elements; it can be used in either of two crossed polarizations for different conditions. The simulator in triangular waveguide (b) has yielded much useful information for E polarization far-from-broadside; it may be historically the first use of triangular waveguide for any purpose.

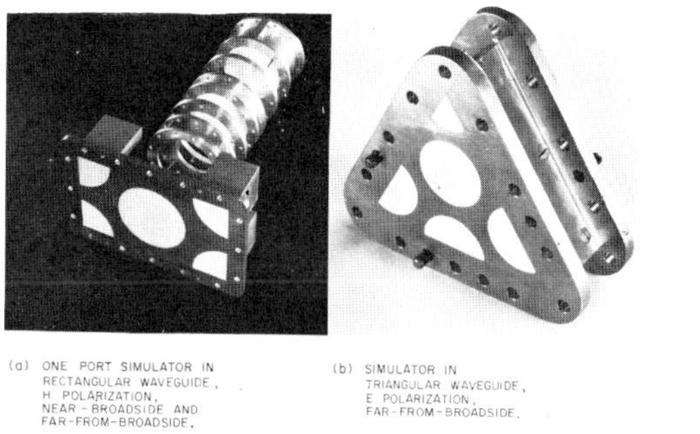

Fig. 17 Photos of the two simulators most used by Wheeler Laboratories.

Fig. 18 shows many configurations that have been constructed and used for various purposes in the author's laboratory. Those for a square grid

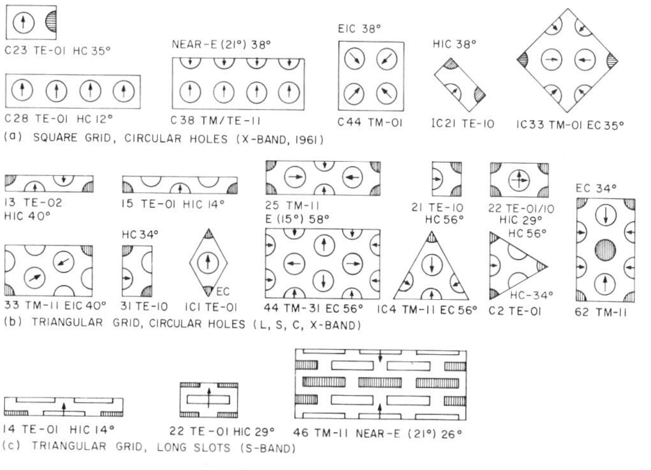

Fig. 18 Use of simulators (Wheeler Laboratories).

(a) were made for early work on a "hard radome" (2) (8) and provided an introduction to the versatility of simulators. The C38 is the one in which the surface-wave phenomenon was discovered; see Fig. 15(c). The simulators for circular holes in a triangular grid (b) show great variety for various purposes, including the examples pictured in Fig. 17 and other figures. The remaining simulators are adapted for long slots in a triangular grid (c); the large one (using the one mode indicated) is strictly not a valid simulator for a linear element, but gave useful information as an approximation. The function of this display is to emphasize the variety of configurations that have been constructed to serve some purposes.

Fig. 19(a) shows some simulators used by the General Electric Co. (courtesy of D. Kuhn) for circular holes in a square grid; it appears that they developed independently the simulator concept and applied it to this case.

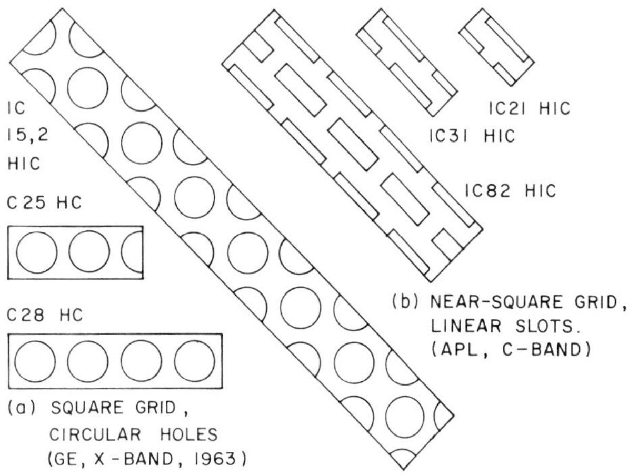

Fig. 19 Use of simulators (GE and APL).

Fig. 19(b) shows some simulators mentioned in papers from the Applied Physics Laboratory of Johns Hopkins University (courtesy of T.C. Cheston) for use in some interesting studies (14) (15).

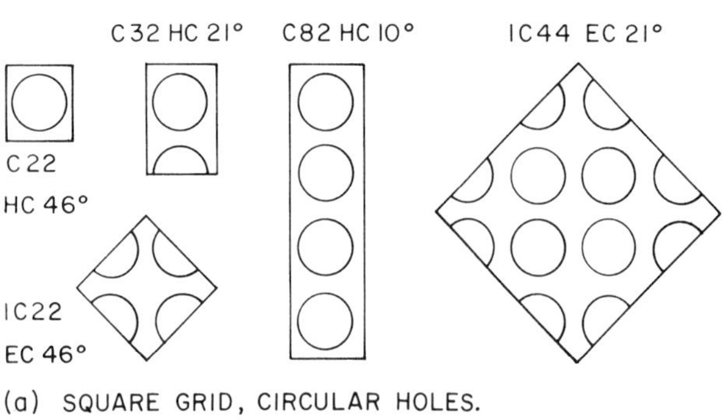

Fig. 20a Use of simulators (Hughes, C-band).

Fig. 20 shows an interesting variety of simulators whose use is reported by the Hughes Aircraft Co. (courtesy of L. Stark).

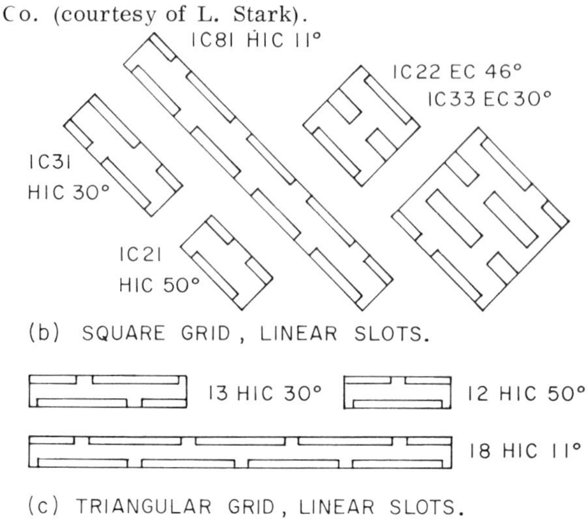

Fig. 20b and c Use of simulators (Hughes, C-band).

Fig. 21 shows some unusual applications of simulators by Texas Instruments, Inc. (courtesy of C. O. Hemmi and P. Montgomery). One is a "one-port" for a linear (electric dipole) element in a square grid (a). The others are for "short slots" in the C plane of a triangular grid (b).

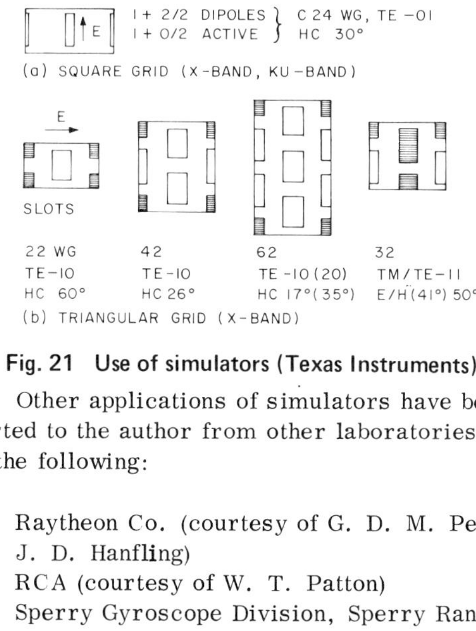

Fig. 21 Use of simulators (Texas Instruments).

Other applications of simulators have been reported to the author from other laboratories, notably the following:

Raytheon Co. (courtesy of G. D. M. Peeler and J. D. Hanfling)
RCA (courtesy of W. T. Patton)
Sperry Gyroscope Division, Sperry Rand Corp. (courtesy of L. Schwartzman, J. J. Stangel, and A. E. Ehlenberger)
Polytechnic Institute of Brooklyn (courtesy of G. H. Knittel)
Lincoln Laboratory (courtesy of B. L. Diamond and G. H. Knittel)
Bell Telephone Labs. (courtesy of N. Amitay)

Most of these applications are similar to some examples illustrated elsewhere herein.

In general, there appears to be an increasing appreciation and use of simulators, with due regard for their versatility and limitations.

X. Summary of Opportunities and Limitations.

In general, the future of the simulator lies in its wider understanding and utilization. It enables measurements of a radiating element in the environment of an infinite array, which cannot be accomplished in any other way. Its use is restricted to an array grid and an element configuration having the required symmetry planes. The operation of one simulator represents only one unique set of scan conditions. The following specific comments are indicative of some specific opportunities and limitations, as seen by this author at this time.

(1) The greatest opportunity seems to lie in the coordinated use of computations of reflection variation by the grating-lobe series with measurements in a "one-port" simulator, along the lines of the design procedure described herein.

(2) The synthesis of a simulator for specified scan conditions is aided by the explicit formulas for this purpose, and by the sector charts showing the available configurations.

(3) The equivalent circuit, evaluated from one set of reflection measurements, is helpful in studying the element behavior from other viewpoints.

(4) The "two-half-port" simulator is perhaps the most enticing of the areas that remain to be explored, especially for the evaluation of a linear element.

(5) The extremes of the variations with scan angle are usually (a) broadside, (b) the greatest angle with E or H polarization, and (c) the scan nearest to incipient GL or surface wave. These are commonly in regions where a simulator requires a moderately large number of elements (a) or the choice is critical to spacing and frequency, (b) and (c).

(6) In the case of an element configuration that is strictly lacking the symmetry required for a simulator, there are various opportunities for devising an equivalent element or half-element which will represent the essential properties of the element.

(7) The choice of an array grid and an element configuration may be influenced by the ability to make computations and simulator tests. This consideration relates to the ability to arrive at a design more expeditiously and to test it more thoroughly.

(8) It is reasonable to suppose that the technology of the simulator in waveguide will continue to advance, even as it has in the recent years of the decade since its first known practical application.

Here it should be mentioned that the alternative measurement technique is the "small array" (10) (18). It contains a moderate number of elements, typically two or three rings around a center element, so the number of elements may be 19 or 37 on a triangular grid. Measurements of an element in a small array give a crude indication of its behavior in a large array, but can be made for all scan conditions. The following is a qualitative comparison between the techniques of small array and simulator

	Small Array	Simulator
Number of elements	moderate	few
Environment of infinite array	no	yes
Close approximation to large-array environment	no	yes
Versatility in scan conditions	all	samples
Freedom from restriction to image symmetry of grid and element	yes	no
Versatility in scan conditions	all	samples

XI. Conclusion.

For testing a radiating element in the environment of an infinite array, the simulator in a waveguide is an indispensable technique. Its application is restricted in respect to the array grid and the element configuration. It enables the simulation only of discrete sets of scan conditions so it should be supplemented by computations of the impedance variation with scan angle over the entire range of operations. On this basis, one useful design procedure has been presented. During a decade of experience, many examples of simulators have been used and much progress has been made in understanding its versatility as well as its limitations.

XII. Acknowledgments.

The author and his associates are most grateful for the opportunities to develop and to utilize this technique, made possible by the Bell Telephone Laboratories and Raytheon Co. under contracts with the

Departments of the Air Force and the Army; special mention is due to D. A. Alsberg, P. H. Smith and R. L. Mattingly of BTL. It is a pleasure also to acknowledge the helpful advice and comments of P. W. Hannan and G. H. Knittel in the preparation of this paper. The many contributions by the author's associates may be inferred from the sampling found in the references herewith; this work was performed under the guidance of N. A. Spencer and G. E. Vaupel, formerly and presently Chief Engineer of Wheeler Laboratories.

XIII. References.

A. Selected Papers by the Author and his Associates.

(1) Wheeler, H. A.: Radiation resistance of an Antenna in an Infinite Array or Waveguide, Proc. IRE, vol. 36, pp. 478-487; Apr. 1948.

(2) Hannan, P. W., Balfour, M. A.: Simulation of a Phased Array Antenna in Waveguide, IEEE Trans., Vol. AP-13, pp. 342-353; May 1965.

(3) Balfour, M. A.: Phased Array Simulators in Waveguide for a Triangular Arrangement of Elements, IEEE Trans., Vol. AP-13, Pp. 475-476; May 1965.

(4) Frazita, R. F.: Properties of Modes in Equilateral-Triangle Waveguides , Wheeler Labs report, to be published; Sept. 1966.

(5) Wheeler, H. A.: The Grating-lobe Series for the Impedance Variation in a Planar Phased-array Antenna , IEEE Trans., vol. AP-14, pp. 707-714; Nov. 1966. (Correction, vol. AP-15, p. 235; Mar. 1967.)

(6) Hannan, P. W.: Simulation of Phased Arrays in Waveguide, P. E. Mayer (ed.), Electromagnetics and Antennas, 1967 Symposium at U. of Ill., pp. 595-651; Jan. 1967.

(7) Balfour, M. A.: Active Impedance of a Phased Array Antenna Element Simulated by a Single Element in Waveguide, IEEE Trans., Vol. AP-15, pp. 313-314; Mar. 1967.

(8) Hannan, P. W.: Discovery of an Array Surface Wave in a Simulator, IEEE Trans., Vol. AP-15, pp. 574-576; Jul. 1967.

(9) Charlton, G. G.: Application of the Grating-lobe Series for Calculating Admittance Variation of a Phased-array Antenna, IEEE Trans., Vol. Ap-15, pp. 818-820; Nov. 1967.

(10) Wheeler, H. A.: A Systematic Approach to the Design of a Radiator Element for a Phased-array Antenna, Proc. IEEE, Vol. 56, pp. 1940-1951; Nov. 1968.

(11) Wheeler, H. A.: A Survey of the Simulator Technique for Designing a Radiating Element, 1970 Phased-Array Antenna Symposium, Digest, pp. 56-59; Jun. 1970. (Correction: Interchange pp. 57 and 58.)

(B) Selected Papers by Others.

(12) Galindo, V., Wu, C. P.: Integral Equations and Variational Expressions for Arbitrary Scanning of Regular Infinite Arrays, IEEE Trans., vol. AP-14, pp. 392-394; May 1966.

(13) Oliner, A. A., and Malech, R. G.: Simulation of Infinite Arrays by Waveguide, Hansen (ed.) Microwave Scanning Antennas, vol. 2, Array Theory and Practice, Academic Press, pp. 322-335; 1966.

(14) Byron, E. V., and Frank, J.: Lost Beams from a Dielectric Covered Phased-array Aperture, IEEE Trans., vol. AP-16, pp. 496-499; Jul. 1968.

(15) Byron, E. V., Frank, J.: On the Correlation Between Wide-band Arrays and Array Simulators, IEEE Trans., vol. AP-16, pp. 601-603; Sept. 1968.

(16) Gregorivich, W. S., Hessel, A., Knittel, G. H., Oliner, A. A.: A Waveguide Simulator for the Determination of a Phased-array Resonance, IEEE 1968 G-AP International Symposium, Program and Digest, pp. 134-141; Sept. 1968.

(17) Hessel, A., Knittel, G. H.: On the Prediction of Phased-array Resonances by Extrapolation from Simulator Measurements, IEEE Trans., vol. AP-18, pp. 121-123; Jan. 1970.

(18) Diamond, B. L.: Small Arrays - Their Analysis and Their Use for the Design of Array Elements, 1970 phased-Array Antenna Symposium, Digest, pp. 52-55; June 1970.

(19) Diamond, B. L., Knittel, G. H.: A New Procedure for the Design of a Waveguide Element for A Phased-array Antenna, 1970 Phased-Array Antenna Symposium, Digest, pp. 60-61; June, 1970. (Entirely by Computation.)

A New Procedure for the Design of a Waveguide Element for a Phased-Array Antenna*

B. L. DIAMOND

Lincoln Laboratory
Massachusetts Institute of Technology
Lexington, Massachusetts

GEORGE H. KNITTEL

Polytechnic Institute of Brooklyn
Graduate Center
Farmingdale, New York

ABSTRACT

In the past, two procedures have been used to design waveguide elements for phased-array antennas; the small-array procedure and the simulator procedure. These procedures rely on measurements to design the element and its matching network, and they have consequent disadvantages.

Recent advances in the ability to accurately compute the active reflection coefficient of waveguide elements have made possible a new, reliable and more-efficient procedure for element design. This procedure relies on computations, and utilizes a waveguide simulator only as the final step to confirm the design.

A detailed discussion of the computational procedure is presented here, delineating its advantages and its limitations. An example is presented of a sophisticated waveguide element designed with this procedure.

I. INTRODUCTION

A phased array of waveguide elements in a conducting plane exhibits an active reflection coefficient, Γ, which varies substantially with scan angle, θ, and frequency, f. The primary problem in element design is to sufficiently minimize $\Gamma(\theta, f)$ over the required ranges of scan angle and frequency, taking into account the imposed non-electrical constraints.

Until recently, two procedures have been used to design waveguide elements, the small-array procedure (Amitay, Butzien, and Heidt, 1968 (1); Hemmi, Hayes, and Voges, 1968 (2); Diamond, 1970 (3)) and the simulator procedure (Wheeler, 1968 (4)**; Wheeler, 1970 (5)).

The small array procedure requires the construction of a small array of elements and the measurement of the coupling coefficients. Next $\Gamma(\theta, f)$ is computed for the center element. A matching network is then chosen and inserted into each element, after which the entire procedure is repeated to verify the results. Details of the small array procedure are given in the references and will therefore not be discussed in this paper. It is worth noting, however, that the main characteristics of this procedure is the use of measurements in a small array to determine $\Gamma(\theta, f)$.

The simulator procedure utilizes measurements of $\Gamma(\theta, f)$ in several waveguide simulators (6) as the basis for choosing a matching network. Measurements are made with the final matching network inserted in each simulator to verify the results. The grating-lobe series (7) is used to interpolate and extrapolate the measured $\Gamma(\theta, f)$ (8). A detailed discussion of the simulator procedure is given in Wheeler's paper (4). The main characteristic of this procedure is the use of measurements in waveguide simulators to determine the $\Gamma(\theta, f)$ on which the impedance matching is based.

Both of the above element-design procedures have certain disadvantages. Since both rely on measurements to determine $\Gamma(\theta, f)$ much time and manpower are required. In the small-array procedure, any change must be made in every element. In the simulator procedure, measurements

*This work was sponsored by the Department of the Army.

**This paper has a great deal of information on element design considerations; it also has many references.

in non-H-plane simulators are very difficult. In addition, both procedures yield values for $\Gamma(\theta, f)$ which are "incomplete" in some sense. The $\Gamma(\theta, f)$ from the small-array procedure is in error for scan angles at which either the main lobe or a grating lobe is near endfire, if the array is too small. The $\Gamma(\theta, f)$ in the simulator procedure is only measurable at a few scan angles and frequencies, and the grating-lobe series is not adequate for all element types (9).

In this paper, we discuss a third element design procedure which is a better choice than the two procedures described above for a class of waveguide elements. This procedure relies on computations to determine array performance; it therefore eliminates the time consuming measurements and also yields the array performance at all scan angles and frequencies. The class of elements for which the procedure is useful is comprised of those which are "computable". Only waveguide elements are discussed here, but the procedure could be used for dipole elements as well.

Section III contains a detailed discussion of the computational procedure and the kind of waveguide elements for which it is valid. Section III presents an example of a linearly-polarized waveguide element designed with this procedure and compares the computed performance with that measured in a waveguide simulator. Section IV contains a discussion of some details relevant to this element design procedure and its extension to other element types.

II. THE NEW PROCEDURE FOR WAVEGUIDE ELEMENT DESIGN

In 1966, Farrell and Kuhn (10) published computations of the performance of an array of rectangular waveguides in a triangular lattice. Sometime afterward Amitay and Galindo (11), Borgiotti (12), Diamond (13), and Schaedla (14) published theoretical solutions which give the performance of circular and rectangular waveguide elements in either rectangular or triangular lattices. These important developments have made possible a new, reliable, and efficient procedure for element design which relies on the computation of $\Gamma(\theta, f)$ (15). This new procedure is outlined in Table I.

TABLE I: A NEW PROCEDURE FOR ELEMENT DESIGN.

1. Choose an element lattice (triangular or rectangular) including dimensions, and an element aperture shape (circular or rectangular) including dimensions.
2. Compute $\Gamma(\theta, f)$ and optimize the element aperture dimensions within the imposed non-electrical constraints.
3. Determine an element matching network by computation. (Alternative: determine the matching network by measurement, representing the computed $\Gamma(\theta, f)$ with a slide-screw tuner).
4. Confirm the final design by measurement in a single simple H-plane simulator.

A detailed discussion of step (1) is beyond the scope of this paper; however, the following guidelines will be given. One usually begins with the element lattice which maximizes the area per element (which has the closest "packing" of grating-lobes) for the specified scan volume. The lattice dimensions should then be chosen so that the area per element is about 5% smaller than that which would permit a grating-lobe at the edge of "real space" for some maximum main-lobe scan angle.* The element aperture shape should usually be taken as rectangular for a linearly-polarized array (to eliminate a propagating cross-mode in the waveguide), and as either square or circular for a circularly-polarized array.

In step (2), one may choose the waveguide aperture size to minimize the variation of Γ with θ and f, or on some other basis. In designing an element for a linearly-polarized array, it is usually best to avoid more than one propagating mode in the waveguide. For some cases in which a dielectric window or plug is used in the waveguide one cannot avoid more than one propagating mode. For other cases, higher waveguide modes are deliberately allowed to propagate in the aperture region of the waveguide to improve scan performance (16). In all of these cases in which higher waveguide modes propagate, the impedance of each waveguide mode looking back into the waveguide element from the aperture plane must be carefully controlled in order to avoid a "resonant value" of impedance, i.e., a value which causes $|\Gamma|=1$. It is easy with the new procedure to determine and avoid these undesired conditions.

Once the waveguide aperture region (including the dielectric plug, if present) has been determined and the resulting array reflection coefficient $\Gamma_A(\theta, f)$ computed, the impedance matching network may be designed (step 3). There is generally no straightforward way to do this; it requires experience and ingen-

*This is based upon a study by the authors which is not yet published.

uity, and each case is different. One usually examines a plot of $\Gamma_A(\theta, f)$ on the reflection chart and, with a knowledge of the available and computable matching structures, then decides upon a particular matching scheme. If the equivalent circuits for the chosen matching structures are either known or computable, one may determine the entire matching-network design by computation. In this case, one may choose to allow interaction between the aperture and the matching structures in one or more higher-order waveguide modes. Alternatively, if the equivalent circuits for the chosen matching structures are not known and not easily computed, one must depend on a measurement to determine the matching network. A slide-screw tuner in a terminated waveguide of the element-guide size may be used to represent the computed $\Gamma_A(\theta, f)$ at each value of θ and f. In this case, the matching structures must be far enough from the aperture so that there is no interaction in higher-order waveguide modes, since higher-order modes at the aperture plane are not correctly excited by the slide-screw tuner.

After completion of the design, it should be checked by measurement in a waveguide simulator (step 4). If the measured matched reflection coefficient in the simulator agrees with the computed matched reflection coefficient, then it is assumed that the computations are accurate, and that they may be confidently used to predict the array performance at all values of θ and f. The simulators (6) suggested for confirmation of the element design are shown in Fig. 1. These are the simplest useful waveguide simulators. Each has only one active element and each represents the infinite array at scan angles near 40° in the H-plane for typical array dimensions.

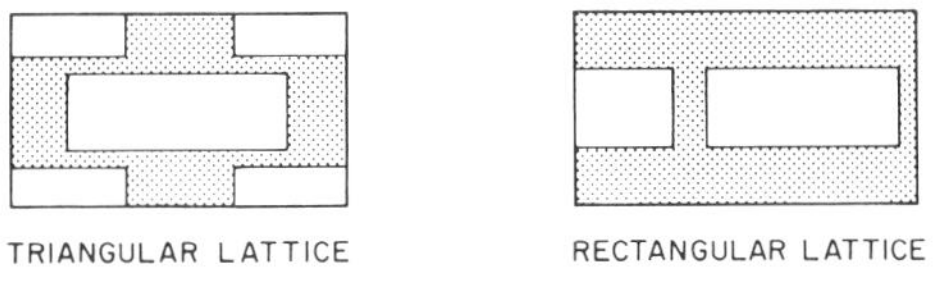

Fig. 1 The simplest useful simulators. Each simulates a single-radiating-lobe scan condition at a scan angle of about 40° in the H-plane.

The conditions sufficient to compute an element design are: 1) the existence of modal representations for the fields in the waveguide region and in the free-space region, and 2) the knowledge of the admittance seen by each mode from the aperture plane. These conditions place minimum constraints on the choice of the element and its matching structures. Rectangular or circular waveguides may be used in rectangular or triangular lattices. Any material or structure may be used in the waveguide region providing its equivalent circuit in the waveguide is known. Any material or structure may be used in the free-space region providing one knows its equivalent circuit for plane-wave incidence.

The accuracy of computation of the array reflection coefficient has been confirmed by measurements in waveguide simulators. Figure 2 is a comparison of measurements and computations for an array of dielectric-filled rectangular waveguides covered by a dielectric slab. This is the array studied by Gregorwich, Hessel, Knittel, and Oliner (17).* Notice the excellent agreement between measurements and computations even in the presence of the resonance.

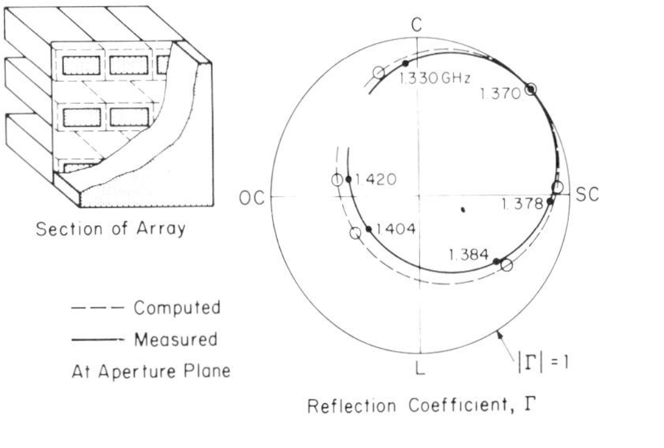

Fig. 2 Comparison of computed Γ with that measured in a simulator for an array exhibiting a resonance.

Figure 3 also shows excellent agreement between measurements and computations made at MIT Lincoln

*The unit cell in this array was 4.46" x 4.46". The waveguides were 4.30"x 2.15" (internal dimensions) and were filled with rexolite (relative permittivity 2.56). A rexolite slab 3.032" thick was placed adjacent to the aperture plane.

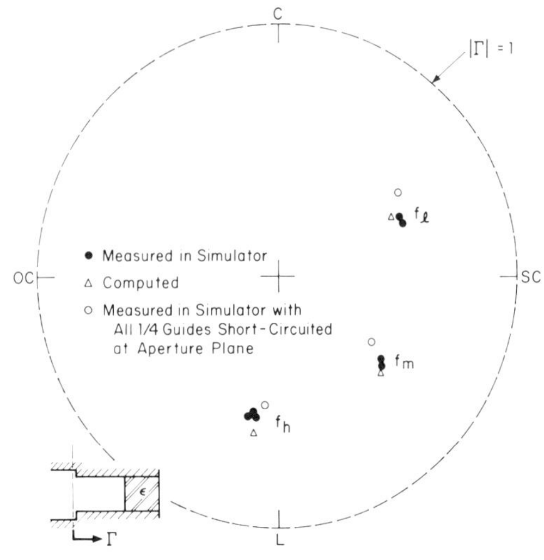

Fig. 3 Comparison of computed and measured Γ for a triangular-lattice array with two waveguide modes above cutoff in the dielectric plug.

Laboratory for a triangular-lattice array of rectangular waveguides fitted with dielectric plugs. Both the TE_{10} and TE_{20} waveguide modes are above cutoff in the plug, but only the TE_{10} mode is above cutoff in the air-filled waveguide. Measurements were made at both ends and at the middle of a 10% frequency band in the triangular-lattice simulator shown in Fig. 1. The importance of the four 1/4 waveguides (required by the simulation) in the simulator was determined by short-circuiting them at the aperture plane with metal tape. It is observed that eliminating the 1/4 waveguides has only a minor effect on the measured reflection coefficient even though a propagating mode exits in the 1/4-waveguide dielectric plugs. *

*In this case the TE_{20} mode of the full waveguide propagates in the 1/4-waveguide plug. It has been our experience that this mode is relatively unimportant in determining the active array impedance, providing that it is not near resonance in the plug. The insensitivity of the measured impedance to the presence of the 1/4-waveguides confirms this.

The advantages of the new element-design procedure are given in Table II.

TABLE II: ADVANTAGES OF THE NEW ELEMENT DESIGN PROCEDURE

1. It relies on computations of the reflection coefficient, $\Gamma(\theta, f)$.

   a. More efficient of time and manpower.

   b. Permits determination of optimum element spacing and aperture size in a simple manner.

2. It yields values of the reflection and transmission coefficients at all scan angles and frequencies for the infinite array. This increases the predictability of the element performance and ensures the absence of any unexpected blindness or surface-wave effects.

3. It can be extended to compute the element coupling coefficients and thence the reflection coefficients and radiation pattern of a finite array with an arbitrary magnitude and phase of excitation.

It seems to us that these advantages relative to the usual procedures are substantial, and that the new procedure should therefore become very popular as the design procedure for the class of waveguide elements which are computable.

III. EXAMPLE OF AN ELEMENT DESIGN WITH THE NEW PROCEDURE

In order to demonstrate the benefit of the new procedure and to gain experience in its use, a practical waveguide element was designed. The element aperture, lattice, and matching network are shown in Fig. 4 (a); the design scan volume is shown in Fig. 4 (b).

The first step in the design was to choose the angle of the array normal (relative to the horizon) and the element lattice to approximately maximize the area per element (within certain non-electrical constraints) and thus minimize the required number of elements. Grating-lobes very near endfire were permitted for scan angles along the periphery of the scan volume at the high-frequency edge of the 10%

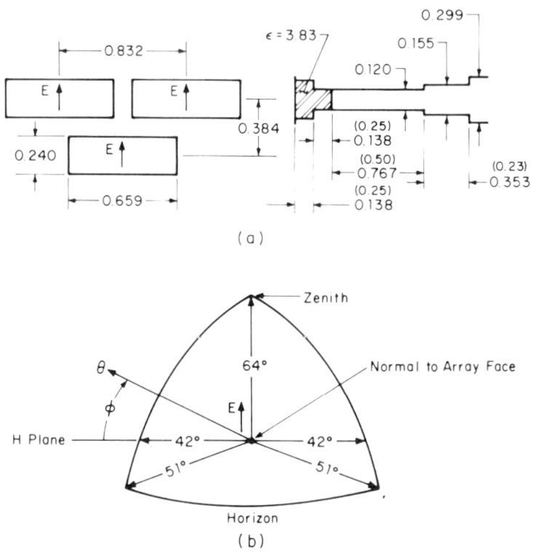

Fig. 4 **A waveguide element designed with the new element-design procedure.**

(a) Element lattice and matching network (all dimensions in free-space wavelengths except numbers in parentheses, which are in guide wavelengths).
(b) Scan volume, 1/4 hemisphere.

band.* An almost -45° isosceles-triangular lattice was obtained in this manner, as shown in the figure.

Since the element is to connect to a phase shifter whose ends are standard air-filled waveguide 0.659** wide by 0.299 high, the width of the element at its aperture was chosen equal to 0.659. It was decided to use a thick quartz plug in the element aperture to protect the phase shifter from the environment. Since the quartz had a relative permittivity of 3.83 (measured), the height of the element at its aperture was reduced to 0.240 in order to limit the propagating modes in the quartz plug to only the TE_{10} and TE_{20} modes. The length of the plug can be easily determined to avoid resonances in the TE_{20} mode, but it is much more difficult to do this over a moderate frequency band when more than one higher-order waveguide mode propagates in the plug.

The next step in the design procedure was to compute $\Gamma(\theta, f)$ at the aperture plane for the element with the same lattice and aperture dimensions shown in Fig. 4, and with the thickness of the dielectric plug chosen so that the impedance seen by the TE_{20} mode from the aperture plane in the direction of the generators is zero. This assures the nonimportance of the TE_{20} mode in the computation of $\Gamma(\theta, f)$; it is as if the TE_{20} mode were well below cutoff in the dielectric plug. The computed variation of Γ with θ, ϕ, and f is shown in Fig. 5;

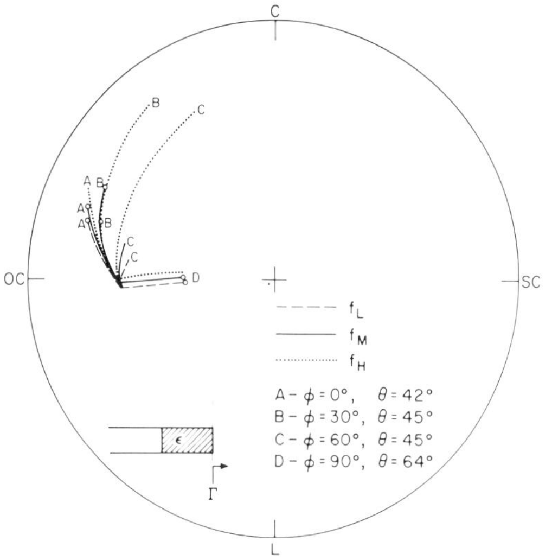

Fig. 5 **Computed Γ at aperture for element with lattice and aperture dimensions as shown in Fig. 4 and filled with a dielectric plug as described in text.**

by examining it, one determines the particular matching concept to be used. Matching structures are initially determined based upon their effect on the TE_{10} mode only, ignoring possible resonances and possible changes in Γ because of the effect of these structures on the TE_{20} mode. After the matching structures are so determined, their ef-

*One would usually keep grating-lobes beyond the edge of "real space." However, when this design was developed, the 5% rule mentioned earlier had not yet been determined; hence, the choice was made, for the purpose of this demonstration, to allow them very near endfire.

**All dimensions are given in free-space wavelengths at midband.

fect on the TE_{20} mode is included in the computations of performance and (generally) found to be small or negligible, except when the TE_{20} mode is placed near resonance by the matching structure. If this mode is near resonance and strongly affects the performance, the matching structure (and sometimes the matching concept) must be altered. If the inclusion of the TE_{20} mode has only a small effect on performance, the matching structure is modified slightly to reoptimize the performance. This procedure for matching the element in the presence of more than one propagating waveguide mode works because of the dominant role played by the TE_{10} mode in the determination of the reflection coefficient.

Notice that as the reference plane in Fig. 5 is moved further back in the waveguide, the Γ loci in the figure spread out more with frequency, since the high-frequency locus moves more rapidly around the reflection chart than the low-frequency locus. However, if the loci are normalized to a different impedance, so that they lie on the other side of the chart, moving back in the waveguide would cause the loci to come closer together or wrap up. This is the matching concept to be employed. The reduction in height of the waveguide and the ending of the dielectric plug (Fig. 4) effectively renormalize the Γ loci so as to place them on the opposite side of the chart from their usual location. The length of reduced-height waveguide then wraps up the loci. Finally, the increases in waveguide height to 0.299 move the loci into the center of the chart.

The equivalent circuit for the matching structures has been determined by analyzing the structures in waveguide via a modal analysis and numerically computing the performance parameters. In the case of the change in waveguide height near the aperture, the equivalent circuits for the TE_{10}, TE_{20}, and TE_{30} modes were determined so that interaction between the aperture and this step could be included for these modes. It was found that such interaction in the TE_{30} mode was negligible, as expected, since this mode is cutoff in the plug. There is interaction in the TE_{20} mode, but this mode is not important except near its resonance; nevertheless, it is important to properly include this mode to ensure that it is not near resonance.

The final matched $\Gamma(\theta, f)$ is shown in Fig. 6. Note that the VSWR is generally less than 3:1 except for a small range of scan angles near the endfire-grating-lobe condition.* The figure also shows that the impedance is especially well matched near broadside scan; a different compromise could have reduced the reflection somewhat for the peripheral scan angles at the expense of a greater reflection near broadside.

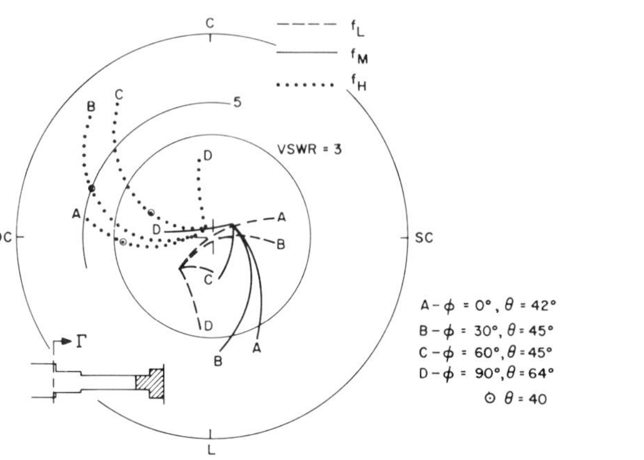

Fig. 6 Computed Γ versus θ, ϕ, and f for element in Fig. 4.

Figure 7 is a comparison of the computed $\Gamma(\theta, f)$

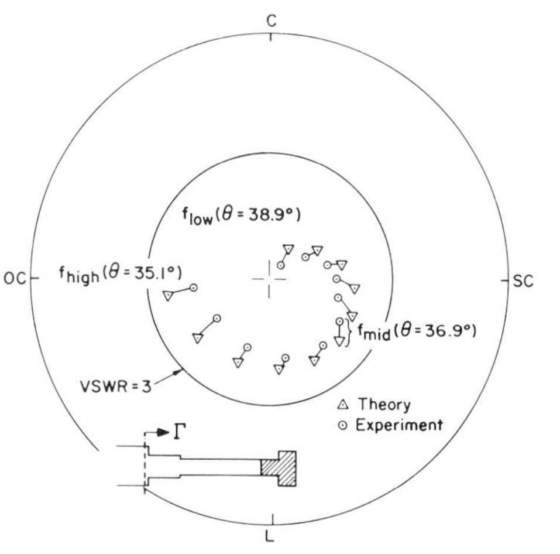

Fig. 7 Comparison of computed Γ with that measured in an H-plane simulator for the element shown in Fig. 4.

*As pointed out earlier, it was not the objective of this study to optimize the element matching. It is expected that a peak VSWR near 3:1 could be attained with this design, if the 5% rule on element area were to be used.

and that measured in an H-plane waveguide simulator like the one on the left side of Fig. 1.* Again we observe close agreement, although not quite as close as that in Fig. 3, perhaps because this element is more complicated than the one displayed in Fig. 3. This agreement confirms the computed design shown in Figure 4 and gives additional confidence in the computed performance presented in Figure 6.

Before leaving this section, it is perhaps useful to point out a few features of the stepped-dielectric-plug matching structure which we have used. First of all, the plug had a length of one-half loaded-guide wavelengths at the center frequency in order to properly locate the admittance (versus scan angle and frequency) loci on the Reflection Chart when normalized to the empty guide admittance. The waveguide width was adjusted to give a very high susceptance (approaching a short circuit) for the TE_{20} mode admittance "looking back" into the matching structure while maintaining the same electrical length for the TE_{10} mode. In order to satisfy these two conditions in other similar designs, the optimum width of the loaded waveguide will depend on the dielectric constant of the loading material, on the relative height of the step in the plug, and on the width of the empty waveguide following the plug. (For high-dielectric-constant materials, this width may considerably exceed that of the loaded waveguide to permit propagation of the dominant mode). After the loci on the reflection chart had been "wrapped up" by the stepped dielectric waveguide and the length of empty reduced-height waveguide, additional steps in waveguide height and/or width were used to center the loci on the reflection chart and to transform to the required waveguide cross-section. Sometimes this can be accomplished with one step; however, it is more likely to require two steps (as in our design) or perhaps a centering step followed by a quarter-wave transformer (three steps). Unfortunately, no straightforward, step-by-step design procedure has been formulated for this type of matching structure, so every change of parameters requires a new computer design by iterative methods.

IV. DISCUSSION

The new element-design procedure discussed here relies on computations to determine $\Gamma(\theta, f)$ rather than on measurements as do the small-array procedure and the simulator procedure. The matching network may be determined by computation or by measurement, as mentioned; in the example presented, the entire design was done by computation so as to emphasize the difference between this procedure and the other procedures. Measurements were made only after the design was completed, and only in a single simulator, to confirm the computed design.

It seems to us that this new design procedure offers substantial advantages over the usual procedures, and that one should make use of it wherever possible. In fact, one should probably choose his element so as to permit use of the computational design procedure. This imposes no substantial restriction on the design, as pointed out earlier. There is still sufficient flexibility in the choice of structure and materials to accommodate nearly any design requirements. The value of this element design procedure has already been recognized at several companies where it has been in use for the past year or two. However, to the best of our knowledge, this is the first detailed description of the procedure and its advantages and limitations relative to the two usual element design procedures. In addition, this paper and another paper in this Proceedings (18)are the first published examples of radiating elements designed with this procedure.

There may, of course, occur from time to time certain unusual cases for which it is necessary or desirable to use a non-computable element. In such cases, one must revert to either the small-array procedure or the simulator procedure and rely on measurements to determine $\Gamma(\theta, f)$ and to design the matching network.

Although only linearly-polarized elements have been discussed in this paper, the computational procedure can be used for the design of circularly-polarized elements as well. Details of such a procedure will not be given here, since they can be deduced from those of the linearly-polarized element design procedure without much difficulty. The principal difference is that for a circularly-polarized element one must be concerned with the element transmission phase as well as the reflection in each of two orthogonal linear polarizations.

We have described the computational design procedure in this paper for waveguide elements exclusively. However, it should be applicable to dipole elements as well. What is needed is a theoretical solution to the dipole-array problem which includes the effects of the stubs supporting the dipoles over the ground plane, since these could play an important role. Once such a solution is available, the same simplification in element design can be realized for arrays of dipoles.

---

*Actually, the data shown in Fig. 7 is for slightly different dimensions than those in Fig. 4 because of an error in construction of the simulator.

V. REFERENCES

1. N. Amitay, P.E. Butzien, and R.C. Heidt, "Match optimization of a two-port phased array antenna element," IEEE Trans. Antennas and Propagation, vol. AP-16, pp. 47-57; January 1968.

2. C.O. Hemmi, W.F. Hayes, and R.C. Voges, "Practical impedance matching of a phased array radiating element," Proc. IEEE, vol. 56, pp.1963-1967; November 1968.

3. B. L. Diamond, "Small arrays -- their analysis and their use for the design of array elements," this Proceedings.

4. H.A. Wheeler, "A systematic approach to the design of a radiator element for a phased-array antenna," Proc. IEEE, vol 56, pp. 1940-1951; November 1968.

5. H.A. Wheeler, "A survey of the simulator technique for designing a radiating element," this Proceedings.

6. P. W. Hannan and M.A. Balfour, "Simulation of a phased-array antenna in waveguide," IEEE Trans. Antennas and Propagation, vol. AP-13, pp. 342-353; May 1965.

7. H. A. Wheeler, "The grating-lobe series for the impedance variation in a planar phased-array antenna," IEEE Trans. Antennas and Propagation, vol. AP-14, pp. 707-714; November 1966. (Correction, vol. AP-15, p. 235, March 1967).

8. G.G. Charlton, "Application of the grating-lobe series for calculating admittance variation of a phased-array antenna," IEEE Trans. Antennas and Propagation, vol. AP-15, pp. 818-820; November 1967.

9. B. L. Diamond, "Resonance phenomena in waveguide arrays," IEEE GAP International Symposium Program and Digest, pp. 110-115. October 1967.

10. G. F. Farrell, Jr., and D. H. Kuhn, "Mutual coupling in infinite planar arrays of rectangular waveguide horns," IEEE Trans. Antennas and Propagation, vol. AP-16; pp. 405-414; July 1968.

11. N. Amitay and V. Galindo, "The analysis of circular waveguide phased arrays," The Bell System Technical Journal, vol. 47, pp. 1903-1932, November 1968.

12. G. V. Borgiotti, "Modal analysis of periodic planar phased arrays of apertures," Proc. IEEE, vol. 56, pp. 1881-1892; November 1968.

13. B. L. Diamond, "A generalized approach to the analysis of infinite planar array antennas," Proc.IEEE, vol. 56 pp. 1837-1851; November

14. W. H. Schaedla, "Equivalent circuit formulation for an array of phased waveguide apertures," IEEE Trans. Antennas and Propagation, vol. AP-18, pp. 28-33; January 1970.

15. G. H. Knittel, "The Design of a radiator element for a phased array antenna - a new and efficient procedure," NEREM RECORD 1969, pp. 158-159; November 1969.

16. N. Wong, R. Tang, S.W. Lee, and W. R. Jones, "Multimode phased array element for wide scan angle impedance matching," this Proceedings.

17. W.S. Gregorwich, A. Hessel, G. H. Knittel, and A. A. Oliner, "A waveguide simulator for the determination of a phased-array resonance," IEEE GAP International Symposium Program and Digest, pp. 134-141, September 1968.

18. G. A. Arredondo, P.E. Butzien, and R.W. Humes, "The design of a wide-band, wide-scan angle waveguide radiating element," this Proceedings.

Wide-Angle Impedance Matching of Phased-Array Antennas - a Survey of Theory and Practice

GEORGE H. KNITTEL

Polytechnic Institute of Brooklyn
Graduate Center
Farmingdale, New York

ABSTRACT

During the past 10 years, many schemes have been proposed for wide-angle impedance matching (WAIM) phased-array antennas. These schemes are designed to minimize the loss in the array realized gain because of the variation of the array reflection coefficient with scan angle.

The various proposed WAIM schemes can be conveniently separated into two categories depending on their location relative to the array "electrical interface," separating the free-space region from the transmission-line region. Substantial improvement in WAIM has been demonstrated for some of the proposed schemes; this is illustrated by "before" and "after" plots on reflection charts. The most useful schemes in practice are conducting baffles in the H-plane of a dipole array, a thin high- ϵ dielectric sheet in front of the aperture plane either in a dipole array or in a waveguide array, and a dielectric slab adjacent to the aperture plane of a waveguide array scanned only in its H plane.

The conflicting statements in the literature regarding the possibility of perfect WAIM in principle are examined. It is concluded that in an infinite phased array one can, in principle, approach arbitrarily close to perfect WAIM, but not attain it exactly.

Areas for additional WAIM research and development are outlined and some desirable objectives for further WAIM study are given.

I. INTRODUCTION

Because of mutual coupling between the radiating elements of a phased-array antenna, the active admittance of the array varies substantially with scan angle and with frequency. During the past decade, a great deal of attention has been given to the problem of matching the array admittance to that of the transmission lines or waveguides feeding the elements, over a wide scan range. It is desirable to obtain the best possible matching over the required range of scan angles and frequencies in order to maximize the array realized gain (the power gain with respect to the power <u>available</u> from the generators). In addition, good matching reduces the harmful effect of aperture reflections on amplifiers which may be used behind each radiating element; possible reradiated multiple reflections are also minimized by sufficiently matching the array.

There are two common ways to view and examine phased-array antennas. These are shown in Fig. 1 for the case of infinite array models. On the

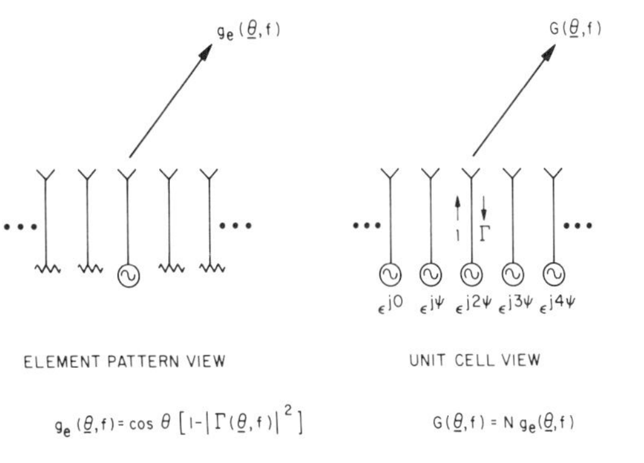

Fig. 1 Two ways to view phased-array antennas.

left side of the figure is the "element-pattern view" in which one thinks in terms of the power or field radiated versus angle from broadside, $\underline{\theta} = \hat{x}\theta_x + \hat{y}\theta_y$ and frequency, f, when only one element is excited and all the others are terminated in their generator admittances.* On the right side of the figure is the "unit-cell view" in which one thinks in terms of the array reflection coefficient, Γ, and the array realized gain, G, versus scan angle from broadside, $\underline{\theta}$, and frequency, f. The unit-cell view is for the case in which all the elements are excited with equal-magnitude generators and with linearly-progressive phase. The two views are related through the equations shown in the figure, via superposition (N is the number of elements). The unit-cell view will be used throughout, except where it is explicitly stated otherwise. The unit-cell view is the one most nearly representative of the array performance during array system operation.

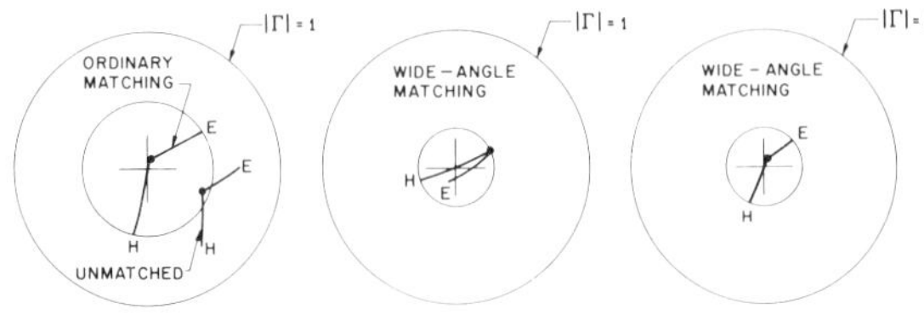

Fig. 2 Definition of wide-angle impedance matching (WAIM); E = E-plane scan, H = H-plane scan.
(a) Ordinary matching with scan-independent structures in element lines.
(b) WAIM by moving the scan loci closer with a scan-dependent matching structure.
(c) WAIM by reducing the size of the scan loci with a scan-dependent matching structure.

The array reflection coefficient, $\Gamma(\theta,f)$ ** is that seen at the array aperture in the mode which carries the power from the generator to the element aperture (this is usually the dominant mode of the element transmission line or waveguide). The reflection coefficient thus defined is sometimes referred to as the active array reflection coefficient; it is related to the (active) array admittance by the usual relationship between load admittance, transmission-line characteristic admittance, and reflection coefficient. Although Γ depends on both scan angle and frequency, its variation with θ is usually much greater than its variation with f.*** This paper is restricted to a survey of the proposed techniques for reducing the variation of Γ with θ, i.e., wide-angle impedance matching (WAIM) an array antenna. It is usually assumed that the array is operating at a single frequency, and is designed to radiate linear polarization. Very little discussion is given here concerning the operation of arrays over wide bandwidths or with circular polarization.

In Fig. 2 are three sketches which define what is meant by wide-angle impedance matching. Figure 2(a) shows a typical Γ-versus-scan locus for a planar array scanned in its E plane and in its H plane. The locus is labeled "unmatched" since it is what one would observe at the array aperture plane. With the addition of identical, scan-independent matching structures in each element line (ordinary matching), one can only design for some compromise match (1, 2), as shown. However, by the addition of a scan-dependent matching structure, one can change the size and/or the shape of the Γ locus, as shown in Figs. 2(b) and 2(c), and thereby substantially improve the wide-angle match of the array. This procedure is known as WAIM, and is the principle subject of this paper.

The benefit from WAIM can be significant. For example, a typical planar array designed to scan throughout 1/3 hemisphere of space at one frequency can be impedance matched with ordinary matching so that $|\Gamma| < 0.55$ (VSWR < 3.5). The maximum loss in transmission for such an array because of mismatch is about 1.6 dB. With an appropriate WAIM scheme, one could expect to reduce the maximum transmission loss to roughly half, or to 0.8 dB, in practice. This is equivalent to increasing the antenna gain by 0.8 dB or to enlarging the aperture area by about 20%, in its effect on system sensitivity.

In the following sections is a list of all of the published WAIM schemes, and (where possible) a comparison of the improvement in WAIM realized. Some of the schemes (those reported before 1966) have been mentioned in a previous survey. (7) These schemes are included here, as well, together with a discussion in much greater depth. Data pertinent

*The terminology throughout will be in terms of generators feeding the elements, and the array performance from the transmission approach (except where explicity stated otherwise). There are, of course, comparable performance parameters for the array in reception, by reciprocity.

**For simplicity, the bar under θ will be dropped from here on; it is to be understood that $\theta = |\theta|$ and that $\Gamma(\theta,f)$ is in general different for different scan planes, at the same value of θ.

***This is the case for a planar array required to scan throughout, say, 1/4 hemisphere over a 10% frequency band.

to these schemes has been replotted, so as to be presentable on a reflection chart for ease of comparison. Most of the theoretical studies surveyed here are based on the model of an infinite array antenna. The experimental results have been obtained in some cases with an infinite array (with the use of waveguide simulators)* and in other cases with a finite array (by the construction and measurement of an actual small array).

Section II of the paper lists all of the schemes proposed for WAIM in the literature. These are conveniently divided into two categories, based upon their location relative to the array "electrical interface", which is defined. A brief discussion of the schemes which are lossy, or for which no substantial WAIM benefit has been demonstrated is given in this section.

Section III discusses in detail those lossless schemes for which a substantial WAIM benefit has been demonstrated, either theoretically or experimentally. A plot of the "before" and "after" array performance is presented for most of the schemes, and a quantitative measure of the match obtainable is given. In each case, an attempt is made to explain the concept behind the scheme, and to put forth a simple physical view with which one can predict the WAIM benefit.

In Section IV, the contradictory statements in the literature concerning the question of perfect WAIM in principle are examined. It is concluded that in an infinite phased array one can in principle approach arbitrarily close to perfect wide-angle match, but never attain it. The question of whether it is possible, in principle, to approach or attain perfect WAIM is important since it bears on whether effort should be expended to find better practical WAIM schemes.

Section V discusses the most practical WAIM schemes and the schemes which are likely to be narrow band. The areas for additional research and development are outlined there and some desirable objectives for further WAIM study are given.

II. SCHEMES PROPOSED FOR WIDE–ANGLE IMPEDANCE MATCHING (WAIM)

It is useful to define an "electrical interface" in phased-array antennas, as indicated in Fig. 3 , which separates the free-space region from the waveguide or transmission-line region. In arrays of waveguides, the electrical interface is composed of the portions of the aperture plane which include the open ends of the waveguides. In arrays of dipoles, the electrical interface is composed of the terminals of the dipoles. In each case, the electrical interface separates two regions for which different modal representations of the fields are appropriate; in neither case is the electrical interface identical with a physical interface.

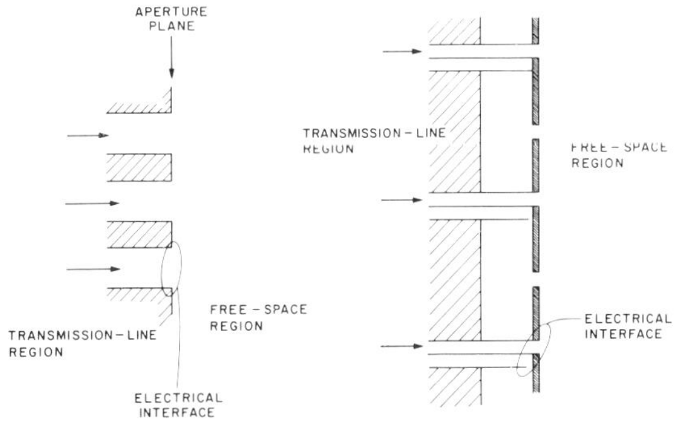

Fig. 3 Definition of array electrical interface
(a) Waveguide array (b) Dipole array

The definition of an electrical interface in phased arrays is useful since all of the published WAIM schemes can then be conveniently classified under two types, those which modify the transmission-line region on the one side of the interface (3-14), and those which modify the free-space region on the other side of the interface (15-33). Further discussion of the various schemes will make use of this classification.

A. TRANSMISSION–LINE–REGION WAIM SCHEMES

There are five distinct transmission-line-region WAIM schemes proposed in the literature. They are as follows.

1. Connecting Circuits (3-6, 8).
   Susceptances or transmission lines are used to interconnect the element transmission lines, thereby introducing scan-dependent susceptances which can partially cancel the variation in the array admittance with scan angle.

2. Dielectric-Loaded Waveguides (9, 12)
   The waveguides in a waveguide array are loaded with dielectric plugs, or homogeneously with dielectric. There is usually only a single propagating mode in the waveguide. The effect of the dielectric on the characteristic admittances of the higher-order cutoff waveguide modes excited at the aper-

*See the excellent survey paper by H. A. Wheeler in this Proceedings.

ture plane is used to improve the wide-angle impedance match.

3. Multimode Waveguides (10, 11, 13, 14).
The waveguides in a waveguide array are designed to propagate more than one mode. The excitation of a higher-order mode is adjusted so that the admittance seen by the higher mode will partially cancel the admittance seen by the dominant mode.

4. Electronically - Tunable Matching (7)
A device is inserted into each element transmission line which introduces an electronically-controlled susceptance. The susceptance is varied with scan angle to effect a wide-angle match.

5. Lossy Matching (7).
A circulator with a matched load is used between the radiating element and its generator or receiver. Reflected power is dissipated in the load.

Of the five schemes, the only lossless ones for which a substantial WAIM benefit has been demonstrated are Connecting Circuits and Multimode Waveguides; hence a detailed discussion of these schemes will be deferred to Section III. Some further comments on the other three schemes are given below.

The reflection coefficient versus scan angle has been computed by Galindo and Wu (9) for arrays of parallel-plate waveguides scanned in one plane and loaded homogeneously with dielectric or with dielectric plugs.* (The plugs extend from the aperture plane back a short distance into the waveguides.) From the results reported by Galindo and Wu, it is clear that the dielectric loading does indeed have a WAIM capability. However, there is generally insufficient data to permit a clear "before" and "after" comparison of the array match or to determine the optimum loading, since that was not a principal objective of their study.

In another paper, Amitay and Galindo (12) have shown an improvement in wide-angle match for an array of circular waveguides in a triangular lattice, scanned in the H-plane. Two cases are compared (their Fig. 5): the case with an air-filled 0.69λ diameter waveguide, and the case with a dielectric-filled (relative permittivity 4.1) 0.34λ circular waveguide. Although the latter case has almost no variation of Γ with θ out to the vicinity of the endfire-grating-lobe condition, an evaluation of the improvement in WAIM is not possible because the former case exhibits a resonance ($|\Gamma| = 1$) before the endfire-grating-lobe condition. However, the work shows that reducing the element size (and loading with dielectric so as to propagate the dominant waveguide mode) can improve the wide-angle match of the array in the H-plane.

The improvement in WAIM found by Amitay and Galindo is probably due partly to the dielectric loading, and partly to the reduction in element size. For small elements, the difference between the isolated element pattern and the ideal $\cos\theta$ power pattern is a measure of the array mismatch. A very small waveguide has an H-plane isolated element pattern which is that of a magnetic dipole in its H-plane ($\cos^2\theta$), and which is narrower than the ideal pattern. A larger waveguide would have an even narrower H-plane isolated element pattern. Hence reducing the waveguide size would reduce the array mismatch versus scan angle for H-plane scan.**

Electronically-tunable matching, although a legitimate WAIM scheme, is probably unwarranted. It requires an additional active device in each element line, thereby substantially increasing the phased-array cost and decreasing its reliability.

Lossy matching usually takes the form of a circulator with a terminated port inserted between the radiating element and its generator. Reflections from the element are dissipated in the circulator termination, and the generator sees a good match at all scan angles and frequencies. This method of treating the symptom rather than the disease does elminate the undesired effect of the mismatch on the generator, but it does not improve the reduced array realized gain. An alternative form of lossy matching utilizes series padding between the element and its generator to reduce the effect of mismatch on the generator. However, this

*If only one waveguide mode is propagating, and if the higher waveguide modes excited at the aperture plane can not see the rear of the plug, the WAIM potential of plugs and homogeneous loading is the same. This is because the higher modes see the same admittance in either case, and because the dominant-mode admittance $Y(\theta, f)$ is independent of the admittance seen in the dominant mode looking toward the generator from the aperture plane.

**The opposite is true for E-plane scan. A very small waveguide has an E-plane isolated element pattern which is that of a magnetic dipole in its E-plane (constant in θ), and which is therefore broader than the ideal pattern. A larger waveguide would have a narrower E-plane isolated element pattern. Hence reducing the waveguide size would generally increase the array mismatch versus scan angle for E-plane scan.

causes a substantial additional loss in the transmitted power, and is therefore generally unacceptable.

D. FREE-SPACE REGION WAIM SCHEMES

There are five distinct free-space-region WAIM schemes. They are as follows.

1. Baffles Between Dipole Elements (15,16). Conducting baffles are inserted between the elements in the H-plane of an array of dipoles over a conducting plane. The baffles substantially reduce the variation in $\Gamma(\theta, f)$ for E-plane scan without affecting that for H-plane scan, thereby improving WAIM.

2. Dielectric Slab Adjacent to Aperture Plane (9,18). A dielectric slab is placed on the aperture plane in a waveguide array to improve the wide-angle match for H-plane scan. Reflection from the free-space-slab interface is used to partially cancel the array aperture reflection versus scan angle. Alternatively, one might think of the slab as broadening the H-plane isolated element pattern.
3. Thin High - ϵ Dielectric Sheet (19-25). A thin high-permittivity dielectric sheet is positioned a short distance in front of the aperture plane in a waveguide array. The sheet acts as a capacitive susceptance which varies with E-plane scan in a way different from its variation with H-plane scan. This scan-dependent susceptance partially cancels that of the array aperture.

4. Loaded Ground Plane (26-29). The ground plane (conducting plane) between the element apertures in a waveguide array is appropriately modified or "loaded." This produces a non-zero tangential E field where the ground plane was, thereby weakening and reducing the period of the field in the aperture plane. The weakened and reduced periodicity weakens the array grating lobes and moves them further from "real space" thus improving WAIM.

5. Closely- Spaced Radiating Elements (9,30-32). The elements of a waveguide array or a dipole array are spaced more closely than is usual. The closer spacing moves the grating lobes further from real space thus reducing the variation of the array susceptance with scan angle and reducing $\Gamma(\theta, f)$.

All of these schemes are lossless, and it is possible to demonstrate a substantial improvement in WAIM for each one; hence a detailed description of each will be given in the next section.

III. SCHEMES WITH A DEMONSTRATED WAIM BENEFIT

Some of the schemes proposed for WAIM have been shown, theoretically or experimentally, to provide substantial benefits. These are discussed in detail in this section (lossy schemes are excluded).

A. TRANSMISSION-LINE-REGION WAIM SCHEMES

In 1963, Hannan, Lerner, and Knittel (3,5) proposed the use of connecting circuits between the element transmission lines to improve WAIM. Figure 4(a) shows the actual circuits (susceptances in this case) interconnecting the element lines. Both first-order circuits (B_1) and second-order circuits (B_2) are shown in an infinite linear array. The actual circuits produce an identical equivalent scan-dependent susceptance in each element line as shown in Fig. 4(b) The first-order circuits produce one equivalent susceptance, $B_{eq,1}(\alpha)$ while the second-order circuits produce a second equivalent susceptance, $B_{eq,2}(\alpha)$, with a different α dependence. In a planar array, the connecting circuits may be used in different scan planes. The resulting equivalent scan-dependent susceptances in either a linear or planar array may be used to partially cancel the variation of the array admittance with scan angle. It is important to note that this is not the same as attempting to cancel the mutual coupling in an array, since that is an entirely different approach.

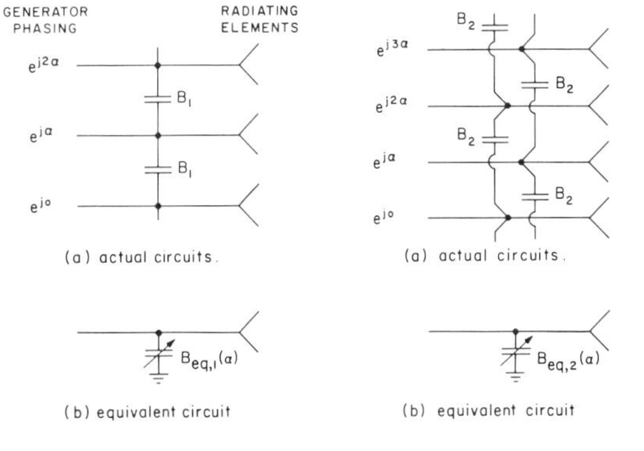

Fig. 4 Scan-dependent connecting circuit susceptances (from Hannan [35]).

The improvement in WAIM calculated by Hannan, Lerner, and Knittel is shown in Fig. 5. Single sets of first-order connecting circuits were used in the E-plane and in the H-plane of an array of dipoles over a conducting plane. For scanning within a $\theta_{max.} = 60^\circ$ conical volume, the array

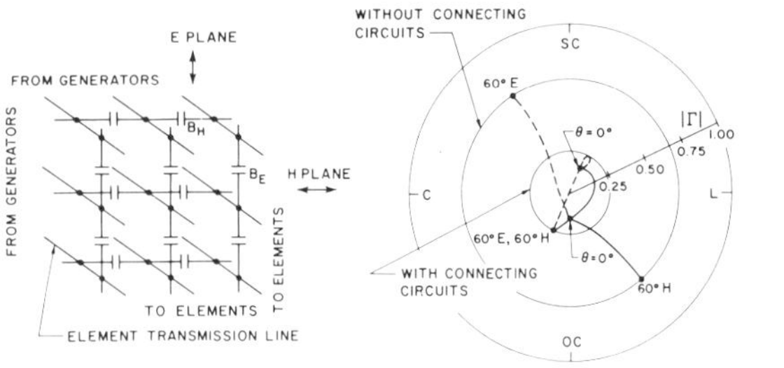

Fig. 5 Connecting circuits in an array of dipoles above a ground plane (from Hannan, Lerner, and Knittel [5]). (a) Actual interconnections (b) Γ versus sean angle

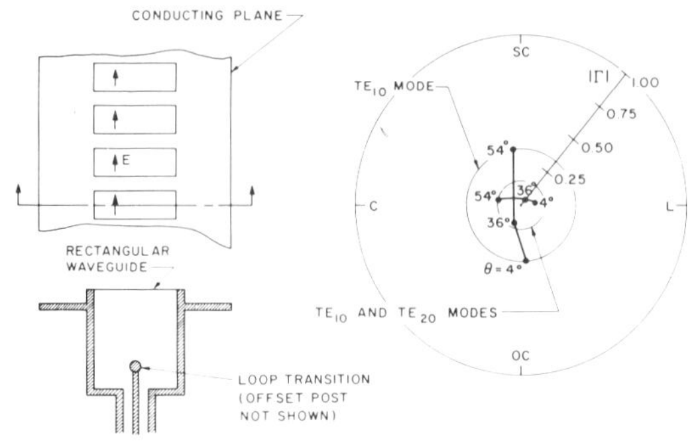

Fig. 6 Excitation of two waveguide modes in a linear array (from Stark, Tang, and Wong [11]). (a) Array (b) Γ versus scan angle

without connecting circuits could be matched with ordinary matching so that $|\Gamma| \leqslant 0.67$ (VSWR $\leqslant 5.1$). With the connecting circuits, and appropriate ordinary matching, the array could be matched so that $|\Gamma| \leqslant 0.26$ (VSWR $\leqslant 1.7$), a great improvement. Amitay (8) has published a systemic procedure for matching an array with connecting circuits.

In 1963 Cook and Pecina (4) proposed a variation of the connecting-circuit scheme which they called "compensation coupling." Their idea was to partially cancel the array mutual coupling by appropriate interconnections of the element transmission lines. Amitay, Pecina, and Wu (6) computed the benefit from this scheme. Element coupling coefficients were measured in a small array. The measured coupling coefficients between the center element and its immediate neighbors were analytically modified to minimize the power returned to the center element over a $\theta_{max.} = 45°$ conical scan volume.

The connecting-circuit scheme is applicable to any phased array since it does not rely on a particular element type. Connnections can be made between element feed lines which are either coaxial cables or waveguides. However, the scheme has never been demonstrated in practice. * It has never been used in an operational array because of the difficulty of interconnecting all of the element lines, and for this reason it will probably remain as only a theoretical technique (although a very useful one, as will be discussed later).

In 1968, Tang and Wong (10) presented a multimode-waveguide technique for WAIM. They worked with a linear E-plane array of open-ended waveguides in a conducting plane, as shown in Fig. 6(a). (The data in this figure are taken from Stark, Tang, and Wong (11).) Both the TE_{10} and TE_{20} modes are above cutoff in the waveguides. These modes are excited in an appropriate ratio by an offset inductive post so as to minimize the reflection-versus-scan locus seen by the dominant coaxial-cable mode at the input to the waveguide. This is accomplished by designing so the aperture reflections in the two modes partially cancel. The scheme was studied both theoretically and experimentally. The measured reflections are shown in Fig. 6(b) for the usual case in which only the TE_{10} mode is excited, and for optimum excitation of both the TE_{10} and TE_{20} modes. For the indicated scan range, $\theta \leqslant 54°$, the respective reflection coefficients are $|\Gamma| \leqslant 0.23$ (VSWR $\leqslant 1.6$), and $|\Gamma| \leqslant 0.09$ (VSWR $\leqslant 1.2$).

One of the principal undesired effects of this

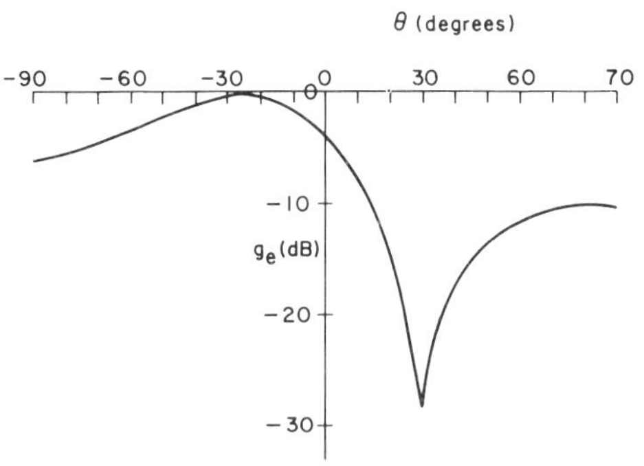

Fig. 7 Element-pattern null in H plane of linear array with TE_{10}, TE_{20}, multimode matching (from Stark, Tang, and Wong [11]). Scanning is in the E plane.

*A few measurements have been made in waveguide simulators at Wheeler Laboratories confirming the benefit, but these were not published (39).

WAIM scheme is the null in the H-plane element pattern, as shown in Fig. 7. The null and the pattern asymmetry are caused by the strong excitation of the TE_{20} mode in the waveguides. Depending upon the application, the null and asymmetry may or may not be important.

Multimode elements for WAIM in a planar array have recently been proposed by Wong, Tang, Lee, and Jones (13). Two designs described by these authors are shown in Fig. 8 (14). In both designs, a dielectric plug is used in which the TE_{10}, TE_{20}, and TE_{30} wave-

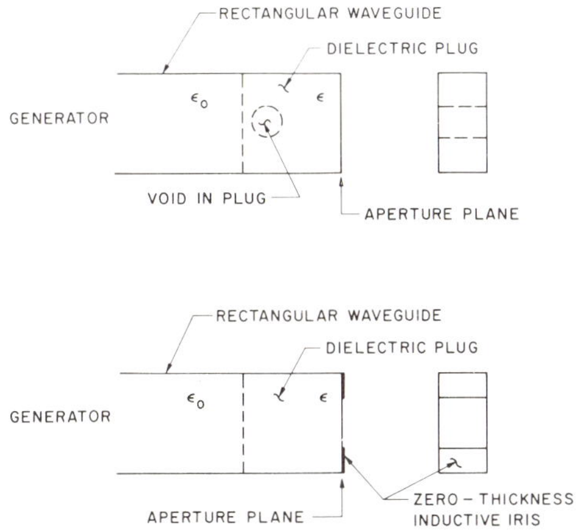

Fig. 8 Element design for use of TE_{10}, TE_{30} multimode matching in a planar array (from Wong [14]).
(a) TE_{30} mode excited by void in dielectric plug
(b) TE_{30} mode excited by iris at aperture plane

guide modes can propagate. In one element, a cylindrical void in the plug couples some of the TE_{10} mode power incident from the generator into the TE_{30} mode. The various parameters are then adjusted to minimize the reflection versus scan angle seen at the generator through a partial cancellation of the reflections which come from the TE_{10} and TE_{30} modes at the aperture plane. In the other element, a zero-thickness iris at the aperture is used to properly excite the TE_{30} mode, and the cylindrical void is unnecessary.

Measurements made with a 256-element planar array have shown that a wide-angle match of $|\Gamma| \leq 0.20$ (VSWR ≤ 1.5) is obtained for a conical scan volume with $\theta_{max.} = 47°$ for the cylindrical-void element. The same element has $|\Gamma| \leq 0.43$ (VSWR ≤ 2.5) for a $\theta_{max.} = 60°$ conical scan volume. A similar match has been computed for the iris element.*

B. FREE-SPACE-REGION WAIM SCHEMES

Edelberg and Oliner put forth a scheme as far back as 1960 which improves the WAIM in dipole arrays (15). It adds conducting baffles between the elements of the array in the H-plane, as shown in Fig. 9(a). In Fig. 9(b) is shown the benefit realized

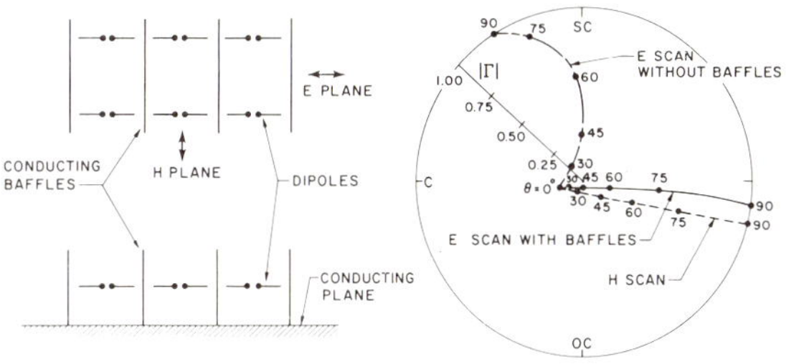

Fig. 9 Baffles in the H plane of a dipole array (from Edelberg and Oliner [15]).
(a) Array (b) Γ versus scan angle

with the baffles; this data has been determined from the formulas and graphs given by Edelberg and Oliner. If the impedance loci in Fig. 9(b) are centered on the reflection chart by the addition of ordinary matching in the element lines, one can realize $|\Gamma| \leq 0.45$ (VSWR ≤ 2.6) without baffles, and $|\Gamma| \leq 0.23$ (VSWR ≤ 1.6) with baffles, for a $\theta_{max} = 60°$ conical scan volume. This work of Edelberg and Oliner is limited to scan angles such that only the main lobe is radiating, and no grating lobe is near endfire. Hence it is valid for dipole spacings less than a half wavelength.

Conceptually, the H-plane baffles are helpful because they modify the E-plane element pattern of the dipole (which is much narrower than the ideal $\cos\theta$ pattern) to be more like the E-plane element pattern of waveguide (which is closer to the ideal pattern). The H-plane element pattern of the dipole is not affected by the baffles; that is what is

*An even smaller reflection, VSWR ≤ 1.67 over a $\theta_{max.} = 60°$ conical volume, has been computed for the iris element. (33) However in that case, the TE_{30} mode is below cutoff in the plug, and a dielectric sheet has been added a short distance in front of the aperture plane. It would seem that the dielectric sheet is therefore primarily responsible for this WAIM benefit (to be discussed later).

desired since it is like the E-plane pattern of a waveguide, closer to the ideal pattern. This reasoning is easily comprehended from the viewpoint of the isolated element pattern which, as mentioned earlier, is a useful indication of WAIM for small elements.

In 1964, Brennecke and Moule (16) experimentally confirmed the WAIM benefit of H-plane baffles in a dipole array. They measured the element coupling coefficients in a square-lattice 9x9 dipole array, with element spacing of about 0.56λ. They then computed the reflection coefficient in the center element versus scan angle for equal-magnitude voltages incident in the element lines. Their computations are plotted in Fig. 10; an endfire grating lobe occurs at $\sin\theta \approx 0.78$, and the main lobe and grating lobe are at equal angles from broadside for $\sin\theta \approx 0.89$. If the loci are centered on the chart for $\sin\theta_{max.} = 0.8$ for a conical scan volume by the addition of ordinary matching, then $|\Gamma| \leq 0.57$ (VSWR ≤ 3.7) without baffles, and $|\Gamma| \leq 0.36$ (VSWR ≤ 2.1) with baffles.

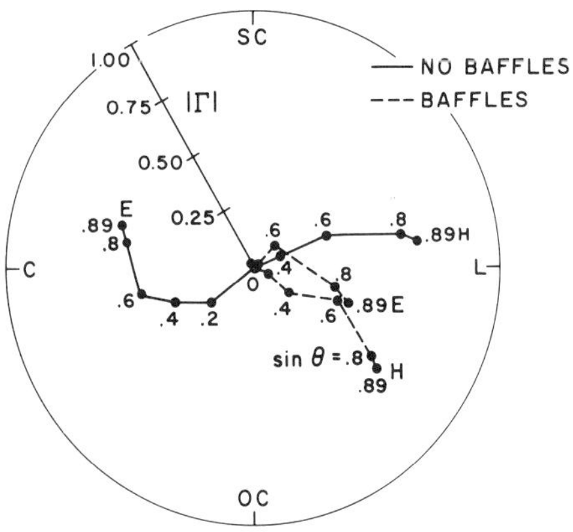

Fig. 10. Experimental confirmation of WAIM for baffles in the H plane of a dipole array (from Brennecke and Moule [16]).

It is clear from the preceding that H-plane baffles in a dipole array offer a substantial WAIM benefit both theoretically and in practice. Dufort (17) has studied the use of H-plane baffles for WAIM in a triangular-lattice array of rectangular waveguides. A sketch of the structure is shown in Fig. 15(b). By an appropriate choice of baffle height, and by spacing the elements much closer than is necessary for the avoidance of endfire grating lobes, he was able to realize $|\Gamma| \leq 0.23$ (VSWR ≤ 1.6) for a $\theta_{max.} = 45°$ conical scan volume. The close element spacing was necessary to avoid blindness or surface-wave effects which otherwise occur for H-plane scan. Brennecke and Moule(16) made measurements on this kind of waveguide array with H-plane baffles, but were unable to realize any WAIM benefit with usual element spacings. Hence, the WAIM potential of H-plane baffles in a waveguide array is not promising.

In 1968, Galindo and Wu (9) and Lee and Mittra (18) demonstrated the WAIM benefit from a dielectric slab adjacent to the aperture plane in a two-dimensional array of parallel-plate waveguides scanned in the H-plane. The waveguides are excited in their TE_1 modes. Figure 11 shows a case computed by Lee and Mittra for the waveguides spaced 0.45λ apart and filled with dielectric (relative permittivity 2.0). Without a slab, the array may be matched to $|\Gamma| \leq 0.37$ (VSWR ≤ 2.2) for $\sin\theta_{max.} \leq 0.95$ ($\theta \leq 72°$) with ordinary matching. With a slab of thickness 0.25λ and relative permittivity 2.0, and with appropriate ordinary matching, $|\Gamma| \leq 0.07$ (VSWR ≤ 1.15).

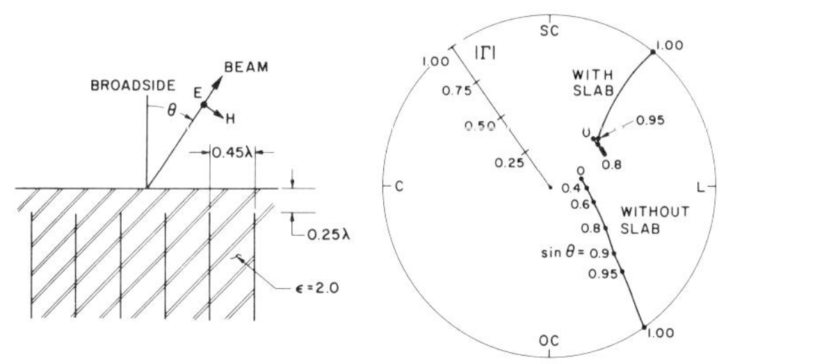

Fig. 11 Dielectric slab adjacent to the aperture plane in a parallel-plate array scanned in the H plane (from Lee and Mittra [18]).
(a) Array (b) Γ versus scan angle

This improvement with the dielectric slab has yet to be confirmed experimentally. However, it would seem that there is no reason it could not be confirmed; hence this scheme is probably of practical importance. It would no doubt work for arrays of rectangular or circular waveguides in a rectangular lattice, and probably for a triangular lattice as well, except near endfire-grating-lobe-conditions. It will not work well for E-plane scan since it then tends to cause resonances or large reflections. (34)

One of the most useful WAIM schemes was proposed by Magill and Wheeler in 1965. (19, 20) Their scheme utilizes a thin, high-permittivity dielectric sheet a short distance in front of the aperture plane in a phased array. The scan-dependent equivalent susceptance introduced by the sheet is used to partially cancel the admittance variation at the array aperture.

Figure 12 is a plot of the variation of the equivalent

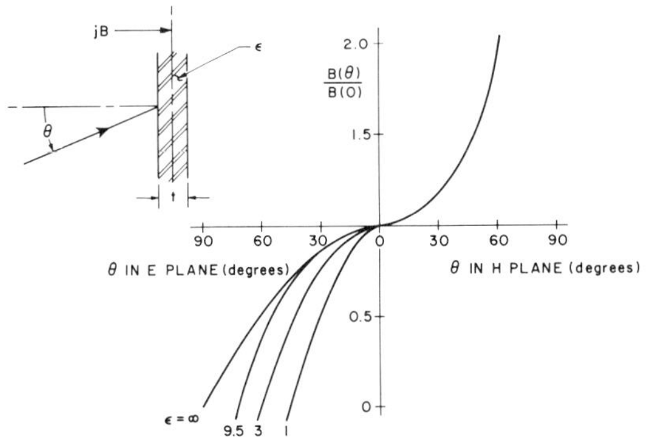

Fig. 12 Relative equivalent susceptance of a thin dielectric sheet for plane-wave incidence (from Magill and Wheeler [20]).

susceptance of a thin dielectric sheet (at its midplane) for a plane wave incident in the E-plane and in the H-plane. The susceptance is normalized to its value at normal incidence, and so the thickness of the sheet, t, drops out. Notice that the susceptance increases with angle of incidence, θ in the H-plane, but decreases with θ in the E-plane. Also, the variation is independent of ϵ in the H-plane, but dependent on ϵ in the E-plane.

The dielectric sheet is spaced a short distance in front of the array aperture plane as shown in Fig. 13.

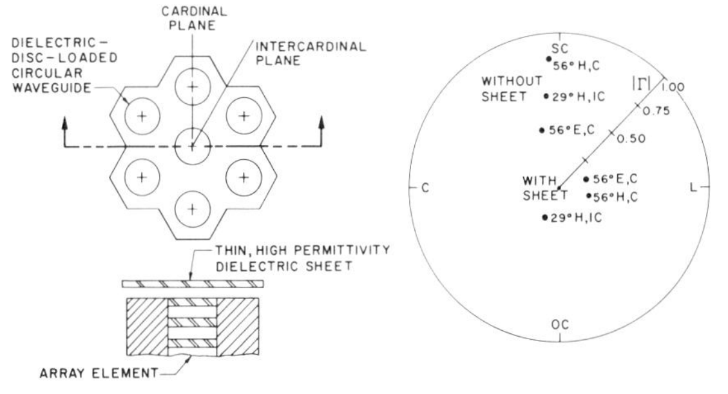

Fig. 13 WAIM benefit with a thin, high-ϵ dielectric sheet, from simulator measurements (from Magill and Wheeler [20]).

Waveguide simulators were constructed, representing scan angles in the cardinal plane with the H field in the scan plane (56° H, C) and with the E field in the scan plane (56° E, C) and in the intercardinal plane with the H field in the scan plane (29° H, IC). The array reflection coefficient was measured in the simulators without the sheet and with the sheet, and the results are shown in the figure.* If the impedance at the three scan conditions are centered on the reflection chart via ordinary matching, one obtains $|\Gamma| \leqslant 0.38$ (VSWR $\leqslant 2.2$) without the sheet, and $|\Gamma| \leqslant 0.20$ (VSWR $\leqslant 1.5$) with the sheet. Magill and Wheeler and Kelly (21) have, computed the WAIM benefit shown by the simulator measurements. Their analysis utilizes only a single mode or plane wave of the free-space region (the main lobe in a finite array) but they show good agreement with the measurements.

A disadvantage of the dielectric-sheet WAIM is that it tends to reduce the allowable scan range in the E-plane (and in scan planes for which a near-endfire grating-lobe condition is approached in a non-H-plane). A case in point has been computed by Schaedla (23) for an infinite array shown in Fig. 14. In the $\varphi = 61°$ scan plane, a grating lobe is at endfire at a scan angle $\theta \approx 70°$. Without the sheet,

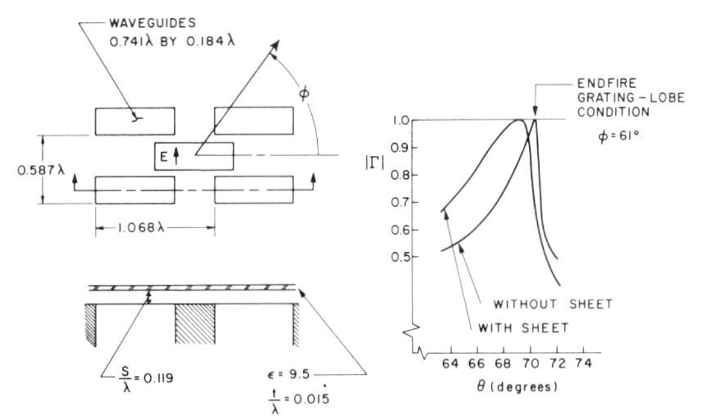

Fig. 14 Computation of a resonance caused by the dielectric-sheet WAIM (Courtesy of W. H. Schaedla and W. T. Patton [23]).

*Magill and Wheeler assume that the 29° H, IC condition is near broadside in performance. Even so, one would be concerned with both their other scan conditions only for a circularly-polarized design. Nevertheless, their data does give a quantitative measure of the WAIM benefit of a thin dielectric sheet.

$|\Gamma| = 1$ at the endfire grating lobe condition, as is usual. With the sheet shown, $|\Gamma| = 1$ at $\theta \approx 69°$ i.e., the sheet has caused a blindness or surface-wave effect at a scan angle somewhat before that for an endfire grating lobe. However, this effect is small for a thin sheet, and tends to zero as $t \to 0$ and $\epsilon \to \infty$, the product of ϵ and t remaining finite.* In the design computed by Schaedla, the element spacing was decreased somewhat so that the blindness effect occurred well beyond the maximum required scan angle. This made possible the WAIM benefit of the sheet over the entire design scan volume.

The benefit from dielectric-sheet WAIM has also been computed by Lee (22) for a two-dimensional array of parallel-plate waveguides scanned in the H-plane. In addition, this scheme has been utilized in a number of practical phased-array designs. (24, 25)

Dufort (1968) has proposed corrugating or loading the ground plane between the radiating elements to improve the WAIM of a waveguide array. (26, 28) The geometries studied by Dufort in this regard are shown in Fig. 15(a) and (c). In the two-dimensional parallel-plate array in Fig. 15(a), the waveguides are fed in their TEM

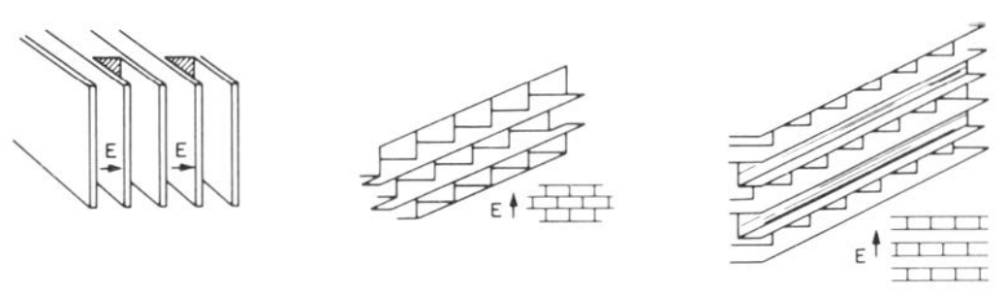

Fig. 15 Additional structures proposed for WAIM (from Dufort [17,26,28]).
(a) Parallel-plate array with one corrugation per period.
(b) Triangular-lattice array of rectangular waveguides with H-plane baffles.
(c) Triangular-lattice array of rectangular waveguides with H-plane baffles and one E-plane corrugation per period.

modes, whose angle of incidence to the aperture plane is variable so as to permit scanning throughout a volume in space; the plate thickness was zero. (26) Every

*This was pointed out by Wheeler and confirmed numerically by Schaedla. It could, perhaps, have been guessed from Fig. 12 from which one observes that $B(\theta) \to 0$ at $\theta = 90°$ E-plane incidence for $\epsilon = \infty$; i.e., that a zero-thickness sheet is invisible to a plane wave traveling along the array (an endfire grating lobe).

other waveguide is short circuited an appropriate distance from the aperture plane. This distance is adjusted (the ground plane is loaded) so as to improve WAIM. With appropriate internal matching; $|\Gamma| \leqslant 0.46$ (VSWR $\leqslant 2.7$) within a quarter of a hemisphere scan volume. A great deal of information is also given by Dufort regarding the WAIM capability of this scheme for various scan ranges. In each case, a grating lobe is allowed at endfire at one extreme of the scan range; the ground plane loading apparently eliminated the $|\Gamma| = 1$ condition usually associated with the endfire grating lobe in a non-H-plane.

For the structure in Fig. 15(c)*, the rectangular waveguides are fed in their TE_{10} modes and scanning is done throughout a volume of space (28). Endfire grating lobes are not permitted (the element area is reduced by 2% from the area which permits endfire grating lobes at the scan extremes). For scanning within a quarter hemisphere of space, $|\Gamma| \leqslant 0.5$ (VSWR $\leqslant 3.0$) with appropriate ordinary matching.

Hannan and Litt (27) and Hessel and Knittel (29) have also proposed modifying or loading the ground plane of an array, but for the specific purpose of eliminating the $|\Gamma| = 1$ condition usually associated with a non-H-plane endfire grating lobe and with a blindness or surface-wave effect. In both of these publications, clear "before" and "after" cases have been shown so as to make evident quantitatively the benefit realized. These are genuine WAIM schemes in that they permit scanning out to and beyond the endifre grating-lobe condition, which would be otherwise impossible because of large mismatch.

The loading scheme studied by Hessel and Knit-

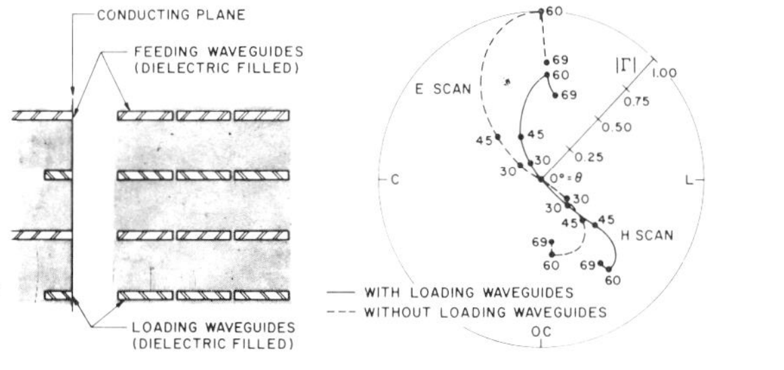

Fig. 16 Loaded ground plane to eliminate $|\Gamma| = 1$ condition (from Hessel and Knittel [29]).
(a) Array structure (b) Γ versus scan angle

*The WAIM scheme in this array is actually a combination of the loaded-ground-plane scheme and the H-plane-baffle scheme.

tel is shown in Fig. 16(a).* The feeding waveguides are excited in their TE_{10} modes; short-circuited loading waveguides are interspaced between the feeding waveguides as shown. The heights of the waveguides were made small so that a simple analysis, utilizing only the TE_{10} waveguide modes, would be valid. Without loading waveguides, $|\Gamma| \leq 1.0$ (VSWR $\leq \infty$) since the $\theta_{max.} = 60°$ conical scan volume includes a non-H-plane endfire grating lobe. With optimum loading from the loading waveguides, $|\Gamma| \leq 0.68$ (VSWR ≤ 5.2). In each case, appropriate ordinary matching is included to perfectly match the array for broadside scan. Figure 17 is a plot of the gain of the grating lobe rela-

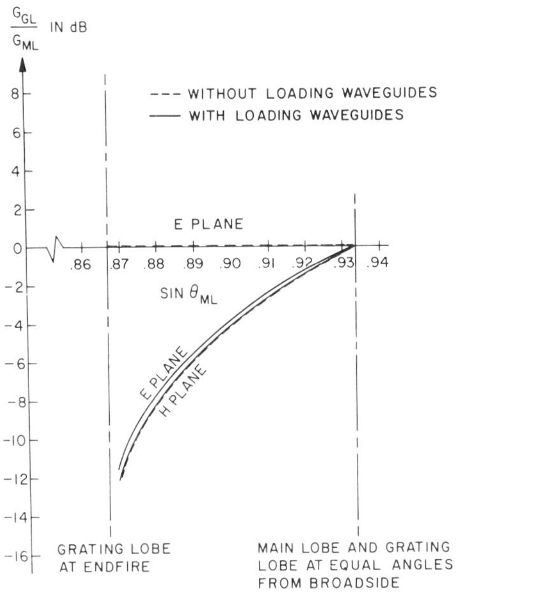

Fig. 17 Gain of grating lobe relative to that of main lobe for loaded-ground-plane array (adapted from Hessel and Knittel [29]).

tive to that of the main lobe for E-plane scan and for H-plane scan. Notice that the loading waveguides substantially reduce the gain of the E-plane grating lobe, thus permitting scanning out to or slightly beyond the endfire grating-lobe condition.

All of the loaded-ground-plane studies reported above were theoretical studies with an infinite-array model. No experimental confirmation of the concept has yet been attempted. Tolerances on the dimensions in a loaded-ground-plane design are likely to be quite critical, and the design may be relatively narrow band.

Perhaps the simplest WAIM scheme is more-closely spacing the radiating elements, as shown for a waveguide array in Fig. 18(a). Starting with an ordinary array of open-ended waveguides spaced to permit endfire grating lobes at the extremes of the scan range, the closer spacing of the elements

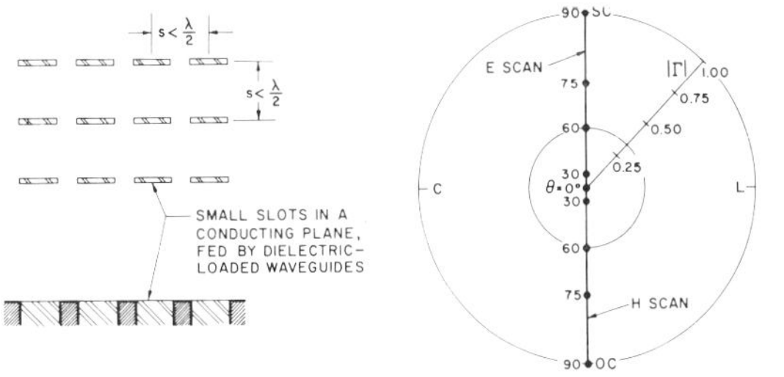

Fig. 18 Closely-spaced elements for improving WAIM (from Wheeler [30]).

monotonically reduces the variation of the imaginary part of the active array admittance with scan by moving the grating lobes further away from real space. In the limit of zero spacing, which also means infinitesimal elements, one has Wheeler's current-sheet array (30), the scan performance of which is shown in Fig. 18(b). The mismatch versus scan angle for this array is caused entirely by the variation of the conductive part of the active array admittance; the variation is now the same in the E-plane and in the H-plane.** For a $\theta_{max.} = 60°$ conical scan volume $|\Gamma| \leq 0.33$ (VSWR ≤ 2.0). The same result holds for a waveguide array in a ground plane or a dipole array in free space in either a rectangular or triangular lattice.

In practice, it is likely that one would want to decrease the element area by only 5% from that determined for endfire grating lobes at the extreme scan angles since the WAIM benefit comes slowly with closer spacing beyond that point. (31)

*This scheme is shown rather than that of Hannan and Litt since it has a more practical geometry. Theoretically, both schemes yield comparable performance.

**This is because the E-plane and H-plane isolated element patterns differ by the same amount from the ideal $\cos\theta$ pattern. Note that the WAIM could be improved by increasing the E-plane waveguide dimension somewhat, since this would narrow the isolated E-plane element pattern thereby bringing it nearer to the ideal pattern.

IV. POSSIBILITY OF PERFECT WAIM

There are conflicting statements appearing in the literature on the question of whether it is possible, in principle, to perfectly impedance match a phased-array antenna over a continuous scan range. Hannan (35) has stated that "... perfect impedance match can theoretically be obtained for all scan angles ..." for which the array reflection coefficient Γ is less than unity in magnitude. Varon and Zysman (36) have stated that "... although an array match can be considerably improved, a perfect match over a continuous scanning range cannot be obtained in general...". Both statements pertain to infinite linear arrays and to infinite planar arrays with a rectangular or triangular element lattice; the elements are fed with uniform amplitude and linearly-progressive phase from generators with an arbitrary internal impedance. Single-frequency operation is assumed.

Hannan's proof utilizes several sets of an infinite number of connecting circuits (see Fig. 4). Each set of connecting circuits is composed of an infinite number of circuits of all orders (only first and second order circuits are shown in Fig. 4). The scan-dependent susceptance $B(\alpha)$ introduced by each set is therefore composed of an infinite number of terms, each term being one component of a complete Fourier-series representation of the $B(\alpha)$ and corresponding to a particular-order circuit. Any required $B(\alpha)$, it is reasoned, is consequently realizable. Several $B(\alpha)$'s, corresponding to several sets of an infinite number of connecting circuits, are then positioned appropriately in the element lines to perfectly match the array impedance over all scan angles for which $|\Gamma| < 1.0$.

Varon and Zysman (36) have shown that for an array with smooth boundaries, the assumption of perfect impedance match over a continuous scanning range contradicts the differentiability requirement inherent in Maxwell's equations and that perfect match is therefore not possible. The periodic nature of the field in front of the array structure is used to show that for a perfectly matched array the aperture electric field is not of bounded variation and hence is not differentiable. In an alternative proof, Varon and Zysman (37) show for an infinite array of uniform, scan-invariant current sources radiating into an obstacle-free space, the active array impedance is an analytic function of the complex scan angle. They then demonstrate that perfect match over a continuous scan range (constant array impedance) contradicts the analytic properties of the impedance and that therefore perfect match is not possible.*

Although not done by Hannan, it can be proven that his connecting-circuit Fourier-series representations of the element-line susceptances, $B(\alpha)$'s, will converge pointwise to the desired susceptances as the number of connecting circuits is made infinite. Hence it follows that the array can be matched as closely as desired over the scan range for which $|\Gamma| < 1.0$. However, it is incorrect to actually pass to the limit of an infinite number of connecting circuits in each set, since the resulting perfect match then contradicts some of the necessary fundamental characteristics of the array fields and the array impedance, as shown by Varon and Zysman.

The degree to which perfect matching can be approached, in principle, by utilizing a large number of connecting circuits is illustrated for an infinite linear array in Fig. 19. In Fig. 19(a) is shown an assumed Γ-versus-scan locus on a reflection chart for a linear array with element spacing of one-half wavelength or less. The different generator phasing, α, correspond to various scan angles; α_0 is broadside scan, and α_4 is endfire scan (for which there is complete reflection of the incident power). With the addition of several sets of a sufficient number of connecting circuits, it is possible

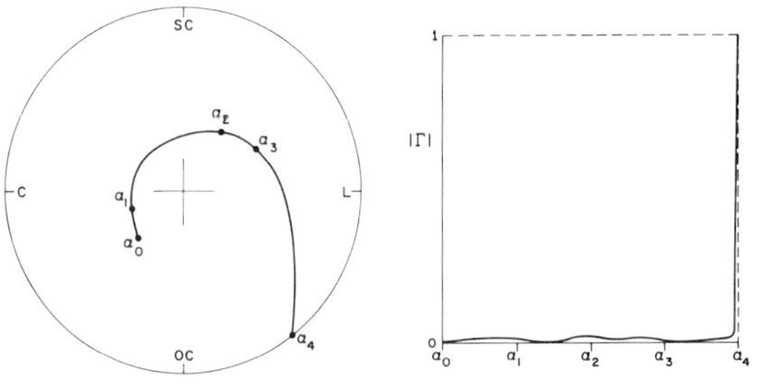

Fig. 19 Demonstration for a linear array, of near-perfect WAIM possible with a sufficient number of connecting circuits.
(a) Γ - versus-scan-angle locus before connecting circuits added.
(b) $|\Gamma|$ - versus-scan-angle locus possible with many connecting circuits.

*The assumption of scan-invariant current sources or forced excitation is equivalent to assuming infinite impedance generators. Since in an infinite array the array impedance is independent of the generator impedance, the proof is valid for finite-impedance generators or free excitation as well.

to obtain the $|\Gamma|$-versus-scan curve shown in Fig. 19(b). At α_4, there is still complete reflection since no amount of matching can change that. For $\alpha_0 \leqslant \alpha < \alpha_4$, $|\Gamma|$ is arbitrarily small, but never constant or equal to zero over any continuous range of α.

The fact that it is possible to approach perfect match arbitrarily closely but not to realize it exactly in phased arrays is consistent with similar phenomena in other disciplines. In network theory, for instance, it is well known that one can realize an arbitrarily-close impedance match over a continuous frequency band between a resistive generator and a (different resistance) resistive load, utilizing a network composed of a sufficient number of linear, passive, lossless, time-invariant structures. However to pass to the limit of perfect match, utilizing an infinite number of structures, contradicts the requirement that the network transfer function be an analytic function of the complex frequency (38). (This requirement is a consequence of the casual nature of the structures, which is itself a consequence of the linear, passive qualities of the structures.) In infinite phased arrays, the assumption of perfect match over a continuous scan range contradicts the requirements of analyticity of the array impedance, and differentiability of the array fields.

To summarize, Hannan's statement that perfect match can be obtained at all scan angles for which $|\Gamma| < 1.0$ is, strictly speaking, incorrect as Varon and Zysman have demonstrated. Nevertheless, Hannan's work shows that infinite phased arrays can be theoretically matched as closely as desired over all scan angles for which $|\Gamma| < 1.0$ (at a single frequency).* This is a significant result; it means that the search for a practical means to achieve near-perfect match over wide scan ranges can be pursued with optimism and expectation.

Before leaving this section, a brief review will be given concerning the question of perfect impedance matching a finite phased array over a continuous scan range. Varon and Zysman (37) have shown that in a finite array excited by infinite-impedance generators, the impedance seen by each generator cannot be perfectly matched over a continuous scan range using lossless, linear, passive, time-invariant matching structures. However, there is no discussion in their paper concerning the restriction of the conclusion to arrays with forced excitation (the array impedance in each element of a finite array does depend upon the generator impedances), or concerning the degree to which perfect match may be approached. It would seem that a finite array fed by finite-impedance generators (free excitation) can at least be matched as closely as desired from a consideration of known results for the two limiting cases, the smallest array (2 elements) and the largest (infinite) array.

*This is also Hannan's present position (39).

V. DISCUSSION

The subject matter covered in this paper has been limited to wide-angle impedance matching of phased-array antennas. In practice, of course, one is concerned with scanning throughout an angular range over a band of frequencies. Although no data has been presented showing the matching versus frequency achievable in phased arrays, some general comments regarding the frequency sensitivity of the various WAIM schemes can be made. It is likely that the performance of some of the schemes will be more sensitive to changes in frequency than that of others since some are highly-tuned or high-Q schemes. Those in this class are the ones with multimode waveguides or with a loaded ground plane.

For a practical phased-array antenna which must scan throughout a volume of space over a band of frequencies, the most useful WAIM schemes are probably closer element spacing, the thin high-ϵ dielectric sheet, and conducting baffles between the elements. An element area about 5% less than that which permits endfire grating lobes at the scan periphery is probably justified in every array design. The additional improvement in array performance for even closer spacing probably does not justify the additional number of elements required (31). The thin high-ϵ dielectric sheet should provide a substantial matching benefit for either waveguide arrays or dipole arrays with either a rectangular or a triangular-lattice. Conducting baffles between the elements in the H plane of a dipole array should also produce a substantial matching benefit in either a rectangular-lattice or a triangular-lattice array. (An array of dipoles above a ground plane utilizing all three of these schemes should be capable of substantially better performance (impedance matching) than what has been obtained to date).

For a practical phased-array antenna which must scan in only one plane but over a band of frequencies, the WAIM scheme consisting of a dielectric slab adjacent to the aperture plane of a waveguide array (H-plane scan) would seem to be an excellent choice. This scheme would likely also be useful for dipole arrays which are required to scan in only the E plane.

As a consequence of the survey given here, the areas for additional work on impedance matching phased-array antennas have become apparent. Suggestions for additional research and development are given below. It would be helpful if in all future publications clear "before" and "after" cases are determined so as to establish the real benefit of the scheme being proposed. Where this is not practical, a comparison of the matching obtained with known cases would be useful. A plot of the array reflection versus scan on a reflection chart (Smith chart) is usually of great value since it permits an immediate determination of whether the reflection locus is"centered", and what additional benefit ordinary matching might provide.

It would be desirable to extend the theoretical study of the conducting baffles to include many unit-cell modes, in order to evaluate its benefit in the vicinity of endfire grating lobes. A systematic theoretical study of the benefit from a thin high- ϵ dielectric sheet should be made. The existing data is rather sparse; computer programs necessary for the study are presently in existence. A way should be found to reduce the connecting-circuit scheme to practice and experimentally demonstrate its benefit. The improvement offered by a dielectric slab adjacent to the aperture plane and by the loaded-ground-plane scheme should be confirmed experimentally. The use of some of the WAIM shemes to optimize the radiated polarization in a circularly-polarized array, rather than to optimize the impedance match, could be examined.

Very little has been done on the design of phased arrays for extremely wide-band performance, i.e. for bandwidths with edge frequencies in the ratio of 2:1 or more. One publication, by Byron and Frank (32), discusses a design in which the elements are spaced closely enough to operate over a 2:1 bandwidth without forming radiating grating lobes; this means element spacings on the order of 1/4 free-space wavelength at the lower frequency. Better ways are needed for designing arrays for extremely wide-band operation. It would be helpful to find a way to eliminate the formation of grating lobes for both extremely wide-band operation and for ordinary 10% - 20% bandwidth operation. The important question of whether there is a fundamental limitation on the ability to theoretically wide-angle wide-band impedance match a phased array needs further examination. (It has already been demonstrated that there is no limitation on the ability to theoretically approach arbitrarily-closely perfect wide-angle match in an infinite array at one frequency.)

The theoretical study of perfect wide-angle matching an infinite phased array deserves more attention. The nature of the failure of Hannan's connecting-circuit model as the number of circuits becomes infinite could be examined. Computations could be made of the shape of the Γ-versus-scan locus as more and more circuits are added. A study should be made of the feasability of perfectly matching the impedance in an arbitrary finite array, including any difference in the conclusion because of forced excitation and free excitation.

Much has been published in the past 10 years on the determination and matching of the impedance of phased-array antennas. It would seem that there is still considerable matter for additional investigation in this fascinating field.

VI. ACKNOWLEDGEMENT

The research reported in this paper was sponsored by the U.S. Army Advanced Ballistic Missile Defense Agency under Contract No. DAHC 60-69-C-0110. Thanks are due to N. Amitay for reviewing the manuscript, and to D. C. Youla for discussions on the question of perfect wide-angle matching. The author is also indebted to H. A. Wheeler and P. W. Hannan for many helpful suggestions and comments.

VII. REFERENCES

(1) Amitay, N., Butzien, P. and Heidt, R.C.: Match Optimization of a Two-Port Phased Array Antenna Element, IEEE. Trans. Antennas and Propagation, vol. AP-16, no. 1, pp. 47-57, January 1968.

(2) Hemmi, C.O., Hayes, W.F., and Voges, R. C.: Practical Impedance Matching of a Phased Array Radiating Element, Proc. IEEE (Special Issue on Electronic Scanning), vol. 56, no. 11, pp. 1963-1967, November 1968.

(3) Hannan, P.W., Lerner, D.S., and Knittel, G.H.: Wide-angle Impedance Matching Calculated for a Phased-Array Antenna, 1963 IEEE Antennas and Propagation International Symposium Digest, pp. 228-233.

(4) Cook, J.S. and Pecina, R.G.: Compensation Coupling Between Elements in Array Antennas, 1963 IEEE Antennas and Propagation International Symposium Digest, pp. 234-236.

(5) Hannan, P.W., Lerner, D.S. and Knittel, G. H.: Impedance Matching a Phased-Array Antenna Over Wide Scan Angles by Connecting Circuits, IEEE Trans. Antennas and Propagation, vol. AP-13, no. 1, pp. 28-34, January 1965.

(6) Amitay, N., Pecina, R.G. and Wu, C. P.: Radiation Properties of Large Planar Arrays, BTL Monograph No. 5047, February 1965.

(7) Oliner, A.A and Malech, R.G.: Methods of Compensation in Microwave Scanning Antennas, vol. 2, Hansen, R.C., ed., New York: Academic Press, 1966, pp. 368-391.

(8) Amitay, N.: Improvement of Planar Array Match by Compensation Through Contiguous Element Coupling, IEEE Trans. Antennas and Propagation, vol. AP-14, no. 5, pp. 580-586, September 1966.

(9) Galindo, V. and Wu, C.P.: Dielectric Loaded and Covered Rectangular Waveguide Phased Arrays, The Bell System Technical J., vol. 47, no. 1, pp. 93-116, January 1968.

(10) Tang, R. and Wong, N.S.: Multimode Phased Array Element for Wide Scan Angle Impedance Matching, Proc. IEEE (Special Issue on Electronic-Scanning), vol. 56, no. 11, pp. 1951-1959, November 1968.

(11) Stark, L., Tang, R., and Wong, N.S.: Multi-mode phased Array Element for Scan Angle Compensation of Element Impedance, Alta Frequenza, vol. 38, 1969 (Special Issue), pp. 249-254.

(12) Amitay, N. and Galindo, V.: Characteristics of Dielectric Loaded and Covered Circular Waveguide Phased Arrays, IEEE Trans. Antennas and Propagation, vol. AP-17, no. 6, pp. 722-729, November 1969.

(13) Wong, N., Tang, R., Lee, S.W., and Jones, W. R.: Multimode Phased Array Element for Wide Scan Angle Impedance Matching this Proceedings.

(14) Wong, N.: Private Communication.

(15) Edelberg, S. and Oliner, A.A.: Mutual Coupling Effects in Large Antenna Arrays II: Compensation Effects, IRE Trans. Antennas and Propagation, vol. AP-8, no. 4, pp. 360-367, July 1960.

(16) Brennecke, N.R. and Moule, W.N.: Use of Fences to Optimize Operating Impedance of Phased Arrays, Using an Improved Measuring Technique, 1964 IEEE Antennas and Propagation International Symposium Digest, pp. 134-142.

(17) Dufort, E.C.: A Scattering Matrix Method for Solving Waveguide Array Impedance Problems, Radio Science, vol. 3 (New Series), no. 5, pp. 475-485, May 1968.

(18) Lee, S.W. and Mittra, R.: Radiation from Dielectric-Loaded Arrays of Parallel-Plate Waveguides, IEEE Trans. Antennas and Propagation, vol. AP 16, no. 5, pp. 513-519, September 1968.

(19) Magill, E.G. and Wheeler, H.A.: Wide-angle Impedance Matching of a Planar Array Antenna by a Dielectric Sheet, 1965 IEEE Antennas and Propagation International Symposium Digest, pp. 164-169.

(20) Magill, E.G. and Wheeler, H.A.: Wide-angle Impedance Matching of a Planar Array Antenna by a Dielectric Sheet, IEEE. Trans. Antenna and Propagation, vol. AP-14, no. 1, pp. 49-53, January 1966.

(21) Kelly, A.J.: Comments on Wide-angle Impedance Matching of a Planar Array Antenna by a Dielectric Sheet, IEEE Trans. Antennas and Propagation, vol. AP-14, no. 5, pp. 636-637, September 1966. Author's reply follows this communication.

(22) Lee, S.W.: Impedance Matching of an Infinite Phased Array by Dielectric Sheets, Electronics Letters, vol. 2, no. 10, pp. 366-368, October 1966.

(23) Schaedla, W.H. and Patton, W.T.: Private Communication.

(24) Stark, L.: Private Communication.

(25) Frank, J.: Phased Array Antenna Development, Technical Memorandum TG-882, Applied Physics Laboratory, The Johns Hopkins University, March 1967.

(26) Dufort, E.C.: Design of Corrugated Plates for Phased Array Matching, IEEE Trans. Antennas and Propagation, vol. AP-16, no. 1, pp. 37-46, January 1968.

(27) Hannan, P.W. and Litt, S.P.: Capacitive ground Plane for Phased Array Antenna, 1968 IEEE Antennas and Propagation International Symposium Digest, pp. 115-123.

(28) Dufort, E.C.: A Design Procedure for Matching Volumetrically Scanned Waveguide Arrays, Proc. IEEE (Special Issue on Electronic Scanning), vol. 56, no. 11, pp. 1851-1860, November 1968.

(29) Hessel, A. and Knittel, G.H.: A Loaded Ground Plane for the Elimination of Blindness in a Phased-Array Antenna, 1969 IEEE Antennas and Propagation International Symposium Digest, pp. 163-169.

(30) Wheeler, H.A.: Simple Relations Derived From a Phased-Array Antenna Made of an Infinite Current Sheet, IEEE Trans. Antennas and Propagation, vol. AP-13, no. 4, pp. 506-514, July 1965.

(31) Diamond, B.L. and Knittel, G.H.: Private Communication.

(32) Byron, E.V. and Frank, J.: On the Correlation Between Wide-Band Arrays and Array Simulators, IEEE Trans. Antennas and Propagation, vol. AP-16, no. 5, pp. 601-603, September 1968.

(33) Lee, S.W. and Jones, W.R.: On the Suppression of Radiation Nulls and Broadband Impedance Matching of Rectangular Waveguide Phased Arrays, Report No. FR 70-14-51, Hughes Aircraft Co., Fullerton, CA.,

(34) Knittel, G.H., Hessel, A. and Oliner, A.A., Element Pattern Nulls in Phased Arrays and Their Relation to Guided Waves, Proc. IEEE, vol. 56, no. 11, pp. 1822-1836, November 1968.

(35) Hannan, P.W.: Proof That a Phased-Array Antenna Can Be Impedance Matched for all Scan Angles, Radio Science vol. 2, (New Series), no. 3, pp. 361-369, March 1967.

(36) Varon, D. and Zysman, G.I.: On the Mismatch of Electronically Steerable Phased Array Antennas, Radio Science, vol. 3, (New Series), no. 5, pp. 487-489, May 1968.

(37) Varon, D. and Zysman, G.I.: Some Properties and Limitations of Electronically Steerable Phased Array Antennas, The Bell System Technical J., vol. 46, no. 9, pp. 1561-1586, September 1967.

(38) Youla, D.C., Castriota, L.J. and Carlin, H.J.: Bounded Real Scattering Matrices and the Foundations of Linear Passive Network Theory, IRE Trans. Circuit Theory, vol. CT- , pp. 102-124, March 1959.

(39) Hannan, P.W.: Private Communications.

The Design of a Wide Band Wide Scan-Angle Waveguide Radiating Element

G. A. ARREDONDO, P. E. BUTZIEN
and R. W. HUMES

Bell Telephone Laboratories, Inc.
Whippany, New Jersey

ABSTRACT

An efficient procedure for designing the radiating elements of a phased array over a large frequency band and wide scan angle is delineated with a design of circular waveguide type radiators. The key factor is to have a complete mathematical characterization of the element so that the active radiation impedance may be calculated accurately as a function of frequency and angle of scan. The element design, including the match, can then be done entirely on paper. The design presented here has been verified with an array waveguide simulator measurement.

I. Introduction

The design of the radiating element of a phased array has been done in the past using primarily two procedures, both of which are essentially experimental. One procedure is to construct a small test array using a tentative element design. The coefficients of coupling are then measured and the active impedance of the element is computed from these measurements as a function of frequency and scan angle. (1, 2) The other procedure is to use a small number of infinite array waveguide simulators (3) to measure as a function of frequency the active impedance of the element at a small number of scan angles. The "grating-lobe series" is then used to calculate the active impedance of the element at other scan angles. (4)

A more efficient procedure for designing the radiating element, in the sense that it does not rely as much on experimental data, is to calculate accurately the element's active radiation impedance, (5, 6, 7) and then to design a network to match the element to the feed line. (8) In this paper we illustrate a procedure for matching an array element over a wide frequency band and scan angle by describing the design of a circular waveguide radiator.

II. Element Specifications

The design objectives chosen are considered typical for a wide frequency band and wide angle coverage application. The element designed is a linearly polarized dielectric-loaded circular waveguide in an equilateral triangular grid array. The geometry of the array is shown in Figure 1, where the arrows indicate the sense of polarization. As can be seen in Figure 1, the element's E-plane is along a cardinal plane of minimum spacing in the array. The types of element, polarization, and array grid were not allowed to change.

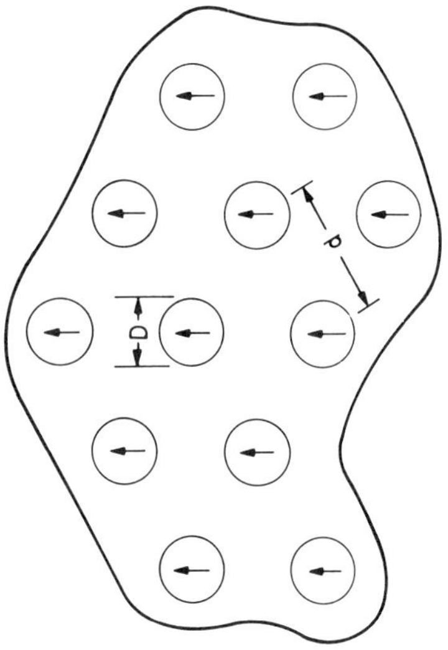

Fig. 1 Array Geometry.

The element was to operate over an 8 to 10 percent frequency band and a scanning cone of approximately 120° with the following specifications: 1) Average reflected power (as a function of frequency and beam direction) of 10 percent; 2) Peak reflected power of 35 percent; and 3) Polarization quality of, at least, 20 dB axial ratio.

One dominant TE_{11} travelling mode was excited in each waveguide. The parameters that were allowed to vary were the dielectric constant of the loading material, the diameter of the waveguide, and the inter-element spacing in the array grid. These variations were constrained to the use of permittivities of practical dielectrics. In addition, no multi-moding in the guides was allowed, and grating lobes were excluded from the 120° scanning cone over the frequency band.

III. Mathematical Characterization of the Element

The active array attributes were calculated using the procedure and computer programs developed by Amitay and Galindo. (5) They formulated the boundary value problem in terms of a vector two-dimensional integral equation for an infinite array of circular waveguide elements arranged in a doubly periodic grid along two skewed (non-orthogonal, in general) coordinates. This general geometry is shown in Figure 2. The formulation accounts for dielectric loading, plugs, sheaths, as well as thin irises at the aperture, but does not allow for obstacles, such as irises, inside the waveguides. (See Reference 5 for the details of the mathematical formulation.)

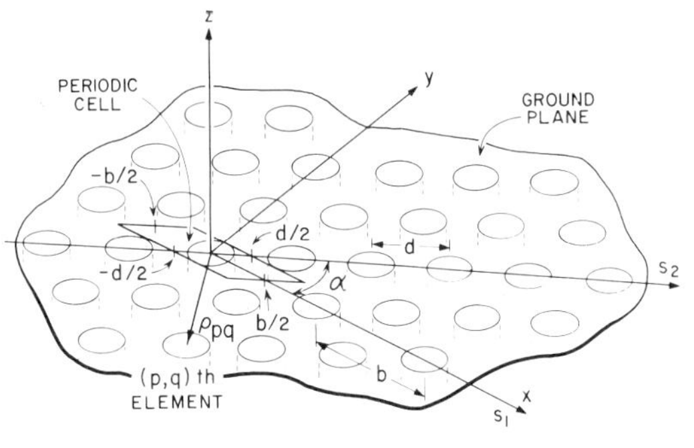

Fig. 2 General Circular Waveguide Array Geometry.

Numerical solutions are obtained by using the Ritz-Galerkin method to solve the integral equation. Amitay and Galindo's computer program was implemented in a time-shared computer. A graphics terminal was employed in an interactive mode which simplified the variation of parameters. Every useful solution was checked for convergence by varying the number of free space and waveguide modes used in the numerical solution. Some solutions were also tested in infinite array waveguide simulators and the agreement in reflection coefficient was found to be excellent. One set of measurements is given in Figure 8 and will be discussed later in this paper.

The dielectric constants of the loading materials used were 9.6, 8.9, 6.7, 4.1, and 2.0. For every dielectric constant the waveguide diameter was so chosen to minimize frequency dispersion, i.e., such that the highest frequency in the band fell just below the cutoff frequency of the next higher order mode.

Many configurations were tried using different types of dielectric loading and combinations of dielectric loading and dielectric plugs. We shall present here only the cases that were eventually matched to meet the specifications. All dimensions are normalized to the mid-band free space wavelength λo.

The results we shall show are for waveguides fully loaded with a 4.1 dielectric material and a diameter $D=0.3564\lambda o$. A spherical coordinate system is used in which θ is the angle off the array normal and ϕ is measured on the face of the array with respect to a plane parallel to the element's H-plane. For each frequency, the radiation admittance was generally found to occupy a fan-shaped area on a Smith chart as a function of pointing direction. The scan contour considered was a quarter of a 120° cone, i.e., starting at boresight ($\theta = 0°$, $\phi = 0°$) scan to 60° along the H-plane ($\theta = 60°$, $\phi = 0°$), then keeping $\theta = 60°$ constant scan in azimuth to the E-plane ($\theta = 60°$, $\phi = 90°$), and finally return to boresight along the E-plane.

The active radiation admittance referred to the aperture plane calculated for a 10% frequency bandwidth and an inter-element spacing $d = 0.6\ \lambda o$ (see Figure 1) is shown in Figure 3. Notice the rap-

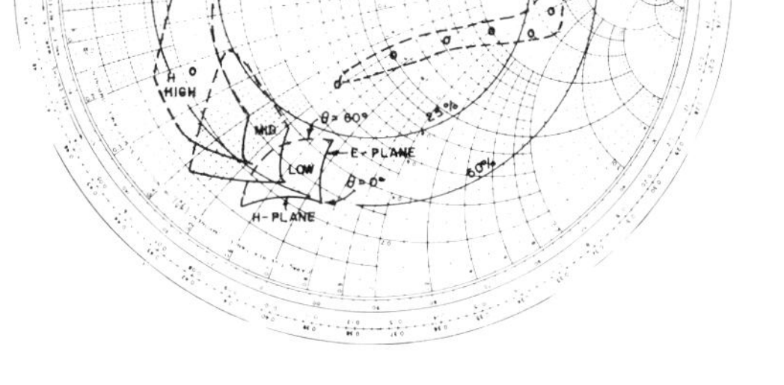

Fig. 3 Active Radiation Admittance Referred to Aperture Plane.

id phase variation in the upper half of the frequency band. This occurred for scan angles of $\theta = 45^\circ$ and above and azimuth cuts of $25^\circ < \phi < 45^\circ$. In addition, the orthogonal port reflection (note that the TE_{11} mode can have two orthogonal orientations) increased in this range so that the total returned power was very high. It would be nearly impossible to match this impedance with a lossless network.

Many variations of the available parameters were tried to eliminate the high total power mismatch and phase variation, but the most effective remedy seemed to be a decrease in relative element spacing. The case of $d = 0.558 \lambda_o$ spacing (3.71 elements per square wavelength as compared to 3.20 for $d = 0.6\lambda_o$ for which the performance improved considerably within the desired scan region is shown in Figure 4. The match for this configuration was obtained for different

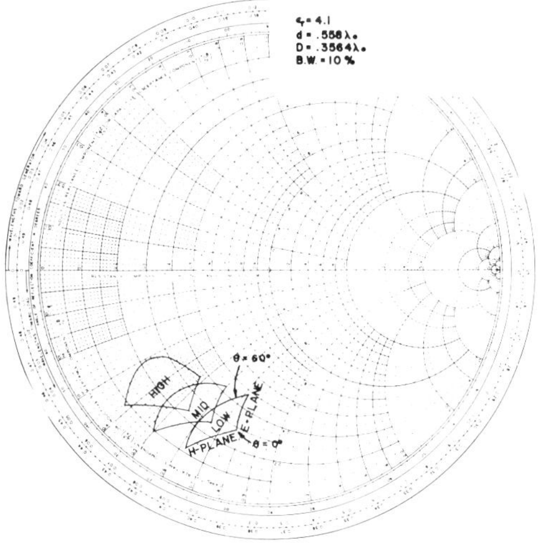

Fig. 4 Active Radiation Admittance Referred to Aperture Plane with Reduced Interelement Spacings.

scan regions and frequency bandwidth, as discussed in the next section.

The far-field element pattern was also calculated. The far-field polarization purity was found to be sufficient to satisfy the specification of at least a 20 dB axial ratio. The highest percentage of returned power coupled to the orthogonal TE_{11} mode was 3 percent.

IV. Matching Network

The practicality of any proposed design was considered to be a critically important aspect of this study. That is, since a large number of these elements would be required in any typical application, the cost is a very important factor and simplicity and low cost go hand in hand. For this reason, only those matching structures that were simple mechanically were considered. Furthermore, the matching network should be purely reactive for maximum efficiency.

The problem of broad-band matching can be treated as a network problem, i.e., match the load admittance given in Figure 4 to a 1 ohm (normalized) transmission line. It is in the nature of the boundary value formulation that if one inserts matching networks in one waveguide to match the active radiation admittance given by Figure 4, one is in fact inserting the same network in all the elements of the infinite array.

The active radiation admittance given in Figure 4 could be readily represented by various equivalent circuits, such as a length of transmission line terminated in a scan-dependent resistive load, or a shunt RLC circuit. Since we had calculated the radation admittance for all the angles and frequencies of interest, we chose to work graphically with all these data in a Smith Chart plot as shown in Figure 4.

Inspection of Figure 4 will show that the best wide-band match can be obtained by a combination of shunt capacitive susceptances. Our procedure was to separately match the center and edges of the band. First a match at mid-band was obtained with one shunt susceptance. Then the low and high edges of the band were matched with a network that was resonant at a frequency close to mid-band, so as not to change the match at mid-band. This procedure can be easily implemented in a computer.

The matching technique leads to many alternative networks. A typical network and the admittance as seen at the reference plane T for an 8 percent bandwidth and a conical scanning region of 60° half angle are shown in Figure 5, where λ_g is the guided wavelength and λ_{go} is that at the center of the band. The average reflected power is 7.6 percent, while the peak is less than 40 percent. For a conical scanning region of 45° half angle, this same structure will provide an acceptable match over a 10 percent bandwidth. This case is shown in Figure 6. A comparison of these admittances with that of the unmatched element illustrates the improvement provided by the matching networks.

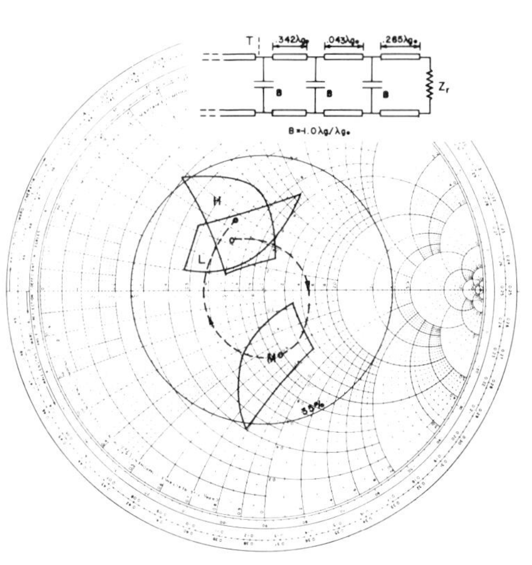

Fig. 5 Admittance of Matched Element Referred to Plane T for a Scan Cone of 120°. (BW = 8%)

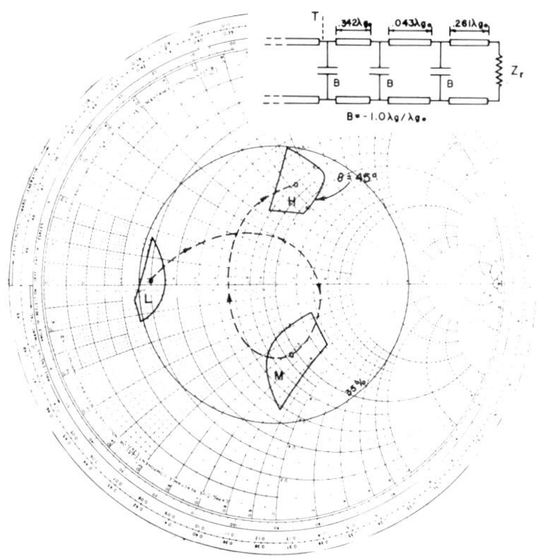

Fig. 6. Admittance of Matched Element Referred to Plane T for a Scan Cone of 90°. (BW = 10%)

V. Experimental Verification

In order to verify the calculations, a simulator was designed with the same array grid dimensions used in the determination of the admittance shown in Figure 4. This simulator is shown disassembled in Figure 7. It corresponds to a scan of approximately 30° from the array normal along the H-plane. (3) A circular waveguide-to-TEM coaxial line transition was also designed. This transition has a VSWR better than 1.15 across the 10 percent band of interest. All measurements were made in the coaxial line.

Figure 8 shows two measurements obtained in the simulator. The return loss without a matching

Fig. 7 Simulator Used for the Experimental Verification.

structure is given by curve A, and that with the matching structure is given by curve B. Also shown in this figure are a few calculated points. As can be seen, the agreement found is excellent.

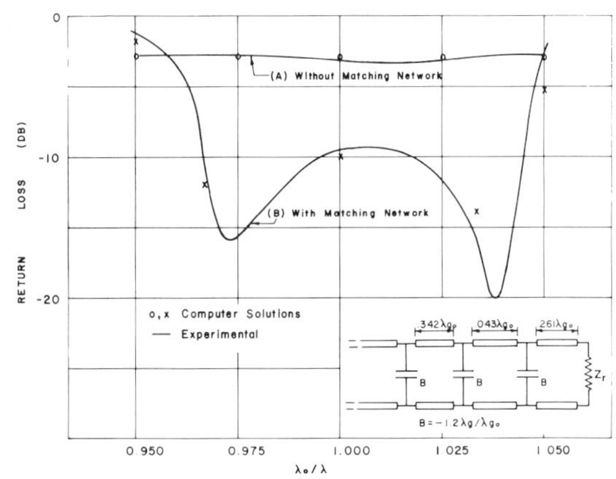

Fig. 8 Results of Simulator Measurements.

VI. Conclusion

A procedure for the design of wide-band, wide-angle radiators for a phased array antenna was illustrated with the design of a circular waveguide element. It was shown that having a complete mathematical characterization of the element allows the designer to choose the antenna parameters such that an acceptable match is readily found. The calculated match was verified for a few angles and a 10 percent frequency bandwidth with a simulator.

VII. Acknowledgements

The firm foundations built by Drs. N. Amitay, V. Galindo and E. Nagelberg and their helpful discussions are gratefully acknowledged. The authors also wish to thank G. W. R. Luderer and W. H. Ierley for their assistance in programming the interactive mode, E. F. Laundry for the design of the circular waveguide-to-coaxial line transition, and L. H. Hendler, J. W. Brown, Miss D. Czahla and Miss A. Russell for their help in many aspects of this work.

REFERENCES

1. Amitay, N., Butzien, P. E., and Heidt, R. C. "Match Optimization of a Two-Port Array Antenna Element," IEEE Trans., AP-16, 47 (1968).

2. Butzien, P. E., "Antenna Element Match Design for Planar Phased Array Antennas," 18th Annual Symposium, USAF Antenna Research and Development Program, Oct. 1968.

3. Hannan, P. W., and Balfour, M. A., "Simulation of a Phased-Array Antenna in Waveguide," IEEE Trans., AP-13, 342 (1965).

4. Wheeler, H. A., "The Grating-Lobe Series for the Impedance Variation in a Planar Phased Array Antenna," IEEE Trans., AP-14, 707 (1966)

5. Amitay, N., and Galindo, V., "The Analysis of Circular Waveguide Phased Arrays," BSTJ, 47, 1903 (1968).

6. Diamond, B. L., "A Generalized Approach to the Analysis of Infinite Planar Array Antennas," Proc IEEE, 56, 1937 (1968).

7. Borgiotti, G. V., "Modal Analysis of Periodic Planar Phased Arrays of Apertures," Proc IEEE, 56, 1881 (1968)

8. Knittel, G. H., "The Design of a Radiator Element for a Phased Array Antenna - A New and Efficient Procedure," 1969 NEREM Record, 11, 158 (1969).

Multimode Phased Array Element for Wide Scan Angle Impedance Matching

N. S. WONG, R. TANG, S. W. LEE
AND W. R. JONES

Hughes Aircraft Company
Ground Systems Group
Fullerton, California

SYNOPSIS

A method of achieving a matched aperture in phased arrays over a wide range of scan angle has been developed. This method has been shown to be effective for two dimensional phased arrays as well as for linear phased arrays. This basic concept involves the excitation of two or more waveguide modes in each radiating element such that the reflections of these modes from the array aperture produce a cancellation effect at the element driving terminal. The implementation of this impedance matching technique in linear arrays of open ended waveguide radiators has been successfully demonstrated (1). The technique has been extended to the two dimensional arrays in the present paper. In the case of the linear array, the dominant TE_{10} and TE_{20} modes are excited in the waveguide aperture by off-set inductive posts, and the input VSWR looking into each element of the linear array has been matched to less than 1.2 over a 60° scan angle. In the case of a two-dimensional array, the open ended waveguide radiators are dielectrically loaded in the immediate vicinity of the radiating aperture so as to support only the TE_{10}, TE_{20}, and TE_{30} modes. Among all the modes at the radiating aperture of the waveguide radiator, the feeding end of each element is unloaded and supports only the TE_{10} mode. Modal amplitude control of the higher order modes is accomplished by means of either a perforated dielectric plug or an inductive iris at the aperture, together with the appropriate choice of length of the dielectric plug (see Figures 1 and 2). The perforated dielectric plug approach was implemented in a two-dimensional array of 256 waveguide elements. The measured input VSWR for this experimental array is 1.5 or less over a scan angle of ±47 degrees from broadside in all scan planes. The input VSWR increases to 2.5 at scan angle ±60 degrees in the diagonal plane of scan. Analytical calculations show that the input VSWR for the experimental two dimensional array with an inductive iris possesses characteristics similar to the perforated dielectric plug where the dimensions of the waveguide element and overall dimensions of the dielectric plug are identical. Calculations also show that the input VSWR of the two dimensional array with

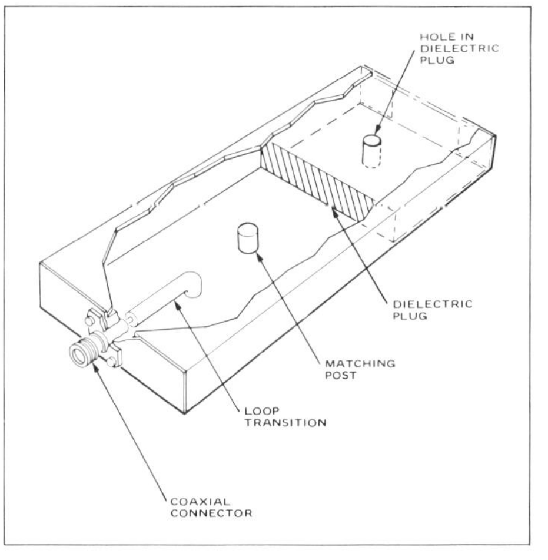

Fig. 1 A Multi-Mode Open Ended Waveguide Radiator with a Perforated Dielectric Slab.

inductive iris can be reduced to a maximum value of 1.67 over ±60 degree scan range, if the element spacing is reduced by 10%. It is expected that similar performance can be obtained with the perforated dielectric plug.

The methods of implementing the multi-mode phased array element described above are practical for most applications where linearly polarized rectangular waveguide radiators are used. When bandwidth is an important factor, the inductive iris method is preferred over the perforated dielectric plug method, because the reduction in transmission path length between the mode transducer and the waveguide aperture allows more broadband operation. In the event that circular polarization

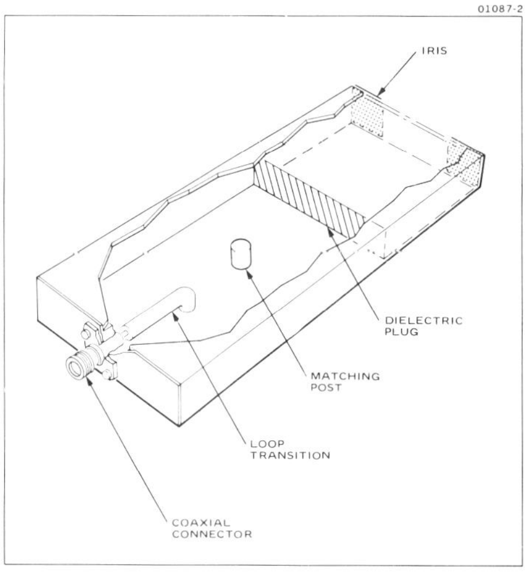

Fig. 2 A Multi-Mode Open-Ended Waveguide Radiator with an Inductive Iris at the Aperture.

is required, circular waveguide with inductive iris may be a possible method of implementing the multi-mode scheme.

I. INTRODUCTION

The recent increase in the application of phased arrays in radar and communication systems has stimulated interest in the study of impedance matching of array apertures over wide scan angles. Various methods have been advanced such as the placement of dielectric sheets in front of the array aperture (2), the introduction of coupling circuits between feed lines to the adjacent elements (3), and the use of baffles (4) and chokes (5). The basic idea behind these matching techniques involves either the modification of the mutual coupling between elements in the array or the modification of the radiation impedance directly. All of the above methods have met with varying degrees of success and all have the disadvantage that they are difficult and/or costly to implement or, as in the case of the first method, may lead to resonance nulls in the active element pattern of the array.

The technique described in this paper evolves from the concept of resolving the mutual coupling coefficients into two or more components with each component corresponding to that of a waveguide mode. One can synthesize the desired mutual coupling coefficients for a matched aperture by means of superpositioning two or more of these wave components. As a result of this superpositioning of modes, flexibility can be gained in the control of the aperture field in the open ended waveguide; therefore, this technique can also be identified as a method of improving the radiation impedance characteristics by means of adjusting the modal content of the various waveguide modes over the aperture of the radiating element.

II. DESCRIPTION OF THE EXPERIMENTAL ARRAY

An experimental two dimensional array has been designed and fabricated to demonstrate the multimode radiator design concept. In order to achieve better impedance match for the broadside beam, a dielectric sheet is first introduced in front of the aperture. The 256-element array is shown in Figure 3. The waveguide element with the perforated

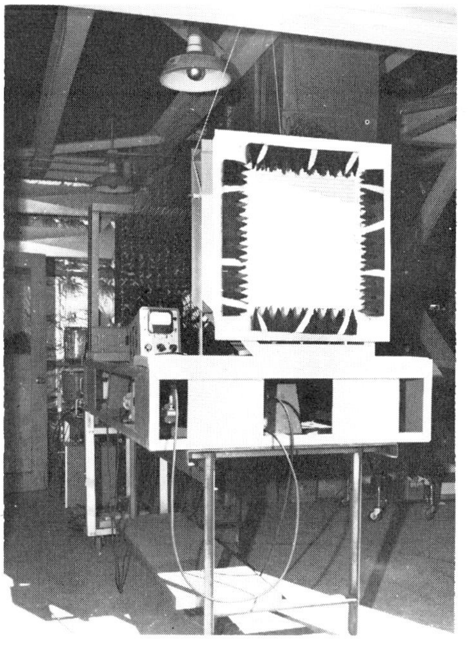

Fig. 3 Experimental Two-Dimensional Array of Open Ended Waveguides.

dielectric plug is shown in Figure 4. An inductive iris version of the element which has a broader bandwidth is shown in Figure 5. The entire array consists of 32 rows of elements with 8 elements in each row. The plexiglas sheet is located in front of the array aperture. The corporate feed network, which consists of 17 of the 16-element, strip-line, power dividers, is located in the back of the experimental array as shown in Figure 6. The input power to the feed network is distributed equally by the first 16-element power divider into the remaining 16 power dividers through a set of coaxial cables.

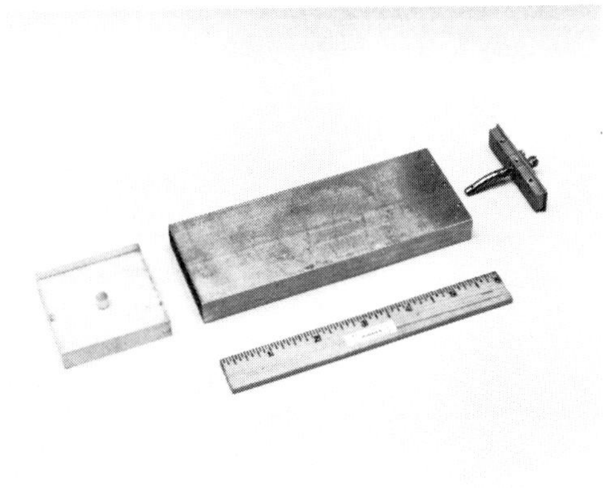

Fig. 4 Open-end Waveguide Radiating Element with Perforated Dielectric Slab.

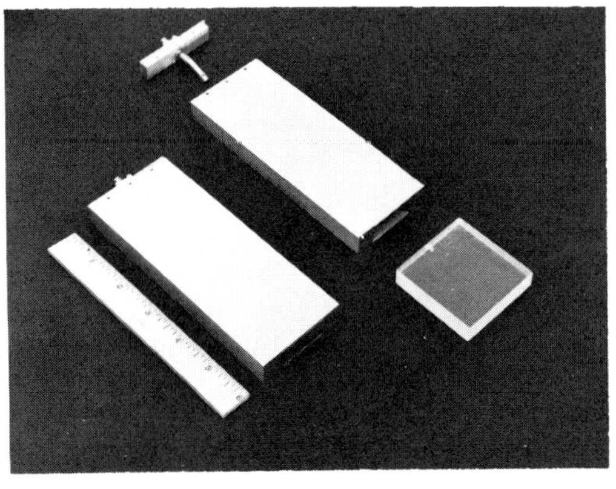

Fig. 5 Open-ended Waveguide Radiating Element with Inductive Irises.

The length of these coaxial cables can be selected to obtain the desired scan angles in a given scan plane. The outputs from the second level of power dividers are connected to the array elements through equal lengths of .141 inch semi-rigid coaxial cables. Impedance measurement on each radiator under test can be performed by replacing the coaxial cable to the radiator by another one of the same length with a slotted line as shown in Figure 6.

This experimental array can be used to perform impedance measurement on the unmatched array simply by omitting the dielectric insert. In this case, only the TE_{10} mode is incident on the waveguide aperture from the driving terminals.

Fig. 6 The Feed Network of the Experimental Two Dimensional Array.

III. IMPEDANCE CHARACTERISTICS OF THE UNMATCHED ARRAY

In order to show the improvement of the matching technique, the radiation impedance before matching was determined analytically and further checked experimentally.

The waveguide radiators were adjusted for the TE_{10} mode excitation and radiation impedance was measured. The input VSWR without dielectric plug is given in Figure 7, and with dielectric plug is given in Figure 8. In the first case, the dielectric plug is removed from the waveguide. In the latter case, the dielectric plug is as shown in Figure 1 except for the absence of the hole. In both cases, a dielectric sheet as shown in Figure 3 is placed in front of the aperture. Measurements of radiation impedance were performed by inserting a slotted line in front of the waveguide element as shown in Figure 6. An analysis (6) was performed for the computation of radiation impedance of open-ended rectangular waveguide arrays of fairly general geometrical configuration. This analysis was programmed for numerical computations and data were obtained for the above cases. The correlation

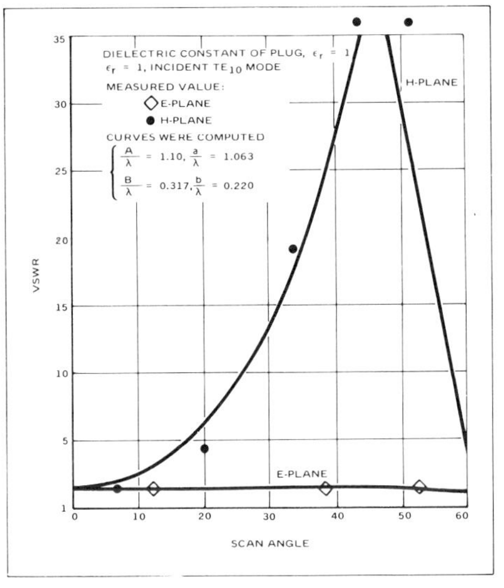

Fig. 7 Input VSWR of Two-Dimensional Waveguide Array (Actual Wall Thickness).

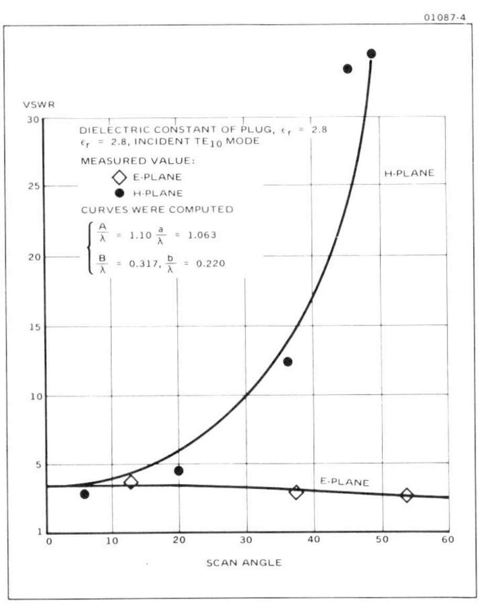

Fig. 8 Input VSWR of Two-Dimensional Dielectrically Loaded Waveguide Array.

of the measured points with the computed curves is quite good, cf. Figures 7 and 8. This comparison establishes confidence in the accuracy of the analytical result for multi-mode excitation. It should be noted that the radiation impedance varies very rapidly with scan in the H plane and a resonance null occurs at approximately 47° scan for the unloaded waveguide array. The resonance null is absent for the dielectrically loaded case although the impedance mismatch is also very large.

IV. RADIATION CHARACTERISTICS OF A MATCHED ARRAY (PERFORATED DIELECTRIC SLAB VERSION)

A. Radiation Impedance

The preliminary optimization of the mode transducer as shown in Figure 1 was performed experimentally. Waveguide simulators were developed to reduce the measurement time. Three waveguide simulators were constructed to simulate scan angles of 9°, 29°, and 40° in the H plane. Since impedance variation for the present array is small for the E plane scan, these preliminary measurements are not required for this scan plane.

The following steps were followed in the optimization procedure. The lengths of the dielectric plug and the thickness of the dielectric sheet (dielectric sheet in the front of the radiating aperture) were first selected to yield reasonable aperture match. Next, the diameter and location of the matching void in the dielectric plug were varied to yield the best impedance characteristics. Essentially, the hole in the dielectric plug couples the incident TE_{10} mode to the required TE_{30} mode with the appropriate amplitude and phase in order to yield the desired cancellation effect.

The experimental array was adjusted according to the findings from the waveguide simulators. The radiation impedance was measured and some minor discrepancy was found to exist between these data and those obtained in the waveguide simulators. After a slight trimming of the waveguide radiation element, data for the curves of the impedance plot as shown in Figure 9a for the E, H, and the diagonal plane respectively were measured. Each impedance point value is an average taken from 6 or more elements. The dashed curve connecting these points is an estimated curve of how the input impedance varies with scan. In general, the measured impedance points showed a VSWR less than 1.4 from broadside to 47° scan in the E, H and D planes, and it increases to 2.5 at 60° scan in the diagonal plane. For comparison purposes, the measured impedance before matching is shown in Figure 9b.

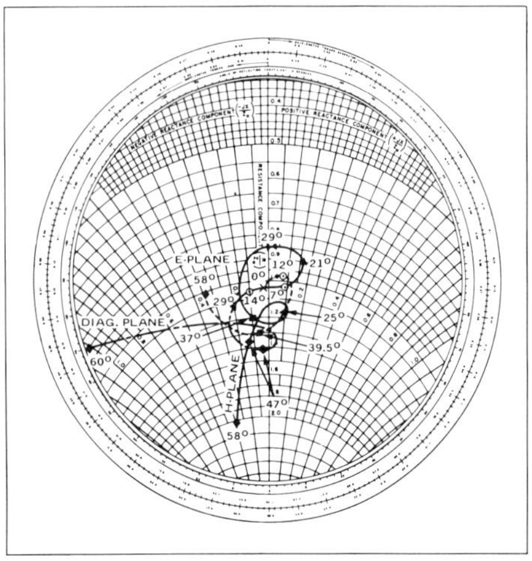

Fig. 9a Measured Impedance Characteristics of Radiating Element Shown in Figure 4.

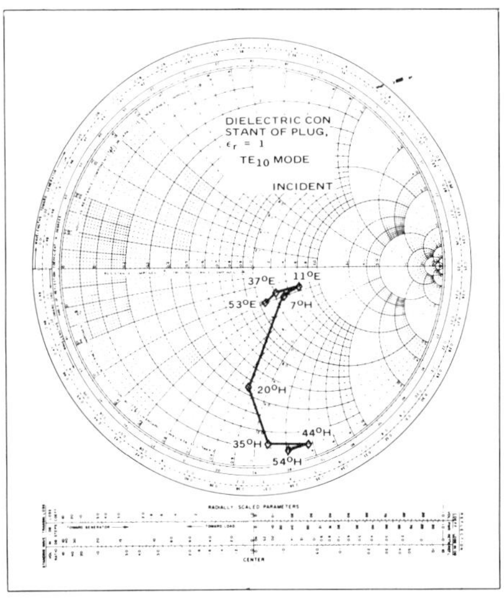

Fig. 9b Measured Radiation Impedance of Experimental Array with Incident TE_{10} Mode.

B. Active Element Pattern

The active element patterns of a matched element were taken along E, H, and D planes. The results are shown in Figure 10. The active element pattern sub-

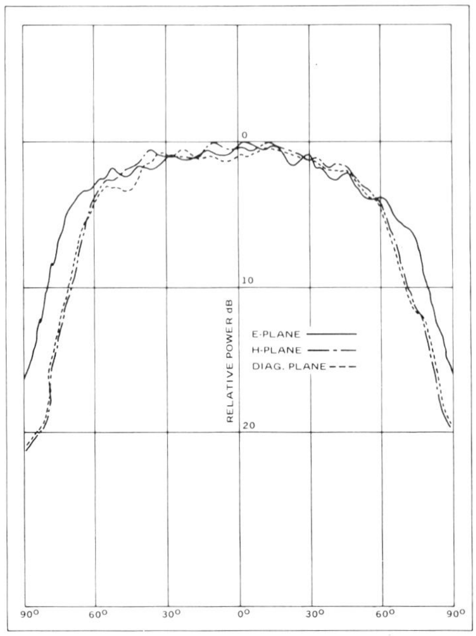

Fig. 10 Active Element Patterns of Radiating Element Shown in Figure 4.

stantiates the impedance data fairly well. The fine ripples in the patterns were caused by edge effects in the experimental array. These measured active element patterns resemble closely the ideal cosine pattern.

C. Array Patterns

Array patterns were taken for the 256-element experimental array. A typical pattern is shown in Figure 11. The array pattern serves the function of evaluating the effect of amplitude and phase errors across the array aperture. The deep nulls on both sides of the array main beam showed that the amount of phase error across the array aperture is very small. The measured peak sidelobe level is 13 dB, indicating that the amplitude distribution is fairly uniform.

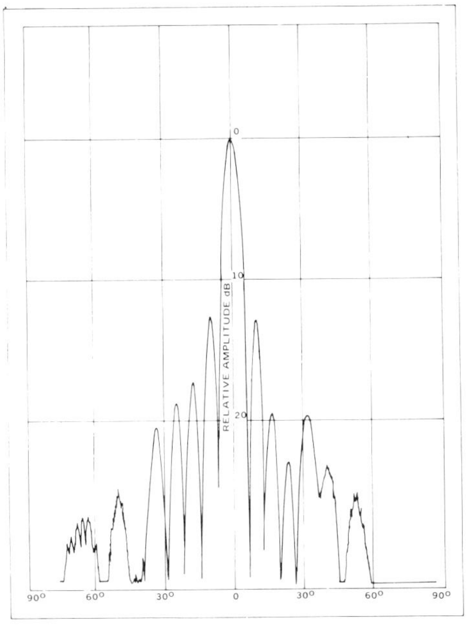

Fig. 11 Radiation Patterns of Experimental Array.

V. RADIATION IMPEDANCE OF THE MATCHED ARRAY (INDUCTIVE IRISES VERSION)

Even though a hole within the dielectrically loaded waveguide accomplishes the task of exciting the necessary higher order modes for the optimal impedance matching of the two dimensional array aperture, this configuration has bandwidth limitation because of the appreciable line length between the hole and the radiating aperture. Bandwidth consideration would require that this path length be minimized. The ultimate limit is realized when the mode transducer is physically located in the aperture. Consequently, a mode transducer should be placed in the aperture directly. An inductive iris is the preferred structure for the purpose of mathematical analysis. The configuration shown in Figure 2 was selected for possible improvement of bandwidth. An analysis on this structure was performed for a very general configuration which may find other applications in addition to the present one. The geometry of the array (cf. Figure 12) has the following flexibility:

(1) The array lattice is arbitrary.
(2) The waveguide wall thickness is arbitrary.
(3) The radiating aperture is arbitrarily located within a diaphragm at the end of the waveguide.
(4) Multiple dielectric layers are allowed.

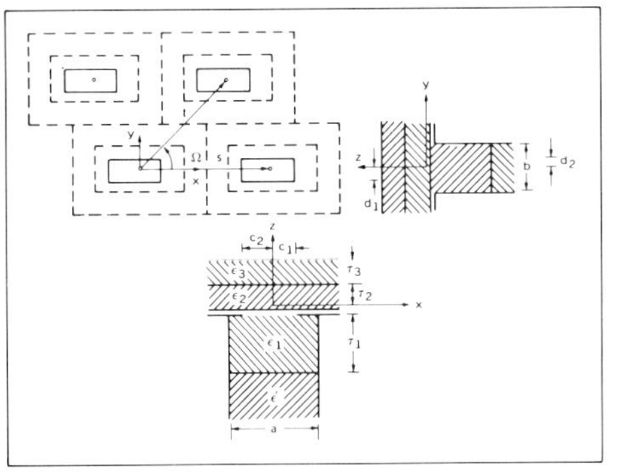

Fig. 12 The Geometry of an Infinite Array of Rectangular Waveguides with Iris at Waveguide Aperture.

The approach to the present problem is based on a straightforward mode matching technique. The equations resulted from the matching of the tangential components of the field are subsequently decoupled or Fourier analyzed in the Galerkin's sense. This procedure in turn yields four infinite sets of linear equations which can be solved numerically with a computer. A computer program of this analysis has been prepared and applied to obtain numerical results on the waveguide element as shown in Figure 2. A detailed description of the analysis was documented in reference (6).

The experimental array under consideration has a triangular lattice and is loaded with a dielectric slab inside the rectangular waveguide. A dielectric sheet is placed in front of the waveguide aperture. The parameters of this array are:

Lattice:	$k_o t = 3.72$, $k_o s = 6.44$, $\Omega = 30^o$
Dielectric:	$\epsilon' = 1$, $\epsilon_1 = 2.56$, $\epsilon_2 = 1$, $\epsilon_3 = 2.6$
	$k_o \tau_1 = 6.24$, $k_o \tau_2 = .426$, $k_o \tau_3 = .708$
Waveguides:	$k_o a = 6.314$, $k_o b = 1.305$, $k_o d_1 = k_o d_2 = .6525$
	$k_o c_1 = k_o c_2$: Varying from zero iris to 60%
Incident Wave:	TE_{10} from the waveguides.

The computed results for the power transmission in the main beam are presented in Figs. 13, 14 and 15 for various planes of scan.

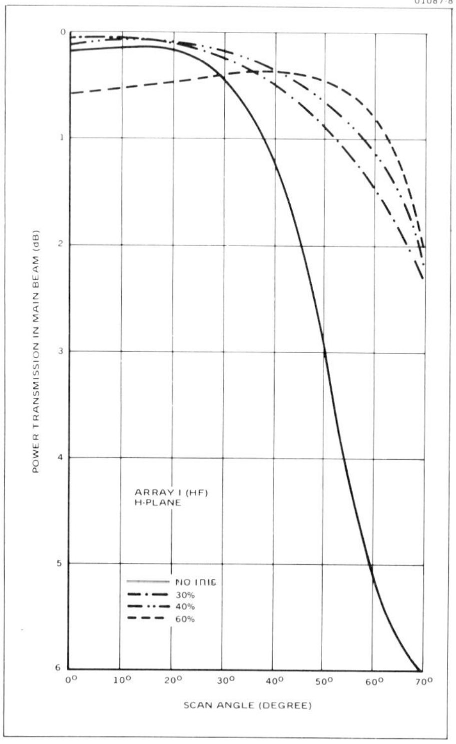

Fig. 13 Power Transmission Coefficients of Two Dimensional Array with Iris at Aperture.

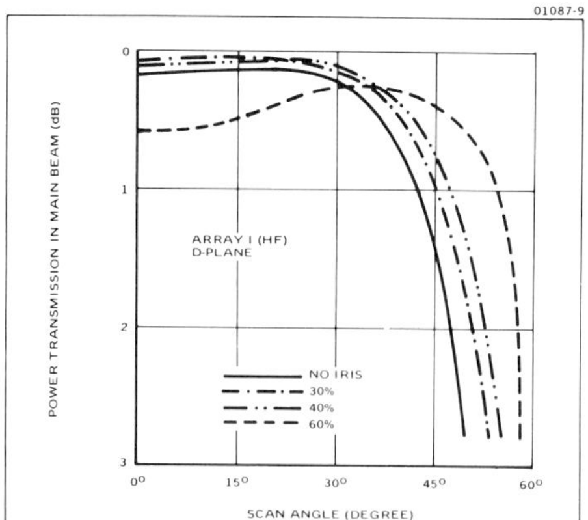

Fig. 14 Power Transmission Coefficients of Two Dimensional Array with Iris at Aperture.

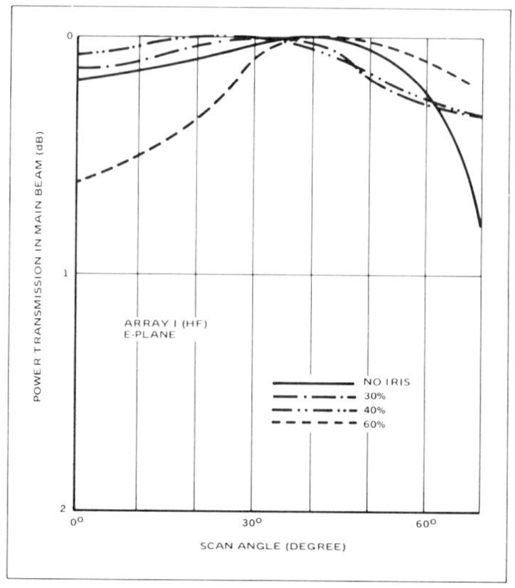

Fig. 15 Power Transmission Coefficients of Two Dimensional Array with Iris at Aperture.

It should be noted that in the absence of the iris, the aperture mismatch increases rapidly as the array is scanned away from broadside along the H and D planes. After the inductive iris has been introduced, the improvement of the aperture mismatch is significant. When the iris width is 40% of the waveguide aperture opening, the power transmission coefficient is about 1.2 dB down at 60° scan along the H plane. The variation in impedance is more rapid in the D plane of scan. In any case, the impedance characteristics after matching is quite similar to the results from the experimental array where a matching void in the dielectric is used. Further improvement in impedance at large scan angles can be realized at some sacrifice to the impedance at broadside if the iris width is increased to 60%.

The impedance characteristic of the above array design at a lower frequency was investigated to determine the bandwidth of the array. Figures 16 to 18 show the transmitted power at H, D, and E planes of scan respectively. The impedance appears to be more constant with respect to scanning in all planes of scan. The reflection loss is less than 1.5 dB over all planes of scan. It may be concluded that the reflection loss of this array over scan is less than 1.5 dB within a 10% band.

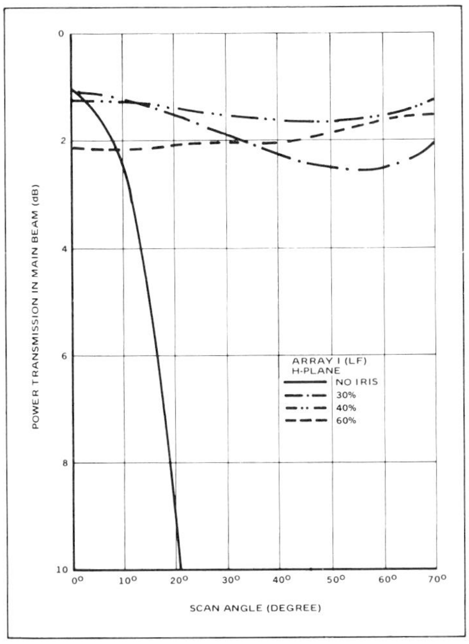

Fig. 16 Power Transmission Coefficient of Two Dimensional Array with Iris at Aperture at 10% Lower Frequency.

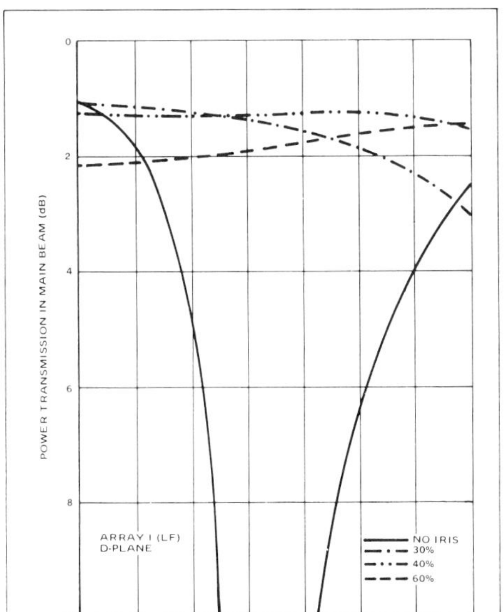

Fig. 17 Power Transmission Coefficient of Two Dimensional Array with Iris at Aperture at 10% Lower Frequency.

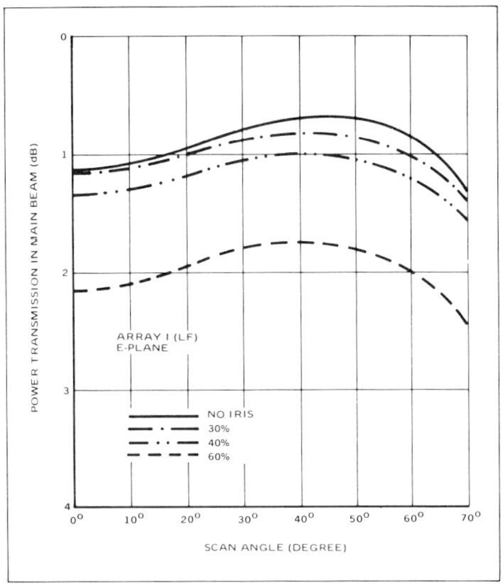

Fig. 18. Power Transmission Coefficient of Two Dimensional Array with Iris at Aperture at 10% Lower Frequency.

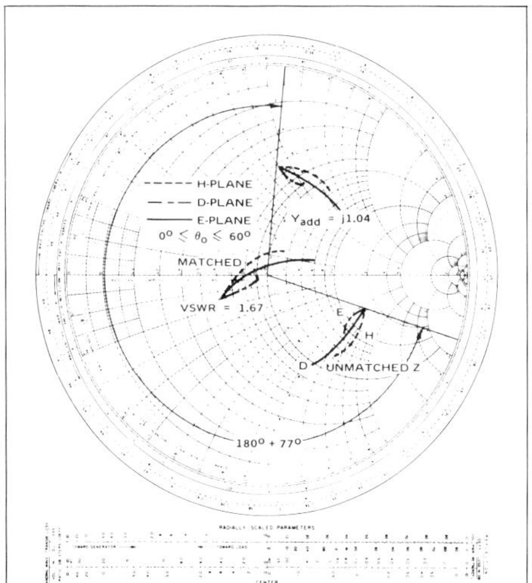

Fig. 19 Calculated Radiation Impedance of Radiating Element Shown in Figure 5 After Final Matching.

It appears from the impedance characteristics shown in Figures 13 to 18 that the element spacing should be reduced if better impedance behavior is required. If 10% reduction in element spacing is assumed, the power transmission coefficients shown in Figures 16 to 18 can be improved by additional matching inside the waveguides in a conventional manner. After matching by either impedance transformers or shunt susceptances, the impedance characteristics are improved to those shown in Figure 19, where the maximum VSWR is 1.67 over scan range of ±60° in all planes.

From the above results, the following observation on the impedance characteristics between the waveguide element as shown in Figure 1 and that of Figure 2 where the matching void is replaced by the inductive iris can be made. For the same array lattice and element spacing, the input VSWR for the case of the inductive iris is essentially the same as that of the perforated dielectric case. However, the calculated input VSWR for the iris case with internal matching has a maximum value of 1.67 over ±60 degrees scan range when all the array dimensions are scaled down 10% from those dimensions used in the perforated dielectrically loaded array. This fact means that the broadband radiator can operate over a 10% bandwidth with a maximum input VSWR of 1.5 over ±47 degrees scan and 2.5 over ±60 degrees scan.

VI. CONCLUSIONS AND DISCUSSION

The technique of achieving a matched radiating aperture for an element in a phased array antenna over a wide scan range has been demonstrated successfully for planar arrays as well as for linear arrays. A maximum VSWR of 1.67 over a ±60° scan range from broadside and overall scan planes is achieved with this technique. It should be pointed out that the experimental waveguide array without dielectric loading and before matching exhibits a resonance null in the H plane scan. This resonance null, however, is completely eliminated after matching over the scan range of ±60°. The capability of this technique in eliminating radiating nulls within a required scan range is very significant; in fact, the numerical calculation on the iris-loaded waveguide array shows clearly that the radiation null is shifted toward the endfire direction as the inductive iris is introduced at the aperture.

Even though the primary goal of this program is to demonstrate the feasibility of impedance matching by superpositioning of modes, other useful information was obtained in the course of the investigation.

The analysis and computer program prepared on this program enables the evaluation of the effects of waveguide size, waveguide wall thickness, element lattice geometry, and dielectric loading inside the waveguides on the radiation impedance of open-ended waveguide arrays. These tools enable the phased array antenna engineer to select and optimize element designs.

Finally, the analytical formulation of the two dimensional array problem is sufficiently general to permit the calculation of radiation impedance of most open-ended rectangular waveguide arrays. The close agreement between the theoretical and experimental results means that this type of phased array can be designed with the aid of the computer program, thus, for the most part, circumventing expensive experimental measurements.

ACKNOWLEDGEMENTS

The authors wish to thank Dr. Richard Mack of AFCRL for helpful comments and suggestions in regard to this program.

REFERENCES

(1) Tang, R. and Wong, N.S.: Multi-Mode Phased Array Element for Wide Scan Angle Impedance Matching, pp. 1951-1959, Special Issue of Proceedings of IEEE on Electronic Scanning, November 1968.

(2) Magill, E.G. and Wheeler, H.A.: Wide-Angle Impedance Matching of a Planar Array Antenna by a Dielectric Sheet, IEEE Trans. on Antennas and Propagation, Vol. AP-14, pp. 49-53, January 1966.

(3) Hannan, P.W., Lerner, D.S. and Knittel, G. H.: Impedance Matching a Phased Array Antenna Over Wide Scan Angles by Connecting Circuits, IEEE Trans. on Antennas and Propagation, Vol. AP-13, pp. 28-34, January 1965.

(4) Brennecke, N.R. and Moule, W.N.: Use of Fences to Optimize Operating Impedance of Phased Arrays Using an Improved Measuring Technique, IEEE G-AP International Symposium, pp. 134-142, 1964.

(5) DuFort, E.C.: Design of Corrugated Plates for Phased Array Matching, IEEE Trans. on Antennas and Propagation, Vol. AP-16, pp. 37-46, January 1968.

(6) Lee, S.W. and Jones, W.R.: On the Suppression of Radiation Nulls and Broadband Impedance Matching of Rectangular Waveguide Phased Arrays, FR 70-14-51, Hughes Aircraft Company, Fullerton, California (under Contract AF19628-68-C-0185).

A New Flush Mounted Antenna Element for Phased Array Application

E. V. BYRON

The Johns Hopkins University
Applied Physics Laboratory
Silver Spring, Maryland

ABSTRACT

A new antenna element is described that can radiate efficiently when arrayed above a closely spaced conducting ground plane. Various forms of the element are shown. Array simulator results, and the element patterns of several configurations are discussed.

INTRODUCTION

Small antennas such as half wave dipoles will not radiate efficiently when mounted parallel to a closely spaced perfectly conducting ground plane. This can be understood by considering the currents on the dipole and on its "image" in the ground plane. The currents are equal in magnitude but 180° out of phase, and as the separation tends to zero their contribution to the far field radiation pattern will cancel.

In this paper a new antenna element (the Collings radiator (l)) is described that can radiate efficiently when spaced above and in very close proximity to a conducting ground plane. This element appears well suited for phased array application with either triangular or rectangular spacing in that it is simple, easy to manufacture, inexpensive, and can be easily hardened. The resulting aperture is flush (i.e. there are no protruding elements) and requires little volume. Forms of the basic element are shown, along with feed methods possible for linear and circular polarization. Simulated aperture impedance measurements and the measured element patterns of several configurations are discussed. Reasonable results have been obtained for elements spaced above a ground plane by as close as one-twentieth of a wavelength.

BACKGROUND

In its simplest form, the Collings radiator consists of a flat circular conducting disk, $\lambda/2$ in diameter, mounted parallel to, and very closely spaced above a conducting ground plane. If the disk is excited from the underside of the ground plane with two probes fed in anti-phase (e.g. at points A_1 and A_2 in Fig. la)

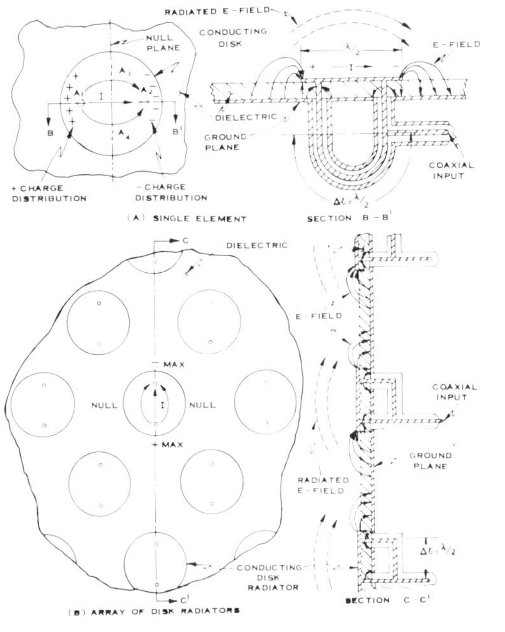

Fig. 1 Disk Radiator

then current will flow across the surface of the disk as indicated. The electromagnetic fields resulting from the charge distribution on the edge of the disk are the primary contributors to the far field radiation pattern, and as shown in Figure 1, they will add to a maximum in a direction normal to the disk. This will be more clearly seen when we subsequently consider a "strip array". Thus the element radiation pattern is substantially that of a half-wave dipole spaced $\lambda/4$ above a ground plane.

If points A_3 and A_4 of Fig. la are fed in phase quadrature with respect to points A_1 and A_2, then circular polarization will result. (For this case the null plane shown in Fig. 1a would not exist.) Fig. 1b depicts a triangular array of disk radiators.

With the probe feeds at points A_1 and A_2 (linear polarization) a null in the electric field distribution will exist as shown in Fig. 1. Thus the elements can be joined in this direction to form the "strip array" configuration shown in Fig. 2a. In Fig. 2b we show a single element cell fed by a waveguide (or a parallel plate transmission line) to achieve the required anti-phase field distribution which would be identically achieved by the balun shown in Fig. 2a. From this figure it can be more readily seen that the feed network induces an equal and opposite charge distribution on each edge of the strip which is the prime contributor to the far field radiation pattern.

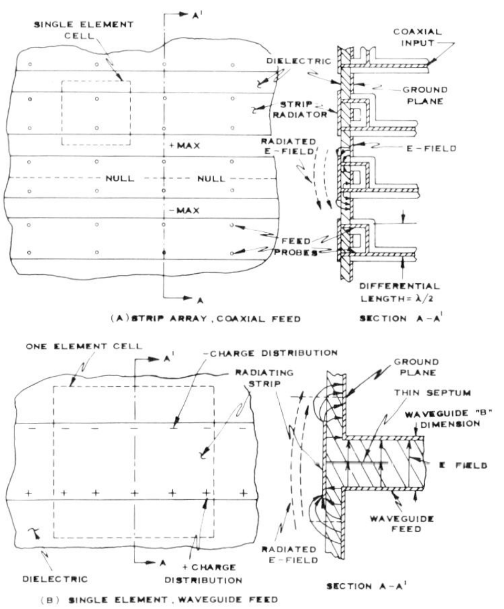

Fig. 2 Strip Radiator.

Figure 3 shows a modification of the strip radiator in which one of the probes in each element is eliminated, thus obviating the need for the two-probe power combiner. This modification (called the single probe strip radiator) requires grounding screws between each radiating segment. In a practical situation the screws could be replaced by eyelets or any other convenient means of obtaining a grounding plane between radiating segments. With reference to Figure 3, the grounding screws form what is equivalent to a dielectric loaded waveguide section fed at one end with a probe which radiates in both directions. The strip width is made one half guide wavelength long (in the dielectric); therefore the electric

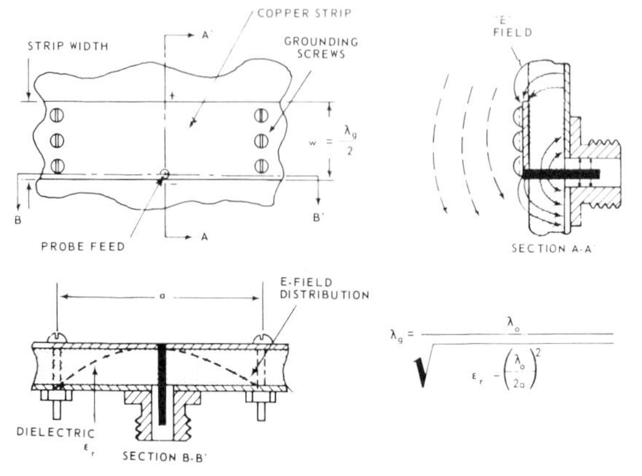

Fig. 3 Single-Probe Strip Radiator.

field vectors at the strip edges are 180° out of phase as shown. The addition of the field vectors in free space is depicted as the dotted lines in this figure; the resulting antenna radiation pattern has a maximum in a direction normal to the aperture.

TWO PROBE DISK AND STRIP RADIATORS

ARRAY SIMULATORS

Waveguide array simulators (2) were built to measure the characteristics of arrays of disk and strip radiators in various triangular and rectangular configurations. Measured transmission coefficients in the simulators indicated that the various elements were essentially lossless. For the strip array the VSWR is a minimum when the strip width is a half-wavelength in the dielectric. For the simulated disk array, the disk diameter is approximately .6 wavelengths in the dielectric. In all of the simulators the aperture impedance was measured with each probe terminated in 50 ohms.

A reasonable impedance match was obtained for a simulated strip array in which the probes were placed at the strip edges. The measured aperture impedance of this model is shown in Fig. 4. Note that no matching was attempted other than by the variation of the probe spacing. The simulator geometry is shown in the inset of this figure.

ELEMENT PATTERNS

A 29 element array of strip radiators with the

element configuration shown in the inset of Fig. 4 was

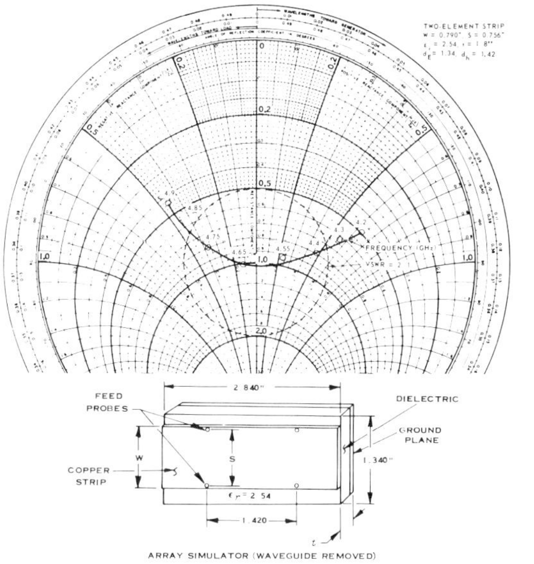

Fig. 4 Measured Aperture Impedance in Rectangular Array Simulator.

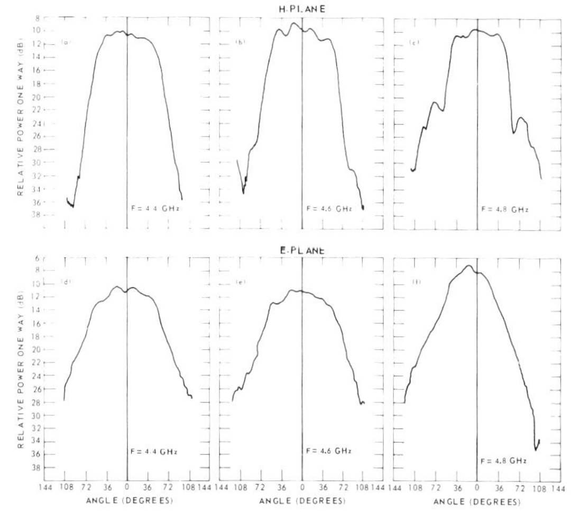

Fig. 6 29-Element Array Element Patterns.

constructed to determine its element pattern. The element pattern (the radiation pattern of the central element with the remaining elements terminated) is indicative of the behavior of a phased array of identical elements under various conditions of scan. In particular, any undesirable aperture resonances that may occur at some scan angles are revealed. (3)

The array is shown in Fig. 5. No anomalies were observed in the measured element patterns in all scan planes. Typical E. and H-plane patterns are shown in Fig. 6. It should be noted that for this geometry, the grating lobe is evanescent at 4.8 GHz for a 45° H-plane scan angle, and for a 55° E-plane scan angle. Thus the distortions in the 4.8 GHz patterns may be indicative of the appearance of the grating lobe in real space. Results similar to the E-plane element patterns were obtained in the 45-degree scan plane.

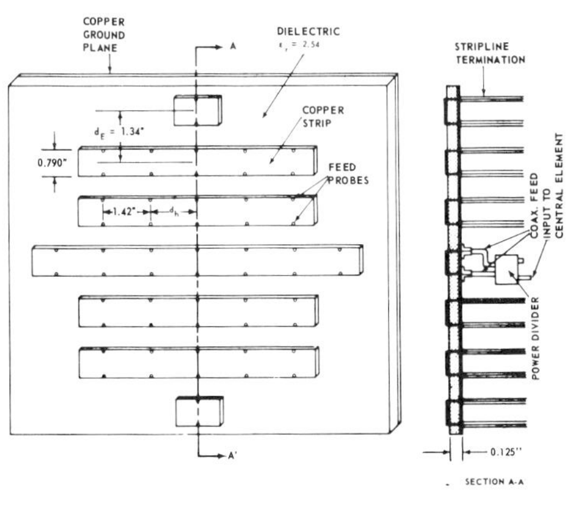

Fig. 5 29-Element Dummy Array Configuration.

Although caution must be used in inferring aperture match from element patterns measured on small arrays, these results (along with the results from array simulator measurements) indicate that at center frequency this array is matched to a VSWR of approximately 3:1 or less for scan angles up to 45° in all scan planes. A larger array can generally be expected to give better results since the effects of incipient grating lobes at scan angles beyond 45° and "edge effects" would be less pronounced. It should be noted that, as stated, no matching was attempted other than by the variation in probe spacing, and no scan compensating matching networks such as a dielectric sheet on the aperture have been employed.

In this array, the elements were spaced above the ground plane by approximately $1/20\ \lambda_o$. These results, imply that a phased array of elements of this type, closely spaced above a ground plane, can be reasonably matched over moderate bandwidths and scan angles.

Of various configurations examined, one model using radiators spaced above a ground plane by a fairly thick dielectric sheet ($\epsilon_r = 4.0$; $t/\lambda_\epsilon = .125$) was more easily matched in the array simulator

(VSWR ≤ 1.5 over a 12% bandwidth), and a 50 element array was constructed.

Subsequent measurements of the H-plane element patterns revealed the existance of unwanted aperture resonances. Experiments verified that these resonances do not exist when the dielectric thickness is reduced. It was further determined that resonances caused by "thick" dielectric apertures could be eliminated by inserting grounding screws between radiating segments, in the manner described for the single probe strip radiator (Fig. 3).

SINGLE PROBE STRIP RADIATOR

The single probe strip radiator retains the desirable features of the two probe strip radiator and has the obvious advantage of requiring half as many probes while eliminating the need for the two probe power combiner. This should result in a simpler and less costly element. It has the disadvantage of requiring grounding screws between segments. It also requires a slightly thicker dielectric (than the two probe element) for a given aperture impedance.

ARRAY SIMULATION PROBLEMS

The single probe modification results in an antenna element which does not possess a symmetry plane through its center in the E-plane (i.e. it cannot be mirrored through its center in the E-plane); therefore, exact simulation of an array of single probe strip radiators is not possible in simple waveguide simulators. This can be seen with reference to the following figures. Fig. 7 shows the desired geometry for a rectangular spaced array of one probe strip radiators. The relative phase at the strip edges (due to the $\lambda/2$ strip width) is depicted by the plus and minus signs in this figure. Also shown is the E-field distribution on the strip edges and the radiated E-field. Choosing the 2 element cell shown by the dashed lines in this figure as the simulator (as previously chosen for the simulation of the 2 probe/element arrays), we have the simulated array shown in Fig. 8. This

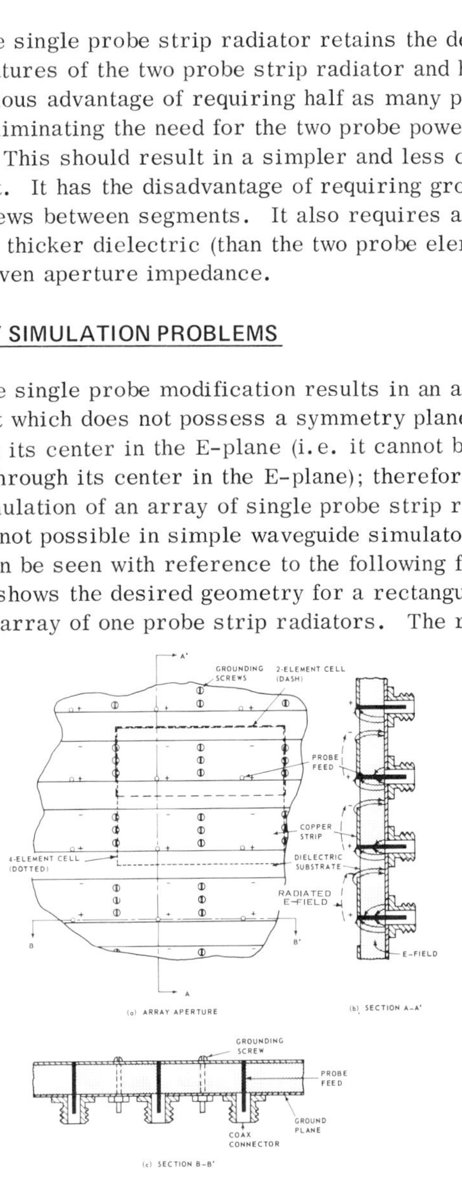

Fig. 7 Array of Single Probe Strip Radiators.

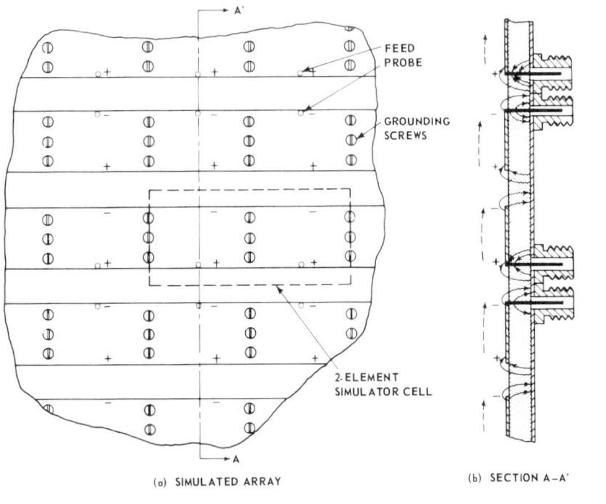

Fig. 8 Simulated Array Using 2- Element Cell Simulator.

simulated array is different from the desired array in that it has probes on alternate edges of each strip (i.e., each strip is mirrored in the E-plane) and these are fed alternately in and out of phase as depicted by the plus and minus signs at the strip edges, (ignoring the phasing required for the simulated 30° H-plane scan angle). However, the resultant radiated field from each strip is as desired.

To determine if a simple waveguide simulator adequately simulates the array of one probe strip radiators shown in Fig. 7 a simulator as shown by the two element cell - the dashed lines - in Fig. 7 was constructed, and the aperture impedance measured. This represents the impedance (for a 30° H-plane scan angle) of the aperture shown in Fig. 8 which was described above.

Another simulator as shown by the four element cell - the dotted lines - in Fig. 7 was then constructed and the aperture impedance measured. This represents an aperture (scanned to 30° in the H-plane) in which the feed probes are on alternate edges of each pair of strips (i.e. each pair of strips is mirrored in the E-plane). Each pair of strips is then fed alternately in and out of phase, resulting in the desired in-phase radiation.

The measured aperture impedance in the above two simulators was identical in every respect. It is reasonable to assume, therefore, that simulators with still additional elements in the E-plane will have the same aperture impedance and thus, in the limit, the desired aperture (Fig. 7) is simulated by the simple "2 element cell" simulator.

ARRAY SIMULATOR RESULTS

Several simulators, represented by the two element cell in Fig. 7, were constructed using materials with dielectric constants of 2.54 and 4.0. In all of the simulators the VSWR was a minimum when the strip width was approximately $\lambda_g/2$ where λ_g is defined in Fig. 3. As in the two probe/element case, simulator transmission measurements determined that the element is essentially lossless.

The solid curve in Fig. 9 shows the measured aperture impedance (with no matching) of the most promising configuration to date. The simulator geometry

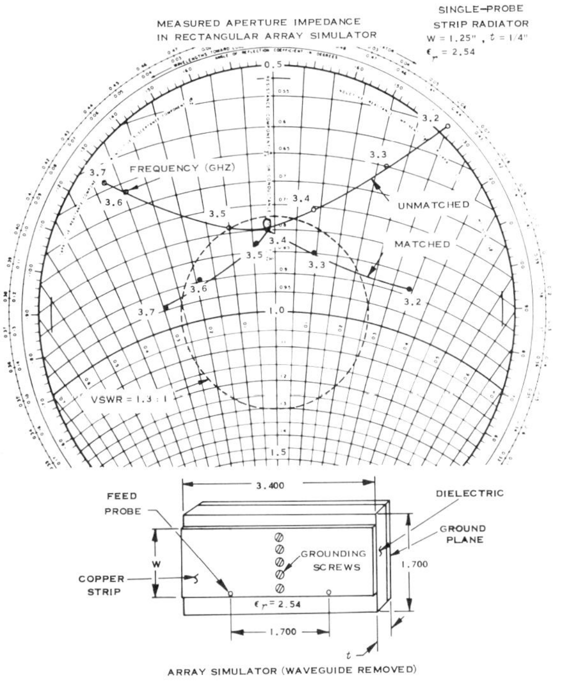

Fig. 9 Measured Aperture Impedance in Rectangular Array Simulator.

is shown in the inset of this figure. Using a simple matching network behind the aperture (i.e., two thin dielectric beads in the coaxial line attached to the terminations) the aperture VSWR was reduced to less than 1.3 to 1 over a 12% bandwidth. These results are shown by the dashed curve in Fig. 9. This element and matching structure is being incorporated in a 50 element test array.

ELEMENT PATTERNS

The twenty-nine element array shown in Fig. 5 was converted to an array of single probe strip radiators by removing one probe from each element and inserting grounding screws between the elements. Measured element patterns of this converted array were very similar to those shown in Fig. 6 for the 2 probe/element array.

A preliminary test was conducted to determine if the 1/4" thick ϵ_r = 2.54 dielectric used in the simulators described above causes any undesirable aperture resonances when used in an array configuration. This consisted of refitting the previously built 50 element array with an aperture of one probe/element strip radiators on 1/4" thick ϵ_r = 2.54 dielectric sheet. No anomalies were observed in the measured element patterns. E and H-plane patterns are shown in Fig. 10.

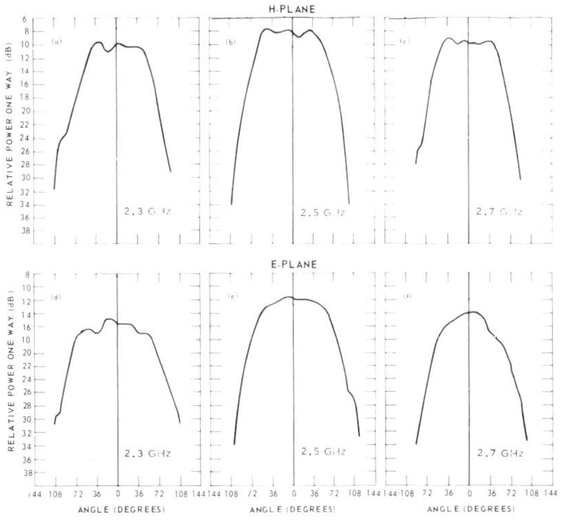

FIG. 10 50 Element Array, Element Patterns with Single Probe Strip Radiators on 1/4" Thick, ϵ_r = 2.54 Dielectric Sheet.

It should be noted that in this frequency range, deep notches were observed in the measured H-plane element pattern of this array with an aperture of 2 probe/element strip radiators on an identical dielectric sheet. These resonances were eliminated when shorting screws were placed between each radiating segment.

For these tests neither of the above arrays were

particularly well matched because they were adapted from previously constructed "two-probe" test arrays. (The near broadside VSWR was approximately 3.5 to 1 or less over their operating bandwidths, as determined in array simulators.)

CONCLUSIONS

Various configurations of a new antenna element have been described. The results obtained in array simulators and from the measured element patterns imply that an aperture using Collings-type radiators can be built that is reasonably well-matched and has good radiation characteristics. The aperture offers considerable potential for phased array application in that it is simple, easy and inexpensive to manufacture, and can be easily hardened. Since the elements are very closely spaced above a conducting ground plane, the resulting aperture is flush and requires little volume; its applicability to conformal arrays is readily apparent.

ACKNOWLEDGEMENTS

Many of the results reported are the work of R. Sydnor. The work was greatly facilitated through the use of aperture matching procedures and computer programs developed by G. Laughlin. T. C. Cheston first recognized that the Collings radiator was well suited to phased array application, and suggested that the disks could be joined to form the strip array. His continued support and encouragement is gratefully acknowledged.

This work was supported by the Strategic Technologies Office of the Advanced Research Project Agency under ARPA Task No. STO AO805.

References

1.) Collings, R.H.P.: U. S. Patent Application No. 807,303; March 14, 1969.

2.) Hannan, P.W., and Balfour, M.A.: Simulation of a Phased Array Antenna in Waveguide, IEEE Trans. on Antennas and Propagation, Vol. AP-13, No. 3, pp. 342-353, May 1965.

3.) Byron, E. V. and Frank, J.: Lost Beams from a Dielectric Covered Phased-Array Aperture, IEEE Trans. on Antennas and Propagation, Vol. AP-16, No. 4, July 1968, pp. 496-499.

Design of a Phased Array Radiating Face for Prevention of Performance Degradation in the Presence of Rain

J. D. HANFLING AND V. A. GIARRIZZO

Raytheon Co., Bedford, Mass.

ABSTRACT

The accumulation of rain in the form of a sheet of water on the face of phased array antennas causes serious array mismatch loss and boresight errors. This can be prevented by the use of a perforated-face design and a non-wetting agent. Rain water will bead up on the face of the radiating element, flow into the perforations and be collected and removed by means of drainage ditches under the radiating face.

The effects of the perforations, drainage ditches, and water thickness on array impedance and surface-wave slowing has been evaluated theoretically and experimentally. Also, a thorough experimental study has been carried out to correlate rain rates with the effective water-sheet thickness that is obtained on several face configurations.

I. Introduction

The accumulation of rain in the form of a thin sheet of water on the face of phased-array antennas causes serious array mismatch loss and boresight errors. This can be prevented by the use of a perforated-ground plane design and a nonwetting agent. Rain water will bead up on the window of the radiating elements, flow into perforations in the ground plane, and be collected and removed by means of drainage ditches under the radiating face.

The effects of the water, performations, and drainage ditches on impedance, surface-wave slowing, and boresight errors have been evaluated theoretically and experimentally. Also, an experimental study has been carried out to correlate rain rates on several face configurations with effective water-sheet thicknesses.

II. Effects on Array Performance

Most array radiators use a ceramic window ($K \approx 6.6$) as part of the element matching network and as a weather seal. The accumulation of rain water ($K \approx 76.7$) on the face of the ceramic detunes the impedance, increases the dissipative loss of the element, and decreases the maximum scan coverage of the array. Also any gradient in the water film as a result of runoff from the top to the bottom of the array will cause boresight errors.

The effect on array impedance and loss at broadside scan caused by a thin sheet of water on the face of an array of flush elements is shown in Figure 1.

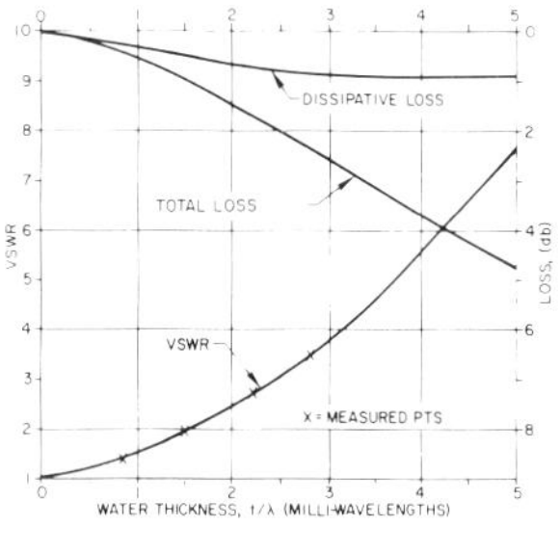

Fig. 1 Performance Degradation Caused By a Water Sheet Over a Matched Array of Flush Elements.

The array has been assumed to be matched to free space under dry conditions. The curves for the VSWR and reflected power have been determined from

$$\text{VSWR} = \frac{1 + \dfrac{B/G_O}{\sqrt{4 + (B/G_O)^2}}}{1 - \dfrac{B/G_O}{\sqrt{4 + (B/G_O)^2}}}$$

where the susceptance for a thin layer of water ($t/\lambda < 0.003$) is

$$\frac{B}{G_O} \approx 2\pi \frac{t}{\lambda} (K - 1)$$

The VSWR curve has been verified experimentally as indicated. The curve for the dissipative loss of a thin layer of water has been calculated from Ref. 1

and for $(t/\lambda < 0.003)$ is

$$L \approx 2\pi \frac{t}{\lambda} K \tan \delta$$

The wave slowing as a result of the water sheet can be computed from

$$1 - \frac{v}{c} \approx \frac{1}{2} \left(\frac{B}{G_O}\right)^2 \frac{1}{K^2}$$

and the corresponding reduction in the maximum scan angle before a grating lobe appears at end fire from

$$\Delta\theta \approx - \sec\theta \left(1 - \frac{v}{c}\right)$$

For an array inclined at 51.5 degrees from the horizontal, the water-sheet thickness versus rain rate at the center element of an array of flush elements is shown in Figure 2. This curve indicates that there is sufficient water buildup for modest rain rates to degrade performance and reouire a rain-removal scheme.

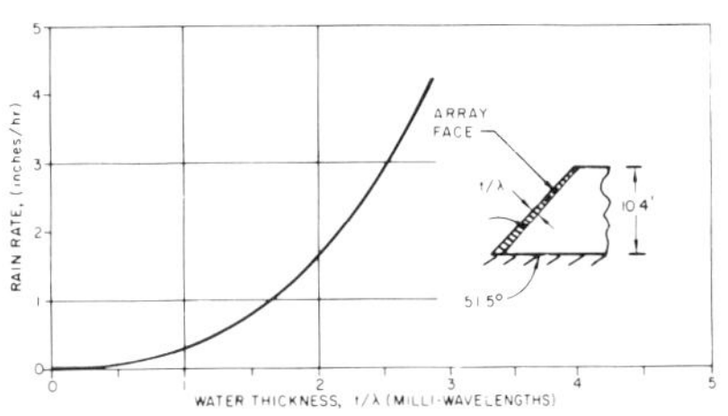

Fig. 2 Calculated Water-Thickness vs. Rain Rate at the Center of an Array of Flush Elements.

III. Perforated Ground-Plane Concept

The formation of a sheet of water over the face of an array, Figure 3(a), can be prevented by setting back the physical ground plane, Figure 3(b). This results in an array of protruding elements with ditches between them for water flow and removal. The perforated ground-plane concept, shown in Figure 3(c), incorporates both the water-removal properties of an array of protruding elements and the electrical ground-plane properties of a flush array. Here, in addition to the ditches, the elements are bridged by a flush conducting-sheet with perforations so that the electrical ground-plane position can be maintained while water is collected and drained.

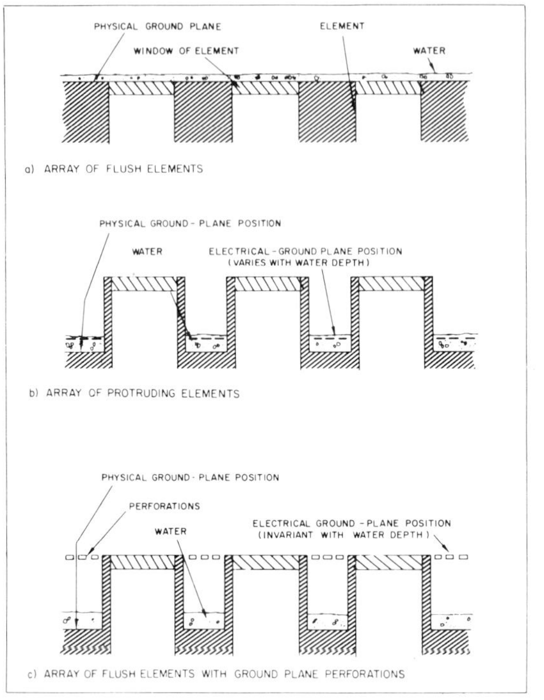

Fig. 3 Development of the Perforated Ground-Plane Concept.

IV. Design of the Perforated Ground Plane

For a triangular grid of elements, a mechanical configuration of a typical perforated face design is shown in Figure 4. The conducting ground plane is formed by means of bars between elements; the bars are situated in cardinal (C) planes, with perforations in intercardinal (IC) planes. This arrangement permits access to the element mounting bolts. The bars are recessed approximately 0.01 wavelength which aids in directing water into the perforations.

The effect of the ground-plane design on array impedance can be determined from computing the series reactance introduced by perforating and recessing the ground plane. The wave slowing, as compared to a flush face design, is determined by

$$1 - \frac{v}{c} \approx \frac{1}{2} \left(\frac{X_S}{R_O}\right)^2$$

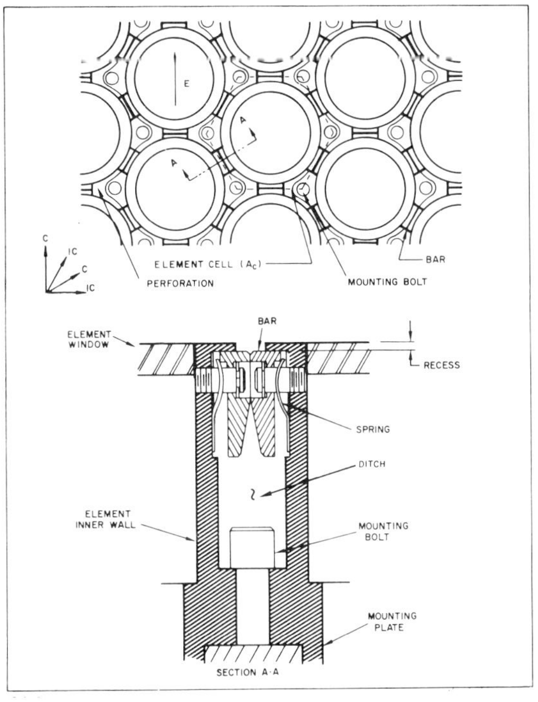

Fig. 4 Mechanical Configuration of Typical Perforated Ground Plane.

where X_S/R_O is the total reactance in series with the element (Ref. 2.). Since it is desirable to minimize the wave slowing, the face design therefore becomes tradeoff among wide-angle scan requirements, performance in rain, and the complexity of the interface between the element and the array structure. A reasonable rule is to permit a maximum wave slowing of 1 to 2 percent.

V. Rain Experiments and Test Results

To confirm the theoretical predictions and to evaluate the proposed design, an experimental program was undertaken. For this program, a mechanical mockup of a vertical section of the array was built that was fourteen feet long and three feet wide and inclined at 51.5 degrees, as shown in Figure 5. A plumbing system was devised which would spray this mockup with the proper density and size of drops to simulate rain. The rate of simulated rain was measured by rain gages. The drop sizes were measured by catching drops in a shallow dish of silicone oil and photographing them as shown in Figure 6. The amount of water on an element was determined by measuring the change in the VSWR from a matched, dry element.

The results of these tests are summarized in Figure 7. Using the perforations and the recess improves the

Fig. 5 Array Face Simulation in Rain Chamber.

Fig. 6 Simulated Rain at 1.8"/HR Captured in Silicone Oil.

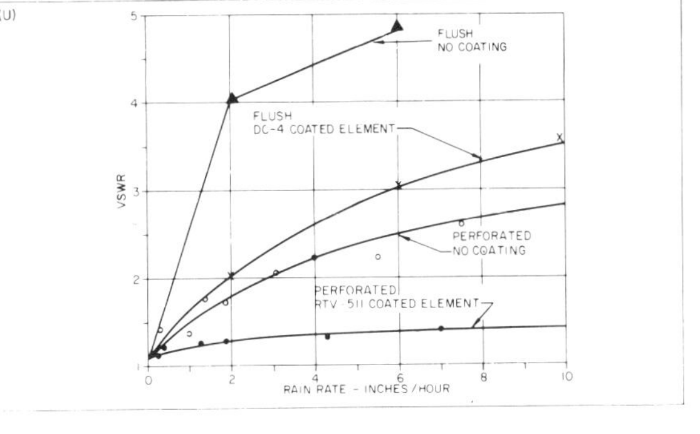

Fig. 7 Measured VSWR vs. Measured Rain Rate for Several Face Conditions.

performance about 2:1 over a flush face design. Coating the element face with a nonwetting agent also results in about a 2:1 improvement. The implementation of the perforated ground plane and a RTV-511 coating results in a VSWR of 1.5 for rain rates up to 15 inches/hr.

VI. Acknowledgments

This work was performed under contract between the Army and Western Electric. The rain problem was part of an advanced array-face design study carried out by engineers from Bell Telephone Laboratories, Raytheon, and Wheeler Laboratories.

VII. References

1) Blevis, B.C.: Losses Due to Rain on Radomes and Antenna Reflecting Surfaces, IEEE Transactions, Vol. AP-13, pp. 175-176, January 1965.

2) Jasik, H.: Antenna Engineering Handbook, First Edition, Ch. 32, pp. 4-12, Ch. 16, pp. 4-28, McGraw-Hill, 1961.

Phased Array Feed Systems, a Survey

ROBERT T. HILL

Naval Ship Engineering Center
Hyattsville, Maryland

Abstract

The "feed system" in phased arrays is surprisingly hard to isolate from the rest of the system. It is rare that the feed is exclusively a power distribution or collection device. Although this function is primary, feed development has in this last decade embraced the functions of formation and steering of single and multiple beams, as well. Then, with the wider availability of phase shifters for individual elements, feed design has more recently emphasized the distributive role alone. Therefore, a categorization along the line of the inclusion and exclusion of beam steering is presented. Accomplishments of the 1960's are reviewed, and finally a discussion of relative merit, problems and design choices is presented. The survey is, for a variety of reasons, restricted primarily to radars of a civilian or military air-surveillance sort in the microwave region.

Introduction and Scope of Survey

The "feed system" in phased arrays is surprisingly hard to isolate from the rest of the radar, much more so than in more conventional radars typically involving tower or mast mounted antennas of a feed horn and reflector type. These classically involve a clear separation of transmitter and receiver from the antenna system, quite foreign to many of today's phased array systems. The replacement of the mechanically rotating antenna structure with inertialess beam steering techniques promoted a blending of transmitter and receiver components into the antenna structure, with the promise of increased efficiency by the proximity to the antenna of both the transmitter and receiver r.f. amplifiers. The elimination of the "rotating joint" in a feed line brought a dispersive or modular potential to the amplifier and array relationship, as in systems involving subarray amplification, which further counters a clear distinction of just what is, in fact, solely the "feed system". As we shall see in several of the systems discussed, this blending is not without exception; for particular applications, phased arrays may still be mounted on movable structure, and a clearer "feed system" is evident.

The opinions or assertions contained herein are those of the writer and are not to be construed as official or reflecting the views of the Navy Department.

It is, of course, impossible to survey absolutely all phased arrays developed to date. There should be no question, then, that this paper, largely because of the author's own background, responsibilities and, hence, contacts in the field, is concerned primarily with radars of a military or civilian nature wherein aircraft or to a lesser degree orbiting or re-entering bodies are the targets and which operate in or near the microwave region. Large arrays of other types, as in radio astronomy, and those operating well outside the microwave region simply could not be surveyed.

There is sufficient interest in the multiple purpose potential of the phased arrays surveyed, such as the detection of targets and the more accurate estimation of their position, to warrant attention to difference pattern characteristics as well as sum. Monopulse extraction is, therefore, discussed.

Finally, this paper is being prepared for presentation with a number of others in which particular feed systems are discussed. I will attempt to avoid these particular ones, and dwell instead on others. I hope to present a "second level" of information in some of these, data not previously reported, and the comments of the experimenters themselves regarding development difficulties, iteration and the like that led finally to the "published" data.

We proceed now with discussion of, first, a categorization of array feed types, and, second, review of the accomplishments of the last decade, and some of the recent proposed systems. Then we will review some of the arguments involved in design choices.

Categorization

Categorization is valuable because it establishes a language. Phased array radars are most often described as being of a particular feed type before any further description is attempted. This survey confirmed, however, that the last two decades have produced virtually tailor-made phased arrays with a resulting variety that nearly defies categorizing. A level or two of classification is possible, but further subclassification is very difficult indeed. I have chosen to describe first phased arrays in which the feed system itself dominates in the steering of the beam and second those in which the feed may play a part in forming the beam but does not beam steer. The wide-spread acceptance of discrete phase shifters for use at each element of an array took the field from the former to the latter type. I find it convenient to discuss the second type, those feeds which do not beam steer, in two further categories, space feeds and constrained feeds. Figure 1 illustrates this categorization with a few systems familiar to many named as examples.

BEAM STEERING INCLUDED		BEAM STEERING EXCLUDED	
FREQUENCY STG	MULTIPLE BEAMS	SPACE FEED	CONSTRAINED FEED
SPS-32,33,48,52	LENSES	REFLECT	MAR
FREQ BY FREQ	AN/SPG-59 (LUNEBERG)	RAR (NRL)	PAR
	TIN HAT	MTR (NWC)	ESAR (FPS-85
	R-2R		GE ON ASMS
			RCA ON ASMS
	MATRICES	FEEDTHROUGH	SOLID STATE ARRAYS
	FAA HEIGHT FDR	SAM-D	
	BUTLER MATRIX	HAPDAR	
		MSR	
		HYBRIDS	
		HUGHES ON ASMS	
		REST	
		HIPSAF	

Fig. 1 Array Feed Systems (Table).

The categorization could, of course, be inverted - that is, a first level might have been Space Feed and Constrained Feed, and Frequency Steering Feeds could have been embraced under Constrained Feed. But the major distinction of, essentially, with and without element phase shifters better addresses the evolution of phased array development. It should be noted that in John Allen's 1962 survey of array radars (1), categorization follows slightly different lines, although the individual types can be found both in that work and here. The intent here is to classify feed systems alone, not total radars as was being done in that work.

Accomplishments of the 1960's

Using this categorization as an outline, we shall review some of the feeds developed in the last decade, trusting that, while it is not a complete survey, each major feed type is represented typically.

Beam Steering Included

Frequency Steering. Generally familiar to radar engineers is the role of the "serpentine" waveguide feed of many phased arrays. These feeds are generally of a series tapped-line type, illustrated most simply in Figure 2(a). From the drawing we

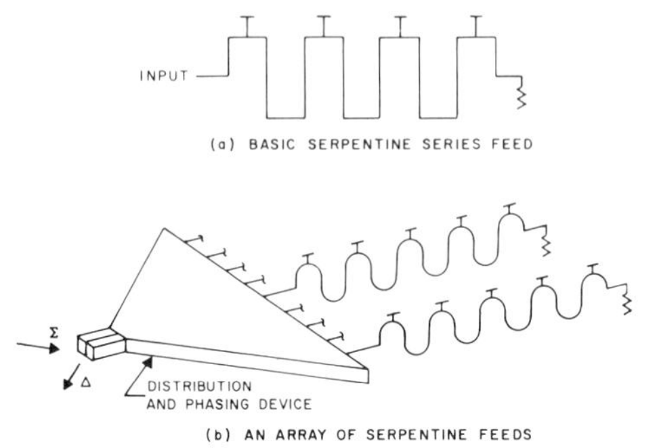

Fig. 2 Frequency Steering.

see that the taps (the radiating elements) bear a phase difference, one to the next, that depends on the number of guide wavelengths, measured in the feed, between them. This is, of course, determined by the frequency of the signal. As in any phased array, the phase difference between the elements determines the beam direction. The ratio of the path length in the feed line between the elements to the distance in space between the elements determines the spatial sensitivity of the frequency steering. This factor has been identified in the literature as the "wrap-up" factor (2), the bigger the wrap-up factor the greater the angle scanned for a given change in frequency. Several radars of significance use this feed principle. Figure 2(b) schematically illustrates the approach for the AN/SPS-33, the Hughes Aircraft built S-band 3-D radar for search and track, which has been operational on board the aircraft carrier Enterprise and the cruiser Long Beach for a number of years. In that system analog ferrite phase shifters are used between a parallel plate distribution device and the serpentine waveguides for steering in the direction orthogonal to the frequency steering. The illustration shows a hybrid feed at the input, providing a difference pattern port for the phase steering direction. Angle monopulse is in one plane only; sequential observations of a target at slightly different frequencies provide angular interpolation in the frequency steering direction. The serpentine feeds were made of quite light weight reduced height waveguide, "snaked" in the H-plane, that is using all H-plane bends. Broadwall multihole waveguide couplers constituted the "taps" for the elements. Making the couplers alternately forward and backward couplers allowed the open end of the auxiliary arms themselves to be the radiators, forming a planar array of open end waveguide elements, backed up immediately by the serpentines. The structure is extremely strong; the unfortunate explosions aboard the Enterprise last year sent debris into the arrays, requiring serpentine replacement, of course, but causing no interruption of the radar functions at the time. The steering sensitivity of the feed in the near normal region is approximately 3 MHz per degree, and the lines have approximately 3 dB loss.

Figure 3 shows the AN/SPS-48 radar antenna, in which a single serpentine is used to feed horizontal straight waveguides with slot radiators. The antenna rotates in azimuth, frequency steering is used only in elevation. This Navy radar (and the Marine Corps counterpart AN/TPS-32) is built by ITT-Gilfillan. Operating in the S-band, it uses a serpentine with all E-plane bends and has a steering sensitivity comparable to the AN/SPS-33 and loss in the 1 to 2 dB region. The taps to feed the horizontal straight waveguides are simply slot couplers in the narrow wall of the serpentine, feeding at a right angle the butted horizontal waveguides. An iris structure in the serpentine opposite the slot contributes to proper match. With this feed arrangement, by the way, comes a peculiarity, the "squinting" of the beam in azimuth as it is frequency steered in elevation. This results from the end feeding of the horizontal straight waveguides. The straight waveguides have, of course, a greatly reduced wrap-up factor and very little steering sensitivity; nonetheless, the azimuth shift of the beam is about one twentieth of the elevation steering (a rough average over the elevation scan region).

Fig. 3 AN/SPS-48 Array Feed.

Another E-plane serpentine guide was designed and produced for an experimental phased array for the Navy several years ago by the Aero Geo Astro Corporation. It also used multihole broadwall couplers after every bend, but used only forward couplers and used coaxial cables to connect the coupler to the radiating element. This serpentine was designed by Dr. Judd Blass, of Terry Microwave at the time, under subcontract.

An array feed worth mentioning here, but not shown in the categorization of Figure 1, is in the class of a "low inertia" steering feed. ITT-Gilfillan, with Marine Corps sponsorship, has developed a phased array which is frequency scanned (by serpentine columns) in elevation and phase scanned in azimuth by a "Delta a" scanner and is illustrated in Figure 4. Essentially a tapped waveguide run in which the "a" dimension can be mechanically varied, the device is certainly appropriate in cases involving a regular and uninterrupted scanning program. The cyclic modulation of the "a" dimension modulates the guide wavelength, of course, changing the phase gradient in azimuth (tap to tap) and resulting in a cyclic and very rapid azimuth scan. The figure shows the linkage assembly to perform the scan. The serpentines for this system are constructed as machined halves of entire serpentines (E-plane) and

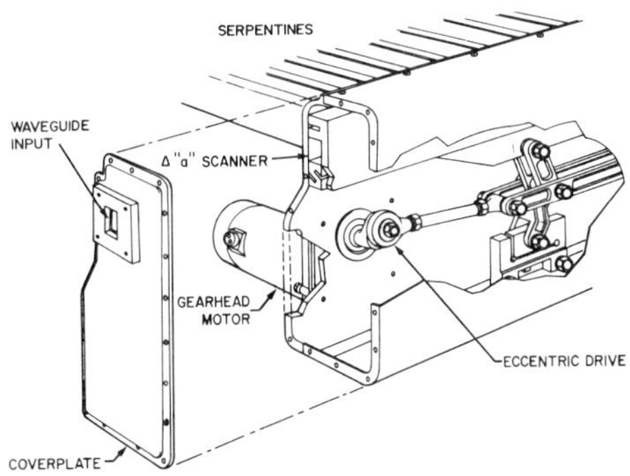

Fig. 4 A Low Inertia Scanning Feed.

fitted together as shown in Figure 5.

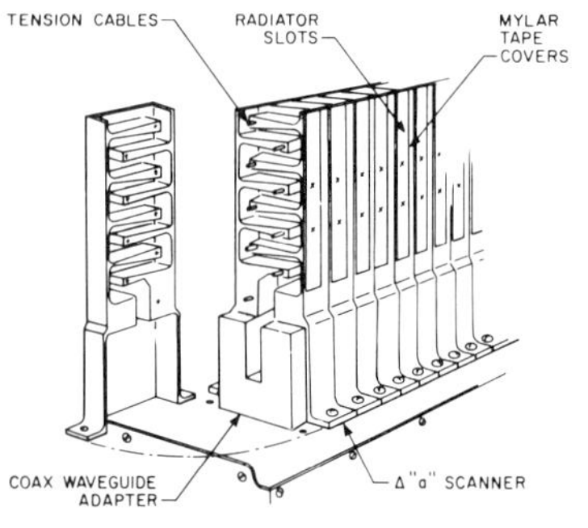

Fig. 5 A Construction of X-band Serpentines.

Frequency steering has been considered for scanning in both dimensions; Figure 6(a) shows such a scheme. If the helix has a wrap-up factor far in excess of that of the serpentines, a monotonic change of

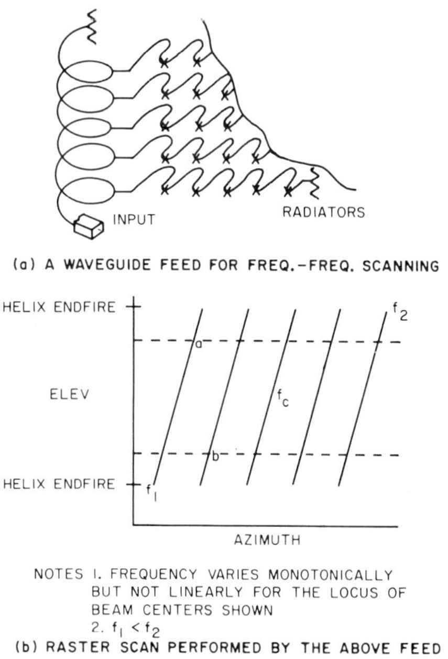

Fig. 6 Frequency by Frequency Steering.

frequency results in the raster scan pattern shown in Figure 6(b). The scheme depends upon the helix "cycling through" conditions of one elevation extreme, through normal, through the other extreme, through end-fire in that direction (and the opposite end-fire), back to the first elevation limit and so on as the frequency is monotonically changed - this cycling has to occur while the serpentines go through only a part of one such cycle in the azimuth steering. In use, the frequencies associated with undesired and poorly performing scan directions would not be transmitted (the scan region outside the dotted lines in Figure 6(b), for example). In this case the wrap-up factors must bear such a ratio that the movement of the beam center in azimuth from point "a" to point "b" is less than an azimuth beamwidth to assure continuous solid angle coverage. Waveguide structure was developed along this line several years ago by the Autonetics Division of North American Rockwell, Anaheim, California; the scheme is similar to that pursued by Dr. Croney at the Admiralty Surface Weapons Establishment in England, as well.

<ins>Multiple Beam Feeds.</ins> Both two and three dimensional lenses or "analog computer" devices have been developed for beam switching - these certainly belong to the category of multiple beam feeds which perform beam steering. The Navy's aborted AN/-SPG-59 radar development used a spherical Luneberg lens to transform a point feed into an appropriately phased set of signals which, when radiated from a spherical array, results in an equi-phase planar front. A two dimensional representation of such a lens is shown in Figure 7, from the basic examination given in Skolnik (3). The lens action demands a change of dielectric constant with radius from a relative value of unity at the circumference to a value of 2 at the center. In the AN/SPG-59 lens, more than 40 concentric shells were used

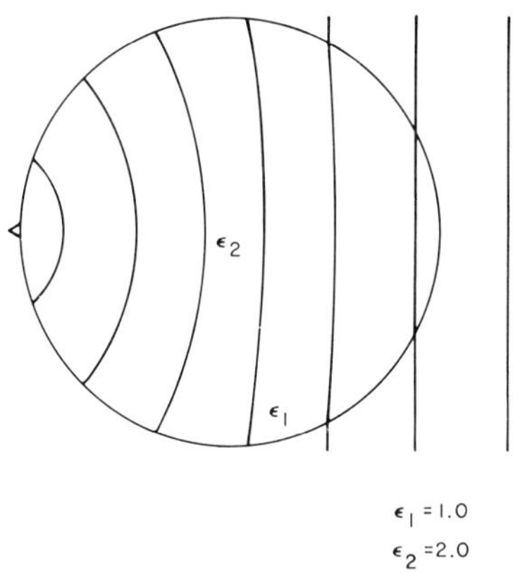

Fig. 7 Two Dimensional Representation of a Luneberg Lens.

in the transmit lens, the dielectric constant of each being controlled by the material density. The receive lenses were made by assembling small cubes, with inserted metal threads controlling the dielectric constant.

An array feed scheme using such a lens is shown in Figure 8. The switches permit each feed point on

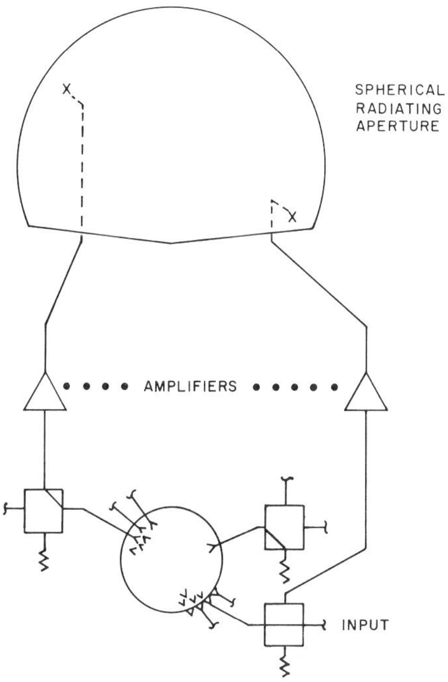

Fig. 8 Feed System Using Luneberg Lens.

the lens to be used as both as excitation point and a collector point. The general possibilities for such a feed include cabling to amplifiers in the one per element manner shown, or connection to include coherent mixing as well so that final amplification and radiation might be at a frequency other than the "phase computation" frequency of the lens. More switching could be included to take advantage of the fact that for any one beam position, less than half of the amplifiers shown in Figure 8 would be used; for example, the amplifier shown at the "input" element in Figure 8 certainly could be used for the diametrically opposite element. All of the amplifiers could by switching at both the lens and the aperture, be shared between diametrically opposite elements - and the same control that causes excitation of a particular input could switch amplifiers so that the hemisphere opposite the feed point is provided with amplification. Most radar applications would define regions on the lens for which excitation as beam inputs is never required and/or for which there is no diametrically opposite element. These considerations complicate the illustrative "half of the amplifiers" and "hemispheric" schemes described here.

A lens similar in function to the two dimensional Luneberg lens is the so called "tin hat" or helmet device (4), a dome shaped parallel plate structure wherein the dome provides a path-length variation to the collector rim ports when one rim port is excited. This path-length variation has the same front forming effect as the dielectric gradient in the Luneberg lens. Figure 9 is a photo of a C-band

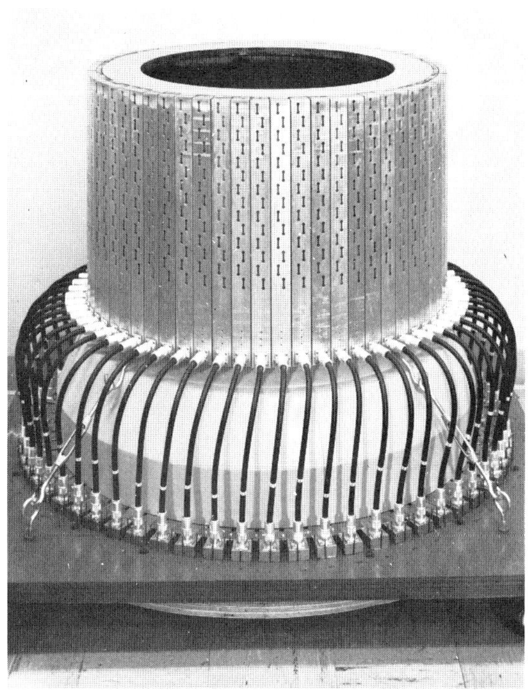

Fig. 9 A Tin Hat Feed.

cylindrical array in which each waveguide column is fed by an output port of the tin hat device at the base of the cylinder. This array was developed by the Hughes Aircraft Company several years ago.

Another feed of this type, that is of the switched beam lens type, is the R-2R lens. Convenient for the feeding of cylindrical or circular arrangements of elements, the lens makes use of the fact that if the elements on a circle of 2R radius are fed through equal line lengths from collector ports on a circle of R radius excited from one port on its circumference, a linear wavefront is formed, providing the ports on the R circle are spaced at twice the central angle of those on the 2R circle. The scheme is shown in Figure 10. It is widely known; the use of it as one of the circular array alternatives at the Naval Electronics Laboratory Center is one of the more recently reported cases (5), and involves a much more elaborate switching scheme to achieve both multi-port excitation for desired weighting and also 360 degree coverage.

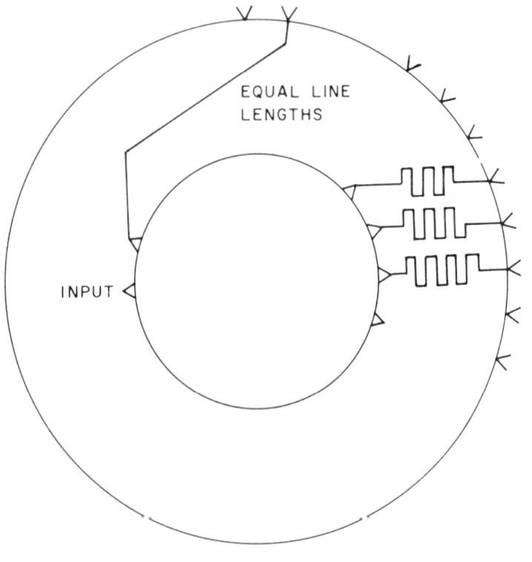

Fig. 10 The R-2R Scheme.

A multiple beam forming and steering feed device that has been used is the matrix of hybrid networks, or the Butler matrix. Such networks were reported nearly ten years ago; the Allen survey (1) and Skolnik (3) both cite several references (6, 7). Figure 11 shows a simple four beam device, composed of variations of the

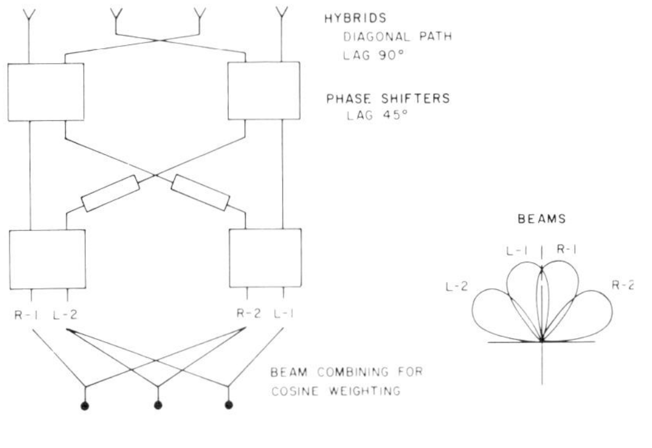

Fig. 11 Four Beam Matrix.

fundamental building block of a quadrature hybrid and a fixed phase shift. The Navy sponsored significant development of such networks at the Naval Research Laboratory (8); an L-band system forming on receive eight beams in azimuth and three in elevation was constructed using eight element matrices in a stripline construction, stacked and crossed to form the pencil beams. Diode controlled switches in a similar construction were used for beam selection. A common practice (as in the NRL program) is to combine the ports of adjacent beams. A cosine aperture weighting then results, rather than uniform, with the often more desirable pattern characteristics. With this arrangement, monopulse processing of the squinted beam type is possible; that is, a magic tee would produce a signal proportional to the amplitude difference of the two inputs. The development of such monopulse switches in three-level stripline was a part of the NRL program. Also included was the specification and acquisition of a 64-element matrix, considered an essential part of any ultimately useful system. Built by Advanced Development Laboratories, Inc., Nashua, N. H., this low power L-band device (1200 to 1320 MHz) was measured by ADL and rechecked, including pattern measurement from a 64-element line array, by NRL. These measurements indicated a phase gradient error of less than 0.2 degrees (with element standard deviation of 6 or 7 degrees for some beam positions), insertion loss around 2 dB typically, beam port isolation greater than 40 dB and VSWR generally less than 1.5 to 1. The beam phase centering requirement for monopulse was treated with compensating fixed phase shift; for at least one beam pair null depth was still limited to approximately 18 dB.

The NRL developments required great patience and iteration in the network construction. The performance was quite sensitive to the tightness of screws holding the stripline plates together, etc. The construction itself was quite heavy, particularly so until crossovers in single layer stripline were developed. A number of concerns have successfully designed and produced crossovers that contribute to this type of feed and, of course, to many other stripline assemblies.

The Butler matrix has been used as a part of more complex array feeds as well, as beam forming networks combining the outputs of subarrays, for example.

A feed system that Hughes Aircraft has developed is the parallel plate radial feed, shown in Figure 12 as it might be used to feed a line array. It is included here in the multiple beam category insofar as higher order TE modes in the circular waveguide in the center correspond to greater phase gradients

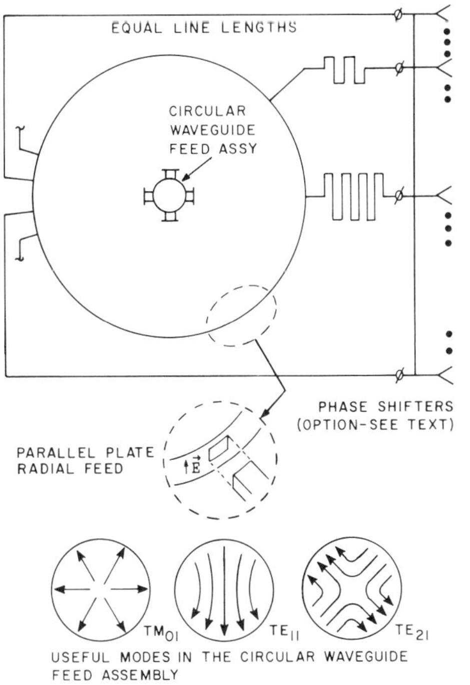

Fig. 12 Parallel Plate Radial Feed.

from port to port at the periphery. Hughes in the early sixties built an S-band radial feed and an appropriate waveguide feed in the center for establishing seven beams, the equivalent of an 8-port Butler matrix. It can be seen that the TM_{01} mode in the circular guide results in a uniform and in-phase excitation at the periphery. The TE_{11} mode results in a complete cosine cycle of amplitude weighting around the periphery. Proper phasing and weighting of the two modes, then, result in an aperture illumination of cosine squared on a pedestal, equiphased. With this excitation, phase shifters as shown could be used for beam steering. On the other hand, the TE_{11} mode and higher order $TE_{n,1}$ modes can be excited in the circular guide feed (with 90-degree phased and oriented pairs, for instance, to create rotating fields in the circular guide and spiral TEM fronts in the parallel plate region, where the pitch of the spiral and, hence, steering angle of the beam, is determined by the particular order of the TE mode) to form multiple beams with no phase shifters.

Hughes engineers report good performance with a simple rectangular waveguide transformation at the periphery as shown in the insert of Figure 12. A version of the feed that could be used to form monopulse squinted beams at the feed structure but with which the external phase shifters would steer the beams has been proposed by Hughes as an improvement in the AN/SPS-33 and is under procurement by the Navy.

Similar parallel plate feed devices of the radial type used for distribution rather than for forming multiple beams have been reported to me by both Sylvania and Raytheon. Raytheon reports an X-band air dielectric device with 4-port contradirectional couplers (small stripline inserts) placed along radii with no wasting at the periphery and claims about 0.5 dB insertion loss. The 324 couplers are arranged in concentric hexagons in this particular experimental device.

The subject of feeds which form multiple beams cannot be left without mentioning the waveguide beam forming matrix (forming 111 receive beams), three of which are used in the AHSR-1 Height Finder of the Federal Aviation Agency and developed by the Maxson Corporation. Each antenna element of a vertical line array is connected to a waveguide line extending behind it. Another waveguide, a "beam forming" guide, taps to these element waveguides with crossed guide couplers. The beam forming guide in which the taps are equi-distant from the elements forms a beam normal to the array. Other beam forming guides tap at a variety of "slants" across the element waveguides, forming beams at various elevation angles. Skolnik (3) summarizes this radar, as well, with a diagram, photo and radar characteristics. A frequent first reaction is to contemplate why the beam forming waveguide using the <u>last</u> tap on each element waveguide is not "cheated" by all the directional couplers that precede it, implying that so little energy is left. A <u>coherent</u> analysis of the <u>entire</u> matrix, of course, discloses that the energy arriving in a plane front at a particular angle reinforces in the beam forming guide "parallel" to the arriving front, with only sidelobe response in the other beam forming waveguides.

Beam Steering Excluded

Relatively "safe" is the classification of feeds for arrays of phase shifting elements into "space" feeds and "constrained" feeds. I suggest "space" is a term slightly preferable to "optical" insofar as fields such as fiber or channeled optics attack the synonomy. Also, I prefer "constrained" as the general class of feeds that route in a manifold structure signal from transmitter or receiver to the array. "Corporate" I think of as a feeding of elements "in a body", i.e. in phase, as an equal line length type of constrained feeding. Corporate feeds are (usually) constrained feeds, but all constrained feeds are not corporate. Finally, and inevitably, there are hybrids of these that I shall describe.

Space Feeds. The paper by Kahrilas (9) in the 1968 special issue of the Proceedings of the IEEE on Electronic Scanning reviews the well-known HAPDAR radar, developed at Sperry, Great Neck, for ARPA and operating at White Sands for the last several years. This well designed space feed array involves the feeding of a circular aperture from the rear, a "feed-through" approach, and, as illustrated in Figure 13, uses 2:1 and 1:1 feedthrough devices in stripline to distribute the desired illumination function across the array and yet preserve collection efficiency. This structure, Kahrilas reports, is called the TACOL for Thinned Aperture Computed Lens; the referenced article provides design and measured data and a number of photographs of the components. A five-horn cluster is used to provide independent sum and difference illumination of the collector surface, and is positioned behind the lens at a distance equal to the lens diameter, i.e. F/D=1. The horn cluster-TACOL combination provides about as much design freedom as has been seen in a space feed system.

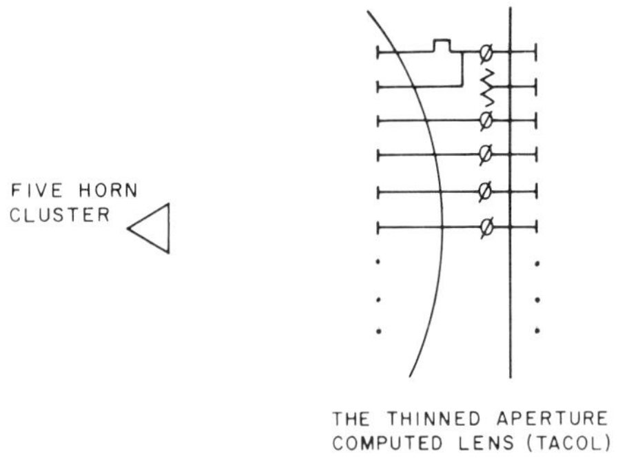

Fig. 13 A Representation of HAPDAR.

The SAM-D radar also uses a feedthrough space fed array, mounted on a tracked vehicle. In this Raytheon development, the array is illuminated by either one of two transmit feedhorns (each connected to a separate transmitter for redundancy) or by the receive horn cluster, depending upon which feed the array is commanded to focus on. In this case we see, then, two interesting features: first, a space duplexing feature involving the refocusing of the array between transmit and receive functions, and, second, a control of which of the two transmitter chains will be employed. The horns are positioned such that the F/D is unity, accommodating the stowage of the array by folding it onto the top of the vehicle. The receive horn cluster is a vertical stack of four multimode horns (10); that is, the elevation patterns are shaped by the four horns combined in the one plane, the azimuth patterns use the multimode sum and difference pattern extraction of each horn in that plane. Raytheon reports they are at this writing approximately 75% completed with the Army Missile Command specified tests on the array, installed in the MSR test facility at Bedford, Mass., since last December. Figure 14 shows an antenna mock-up. Another radar to use the space duplexing feature of feedthrough space feeding is the S-band Missile Site Radar (MSR) (11), another Raytheon development and part of the Ballistic Missile Defense program.

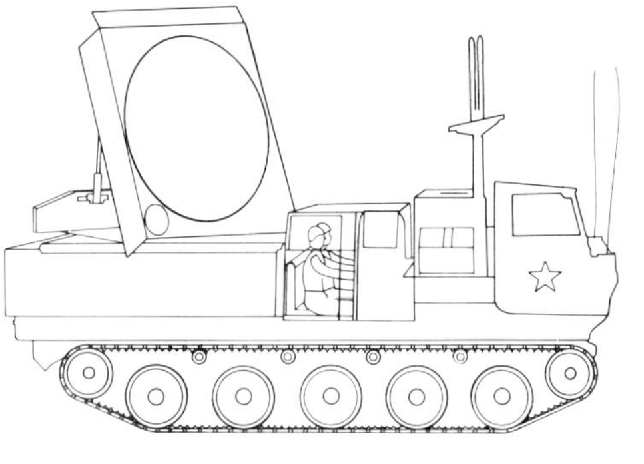

Fig. 14 SAM-D Antenna Mock-Up.

The use of horn clusters to illuminate feedthrough space fed arrays was examined at some length at Westinghouse under the investigation of planar array alternatives late in the Navy's AN/-SPG-59 program. Clusters of a dozen or more horns were examined with rather elaborate coupler and divider assemblies behind the horns to achieve the desired sum and difference illumination functions. Among the alternatives proposed by Westinghouse was a plane-mirror folding of the cone of the feedthrough area behind the array, so that four planar arrays could be fed, individually, in a rather entwined vertical structure, more compatible with the ship architecture at the time. This is a variant I've not seen elsewhere. The work on these feeds was done by Mr. H. Querido , to the best of my knowledge; reports are not known to me.

The other type of space feeding obviously involves illumination of the aperture from the front rather than from the back, so-called "reflect" feeding. While the feedthrough approach offers greater freedom in designing the feedhorn assembly (without blocking the aperture), the reflect array permits the area immediately behind the array to be used for phase shifter control and drive assemblies, structural members, heat removal apparatus and the like. The major disadvantage is that the suspended horn not only has an aperture limitation, but positioning and bracing structure must also be minimized, making it susceptible to inadvertent defocusing in the field.

As in the feedthrough approach, we can look to Raytheon for experience with the reflect approach as well. In the middle sixties, an independent research and development program there involved various array sizes leading to a 1300 element reflect array, circularly polarized. This work led to the X-band reflect array radar, developed by Raytheon and now operating at the Naval Weapons Center, Chine Lake, California. Figure 15(a) shows the array, Figure 15(b) its horn cluster and monopulse assembly. NWC personnel (A. J. Paulsen and others) report that one of the difficulties encountered in the early operation was the alignment of the feed. The system was carefully aligned in June 1969. By September about 6 dB had been lost in monopulse null depth (an average with scan). Some failed phase shifters, a fairly localized set, were replaced and a dB or two was recovered. Moving the feed was not attempted at that time, due to the structural difficulty - it is suggested that a better arrangement of adjustable members would have been possible in design.

A later development at Raytheon is in their RARF radar, a reflect array at K_u-band featuring a polarization agile feed cluster of four horns. Each horn has a pair of orthogonal exciters, but rather than route eight waveguides to this assembly, the horns are summed and the two difference patterns formed for each of the two sets of exciters, so that six waveguides feed this assembly. The weighting and phasing of these two sets of three inputs determine the polarization degree of circularity and sense. The antenna is shown in Figure 16.

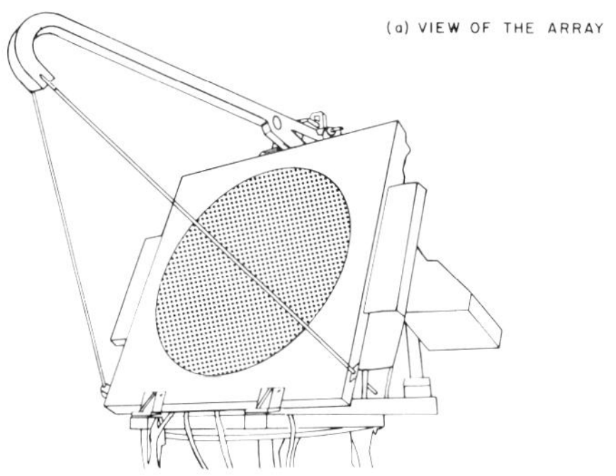

Fig. 15 a Multitrack Radar (MTR)

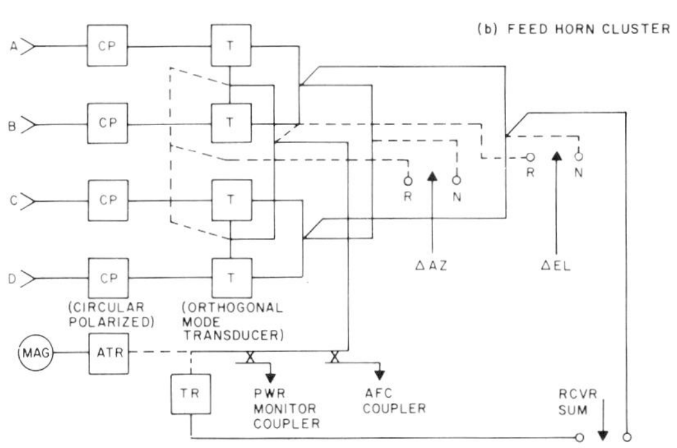

Fig. 15 b Multitrack Radar (MTR)

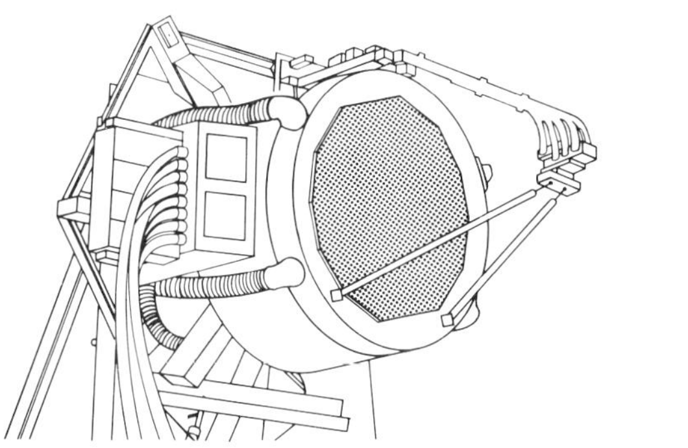

Fig. 16 Raytheon RARF Array.

A curious part of Raytheon's work in reflect arrays is that the use there of row and column phase control (wherein spherical correction is approximated by the product of two "crossed cylinder" factors) is somewhat incompatible with offset feeds. With the exception of some early FAD work, the centered feeds have been used. My recent discussions with Raytheon personnel indicate that if a different control scheme had been used, they might have exploited the offset feed to a greater extent.

At the Naval Research Laboratory, an S-band reflect array radar (RAR) has been operated for a number of years following its development by the Blass Antenna Electronics Company. It has employed a centered feed horn assembly which was a vertical stack of two multimode horns; that is, a horn pair was used to form the elevation difference pattern, the multimode difference pattern extraction was used in azimuth. More recently, a single multimode aperture has been used to form difference patterns in both planes.

Constrained Feeds. The Bendix-built AN/FPS-85 radar (12) at Eglin Field, Florida, is an example of one kind of constrained or channeled feed system, one in which properly phased signals are generated at HF to be fed to each element after up-conversion and amplification. I've considered the beam steering mechanism as not being a part of the feed, although a feed-like structure (a clever tapped coaxial cable structure not unlike a set of serpentine lines in principle) was used in the ESAR predecessor at Bendix, and was proposed in the Navy's SEADAR studies there in the early 1960's, to generate at HF the required row and column phasing. SEADAR was a thorough study, carried well into component and materials selection phases with detailed full-scale mock-ups of several alternatives in the conversion, distribution and amplification region (the "feed") behind the array face. Many useful specifications came from this ambitious program, but finally the sheer size and weight of the potential system precluded its further development for shipboard use. This, like the ESAR and AN/FPS-85, used clusters of simultaneous beams to search for, acquire and track targets. The MAR radar, a predecessor of the PAR (11) under development by General Electric as a part of the Safeguard Program, and the PAR itself similarly use constrained feeds with amplification on either a per-element basis or a subarray basis, and form clusters of beams from which monopulse information is extracted. Some of these systems result in remarkable entanglements of semi-rigid coaxial cable behind the array face, to a degree that cannot be appreciated from the inevitably simplified notional drawings. I have found personal examination of the MAR, the AN/SPG-59, the ESAR and other cabled feeds most sobering indeed and have expressed concern that this "behind the face" area can be too easily slighted in early design. In more recent work under Provencher at the Naval Electronics Laboratory Center in San Diego, feed system hardware has received an impressive amount of early attention. Figure 17 shows this feed area for a part of a multi-band array, with both coaxial cable and stripline power dividers evident. Careful examination reveals the L-band 1:2 divider at the center foreground, the 1:4 S-band divider at the right feeding further dividers in stripline and the two levels of C-band division (at right and in the center stacks) both in stripline. Constrained feeding with coaxial cables forms a part of many other feed systems, such as connecting lenses to arrays, etc. Problems encountered in Navy work have included the flow of center conductor support material; in the AN/SPG-59 program, a spiral type teflon support was found unsatisfactory even after a holding-finger design at each connector was tried. Finally, a five-ribbed spline support was developed and found satisfactory.

Fig. 17 NELC Three Band Feed.

Another type of constrained feed, and one of the more attractive for many applications, is the waveguide series feed. The center-fed series feed with cross-guide couplers at each element was carried into a rather advanced state of development by the General Electric Company, Syracuse, N. Y., under funding by all three services. This feed is illustrated schematically in Figure 18. The G. E.

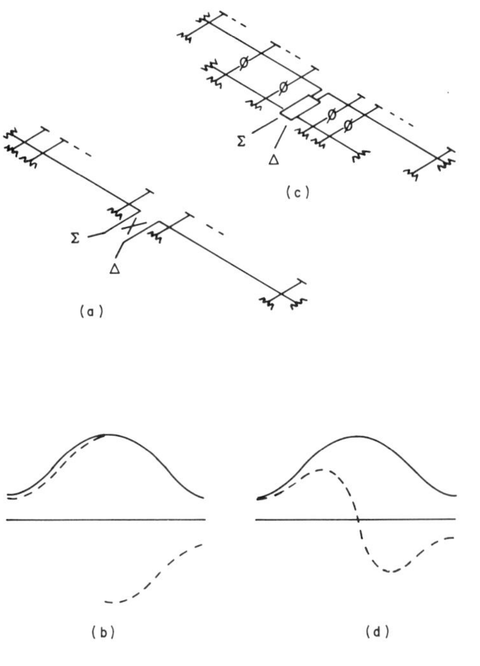

Fig. 18 Waveguide Series Feeds.

work included 1) design procedure, wherein coupler values could be upper bounded for acceptable broad band properties and all coupler values determined for desired illumination functions, 2) construction techniques, resulting in very light weight and inexpensive construction, 3) system trades, wherein illumination functions were investigated in an attempt to meet in one design both sum (as in search) and difference (as in track) pattern requirements, and 4) transient analysis, wherein "response function", "instantaneous bandwidth" or "signal distortion" properties of such a feed were investigated.

The Navy showed considerable interest in the earlier variation, Figure 18(a); in the ASMS program, as did the Army in the AADS-70 and SAM-D programs. To the Navy it represented a very controllable (in design) distribution device which could be used in feeding the rows, say, of a planar array, with a single vertical feed to connect the rows. The form factor, resulting in a more slab-like array structure than space feeding would allow, was attractive. The middle 1960's brought advancement in experimentation that was impressive. In considering it for a multi-function radar application, the Navy was uncertain as to the specific illumination function that would be optimum. While the center feeding with the hybrid allows a delightfully simple extraction of both sum and an interferometer type of difference pattern (in addition to the self duplexing feature through phase shifter control between transmit and receive), it is demanded that one illumination function be built into the device, and treated in halves in difference, as shown in Figure 18(b). This, it was recognized, was a "single degree of freedom" (one illumination function) in the design of a multiple purpose device. G. E. and others (15) dealt briefly with this dilemma, and, just as compromising looked the bleakest, a modified version offering two degrees of freedom in illumination function determination was made available. This author's exposure to it was in association with SAM-D and ASMS joint proceedings in 1967 and, in that effort at least, it was referred to as a "Lopez" feed, crediting A. Lopez of the Wheeler Laboratories with its development. Figure 18(c) shows the basic scheme, and Figure 18(d) the design freedom afforded. G. E. proceeded with this refinement of their previous work and refer to the feed as a "tandem" feed. It should be noted that in forming the sum and difference patterns at the combining network in the center, one need not associate either row of couplers uniquely with the sum or the difference pattern, although such combining is a special case, of course. In general, the two couplers per element and the combining or hybrid network assembly in the center allow individual illumination functions to be achieved for sum and difference pattern formation. The truncation of the back row is indicative of the fact that, in designing desirable sum and difference patterns, the illumination functions need not differ in the edge region, but the central region is vital. A further design choice, the fixed phase shift between coupler rows, is handled in the G. E. work by a positioning or "slant" of one row relative to the other. A program of transient analysis was sponsored by the Navy at G. E. (contract N00024-68-C-1093) in which it was desired to synthesize the impulse response function (itself a function of steering angle) of a "tandem feed" through c. w. sampling. A computer simulation of the feed was made, involving c. w. excitation and a recording of amplitude and phase in the far field for various beam steering angles. This was spot checked with c. w. operation of an experimental array with good agreement. From these data, a convolution with any input waveform is possible to determine the distortion in the signal as might be received at any far field point.

Recent discussions with the Hughes Aircraft Company, Fullerton, disclose their development of a tandem feed in light weight reduced height waveguide and their development of design procedures. An S-band 24 element (12 per half) feed has been built in which no truncation is used in the back row. The particular design involves the sum pattern being uniquely determined by the front row. The design was iterated to balance coupler limitations with efficiency (waster load loss at the end of the feed) for a particular desired pair of illumination functions. In this design, the maximum coupler value is between 6 and 7 dB, and an efficiency of about 83 percent is achieved (power radiated to power input) for patterns of about 28 dB maximum sidelobes, sum and difference. The design process developed at Hughes by DuFort and Jones and soon to be reported in the literature, shows this achievement to be very nearly the upper bound on efficiency for the particular constraints.

Row feeding with such a waveguide structure can, of course, be regarded as "subarray" feeding, but certainly more familiar in the subarray concept is feeding in rectangular and nearly square subarrays, then forming an array by a "brick wall" assembly of these modules. The development of appropriate power dividers has been advanced at RCA, Moorestown, N. J., and at the Applied Physics Laboratory of the Johns Hopkins University. RCA has proposed such an array feed for the Multi-Function Array Radar (MFAR) of the AEGIS weapon system (formerly the ASMS) and are proceeding with that development. Both the RCA approach and the APL work supporting the Navy program are corporate feeds in a stripline type assembly, and are of the "reactive" type, that is at each junction there is no load device, as there is in a "matched" corporate assembly. The 1968 paper by Stark et al (14) describes a subarray feed assembly of the "matched" type, an assembly of waveguide hybrid devices providing 1:24 division. The article points out the salient advantage of the matched type as reducing the effects of reflections which might otherwise cause spurious sidelobes. Recent discussions with Raytheon have disclosed in-house work on an X-band corporate feed in which a resistive film is used with a "tuning fork" power divider in a way that 4-port hybrid performance is achieved, providing a matched corporate feed in stripline.

Figure 19 shows a view of the 1:16 reactive feed (a) and the subarray module (b) with which it is used. The entire module has been power tested with both radiate and short circuit conditions at the elements in the region 25 to 40 kw peak. Incidentally, APL has developed a power testing technique in which a radioactive source is used in proximity to stimulate free charges and resultant arcing. Proper control and statistical interpretation can shorten the time required for power testing markedly. A recent article by Robinson of RCA (15) illustrates their network.

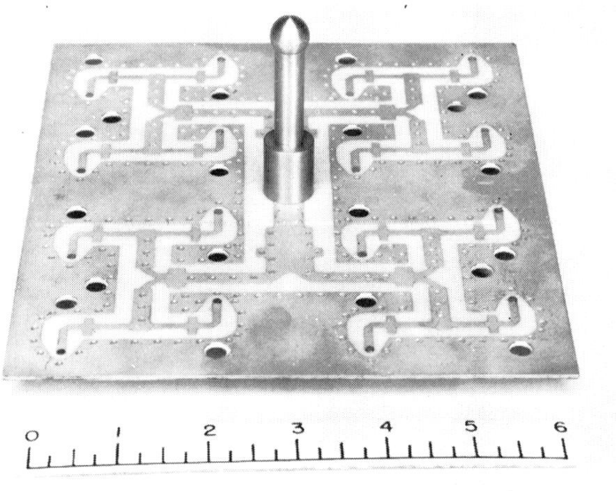

Fig. 19a A Corporate Feed

Fig. 19b A Corporate Feed

Much of the community is involved in the race to "solid state arrays". Proposals generally involve a form of constrained feeding of these amplifiers (or locked oscillators), although not without exception. Meads, Harper and Hsiao at NRL have covered some aspects of feeding (multiple band conversion schemes, etc.) in recent memoranda. Other work involves very compact printed or stripline structure at each element, such as the experimental system at Ryan in San Diego, seen by myself about a year ago.

Hybrid Systems. The Hughes proposal in the recent AEGIS Contract Definition (under the General Dynamics, Pomona, team) is an example of a hybrid system and is illustrated in Figure 20. It

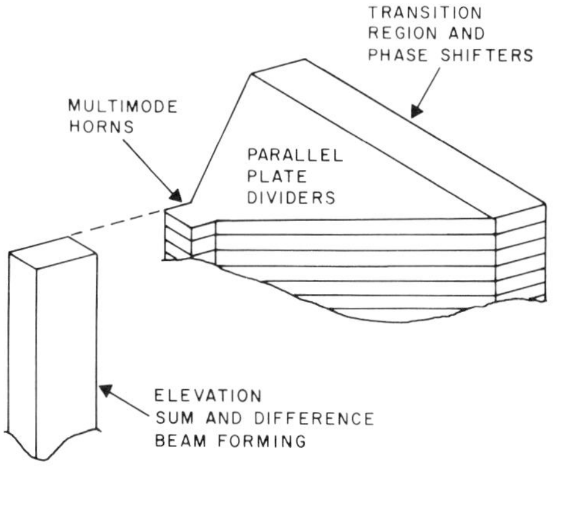

Fig. 20 A Hybrid Feed - Hughes on ASMS.

involves a stack of parallel plate TEM power dividers each feeding a row of elements in a one-dimensional space feed manner. Each of these dividers is fed by a multimode horn, using weighting of the TE_{10}, TE_{20}, TE_{30}, and TE_{40} modes for independent sum and difference pattern determination. (Later work, Hughes reports, has indicated that the TE_{40} contribution is probably not required.) The transition from the parallel plate device to the waveguide housed phase shifters has been studied at Hughes by DuFort and reported in the 1968 paper (16).

A hybrid feed for an array intended to scan only slight angles has been developed by RCA and is the subject of the paper by W. Patton in these proceedings, as well as an earlier report (17). The approach, under an Air Force program called REST, involves illuminating from a point feed a spherical collector surface, then cabling these signals through phase shifters to groups of radiating elements on the array. A notional sketch, Figure 21, illustrates this hybrid scheme.

Subarray feeding has previously been described, both with rows and also the more conventional rectangular subset of elements forming the subarrays. It was quickly established that the rectangular sub-

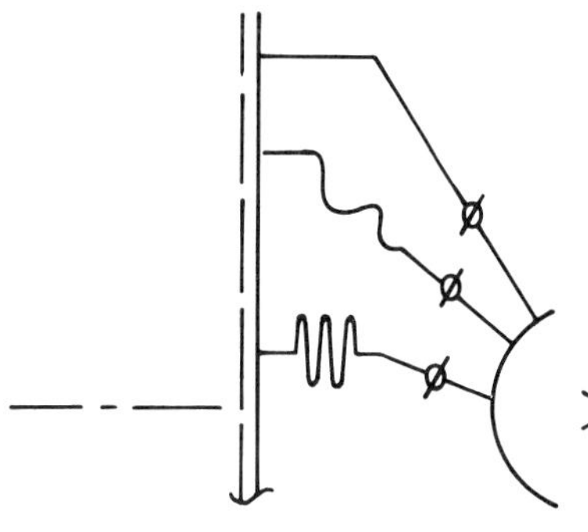

Fig. 21 A Hybrid Feed - RCA REST.

arrays should be stacked with some staggering (as in a brick wall) to reduce illumination quantization effects. This started an evolution of subarray feeding as follows, that leads to our final system surveyed. APL, and probably others, has studied subarrays, using the stripline type feed, that are not rectangular but rather "windmill" shaped so that adjacent subarrays are interlocked, reducing the spatial granularity in the complex illumination function. Next in the evolution, I have discussed with various concerns the interlacing of cabled subarrays, wherein a "subarray" may have a central density of elements but a few elements in extreme positions, further smearing aperture quantization effects by this interlacing. In each of these steps, in this evolution, elements are uniquely identified with a subarray (and, hence, with one amplifier or time delay device, etc.). The next evolutionary step is that of the HIPSAF system at Hughes a hybrid system wherein each element receives a contribution from each constituent "subarray" illumination function. These functions are space fed to the array and are shown as the beams in Figure 22, having been formed by an 8 by 8 crossed Butler matrix assembly. These devices are fed by a 1:32 stripline corporate feed. The Butler assembly has a spherical aperture for focusing at the collector surface of the feedthrough array, F/D of 0.5 is used, with a resulting high illumination efficiency - about 0.2 dB loss in the space feeding is claimed. Figure 23 shows the assembly, the crossed Butler is in waveguide (0.4 dB loss each way is claimed), the feedthrough array (not shown) uses open waveguide elements on both surfaces; the time delay mechanism is by cable lengths in the present breadboard. The entire feed promises very broadband operation, the Butler assembly providing a feed that exactly compensates for the migrating focus, with signal spectral (sideband) constituents, of the feedthrough array. Certainly a feed system offering the proverbial "something for everybody", and a fitting close to the survey.

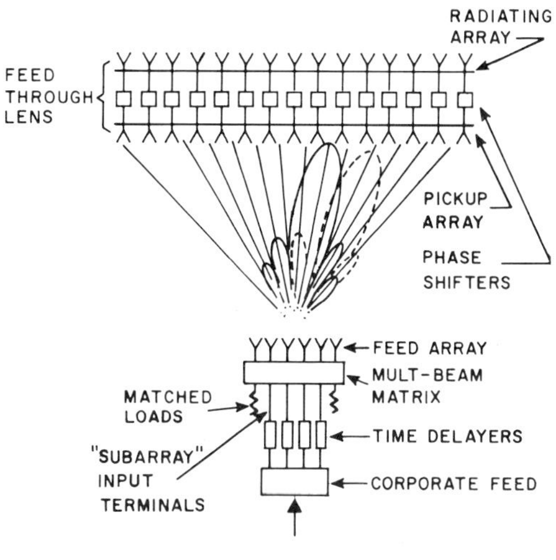

Fig. 22 HIPSAF Diagram.

Fig. 23 HIPSAF Construction.

Design Considerations

This author's participation in a number of the recent evaluations of proposed systems has revealed this fact that subtle design differences of fractions of a dB in loss (as in spillover versus insertion loss) and the like are certainly involved in setting the course of exploratory development, but in engineering development the more mundane matters of form factor, accessibility and cost quickly dominate. I don't find this inappropriate at all; it's simply ironic to devote one's attention to signal distortion in line feeds one minute and hail the rigidity the waveguide runs afford the array the next. A case from the SAM-D/ASMS joint design explorations in 1967 illustrates. One contribution involved a feedthrough array at C-band for the SAM-D use; but to meet the Navy requirement for S-band operation and the Navy's greater need for over-pressure, the peripheral mounting of the larger array simply would not suffice, and a waveguide constrained feed structure was offered instead. Other form factor matters are important, particularly those that bear on maintainability. The choice of feed structure must be jointly considered with phase shifter drive and control circuitry location, drive cabling, air flow for heat removal and the like.

I have previously mentioned "matched" and "reactive" feed types in the context of the constrained power dividers, and implied a concern with the energy reflected within the feed or at the elements. Reflections can cause VSWR sidelobes, secondary beams or a general pattern degradation if not absorbed in many types of systems, not only in the corporate feeds. In reflect space feeding, the effect of the image of the feed horn, in the plane surface of the array face has been observed by pattern measurements. It has come to my attention that in the HAPDAR a slight ripple on the signal due to reflective paths within the chamber behind this feedthrough space fed array can be observed, although no pattern effects are evident. I recall attention in the design of serpentine feeds with reactive taps to slightly staggering the tap points, accepting slight loss in gain and other effects, to avoid the broadside resonance phenomenon in which reflective contributions from each tap add in phase at the input for regularly spaced taps, giving high VSWR and sudden gain loss.

The space feed versus constrained feed argument concerning loss deserves comment. The reflect approach does limit horn cluster design and taper, so that perhaps it remains least efficient with a 2 or 3 dB loss (spillover and face reflections), whereas the feedthrough approach provides considerably more opportunity to shape the illumination for edge efficiency, 1.0 dB is possible. Most constrained feeds must run close to 1.0 dB in loss. The radial parallel plate devices perhaps offer the greatest efficiency, with total insertion loss (taps included) of as little as half a dB, excluding admittedly any necessary cabling to the aperture.

Of course, the feed is one part of a signal filter and hence its contribution to distortion, its transfer or transient characteristics, are of interest. These topics are explored in early M.I.T. reports (18) and more recently by Adams (19) and Kinsey who, with Horvah, is contributing applicable material in these proceedings. Attention was given to feed and component transient characteristics by Ross (20) at Sperry, as well. For most applications of my association, signal bandwidths required do not rival the bandwidths of candidate array feeds for most reasonable criteria for band limits. However, just as need for more range resolution took us away from frequency steering feeds, demand for even greater information content (bandwidth) in a single sample will continue to drive us toward time delay or frequency insensitive array geometries and feed systems.

References

1. Allen, J.L.: Array Radars - A Survey of Their Potential and Their Limitations, The Microwave Journal, May 1962.

2. Allen, J.L.: The Theory of Array Antennas, Lincoln Laboratory Technical Report No. 323, July 25, 1963.

3. Skolnik, M.I.: Introduction to Radar Systems, McGraw-Hill, New York, 1962.

4. Johnson, R.C.: The Geodesic Luneberg Lens, The Microwave Journal, August 1962.

5. Boyns, J.E.: Circular-Array Radar Antenna, NELC Report 1535, February 2, 1968.

6. Butler, J. and Lowe, R.: Beam Forming Matrix Simplifies Design of Electronically Scanned Antennas, Electronic Design, Vol. 9, April 12, 1961.

7. Shelton, J.P. and Kelleher, K.S.: Multiple Beams from Linear Arrays, IRE Transactions, Vol. AP-9, March 1961.

8. Harper, W.H.: A Matrix-Steered Radar Concept Final Report, NRL Report 6863, April 4, 1969.

9. Kahrilas, J.R.: HAPDAR - An Operational Phased Array Radar, Proceedings of the IEEE, November 1968.

10. Howard, D.D.: A New Broadband, Efficient, and Versatile Feed for the Monopulse Tracking Radar, NRL Report 5559, December 29, 1960 (published Confidential, now Unclassified)

11. Johnson, C.W.: Ballistic Missile Defense Radars, IEEE Spectrum, March 1970.

12. Reed, J.E.: The AN/FPS-85 Radar System, Proceedings of the IEEE, March 1969.

13. Kinsey, R.R.: Line Source Aperture Monopulse Excitations, Final Report of Contract AF19(628)-5905, Air Force Cambridge Research Laboratories Report 68-0254, April 1968.

14. Stark, L. et al: Microwave Components for Wide-Band Phased Arrays, Proceedings of the IEEE, November 1968.

15. Robinson, A.S.: Synthesis of Phased Array Radar Systems, RCA Engineer, Vol. 15, No. 3, October-November 1969.

16. DuFort, E.C.: A Design Procedure for Matching Volumetrically Scanned Waveguide Arrays, Proceedings of the IEEE, November 1968.

17. Patton, W.T.: Analysis of Radar Electronic Scan Technique (REST), Air Force Systems Command Report ESD-TR-66-355, July 1966.

18. Allen, J.L. et al: Phased array Radar Studies, Lincoln Laboratories Technical Report No. 236, November 13, 1961, p. 243.

19. Adams, W.B.: Phased Array Radar Performance with Wideband Signals, EASTCON'67, supplement to IEEE Transactions, AES-3, No. 6, November 1967.

20. Ross, G.: Transient Response of Antenna Feed Networks for Array Systems, Sperry Gyroscope Company Internal Report No. EB-5287-0285, May 20, 1963.

Diode and Ferrite Phaser Technology*

D. H. TEMME

Lincoln Laboratory
Massachusetts Institute of Technology
Lexington, Massachusetts

ABSTRACT

This paper surveys configurations for elemental and module phasers of both diode and ferrite types, including digital and analog versions. Driver considerations for all types and their interrelation with the phaser are outlined. Design trends are also noted.

Introduction

The phased-array antenna system configuration influences the phase shifter design. The phaser can usually be categorized as either elemental or module type. The elemental type is used for space, corporate and constrained feed systems (Fig. 1a). The entire antenna face may share a transmitter with other antenna faces via some sort of switching network. (When several transmitter tubes are properly phased, their output can be combined in a hybrid network so that the power sums at the appropriate port). The contrasting module approach (Fig. 1b) has a power source for every an-

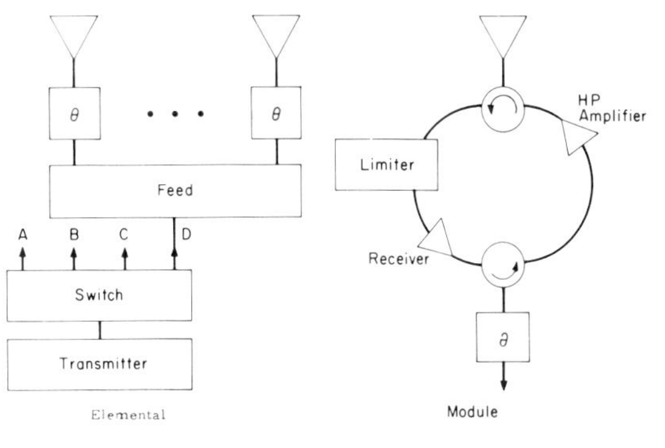

Fig. 1 System Considerations.

tenna element with the phase shifter before the final power stage or transmit and low-noise amplifiers on receive. In the case of the element phase shifter, the key performance requirements are the peak power and average power handling capability, the insertion loss and the cost while in the case of the module phase shifter the key consideration is primarily cost since the power and insertion loss requirements are minimal. A comparison of the elemental and module phase shifters requires a comparison of the two systems. Even a meaningful cost comparison of an elemental phase shifter requires the consideration of many factors. All system costs incurred by the phase shifter should be considered. Besides the cost of the hardware items that should be included are the cost of generating the RF power dissipated by the phase shifter, the cost of flying its weight in case of an airborn system and the cost of spares and maintenance, which is determined by its reliability. All these factors and many others make any general conclusions about phasers very difficult. There are general technological trends which will be discussed. However, cost trends are much less clear; costs are rapidly changing and must be evaluated at each point in time for each particular system and thus are not discussed here.

Diode Phasers

The majority of the diode phasers are of the discrete-step type and consist of a number of cascaded bits. (1-4) Each bit uses one or more diodes which operate as switches biased in forward or reversed polarities. There are three basic ways to employ these switches for discrete phase steps. In each case the electrical length of the signal path is altered by changing the state of the switch between closed and open states. One type depends upon signal reflection from switches incorporated in a transmission line. They are depicted for the shunt switch case in Fig. 2a. In each case the electrical length of the signal path is altered by changing the state of

*This work was sponsored by the Department of the Army.

the switch between closed and open states. In another type, switches are used to switch in and out lengths of transmission lines as shown in Fig. 2b. The loaded line configuration of Fig. 2c shows a pair of switches that alter the reactive loading of a quarter wavelength of transmission line to change the insertion phase. The dual of these circuits can generally be constructed when a shunt switch and shorted stub are replaced by a series switch and open-circuited stub.

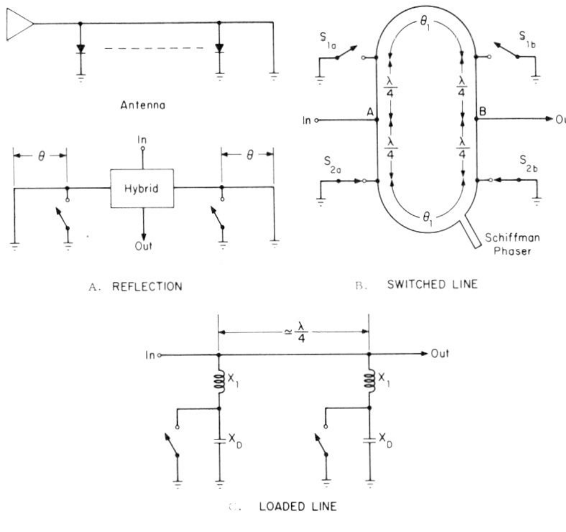

Fig. 2 Diode Phaser Configurations.

The phase shift and power handling capability of these circuits can readily be calculated once a maximum permissible current for the closed switch position and a maximum voltage for the open position are specified. (5, 6) Also, when the ideal switches are replaced by actual diodes, the losses of the circuit can generally be calculated by treating the introduction of losses as a perturbation. The usual practice is to express losses in terms of the ratio of the operating frequency to the characteristic cutoff frequency of the diode.

A few performance generalizations can be given for these circuits. The peak power is inversely proportional to the phase shift increment and is directly proportional to the power handling capability of the diode. (The power handling capability of the diode can be considered as the product of the maximum RF forward current and the maximum RF voltage when reversed biased. Caution must be exercised in establishing the maximum RF current and voltage values. Values probably should be limited by the allowed RF dissipation. This should be checked experimentally at the operating peak power and pulse length for both states). Thus, the phase shift increment of a diode phaser with a large peak power handling capability must be small and the number of diodes required large. The circuit structure with the lowest copper loss for this case is the loaded line structure. An example is Fig. 3

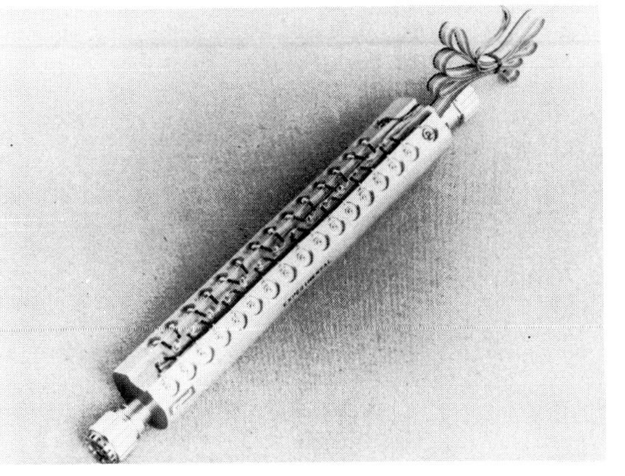

Fig. 3 A Loaded Line S-Band Phaser (Courtesy of Microwave Assoc.).

where a coax construction is used. For a lower power module phaser, a microwave integrated circuit (MIC) phaser would be the most likely choice since it can be easily and compactly fabricated. Fig. 4 shows a switched line implementation. The MIC medium has been very useful in reducing the number of iterative designs required to obtain the desired performance since the parameters are generally better known and the parasitics are small. The use of diode chips eliminates package parasitics.

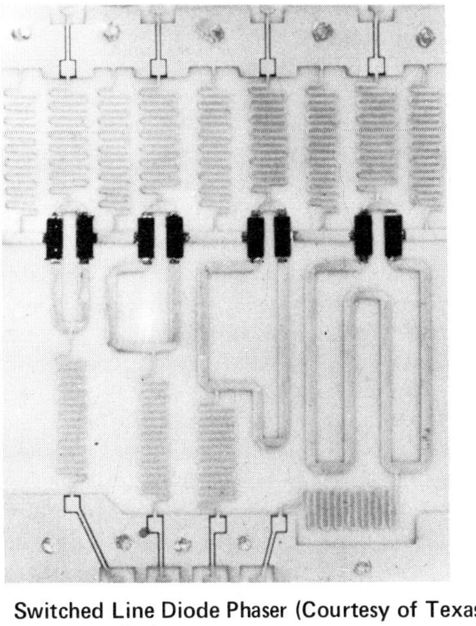

Fig. 4 Switched Line Diode Phaser (Courtesy of Texas Instruments Co.).

The varactor diode analog phased (7, 8) has not been widely used. It is limited in power handling capability and figure of merit. It can be an effective way to provide a number of lower order bits in a low power phaser.

Most diode phaser drivers are the type depicted in Fig. 5a. Many present day diodes are sufficiently uni-

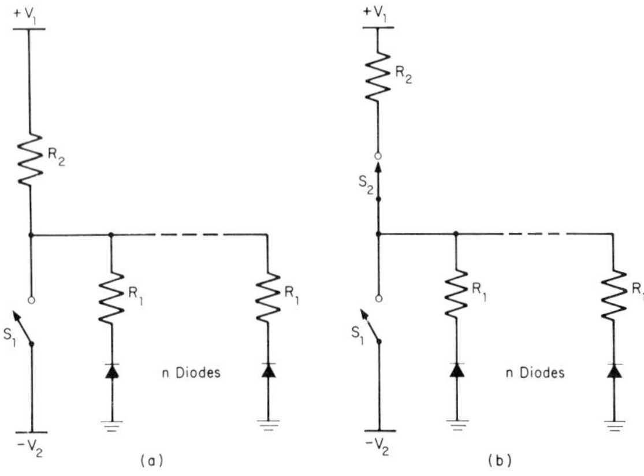

Fig. 5 Driver Circuits for A Diode Phaser.

form so that the current sharing resistors (R_1) are not required. The switching speed to the reverse biased state can be decreased for high power units with large stored charge in the forward biased state with more complex drivers like the one shown in Fig. 5b. However, such speeds are seldom necessary.

Ferrite Phasers

The discussion of ferrite phasers will be restricted to the remanent, or latching, type. The insertion phase of these devices depends upon the state of the magnetization of the ferrite toroid in the phaser. The magnetization is changed by passing a current pulse through the wire which threads the toroid. No holding current is required. An analog phaser (continuous change in phase) can be obtained by partially switching the magnetization to obtain the desired phase shift. Digital devices can be made by cascading ferrite toroids in one structure and switching the magnetization of each toroid between the two remanent states as in the case of a computer memory core. A nonreciprocal phase type of ferrite phaser has been most widely used to date because of its superior performance to the reciprocal type. A recent innovation in reciprocal phaser has markedly improved their figure of merit and is destined to find wider application.

The configuration for the nonreciprocal latching ferrite waveguide phaser is shown in Fig. 6. If the toroid is appropriately placed in the waveguide, a strong gyro-

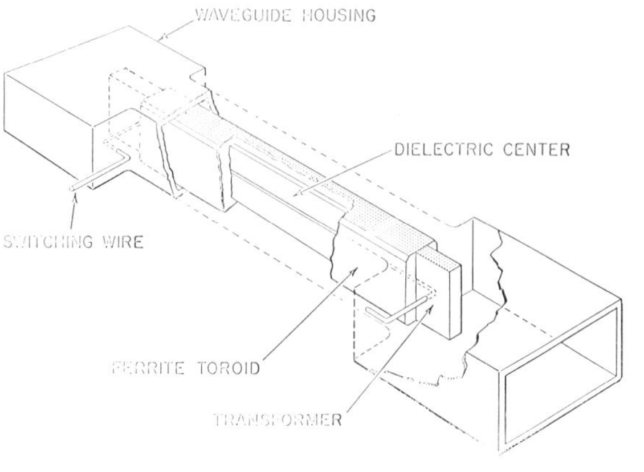

Fig. 6 Nonreciprocal Ferrite Phaser.

magnetic interaction occurs in the region of circular polarization to slow the wave propagating down the guide for one sense of magnetization and only a weak interaction occurs for the opposite sense of magnetization. The calculation of the performance characteristics requires the solution of a boundary value problem. The complexity necessitates a numerical solution with the aid of a computer. (9)

Another popular configuration has been the circular toroid (10) since this configuration could be easily fabricated with ceramic technology existing a few years ago. The performance is good for low power applications. For high power applications the rectangular configuration is preferred since a larger height toroid can be used and still avoid the onset of higher order mode propagation due to dielectric loading of the guide (11) power capability of the phaser increases with guide height since the RF magnetic field intensity can be reduced below the critical RF field intensity where the attenuation of the phaser increases due to spinwave losses. The ability to form and to use "as fired" rectangular toroids (12) make it possible to fabricate low cost rectangular considerations.

Recent work (13) has shown that dielectric loading of a slot of the toroid with very high dielectric constant material can produce better figures of merit, lower switching power, but with a decrease in the peak power capability. An optimum dielectric constant for high power devices may be about 30 to 40; this is a value somewhat greater than previously used and is just becoming available in newly developed materials. Improved measurements of magnetic losses (14) tend to confirm the suspected relation between magnetic loss and the spin wave linewidth (15) a fundamental parameter in calculating the peak power capability of the device. Thus the attainable figure of merit will decrease as the required power handling increases.

The determination (13, 16) that manganese doping of garnet reduces its magnetostriction sensitivity without significantly effecting its microwave properties is particularly significant for S-band devices in which garnet must generally be used because of its reduced temperature sensitivity. A variation in phase shift caused by the variation in remanent flux due to mechanical or temperature stresses via the magnetostriction mechanism has been particularly troublesome when using garnet materials.

A ferrite phaser configuration for microwave integrated circuit applications is the meander line phaser (17) shown in Fig. 7. Its operation depends upon the

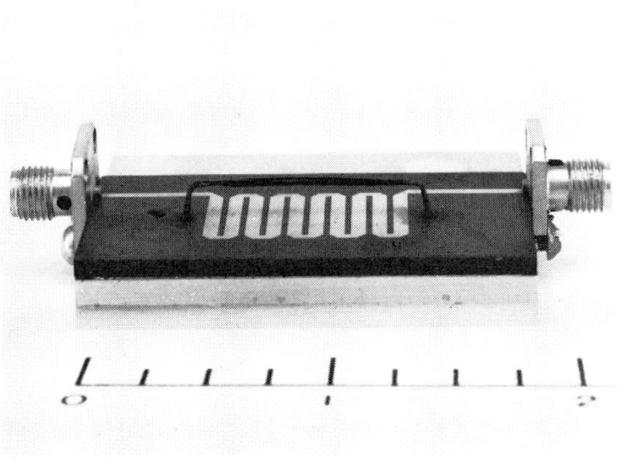

Fig. 7 A Planar Ferrite Meander Line Phaser.

time phasing of the RF magnetic field between adjacent meander lines to produce circular polarization and the attendant gyromagnetic interaction. The strength of the interaction depends upon the direction of the magnetization parallel to the meander line which is switched as a toroid in the plane of the magnetic substrate. These devices are structurally simple and inexpensive. They are low power devices, typically less than 100 watts, and have about a 2 dB insertion loss.

The ferrite phase shifters that have been described are nonreciprocal phase type of devices, that is, their phase shift is different in the two different directions of propagation. This type has been most widely used to date because of their superior loss performance when compared with reciprocal type devices. However, a recent innovation (18) in reciprocal phaser design has markedly reduced their loss and this device is destined to find wider application. The principle of the reciprocal ferrite phaser is depicted in Fig. 8. Two nonreciprocal

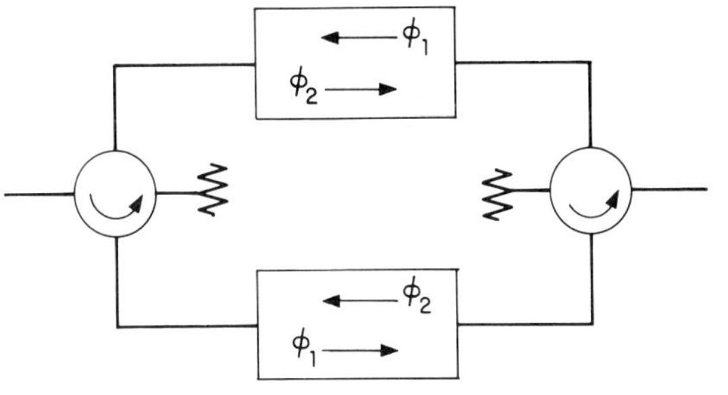

Fig. 8 Recriprocal Phaser Schematic.

phasers connected with two circulators as shown provide reciprocal phase shift between the two ports. The equivalent function is accomplished by cascading a nonreciprocal phase Faraday rotator with a nonreciprocal circular polarizer (19) (ferrite quarter-wave plates) at each end of the rotator (Fig. 9). A

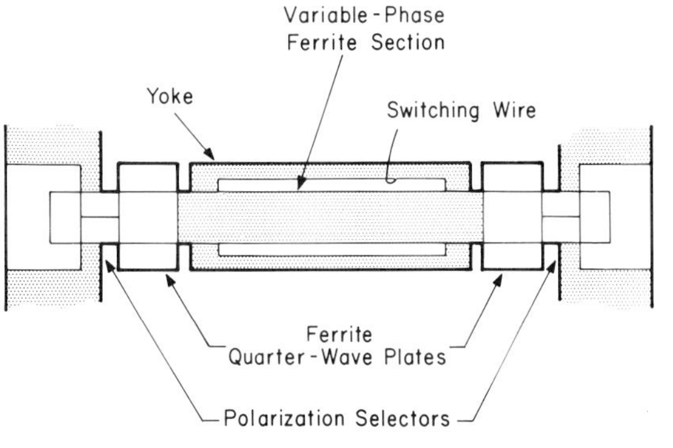

Fig. 9 Reciprocal Phaser Configuration.

ferrite yoke is attached externally on the rotator section to complete the ferrite toroid for latching the magnetization at the appropriate level for the desired phase setting. Phasers with this structure have an insertion loss less than one decibel. Fig. 10 is an X-band unit (20) with a peak power capability of 400 watts and weighs (minus support structure) one ounce. Devices with kilowatts of peak power and a few hundred watts of average power capacity have been built. (21) Their switching speed is limited by eddy currents generated when switching the flux passing through the metallic film waveguide wall of the rotator.

Fig. 10 X-Band Reciprocal Ferrite Phaser (Courtesy of Hughes Aircraft Co.).

There are reciprocal phaser configurations with toroids completely internal to the waveguide housing to avoid switching speed limitations. Since they are more lossy and have a lower peak power capability, they will not be described here.

Two driver circuits for ferrite phasers are schematically shown in Fig. 11. The transistor circuit (Fig. 11a) can be used to switch the toroid between its two

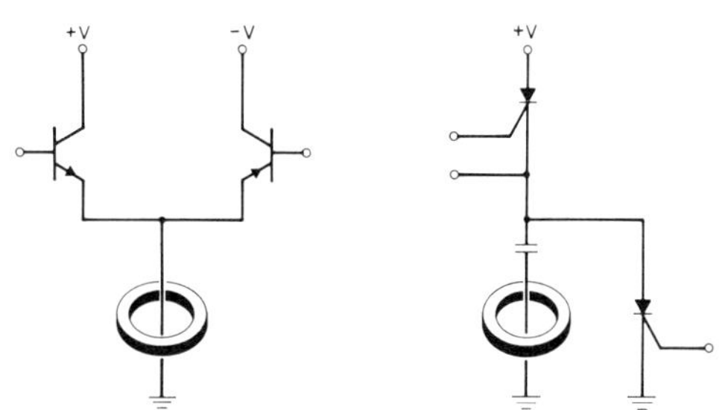

(a) Transistors (b) Silicon-Controlled Rectifiers

Fig. 11 Driver Circuits for Ferrite Phasers.

remanent states for digital phasers. Fig. 11b shows a SCR-capacitor circuit well known because the silicon controlled rectifier (SCR) switches are inexpensive. The recovery time of a low cost SCR limits the switching speed to about thirty microseconds. The switching speed of this type of circuit can be improved by substituting transistors for the SCR. Another technique that is beginning to be used more widely is the use of the transistor amplifier as a flux driver (6) to meter a flux change to obtain the correct phase setting. To a first approximation, setting a phaser toroid at the same flux level, maintains the same phase setting even though the magnetization changes with temperature, and is different from unit to unit due to manufacturing tolerances. A practical switching speed for a flux driven single toroid analog phaser is about ten microseconds. One microsecond switching speed can be obtained by flux driving each bit in a digital configuration.

Phaser Contrasts and Design Trends

Important performance factors for diode and ferrite devices are listed in Table I. Some similari-

IMPORTANT PERFORMANCE CONSIDERATIONS

	Diode	Ferrite
1. Type of Circuit Element	Lumped	Distributed
2. Quality Factors	Cut-off Frequency	Spin Linewidth
	$f_c = \frac{1}{2\pi\sqrt{R_F R_R}\,C}$	Remanence Ratio
3. Power Handling Capability	V_B – Breakdown Voltage	
	I_{RF} – RF Current Limit $\sim I_B \omega \tau$	Spin Linewidth
	Thermal Structure	Thermal Structure
	Pulse Length	Magnetic Moment Temperature Sensitivity
	Peak Internal Temperature Rise	
4. Switching Speed and Control Power	Stored Charge $\sim I_B \tau$	Coercive Force
	Breakdown Voltage	Hysteresis Loss

Table I. Important Performance Considerations

ties and contrasts are noted. That a diode can be represented as a lumped element is a design advantage and permits designing compact devices at the very low microwave frequencies. On the other hand, the distributed nature of ferrite elements helps alleviate mechanical tolerance problems at the higher frequencies. The losses increase and the power handling capability of diodes decreases with increasing frequency. The loss and power handling factors for ferrite, spin linewidth and remanence ratio are frequency independent. Both diode and ferrite phasers have adjustable factors (breakdown voltage, junction area, spin linewidth) for power handling - loss optimization. Reliable values for diode RF voltage and current levels are probably still best determined experimentally. In principle, the average power capability of a ferrite device can be calculated from the magnetic moment-temperature sensitivity, thermal structure and flux driver characteristics, but in practice can more readily be determined experimentally. In both cases, the cost of the driver increases with the increasing power capability of the phaser. The insensitivity of ferrite to transients and nuclear radiation in contrast to that for diodes is sometimes of great importance.

Some contrasts of diode and ferrite phaser device performance are listed in Table II. The table lists what is considered to be the typical range of performance values in devices that have been fabricated;

GENERAL CONTRASTS IN PHASING

	Diode (Step)	Remanent Ferrite (Analog and Step)
Frequency	VHF – X-band	S – K-band
Insertion Loss/2π	0.5 to 2.0 dB	0.5 to 1.0 dB
Temperature Sensitivity	Negligible	< 2 percent
Peak Power	20 kW*	100 kW
Average Power	2000 W	1000 W
Bandwidth	Octave – 10 percent	
Switching Speed	1 nsec – 10 μsec	0.1 μsec – 30 μsec
Control Power Energy	0.1 – 5 W	20 – 2000 μJ

* Pulse Length Dependent

Table II. General Contrasts in Phasing

however, the best values were not all obtained in one device. The lower values of insertion loss are for low power devices. The temperature sensitivity listed for the ferrite phaser is based upon a flux driver. The peak and average power values listed for diode and ferrite devices are considered practical at L and S-bands, respectively. Practical values decrease with frequency in both cases. High switching speeds are usually more easily achieved in diode devices. The control power of the devices are usually comparable for medium power devices for switching rates of 1 to 5 KHz. New cost reduction techniques are continually being found in both technologies; there does not appear to be any fundamental cost considerations that would suggest a trend applicable to all of the many different phased-array antenna applications.

Conclusions

It is apparent that the interrelation of the performance parameters is sufficiently complex (and still rapidly changing) that a carefully detailed examination of alternative approaches is usually required to make an intelligent choice for a particular systems application. At present, diode devices are unique at the lower frequencies and in some functions at the high frequencies. Ferrite devices are unique in some functions at the high frequencies. The area of competitive performance is increasing. This trend is welcome and can be expected to continue.

REFERENCES

1. Temme, D.H.: Phased Array Radar Studies, M.I.T., Lincoln Laboratory Tech. Dept. No. 228, pp. 77-78, DDC249470, August 1960.

2. White, J. F.: High Power, p-i-n Diode Controlled, Microwave Transmission Phase Shifters, IEEE Trans. Microwave Theory Tech. 13, 233-242 (1965).

3. Burns, R.W. and Stark, L.: PIN Diodes Advance High-Power Phase Shifting, Microwaves 11, 38-48 (1965).

4. The Development of Broadband Digital Phase Shifting Networks Using Semiconductor Switches, Hyletronics Corp., Final Rept., Contract No. bsr 91300, February 1966.

5. Hines, M.E.: Fundamental Limitations in RF Switching and Phase Shifting Using Semiconductor Diodes, Proc. IRE 45, 697-708 (1964).

6. Ince, W.J. and D. H. Temme, Advances in Microwaves, Vol. 4 Academic Press, (1969).

7. Hardin, R. H., Downey, E. J.: and Munushian, J.: Electronically Variable Phase Shifter Utilizing Variable Capacitance Diodes, Proc. IRE 48, 944-945 (1960).

8. White, J.F.: Figure of Merit for Varactor Reflection Type Phase Shifters, IEEE NEREM Record pp. 206-207 (1965).

9. Ince, W.J. and Stern, E.: Nonreciprocal Remanence Phase Shifters in Rectangular Waveguide, IEEE Trans. Microwave Theory Tech. 15, 87-95 (1967).

10. Frank, J., et al.: Latching Ferrite Phase Shifter for Phased Arrays, Microwave J. 10, 97 (March 1967).

11. Ince, W.J., et al: A Comparison of Two Nonreciprocal Latching Phaser Configurations, IEEE Trans. Microwave Theory Tech. to be published Jan. 1971.

12. Temme, D.H.: et al; A Low-Cost Latching Ferrite Phaser Fabrication Technique, G-MTT Digest 88-96, (1969).

13. Ince, W.J., et al: The Use of Manganese-Doped Iron Garnets and High Dielectric Constant Loading for Microwave Latching Ferrite Phasers,

G-MTT Digest, 327-331 (1970).

14. Green, J. J.: RADC-TR-69-338, Final Tech. Report (1969).

15. Rodrigue, G. P., et al: Operating Dynamic and Performance Limitations of Ferrite Digital Phase Shifter, IEEE MTT-15, 709 (1967).

16. Dionne, F. G. and Paul, P. J.: Magnetic Anisotrophy and Magnetostriction of $Y_3Mn_{0.1}Fe_{4.9}O_{12}$, Mater, Res. Bull, 4 (Pergamon Press, London, 1969).

17. Roome, G. T. and Hair, H. A.: Thin Ferrite Devices for Microwave Integrated Circuits; IEEE Trans. MTT 16 411 (July, 1968).

18. Boyd, C.R., Jr.: A Dual-Mode Latching, Reciprocal Ferrite Phase Shifter, G-MTT Digest 337-340, (1970).

19. Fox, A.G., Miller, S.E. and Weiss, M.T.: Behavior and Applications of Ferrite in the Microwave Region, B.S.T.J., Vol. 31, No. 1, p. 98, January 1955.

20. Roberts, R.G.: An X-Band Reciprocal Latching Faraday Rotator Phase Shifter, G-MTT Digest, 341-345, 1970.

21. Boyd, C.R., Jr. et al: An S-Band Dual Mode Reciprocal Ferrite Phase for Use at High Power Levels, G-MTT Digest, 346-350, 1970.

Beam Steering of Planar Phased Arrays

T. C. CHESTON

The Johns Hopkins University
Applied Physics Laboratory
Silver Spring, Maryland

Abstract

Methods of steering the beam of a planar phased array are discussed. Special attention is given to the possibility of sharing an electronic driver between several phase shifters to reduce the overall cost.

Introduction

Two dimensional beam steering of planar phased array antennas is obtained by adjusting the phase shifters to give a planar phase front (modulo 2π) normal to the direction of the desired beam. (1-4) In the specific case of microwave ferrite phase shifters the choice of how the phase control is achieved is of special importance since it has a decisive effect on the overall cost of a phased array. The present-day cost of a phase shifter driver with its logic circuitry typically equals or even exceeds the pro-rated total cost of all other components of the antenna system, i.e. phase shifters, feed-network, radiating aperture and structure. Beam steering systems where one driver and logic are shared by two or more phase shifters, therefore, become of interest. Such systems are possible if 720 degrees of phase shift are available rather than 360 degrees and application becomes reasonable if, on transmitting, the aperture is divided into subarrays, each with its own power amplifier.

It is not clear why such a large proportion of the cost is lodged in the driver and learning curves have predicted sharp decreases of this cost. So far, no evidence of such decreases has been seen.

Row and Column Steering

The simplest form of beam steering involves the use of orthogonal row and column steering commands which are in the form mT_{xs} and nT_{ys} for the mth row n^{th} column. T_{xs} and T_{ys} are the computed values for the element-to-element phase increments for the particular beam-pointing direction. To reduce the computations, use may be made of the symmetry planes of the antenna. The row and column commands may additionally contain corrections to compensate for known phase errors such as those caused by optical or series feed systems or other corrections, for example to compensate for changes of phase shift as a function of temperature. At each element the row and column phases must be added by appropriate circuitry to form the command for the phase shifter driver; there will be N drivers for an array of N elements and phase shifters. If two phase shifters giving a total of 720 degrees of phase shift had been in series with each radiating element, one for the row command and one for the column command, then it would have become possible to use, in the limit, as few as one (high power) driver per row and one per column. The total number of drivers for a square array of N elements is then reduced to $2\sqrt{N}$. This saving is offset by the increased phase shifter cost and the increase in insertion loss.

Subarray Steering

Figure 1 shows two methods of beam steering

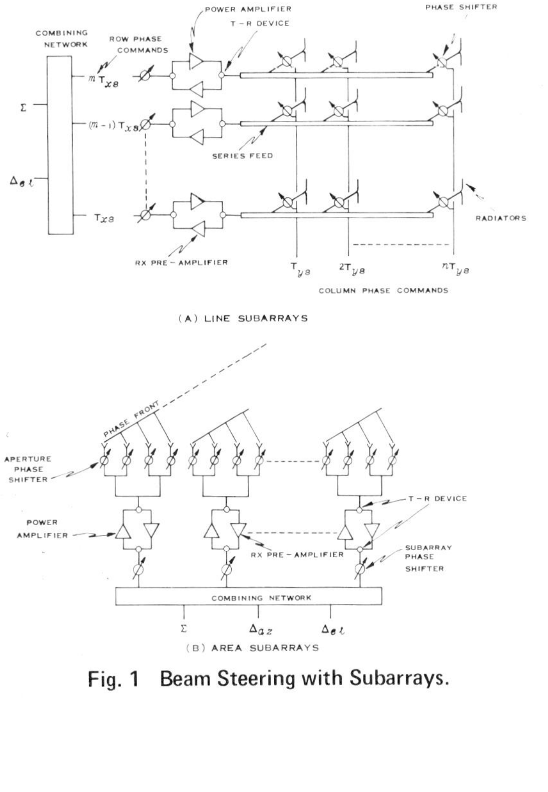

Fig. 1 Beam Steering with Subarrays.

where the aperture is divided into subarrays, each with its own power amplifier on transmitting and preamplifier on receiving. In both cases 720 degrees of phase shift are available. In Fig. 1a the row and column commands have separate phase shifters: a minimum of one driver per row and one per column are possible. The additonal row phase shifters carry only relatively low RF power since they preceed a power amplifier and their insertion loss is therefore compensated. The system of Fig. 1b uses the aperture phase shifters to steer all the subarray beams identically, i.e. phase commands are repeated from subarray to subarray. The system has the potential of providing very wide instantaneous bandwidth by replacing the subarray phase shifters with time delay elements. (5) The subarray phase shifters are set to line up the phase fronts of the subarrays. With N antenna elements and S subarrays, a minimum of $\frac{N}{S}$ + S drivers is possible.

Without distributed amplifers the schemes of Fig. 1 incur disadvantages: The insertion loss is increased (about 1 dB) by the added subarray phase shifters and their design has to be adequate for the higher power that they have to pass when transmitting.

Beam Shaping

Complete control of both amplitude and phase at the aperture is required to obtain a best fit to a desired shape of radiation pattern. It can be shown, however, that phase control alone can give good approximations. In particular, a phase front shaped like a gable or like a circular cylinder may be used to broaden a radiation pattern and more complicated phase distributions can be employed to otherwise shape the pattern or to generate multiple beams (6) (at the cost of 3 dB peak gain every time the number of beams is doubled). This last capability - formation of multiple beams - is readily implemented by row and column systems but runs into trouble with the rough granularity of the subarrays of Fig. 1b. Row and column systems can cope with most phasing requirements including those of optical feeds where the spherical phase correction may be approximated by that of two orthogonal circular cylinders.

Flux-Drive and Digital Phase Shifters (7)

Flux-drive phase shifters are analogue devices and require a fairly complex logic section which not only adds the phase commands for the two planes of scanning, but also limits the flow of flux to values corresponding to the required phase shift. The driver section itself merely provides power amplification and can be relatively simple. In a subarray system the logic can be shared between all corresponding phase shifters. For a given driver, the time taken to set a phase shifter is proportional to the number of phase shifters it serves. The number of phase shifters that can usefully be connected to one driver will be determined, therefore by balancing available power from the drivers against the time budget.

The drivers for digital phase shifters are relatively simple in design, but one driver is required for each of the bits. If each driver is capable of driving one 180 degree bit, then it also can drive two 90 degree bits in the same period of time, or four 45 degree bits and so on. In this way, a given system can use, on the average, significantly less than one driver per phase shifter bit at the cost of more complex wiring. In most cases even a simple driver can accommodate several 180 degree bits and the total number required in a subarray system can be reduced to, at least, an average of one driver per several phase shifter bits.

With analogue phase shifters the beam steering is exact within the accuracy of the devices. With digital phase shifters the beam steering accuracy can be made very high (8-9) by a more complex method of beam steering, not amenable to row and column or subarray computations. In the limit only the smallest increment of one single phase shifter may be actuated to steer the beam through its smallest step.

The most desirable place for the logic and driving circuitry is remote from the phase shifter rather than joined to it to form a unit, for the following reasons:

1. The driver generates heat and should therefore be kept away from the temperature sensitive ferrite.

2.) There are more input leads to the driver than output leads.

3.) The ferrite phase shifter is expected to have very much higher reliability then the driver. The driver should be accessible and capable of being replaced without necessarily disturbing the phase shifter.

4.) More space is available for the driver in a remote position which may allow a design with improved reliability and lower costs.

5.) If a driver is to be used for several phase shifters then remote mounting would appear most convenient.

In the case of an optically-fed lens system the only reasonable available space for the driver is right on the phase shifter.

Phase Quantization Effects

Phase quantization of digital phase shifters can give rise to significant sidelobes (quantization lobes). These may be reduced by decorrelating the phase quantization error (10-13) or by adding a constant phase term in the beam steering command to all phase shifters on successive transmissions or between transmitting and receiving. This will change the digital approximation and hence the value of the quantization lobe.

With subarray or other systems where two phase shifters are in series, the effective number of bits of the total phase shift is reduced. The subarray phase shifter which is used in smaller numbers should, therefore, have more bits giving greater accuracy. This will also reduce grating lobes caused by the repetitive geometry of the subarrays.

References

1. Von Aulock, W.H.: Properties of Phased Arrays, IRE Trans., Vol. AP-9, October 1960.

2. Allen, J.L.: The Theory of Array Antennas, MIT Lincoln Lab. Tech. Report 323, July 1963.

3. Strauss, H.J.: Non-Orthogonal Beam Steering Commands for Array Antennas, Proc. IEEE, Vol. 56, No. 11, November 1968.

4. Hatcher, B.R.: Collimation of Row- and- Column Steered Phased Arrays, Proc. IEEE, Vol. 56, No. 11, November 1968.

5. Tang, R.: Survey of Time-Delay Beam Steering Techniques, This Proceedings.

6. Queen, J.L. and Cheston, T.C.: Generation of Multiple Simultaneous Beams by Aperture Phase Control, Electronic Letters, Vol. 5, No. 20, October 2, 1969.

7. Temme, D. H.: Diode and Ferrite Phaser Technology, This Proceedings.

8. Frank, J. and Ruze, J.: Beam Steering Increments for a Phased Array, IEEE Trans., Vol. AP-15, No. 6, November 1967.

9. Hatcher, B. R.: Granularity of Beam Positions in Digital Phased Arrays, Proc. IEEE. Col. 56, No. 11, November 1968.

10. Miller, C. J.: Minimizing the Effects of Phase Quantization Errors in an Electronically Scanned Array, Proc. 1964 Symp. on Electronically Scanned Array Techniques and Applications, RADC-TDR-64-225, Vol. 1, pp. 17-38, RADC, Griffiss Air Force Base, N.Y.

11. Skolnik, M. I.: Radar Handbook, Section 11, Array Antennas, Cheston T. C. and Frank, J., McGraw Hill, 1970.

12. Valentino, P.: A Technique for Randomizing Phase Errors for a Confined Feed Array, 1960 USNC-URSI Spring Meeting, Commission 5, Session 3.

13. Buck, G. J.: Quantization and Reflection Lobe Dispersion, This Proceedings.

Quantization and Reflection Lobe Dispersion

G. J. BUCK

Aerospace Group
Hughes Aircraft Company
Culver City, California

ABSTRACT

The quantization and reflection lobes that are generated in planar arrays that are digitally scanned in phase can be dispersed by the addition of a phase perturbation into the feed structure. The optimum phase-perturbation function for dispersing the reflection and quantization lobes can be expressed explicitly in terms of the amplitude of the aperture distribution. The amount of reflection lobe dispersion obtained is limited by the directivity of the array and the range of the phase gradients introduced by the phase-perturbation function. A modest increase in the power density of the dispersed quantization lobes occurs if the range of the phase gradients is chosen to optimize the dispersion of reflection lobes rather than the dispersion of quantization lobes.

INTRODUCTION

Quantization lobes are generated by the periodic phase errors present in planar arrays that are digitally scanned in phase. (1) Reflection lobes are also generated in arrays with traveling-wave feeds by the coherent multiple reflections within the regular feed structure. (2), (3), (4) The dispersion of both quantization and reflection lobes can be accomplished by the addition of a nonlinear phase perturbation to the linear phase distribution normally provided by the feed structure. (1), (5) The phase perturbation can be created by the use of a curved traveling-wave feed structure or by the variation of the width of the feed waveguide at different points in the feed structure. Practical limitations upon the curvature of the feed structure and the range of waveguide widths that can be used, place a limit on the phase perturbations that may be obtained. For this and other reasons to be described in this paper, the degree to which the reflection and quantization lobes can be dispersed is consequently limited.

REVIEW OF REFLECTION LOBE DISPERSION

For a given range of the gradient of the phase-perturbation, the phase-perturbation function that causes the greatest dispersion of the reflection lobes can be explicitly described in terms of the amplitude of the aperture distribution. By the method of stationary phase, (6) the optimum phase-perturbation, $\Phi(x)$ for a linear array has been shown (7) to be proportional to the doubly integrated square of the amplitude distribtuion, $A(x)$, that is

$$\Phi(x) \sim \iint A^2(x)\, dx \quad . \tag{1}$$

The dispersion factor, R, as defined in this paper, is the ratio of the maximum power densitites in the reflection or quantization lobes before and after dispersion by the addition of the phase perturbation. As shown in Equation (2), the amount of the dispersion of the reflection lobes that can be obtained in a one-dimensional array is limited by the directivity, D, of a linear array of isotropic radiators, by the range of phase gradients $\Delta \frac{d\Phi}{dx}$, that is introduced by the phase perturbation function, and by the wavelength, λ.

$$R = \frac{4\pi}{\lambda \, \Delta\left(\frac{d\Phi}{dx}\right) D} \quad . \tag{2}$$

In a two-dimensional, rectangular array with a separable aperture distribution, $A(x, y)$, where

$$A(x, y) = A_x(x)\, A_y(y) \quad , \tag{3}$$

the separable components, $\Phi_x(x)$ and $\Phi_y(y)$, of the optimum phase perturbation function for dispersing reflection lobes are related to the aperture distribution in a similar fashion; that is

$$\Phi_x(x) \sim \iint A_x^2(x)\,dx \qquad (4a)$$

and

$$\Phi_y(y) \sim \iint A_y^2(y)\,dy \qquad (4b)$$

The two-dimensional dispersion factor is given by[7]

$$R = \frac{\pi}{D} \; \frac{4\pi}{\lambda L\left(\frac{d\Phi_x}{dx}\right)} \; \frac{4\pi}{\lambda L\left(\frac{d\Phi_y}{dy}\right)} \qquad (5)$$

where D is the directivity of the planar array.

The same technique that has been used for analyzing the dispersion of reflection lobes can be used, with some refinements, to determine a similar optimum phase-perturbation function for dispersing quantization lobes. The limits on the dispersion of quantization lobes can also be described.

QUANTIZATION LOBES

The cause of quantization lobes can be understood if one considers the antenna as a composite array of sub-arrays. Each sub-array consists of a group of elements contained in one period of phase error. The quantization lobes appear as grating lobes of the composite array with an element factor determined by the far-field pattern of the sub-array. As shown by Miller, (1) the phase error, α^ϵ, within the aperture can also be represented by an amplitude function or aperture-error function, A_m^ϵ, that is in phase quadrature with the desired aperture distribution; that is

$$A_m^\epsilon \approx j A_m e^{-jmd_o u_o} \varphi_m^\epsilon \qquad (6)$$

where A_m is the amplitude of the excitation of the mth element, φ_m^ϵ is the phase error, d_o is the element spacing, and the array is scanned to a position given by u_o, where

$$u_o = \frac{2\pi}{\lambda} \cos\theta_o \quad . \qquad (7)$$

The angle, θ_o, is measured from the axis of the array. The far-field pattern of the aperture-error function can then be used to describe the quantization lobes.

The latter representation will be described in this paper as it better demonstrates the dispersion of the quantization lobes by the addition of an aperture-phase perturbation. The representation of the aperture-error function by means of a summation will further simplify the analysis and also allow the direct extension of the results for linear arrays to two-dimensional arrays.

ANALYSIS

First, let us define the periodic quantization function, $G(\sigma)$ as

$$G(\sigma) = \frac{\sigma}{\pi} \quad , \qquad |\sigma| < \pi$$

and

$$G(\sigma + p2\pi) = G(\sigma) \quad , \qquad p = 0, \pm 1, \pm 2 \ldots \quad . \qquad (8)$$

$G(\sigma)$ is illustrated in Figure 1.

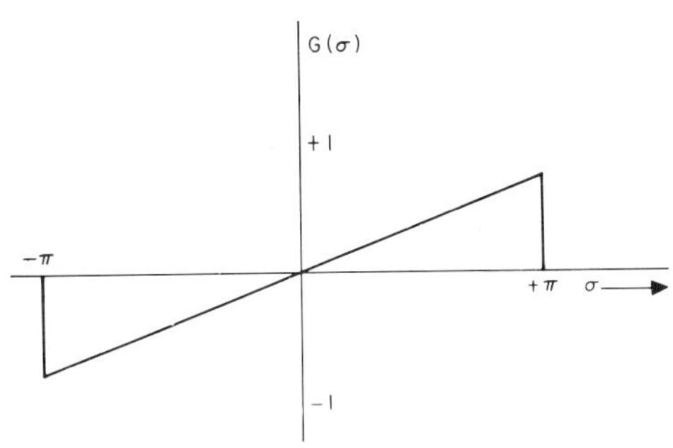

Fig. 1 Quantization Function.

$G(\sigma)$ may be represented by a Fourier series as

$$G(\sigma) = -j \sum_{\substack{n=-\infty \\ n\neq 0}}^{n=+\infty} \frac{(-1)^n}{n\pi} e^{jn\sigma} \tag{9}$$

The phase error and the aperture-error function can be written in terms of the quantization function as

$$\varphi_m^{\epsilon} = \frac{\pi}{2^N} G(2^N m d_o u_o + pm\, 2\pi) \quad , \qquad p = 0, \pm 1, \pm 2 \ldots \tag{10}$$

and

$$A_m^{\epsilon} = j \frac{\pi}{2^N} A_m e^{-jmd_o u_o} \cdot j \sum_{\substack{n=-\infty \\ n\neq 0}}^{+\infty} \frac{(-1)^n}{n\pi} e^{jn(2^N m d_o u_o + pm\, 2\pi)} \tag{11}$$

where N is the number of bits in the digital phase shifter, and the integral values of p represent multiple increments of phase shift between adjacent elements in the array. The far-field radiation pattern, $E^{\epsilon}(u)$, of the aperture-error function in the direction represented by u is given by

$$E^{\epsilon}(u) \sim \sum_{m=-m_o}^{+m_o} A_m^{\epsilon} e^{+jmd_o u} \tag{12}$$

After interchanging the summation of the m elements in the array with the summation of the Fourier series, the far-field radiation from the aperture-error function can be represented as

$$E^{\epsilon}(u) = \sum_{\substack{n=-\infty \\ n\neq 0}}^{+\infty} E_n(u) \tag{13}$$

where

$$E_n(u) \sim -\frac{(-1)^n}{n\, 2^N} \sum_{-m_o}^{+m_o} A_m e^{+jm d_o \left(u - u_o + n\, 2^N u_o + 2\pi \frac{np}{d_o}\right)} \tag{14}$$

The $E_n(u)$ are the quantization lobes and appear in u space at the points for which

$$u = u_o - n\left[2^N u_o + \frac{2\pi p}{d_o}\right] \qquad n = \pm 1, \pm 2, \pm 3 \tag{15}$$

where p is selected to minimize the quantity in brackets. The lobes have an amplitude of

$$E_n\left(u_o - n\, 2^N u_o + p\frac{2\pi}{d_o}\right) \sim -\frac{(-1)^n}{n\, 2^N} \sum_{m=-m_o}^{m=+m_o} A_m \quad . \tag{16}$$

After normalization to the main beam, the amplitude of the quantization lobe has the well known value of

$$E_n = -\frac{(-1)^n}{n\, 2^N} \quad . \tag{17}$$

This representation is applicable for all values of scan angle.

The summation in Equation (14) will also give nonzero values when the entire coefficient of m is a multiple of 2π. The nonzero sums correspond to the higher order lobes generated by an aperture with point sources. These higher order terms can be neglected in arrays with radiating elements of a size comparable to the element spacing.

The addition of a phase perturbation, $\Phi(x)$, into the feed structure will cause the u_o in the argument of the quantization function to vary as a function of its position within the aperture. The expression for the nth lobe becomes

$$E_n(u) \sim \frac{(-1)^n}{n\, 2^N} \sum_{m=-m_o}^{+m_o} A_m e^{-jm d_o \left(u - u_o + n\, 2^N u_p(x) + \frac{2\pi np}{d_o}\right)} \tag{18}$$

where

$$u_p(x) = \frac{d\,\Phi(x)}{dx} \tag{19}$$

If the summation in Equation (18) is approximated by an integral over the aperture, the method of stationary phase (6) can be used to show that

$$E_n(u) \sim \frac{(-1)^n}{n\,2^N} \frac{\sqrt{\pi}\,A(a)}{\sqrt{\frac{1}{2}\,n\,2^N \left.\frac{d^2\Phi}{dx^2}\right|_{x=a}}} , \tag{20}$$

where A_m has been replaced by a continuous function $A(x)$, and the stationary point appears at

$$x = a$$

where

$$\left.\frac{d\Phi}{dx}\right|_{x=a} = u \quad . \tag{21}$$

In order to disperse each quantization lobe uniformly over a range of u, that is for $E_n(u)$ to be constant over the dispersion range of u, the following equality,

$$\left.\frac{d^2\Phi}{dx^2}\right|_{x=a} = C\,A^2(a) \quad , \tag{22}$$

must hold for each value of a, that is the equality,

$$\frac{d^2\Phi}{dx^2} = C\,A^2(x) \tag{23}$$

must hold over the entire aperture. C is a yet undetermined constant.

From Equations (18), (20) and (23) the dispersion, which is defined as the ratio of the dispersed to the undispersed quantization lobe, can be shown to be

$$R = \frac{4\pi\,\lambda}{2n\,\lambda\,2^N} \frac{\left| \int_{-x_o}^{+x_o} A^2(x)\,dx \right|}{\Delta\left(\frac{d\Phi}{dx}\right) \left| \int_{-x_o}^{+x_o} A(x)\,dx \right|^2} , \tag{24a}$$

where

$$\Delta\left(\frac{d\Phi}{dx}\right) \triangleq \int_{-x_o}^{+x_o} \frac{d\Phi^2}{dx^2}\,dx \tag{24b}$$

Because the directivity, D, of a linear array of isotropic elements is given by

$$D = \frac{2 \left| \int_{-x_o}^{+x_o} A(x)\,dx \right|^2}{\lambda \left| \int_{-x_o}^{+x_o} A^2(x)\,dx \right|} , \tag{25}$$

The dispersion can be written as

$$R = \frac{4\pi}{n\,2^N\,\lambda\,\Delta\left(\frac{d\Phi}{dx}\right) D} \tag{26}$$

For there to be but one stationary point within the aperture for each dispersed quantization lobe, the range of the gradient of the phase-perturbation function must be limited to

$$\Delta\left(\frac{d\Phi}{dx}\right) \le \frac{2\pi}{2^N d_o} \tag{27}$$

Although the introduction of a larger range of phase gradients would reduce the contribution from each stationary point, the constructive addition of the contributions from multiple stationary points would be increased. Thus, the maximum dispersion that can be obtained is given by

$$R = \frac{2\frac{d_o}{\lambda}}{nD} \qquad (28)$$

For uniform dispersion over the optimized range, the magnitude of each of the quantization lobes can be determined from Equations (17) and (28) as

$$|E_n|_{\text{dispersed}} = \frac{1}{n2^N}\sqrt{R} = \frac{1}{2^N_n}\sqrt{\frac{2\frac{d_o}{\lambda}}{nD}} \qquad (29)$$

where E_n has been normalized to the field strength at the center of the main beam of the error-free aperture distribution. From Parseval's relation for a Fourier series and Equation (17) the total normalized sidelobe power, P_T, is

$$P_T = \sum_{\substack{n=-\infty \\ n\neq 0}}^{+\infty} |E_n|^2 = \frac{2}{2^{2N}}\sum_{n=1}^{\infty}\frac{1}{n^2} = \frac{\pi^2}{3.2^{2N}} = \frac{3.28}{2^{2N}} \qquad (30)$$

After dispersion the rms power density of the sidelobes, P_{rms}, can be obtained from Equation (29) as,

$$P_{rms} = \sum_{\substack{n=-\infty \\ n\neq 0}}^{+\infty} |E_n|^2 R = \frac{2\frac{d_o}{\lambda}}{2^{2N}D}\cdot 2\sum_{n=1}^{\infty}\frac{1}{n^3}$$

$$= \frac{2\frac{d_o}{\lambda}}{2^{2N}D}\, 2.4 = 2\frac{d_o}{\lambda}\,(0.73)\frac{P_T}{D} \qquad (31)$$

The rms sidelobe power density is defined as the power density averaged over the region of uniform dispersion. If all of the components are orthogonal over the dispersion region, the power densities associated with the various components can be added to give the rms power density. The result shown in Equation (31) is slightly less than Miller's (1) estimate (by a factor of 1.37 for half-wave spacing), because Miller assumes that all of the sidelobe power appears in real space. Actually a significant fraction of the sidelobe power in the higher order terms is reactive and appears in imaginary space.

If all of the components of the dispersed quantization lobes were to add, in phase, in some direction, the peak power density of the dispersed sidelobes, P_p, would be

$$P_p = \left|\sum_{\substack{n=-\infty \\ n\neq 0}}^{+\infty} \frac{1}{2^N_n}\sqrt{\frac{2\frac{d_o}{\lambda}}{nD}}\right|^2 = \frac{2\frac{d_o}{\lambda}}{2^{2N}D}\left|\sum_{n=1}^{\infty}\frac{2}{n^{3/2}}\right|^2$$

$$= \frac{2\frac{d_o}{\lambda}}{2^{2N}D}\cdot 27.3 \qquad (32)$$

Thus, the peak power of the sidelobe that can occur is 11.4 times or 10.55 dB above the rms sidelobe level. The in-phase summation of the first six quantization lobes is 3.98 or 6 dB above P_{rms}.

TWO-DIMENSIONAL ARRAYS

In two dimensions, the phase error in an aperture that is due to quantization errors can also be represented by the amplitude of an aperture error-function, A^{ϵ}, where

$$A^{\epsilon}_m = j\,\frac{A(x_m, y_m)\pi}{2^N}\,G(\sigma_m)\,, \qquad (33)$$

and $G(\sigma)$ is the quantization function that has been defined previously. In two dimensions σ is given by

$$\sigma_m \sim -k \cdot r_m \quad ,$$

where $\bar{k}$ is the propagation vector for the desired beam direction and $\bar{r}_m$ is the position-vector for the mth element relative to the origin of the rectangular coordinate system that lies in the plane of the array. The quantization function can again be expressed by the Fourier series as

$$G(\sigma) = j \sum_{\substack{n=-\infty \\ n\neq 0}}^{n=+\infty} \frac{(-1)^n}{n\pi} e^{jn\sigma} \tag{9}$$

where σ is related to the direction cosines of the propagation vector and the components of the position vector for each element as follows,

$$\sigma_m = 2^N \bar{k} \cdot \bar{r}_m = 2^N (\bar{k}_x x_m + \bar{k}_y y_m) \tag{34}$$

The aperture error-function for an array with a separable aperture distribution can then be written as

$$A_m^{\varepsilon} = -\frac{A_x(x_m)A_y(y_m)}{2^N} \sum_{\substack{n=-\infty \\ n\neq 0}}^{-\infty} \frac{(-1)^n}{n} e^{jn2^N k_x x_m} e^{jn2^N k_y y_m} \tag{35a}$$

where

$$A(x_m, y_m) = A_x(x_m) A_y(y_m) \tag{35b}$$

The addition of a separable phase perturbation function (x, y) to the aperture distribution, where

$$\Phi(x, y) = \Phi_x(x)\, \Phi_y(y) \quad , \tag{36}$$

will effectively cause k_x and k_y to vary over the aperture. If k_x and k_y vary sufficiently slowly, the method of stationary phase can again be used to evaluate the magnitude of the dispersed quantization lobes. Since the series representation of the aperture error-function is separable, the analysis for each of the two variables is identical to the previous analysis of the one-dimensional aperture.

The optimum phase perturbations are accordingly given by

$$\frac{d^2 \Phi_x}{dx^2} \sim A_x^2(x) \tag{37a}$$

and

$$\frac{d^2 \Phi_y}{dy^2} \sim A_y^2(y) \tag{37b}$$

and the optimum range of the gradient of the phase perturbation is given by

$$\lambda \Delta\left(\frac{d\Phi_x(x)}{dx}\right) = \frac{4\pi}{2^N\left(\frac{2d_o}{\lambda}\right)} \tag{38a}$$

and

$$\lambda \Delta\left(\frac{d\Phi_y(y)}{dy}\right) = \frac{4\pi}{2^N\left(\frac{2d_o}{\lambda}\right)} \,. \tag{38b}$$

From Equations (24) and (27) the one-dimensional dispersion with the optimum range of phase gradients can be written as

$$R = \frac{\lambda\left(2\frac{d_o}{\lambda}\right)}{2n} \frac{\left|\int_{-x_o}^{+x_o} A^2(x)\, dx\right|}{\left|\int_{-x_o}^{+x_o} A(x)\right|^2} \tag{39}$$

In two dimensions the equation similar to the above equation is

$$R = \frac{\lambda^2\left(2\frac{d_{ox}}{\lambda}\right)\left(2\frac{d_{oy}}{\lambda}\right)\left|\int_{-x_o}^{+x_o} A_x^2\, dx\right|\left|\int_{-y_o}^{+y_o} A_y^2\, dy\right|}{4n^2\left|\int_{-x_o}^{+x_o} A_x(x)\, dx\right|^2\left|\int_{-y_o}^{+y_o} A_y(y)\, dy\right|^2} \quad (40)$$

where d_{ox} and d_{oy} are the element spacings. Since the directivity of the array can be expressed as

$$D = \frac{4\pi}{\lambda^2}\,\frac{\left|\int_{-x_o}^{+x_o} A_x(x)\, dx\right|^2\left|\int_{-y_o}^{+y_o} A_y(y)dy\right|^2}{\left|\int_{-x_o}^{+x_o} A_x^2\, dx\right|\left|\int_{-y_o}^{+y_o} A_y^2\, dy\right|} \quad (41)$$

the maximum dispersion for each quantization lobe of the two-dimensional array can be written as

$$R = \frac{\left(2\frac{d_{ox}}{\lambda}\right)\left(2\frac{d_{oy}}{\lambda}\right)}{n^2}\,\frac{\pi}{D} \quad (42)$$

For simplicity let

$$2\frac{d_{ox}}{\lambda} = 2\frac{d_{oy}}{\lambda} = 2\frac{d_o}{\lambda} \quad (43)$$

From Equations (17) and (42), the normalized magnitude of the dispersed grating lobe can be expressed as

$$|E_n|_{dispersed} = \frac{2\,d_o}{\lambda}\sqrt{\frac{\pi}{D}}\,\frac{1}{n}\,\frac{1}{2^{2N}} \quad (44)$$

P_{rms} is accordingly given by

$$P_{rms} = \frac{\pi}{D}\cdot\frac{1}{2^{2N}}\left(\frac{2\,d_o}{\lambda}\right)^2\sum_{\substack{n=-\infty\\ n\neq 0}}^{+\infty}\left(\frac{1}{n^2}\right)^2 = \frac{\pi}{D}\,\frac{\left(2\frac{d_o}{\lambda}\right)^2}{2^{2N}}\,(2.165) \quad (45)$$

If all of the dispersed lobes were to add in phase at some point in space, the power density in this peak would be given by

$$P_p = \frac{\pi}{D}\,\frac{\left(2\frac{d_o}{\lambda}\right)^2}{2^{2N}}\left[2\sum_{n=1}^{\infty}\frac{1}{n^2}\right]^2 = \frac{\pi}{D}\,\frac{\left(2\frac{d_o}{\lambda}\right)^2}{2^{2N}}\,\frac{(\pi)^4}{9}, \quad (46)$$

or relative to the rms power density as

$$\frac{P_p}{P_{rms}} = \frac{(\pi)^4/9}{2.165} = 5.0 \text{ or } +7 \text{ db} \quad (47)$$

If just the first six terms of the series add in phase, the power density, P_6, would be

$$\frac{P_6}{P_{rms}} = 3.45 \text{ or } +5.4 \text{ db} \quad (48)$$

OVERDISPERSION OF QUANTIZATION LOBES

The phase perturbation that is introduced into the feed structure of the array disperses both the reflection lobes and the quantization lobes. The range of the phase gradients that is introduced to obtain the greatest practical dispersion of the reflection lobes is normally much larger than the range of phase gradients that provides the optimum dispersion of the quantization lobes. The introduction of the large phase perturbation "overdisperses" the quantization lobes and effectively makes the phase errors in the aperture random. Miller (1) has shown that Rayleigh statistics can be used to describe the overdispersed quantization lobes.

If the quantization lobes are overdispersed or made random, then for, I, independent observations and a confidence level of 90 percent, Rayleigh statistics indicate that the peak level of the sidelobes P_p (90 percent), is given by

$$P_p \text{ (90 percent)} = P_{rms} \ln_e (10 \text{ I}) \qquad (49)$$

If it is assumed, as Miller did, that the number of independent sidelobes in real space is equal to the directivity, D, of the array, then Equation (49) may be rewritten as

$$P_p \text{ (90 percent)} = P_{rms} \ln_e (10 \text{ D}) \qquad (50)$$

The statistical estimate of the peak power density is compared in Figure 2 with the peak power density

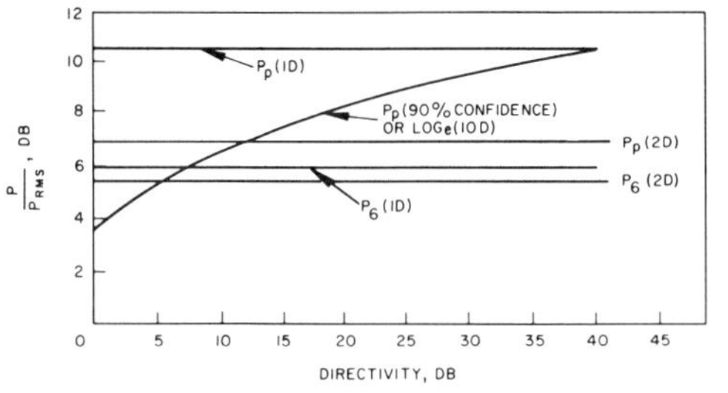

Fig. 2 Peak Power Density Relative to RMS Power Density of Quantization Lobes.

for optimum dispersion and with the value of the peak sidelobe composed of the first six Fourier components for both one- (1D) and two- (2D) dimensional arrays. The peak sidelobe level, P_p, that could occur if all of the dispersed components added together may not necessarily occur. On the other hand, the in-phase addition of the first six components, P_6, is almost certain to occur at some point in the radiation pattern of an array of moderate or large size. The actual peak quantization lobe that would be observed, with optimum uniform dispersion, can be expected to lie between P_p and P_6. The over-dispersed sidelobe level is represented by P_p(90 percent). The comparison in Figure 2 shows that in the two-dimensional arrays, the optimum phase perturbation produces lower quantization lobes than the larger perturbation that overdisperses the lobes. The accuracy of the comparison, however, is open to some question, since the statistical calculation of the sidelobe level is based on an estimate of the number of independent combinations of sidelobes that effectively occur in the radiation pattern of an array. This number has been assumed to be equal to the directivity of the array and may perhaps differ from this number by a factor of two or so. Fortunately, for gains of 20 dB or so, a change in D by a factor of two changes the statistical estimate by less than 0.5 dB. The comparison can, accordingly, be considered to be accurate for the larger antennas.

REFERENCES

1. Miller, C.J.: Minimizing the Effects of Phase Quantization Errors in an Electronically Scanned Array, Proc. of Symposium on Electronically Scanned Array Techniques and Applications, RADC-TDR, Volume 1, July 1964, pp. 17-38.

2. Kurtz, L.A. and Elliott, R.S.: Systematic Errors Caused by the Scanning of Antenna Arrays: Phase Shifters in the Branch Lines, IRE Trans. Antennas Propagation, 4, 619, 1956.

3. Hansen, R.C.: Microwave Scanning Antennas, Vol. III, pp. 40-43, Academic Press, New York, 1966.

4. Spradley, J.L.: Systematic Errors in the Scanning of Traveling-Wave Antenna Arrays, TM 474, Hughes Aircraft Company, Culver City, California, 1957.

5. Valentino, P.: A Technique for Randomizing Quantized Phase Errors for a Confined Feed Array, URSI Spring Meeting, 1969, Wash. D.C. p. 59.

6. Silver, S.: Microwave Antenna Theory and Design, pp. 119-122, Dover Publication, Inc. New York, 1965.

7. Buck, G.J.: Reflection Lobe Dispersion in a Phased-Array Antenna, to be published as a communication in IEEE Transactions on Antennas and Propagation.

The Step Recovery Diode: An Analog Phase Shifter

JAMES P. WEIR, Jr.
RAYMOND C. KENT

Teledyne Ryan Aeronautical Co.
San Diego, California

ABSTRACT

A step recovery diode (SRD) is used frequently in multiplier circuits to generate microwave frequencies. The phase of the multiplier output may be controlled by changing the bias applied to the diode. Thus the circuit performs as a combined multiplier and analog phase shifter. This circuit has many applications in microwave phased array antenna systems. Experimental transmitting and receiving phased arrays are descri bed.

INTRODUCTION

Step recovery diodes are often used as frequency multipliers in solid state microwave circuits. The multiplier output is a harmonic of the input frequency and is phase coherent with the input. The phase of the output may be varied by controlling the bias applied to the diode. Thus the circuit performs as an analog phase shifter as well as an efficient frequency multiplier. This circuit has many applications in microwave phased array antenna systems.

STEP RECOVERY DIODE

The step recovery diode or SRD is a charge-storage diode also known as a snap varactor diode, punch-through varactor or Bimode® diode. These diodes are designed so that most of the minority carriers injected under forward bias are stored near the junction and are immediately available to contribute to reverse conduction. This is accomplished by a doping profile having a steep gradient near the junction or by a *pin* profile with a narrow *i* region (1).

SRD FREQUENCY MULTIPLIER

Step recovery diodes have been extensively studied (2-4), and have found wide applications as simple, efficient and stable high order multipliers.

A step recovery diode has two discrete states of reactance and switches very rapidly between these two states. Figure 1 shows the basic operation

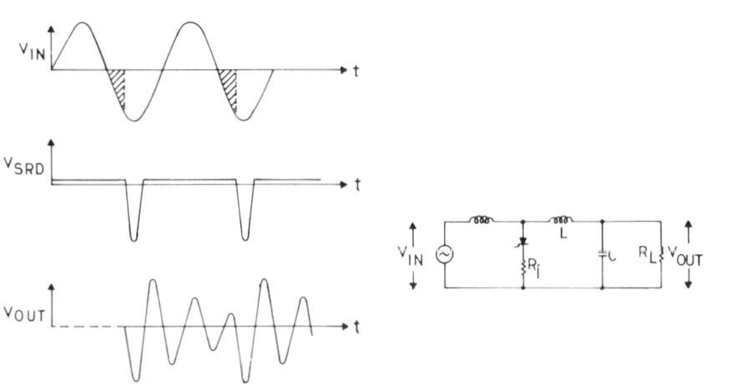

Fig. 1 SRD Multiplier Waveforms.

of an SRD multiplier. During the forward bias of the RF cycle, the impedance is very low and charge is stored near the junction in the form of minority carriers. During the reverse cycle of RF, these minority carriers have been swept out of the junction. At this point the diode steps to a high impedance state.

When the diode steps, a sharp transient is formed across the diode. This transient is allowed to ring down in an output resonator to produce the output which is a harmonic of the input frequency. These multipliers operate with efficiencies greater than 1/N at output frequencies up to 13.5 GHz.

SRD PHASE SHIFTER

An SRD frequency multiplier may also be used as a phase shifter by controlling the bias applied to the diode. Figure 2 shows the multiplier with the addition of bias to the diode. The dashed line is the input voltage minus the bias voltage. Since the diode with bias is driven in the forward cycle for

® Varian Associates

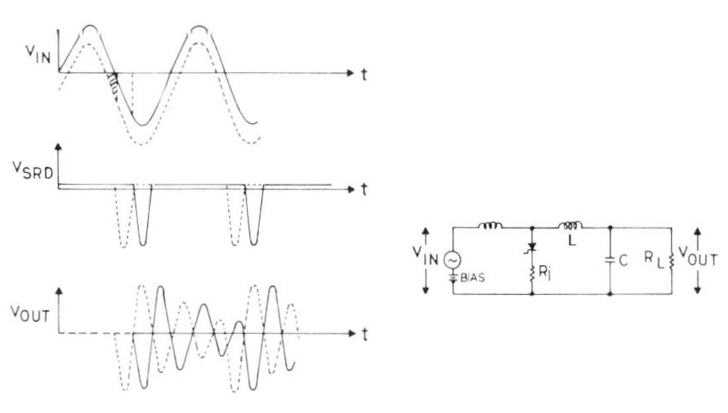

Fig. 2 SRD Multiplier Waveforms with Bias.

a shorter period of time, the stored charge is less and is swept out of the junction at an earlier time in the reversed cycle. Thus the bias changes the point on the input cycle at which the diode steps and advances or delays the time of the sharp transient. This shifts the phase of the output with respect to the input. The amount of phase shift depends on the order of multiplication.

A linear phase shift as a function of bias current may be obtained by proper design of the bias network. The output may also be temperature compensated by a proper choice of the temperature coefficient of the bias network resistors.

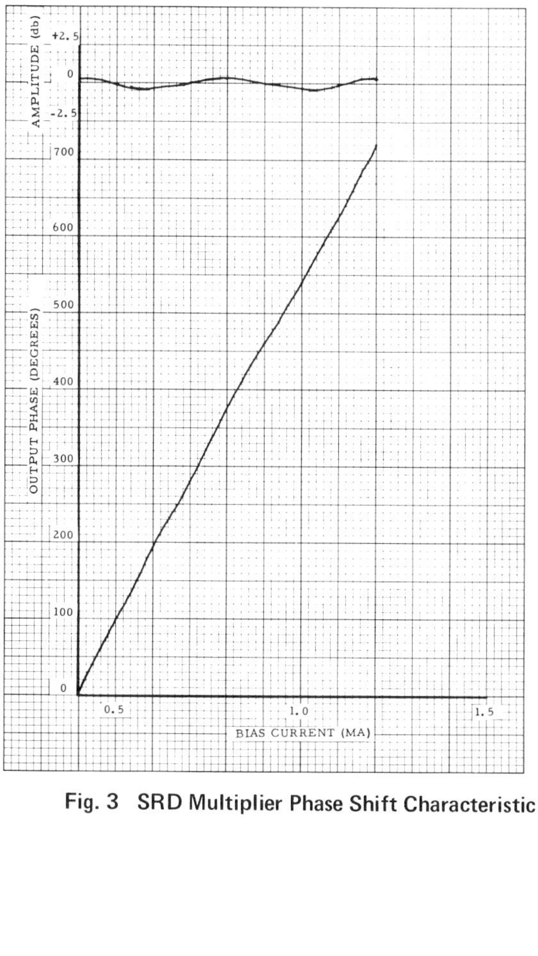

Fig. 3 SRD Multiplier Phase Shift Characteristic.

Figure 3 shows the phase shift characteristic of times ten multiplier with an output frequency of 2230 MHz. As the bias current is increased from 0.4 to 1.2 milliamperes the output phase shifts 720 degrees. The maximum deviation from linear is less than four degrees. The output power shown in the upper curve is approximately four milliwatts and is constant within 0.5 dB during the phase shift. The maximum rate of change in bias current is limited by the capacitance of the diode and time constant of the bias circuit. Phase shifts of plus and minus 360° have been achieved at rates up to 100 KHz. No hysteresis in phase shift occurs as the bias is changed.

PHASED ARRAY CIRCUITS

SRD phase shift multipliers have been utilized in integrated phased arrays. These arrays consist of a number of small microwave receiving or transmitting modules. Each module has an antenna or radiating element and the modules are grouped together to form an array. Beam steering of the array is accomplished by shifting the phase of the signal received or transmitted by the antenna element.

A UHF source is usually used to drive a transmitting array. This UHF power is divided into equal parts to drive the transmitter modules. Each module contains a buffer amplifier and an SRD multiplier as shown in Figure 4. The output of the multiplier is passed through a bandpass filter to suppress the unwanted harmonics. This output may be applied directly to the antenna or through a power amplifier to obtain higher power output.

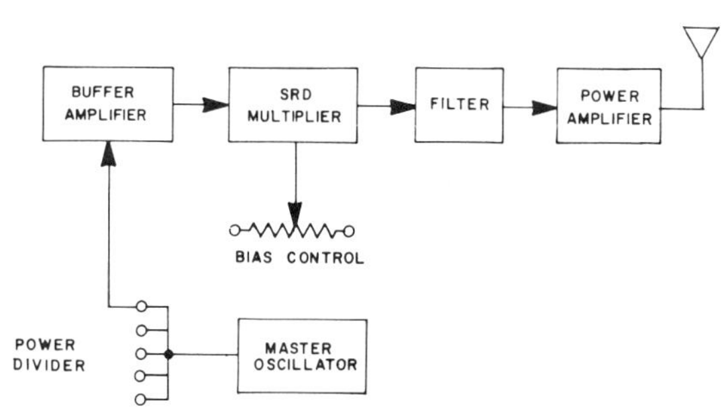

Fig. 4 Transmitter Module Block Diagram.

In receiving arrays the SRD phase shifting multiplier is used as a local oscillator in each receiving module. Shifting the phase of the local oscillator produces a corresponding phase shift in the intermediate frequency. This is equivalent to shifting the phase of the received signal. The block diagram of a receiver module is shown in Figure 5. The output of

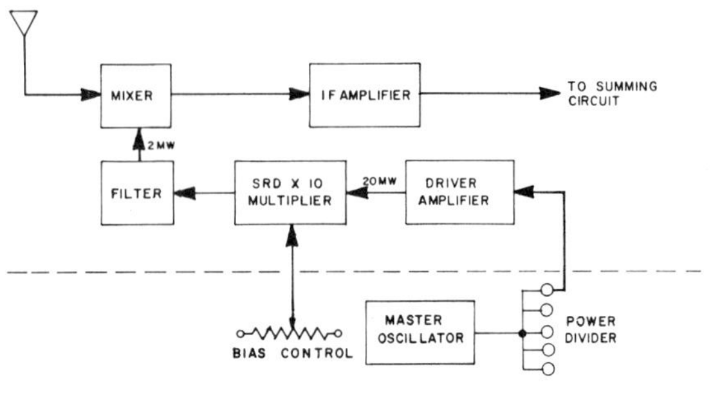

Fig. 5 Receiver Module Block Diagram.

a reference oscillator is divided into equal parts and applied through isolation amplifiers to the input of the SRD multipliers. The output of each multiplier is passed through a bandpass filter to the signal mixer. This filter suppresses the undesired harmonics. Measurements have been conducted to determine the phase noise introduced by SRD multiplier (5). The noise measured is typically lower than single-cavity klystrons.

The bias voltages required to steer the beam in a phased array may be obtained from a simple resistive matrix shown in Figure 6. This matrix is connected for

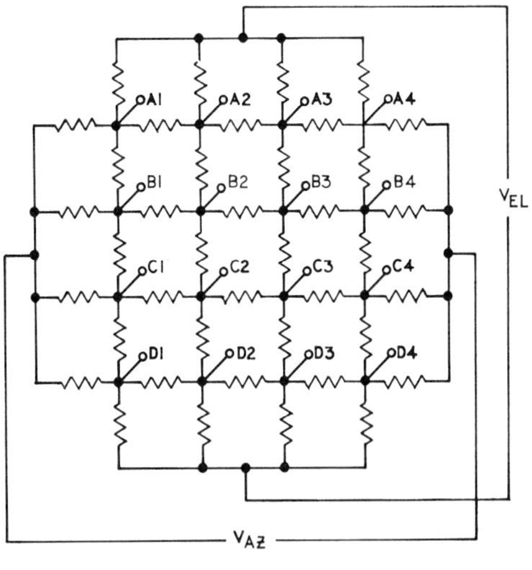

Fig. 6 Steering Matrix.

a sixteen element (4 x 4) planar array. Each resistor junction is connected through a current source to the bias of a corresponding SRD in the array. The azimuth steering voltage is applied across one axis while the elevation steering voltage is applied across the other axis of the matrix. Since the phase is linear with bias, the phase will vary in a $\phi, 2\phi, 3\phi \ldots n\phi$, progression required to steer the beam of the array.

In large arrays, the voltages from the matrix pass through logic circuits that step the bias applied to the multipliers. This steps the phase in 360 degree increments. When the matrix output increases above the voltage required to shift the phase 360°, the logic steps the voltage back to 0° and then on to the equivalent phase shift required. Thus, the SRD multipliers are operated over the most linear 360° portion of their phase shift characteristic.

PHASED ARRAY APPLICATIONS

SRD multipliers have been used as the phase shifting device in receiving and transmitting arrays at S, C, and X-band. Figure 7 shows a stripline

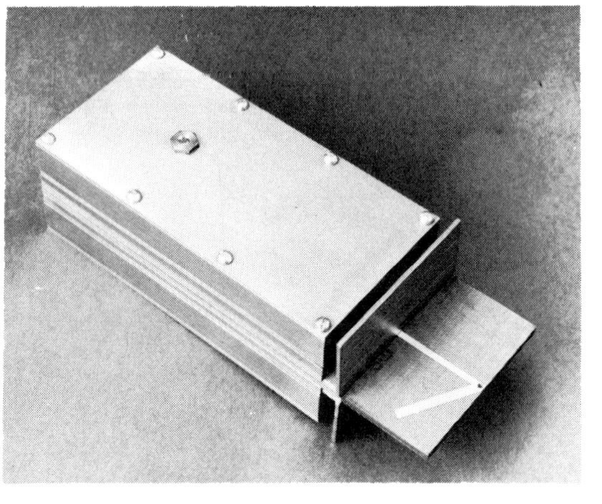

Fig. 7 S-Band Receiver Module.

module of an S-band receiving array. The multiplier from this module is shown in Figure 8. This is a times ten multiplier with an output frequency of 2230 MHz. The input is applied to the large choke at the end of the resonant line. The SRD is operated in the series mode and is located one-third of the line length from the input choke. The resonant line is short circuited at the end by a plated-through hole. The output is coupled across a small gap to the output line. This line is connected to an interdigital bandpass filter which suppresses the spurious harmonics. The output of the filter is connected to the signal mixer on an adjacent stripline board.

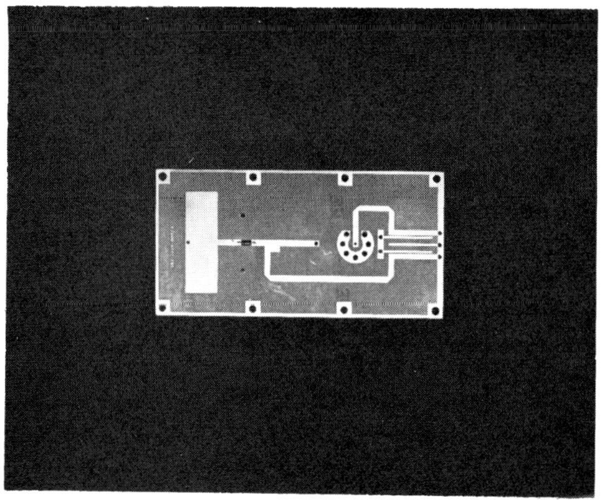

Fig. 8 S-Band SRD Multiplier.

Figure 9 shows the MIC circuits on alumina substrates of a C-band receiver module. This receiver array operates at a frequency of 5748 MHz and an IF output of 60 MHz. The X8 multiplier operates in the shunt mode with the SRD connected between the resonant line and ground. This receiver module is a one inch cube.

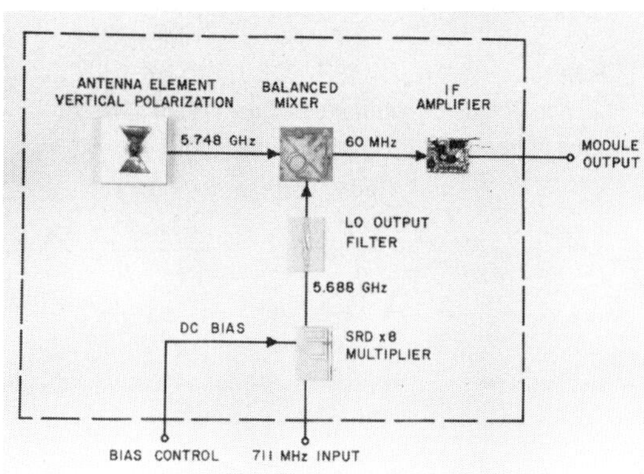

Fig. 9 C-Band Receiver Module.

Figure 10 shows the test fixture with sixteen modules from a C-band Combined Acquisition and Tracking radar. Each module contains a pulsed SRD transmitter and an SRD L. O. receiver duplexed to a common bowtie antenna. The radar transmits a 0. 5 microsecond pulse with a peak power of 300 milliwatts per module.

Figure 11 is a thirty-six element (6 x 6) S-band receiving array. This receiver operates at 2290 MHz with an IF of 60 MHz. The antenna beamwidth is approximately 18 degrees wide in azimuth and elevation. The receiver beam may be scanned ±60° in any axis at scan rates greater than 10 KHz.

Fig. 10 C-Band Combined Acquisition and Tracking Radar.

Fig. 11 S-Band Receiver Array.

Figure 12 shows the receiver antenna patterns at various scan angles with a constant gain setting. These patterns compare very closely with the calculated patterns and demonstrate the precise phase control achieved by SRD phase shifters.

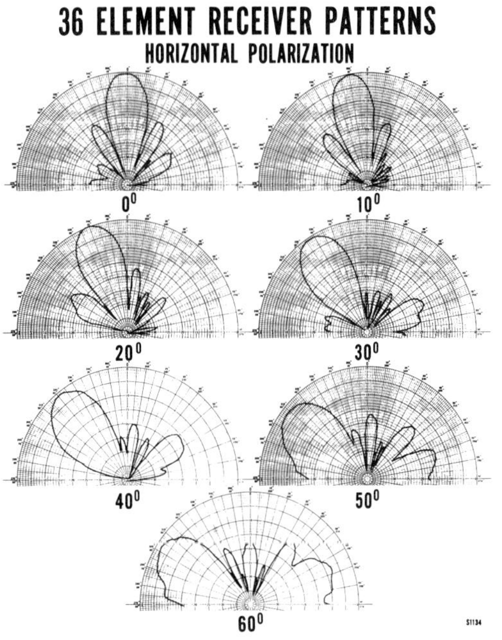

Fig. 12 Phased Array Antenna Patterns.

CONCLUSIONS

The step recovery diode multiplier is a combined analog phase shifter and frequency multiplier. It is an extremely simple, stable and efficient method of generating phase controlled microwave power. A continuous linear phase shift of several hundred degrees may be obtained at very rapid rates with very low scanning power. This solid state device is highly reliable and has many applications in phased array systems.

ACKNOWLEDGMENT

The authors wish to thank John Aasted and Robert Apgar for their helpful comments.

REFERENCES

1. Mall, J., Krakauer, S., and Shen, R.: P-N Junction Charge Storage Diodes, Proc. IRE., January 1962.

2. Krakauer, S.S.: Harmonic Generation, Rectification and Lifetime Evaluation With the Step Recovery Diode, Proc. IRE, July 1962.

3. Interim Report No. 29169-2, SRD Frequency Multipliers, Ryan Electronic and Space Systems, Contract No. NAS 8-20257, July 1966.

4. Harmonic Generation Using Step Recovery Diodes and SRD Modules, Application Note 920, Hewlett-Packard Associates, 1968.

5. Johnson, S. L., Smith, B. H., and Calder, D. A.: Noise Spectrum Characteristics of Low Noise Microwave Tubes and Devices, Proc. IEEE, February 1966.

Time Scanned Array Radar

GEORGE J. VOGEL, B. J. LYONS,
Rome Air Development Center
KEN PARSONS, L. W. BEACHLER
Bendix Communications Division
Towson, Maryland

1.0 ABSTRACT

The time scanned array radar is an electronic beam steering technique for narrow or wideband signals. A time delay or phase gradient is provided on the array for beam steering by control of the radar signal. No beam steering control is required at the element level.

2.0 SUMMARY

2.1 THE TECHNIQUE

Time Scanned Array Radar is an array which is steered by generating a series of pulses which have their relative phase and their interpulse spacing accurately controlled. Each element contains a fixed delay line, the length of which is a linear function of the position of the element in the array and not the direction to which the array is steered. The pulse train is radiated from all elements without any change. The fixed delays are such that at the nth element, the first pulse of the pulse train is radiated at the same time the nth pulse is radiated from the first element. The relative phases of these pulses will determine the array beam direction. If each pulse of the pulse train has a relative phase corresponding to the phase required at each element for a particular direction, then the array forms a beam in the desired direction. Each element of the array radiates a pulse train with the same pulse-to-pulse gradient, but displaced in time. The sum of these pulse trains is another pulse train with the same relative pulse-to-pulse phase gradient. If a receiver array is arranged the same way, with the same fixed delays, then it will automatically sum the received pulse train coherently if it is received from a direction corresponding to the pulse-to-pulse phase gradient. Instead of providing a pulse-to-pulse gradient, the interpulse period can be increased or decreased by an amount equal to the element-to-element time delay required for a particular beam direction. This then, results in a true time delay controlled phased array radar.

The time scanned array radar concept does not use any active control devices at the element level and the receiver forms all the beams simultaneously. All the beams are formed on a single outlet and are identified by a pulse-to-pulse code. This time code is provided by the transmitter and is what controls the beam direction, which is where the name was derived from. This system works both as a true time controlled array and a phase controlled array.

Figure 1 illustrates the concept when used as a

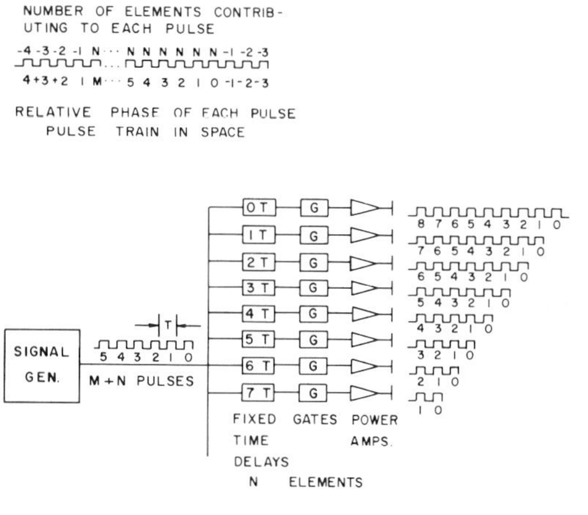

Fig. 1 Phase Control System.

phase controlled array. A fixed delay is in the signal path of each element. This delay has a different value from element to element. A gate is provided in each element after the fixed delay. This gate is not necessary to the operation of this system, but provides some improvement and flexibility. All the gates are switched at the same time. The use of the gate will be explained later.

At the input to the array, a pulse train is provided. The period T, of the pulse train is the same as the T in the fixed delays.

The pulse train is split and sent to each radiating element through the fixed delays. The right side of the figure shows the timing of the pulse trains as they leave each element. In the first element, the pulse train is undelayed. From the second element, the pulse train is delayed one pulse period. Each successive element delays the pulse train one additional pulse period. At a given instant of time, a different pulse in the pulse train is radiating from each element represented by a column of the pulses. To steer the array, a phase gradient must be provided in these pulses. The phase of pulse #1 must be different from each of the other pulses. This can easily be provided by inserting a pulse-to-pulse phase gradient on the pulse train at the input to the array. The pulse train shown below the figure illustrates the summed pulse train in space. The N represents the number of elements in the array. The figures at the top of the pulse train represent the number of elements contributing to that pulse. The figures below the pulse train represent a numbering sequence for the pulse train where M + N is the number of pulses at the input to the array. By activating all the gates at the same time, the radiated pulse train can be started and stopped at any place and any of the pulses can be eliminated. It may not be apparent but this pulse train has the same pulse-to-pulse phase gradient as the original. In the desired direction, all the pulses in a column are in phase. Each successive column changes the phase by one increment. On receive, the process is reversed, and a single pulse train is formed at the output, with the same pulse-to-pulse phase gradient as before.

Figure 2 shows the same system but used as a time delay system. In this one, the pulse period is changed to a new value of $T \pm \Delta t$. The fixed delays do not change. The result is that out in space, the pulses are shifted slightly from before. Again, the first element doesn't delay the signal. The second pulse train is delayed by an amount equal to T, so that these two pulses in any column are separated in time by an amount equal to Δt. The next pulse train is delayed by 2T. The first pulse in this train is shifted from the third pulse in the undelayed train by an amount $2\Delta t$. In any vertical column, a time delay gradient of Δt is provided across the array. The summed pulse train has a pulse-to-pulse period of $T \pm \Delta t$, the same as the input. On receive, the same thing happens in reverse and a single pulse train is formed at the output. The gates are used as before, to eliminate any undesired pulses before being radiated.

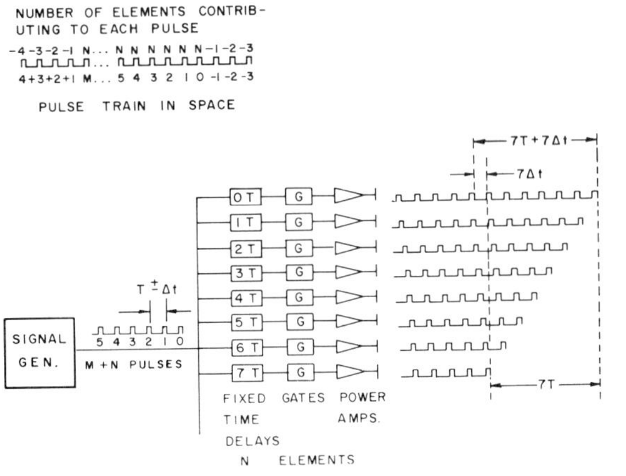

Fig. 2 True Time Control System.

2.2. FEASIBILITY STUDY RESULTS

A twelve-month study concluded that the technique is a feasible beam steering technique for wide or narrow bandwidth signals. TSAR provides phase beam steering when only phase between pulses fed to the array is controlled, and true time delay steering when the interpulse delay is controlled.

Time Scanned Array Radar steering demonstrates comparable flexibility to other beam steering schemes with respect to signal modulation and aperture illumination. Transmit beam switching speeds are limited by the inter-element delays in a way similar to other systems utilizing delays to develop elemental beam steering control. In a practical TSAR system, this time would be about three microseconds or less. In addition, it is possible to intersperse two pulse trains if the individual pulses are short compared to the elemental delay, T.

A significant part of the cost trade between conventional steering on an elemental level and use of TSAR concerns the inter-element delay cost. For real time delay steering, switched time delays at the element level would be replaced by fixed delays between elements or subarrays. The inter-element delay should be held to a low value consistent with the requirements of no overlap between pulses of the transmitted signal. Since TSAR requires a pulse train, a range ambiguity occurs. The usual approach for handling this problem is by using a different PRF on successive pulse bursts. For TSAR this could be an expensive solution because changing the nominal PRF would normally require changing the inter-element delay. However, the PRF can be varied over narrow limits without actually changing the delays. This imposes an error phase gradient on the system which has to be corrected by changing the pulse-to-pulse phase gradient. The PRF can be increased and decreased from the normal on alternate observations to double the reduction in range ambiguities. This variation would not be sufficient to completely resolve all the ambiguities. In many applications, the minimum cost-effective bandwidth of a TSAR approach would be greater than the requirements. This minimum bandwidth for TSAR would be of the order of 10 to 25 MHz depending upon the relative cost of components. This also means that in most applications a true time delay TSAR would be utilized even though comparative systems for the same applications could be of the phase type.

Single unit control for narrow or wide bandwidth beam steering is provided and no beam steering control at the element level is required. Full array gain can be obtained in a completely passive manner on receive, but some form of integration of the received pulse train, as in any system utilizing pulse bursts, must be used. Time Scanned Array Radar beam steering may be utilized in two dimensions. However, large delays are required for the second dimension which would not be practical. Some other method of beam steering should be applied between time scanned rows of array elements of the planar array. Again, in a practical system, even in a linear array, the TSAR system would be combined with phase shifters. It should be readily apparent, that as the delay gets larger, the cost eventually becomes greater than that of a phase shifter. One approach to this problem is to use a relatively small TSAR linear array, space the elements far apart, and intersperse with other identical TSAR arrays with phase shifters between the TSAR arrays. This reverses the usual subarray, i.e., the interelement phase shifters are few in number and the subarray time delay control "units" are many. A planar array would be composed of many of these linear arrays controlled by means of phase shifters and/or time delay units.

The requirements on the accuracy of the phase and/or delay between successive pulses of the train is one of the most demanding requirements in the TSAR system. Experience indicates that errors in the signal generator as well as errors due to the inter-element delay dependence on temperature can be sufficiently controlled if careful design and temperature control for the signal generator circuitry and inter-element delay lines are utilized. This care in signal generator design is reasonable because only one signal generator is required per system. The accuracy requirements of the fixed delay lines and the spacing between pulses are no more critical in terms of phase than in any other corporate feed system. The fixed delays can have random errors and/or be adjusted by means of phase trimmers. The interpulse spacing is more critical as random errors cannot be compensated for. The total random phase errors in TSAR can be allowed to deteriorate as much as in any other system. The major problem is in errors that accumulate. These can be calibrated out but must not be allowed to drift very much between times of calibration. Some approaches for a TSAR signal generator are better than others in this respect.

Space taper on transmit and/or receive can easily be obtained by eliminating equipment behind inactive elements, the same as is done in any other array.

2.3. EXPERIMENTAL TEST PROGRAM

The objective of the experimental test program was to demonstrate the feasibility of the time scanned array radar beam steering principle by performing tests on a breadboard model. An existing government owned 90 element linear wideband phased array located at the Bendix Communications Division plant at Towson, Maryland was used as a test bed. Transmit and receive array testing of the TSAR beam steering technique was conducted separately in a closed system using three sets of 90 delay line cables to simulate the wavefront propagation delay associated with transmitting or receiving a plane wavefront from a linear array at 0°, 30° and 40° from boresight. Four configurations were used to test the TSAR beam steering technique for transmit or receive, and for true time delay or phase control.

Figure 3 shows the real time delay steered simulated transmit array. Shown is a feedback loop pulse train generator feeding a train of 46 Nsec pulses with To equal to 426 Nsec into a series tapped delay line. The delay line provides the fixed delay required for each array element by the TSAR technique. The TSAR real time delay steering is applied to 9 subarrays of 10 elements each. The other three configurations were similar. The TSAR phase controlled configuration used only 9 elements spaced 6.25λ apart instead of the nine subarrays. This resulted in grating lobes.

Figure 4A shows the calculated phase steered transmit array angular response for a simulated target angle of 41°. The lobe at the right side of the figure represents the desired pattern. All others are grating lobes. At 41° the signal with a half voltage pulse pulsewidth of 213 Nsec is a narrow bandwidth signal for the transmit array. The half

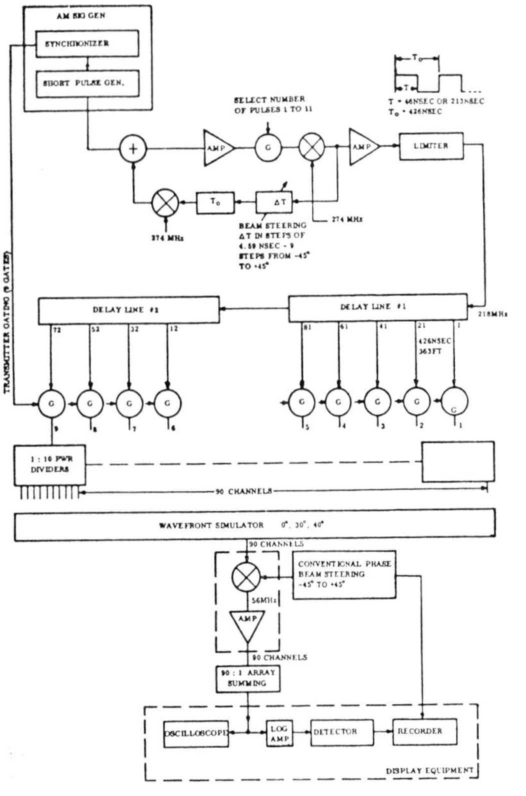

Fig. 3 TSAR Wideband Steering for Simulated Transmit Array.

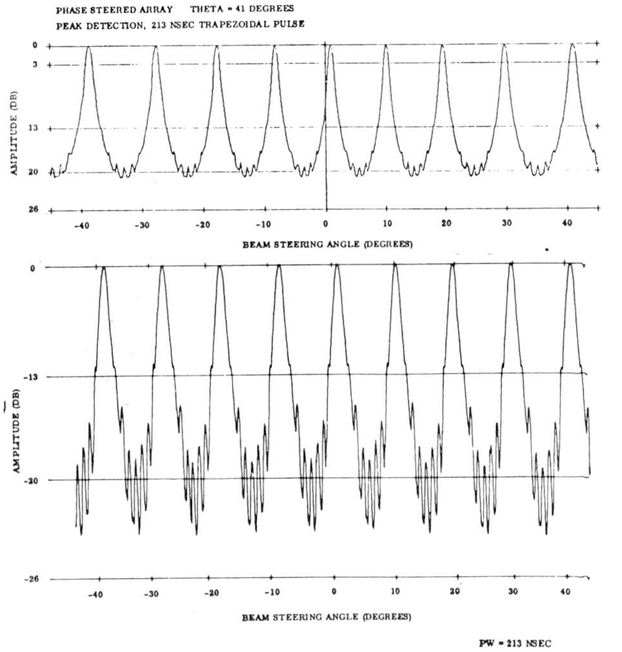

Fig. 4 Narrowband Transmit Array Angular Response 213 Nsec Pulse TSAR Phase Steered Array (θ = 41°).

voltage pulsewidth is 1.5 times the simulated wavefront propagation time between the first and ninth elements. The filling in of the nulls is due to taking the peak voltage response rather than a matched filtered response for the signal out of the array. This effect is a result of the measurement technique rather than the beam steering technique. Figure 4B shows the measured angular pattern at 41°.

Figure 5 shows the calculated and measured beam

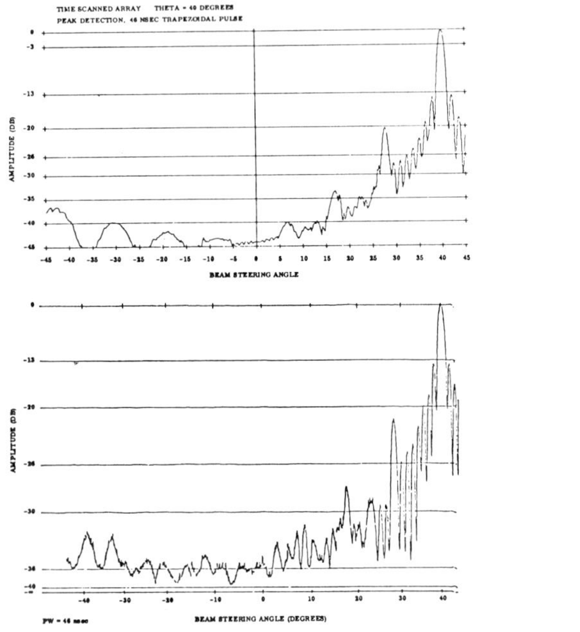

Fig. 5 Wideband Transmit Array Angular Response θ = 40°.

patterns at 40° for the real time delay transmit case. The wavefront propagation time across the array at 40° is more than 3 times the half voltage width of the input pulses. However, with TSAR true time delay steering, no significant loss of peak array response amplitude was calculated or measured. The prominent sidelobes at 28° and 18° are due to the sub-arraying used.

Figure 6 gives the calculated and measured temporal response for the 46 Nsec pulse used as an input to the TSAR real time steered transmit array at boresight. The overshoot is caused by the fast rise time of the input pulse ringing the receiver.

Figure 7 gives the real time delay steered array response at 40°. Comparison of Figure 6 and 7 indicate that the transmit array temporal response is essentially independent of beam steering angle with TSAR real time delay steering. The peak array response was only slightly dependent on the beam steer-

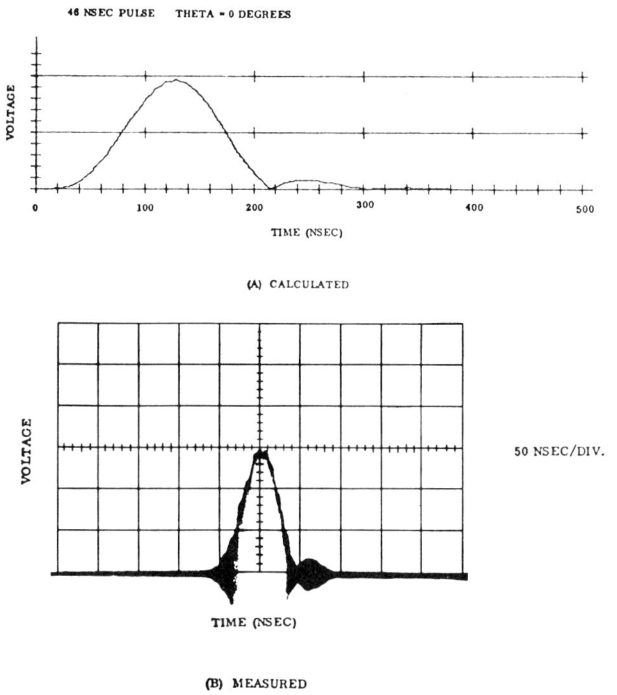

Fig. 6 Temporal Response Wideband Time Scanned Transmit Array (θ = 0°).

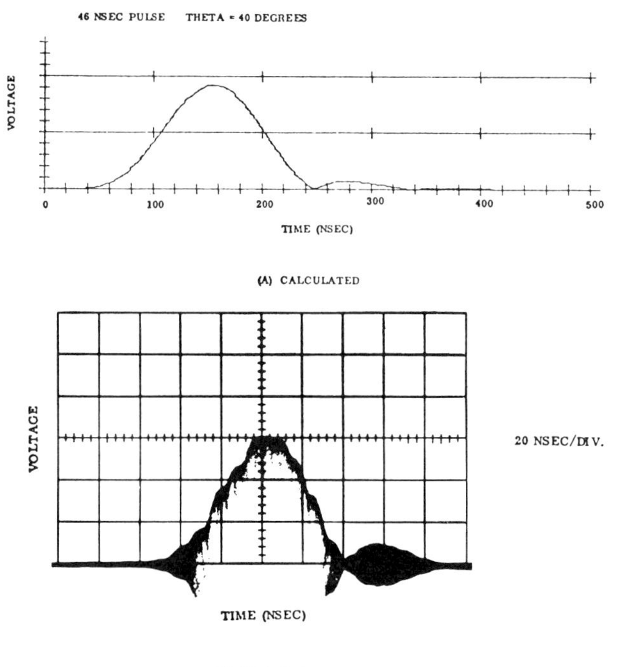

Fig. 7 Temporal Response Wideband Time Scanned Transmit Array (θ = 40°).

ing angle and was attenuated 1 dB. This loss was primarily due to TSAR delay steering being applied in steps rather than continuously.

The transmit array results obtained indicate that TSAR phase steering and real time delay steering can provide satisfactory beam pointing control with acceptable angular and temporal responses.

2.4. CONCLUSION

One of the main advantages of this concept is economics. A second advantage is less complexity. The concept eliminates the need for element phase shifters and their control circuitry. In addition, the feed system and fixed delays can be arranged in a tree network. The preamplifiers and final power amplifiers can then be placed in the most optimum location within the tree. As an example, in one configuration where track and search requirements are both present, there are 8 elements per preamplifier and 2 elements per power amplifier. These ratios are strictly a function of loss and gain considerations. Optimum cost for a given requirement would depend upon relative cost of components. It is estimated that the electronics for an array of this configuration, 64 x 64 receive aperture and a 32 x 32 transmit aperture, using modified AN/FPS-85 components, would cost approximately $\$10^6$. This does not include computers, data processors, etc. The receive aperture for beamwidth considerations would be 4096 elements, provide the gain of about 2000 lossless elements, and require 512 preamplifiers. The transmit aperture would be 1024 elements with 512 power amplifiers having an additional loss of 0.6 dB.

Reference

1. Lyons, B. J. and Buczkowski, L. J.: Time Scanned Array Radar, RADC-TR-67-449, Contract AF30(602)-4232.

The Design of Hybrid Multiple Beam Forming Networks

K. H. HERING

TRW Systems Group
Los Angeles, California

Abstract

Guidelines for the design of the 180-degree hybrid type network are presented which have not been available. A sixteen-element 180-degree hybrid matrix is designed and an indication is given as to how to design higher order matrices using 180-degree hybrids.

Summary

The simplest and lowest order beam forming matrix is shown in Fig. 1 and consists of a 90- or 180-degree hybrid. The simple 180-degree hybrid matrix is better known as a magic tee.

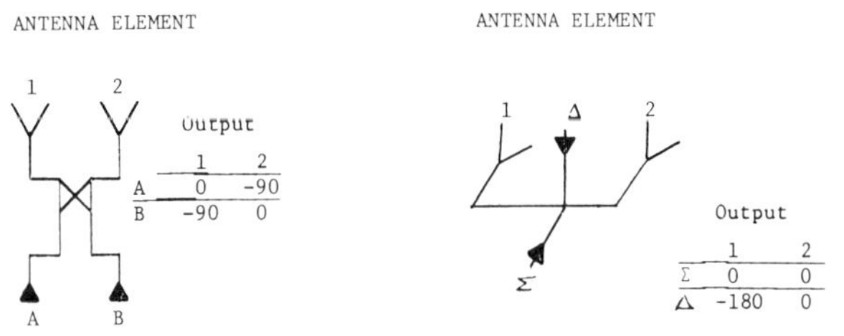

Fig. 1 Two element 90 and 180° Hybrid Matrices and Phase at Outputs – First Order Hybrid Matrices.

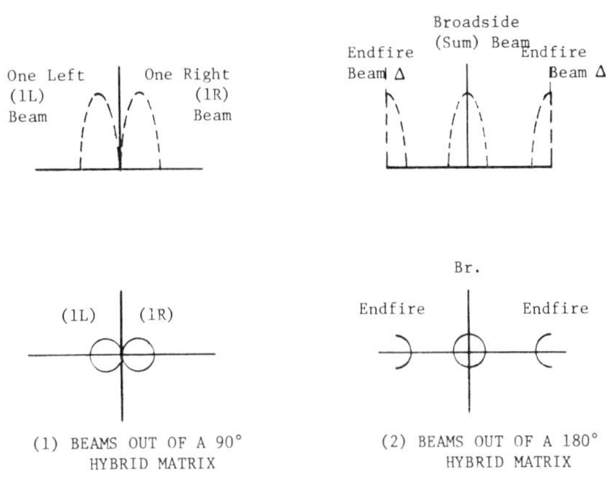

Fig. 2 Beams Generated by a Two-Element Matrix.

Fig. 2 shows representatively the beams for the 90- and 180-degree one-hybrid matrix (1).

The design theory for a 90-degree hybrid multiple beam-forming network is available (2). Such information is not available for the 180-degree hybrid matrix. The characteristic equations can be derived (3) (Table 1).

EQUATIONS FOR MULTIPLE BEAM FORMING MATRICES

TABLE 1

	90° Hybrid Matrix	180° Hybrid Matrix
Far-field Amplitude	$\|E_k\| = \frac{1}{n} \frac{\sin n[\frac{\pi d}{\lambda} \sin \alpha - \frac{(2K-1)}{n} \frac{\pi}{2}]}{\sin[\frac{\pi d}{\lambda} \sin \alpha - \frac{(2K-1)}{n} \frac{\pi}{2}]}$	$\|E_p\| = \frac{1}{n} \frac{\sin n [\frac{\pi d}{\lambda} \sin \alpha - \frac{P\pi}{n}]}{\sin[\frac{\pi d}{\lambda} \sin \alpha - \frac{P\pi}{n}]}$
Position of main beams	$\sin \alpha_B = \frac{\lambda}{nd} (K - \frac{1}{2})$	$\sin \alpha_B = \frac{P\lambda}{nd}$
Sector coverage	$\Theta = 2 \sin^{-1} [\frac{\lambda}{2d}(1 - \frac{1}{n})]$	$\Theta = 2 \sin^{-1} [\frac{\lambda}{nd}]$
Beam crossover angle	$\alpha_c = \sin^{-1} (\frac{K\lambda}{nd})$	$\alpha_c = \sin^{-1} [\frac{\lambda}{2d} (\frac{1}{n} + \frac{2P}{n})]$
Beam crossover level	$\|E\|_c = \frac{1}{n} \frac{1}{\sin (\frac{\pi}{2n})}$	$\|E\|_c = \frac{1}{n} \frac{1}{\sin (\frac{\pi}{2n})}$
Beam movement with frequency	$d\alpha_B = \tan \alpha_B (\frac{d\lambda}{\lambda})$	$d\alpha_B = \tan \alpha_B (\frac{d\lambda}{\lambda})$
Inter element phase shift	$\frac{(2K-1) \pi}{n}$	$\frac{2P \pi}{n}$

For a 90-degree hybrid matrix, the first beams to the right (1R) and to the left (1L) of broadside appear always at the extreme ports. The 2R and 2L beams are

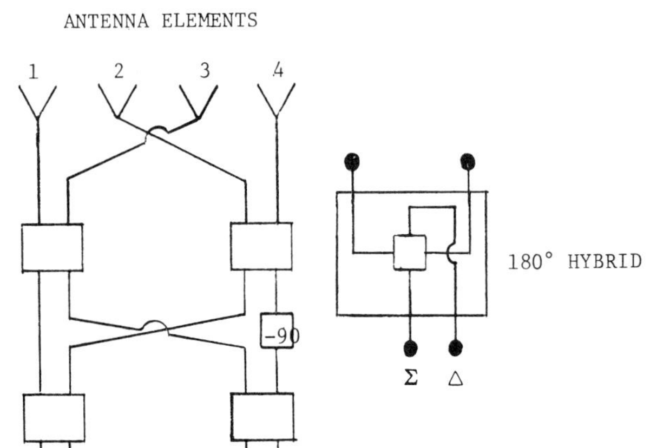

Broadside End-Fire 1R 1L

BEAM OUTPUTS

	Output 1	2	3	4
Br.	0	0	0	0
Endf	0	0	-180	-180
1R	0	-90	-180	-270
1L	0	90	180	270

Fig. 3 Four by Four Beam Forming Matrix. Second Order Matrix of 180° Hybrids.

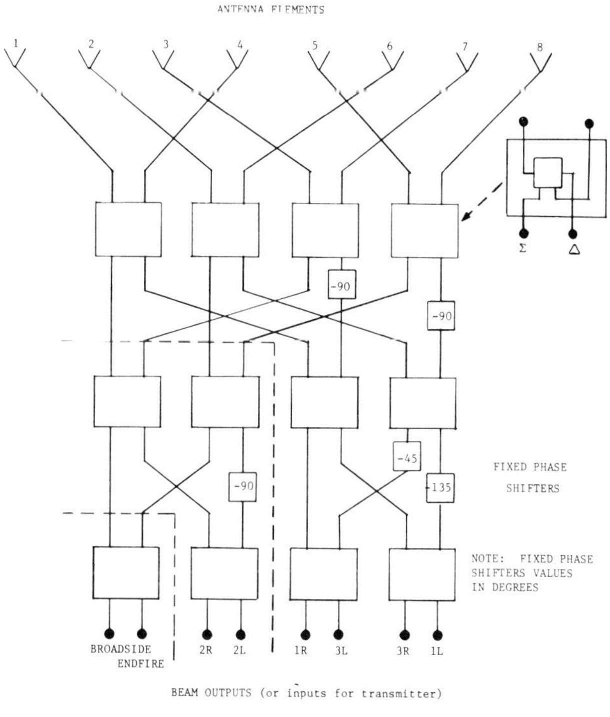

Fig. 4 **Eight by Eight Beam forming Matrix. Third Order Hybrid Matrix Using Sum-and - Difference Devices.**

at the two centrally located output ports. The beam sequence is different for the 180-degree hybrid matrices as seen from Figs. 3 and 4.

The first two ports from the left are always the broadside and endfire beam ports. The remaining $(\frac{n}{2} - 2)$ ports on the left half of the n-element are the ports for beams two times the beam number of the 2^{N-1} element matrix, where N gives the matrix order.

The low left hand corner of a n-element 180-degree hybrid matrix is identical to the complete (2^{N-1}) - element 180-degree hybrid matrix, except for the beam numbers which change; but in a predictable manner.

The phase shifts required in the right hand side in the row just above the beam outputs are calculated by adding to every second length of line of the left half, $-\frac{360}{n}$ degrees of phase shift, where n equals the number of elements (ports) of the matrix. Therefore, for a sixteen-element matrix, -22.5 degrees is added to the second section of line counted from the left. This technique can be applied to design matrices of any order. The linear phase taper which is required to occur across the array of elements to generate a given beam can be calculated systematically as is shown in Figs. 1 and 3.

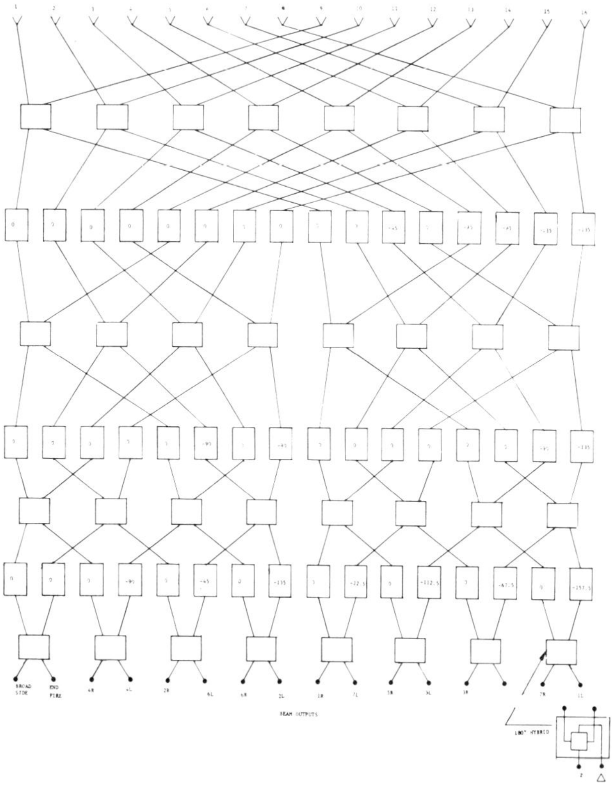

Fig. 5 **Sixteen-Element Butler Hybrid Matrix. A Broadside Beam is Obtained through the Use of 180° Hybrids.**

The four-element matrix, of second order, has phase settings of 0, -90, -180, and -270 degrees at the element ports 1, 2, 3 and 4, respectively for the 1R beam. If a plane wave is incident on the four-element (array) matrix at an angle corresponding to the phase settings as outlined, then all signals are in phase at the beam port labeled, 1R. When the plane wave is incident from a direction which does not relate to any of the phase settings outlined in Fig. 3, then at more than one beam port signals are detected.

Conclusion

Several interesting, and useful, differences between the types of multiple beam forming matrices have been brought out, which have been used to form guidelines for the design of a n-element 180-degree hybrid matrix. In applications, where a beam at broadside is required, a 180-degree hybrid matrix can be used in place of the 90-degree hybrid matrix.

References

(1) Butler, J. L., "Microwave Scanning Antennas," Vol. III Array Systems Academic Press 1966 Chapter 3.

(2) Moody, H. J., "The Systematic Design of the Butler Matrix," IEEE Trans., AP-12, Correspondence November 1964, pp. 786-788.

(3) Delaney, W. P., "An RF Multiple Beam-Forming Technique," IRE Trans. of Military Electronics MIL-6-April 1962, pp. 179-186.

Bandwidth Criteria for Phased Array Antennas

J. FRANK*

Technology Service Corporation
Silver Spring, Maryland

Abstract

Bandwidth criteria are developed for narrow band signals with frequency agility, and for instantaneous broadband signals. The basic mechanisms which degrade array performance as frequency is changed are examined. These include frequency scanning caused by the aperture and by the feed. Several types of feeds are discussed, including both dispersive and non-dispersive feed networks. Finally, the matched filter performance is examined.

Introduction

An array that is steered by phase shift, rather than time delay, will scan as the frequency is changed. This fundamental mechanism limits the bandwidth of a phased array. Only a limited amount of frequency change can be permitted before the beam is no longer pointing in the desired direction. (1) The use of time delays at each element would provide very broad instantaneous bandwidths. For acoustic arrays it is practical to have a true time delay at each element and thereby achieve multi-octave bandwidths. At microwave frequencies time delay at each element is presently too costly and instantaneous bandwidths are more commonly on the order of one or two percent. Most arrays may be tuned over a larger bandwidth (five to ten percent) by resetting the phase shifters when the frequency is changed.

A reasonable first guess at the bandwidth of a scanning array is

$$\text{Bandwidth } (\%) = \text{Beamwidth (degrees)}$$

That is, an array will possess a one percent instantaneous bandwidth, if it has a broadside beamwidth of one degree. The actual bandwidth may be more or less than this value depending on the feed, transmitted waveform, and required region of scan, but this simple formula will quickly provide an estimate that is rarely off by a factor of more than two.

Selection of a more accurate criterion requires consideration of the manner in which the array will be used. For example, one array may be required to transmit a narrow pulse (τ) which is equivalent to transmitting a wide band of frequencies $(\Delta f \approx \frac{1}{\tau})$. In this case, the beam for each spectral component is pointed in a slightly different direction and an average loss will occur. Another array may be used instantaneously over a narrow band of frequencies, but it may be desired to hop from frequency-to-frequency without resetting the phase shifters. Here, the loss at the edges of the band will be severe unless the bandwidth is more limited. The latter case, where only a narrow band signal is used with frequency agility, can be analyzed on a CW basis to determine array bandwidth. In addition to yielding a bandwidth criterion, this analysis provides an insight to the mechanisms which degrade array performance as frequency is changed. These mechanisms will be elaborated upon and several feed systems using dispersive and non-dispersive media will be discussed. The manner in which a single frequency scans the beam will then be extended to the broadband (Pulse) case and similar criteria will be developed. The broadband criteria will also be cast in terms of the aperture fill time.

In both the CW and Pulse cases, the criteria will allow a maximum one-way energy loss of 1 dB or less. This limit is somewhat aribtrary and it is hoped that the development of the criteria will provide sufficient understanding so that the reader may derive a criterion to suit his own requirements. For an antenna designer the one-way loss in gain or change in pattern is meaningful. To a radar designer, the two-way degradation in Signal/Noise after matched filtering may be more important. This will also be discussed and graphs will be presented showing the Signal/Noise degradation as a function of bandwidth (or equivalently, as a function of the aperture fill time).

*Formerly with the Johns Hopkins University, Applied Physics Laboratory.

Bandwidth Limitations

The fundamental limitations placed on bandwidth are caused by two effects:

1. Aperture Effects - an aperture steered by phase, rather than time delay, will scan the beam as frequency is changed.

2. Feed Effects - most feed networks (with the exception of an equal line length feed) provide an aperture phase excitation that varies as the frequency is changed, thereby scanning the beam.

The two effects are additive so that if the aperture causes the beam to scan by $\Delta_a(\sin\theta_o) = X$, and the feed causes the beam to scan by $\Delta_f(\sin\theta_o) = Y$, then the beam will scan by $\Delta_{a+f}(\sin\theta_o) = X + Y$. For most practical cases, where the scanning due to frequency is small an excellent approximation is given by

$$\Delta_{a+f}(\theta_o) = \Delta_a(\theta_o) + \Delta_f(\theta_o)$$

For example, if the beam is frequency scanned one degree by the aperture and two degrees by the feed, the total scanning due to frequency is three degrees. Therefore, the effects can be examined separately for ease in analysis and combined after each is understood.

Aperture Effects

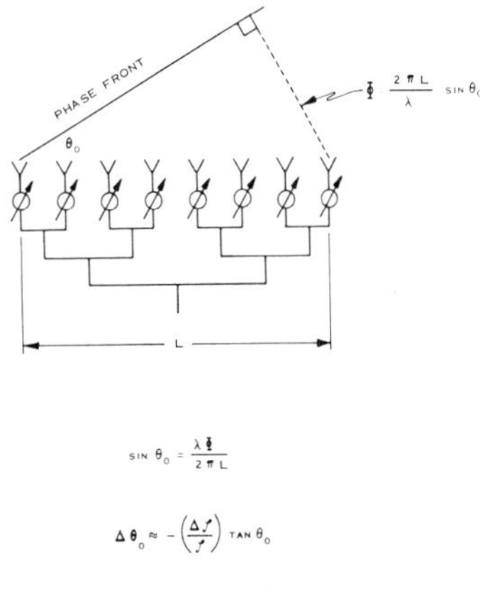

Fig. 1 Scanning Due to Aperture Effects.

Figure 1 shows a parallel (equal line length) fed array that is scanned to an angle, θ_o. This type of feed is used as an illustration since it exhibits no Feed Effects (i.e., a signal at the input illuminates each of the radiating elements with the same phase, regardless of the frequency). A linear phase slope across the array is required to scan the beam to θ_o and the required phase at the end element is $\Phi = \frac{2\pi L}{\lambda}\sin\theta_o$ (radians) with respect to the first element. This equation can be rewritten

$$\sin\theta_o = \frac{\Phi\lambda}{2\pi L}$$

It is apparent that if Φ and L are fixed, then a change in wavelength will produce a proportional change in scan angle. This is the aperture effect. A little mathematical manipulation (Appendix A) shows that the change in scan angle is given by

$$\Delta\theta_o \approx -\left(\frac{\Delta f}{f}\right)\tan\theta_o \qquad \text{(radians)}$$

$$\frac{\Delta f}{f} = \text{Fractional Bandwidth}$$

Several points are worth noting:

1. Aperture size and beamwidth have no effect on the change in scan angle.

2. The amount of scanning is dependent on the original scan angle and the effect is most noticeable at large scan angles. At $\theta_o = 0$, there is no scanning regardless of the amount of change in frequency, and the antenna has infinite bandwidth. It should therefore be apparent that before a bandwidth criterion is specified, the maximum scan angle of the array must be specified.

3. As frequency is increased, the scan angle will decrease.

It might appear from 1. above that the aperture size and beamwidth are of little importance. This is not the case. Stating that the beam scans by one degree or two degrees does not say much about the degradation in performance. However, stating that the beam is scanned by one quarter or one half beamwidth gives a good indication of the degradation. As shown in Figure 2, a change in frequency will cause the entire antenna pattern to be translated and undistorted if the abscissa is plotted in $\sin\theta$. The pow-

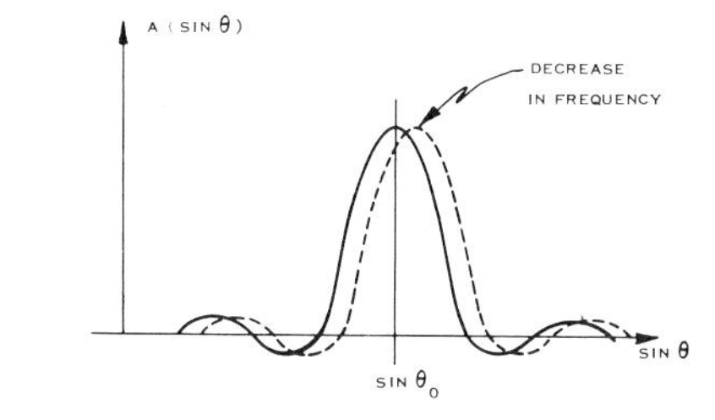

CRITERION (CW): SCAN $\pm \frac{1}{4}$ BEAMWIDTH; MAXIMUM LOSS = 0.7 db

CRITERION (PULSE): SCAN $\pm \frac{1}{2}$ BEAMWIDTH; MAXIMUM LOSS ≈ 0.8 db

Fig. 2 Frequency Scanned Beam.

er radiated in the original direction $(\sin \theta_0)$ is decreased because of this translation. If it is known that the beam has been frequency scanned by a half beamwidth, it is immediately known that the power radiated in the original direction is down by 3 dB. For the CW case, the criterion developed here will permit the beam to scan by a maximum of one quarter beamwidth away from the desired direction, thereby incurring a maximum one-way loss of 0.7 dB. This criterion actually permits frequency scanning of one half beamwidth, since the beam is allowed to scan from one quarter beamwidth less than the desired direction to one quarter beamwidth more than the desired direction. A plot of the loss in gain as a function of the fractional beamwidth scanned away from desired is shown in Figure 3.

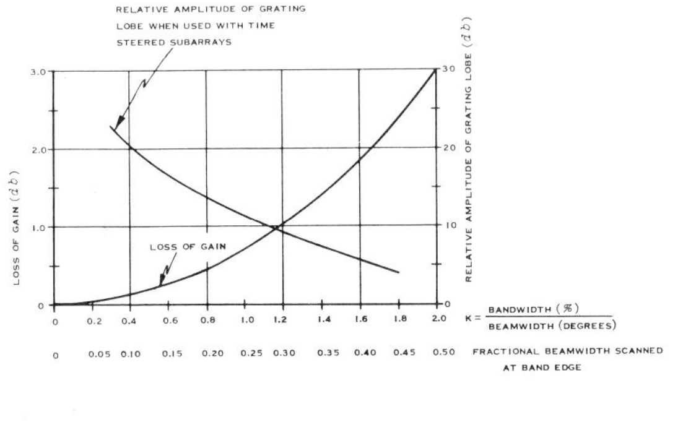

Fig. 3 Loss with Scanning Due to Aperture; Grating Lobe Levels when used with Time Steered Subarrays (CW) (After Cheston and Frank [1]).

As stated previously, the frequency scanning will be greatest at the largest scan angles. For many antennas, a scan angle of 60° is required and this will be the scan angle used in the development of all criteria in this paper. For this case, Appendix B derives the bandwidth for a given broadside beamwidth. The result is

$$\text{Bandwidth } (\%) = \text{Beamwidth (deg)}. \quad \text{(CW)}$$

It is interesting to note that although this result has been derived in Appendix B for a given illumination, the result holds up quite well for a wide range of aperture illuminations.

Pulsed Spectra

When a pulse is transmitted, many frequency components are transmitted simultaneously. (2, 3) It will be assumed that the central spectral component is scanned to the desired direction and that the symmetric spec tral components are frequency scanned symmetrically about the desired direction. In this case, the 'broadband' (Pulse) antenna pattern is the sum of the CW antenna patterns $A(\omega,\theta)$, weighted by the signal spectrum $S(\omega)$ and is given by

$$A(\theta) = \int_{-\infty}^{\infty} A(\omega,\theta)\, S(\omega)\, d\omega$$

This can also be approached from the frequency domain where the transmitted spectrum $S(\omega)$ is modified by the antenna pattern so that the spectrum at a point in space is $S(\omega)\, A(\omega,\theta)$. The time waveform is the fourier transform of the spectrum

$$S(t,\theta) = \int_{-\infty}^{\infty} S(\omega)\, A(\omega,\theta)\, e^{j\omega t}\, d\omega$$

Evaluating at t = 0, so that the time waveform is centered at the target yields the previous equation.

Of interest is the energy contained in the spectrum which reaches the target at θ_0. This is just a summation of the energy in each spectral component in the band of interest as given by

$$\text{Energy }(\theta_o) = \int_{f_o-\Delta f/2}^{f_o+\Delta f/2} |S(\omega)\ A(\omega,\theta_o)|^2\ d\omega$$

Computations for the one-way loss in energy at the target have been performed for the spectra shown in Figure 4. The loss in energy is plotted in Figure 5 as a

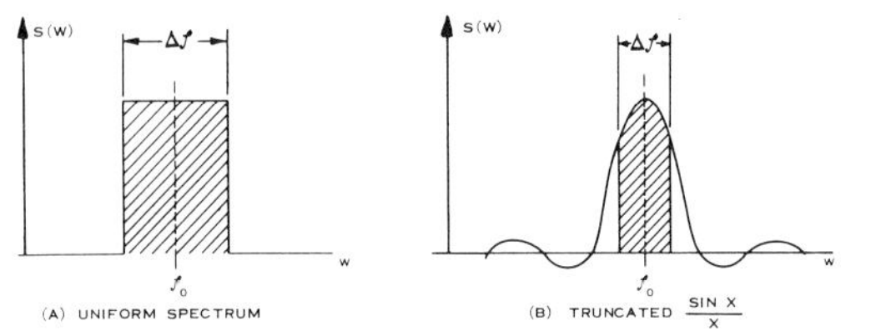

Fig. 4 Spectra used in Computing Energy Loss for Pulsed Signals.

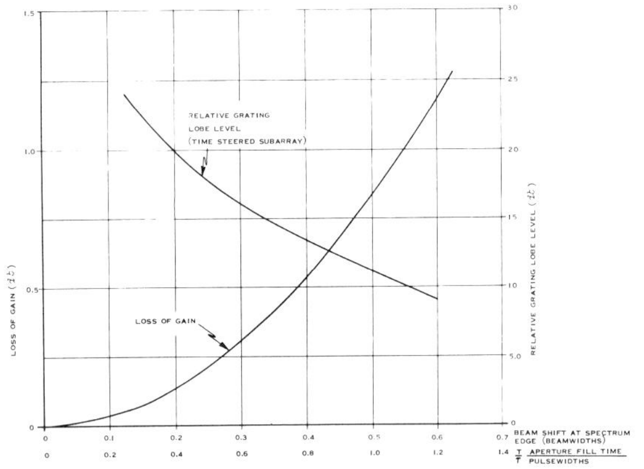

Fig. 5 Loss of Energy due to Aperture Scanning; Energy in Grating Lobe when used with Time Steered Subarrays (Rectangular Spectrum).

function of the maximum scanning by the edge spectral component. The broadband criterion will allow the beam to scan from $-\frac{1}{2}$ Beamwidth* to $+\frac{1}{2}$ Beamwidth* (where Beamwidth* is a local beamwidth) as the spectrum (Fig. 4a) is traversed from one edge to the other edge. This provides exactly twice the bandwidth as the CW case and is given by

$$\text{Bandwidth (\%)} = 2\ \text{Beamwidth (deg)} \quad \text{(Pulse)}$$

Using this criterion, the maxiumum loss is approximately 0.8 dB. It is seen that although twice as much frequency scanning is permitted for the broad band (Pulse) case, the loss is approximately the same as in the CW case. The $\frac{\sin x}{x}$ spectrum (rectangular pulse) as shown in Figure 4b has also been examined. Here only the portion of the spectrum between 3.9 dB points was considered, since a receiver designed to accept only the energy in a band of $\Delta f \approx \frac{1}{\tau}$ would not accept the spectral components outside of this range. The computed values for loss in energy associated with the truncated $\frac{\sin x}{x}$ spectrum were so close to those computed for the uniform spectrum that it was not felt worthwhile to include them separately. For the broadband criterion defined here the loss in gain for the truncated $\frac{\sin x}{x}$ was 0.7 dB as compared to 0.8 dB loss for the uniform spectrum.

Hammer (4) has also calculated the loss in energy due to transmitting a pulsed signal. His results differ from those given here (he shows higher losses) since he attributes the loss in all spectral components to the antenna. The calculations here attribute only the losses in the frequency band which would be accepted in a receiver possessing a flat frequency response over the bandwidth Δf. In either case, the loss is dependent on the particular spectrum chosen. For example, if Hammer had chosen the uniform spectrum (Fig. 4a), his results would be in agreement with those given here. It does not seem reasonable to penalize the antenna alone for the choice of spectrum especially since perfect matched filter receivers are not commonplace. This topic will be further discussed later in this paper under matched filter performance.

Aperture Fill Time

A convenient way of thinking about the bandwidth of an array is in terms of the aperture fill time. As shown in Figure 6, if a signal is incident from θ_o, the signal travels a distance $L \sin\theta_o$ farther to the last element than it does to the first element. The time it takes for the signal to be present in all elements is just $T = L/C \sin\theta_o$; the aperture fill time for a parallel fed array. The bandwidth criterion (Pulse) discussed above can be simply related to the ratio of the r.f. pulse width τ, to the aperture fill time T. Appendix C demonstrates that for $\tau = T$, the frequency band defined by $\Delta f = \frac{1}{\tau}$

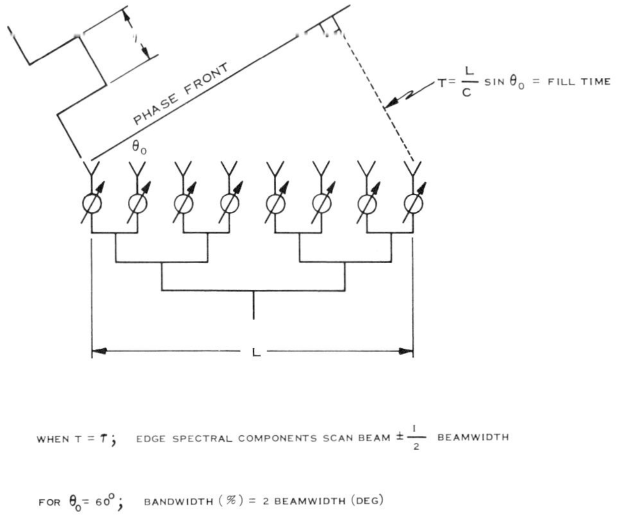

Fig. 6 Aperture Fill Time.

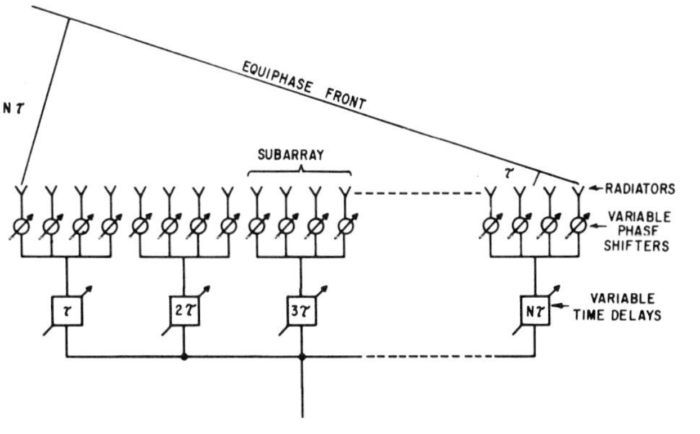

Fig. 7 Time Steered Subarrays: Phase Steered Aperture.

is just sufficient to scan the beam through one local beamwidth. Thus the criterion

$$\text{Bandwidth (\%)} = 2 \text{ Beamwidth (deg)} \quad \text{(Pulse)}$$

is equivalent to saying

$$\text{Pulse Width } (\tau) = \text{Aperture Fill Time (T)}$$

Figure 5 can therefore be used with variable pulse widths as indicated, since the reciprocal of the pulse width ($\Delta f = \frac{1}{\tau}$) is linearly related to the fractional beamwidth scanned.

Time Delays and Subarrays

Although time-delay at each element is prohibitively costly with present microwave techniques, a marked increase in instantaneous bandwidth can be obtained using subarrays with time delay (5). As shown in Figure 7 the aperture may be divided into subapertures (termed subarrays) and a time delay network may be placed behind each subarray. In this fashion the instantaneous bandwidth is doubled by the use of only two time delay networks for a linear array (four networks for a planar array). In general, a planar array will require at most N^2 time delay networks to increase the bandwidth by a factor of N, and this can be reduced if the symmetry of the array is used to advantage.

The loss in gain can be understood by examination of Figure 8. At band center, the subarray pattern and the array factor attain maximum gain coincidentally.

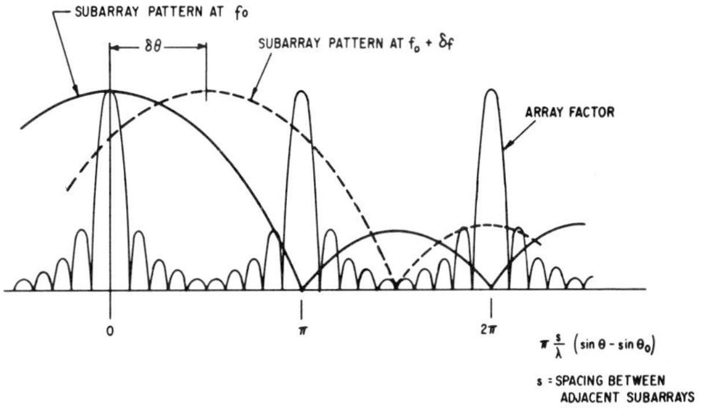

Fig. 8 Grating Lobes Induced by Time Steered Subarrays.

As the frequency is changed, the array factor stays fixed in space since it is time steered. The subarray pattern scans with frequency, since it is phase steered. Therefore the loss in gain with a change in frequency is the same as though the aperture were just that of the subarray, and this has been plotted in Figs. 3 and 5. With time steered subarrays, however, there is a difference in the pattern. In this case a grating lobe will form at the intersubarray spacing. With no frequency scanning, the peak of the grating lobe falls just at the null of the subarray pattern (assuming uniform illumination). With frequency scanning, the subarray pattern moves over the array factor grating lobe and produces a grating lobe in the pattern. The amplitude of the grating lobe has been calculated and the results plotted in Figs. 3 and 5 for the CW and Pulse cases respectively. It is seen that for the criteria set up in this paper, the grating lobe at worst will be down by 11 dB from the main beam in both the CW and Pulse cases.

Feed Effects

When a feed other than a parallel feed is used, phase errors are produced across the aperture due to the feed alone. An example is the end fed series feed shown in Figure 9. The total path length across the feed is $\Phi = \frac{2\pi L}{\lambda}$ radians assuming free space propagation.

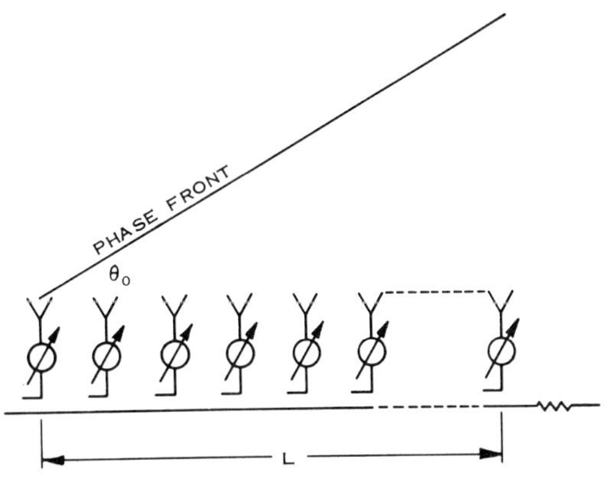

$$\Phi_f = \frac{2\pi L}{\lambda}$$

$$\Delta\theta = \left(\frac{\Delta f}{f}\right) \frac{1}{\cos\theta_0} \frac{\lambda_g}{\lambda} \quad \text{WITH WAVEGUIDE}$$

$$T = \frac{L}{C}\sin\theta_0 \pm \frac{L}{C}\frac{\lambda_g}{\lambda} \quad \text{APERTURE FILL TIME}$$

Fig. 9 End Fed Series Array.

When the frequency is changed the change in phase across the aperture will be

$$\Delta\Phi = \left(\frac{\Delta f}{f}\right) \frac{2\pi L}{\lambda} \quad \text{(radians)}$$

and this change in phase will tilt the beam just as phase shifters would. The angular change for a change in frequency is shown in Appendix D to be

$$\Delta\theta_0 \approx \left(\frac{\Delta f}{f}\right) \frac{1}{\cos\theta_0} \quad \text{(radians)}$$

This effect is similar to the aperture effect in that it results in a pure translation of the beam with no distortion. For one direction of scanning (towards the feed load) the feed effect tends to cancel the aperture effect. In fact for end-fire (toward load) a non-dispersive series feed is an equal line length feed and the array has no bandwidth limitations. This, however, is not a common use and most arrays are required to scan in both directions from broadside. In the direction of the feed input, the aperture and feed effects reinforce and the antenna scans rapidly with frequency. For a 60° scan angle, the scanning caused by the feed alone is slightly greater than caused by the aperture. Therefore, the bandwidth of the series feed is essentially half that of the equal line length feed.

In waveguide, the signal propagates more slowly which is equivalent to having a longer feed line. This causes the feed to scan more rapidly and the series feed performance degrades accordingly. The slowing in waveguide is proportional to λ/λ_g and the array now scans by

$$\Delta\theta \approx \frac{\lambda_g}{\lambda} \left(\frac{\Delta f}{f}\right) \frac{1}{\cos\theta_0} \quad \text{(radians)}$$

For this case the bandwidth criteria which assume $\pm 60^{\circ}$ scanning from broadside are:

$$\text{Bandwidth (\%)} = \frac{1}{\left(1 + \frac{\lambda_g}{\lambda}\right)} \text{Beamwidth (deg)} \quad \text{(CW)}$$

$$\text{Bandwidth (\%)} = \frac{2}{\left(1 + \frac{\lambda_g}{\lambda}\right)} \text{Beamwidth (deg)} \quad \text{(Pulse)}$$

Once again the pulse criterion is equivalent to stating that the aperture fill time is equal to the width of the r.f. pulse. Here the aperture fill time $T = L/C \sin\theta_0 + L/C \frac{\lambda_g}{\lambda}$ is seen to be the time required for the signal to traverse the array and then return to the input.

A center fed series array (6) as shown in Figure 10a may be thought of as two end fed series arrays. At band center the phase shifters are set so that all ele-

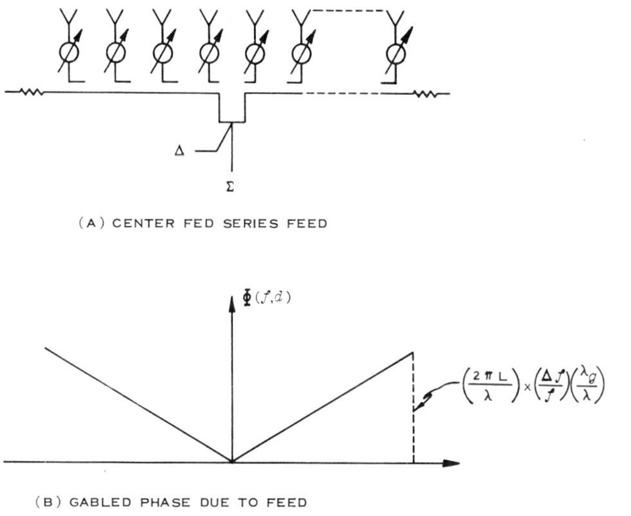

Fig. 10 Center Fed Series Array.

ments radiate in phase. As the frequency is changed, each half scans in the opposite direction due to the gabled phase slope shown in Figure 10b. This results in a broadening of the beam with no change in direction. If the two beams (Fig. 11a) were to move far enough apart, the beams would split. However, by restricting the bandwidth to reasonable values, the result is a loss in gain caused by the beam broadening. It should be noted that this feed does have bandwidth limitations at broadside.

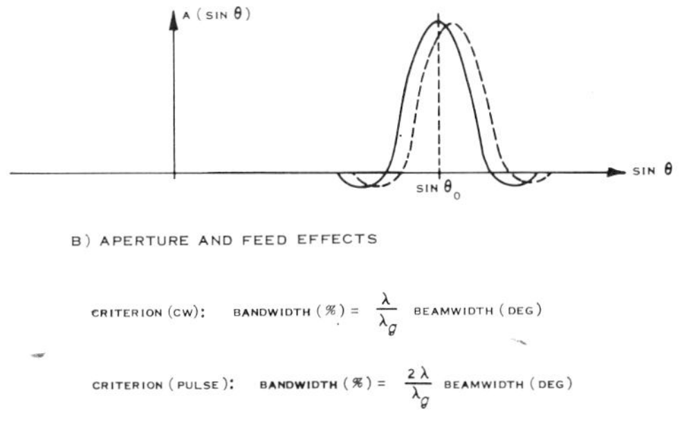

Fig. 11 Aperture Effects Superimposed on Feed Effects: Center Fed Array.

As the array is scanned aperture effects are also introduced, and are superimposed on the feed effects. As shown in Figure 11b, for one half of the array (solid curve) the aperture and feed effects tend to cancel, while for the other half (dashed curve) they tend to reinforce. The performance of the center feed is worse than the parallel feed at broadside but quite comparable at large scan angles. At a scan angle of 60°, the center fed array will have approximately $\frac{1}{4}$ dB more loss than the parallel fed array. For simplicity, the same criteria will be used for both feeds.

The discussion above has assumed that the center feed used non-dispersive transmission with free space propagation. If waveguide is used, the center fed array performance degrades and the criteria are:

$$\text{Bandwidth (\%)} = \frac{\lambda}{\lambda_g}\ \text{Beamwidth (deg)} \quad \text{(CW)}$$

$$\text{Bandwidth (\%)} = 2\frac{\lambda}{\lambda_g}\ \text{Beamwidth (deg)} \quad \text{(Pulse)}$$

The performance of a space fed array (Figure 12) can be analyzed by first thinking of a space feed with a very long focal length (e.g., an infinite focal length). In this case the performance is identical to that of an equal line length feed. As the feed is brought closer to the aperture a spherical phase front is imposed on the aperture. This phase error normally compensated for by the phase shifters. As frequency is changed a symmetric phase error (spherical) is imposed on the array which can be approximated by a parabolic phase error (for a line array). The phase error at the edges is never as great as that of the center fed series array. As the focal length decreases (Fig. 12b) the phase error approaches the gabled phase error of the center fed array and in the limit of zero focal length the performance is identical to that of the center fed array. Since the bandwidth criterion for an equal line length array is the same as the criterion for a center fed array (free space propagation), the bandwidth criterion for a space feed is also the same as that for an equal line length feed. Once again, the performance over the scan region is not quite as good as the equal line length feed. However, at a 60° scan angle the additional loss in g ain (over the equal length feed) will be less than $\frac{1}{4}$ dB, and for simplicity the same criterion will be applied.

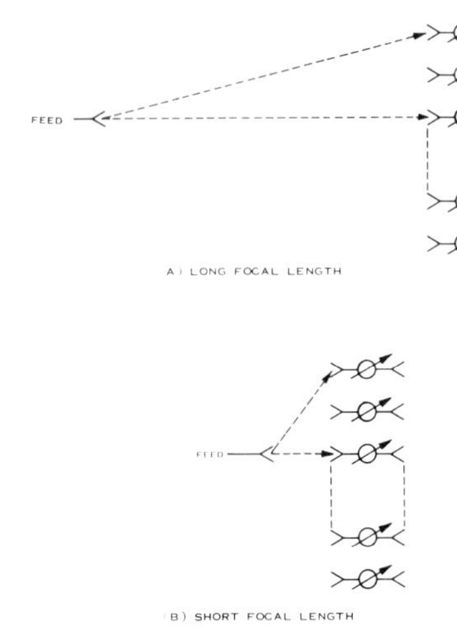

Fig. 12 Optical (Space) Feed.

Matched Filter Performance

The discussions above dealt only with the loss in the one-way antenna pattern. This provides physical insight from the viewpoint of the antenna designer. For the radar designer more information is needed. This was touched on briefly in the discussion of the spectrum of Fig. 4b. It was then (somewhat arbitrarily) stated that the loss in energy from the unshaded portion of the spectrum should not be attributed to the antenna, since the receiver would not accept the energy if it were available. A true matched filter receiver could actually recover some of this energy, and it is of interest to examine the degradation in terms of the matched filter. When a spectrum $S(\omega)$ is transmitted, the radar designer normally builds (or approximates) a matched filter with a conjugate response $S^*(\omega)$. The antenna itself can be thought of as a filter which is sensitive to scan angle and has an impulse response of $A(\omega,\theta_o)$. The original spectrum $S(\omega)$ is therefore modified both on transmit and on receive with a resultant spectrum $S(\omega)\ A_T(\omega,\theta_o)\ A_R(\omega,\theta_o)$ incident to the receiver. An appropriate matched filter for this spectrum would have the impulse response $[S(\omega)\ A_T(\omega,\theta_o)\ A_R(\omega,\theta_o)]^*$. This is seen to be a matched filter which varies with scan angle and is difficult to implement. This problem has been analyzed by Sklar (7) and by Rothenberg and Schwartzman. (8) Rothenberg and Schwartzman (9) have provided a comparison between the filter being matched to the input (transmitted) spectrum $S(\omega)$ and the filter being matched to the output (received) spectrum $S(\omega)\ A_T(\omega,\theta_o)\ A_R(\omega,\theta_o)$. Their results are shown in Figure 13. The results are given in terms of the

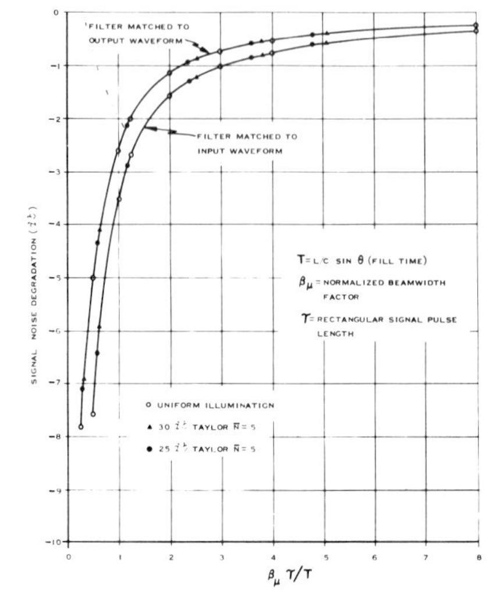

Fig. 13 Two-Way Loss in Signal/Noise (Equal Line Length Feed) (From Rothenberg and Schwartzman [9]).

ratio of pulse width to aperture fill time. The factor B_u is the ratio of beamwidth of the antenna to the beamwidth of a uniformly illuminated aperture of the same size. Several items are worth noting:

1. The beamwidth or normalized aperture fill time (T/B_u) determines performance rather than the aperture fill time itself. This is understandable since the outer elements in a tapered illumination do not carry the weight of the inner elements and could not be expected to have as much affect.

2. The two-way loss for the criteria developed in this paper $(\tau = T)$ is seen to be 2.5 dB or 3.5 dB depending on the filter chosen. This is more than indicated in Figure 5 where a one-way loss of 0.8 dB was shown. The discrepancy occurs because the matched filter analysis has been applied to a particular waveform; a rectangular pulse using an entire spectrum given in Fig. 4b. If the spectrum of Figure 4a had been chosen, the loss would be as given in Figure 5.

3. The matched filter technique (using either spectrum) is a very reasonable way of performing the analysis to understand the theoretical limit of performance. However, a practical receiver design only approximates a matched filter and a loss is normally attributed to the receiver. In a practical receiver the far out spectral components are

often heavily attenuated, and in the opinion of the author the energy losses shown in Figure 5 come closest to describing the loss which should be attributed to the antenna.

For the radar designer, the losses are very real regardless of where he choses to allocate them. It is his job to tailor the waveform, antenna, and receiver to minimize these losses.

A comparison among the different feed networks is shown in Figure 14 for a typical array. Several pulse

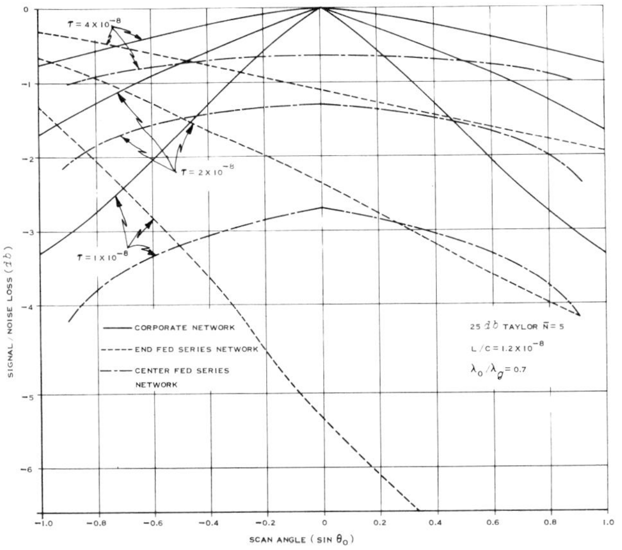

Fig. 14. Two-Way Loss in Signal/Noise (Filter Matched to Output Waveform) (From Rothenberg and Schwartzman (9)).

lengths are considered. It is seen that the corporate (equal line length) network performs best over most scan regions. This is especially true since waveguide networks have been assumed for the end and center fed series networks, as would be the case in most practical applications.

Conclusions

A summary of criteria is given in Figure 15. It should be remembered that all criteria were developed for a worst case situation (i.e., for an antenna scanned to 60° from broadside). This may not be very reassuring since the antenna has also experienced its maximum loss in gain (at least 3 dB) at a 60° angle of scan. It might be reasonable to adapt the bandwidth to the scan angle or design the matched filter for an angle close to 60° rather than broadside. In any case, the radar designer should appreciate the interaction among the wave-

USE OF THESE BANDWIDTH CRITERIA WILL PERMIT SCANNING TO 60° FROM BROADSIDE WITH A MAXIMUM ONE-WAY ENERGY LOSS OF APPROXIMATELY 3/4 db

FEED	CW BANDWIDTH (%)	PULSE BANDWIDTH (%)
EQUAL LINE LENGTH	BEAMWIDTH (DEG)	2 X BEAMWIDTH (DEG)
END FED SERIES	$\frac{1}{\left(1 + \frac{\lambda_g}{\lambda}\right)}$ X BEAMWIDTH (DEG)	$\frac{2}{\left(1 + \frac{\lambda_g}{\lambda}\right)}$ X BEAMWIDTH (DEG)
CENTER FED SERIES	$\frac{\lambda}{\lambda_g}$ X BEAMWIDTH (DEG)	$2\,\frac{\lambda}{\lambda_g}$ X BEAMWIDTH (DEG)
SPACE FED (OPTICAL)	BEAMWIDTH (DEG)	2 X BEAMWIDTH (DEG)

IN THE ABOVE CRITERIA ALL BEAMWIDTHS REFER TO THE BROADSIDE BEAMWIDTH

λ_g = GUIDE WAVELENGTH

λ = FREE SPACE WAVELENGTH

FOR SERIES FEEDS WITH FEED PROPAGATION AT VELOCITY OF LIGHT USE $\frac{\lambda_g}{\lambda} = 1$

Fig. 15. Summary of Bandwidth Criteria

form, antenna and receiver, and then design for maximum performance.

The criteria developed here are not strict but are intended as a means of comparison between the different feed networks. It is hoped that this paper will help the reader provide his own criteria based on the system requirements.

Acknowledgements

The author gratefully acknowleges the valuable contributions made by T.C. Cheston, G.J. Laughling and C. A. Shipley.

References

(1) Skolnik, M.I.: Radar Handbook, Section 11, Array Antennas by T. C. Cheston and J. Frank, McGraw Hill, 1970.

(2) Adams, W. B.: Phased Array Radar Performance with Wideband Signals, Aerospace and Elec. System, Technical Convention Record, pp. 257-271, November, 1967.

(3) Sharpe, C. B. and Crane, R. B.: Optimization of Linear Arrays for Broadband Signals, IEEE Trans. Vol. AP-14, No. 4, pp. 422-427, July 1966.

(4) Skolnik, M.I.: Radar Handbook, Section 13, Frequency-Scanned Arrays, by I.W. Hammer, McGraw Hill, 1970.

(5) Tang, R.: Survey of Time-Delay Beam Steering Techniques, Digest of Phased-Array Antenna Symposium, PIB, Long Island Graduate Center, June, 1970.

(6) Kinsey, R.R. and Horvath, A. L.: Transient Response of Center-Series Fed Array, Digest of Phased-Array Antenna Symposium, PIB, Long Island Graduate Center, June, 1970.

(7) Sklar, J.R.: Short Pulse Limitations of Phased Arrays, Chapter II of Lincoln Laboratory Report #TR-236, November, 1961.

(8) Rothenberg, C. and Schwartzman, L.: Phased Array Signal Bandwidth, Digest 1969 IEEE Internat'l Symposium on Antennas and Propagation, pp. 116-123, December 1969.

(9) Rothenberg, C. and Schwartzman, L.: Private Communication.

Appendix A

Aperture effects cause a change in scan angle with frequency of

$$\Delta\theta_o \approx -\frac{\Delta f}{f}\tan\theta_o \quad \text{(radians)}$$

Proof: As shown in figure 1, an array scanned to θ_o has a linear phase taper with a maximum value of

$$\Phi = \frac{2\pi L}{\lambda}\sin\theta_o \quad \text{(radians)}$$

or

$$\sin\theta_o = \frac{\Phi\lambda}{2\pi L}$$

For a differential change in wavelength

$$\frac{d}{d\lambda}(\sin\theta_o) = \frac{d}{d\lambda}\left(\frac{\Phi\lambda}{2\pi L}\right)$$

or

$$\cos\theta_o \frac{d\theta_o}{d\lambda} = \frac{\Phi}{2\pi L}$$

then

$$d\theta_o = \frac{\Phi \quad d\lambda}{2\pi L\cos\theta_o}.$$

Since

$$\frac{\Phi}{2\pi L} = \frac{\sin\theta_o}{\lambda},$$

$$d\theta_o = \frac{d\lambda}{\lambda}\tan\theta_o = -\frac{df}{f}\tan\theta_o$$

And for small finite changes in frequency

$$\Delta\theta_o \approx -\frac{\Delta f}{f}\tan\theta_o. \quad \text{(radians)}$$

The minus sign indicates that as frequency is increased the scan angle decreases.

Appendix B

For an array scanned to 60°, aperture effects will cause the edge spectral components to scan the beam from -1/4 Beamwidth* to +1/4 Beamwidth* (where Beamwidth* is a local beamwidth) when the bandwidth is defined by

Criterion (CW): Bandwidth (%) = Beamwidth (deg)

Proof: To move the beam from -1/4 Beamwidth* to +1/4 Beamwidth* requires a total angular change as given by Appendix A of

$$\Delta\theta_o = \frac{\text{Beamwidth}^*}{2} = -\left(\frac{\Delta f}{f}\right)\tan\theta_o \,\text{(radians)}$$

It is well known that an antenna beam broadens as it is scanned from broadside, so that

$$\text{Beamwidth}^* = \frac{\text{Beamwidth}}{\cos\theta_o} = \frac{\lambda/L}{\cos\theta_o} \quad \text{(radians)}$$

therefore

$$\frac{\lambda}{2L\cos\theta_o} = \frac{\Delta f}{f}\tan\theta_o.$$

letting

$$\text{Bandwidth (\%)} = 100\left(\frac{\Delta f}{f}\right)$$

$$\text{Beamwidth (deg)} = 57.29\,(\lambda/L)$$

and

$$\theta_o = 60^\circ$$

yields

$$\text{Beamwidth (deg)} \approx \text{Bandwidth (\%)}$$

Appendix C

If the r. f. pulse width (τ) is equal to the aperture fill time ($T = L/C \sin \theta_o$), then the change in frequency defined by $\Delta f = \frac{1}{\tau}$ scans the antenna beam by one local beamwidth (Beamwidth*).

Proof: Referring again to Figure 1, the phase at the edge of the aperture is given by

$$\Phi = \frac{2\pi L}{\lambda} \sin \theta_o \quad \text{(radians)}$$

or

$$\Phi = \frac{2\pi L}{C} f \sin \theta_o \quad \text{(radians)}$$

For a change in frequency, the change in phase is

$$\Delta\Phi = \frac{2\pi L}{C} \Delta f \sin \theta_o \quad \text{(radians)}$$

letting $\frac{L}{C} \sin \theta_o = T$

and $\Delta f = \frac{1}{\tau}$

yields $\Delta\Phi = 2\pi \frac{T}{\tau}$

for $T = \tau$ $\Delta\Phi = 2\pi$

A linear phase change of 2π across an aperture will scan the beam by

$$\frac{\lambda}{L \cos \theta_o} = \text{Beamwidth}^*$$

Appendix D

Feed Effects cause a change in scan angle with frequency of

$$\Delta\theta \approx \frac{\Delta f}{f} \frac{1}{\cos \theta_o} \quad \text{(radians)}$$

Proof: Referring to Figure 9, the total feed path length from first to last radiating element is (assuming free space propagation in the feed)

$$\varphi = \frac{2\pi L}{\lambda} \quad \text{(radians)}$$

or

$$\varphi = \frac{2\pi L}{C} f$$

The change in φ for a change in frequency is

$$\Delta\varphi = \frac{2\pi L}{C} \Delta f$$

This linear change in phase across the aperture caused by the change in frequency scans the beam. To observe just how far the beam is scanned, examine the way in which an aperture is scanned with phase. For a given scan angle θ_o, the required phase across the array is

$$\Phi = \frac{2\pi L}{\lambda} \sin \theta_o$$

the required change in Φ for a change in scan angle is

$$\frac{d\Phi}{d\theta} = \frac{2\pi L}{\lambda} \cos \theta_o$$

or

$$\Delta\Phi \approx \frac{2\pi L}{\lambda} \cos \theta_o \, \Delta\theta_o = \frac{2\pi L f}{C} \cos \theta_o \, \Delta\theta_o$$

When the change in phase across the array is induced by the feed

$$\Delta\varphi = \Delta\Phi$$

or

$$\frac{2\pi L}{C} \Delta f \approx \frac{2\pi L f}{C} \cos \theta_o \, \Delta\theta_o$$

and

$$\Delta\theta_o \approx \frac{\Delta f}{f} \frac{1}{\cos \theta_o} \quad \text{(radians)}$$

Survey of Time-Delay Beam Steering Techniques

R. TANG

Hughes Aircraft Company
Ground Systems Group
Fullerton, California

Synopsis

This paper describes the concept of time-delay beam steering and consists of a general survey of such techniques. In particular, the emphasis is placed upon the various subarray techniques since these techniques meet all the performance requirements of practical radar systems at a reasonable cost. The bandwidth limitations of the subarray systems are described. The various methods of designing time-delay circuits to minimize the number of time-delay bits or to simplify driver control are also discussed. Problems in the design of time-delay circuits are briefly described.

Introduction

In order to achieve high range resolution in a phased array radar, the antenna must be able to transmit and receive short pulses or wide band signals. When the time duration of the pulsed signal is less than the "fill-up" time of the array aperture as shown in Figure 1, only a portion of the array aperture is illuminated at a given instant. The "fillup" time is defined by the following relationship:

$$\text{Fillup Time} = \frac{L \sin \theta}{C}$$

Where L is the length of the array

θ is the scan angle from array broadside

C is the velocity of light in free space

This partial illumination results in a reduction of signal amplitude and a spreading of the pulse width. The reduction in signal amplitude as a function of array size for a sin x/x pulse width with 10% signal bandwidth is shown in Figure 2. In order to

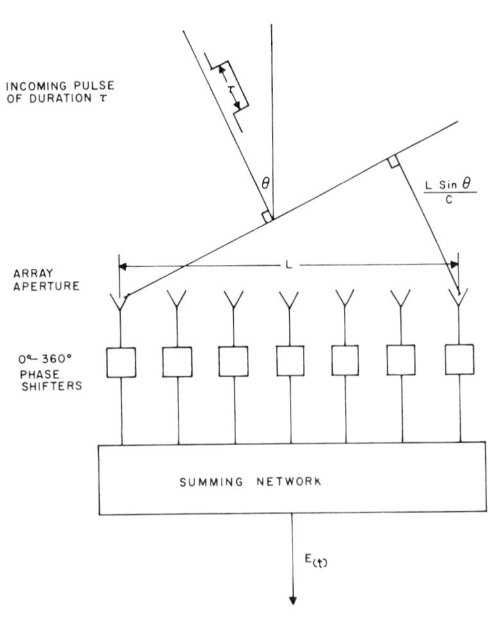

Fig. 1 Aperture size vs. pulse length.

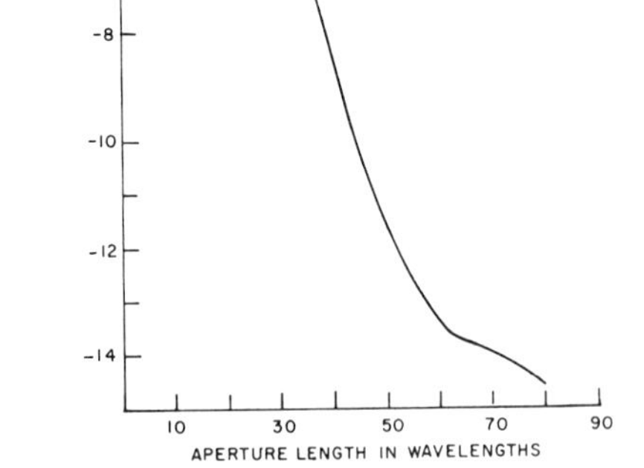

Fig. 2 Relative amplitude of a wide band pulse vs. aperture size.

eliminate these dispersive effects, a time-delay beam-forming system is then necessary to provide an addition of the signals from the various portions of the array aperture at roughly the same time.

In general, the various types of time-delay beam-forming systems can be grouped into two categories; namely, the time-delayer-per-element-type systems and the time-delayer-per-group-of-element-type systems, commonly known as subarray systems. The beam forming concepts in the former type of systems are in general true time-delay except for the bandwidth limitation of the components used in the system. The beam-forming concepts in the latter type of system are only a partial realization of a true time-delay system. However, for most practical radars, the subarray-type systems are capable of achieving the required instantaneous bandwidth at a more reasonable cost. Therefore, the emphasis of this paper is placed on the subarray-type systems.

Time-Delayer Per Element Systems

The various types of time-delayer-per-element systems can be grouped into optical systems and constrained-feed systems. The optical systems include the Luneberg lens, spherical reflector with multiple duplexing feed horns (1), constrained lenses (2), and other forms of geodesic lenses. (3) In these optical systems, time delay beam-forming is accomplished by designing the lens to provide equal path lengths from the feed horn to a plane wavefront in the radiating aperture. Typical performance characteristics for the optical systems are their extremely wide instantaneous bandwidth capability (in the order of 20% to 50%) and their multiple simultaneous beam capability (as many beams as there are feed horns). For a narrow beam system, the number of beams (which is equal to the number of feed horns) required to fill a specified scan coverage could be in the order of hundreds or thousands. Beam scanning of these systems is accomplished by switching from one feed horn to another. In system applications where the number of beams required to cover a specified scan coverage is much less than the number of array elements in the radiating aperture, the number of switches required for beam steering is correspondingly much fewer than the number of phase shifters in a phase scanned approach. Therefore, in these applications, the optical approach has a cost advantage over the phase scanned approach. The scan coverage of some of these systems has been demonstrated over a conical angular sector of 120 degrees (±60 degrees from broadside). Since there are no active components in the beam forming circuits of these systems, the random errors in the radiating aperture are minimized.

Another type of optical system is the prismatic compensation technique, (4) which resolves the incoming signal into its frequency spectral components and each frequency component is then phase shifted and recombined into the required pulse output. This technique uses a matrix of filters and phase shifters instead of time delayers to achieve time delay beam steering. In a space feed system of either the feed-through or the reflective array configuration, only a relatively small matrix is needed behind the primary feed array to provide wide band operation. This system has been demonstrated to operate over approximately a 8% instantaneous bandwidth.

The second category of time-delayer-per-element type of system is the constrained feed type of system. In the constrained-feed systems, there are the Blass Matrix (Brighan's Version), (5) the pulse-to-pulse controlling technique, (6) and the I-F time-delay steering technique. (7) The latter two techniques can be used in the subarray concept. In the Blass Matrix technique as shown in Figure 3A, the required progressive time delay between elements for a given beam scan direction is accomplished by the uniform increments in the path lengths from the beam input terminal to the radiating elements. However, the bandwidth of the system is limited by the

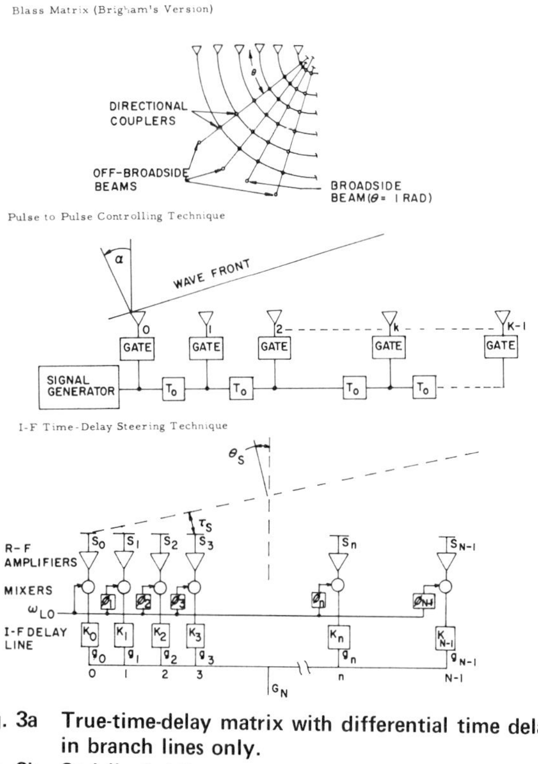

Fig. 3a True-time-delay matrix with differential time delays in branch lines only.

Fig. 3b Serially fed linear array.

Fig. 3c I-F time delay steered array antenna with phase correction.

performance of the directional couplers. Deterioration in the isolation of the couplers would result in cross coupling between the beam terminals. In the pulse to pulse controlling technique as shown in Figure 3B, time delay beam steering for an end-fed linear array is accomplished by controlling the relative time delay between successive pulses in a uniform pulse train. This technique eliminates the requirement of a time delay phase shifter per element. However, in place of thetime delay phase shifter, there is a time gate device at each element level and a long inter-element delay line between adjacent elements. The gating at all the element channels is required in order to avoid any loss in the signal to noise ratio. This loss occurs because of the partial illumination of the array aperture in the transient time of filling the long tapped delay line. The error effects in the system are also cumulative due to the serial feeding of the array elements. In the I-F time delay technique as shown in Figure 3C, I-F signal delay lines are used to provide time coincidence of the signal envelopes at the input to the summing network and phase shifters are used in the local oscillators of each mixer to provide in-phase summation of the signals within the envelopes. This technique relaxes the tolerance requirements on the delay lines due to the lower frequency, however, a preamplifier may be required in front of each mixer to overcome the mixer losses.

Subarray Systems

The various types of subarray systems are the contiguous subarray coordinate system (8) (See Figure 4),

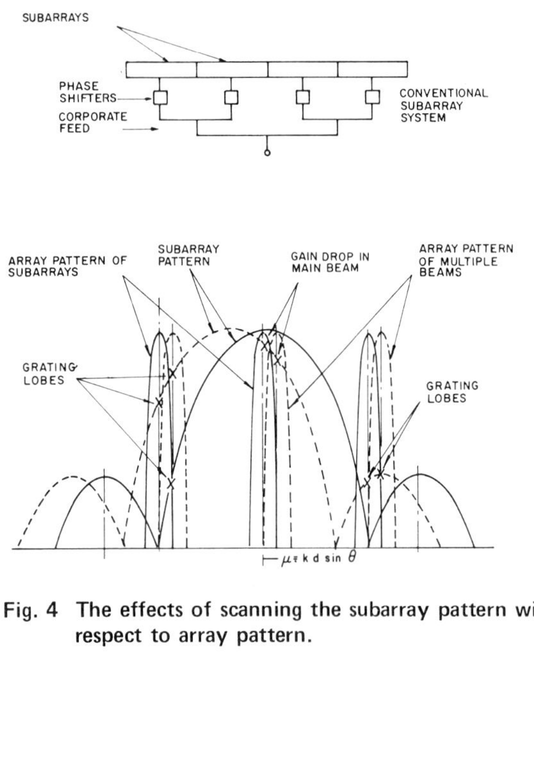

Fig. 4 The effects of scanning the subarray pattern with respect to array pattern.

the interlaced (9) or partially overlapped subarray (10) (See Figure 7) and the completely overlapped subarrays. (11) In these subarray schemes, the instantaneous bandwidth of the system increases with the increasing amount of overlapping of the subarray. For a given operating bandwidth, the completely overlapped subarray system requires the minimum number of subarrays or time delay phase shifters, thus it minimizes the cost.

The contiguous subarray system is the simplest subarraying concept wherein the radiating aperture is divided into a number of smaller subarrays which border each other at the edges (see Figure 4). The length of each subarray is made small compared to the length of the pulsed signal so that the entire subarray is illuminated during the duration of the pulsed signal. Each subarray is then connnected to a common feed point via time delay type phase shifters. These time-delay phase shifters steer the array pattern of the array of subarrays to the desired scan direction in space. Each subarray is steered to the same direction by means of the 360 degree phase limited phase shifters at the element level. Because of the 360 degree limitation, the phase shifters, in general, must be recycled, i.e., differ from the proper phase by an integral number of 360 degrees at the center frequency. This results in a linear progressive phase across the subarray that is essentially independent of frequency which means that the pattern of the subarray, whose elements are spaced by "d" distance apart, scans with frequency according to the relationship:

$$k\ d\ \sin\ \theta = \varphi \quad = \text{inter-element phase shift}$$

Since k is proportional to frequency and φ is independent of frequency, the scan angle "θ" must change with frequency.

The frequency scanning of the subarray pattern with respect to the array pattern of the subarrays leads to degradation in array gain and in the formulation of grating lobes as shown in Figure 4. These effects are greatly magnified when multiple simultaneous beams are generated in the time delayed portion of the feed system because the same subarray pattern must serve for all of the array patterns. The multiple simultaneous beams can be formulated by splitting each subarray terminal into multiple outputs with a power divider network. The corresponding output of the power divider networks are connected to a time-delay feed network for each one of the multiple beams. The amount of gain degradation and the grating lobe level are shown as a function of the ratio of bandwidth over the subarray beamwidth in Figure 5. For a subarray of approxi-

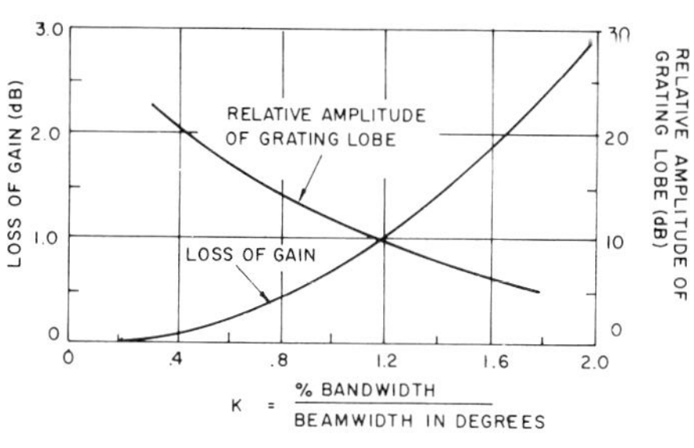

Fig. 5 Loss of gain and grating lobe amplitude as a function of bandwidth when scanned 60°.

mately 5 wavelengths long (~10° beamwidth) operating over a 10% bandwidth, the array gain is reduced by approximately 1 dB at the band edges when the main beam is scanned to 60 degrees from broadside. For the same conditions, the grating lobe level is approximately -9 dB down from the peak of the main beam. These results ignor the "cosine" active element factor. In actual values, the above results should be modified by a cosine element factor. The effect of subarray size on the width of the main lobe and the peak sidelobe level of a weighted, 500 MHz compressed pulse is shown in Figure 6. The peak time sidelobe

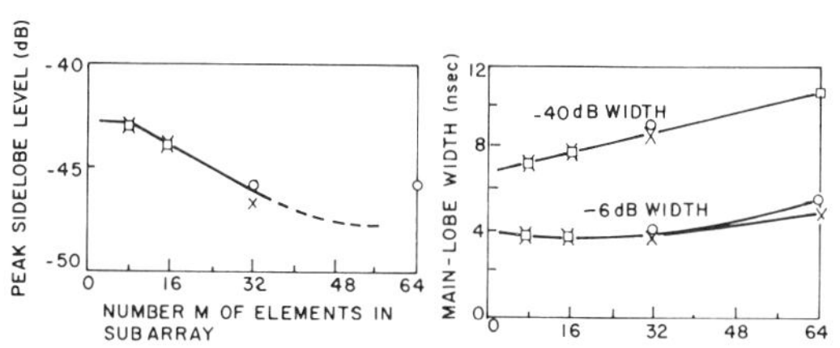

Fig. 6 Effect of subarray size on width of main lobe and peak sidelobe level of weighted, 500 MHz compressed pulse. Scan angle = 45°, $\eta = d/\lambda_0 = 0.559$. Crosses are for target at center of beam. Circles are for target displaced by 1.1° from beam center.

is not affected by the subarray size as significantly as the pulse width or the peak array gain (See Ref. 12). If fact, the peak time sidelobe level decreases somewhat with increasing subarray size.

The interlaced and partially overlapped subarray schemes, as shown in Figures 7A and 7B, have the size of the subarray aperture substantially larger than the spacing between the subarray centers. In so doing,

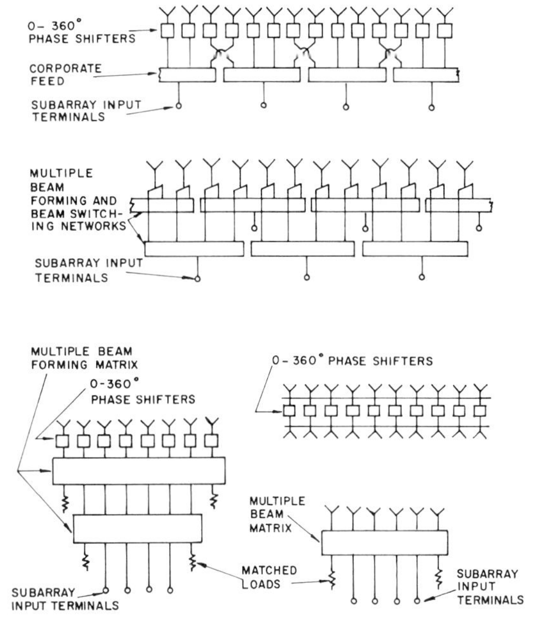

Fig. 7 Subarray schemes for minimizing gain reduction and grating lobe level.

Fig. 7a Interlaced subarray scheme.

Fig. 7b Overlapped subarray scheme.

Fig. 7c Completely overlapped subarray scheme. (Constrained feed approach).

Fig. 7d Completely overlapped subarray scheme. (Space feed approach).

the subarray null width becomes narrower than the spacing between the grating lobes; hence, the grating lobes are in the low sidelobe region of the subarray pattern. This in turn allows for better grating lobe suppression over a larger frequency band. As an example, when the width of the subarray apertures is twice their phase center spacing, the grating lobes are moved to approximately the second null of the subarray pattern. Amplitude and phase weighting can be applied to the subarray elements in order to control to some extent the shape and sidelobe structure of the subarray pattern. The interleaved subarray has the signal from each element fed into only one subarray; whereas, the overlapped subarray has the signal from each element divided and fed into two or more subarrays. In the case of the interleaved subarray, the effect of overlapping is obtained by interleaving the elements belonging to one subarray with the elements belonging to adjacent subarrays, so that adjacent subarrays share the elements of a common area on the array.

In order to minimize the gain reduction and grating lobe level over a wide bandwidth in a subarrayed aperture, an ideal subarray pattern would be a rectangularly shaped pattern which is narrower than the angular separation between the first grating lobes. This ideal rectangular subarray pattern has no reduction in gain as the main beam is scanned within the subarray pattern and it has complete grating lobe suppression outside the subarray pattern. These characteristics are illustrated in Figure 8. The completely overlapped subarray schemes, as shown in Figures 7C and 7D achieve the closest approximation to the ideal rectangular subarray pattern. The amplitude distribution across the radiating aperture corresponding to each subarray input terminal is a truncated Sin u/u distribution; therefore, its radiation pattern closely approximates a rectangular pulse-shaped pattern. For a narrow beam system, it is not practical to build a large multiple beam matrix system as shown in Figure 7C. The space fed version of this completely overlapped subarray system as shown in Figure 7D is a simple and efficient approach in accomplishing the overlapping of the subarrays. Since the space-fed system is the optimum subarray design in terms of performance and cost, a detailed description of this system is given below.

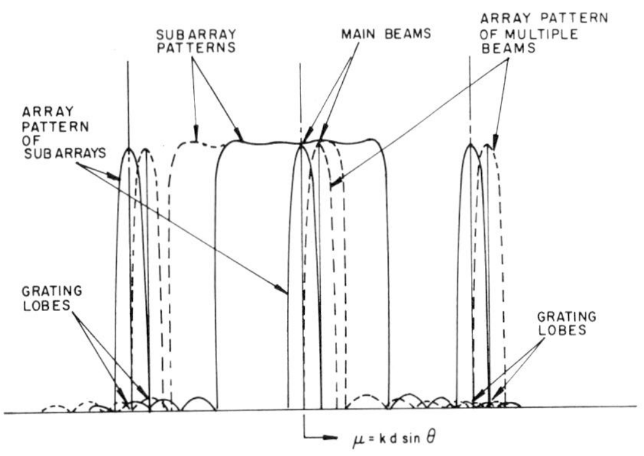

Fig. 8 The effect of scanning the ideal "rectangular" subarray pattern with respect to array pattern.

The space fed system consists of a conventional feedthrough lens fed by a Butler matrix (or a Blass matrix . The basic configuration of this beam-forming system is shown in Figure 9. In the feed-through lens system; there is a constant phase-type phase shifter between each pair of radiating and pickup elements. These phase shifters will have a maximum phase shift of 360 degrees and their phase characteristics remain essentially constant with frequency. The multiple feed horns in the feed array are connected to the outputs of a Butler matrix. Each input port of the Butler matrix represents a subarray input terminal. The reason for this identification is the fact that the feed network to the input ports of the Butler matrix is exactly the same as that for the feeding of the subarray input terminals of a conventional subarray system. The difference between the two systems lies in the fact that the radiating apertures of the subarrays of the present system overlap each other and each extends completely across the entire radiating aperture of the antenna, whereas the radiating apertures of the subarrays of the conventional subarray system only occupy a small portion of the total radiating aperture of the antenna and they do not overlap each other. With signal applied at a subarray input terminal, the excitations at the feed horns are uniform in amplitude and progressive in phase across the radiating aperture of the feed array (see Figure 9). The illumination pattern of the feed array corresponding to this aperture distribution is therefore a sin u/u pattern. The sin u/u illumination pattern is picked up by the receiving aperture of the pickup array. The spherical phase distribution of the signals at the pickup array are compensated by the phase shifters in the lens. Hence, the illumination across the final radiating aperture of the feed-through lens is still approximately a sin u/u distribution (See Figure 9 . The sin u/u distribution in turn provides a subarray radiation pattern which is approximately rectangular in shape.

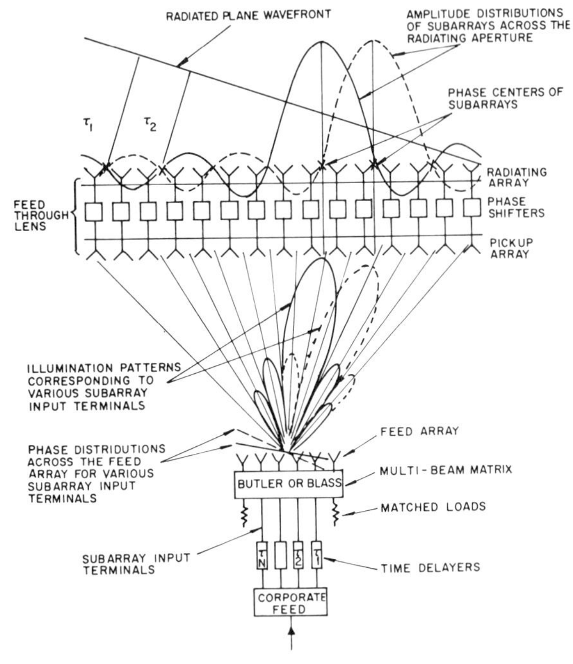

Fig. 9 Completely Overlapped Subarray Antenna System.

Since the excitation of each subarray input terminal yields an aperture distribution across the feed array which is uniform in amplitude and progressive in phase, the excitations of all the subarray input terminals with the corporate feed will yield a sin u/u distribution across the feed array. The primary illumination pattern of the feed array when all the subarray input terminals are excited is therefore a sector beam pattern with steep drop-offs at the edges of the feed-through lens. Hence, the spillover loss of this feed system is extremely small. For example, for a $1^\circ \times 1^\circ$ circular planar array system with a $f/D = 1/2$, the spillover loss is approximately 0.1 dB. The sector beam primary pattern also provides a fairly uniform illumination across the central portion of the radiating aperture with a small amount of amplitude taper near the edges of the lens due to the inverse square law spreading of the signal. The aperture efficiency of the final radiating aperture as all the subarray input terminals are uniformly excited is approximately 90 percent. The peak sidelobe level corresponding to this aperture distribution is approximately 23 dB. For the purpose of comparison, the gain reduction of the completely overlapped subarray system as a function of the ratio of bandwidth over subarray beamwidth is plotted in Figure 10. For a 10% bandwidth and 10° subarray beamwidth, the gain reduction of the completely overlapped subarray antenna is only half of that of the conventional subarray antenna. In other

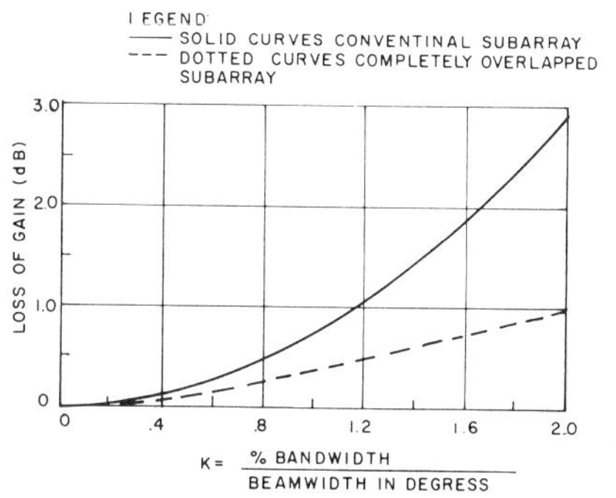

Fig. 10 Loss of gain as a function of bandwidth when scanned 60°. Note: There are no well defined grating lobes with the rectangular subarray pattern; however, some degradation of sidelobe structure occurs if array is operating beyond K = 1.2.

words, for the same allowable gain reduction and bandwidth, the completely overlapped subarray antenna minimizes the number of required subarrays and time delayers.

Time-Delay Network Design

The three time-delay feed configurations which deserve prime consideration are shown in Figure 11.

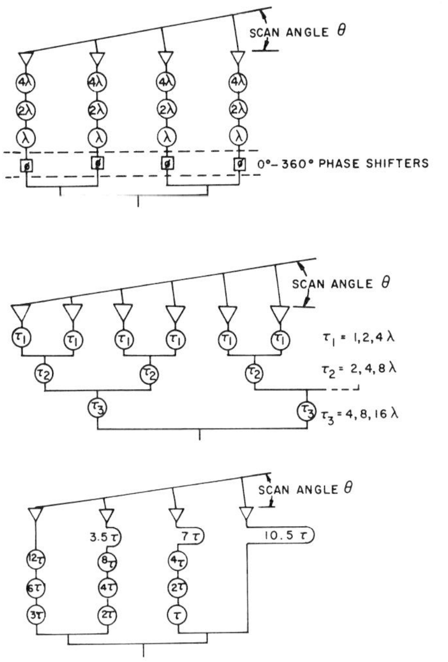

Fig. 11a Antenna feed system using identical time delayers for each subarray.

Fig. 11b Antenna feed system using time delayers placed at various levels of the corporate feed network. The longer time delay bits feed large blocks of the array.

Fig. 11c Antenna feed system using time delayers permitting simplified computation and control circuits.

The first method, Figure 11A uses identical binary, digital time-delayers for each subarray in order to provide the required time delay for each beam position. The configuration of 11B is evolved by recognizing that the long time-delay bits in Figure 11A need not be located in all subarray circuits, but can be used instead, to feed blocks of subarrays. This configuration reduces the total number of time-delay bits required and does so by reducing the number of larger bits significantly. Figure 11C shows a configuration which reduces the amount of computation and control circuits for the time-delayers. By energizing all time delayers of each horizontal level simultaneously, the beam is scanned in an amount corresponding to the inter-subarray time delay setting "τ" Fixed scan angle biasing lines are shown in the figure to produce limits which are symmetrical about broadside.

The circuit design of the time-delayer consists of switches and appropriate lengths of TEM line. The two basic configurations of Figures 12A and 12B use serial type switching circuits employing SPDT switches or switchable circulators. Figure 12C shows a parallel

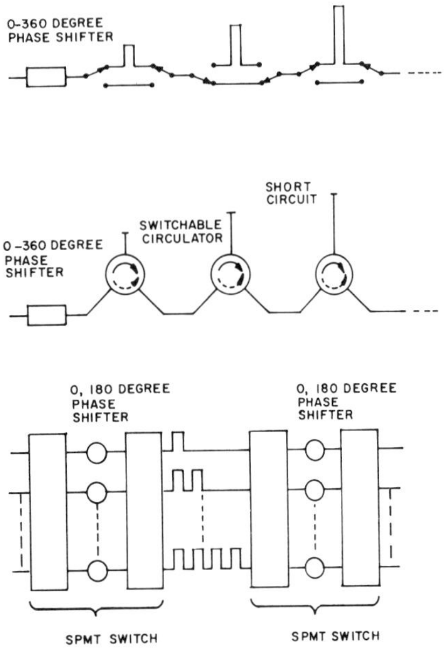

Fig. 12 Circuit configurations for time delayers.

Fig. 12a Time delay by SPDT switches.

Fig. 12b Time delay by switchable circulators.

Fig. 12c Time delay by switched parallel lines.

switched line circuit using two back-to-back Butler Matrices with one bit phase shifters in between as single pole multi-throw switches. This parallel line switching circuit is the lowest loss time delay circuit since it has only two cascaded switches. It is also a relatively high power circuit.

Conclusions

Among all the time-delay beam steering techniques covered by this survey, it is this writer's opinion that the completely overlapped subarray system as shown in Figure 8 offers the best trade-off in terms of performance and cost within the performance requirements of most high resolution radar systems, the completely overlapped subarray system can perform almost equally as well as a true time delay system.

REFERENCES

1. Tang, R. and Stark, L.: High Resolution Hemispherical Reflector Antenna Technique, Proceedings of Symposium on Electronically Scanned Array Techniques and Applications, RADC-TDR-64-225, Vol. II, pp. 73-97, July 1964, Secret.

2. McFarland J. L. and Ajioka, J.S.: Multiple-Beam Constrained Lens, Microwaves, pp. 81-89, August 1963.

3. Thiele, G. A. and Rudduck, R.C.: Geodesic Lens Antenna for a Low-Angle Radiation, IEEE Transaction on Antennas and Propagation, July 1965, Vol. AP-13, pp. 514-521.

4. Slate, Matthew W. and Ranghelli, Joseph: Prismatic Effect Compensation Technique, Final Report No. FR-711, RADC-TR-68-16, March 1968.

5. Ricardi, L.J.: Array Beam Forming Networks, Report No. TN-1965-12, April 1965.

6. Lyons, B. J. and Buczkowski, L.J.: Time Scanned Array Radar, Final Technical Report No. RADC-TR-67-449, September 1967.

7. Payne, John B.: Wideband Array Antenna Using I-F Time Delay Steering, Final Technical Report No. RADC-TR-66-289, October 1966.

8. Frank, J.: Phased Array Antenna Development, Applied Physics Laboratory, Johns Hopkins University, Report TG-882-March 1967.

9. Whistler, W.T., Hein, C.W. and Kuhn, D.H.: Array Beam Switching Study, General Electric Co., Final Report No. RADC-TR-67-543, November 1967.

10. Nolen and White, Sub-array Analysis and Design, Technical Report No. RADC-TDR-64-225, Vol. I, July 1964.

11. Tang, R. and Wong, N.S.: High Performance Space Feed, Hughes Aircraft Company Report No. TR-69-384, 25 August 1969, Secret.

12. Radar Discrimination Technology, Lincoln Laboratory, Massachusetts Institute of Technology, 30 November 1967, pp. 103-104.

Transient Response of Center-Series Fed Array

R. R. KINSEY and
A. L. HORVATH

General Electric Company
Heavy Military Electronic Systems
Syracuse, New York

ABSTRACT

The instantaneous frequency response of a linear array antenna fed by a tandem series feed (TSF) network, having independent sum and difference amplitude tapers, has been investigated both analytically and experimentally. Far-field measurements are compared with the computed data and are shown to be in good agreement.

It is found that the instantaneous bandwidth of the TSF network is generally less than the bandwidth of the corporate fed array, but that it varies less rapidly with scan angle.

I. INTRODUCTION

The response of an array antenna to an arbitrary signal can be calculated in the same way that one would calculate the response of a linear filter. It is well known from linear system theory that the output of a linear filter can be expressed as the convolution of the filter impulse response and the input signal. An equivalent expression of the output is given by the product of the filter frequency response and the spectrum of the input signal. The frequency response of the array antenna varies with scan angle and signal angle of arrival. The way in which the frequency response varies is determined by the type of steering employed, the array size and shape, and the frequency response of the feed structure.

II. THEORY OF ARRAY FREQUENCY RESPONSE

Consider an N-element linear array with element spacing d, as shown in Figure 1. The signal at the nth element is multiplied by the complex weight $a_n e^{j\phi_n}$.

The signals at each of the elements are then added together. It is desired to relate the signal at the summer output to the signal which is incident on the array; further, it is desired to find the array impulse response h(t), or equivalently the array frequency response $H(\omega)$ The relationship between the input and output variables is given by the usual expression for

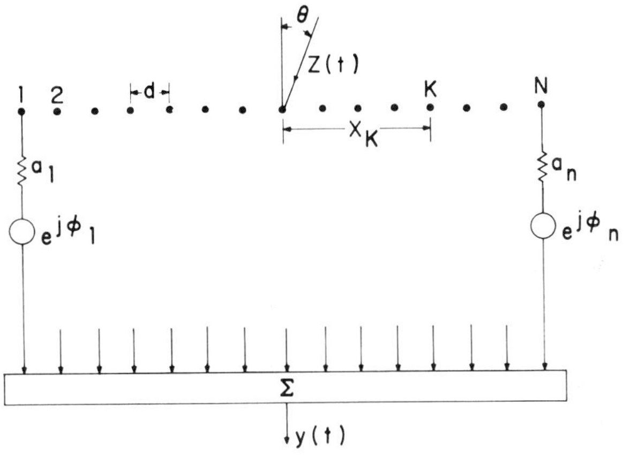

Fig. 1 Linear Array Model.

linear systems

$$y(t) = \int_{\infty}^{\infty} z(\tau)\, h(t-\tau)\, d\tau \qquad (1)$$

$$Y(\omega) = H(\omega)\, Z(\omega)$$

where

$z(\tau)$ = The signal incident on the array
$Z(\omega)$ = Incident signal spectrum
$y(t)$ = The output of the array
$Y(\omega)$ = Output spectrum

There are several ways to derive the system function $H(\omega)$. The most straightforward way is to make use of the fact that if the input of a linear system is an exponential $e^{j\omega t}$ then the output is also an expotential proportional to the input; $y(t) = H(\omega)e^{j\omega t}$. Now if it is assumed that the array is composed of isotropic elements, which have a flat response over the band of interest, the array response due to a CW signal $e^{j\omega_o t}$ incident at an angle θ measured from broadside is given by

$$y(t) = \sum_{n=1}^{N} a_n e^{j\phi_n} e^{j\omega_o (t + \frac{nd}{c}\sin\theta)}$$

$$= e^{j\omega_o t} \sum_{n=1}^{N} a_n e^{j\phi_n} e^{j\frac{\omega_o}{c} nd \sin\theta} \qquad (2)$$

Suppose the incident signal is $e^{j(\omega + \omega_o)t}$, where ω represents a frequency deviation from the carrier, ω_o. The array response is given by

$$y(t) = e^{j(\omega + \omega_o)t} \sum_{n=1}^{N} a_n e^{j\phi_n} e^{j\left(\frac{\omega + \omega_o}{c}\right) nd \sin\theta}$$

$$= H(\omega, \theta)\, e^{j(\omega + \omega_o)t} \qquad (3)$$

Now $H(\omega, \theta)$ is the low-pass equivalent of the array filter function. Actually, $H(\omega, \theta)$ represents a family of filter functions, since the array frequency response depends on the signal angle of arrival. The complex envelope of the array impulse response is simply the Fourier transform of $H(\omega, \theta)$

$$h(t, \theta) = \frac{1}{2\pi} \int_{-\infty}^{\infty} H(\omega, \theta)\, e^{j\omega t}\, d\omega \qquad (4)$$

Knowledge of $H(\omega, \theta)$ or $h(t, \theta)$ allows one to calculate the output of an array for an arbitrary input signal. If an input signal is given by $z(t) = \{Re\ \hat{z}(t) e^{j\omega_o t}\}$, then the array output can be expressed as

$$\hat{y}(t) = \int_{-\infty}^{\infty} h(t - \tau, \theta)\, \hat{z}(\tau)\, d\tau \qquad (5)$$

or

$$\hat{Y}(\omega) = H(\omega, \theta)\, \hat{Z}(\omega)$$

The real output signal becomes

$$y(t) = Re\left\{\hat{y}(t)\, e^{j\omega_o t}\right\} \qquad (6)$$

For an ideal phase-steered array, the elemental phase shifts ϕ_n are constant with frequency. If the array is to be steered to the angle θ_o, the proper elemental phases are given by

$$\phi_n = -\frac{\omega_o}{c} nd \sin\theta_o \qquad (7)$$

The array frequency response now becomes [from Equations (3) and (7)],

$$H(\omega; \theta, \theta_o) = \sum_{n=1}^{N} a_n e^{j\omega_o \frac{nd}{c}\left\{\left(1 + \frac{\omega}{\omega_o}\right)\sin\theta - \sin\theta_o\right\}} \qquad (8)$$

Note that $H(0; \theta, \theta_o)$ is simply the usual CW array pattern, and $H(\omega; \theta, \theta_o)$ is a rescaled version of the CW array angle pattern. It is convenient to discuss the frequency response in terms of amplitude and phase distortion of the incident signal. The filter function, given by Equation (8), can be expressed as a magnitude and phase.

$$H(\omega; \theta, \theta_o) = |H(\omega; \theta, \theta_o)|\, e^{j\Psi(\omega; \theta, \theta_o)} \qquad (9)$$

A distortionless filter has a constant magnitude and linear phase over the frequency band of interest. If the a_n are symmetric about the center of the array, it is seen from Equation (8) that there is no phase distortion of the signal. The boresight (pointing direction) bandwidth can be defined in the same way as the CW beamwidth. Substituting in Equation (8),

$$H(0, \alpha_o + \Delta\alpha, \alpha_o) = H(\omega_B, \alpha_o, \alpha_o)$$

$$\omega_B = \frac{\omega_o \Delta\alpha}{\alpha_o} \qquad (10)$$

so that

$$B = \frac{f_o \Delta\alpha}{|\alpha_o|}$$

$\alpha = \sin\theta$

$\Delta\alpha$ = beamwidth at f_o

f_o = carrier frequency (Hz)

B = bandwidth (Hz)

Suppose the CW beamwidth is defined as $\frac{\lambda_o}{D}$ where D is the aperture length. Equation (10) then becomes

$$B = \frac{f_o \lambda_o}{D|\alpha_o|} = \frac{c}{D|\alpha_o|} = \frac{1}{T} \tag{11}$$

T is called the time-depth of the aperture. For negligible signal distortion, it is required that $BT \leqslant 1$.

In this ideal case, the array frequency response is related to the CW angle response. It is seen from Equation (8) that

$$E(\alpha) = H(0;\alpha,\alpha_o) = \sum_{n=1}^{N} a_n e^{j\frac{\omega_o}{c} nd(\alpha - \alpha_o)} \tag{12}$$

where $E(\alpha)$ is the CW angle pattern. It is seen that $H(\omega;\alpha,\alpha_o)$ can be expressed as follows.

$$H(\omega;\alpha,\alpha_o) = E\left[\alpha\left(1 + \frac{\omega}{\omega_o}\right)\right] \tag{13}$$

Hence, the array frequency response can be found by rescaling the CW angle response. When the effects of the feed system are considered, it may no longer be possible to express the array frequency response in terms of the CW angle pattern.

In general, the array and its associated feed structure can be characterized as shown in Figure 2. The $h_k(t)$ represents the impulse response of the feed structure between the kth element and the summer. It is then easy to show that the array frequency response is given by

$$H(\omega,\alpha) = \sum_{K=1}^{N} H_k(\omega) e^{j\frac{X_k}{c}\alpha(\omega + \omega_o)} \tag{14}$$

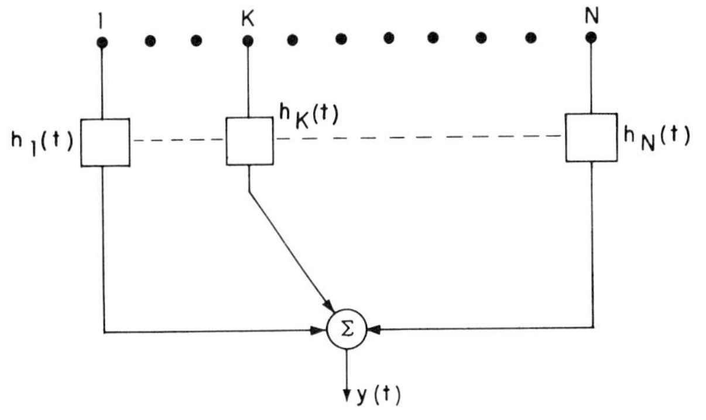

Fig. 2 Linear Array with Feed Structure.

where

$H_k(\omega)$ = The low pass equivalent of the frequency response of the kth path

X_k = The distance of the kth element from the phase reference

In order to analyze the effect of the feed structure, it is necessary to determine the $H_k(\omega)$ for a given feed structure.

It is convenient to rewrite Equation (14) in a form which separates the effect of steering devices from the dispersive effect of the feed. This results in the following characterization of the array frequency response.

$$H(\omega,\alpha) = \sum_{k=1}^{N} \underbrace{A_k(\omega) e^{-j\beta_k(\omega)}}_{\text{Feed Structure}} \underbrace{e^{j\phi_k(\omega)}}_{\text{Steering Devices}} \underbrace{e^{j\omega_o \frac{X_k}{c}\alpha\left[1 + \frac{\omega}{\omega_o}\right]}}_{\text{Incoming Time Shift Effects}} \tag{15}$$

Note that, in general, the effects of the feed structure are not separable from the ideal array response. For the parallel-fed array, the feed structure responses are ideally the same for each element, so that a single compensation filter may, in principle, be used to maintain the following

$$A_k(\omega)\, e^{-j\beta_k(\omega)} = A_k(0) = a_k \qquad (16)$$

and Equation (13) applies.

III. TANDEM SERIES FEED ANALYTICAL MODEL

The tandem series feed network (1) is illustrated schematically in Figure 3. The sum and difference

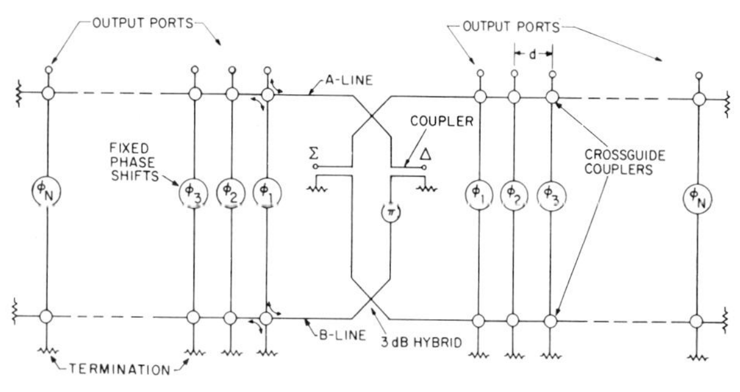

Fig. 3 Tandem Feed Schematic.

excitations are each formed by the combination of the two feedlines, including recoupled waves. The major analytical problem is to relate the amplitude and phase of the element excitation to the design parameters of the feed. It is possible to relate the excitation at each element to the waveguide and coupler parameters, using a closed-form expression. However, an analytical formulation of this relationship is complex and extremely difficult to work with. When using a computer, it is much simpler to calculate progressively the excitation at each element. The signal flow at each element is shown in Figure 4. If the frequency characteristics of the couplers and waveguide are known, the complex excitation at each element can be determined.

It is assumed that the array elements are equally spaced, and that the signal progresses from the ith element to the $(i+1)^{th}$ element through a section of waveguide feedline of length d. The waveguide imposes a phase lag on the signal. Hence, the residual phase due to the waveguide can be expressed in terms of the low-pass equivalent of the waveguide propagation factor. At each coupler, the transmitted wave V_t and the coupled wave V_c may be expressed in terms of the incident wave V_i.

$$\begin{aligned} V_t &= V_i \cos\sigma\, e^{-j\sigma} \\ V_c &= V_i \sin\sigma\, e^{-j\left(\sigma - \frac{\pi}{2}\right)} \end{aligned} \qquad (17)$$

It can be seen from Figure 4 that the element

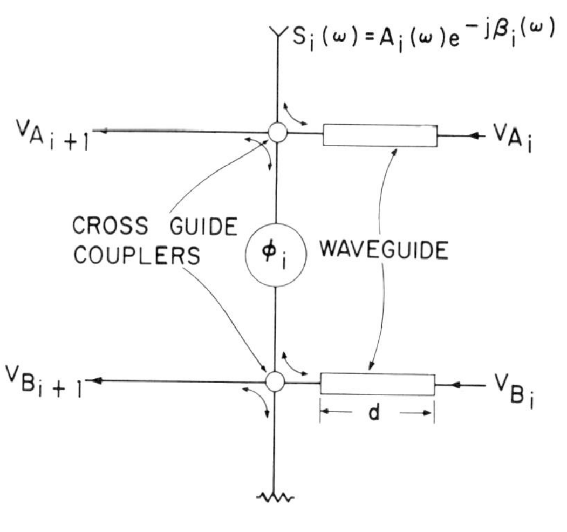

Fig. 4 Signal Flow at i^{th} Element.

excitation $S_i(\omega)$, and the A- and B-line signals $V_{A_{i+1}}$ and $V_{B_{i+1}}$ which are transmitted to the i+1th element, can be calculated from the signals V_{A_i} and V_{B_i}, using the following equations. (Common phases have been deleted in all expressions.)

$$B_i(\omega) = V_{A_i} e^{-j\beta_g(\omega)d} \sin\left[\sigma_{A_i}(\omega)\right] e^{-j\sigma_{A_i}(\omega)} + V_{B_i} e^{-j\beta_g(\omega)d} \sin\left[\sigma_{B_i}(\omega)\right] \cos\left[\sigma_{A_i}(\omega)\right] e^{j\left[\phi_i - \sigma_{B_i}(\omega) - \sigma_{A_i}(\omega)\right]} \quad (18)$$

$$V_{A_{i+1}} = V_{A_i} e^{-j\beta_g(\omega)d} \cos\left[\sigma_{A_i}(\omega)\right] e^{-j\sigma_{A_i}(\omega)} - V_{B_i} e^{-j\beta_g(\omega)d} \sin\left[\sigma_{B_i}(\omega)\right] \sin\left[\sigma_{A_i}(\omega)\right] e^{j\left[\phi_i - \sigma_{B_i}(\omega) - \sigma_{A_i}(\omega)\right]} \quad (19)$$

$$V_{B_{i+1}} = V_{B_i} e^{-j\beta_g(\omega)d} \cos\left[\sigma_{B_i}(\omega)\right] e^{-j\sigma_{B_i}(\omega)} \quad (20)$$

$\beta_g(\omega)$ is the waveguide propagation factor.

$\sigma_{A_i}(\omega)$ and $\sigma_{B_i}(\omega)$ are the A- and B-line coupler phase lags that are related to the coupling coefficient ρ (measured in dB) by the following equation.

$$\sigma = \sin^{-1}\left[10^{\frac{-\rho}{20}}\right] \quad (21)$$

If the frequency dependence of the coupling coefficient is known, then the frequency-dependent element excitations are easily computed through the use of the above equations. For this study it was assumed that $\rho(\omega)$ was linear across the frequency band of interest.

The coupler lag given by Equation (21) was derived on the basis of a coupler with infinite directivity, having coupling apertures symmetrically centered on a diagonal of the region common to two identical crossed guides. Experimental measurements agree well with the computed lag for couplers with directivities of 20 dB or more. Strictly speaking, the couplers employed in the TSF network that was used for the array response measurements depart from the assumption that the feedline and crossed guides be identical. The crossed guides in this particular feed were single-ridged waveguides and of lesser width than the unridged feed guide. The result is that the actual coupler lag is somewhat less than that predicted by Equation (21), the deviation being greatest for tight coupling values. This is a second-order effect in the frequency response computations since, for the most part, the tightest couplers are associated with the elements that have the smallest amplitude weights. The phase-lag errors vary from a fraction of a degree to a maximum of 7 degrees for a 6 dB coupler, and have been ignored in the computation of array frequency response.

IV. EXPERIMENTAL TSF ARRAY ANTENNA

A 48-port TSF network was designed, using the computer program TANDEM (1). Taylor and Bayliss excitations were impulse sampled to obtain the design amplitudes at each element port. The compensating phase leads, required between the front and rear tandem feedlines, were made to progress in a more or less linear manner by a judicious choice in the maximum coupling values in each line. The A-line coupling values range from 6 dB to 15.6 dB and the B-line values from 7 dB to 20 dB at the design frequency. Design parameters also include a feedline attenuation of 0.01 dB between cross-guide couplers, minimum coupler directivity of 20 dB, and typical coupling sensitivity of 0.3 dB/100 MHz. The rear feedline was truncated following the 10th coupler, two couplers before the midway point on either half-line, and a first-order weighted least squares fit to the exact phase compensation was implemented by tilting the rear feedline an appropriate amount. Over the 3.1 to 3.5 GHz band, the computed sum and difference sidelobes remain below -30 dB, taking into account the variation in feed network coupling values and compensating phase leads.

Figure 5 shows the completed array, set up for measurement, in a rotatable Butler building at the General Electric antenna test facility. Precision

Fig. 5 Experimental TSF Array Antenna Setup.

mechanical phase shifters were employed to set the desired phase at each of the element ports. Semi-rigid miniature coax lines connect the mechanical phase shifters to slot dipole radiating elements spaced on 1.90 inch centers. A phase-trimming adjustment in the base of the dipole was used at the time of assembly to trim the electrical length of each coax run to be equal to within 2 degrees rms. Thus, the beam-steering phase shifts could be accurately introduced by disconnecting the coax run from the phase shifter and connecting the automatic phase measurement system. To facilitate this phasing operation, a TNC coax connector was modified to provide a push-on connection between the coax run and the phase shifter.

The dipoles were fabricated by investment casting and were cadmium plated to provide a corrosion-resistant finish. An E-plane horn flare having a 6-inch aperture protrudes through the side of the building. This serves to reduce the elevation HPBW to 30 degrees and to increase elemental gain by some 5 dB.

V. PATTERN RANGE MEASUREMENT OF TSF ARRAY RESPONSE

The experimental data actually employed to determine TSF array response constitute samples of the complex frequency response function $H(\omega; \alpha, \alpha_0)$. Experimental recordings of phase and amplitude were made as the pattern range transmitter was swept through a 200 MHz band centered about each of four frequencies (3.15, 3.25, 3.35 and 3.45 GHz). These recordings were taken with the array phased for scan angles of zero degrees, 30 degrees, and 50 degrees relative to broadside for each of the above center frequencies. Sum and difference port signals were recorded in each case, with the incident wave arriving first from the boresight direction and secondly from an off-boresight direction corresponding to the half-power point of the sum pattern. In addition, for each scan setting of the phase shifters, sum and difference angle patterns were recorded with the transmitter frequency fixed at -100, -50, 0, 50, and 100 MHz relative to the applicable one of the four carrier frequencies.

The 200 MHz band used in the measurement program is sufficient to verify the predictions made with the analytical model. In addition, most practical signal bandwidths are much less than 200 MHz at S-Band. However, it is not proper to use this band-limited data for computing the array impulse response.

Figure 6 is a block diagram of the pattern range equipment. The equipment at the receive site can be identified in the photograph of Figure 5.

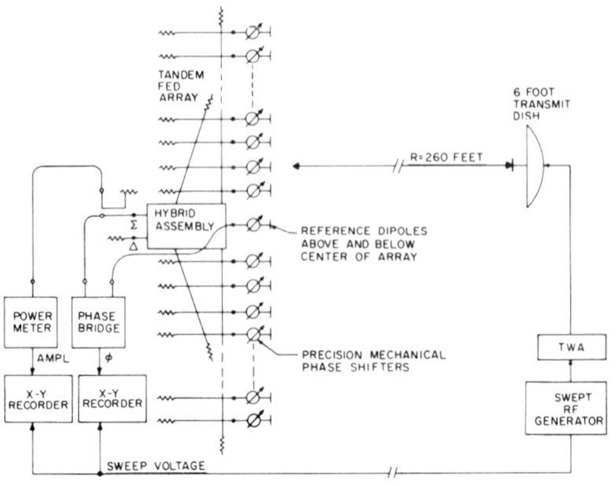

Fig. 6 Block Diagram of Pattern Range Setup.

It is desired that the reference antenna for phase measurements have its phase center located at the center of the array aperture. The reference antenna should not alter the mutual coopling (or effective area) of the innermost elements. These two constraints were satisfied by adding a 5-element section of flare above and below the 48-element array aperture. The center element of each 5-element section was active and the 2 elements on either side were terminated in a matched load. This provided an element pattern for the two reference antennas that was essentially the same as a typical pattern for the elements in the 48-element array. The two active elements from the reference flares were then added together in a 3 dB hybrid coupler to obtain a single reference signal. The reason behind this hybrid summing was an attempt to minimize the effect of ground reflections or differences in path length between the transmit dish phase center and the flares.

To make certain that the measured phase response of the array antenna was not due in part to differences in electrical lengths, or dispersion characteristics, between the array hybrid assembly and the transmission line from the reference antennas, all coax lines from the elements to the phase shifters, except the centermost two, were disconnected and replaced by matched terminations. Line lengths to the phase-measuring system were then adjusted until the phase plot had a minimum slope over the 200 MHz band of interest. The coax lines from the elements were then reconnected to the phase shifters and the recordings of phase for the sum and difference ports were made. In essence, the phase reference became a two-element array having a hybrid assembly identical to the 48 element array. Since the effective time depth of the antenna is known to be proportional to the length of the active aperture, the error introduced by this reference technique is seen to be completely negligible.

The analytical model of the TSF array assumed ideal phase shifters; i.e., the insertion phase settings remain constant over any 200 MHz band of interest. The precision mechanical phase shifters actually employed exhibit a change in insertion phase with frequency that depends on the phase setting. Typically, this change amounted to 30 degrees/100 MHz at the minimum insertion phase setting and 40 degrees/100 MHz at the maximum insertion phase setting. The procedure followed in setting up the reference phase involved the two centermost phase shifters, thereby effectively removing the mean variation in insertion phase with frequency. The errors at each element can thus range up to ±5 degrees/100 MHz for the extreme settings of insertion phase. These phase variations constitute a source of error in the experimental phase plots, albeit a second-order error. This conclusion is justified on the basis that computations of frequency response for 3-bit phase quantization gave nearly identical results to those computed for exact phase-shifter settings. By far the most important factor is the added line length between successive elements.

VI. COMPARISON OF ANALYTICAL AND EXPERIMENTAL DATA

Using the design parameters of the experimental TSF array antenna, and the computer program alluded to earlier, plots of array response were computed for comparison with the pattern range measurements. Because of good agreement between computed and experimental response, the sparse data sampling selected for this paper is sufficient to illustrate array response for all conditions of measurement described in the previous section.

For broadside steering and signals incident at boresight (Figure 7) the difference signal theoretically has a flat frequency response at $-\infty$ dB. When the array is steered off-broadside, however, the difference signal has a null only at the carrier frequency for waves incident from the direction of scan (Figure 8), and at an offset frequency if the angle of arrival does not correspond to the scan direction (Figure 9). In spite of these effects, proper tracking signals can be obtained providing the track sample maintains symmetry in frequency about the carrier. In all cases the magnitude of the sum beam remains reasonably flat for a large band of frequencies.

The phase characteristic is approximately linear for both the sum and difference signals (Figures 10, 11, and 12). Note that the large apparent discrepancy in the Δ phase plots of Figure 12 between +60 and +100 MHz is actually a measurement error due to insufficient signal strength, as can be seen from Figure 9. The slope of the phase for the difference signal is approximately twice the phase slope of the

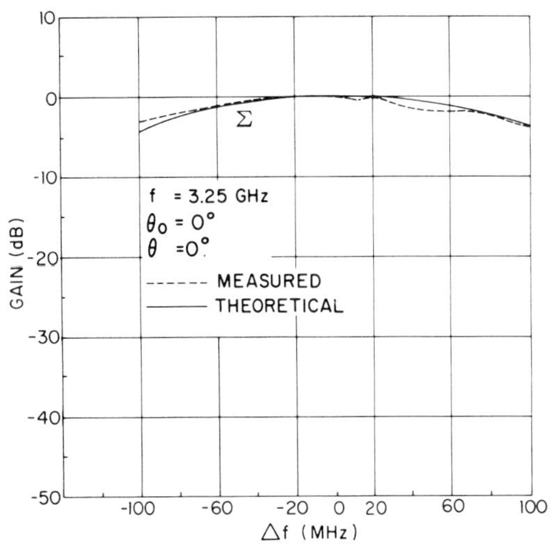

Fig. 7 Experimental Frequency Response Plots (Gain), f_o = 3.25 GHz, $\theta = 0^o$, $\theta_o = 0^o$

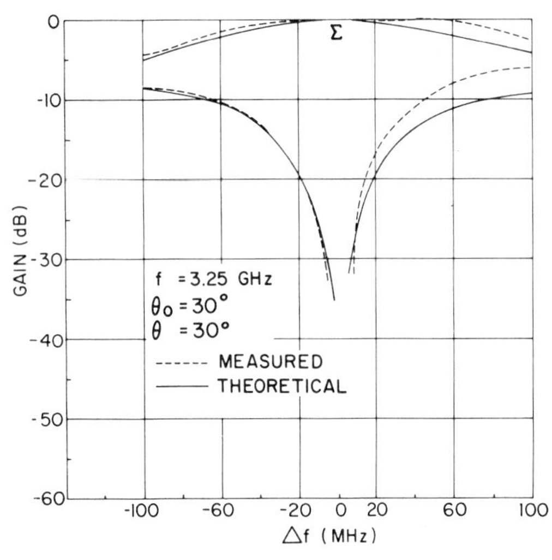

Fig. 8 Experimental Frequency Response Plots (Gain), f_o = 3.25 GHz, $\theta = 30^o$, $\theta_o = 30^o$

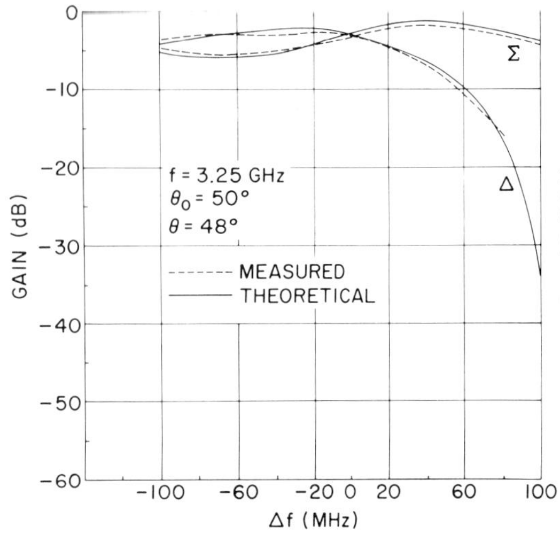

Fig. 9 Experimental Frequency Response Plots (Gain); f_o = 3.25 GHz, θ = 48°, θ_o = 50°

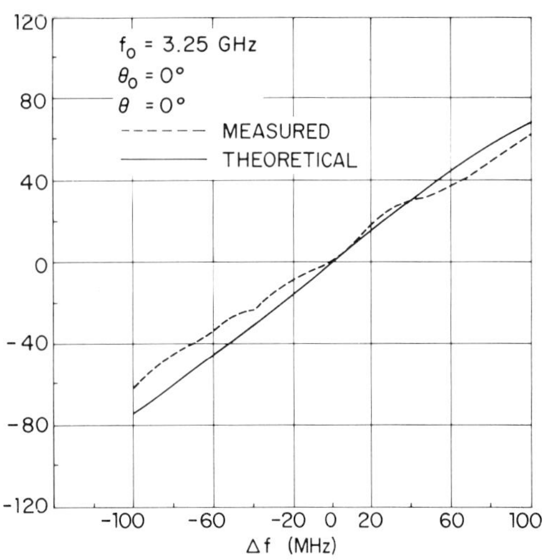

Fig. 10 Experimental Frequency Response Plots (Phase), f_o = 3.25 GHz, θ = 0°, θ_o = 0°

corresponding sum signal. In the time domain, this simply means that the difference signal experiences a time delay in the feed approximately twice that of the sum signal. For this array the time delay amounts to about 2 ns, and hence is not significant. A simple physical explanation for this effect derives from the sum and difference amplitude tapers. The difference amplitude taper peaks up near the middle of each array half,

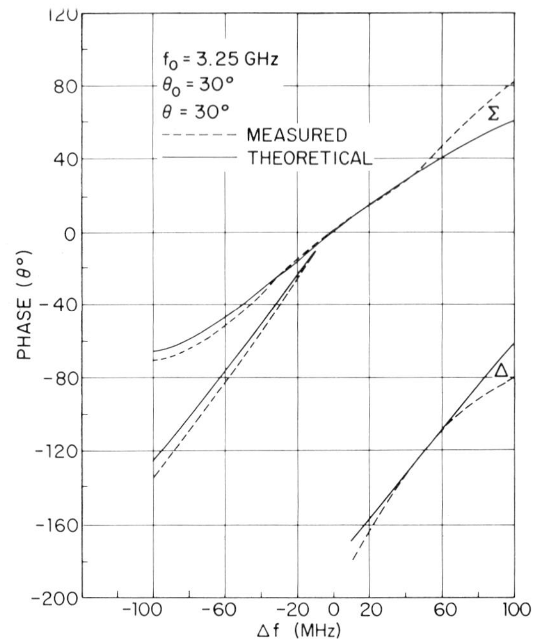

Fig. 11 Experimental Frequency Response Plots (Phase), f_o = 3.25 GHz, θ = 30°, θ_o = 30°

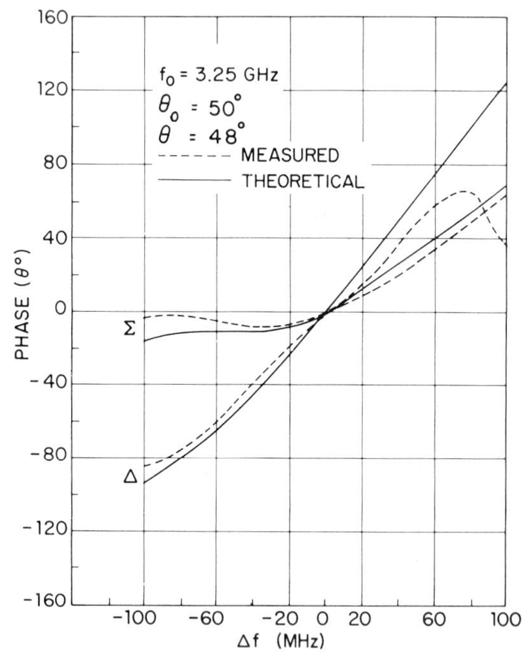

Fig. 12 Experimental Frequency Response Plots (Phase), f_o = 3.25 GHz, θ = 48°, θ_o = 50°

while the sum amplitude taper peaks in the center of the aperture. The effective phase-center separation of the two aperture halves is thus greater for the difference excitation than it is for the sum excitation, which explains the difference in time delay for the two signals.

Figures 13 and 14 are CW angle patterns taken at 50 and 100 MHz above 3.25 GHz, the frequency for which the array had been phased for broadside scan. The center-frequency angle pattern is shown in Figure 15. Examination of these patterns reveals that the near-in sidelobes, particularly the first, merge with the main beam, while the farther-out sidelobes are almost

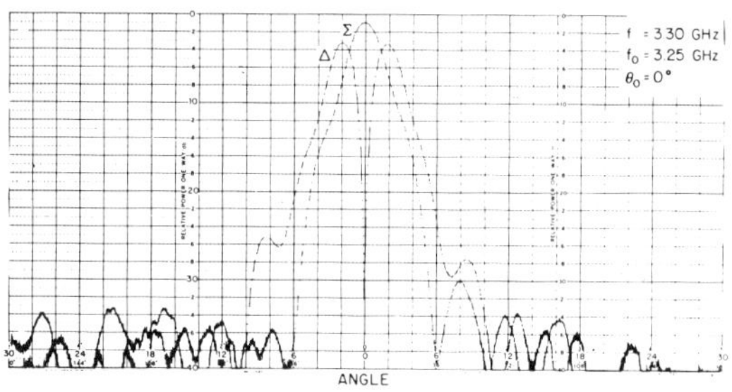

Fig. 13 Experimental CW Pattern, f = 3.30 GHz, f_o = 3.25 GHz, $\theta_o = 0^o$

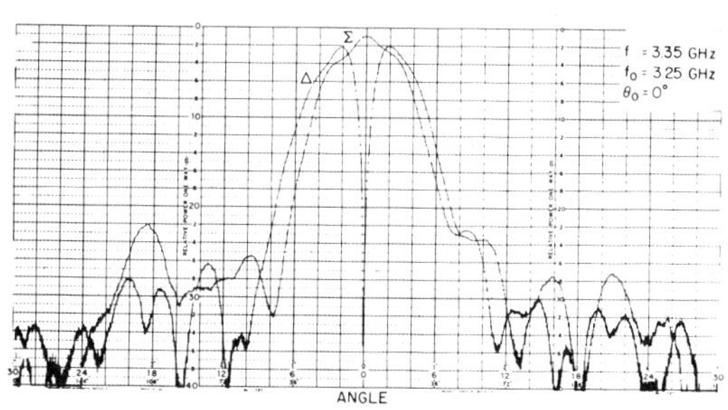

Fig. 14 Experimental CW Pattern, f = 3.35 GHz, f_o = 3.25 GHz, $\theta_o = 0^o$

unaffected for the +50 MHz pattern, but raised several dB for the +100 MHz pattern. The width of the sum and difference main beams is practically unchanged down to the-10 dB level for ± 50 MHz patterns, but broadened considerably for ±100 MHz patterns. Similar pattern sets for other scan angles and other center frequencies exhibit virtually the same behavior.

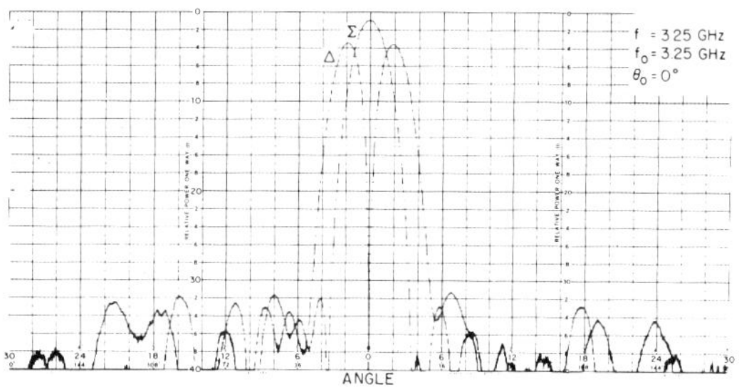

Fig. 15 Experimental CW Pattern, f = 3.25 GHz, f_o = 3.25 GHz, $\theta_o = 0^o$

Figures 16 and 17 are CW angle patterns companion to Figure 15, and represent scan angles of 30 degrees and 50 degrees at a frequency of 3.25 GHz. Sidelobe levels remain low and the pattern beamwidth broadens as the secant of the scan angle, as expected.

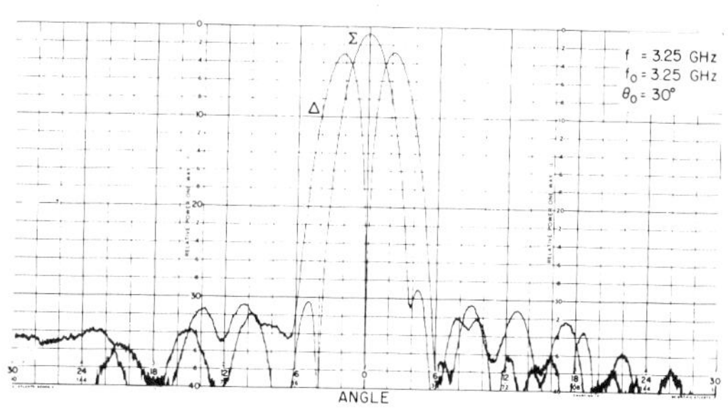

Fig. 16 Experimental CW Pattern, f = 3.25 GHz, f_o = 3.25 GHz, $\theta_o = 30^o$

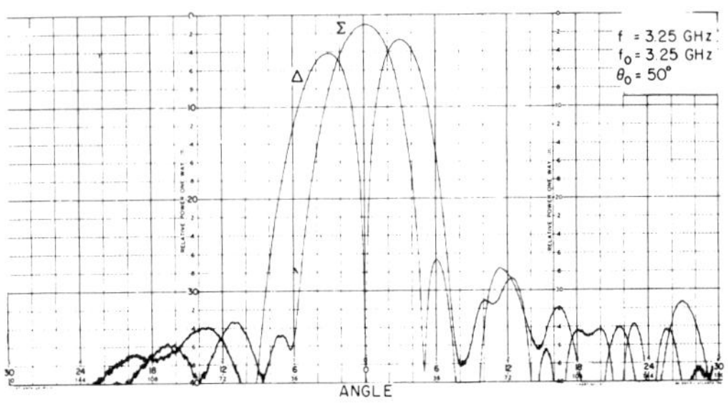

Fig. 17 Experimental CW Pattern, f = 3.25 GHz, f_o = 3.25 GHz, θ_o = 50°

VII. TIME RESPONSE OF THE TSF ARRAY

In general, the array output may be found by computing the inverse Fourier Transform of the product of the input signal spectrum and the array frequency response. A direct time domain approach results in simpler expressions for boresight signals.

If it is assumed that the coupling coefficients are constant over the signal bandwidth, then the residual phase at each element is due to the waveguide sections only. The results of calculations and experimental data indicate that this is a reasonable assumption. The array filter function under these conditions is given by

$$H(\omega;\alpha,\alpha_o) = \sum_{\substack{k=-N \\ k\neq 0}}^{N} A_k e^{-j\frac{|X_k|}{c}\left[\beta_g(\omega)-\beta_g(0)\right]} e^{j\frac{X_k}{c}\omega_o\left[(1+\frac{\omega}{\omega_o})\alpha-\alpha_o\right]} \tag{22}$$

It is also assumed that the signal bandwidth is such that only the linear term in the series expansion of $\beta_g(\omega)$ is required. Hence,

$$H(\omega;\alpha,\alpha_o) = \sum_{\substack{k=-N \\ k\neq 0}}^{N} A_k e^{-j\frac{|X_k|}{c}\left[\frac{\omega_o\omega}{\sqrt{\omega_o^2-\omega_c^2}}\right]} e^{j\frac{X_k}{c}\omega_o\left[(1+\frac{\omega}{\omega_o})\alpha-\alpha_o\right]} \tag{23}$$

The inverse Fourier transform of Equation (23) is the array impulse response.

$$\hat{h}(t,\alpha,\alpha_o) \approx \sum_{\substack{k=-N \\ k\neq 0}}^{N} A_k e^{-j\frac{X_k}{c}\omega_o(\alpha-\alpha_o)} \delta\left[t+\frac{X_k}{c}\alpha-\frac{|X_k|}{c}\frac{\omega_o}{\sqrt{\omega_o^2-\omega_c^2}}\right] \tag{24}$$

where

$\delta(t)$ = The Dirac delta function

The response $\hat{y}_o(t)$ to an arbitrary signal $\hat{s}(t)$ is the convolution of $\hat{s}(t)$ and $\hat{h}(\tau,\alpha,\alpha_o)$.

$$\hat{y}_o(t) = \sum_{\substack{k=-N \\ k\neq 0}}^{N} A_k e^{-j\frac{X_k}{c}\omega_o(\alpha-\alpha_o)} s(t-\tau_k) \tag{25}$$

where

$$\tau_k = \frac{X_k}{c}\left[\alpha + \frac{\omega_o}{\sqrt{\omega_o^2-\omega_c^2}}\right] \quad \text{for } k < 0$$

or

$$\tau_k = \frac{X_k}{c}\left[\frac{\omega_o}{\sqrt{\omega_o^2-\omega_c^2}} - \alpha\right] \quad \text{for } k > 0 \tag{26}$$

The boresight response becomes

$$\hat{y}_o(t)\big|_{\alpha+\alpha_o} = \sum_{\substack{k=-N \\ k\neq 0}}^{N} A_k \hat{s}(t-\tau_k) \tag{27}$$

It is seen from Equation (27) that the array output for boresight signals can be calculated as a weighted sum of time-shifted replicas of the input complex envelope.

If an impulse of energy is incident on the array antenna aperture (from the α_o direction), the hybrid assembly output will consist of the superposition of the contributions from each array element. In general, these contributions are not time-coincident because of time differences in arrival of the wavefront at each element and because of unequal delays within the feed network. The time span between the first and last of the elemental signals is referred to as the time-depth of the antenna. For a center series-fed array, the time-depth can be expressed by

$$T = \frac{D}{2c} |\alpha_o| + \frac{D}{2c} \frac{\omega_o}{\sqrt{\omega_o^2 - \omega_c^2}} \qquad (28)$$

where the first term is the result of incoming time shift effects and the second term is the differential delay introduced by the feed structure.

Comparing this with the time depth of the parallel-fed array $(T = \frac{D}{c} |\alpha_o|)$, it is seen that the bandwidth of the tandem-fed array will be less than the bandwidth of a parallel-fed array. This is so because the quantity multiplying $\frac{D}{2c}$ in the second term of Equation (28) is always greater than unity for air-filled waveguides. Typical numbers indicate that the $\frac{1}{T}$ bandwidth of the TSF array, steered 60 degrees off-broadside, will be about 75 percent of the bandwidth of a parallel-fed array steered to the same angle. It is interesting to note from Equation (28) that although the bandwidth of the TSF array may be less than the parallel-fed array, it is also less sensitive to scan angle. The implication is that it might prove feasible to utilize a fixed corrective filter to compensate for frequency response effects of the TSF array.

ACKNOWLEDGEMENT

A major portion of this work was sponsored by the U. S. Navy (Naval Ship Systems Command) under Contract No. N00024-68-C-1093.

REFERENCES

1. Kinsey, R.R.: Line Source Aperture Monopulse Excitations, AFCRL Contract No. AF19(628)-5905, Final Report, General Electric Company, April 1968 (Available as ASTIA AD 671967).

BIBLIOGRAPHY

1. Lopez, A.R.: Monopulse Networks for Series Feeding an Array Antenna, IEEE Transactions on Antennas and Propagation, Vol. AP-16, No. 4, July 1968, pp. 436-440.

2. Adams, W. B.: Some Consequences of Signal Bandwidth and Scanning Angle on Phased Array Characteristics, MSEE Thesis, Syracuse University, Syracuse, New York, June 1966.

3. Adams, W. B.: Phased-Array Radar Performance with Wideband Signals, Supplement to IEEE Transactions on Aerospace and Electronic Systems, Vol. AES-3, No. 6, November 1967.

Mutual Coupling Effects in Circular Arrays on Cylindrical Surfaces — Aperture Design Implications And Analysis

A. HESSEL

Polytechnic Institute of Brooklyn Graduate Center
Farmingdale, New York

ABSTRACT

This paper discusses mutual interaction effects in uniformly-spaced circular arrays on conducting cylindrical surfaces. The influence of mutual coupling on the element pattern in a match-terminated circular array environment is emphasized. The role of this element pattern in determining the array performance is stressed and the methods of its analysis are reviewed.

I. INTRODUCTION

The recent developments in low-inertia phase shifters and electronic switches have aroused considerable interest in large, circular arrays on conducting cylindrical surfaces for rapid, distortionless 360° scanning and conformal phased-array applications.

A low side-lobe, high gain aperture design of such arrays requires the knowledge and control of mutual coupling parameters in the array environment. An inadequate handling of the mutual interaction in directive circular arrays, may easily result in a reduced gain and a surprisingly high side-lobe level. It should also be noted that mutual coupling becomes increasingly important with the stringency of the design requirements.

It is well known that due to the curvature of conducting surfaces, the magnitudes of the inter-element coupling coefficients in cylindrical arrays are less than in corresponding planar structures. This advantage, however, is offset by other factors to be described below.

In both planar and circular arrays, mutual coupling affects the peak element gain, as well as the amplitude and phase of the element pattern in a match-terminated array environment. This pattern will be referred to as the Array Element Pattern (AEP).

In large planar arrays, the AEP is identical for all elements (save for end effects) and thus factors out of a fully excited array pattern expression. As a result, the phase variation of AEP is of no consequence and only its amplitude is significant, influencing the peak array gain, its scan variation, and the far side lobes.

In circular arrays this is not the case; the various elements point in different directions and the AEP can no longer be factored out. Consequently, both amplitude and phase of a circular AEP are significant, affecting the array gain as well as the <u>near</u> side lobes. In addition, contrasted to the planar case, the AEP plays a major role in synthesis of directive circular arrays. In view of the above, the AEP constitutes an important tool in design of uniformly-spaced circular arrays.

The analysis of these arrays may be considerably simplified with the help of the phase-sequence approach, (1-3 which is a consequence of the rotational symmetry of these antennas and facilitates the decomposition of circular array excitation into simpler modal constituents. The n-th phase sequence ($0 \leq n \leq N-1$) results, when consecutive elements of a uniformly-spaced ring array are driven by uniform amplitude voltage generators with a progressive phase delay of $\frac{2\pi}{N}n$ radians, where N is the total number of elements in the ring. With this excitation, the performance of all elements is identical and the analysis reduces to the solution of a boundary value problem in a typical unit cell.

Until recently, this procedure was applied only to ring arrays of a relatively small number of axial, or circumferential, linear-dipole elements in free space and to ring arrays of monopoles above a ground plane. (2, 3) The elements were assumed driven by prescribed phase-sequence gap voltages (forced excitation) and the current distribution on a typical dipole was obtained, to various orders of approximation, via solution of a single integral equation. This type of antennas and their models have the following disadvantages:

(a) The linear dipole model is too idealized. In a practical array environment, the necessary supports and baluns, disregarded in analysis, may alter the calculated current distribution to a significant degree.

(b) In view of the otherwise excessive backlobe, these arrays require spacings not exceeding 0.4λ. (4-6)

(c) They are not suitable for rugged field applications.

(d) Because practical sources are of the constant power type, the forced gap excitation (requiring zero impedance generators) represents an unrealistic assumption.

Only recently has analysis of mutual coupling in circular arrays on conducting cylinders received attention. (7-17) A metal cylinder, besides providing support for the elements, facilitates, in the case of slots, removal of the feed structure from the active region of radiating elements, with a resultant simplification of geometry and predictability of array performance. In addition, because of the directive element patterns it produces, the cylinder permits larger inter-element spacings than the free-space ring arrays, without generation of an excessive back-lobe. (4)

II. CHARACTERISTICS OF ARRAY ELEMENT PATTERNS IN UNIFORMLY SPACED CIRCULAR ARRAYS ON CONDUCTING CYLINDRICAL SURFACES

A. THEORETICAL RESULTS

The extent to which mutual coupling may affect a circular array AEP is illustrated with the help of a simple model (7, 8, 11) of a ring array of axial slits in a large conducting cylinder (see Fig. 1). The parallel-plate guide TEM-mode-fed slits are matched to free-space, under the n = 0 phase-sequence excitation, by means of identical matching-networks, and the waveguides are terminated in their characteristic impedances.

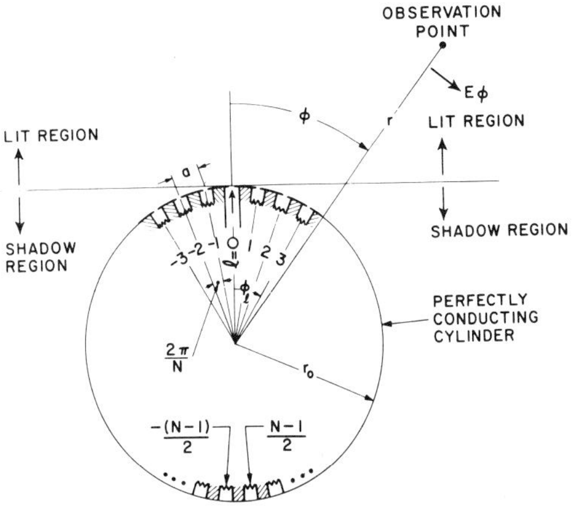

Fig. 1 Circular Array of Parallel-Plate-Guide Fed Axial Slits in a Conducting Cylinder; Array Element Pattern Excitation. (From [11]).

Figure 2 presents a comparison between typical E-plane element gain patterns of the array in Fig. 1, with circumference to wavelength ratio $kr_0 = 105$, slit width $t = 0.1\lambda$ for three values of inter-element spacing $\frac{1}{2} < a/\lambda < 1$, and the E-plane gain function of a corresponding isolated, matched slit. In contrast to the isolated-slit-gain of 3dBI, the peak realized gain for $a/\lambda = 0.525$ and 0.6 approximately equals the unit cell gain $\frac{2\pi a}{\lambda}$, while for $a/\lambda = 0.8$ it is even 3 dB higher than $\frac{2\pi a}{\lambda}$.

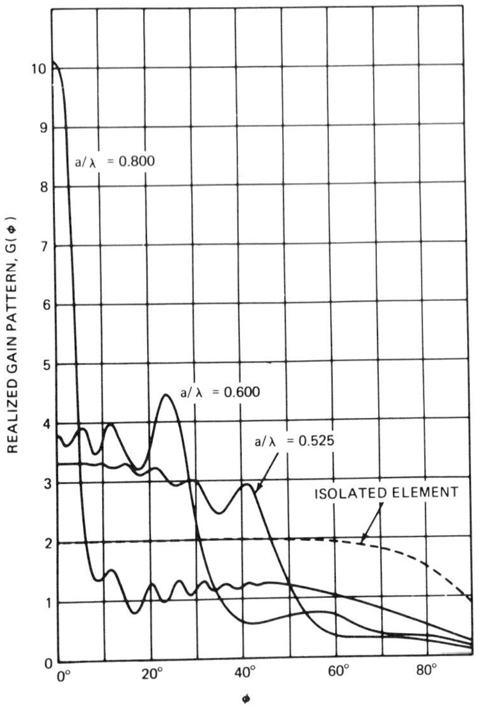

Fig. 2 Effect of Mutual Coupling on the E-plane Element Gain Function for Array in Fig. 1, $kr_0 = 105$, $t/\lambda = 0.1$. (From [10] and [11]).

This gain increase, associated with a considerable narrowing of the AEP as compared with the isolated element pattern, is accompanied by a dip of broadside. For $a/\lambda = 0.525$, the dip appears near 60° and tends, with increasing spacing, towards broadside. Theory predicts appearance of the dip at an angle ϕ_d for which the corresponding planar AEP exhibits in the E-plane an incipient grating lobe null. (7, 11)

Figure 3 shows the phase variation (referenced to the element) associated with the patterns of Fig. 2. Excursions of 30° - 40° are observed, predominantly

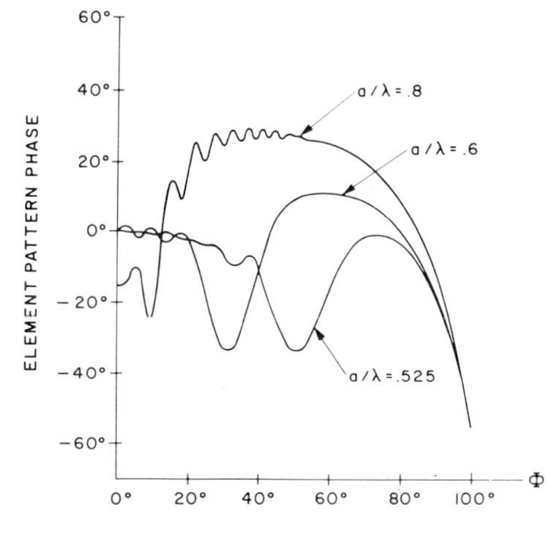

Fig. 3 Array Element Pattern Phase; $kr_o = 105$, $t/\lambda = 0.1$. (From [10]).

in the dip regions, indicating that the phase center of an element in a circular array environment, varies with the observation angle.

Another characteristic feature of AEPs in Fig. 2 is the appearance of a ripple off broadside. The origin of the ripple as well as the physical phenomena associated with the formation of the AEPs and, by implication of the formation of the AEPs for cylindrical arrays of uniformly spaced finite aperture elements on large cylindrical surfaces, (13) are discussed in Refs. (7) and (11) and are summarized below.

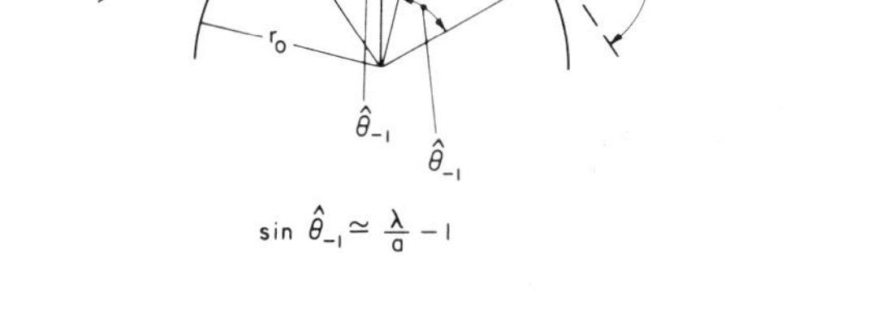

Fig. 4 Wave Contributions in the Forward (Lit) Region of AEPs of Circular Arrays on Conducting Cylindrical Surfaces.

In the forward (lit) region, $|\phi| < 90°$ in Fig. 1, the AEP for medium, or large cylinders, results basically from a superposition of two wave-types (see Fig. 4). The dominant contribution, the space-wave, is analogous to the E-plane AEP of an infinite array of aperture elements in a ground plane; it appears to originate from the driven element and its pattern is proportional to an angle dependent, active element transmission coefficient $T(\nu)$. In the circular array case $T(\nu)$ is the ratio of the total dominant mode voltage to the incident mode voltage in a typical feed guide, under the ν-th phase sequence excitation, and is obtained by essentially replacing the "admittance crater" of a corresponding planar array by a modified "crater" associated with the cylindrical geometry. For an observation angle ϕ, $T(\nu)$ is to be evaluated at $\nu = kr_o \sin\phi$ (see section III). The space-wave is responsible for the "grating-lobe" dips in the AEPs of Fig. 2. The extent of the dips is proportional to $(kr_o)^{-2/3}$ (power) (11).

Other contributions to the AEP in the forward region are due to creeping-waves which are the guided waves of the array structure. *

Creeping-waves are encountered in radiation of isolated apertures on conducting cylindrical surfaces. These waves propagate with nearly free-space velocity in their direction of travel around the cylinder. They radiate off along the local tangent and therefore contribute significantly to the far field pattern only in its shadow region. While traveling on the cylindrical surface they attenuate exponentially.

In contrast, creeping-waves in uniformly-spaced circular arrays on conducting cylindrical surfaces contribute significantly not only in the shadow region, but more importantly in the forward (lit) region. This is a consequence of the angular periodicity of ring arrays which gives rise to grating-lobes (harmonics) of the clockwise and counterclockwise traveling fundamental creeping-waves. While the fundamental creeping waves closely resemble those on a smooth cyliner, their grating-lobes, for spacing $\frac{\lambda}{2} < a < \lambda$, radiate away from the local tangent in the forward direction. The angle of radiation $\hat{\theta}_{-1}$ with respect to the local radius (see Fig. 4) is approximately given by

*Planar arrays do not support creeping-waves, but leaky waves. These do not contribute directly in the far field and therefore, in the planar case the space wave alone contributes the AEP.

$$\sin|\hat{\theta}_{-1}| = \frac{\lambda}{a} - 1 \qquad (1)$$

and corresponds to that of a grating-lobe for a locally planar array of spacing "a", driven with a steering phase delay "ka" of the fundamental creeping-wave.

Superposition of the space-wave and the creeping-wave grating-lobe contributions yields an interference pattern which accounts for the rippled main lobe of the AEP and also for some filling of the space-wave dips.

As seen from the geometry of Fig. 4, the creeping-wave grating-lobes contributing at a pattern angle ϕ, travel around the array face through the angle $\phi \pm \hat{\theta}_{-1}$ and in particular those appearing at broadside, traverse an angle $|\hat{\theta}_{-1}|$ or an arc $r_o|\hat{\theta}_{-1}|$. As "a" approaches λ, this path length decreases, in view of Equation (1), causing a proportional (in dB) reduction of attenuation and consequently, an amplitude increase of the creeping-wave grating-lobes that contribute in the broadside direction. This in turn produces a significant change of the broadside gain (increase or reduction depending on the relative phasing of the space-wave and the creeping-waves). For the particular combination of kr_o and $a/\lambda = 0.8$ in Fig. 2, an in phase addition takes place at broadside. In section IIB a case of destructive interference is encountered.

The effect of a smaller cylinder radius is exhibited in Fig. 5 showing the E-plane AEP of the array in Fig. 1, with kr_o=21. The pattern is distorted and its basic features are not easily recognizable. This is attributed to a reduced grating-lobe dip and strong creeping-wave effects.

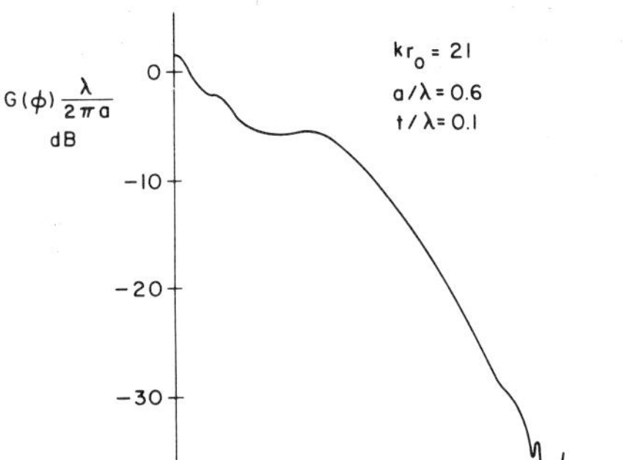

Fig. 5 E-plane Element Gain Pattern for Array in Fig. 1; $kr_o = 21$, $a/\lambda = 0.6$, $t/\lambda = 0.1$. (From [11]).

An analysis of the element gain function in infinite stacked ring arrays of open-ended rectangular waveguides with circumferential polarization, appears in these Proceedings (13) and the reader is referred there for details.

Reference (15) considers the radiation characteristics of infinite, stacked ring arrays of open-ended circumferentially polarized, rectangular waveguide elements in a large conducting cylinder. In particular, for a rectangular and triangular grid arrangement, (15) presents the pattern due to a uniform amplitude and linear phase taper excitation of a single, axial, linear-array-column in the match-terminated circular array environment. The entire field is derived from a single, axial component of the magnetic field. This is an approximation which seems to yield fairly accurate results, as shown in Section IIB.

For axial polarization and wider slits with an assumed sinusoidal aperture electric field, the AEP indicates weaker mutual interaction (Fig. 6); creeping wave effects are subdued and the AEP in the forward region is determined principally by the space-wave and is numerically close to the isolated element pattern.

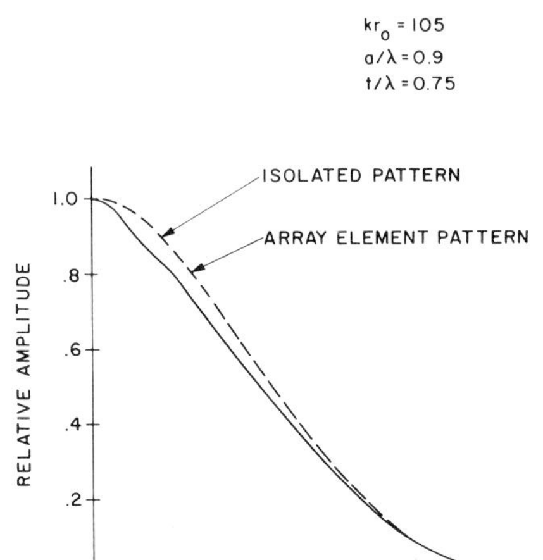

Fig. 6 Element Gain Function for Array in Fig. 1 with Axial Polarization; $kr_o = 105$, $a/\lambda = 0.9$, $t/\lambda = 0.75$. (From [10]).

Mutual coupling in single ring arrays of dipoles in the presence of small to medium size conducting cylinders, is considered in Reference (16). Fig. 7 pre-

CYLINDER RADIUS	RADIUS TO DIPOLES	Z_{11}
1.0λ	1.25λ	85.20+j69.57
2.0λ	2.25λ	85.40+j70.89
3.0λ	3.25λ	85.48+j71.38
4.0λ	4.25λ	85.52+j71.64
5.0λ	5.25λ	85.55+j71.80

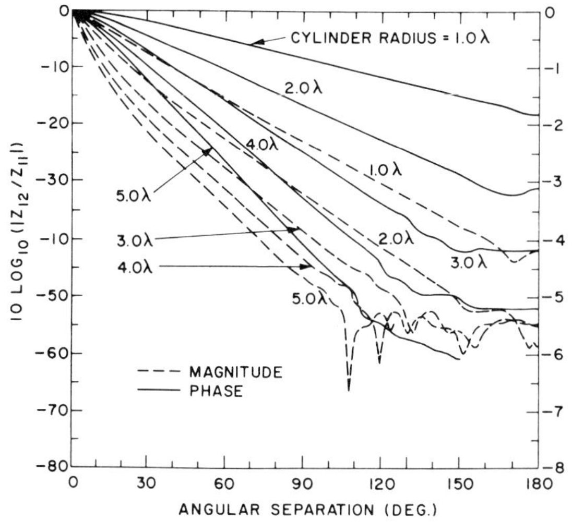

Fig. 7 Mutual Impedance vs. Angular Separation Between Two Side-by-Side, Axial, λ/2 Dipoles in Front of a Conducting Cylinder. (From [16]).

sents, the magnitude and phase plots of the mutual impedance Z_{12}, as a function of the angular separation between two side-by-side, linear, single mode, axial λ/2 dipoles, spaced λ/4 from the surface of a conducting cylinder (radius varying from 1λ to 5λ).

The field of a single dipole, in azimuth, has been represented by a harmonic series and along the axis by a Fourier integral, both appropriately truncated for their numerical evaluation.

For small separations, it was found that Z_{12} may be adequately approximated by the geometric-optical result. Beyond the shadow boundary, the magnitude of Z_{12} decays exponentially (linear in dB) and its phase varies linearly with the dipole spacing, as would be expected on the basis of a creeping-wave interaction mechanism. In the back-lobe region of the excited dipole the interference between the clockwise and counter-clockwise creeping-waves causes a departure from linearity.

Figure 8 shows, the H-plane AEP of a single ring array of uniformly spaced, λ/2 axial dipoles at a distance λ/4 from a 3λ radius cylinder, with the inter-element spacing as parameter. The load (generator) impedances have been chosen to conjugate-match the single dipole self impedance in the presence of the cylinder. Results indicate narrowing of the AEP with increased spacing, but grating lobe dips are not observed (and not anticipated).

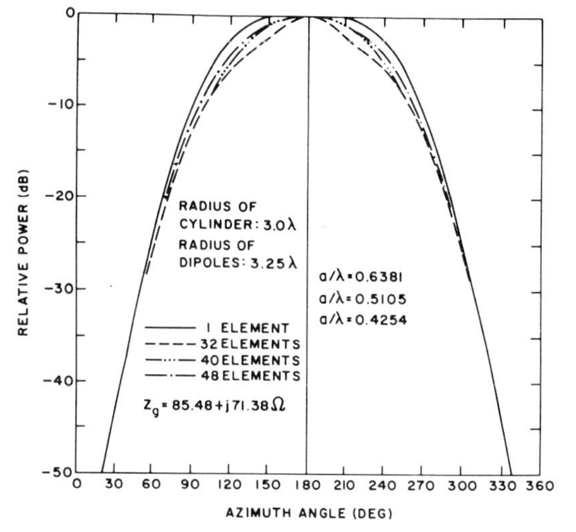

Fig. 8 H-plane AEP for a Single Ring of Axial, λ/2 Dipoles at a Distance λ/4 from the Surface of a Central, 3λ Radius Conducting Cylinder. (From [16]).

CYLINDER RADIUS	Y_{11}
2.0λ	0.9498 + j0.2821
3.0λ	0.9583 + j0.2784
4.0λ	0.9626 + j0.2765
5.0λ	0.9651 + j0.2752

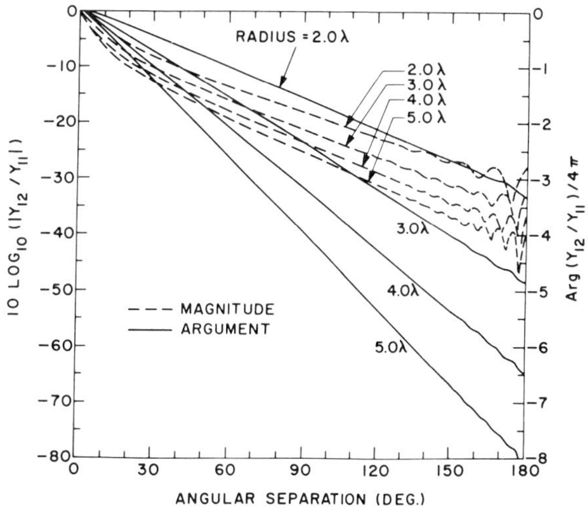

Fig. 9 Mutual Admittance vs. Angular Separation Between Two Side-by-Side, Axial, λ/2 Slots in a Conducting Cylinder. (From [16]).

Plots of mutual admittance between two side-by-side axial, single mode $\lambda/2$ slots in conducting 2λ-5λ radii cylinders, vs. angular separation, appear in Fig. 9. As expected, mutual coupling for the slot case is stronger than for the dipole case.

B. EXPERIMENTAL EVIDENCE

The Naval Electronics Laboratory Center (NELC) antenna group headed by J. H. Provencher, performed measurements of the AEPs on two circumferentially polarized, large, circular arrays. The first antenna (Fig. 10) is a single ring array of 128 sectoral-horn-

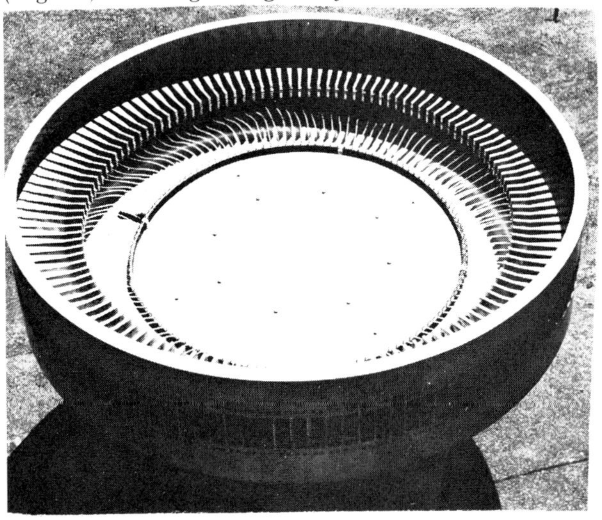

Fig. 10 NELC 4' Radius, 128 Horn, Single Ring Array.

fed rectangular apertures in a 4' radius conducting cylinder. The apertures, $\frac{\lambda}{2}$ in the E-plane and $\frac{3}{2}\lambda$ in the H-plane are 0.65λ apart at the design frequency 3.2 GHz.

Figures 11-19 show the comparison between measured E-plane AEPs and the corresponding isolated element patterns (single horn-fed aperture on the cylinder) in the 2.7-3.9 GHz frequency range. Figure 11

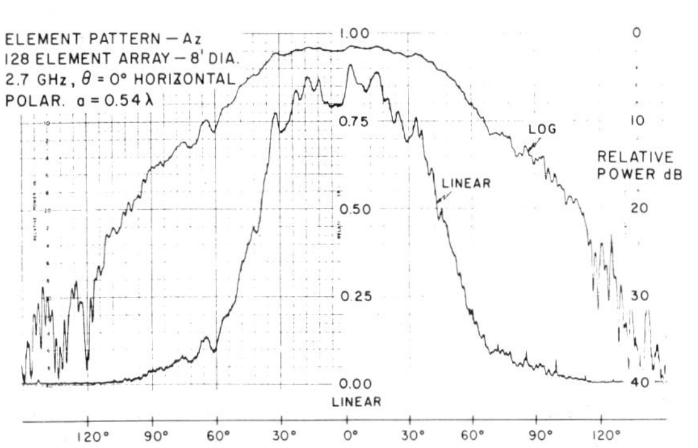

Fig. 11 NELC 4' Radius AEP; f = 2.7 GHz, $a/\lambda = 0.54$.

indicates that, for $a/\lambda = 0.54$, the AEP exhibits a 3 dB beamwidth of $\approx 95°$, against $144°$ of the corresponding isolated element pattern in Fig. 12. The

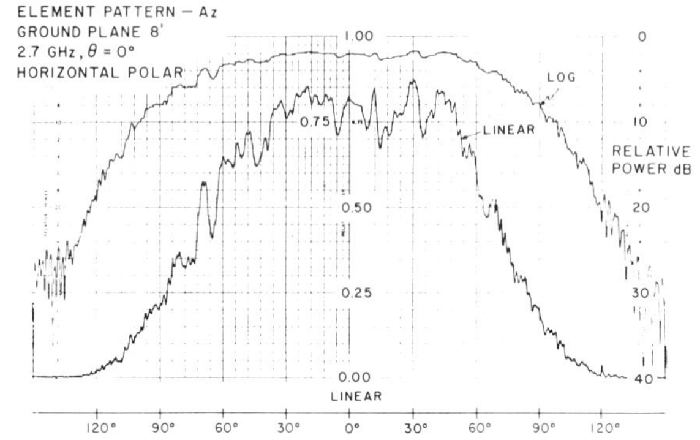

Fig. 12 Isolated Element Pattern - A Single Horn-Fed Aperture on 4' Radius Conducting Cylinder; f = 2.7 GHz.

f = 2.7 GHz.

theoretical grating-lobe dip location for this spacing is 58°. Although the dip does not explicitly appear, its influence manifests itself in a considerable narrowing of the beamwidth in Fig. 11 as compared to Fig. 12. A similar comparison is seen in Fig. 13 and 14 at 2.9 GHz. Figure 13 indicates further nar-

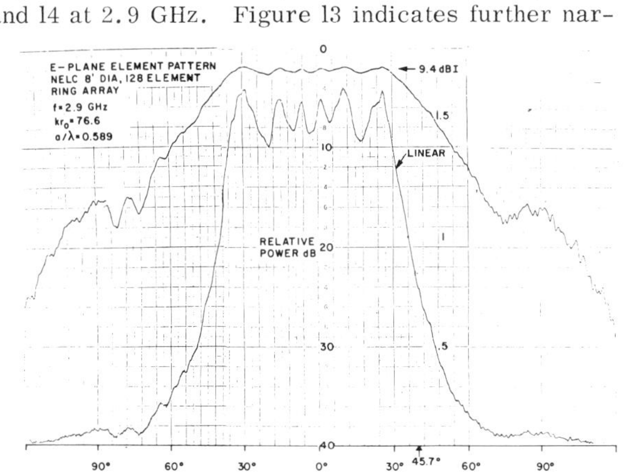

Fig. 13 NELC 4' Radius AEP; f = 2.9 GHz, $a/\lambda = 0.589$.

rowing of the AEP beamwidth for $a/\lambda = 0.589$, the predicted grating-lobe dip location being at 45.7°. The space-wave-creeping-wave interference ripple is clearly visible and accentuated on the linear scale. The peak gain is 9.4dBI. Figure 15 presents the AEP at the design frequency 3.2 GHz ($a/\lambda = 0.65$). Peak element gain increases to 9.9 dBI and the pattern narrows further, with the predicted grating-lobe dip location at 32.5°. Figure 16 shows the AEP at 3.5 GHz ($a/\lambda = 0.70$). The grating-lobe dip is now discernible near the theoretical 25.4°.

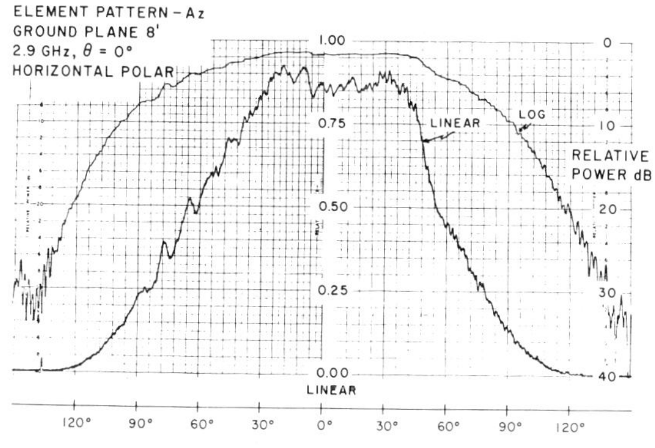

Fig. 14 Isolated Element Pattern (as in Fig. 12); f = 2.9 GHz.

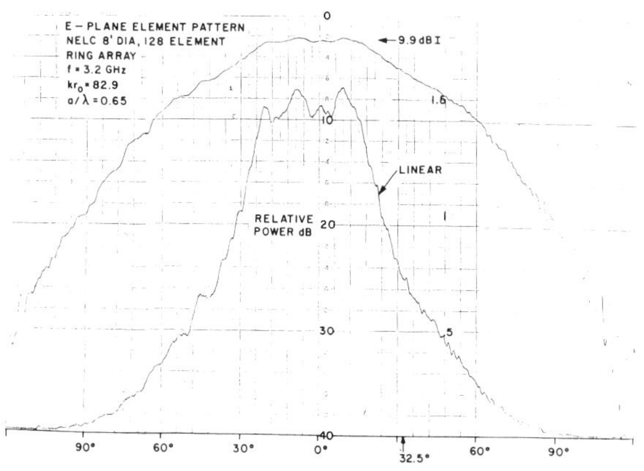

Fig. 15 NELC 4′ Radius AEP; f = 3.2 GHz, a/λ = 0.65.

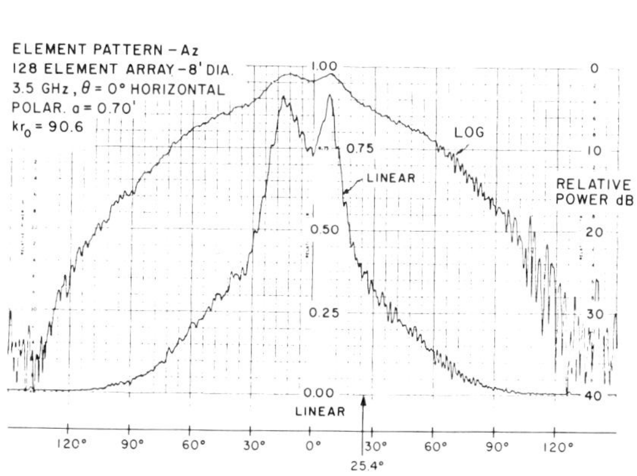

Fig. 16 NELC 4′ Radius AEP; f = 3.5 GHz, a/λ = 0.70.

The corresponding isolated element pattern found in Fig. 17 is very different. The AEP at 3.9 GHz (a/λ = 0.78) in Fig. 18 closely resembles the predicted E-plane AEP of Fig. 2 for a/λ = 0.8. The 3dB

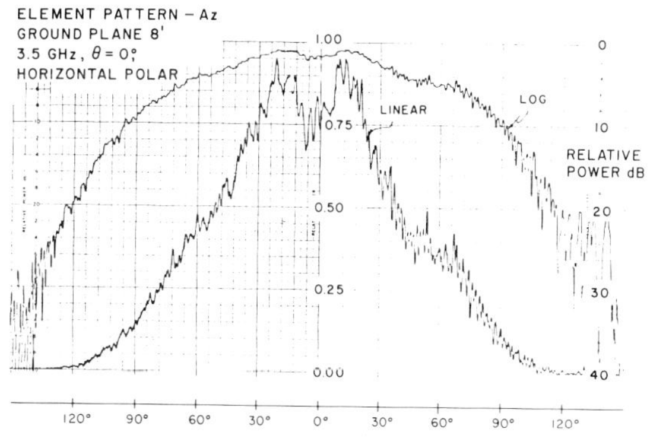

Fig. 17 Isolated Element Pattern (as in Fig. 12); f = 3.5 GHz.

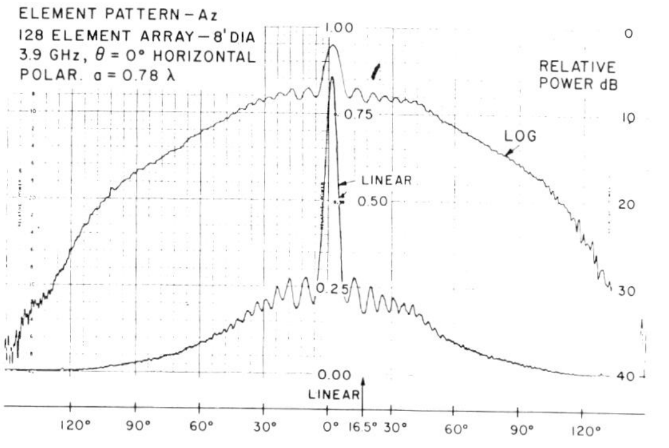

Fig. 18 NELC 4′ Radius AEP; f = 3.9 GHz; a/λ = 0.78.

beamwidth is about 6° and the peak gain 13.6 dBI, (5dB higher than that of the isolated element pattern in Fig. 19. This compares to a 7dB increase in

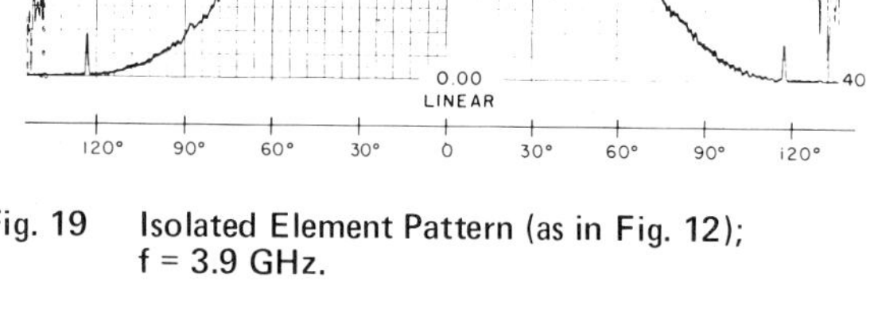

Fig. 19 Isolated Element Pattern (as in Fig. 12); f = 3.9 GHz.

Fig. 2 for the array of slits). The theoretical grating-lobe dip location is 16.5°. The dip however is filled by the pronounced interference ripple. The reduced grating-lobe dip and the 2dB difference in gain-increase are understandable in light of the reduced level of mutual coupling in the single ring horn array, as compared with the two dimensional slit array.

The second antenna is a 60° sector stacked-ring array of open-ended rectangular waveguides in a conducting cylinder (Fig. 20). The array contains 42 staggered, linear arrays (columns) of 32 elements each, forming a triangular grid. Figures 22–34 present the measured and calculated (15) principal E-plane AEP's in the frequency range 2.9–3.9 GHz, along with two measured isolated element patterns. Figure 21 shows the grating-lobe grid for this array, appropriate to 3.3 GHz. The grid, drawn as if the array were planar, (13) indicates that in the E-plane AEP, the grating-lobe dip is caused by scanning the first E-plane grating-lobe to the edge of real space. The predicted angular locations of the grating-lobe dips at ± 31° are well borne out by both the experimental and the calculated patterns.

1344-Element cylindrical array operates at S-band. The 42 columns are mounted on an eight foot radius.

Fig. 20 NELC 8′ Radius, Triangular Grid, Stacked-Ring Array of Open-Ended Rectangular Waveguides.

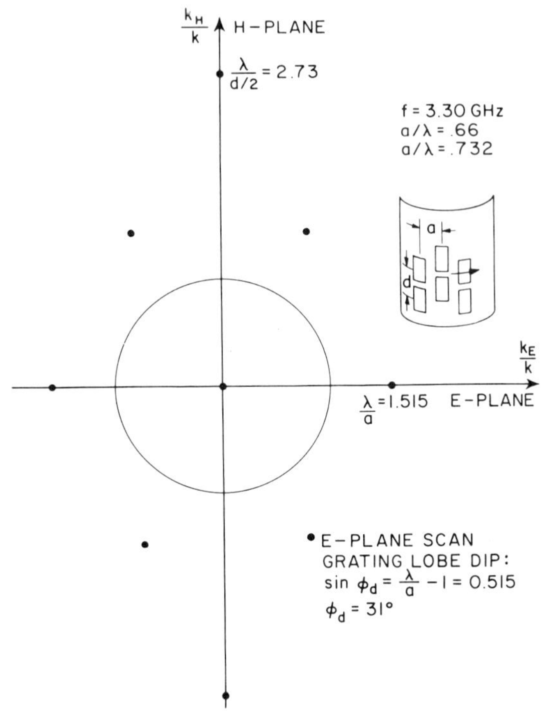

Fig. 21 Grating-Lobe Grid for the Array of Fig. 20; f = 3.3 GHz.

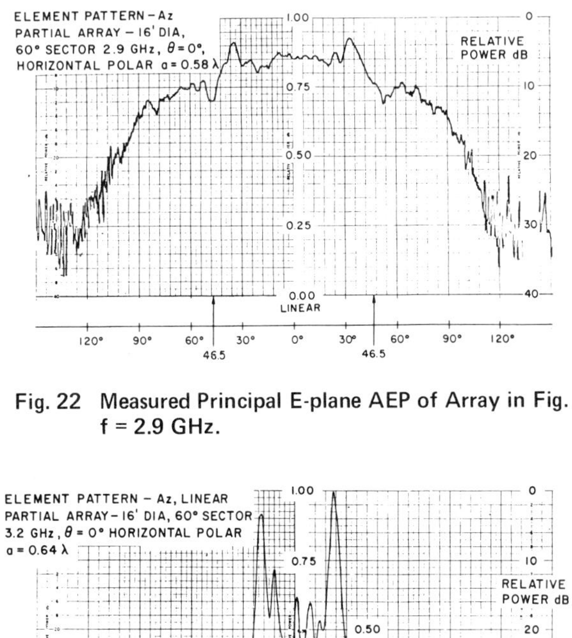

Fig. 22 Measured Principal E-plane AEP of Array in Fig. 20; f = 2.9 GHz.

Figure 22 is the measured principal E-plane AEP at 2.9 GHz. Grating-lobe dips are well formed and follow eq. (1). Figures 23 and 24 show the experimental and theoretical AEP at 3.2 GHz. Grating-lobe dips appear at about ± 30° and, the additive at broadside, space-wave creeping-wave interference, is clearly observed. Except for some distortion, due to asymmetries in the experimental array and end effects, a good qualitative agreement is obtained between theory and experiment.

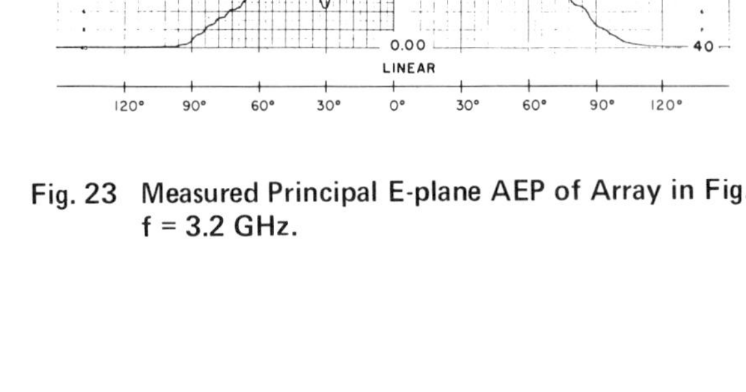

Fig. 23 Measured Principal E-plane AEP of Array in Fig. 20; f = 3.2 GHz.

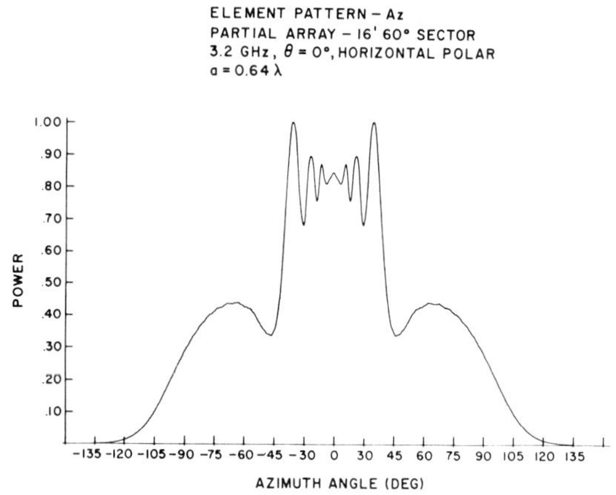

Fig. 24 Calculated AEP of Infinite Array corresponding to Fig. 20; f = 3.2 GHz.

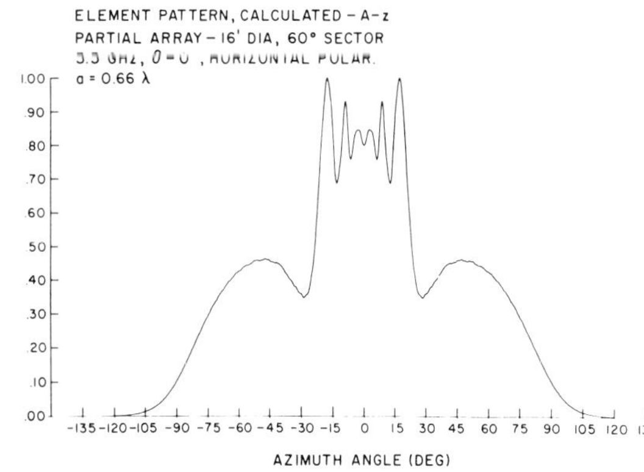

Fig. 26 Calculated Principal E-plane AEP as in Fig. 24; f = 3.3 GHz.

Figures 25 and 26 show the AEP's at 3.3 GHz, in which the creeping-waves are "out of phase" at broadside. Figures 27–29 correspond to 3.4 GHz. Grating-lobe dips move in and the ripple increases. The isolated element pattern in Fig. 29 differs considerably from the AEP. Finally, Figs. 30 and 31 present AEP'S at 3.9 GHz ($a/\lambda = 0.78$). Grating-lobe dips follow their theoretical $\pm 15.7^\circ$ values yielding a narrow beamwidth. Large amplitude creeping-wave grating-lobes interfere destructively with the space-wave at broadside resulting in a 2 dB dip which is presumably accompanied by some phase variation. Such an AEP is obviously disadvantageous and for this reason the E-plane spacing should be kept below, say, 0.7λ. Further reason for spacing limitation is discussed in the subsequent section.

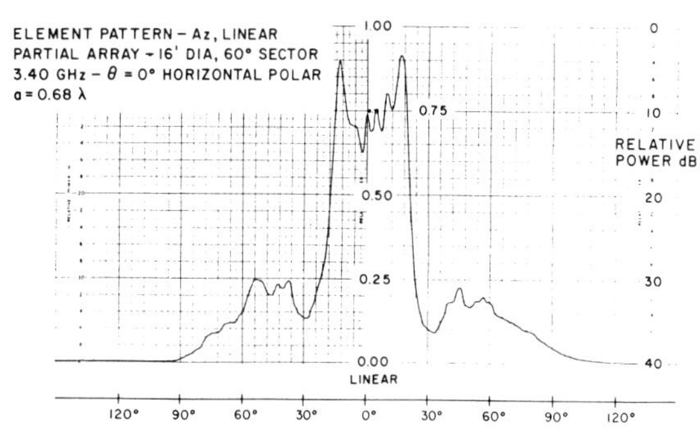

Fig. 27 Measured Principal E-plane AEP of Array in Fig. 20; f = 3.4 GHz.

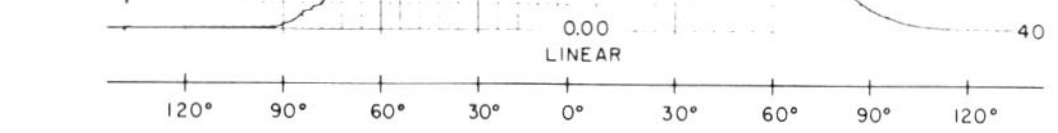

Fig. 25 Measured Principal E-plane AEP of Array in Fig. 20; f = 3.3 GHz.

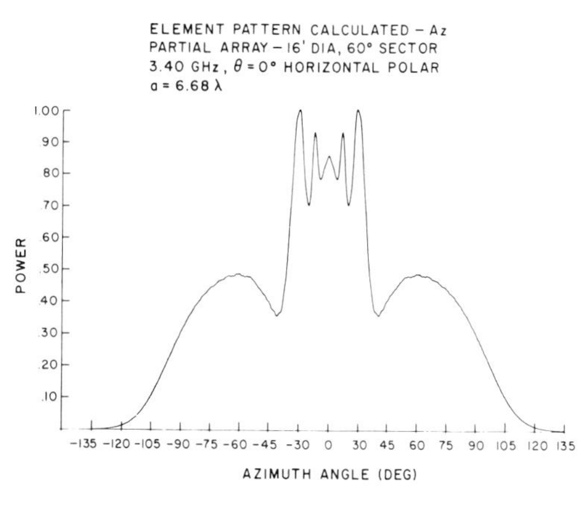

Fig. 28 Calculated Principal E-plane AEP as in Fig. 24; f = 3.4 GHz.

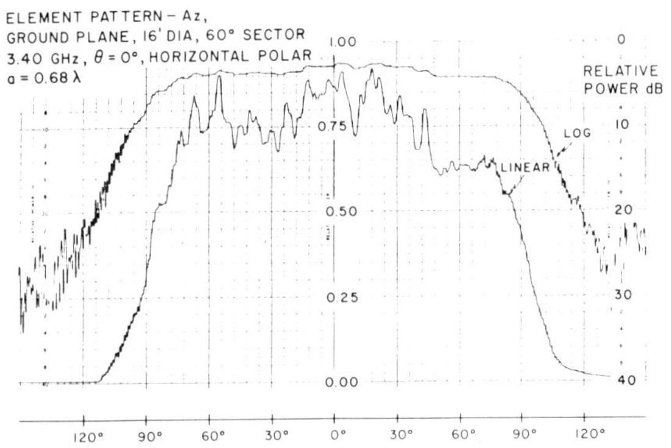

Fig. 29 Measured E-plane Isolated Element Pattern; $r_o = 8'$, f = 3.4 GHz.

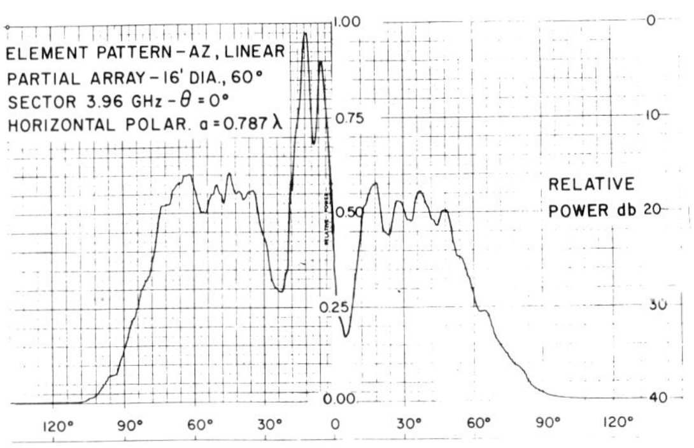

Fig. 30 Measured Principal E-plane AEP of Array in Fig. 20; f = 3.9 GHz.

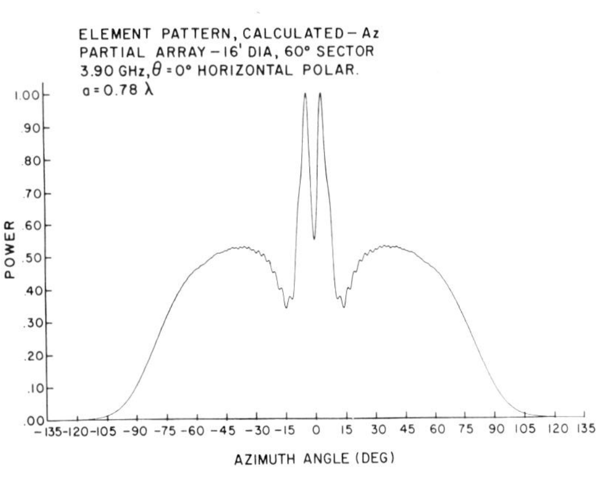

Fig. 31 Calculated Principal E-plane AEP as in Fig. 24; f = 3.9 GHz.

C. THE ROLE OF THE AEP IN CIRCULAR ARRAY PERFORMANCE

When an arc of a uniformly spaced circular array is fully excited for beam formation, mutual coupling enters via the AEP. In particular, denoting $g_i(\theta, \phi)$ the realized gain function of the i-th element, one finds (7-10) that the maximum realized gain $G_{max}(\hat{\theta}, \hat{\phi})$ of the excited arc, in a particular direction $(\hat{\theta}, \hat{\phi})$ is,

$$G_{max}(\hat{\theta}, \hat{\phi}) = \sum_i g_i(\hat{\theta}, \hat{\phi}) = \sum_i g_o(\hat{\theta}, \hat{\phi} - \frac{2\pi}{N} i) \quad , \qquad (2)$$

where the summation extends over the excited elements. The second form in Equation (2) implies that $G_{max}(\hat{\theta}, \hat{\phi})$ is a sum of values of the central element realized gain function, sampled at angles corresponding to positions of the excited elements.

The illumination vector $\underline{a}^o$ that yields $G_{max}(\hat{\theta}, \hat{\phi})$ is given by

$$\left.\begin{aligned} |a_i^o| &= \sqrt{g_i(\hat{\theta}, \hat{\phi})} \\ \angle a_i^o &\equiv \text{total phase correction for the } (\hat{\theta}, \hat{\phi}) \text{ direction} \equiv \\ &\equiv \text{path length correction} + \text{AEP phase correction} \end{aligned}\right\} . \qquad (3)$$

For an arbitrary excitation $\underline{a}$, the realized gain of the arc array is reduced by the taper efficiency η, where, with $|\underline{a}|^2 = 1$ and $|\underline{a}^o|^2 = 1$

$$\eta = \frac{G(\hat{\theta}, \hat{\phi})}{G_{max}(\hat{\theta}, \hat{\phi})} = |\underline{a}^* \cdot \underline{a}^o| \quad . \qquad (4)$$

Equations (2-4) yield the well known planar results, when all g_i become identical.

An interesting equivalence has been established between a large multiple-faced planar phased-array and an equal diameter step-scanned circular array. For the case, when the AEP's of the two arrays are nearly equal, it is shown (10) that the scan average gain of the multi-faced phased-array system is approximately equal to the scan-independent gain of the circular array.

Typical maximum E-plane arc array patterns for the axial, matched, thin slits of Fig. 1, are shown in Figs. 32 and 33. It is seen that doubling the number of excited elements (from 60° arc to 120° arc) yields only 1 dB increase in realized gain. This is due to an inefficient radiation of the edge elements, as seen from the associated element patterns.

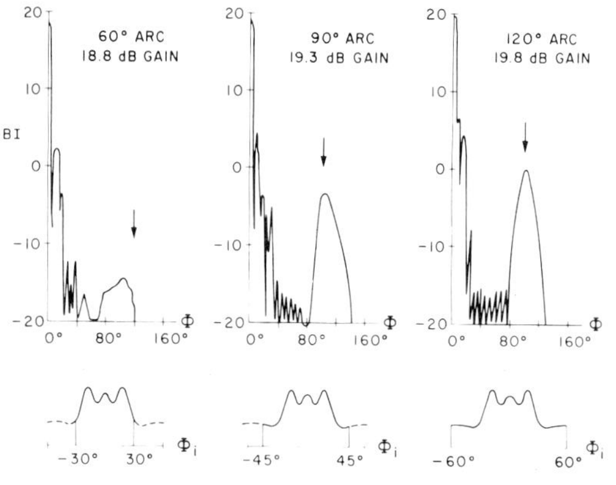

Fig. 32 Maximum Gain E-plane Patterns for Arc Array of Fig. 1; kr_o = 86, a/λ = 0.65. (From [10]).

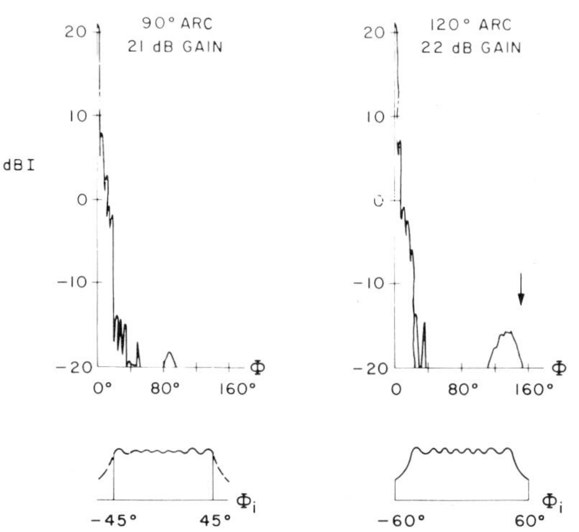

Fig. 33 Maximum Gain E-plane Patterns for Arc Array of Fig. 1; kr_o = 86, a/λ = 0.535. (From [10]).

The patterns for 90° and 120° arc exhibit diffuse, high side lobes at about $|\phi| = 100°$. Tendency to form such lobes is characteristic of large, directive circular arrays. These lobes have been identified as grating-lobes of the nearly planar sub-arrays formed by the far end elements of the excited arc. Since the sub-arrays are phased for broadside of the entire arc array, grating-lobes are formed in a backward direction, corresponding to the phase taper and spacing (Fig. 34). The diffuse character

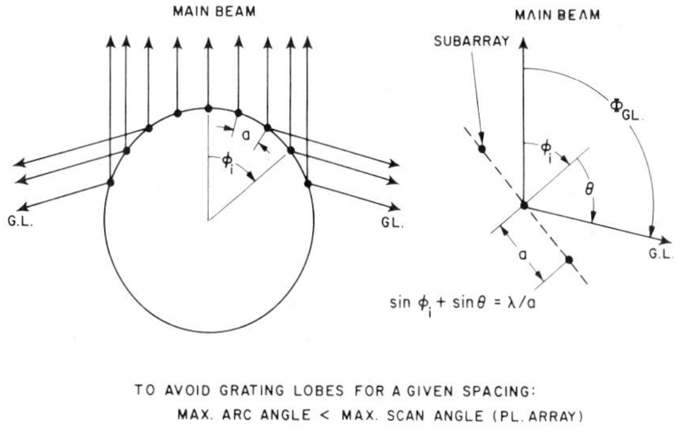

Fig. 34 Grating Lobes in Circular Arc Arrays (From [10]).

of these grating-lobes is due to defocusing associated with the curvature of the cylindrical surface. It should be noted that the relevant subarrays are centered approximately at angular locations numerically equal to the maximum E-plane scan angles of a corresponding planar array. Hence, for a given electrical spacing, an approximate rule for avoiding grating-lobes, is to limit the extent of the excited arc to the maximum scan angle without grating-lobes in a corresponding planar array. Alternatively, for a given arc length, one may have to reduce the spacing.

A more accurate formula reads [17]:

$$\sin|\Phi_{GL}| = \frac{\lambda}{2a} \quad (5)$$

where Φ_{GL} is the grating-lobe emergence angle with respect to the beam pointing direction. The angular location of the centers of the relevant subarrays is

$$\Phi_c = \frac{\Phi_{GL}}{2}\ . \quad (6)$$

To avoid an excessively high grating-lobe level, for a fixed spacing $\frac{\lambda}{2} < a < \lambda$ and with the array phased for beam formation, the excited arc should be kept to less than Φ_c. Alternatively, a/λ may be decreased at an increased cost. An appropriate amplitude taper could also be applied to the relevant subarrays, at the expense of gain.

It should be kept in mind, that when the beam is phase-scanned in azimuth, the grating-lobes, although suppressed at broadside, may reappear for scan angles off broadside.

CONCLUSION

To summarize, it may be asserted that considerable progress has been made towards understanding of mutual coupling effects in linearly polarized, uniformly spaced circular arrays of waveguide-fed apertures on conducting cylinders. Basic theoretical predictions have been borne out by experiment. Mutual coupling analysis has also been carried out in single ring arrays of axial $\frac{\lambda}{2}$ dipoles in front of, and $\lambda/2$ slots on, small to medium size conducting cylinders. Study in this direction remains to be continued.

The Array Element Pattern (AEP) has been found a useful tool for evaluating the circular array element performance and estimating the importance of mutual coupling in array environment. Analysis and synthesis of uniformly-spaced circular arrays are reduced to a linear superposition of sampled values of a single AEP, weighted with appropriate excitation coefficients. (18) The maximum realized gain of a uniformly spaced circular arc array, the corresponding "optimal" free-excitation and the taper efficiency when the illumination departs from the optimal, are related to the sampled values of the AEP.

The analysis has established certain similarities and differences between the AEPs of circular arrays of apertures on large radii conducting cylindrical surfaces and those of corresponding planar arrays. It has been shown that the H-plane AEPs are quite similar and that the differences appear predominantly in the E-plane. In the latter case the curvature reduces the influence of mutual coupling giving rise to the AEP dips instead of nulls at incipient grating-lobe angles. In a circular array, the latter condition corresponds to the "cutoff" of the next higher mode of the unit cell waveguide. (11) The dips and the accompanying phase variation, both of which should be accounted for in a high-gain low-side-lobe design of a circular array, taper off with diminishing kr_0.

Further differences between the planar and circular AEPs, particularly in the E-plane, are caused by creeping-wave effects, absent in planar arrays. In large circular arrays of apertures on conducting cylindrical surfaces, creeping-waves give rise, off broadside, to a ripple of the AEP. This ripple grows with increasing a/λ and tapers off with increasing kr_0. For $a/\lambda = 0.8$ it gives rise to considerable distortion of the circular AEP as compared to the corresponding planar E-plane AEP and may boost or reduce the peak realized element gain by 2 to 3 dB.

It should be noted that a single AEP has significance either in full 360° circular arrays or else, as in large planar phased arrays, for elements in the central portion of a large circular arc array. For near-edge elements AEP loses its simplicity, because each of these elements has a different AEP.

It should also be pointed out that the amplitude and phase variation of AEP make the pattern synthesis of circular arrays more difficult than that of planar arrays.

An experimental procedure of determining the AEP is strongly advocated as a check of analysis and in cases where analysis is too difficult to perform, a small array AEP technique should prove a valuable design tool.

Large, directive circular arc arrays are prone to form pronounced, broad, far side-lobes. These lobes have been identified as grating-lobes due to the nearly planar subarrays of outer elements of the excited arc and arise because of the subarrays orientation and their phasing for beam formation. Means to suppress these grating-lobes are described.

Future investigations may deal with the question of cross-polarization in scanned stacked ring arrays of linearly polarized elements; with circular arrays of inclined slots on, or dipoles in front of a central conducting cylinder; with ring arrays of circularly polarized elements; and with effects of cylindrical dielectric radome shells on mutual coupling in circular arrays.

Further afield lie studies of mutual coupling in arrays on surfaces of revolution e.g. cones, spheres, ellipsoids, and the determination of performance of conformal phased arrays on more complicated curved conducting surfaces.

III. METHODS OF CIRCULAR ARRAY ANALYSIS

As in the case of planar phased arrays, several methods are available for analysis of circular arrays on conducting cylindrical surfaces.

1. An Element by element approach, which is applicable not only to uniformly spaced 360° arrays, but to conformal arrays as well. In the absence of symmetry this approach requires an inversion of a large $(\underline{\underline{1}} + \underline{\underline{Y}})$ or $(1 + \underline{\underline{Z}})$ matrix of order N. In full 360° circular arrays, the presence of rotational symmetry permits a simple inversion via the diagonalization of the $\underline{\underline{Y}}$ $(\underline{\underline{Z}})$ matrix in an eigenbasis of $\underline{\underline{Y}}$ $(\underline{\underline{Z}})$. The eigenvalues and eigenvectors of $\underline{\underline{Y}}$ $(\underline{\underline{Z}})$ are the phase sequence admittances and excitations respectively of the cylindrical array. This approach is well known (see (1-3) and lately (8)).

2. Unit Cell approaches for 360° uniformly spaced circular arrays.

(2a). In the physical space (number of unit cells equal to number of elements N).

The results derived by this method for single mode elements, coincide with those of the element by element approach. Method (2a) has an advantage for multimode elements in that it easily permits setting up of a mode matching procedure or an integral equation over a single element. Both (1) and (2a) however wind up in a harmonic series representation for the far fields with its usual convergence difficulties when the circumference to wavelength ratio kr_o increases.

(2b). In an infinitely extended angular space.

This approach yields, as in the case of a smooth cylinder, (19) a convergent integral representation of the AEP for large values of kr_o without the necessity of performing a Watson transformation (20) of the harmonic series and generalizes the method (2a) to parallel that used in analysis of infinite phased arrays.

In view of the familiarity of the antenna engineers community with the element by element approach, we shall restrict our considerations to the unit cell methods. The exposition is based primarily on the material in reference (11). Certain basic points however, in connection with validity of the approach (2b) which are implicit in (11), will be highlighted below.

Furthermore, in view of the fact that methods of stacked ring array analysis are an extension of techniques to be described, the subsequent considerations for clarity of presentation are restricted to the two dimensional array of Fig. 1.

It is assumed that the feed waveguides propagate only the TEM mode and that the slits are sufficiently narrow to validate a single aperture mode approximation.

The rotational symmetry of uniformly-spaced circular arrays implies a corresponding rotational invariance of the AEP. As a result, the fields radiated due to phase sequence excitation of uniformly-spaced circular arrays, obey Floquet's theorem (see Appendix). At the outset of analysis, it is convenient to assign a reference plane to each feed waveguide. The reference planes are equi-distant and sufficiently far removed from the respective apertures.

(2a). The Unit Cell Method in the Physical Angular Space.

In this method the extent of the angular domain of interest is restricted to the physical space $-\pi \leqslant \Phi \leqslant \pi$ (see Fig. 34a), where, in order to assure the single-valuedness of fields, the requirement of their 2π periodicity is explicitely imposed.

A. Array Excitation

With respect to the chosen set of reference planes, the array is considered an N-port network, whose excitation may be described in terms of an N component, complex, incident-wave vector $\underline{a} = \{a_\ell\}$, $(\ell = 0, 1 \ldots N\text{-}1)$. A unity amplitude excitation of the single $\ell = 0$ element is characterized by $\underline{a} = \underline{a}^\circ = \{\underline{a}_\ell^\circ\} = \{\delta_\ell^\circ\}$, where δ_ℓ° denotes the Kronecker symbol. The vector $\underline{a}^o$ may be expressed as a linear superposition of phase-sequence excitation vectors $\underline{a}^{on}$

$$\underline{a}^o = \sum_{n=0}^{N-1} \underline{a}^{on} = \frac{1}{N} \sum_{n=0}^{N-1} \{e^{-jn\frac{2\pi}{N}\ell}\} \tag{7}$$

where each $\underline{a}^{on}$ represents a uniform amplitude, progressive, inter-element phase delay excitation of $2\pi n/N$ radians ($2\pi n$ radians total phase shift).

Denoting $\underline{E}(r, \Phi, n)$ the electric field resulting from the n-th phase-sequence $\underline{a}^{on}$, the electric field due to the single $\ell=0$ element excitation becomes, in view of Equation (7),

$$\underline{E}(r, \phi) = \frac{1}{N} \sum_{n=0}^{N-1} \underline{E}(r, \phi; n) \quad . \tag{8}$$

B. Electromagnetic Problem for a Phase Sequence Excitation

Due to angular periodicity of a circular array, the electromagnetic problem for the n-th phase-sequence excitation reduces to the field solution within a single unit cell, shown in Fig. 36 with ν replaced by n for the array of Fig. 1. This unit cell consists of a junction between a feed waveguide and a radial waveguide whose walls are characterized by phase-shift (Floquet) boundary conditions.

$$\underline{E}\left(r, \frac{\beta}{2}; n\right) = e^{-jn\beta}\, \underline{E}\left(r, -\frac{\beta}{2}; n\right)$$

$$\underline{H}\left(r, \frac{\beta}{2}; n\right) = e^{-jn\beta}\, \underline{H}\left(r, -\frac{\beta}{2}; n\right)$$

Validity of the unit cell concept and of (9) is a direct consequence of Floquet's theorem (see Appendix). This theorem holds, as long as there is at least one complete ring, i.e., as long as rotational symmetry is present.

In the radial waveguide, the fields are expanded in terms of a complete set of modes of the form $e^{-j(n+\frac{2\pi}{\beta}m)\Phi} = e^{-jn_m\Phi}, (m = 0, \pm 1 \ldots)$, individually satisfying (9)(see Appendix). In this fashion

$$E_{\phi}(r, \phi; n) = \frac{1}{\beta r} \sum_{m=-\infty}^{\infty} V_m(r; n)\, e^{-jn_m\phi}, \quad r > r_0,\ -\frac{\beta}{2} \leqslant \phi \leqslant \frac{\beta}{2}$$

where from Maxwell's equations or from radial transmission line theory and in view of the radiation condition, one has for the LSE polarization

$$\frac{V_m(r; n)}{\beta r} = A_m(n) H_n^{(2)'}(kr)$$

In (11), A_m is independent of r and (') denotes the derivative with respect to the argument of the Hankel function $H_n^{(2)}(kr)$.

From (11)

$$V_m(r; n) = \frac{r}{r_0} \frac{H_{n_m}^{(2)'}(kr)}{H_{n_m}^{(2)'}(kr_0)} V_m(r_0; n)$$

so that

$$E_{\phi}(r, \phi; n) = \frac{1}{\beta r_0} \sum_{m=-\infty}^{\infty} V_m(r_0; n) \frac{H_{n_m}^{(2)'}(kr)}{H_{n_m}^{(2)'}(kr_0)} e^{-jn_m\phi}$$

In Equation (12)

$$V_m(r_0; n) = r_0 \int_{-\beta/2}^{\beta/2} d\phi\, E_{\phi}(r_0, \phi; n)\, e^{jn_m\phi}, \qquad (13)$$

and $E_{\phi}(r_0, \phi; n)$ is the aperture field for the n-th phase-sequence excitation.

The mode functions of the feed guide are conveniently defined with respect to a Cartesian coordinate system shown in Fig. 36. The corresponding modal voltages and currents are denoted V_n^i and I_n^i.

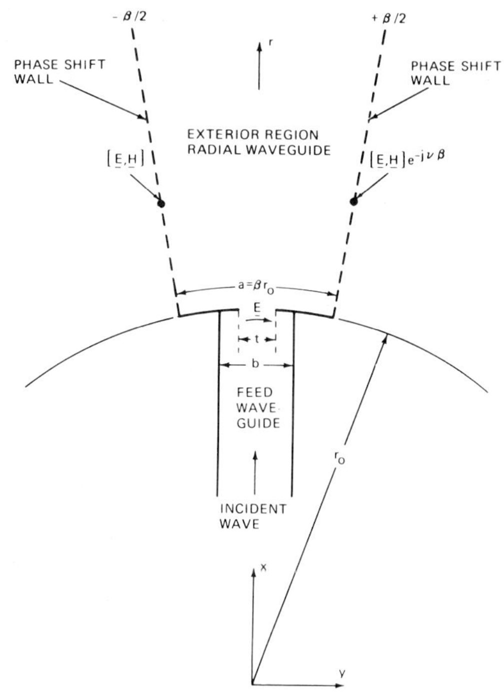

Fig. 36 Unit Cell for Array in Fig. 1. (From [11]).

The electromagnetic problem within a unit cell amounts to the determination of the tangential electric field aperture distribution, from which all other quantities of interest are derivable. For multimode aperture elements, variational methods may be employed, along the lines described in (21). These methods are applicable, upon neglecting the curvature over the extent of the aperture. For large radii cylinders, such an approximation introduces a negligible error.

For single mode elements, e.g. thin slits, the method leads, as in the planar case, to a grating lobe series for the active aperture admittance $Y_a(n) = G_a(n) + j\,B_a(n)$ which describes the effect of the slit on the dominant feed waveguide mode and which is identical for all elements. In forming $Y_a(n)$ it is found that the admittance "crater" for the corresponding planar array of slits is replaced by a different "crater" appropriate to cylindrical geometry. This is accomplished by matching e.g. the complex power on both sides of the aperture, yielding

$$Y_a(n) = \sum_{m=-\infty}^{\infty} \left| \frac{V_o(r_o, n_m)}{V^i(r_o n)} \right|^2 Y(r_o, n_m) + \sum_{m=-1}^{\infty} \left| \frac{V^i_m(r_o, n)}{V^i(r_o, n)} \right|^2 Y^i_m \quad (14)$$

In (14) $Y(r_o, n_m)$ and Y^i_m are, respectively, the various modal admittances of the radial and the TEM feed guide. The dominant radial mode characteristic admittance

$$Y(r_o;n) = \frac{1}{\zeta_o \beta r_o} Y'(r_o, n) \quad (15)$$

where $Y'(r_o, n) = -j \frac{H^{(2)}_n(kr_o)}{H^{(2)'}_n(kr_o)}$ is the normalized dominant mode radial guide modal admittance and ζ_o is the free space impedance. To insure an effective power transfer through the slits, an identical network is incorporated in each waveguide, to match the feed waveguide to $Y_a(n)$ at a particular value $n = \bar{n}$, e.g. $\bar{n} = 0$. The equivalent circuit, as "seen" by the dominant feed waveguide mode, is shown in Fig. (37). With the notation of this figure one has identically

$$V_m(r_o;n) = \frac{V_m(r_o;n)}{V^i_o(r_o, n)} \cdot \frac{V^i_o(r_o;n)}{V^i_o(r_o^-;n)} \cdot \frac{V^i_o(r_o^-;n)}{V^{inc}_o} \cdot V^{inc}_o \quad (16)$$

Here $V^i_o(r_o, n)$ is obtained as in (13) by a scalar product of the aperture field with the dominant mode function of the feed guide.

Defining

$$\frac{V^i_o(r_o^-;n)}{V^{inc}_o} = T(n, \bar{n}) \quad (17)$$

as the active element transmission coefficient (with the dependence on the matching network explicitly indicated), (16) may be rewritten

$$V_m(r_o;n) = \bar{n}\, T(n, \bar{n}) \frac{V_m(r_o;n)}{V^i_o(r_o;n)} V^{inc}_o, \quad (18)$$

where $\bar{n} = \frac{V^i_o(r_o, n)}{V^i_o(r_o^-, n)}$ is defined in Fig. 37. It should be noted that $T(n, \bar{n})$ and V_m / V^i_n depend solely on the shape of the aperture field, but, not on its amplitude.

With the aid of the network in Fig. 37, one finds

$$T(n, \bar{n}) = \frac{2 G_a(\bar{n})}{Y_a(n) + Y_a^*(\bar{n})}, \quad (19)$$

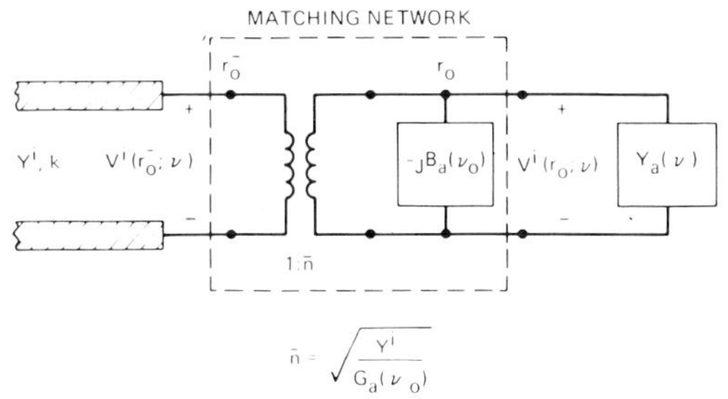

Fig. 37 Equivalent Network for Unit Cell of Fig. 36. (From [11]).

where* denotes complex conjugate. Substituting (18) into (10) yields

$$E_\phi(r, \phi;n) = \frac{V^{inc}_o}{\beta r_o} \bar{n}\, T(n, \bar{n}) \sum_{m=-\infty}^{\infty} \frac{V_m(r_o;n)}{V^i_o(r_o, n)} \frac{H^{(2)'}_{n_m}(kr)}{H^{(2)'}_{n_m}(kr_o)} e^{-jn_m\phi} \quad (20)$$

Since $e^{-jn\frac{2\pi}{N}} = e^{-jn_m\frac{2\pi}{N}}$, it follows by uniqueness that all phase-sequence field quantities are periodic in n with the period of $\frac{2\pi}{\beta} = N$. Hence,

$$E_\phi(r, \phi, n) = E_\phi(r, \phi, n_m) \quad (21)$$

and therefore, in view of (13)

$$V_m(r_o;n) = V_o(r_o;n_m) \quad (22)$$

Also

$$V^i_o(r_o;n) = V^i_o(r_o;n_m) \quad (23)$$

and

$$T(n, \bar{n}) = T(n_m, \bar{n}). \quad (24)$$

Consequently,

$$E_\phi(r, \phi;n) = \frac{V^{inc}_o}{\beta r_o} \bar{n}\, T(n, \bar{n}) \sum_{m=-\infty}^{\infty} \frac{V_o(r_o;n_m)}{V^i_o(r_o, n_m)} \frac{H^{(2)'}_{n_m}(kr)}{H^{(2)'}_{n_m}(kr_o)} e^{-jn_m\phi} \quad (25)$$

C. The Array Element Pattern

Using superposition in accordance with (8), one has for the AEP

$$E_\phi(r,\phi) = \frac{V_{o\,inc}}{2\pi r_o}\,\bar{n}\sum_{n=0}^{N-1} T(n,\tilde{n}) \sum_{m=-\infty}^{\infty} \frac{V_o(r_o;n_m)}{V_o^i(r_o,n_m)}\,\frac{H_{n_m}^{(2)\prime}(kr)}{H_{n_m}^{(2)\prime}(kr_o)}\,e^{-jn_m\phi}\,, \tag{26}$$

where $n_m = n + \frac{2\pi}{\beta}m = n + Nm$.

In view of (28),

$$E_\phi(r,\phi)\ \frac{V_{o\,inc}}{2\pi r_o}\,\bar{n}\sum_{n=0}^{N-1}\sum_{m=-\infty}^{\infty} T(n_m;\tilde{n})\,\frac{V_o(r_o,n_m)}{V_o^i(r_o,n_m)}\,\frac{H_{n_m}^{(2)\prime}(kr)}{H_{n_m}^{(2)\prime}(kr_o)}\,e^{-jn_m\phi} = \frac{V_o^{inc}}{2\pi r_o}\,\bar{n}\sum_{n=-\infty}^{\infty} T(n;\tilde{n})\,\frac{V_o(r_o,n)}{V_o^i(r_o,n)}\,\frac{H_n^{(2)\prime}(kr)}{H_n^{(2)\prime}(kr_o)}\,e^{-jn\phi}\,, \tag{27}$$

where use has been made of the identity

$$\sum_{n=0}^{N-1}\sum_{m=-\infty}^{\infty} f(n+mN) = \sum_{n=-\infty}^{\infty} f(n)\ . \tag{28}$$

In the far field (27) becomes

$$E_\phi(r,\phi) \sim -j\,\frac{V_o^{inc}}{2\pi r_o}\,\bar{n}\,\sqrt{\frac{2}{\pi kr}}\,e^{-j(kr-\frac{\pi}{4})}\sum_{n=-\infty}^{\infty}\frac{V_o(r_o;n)}{V_o^i(r_o;n)}\,T(n,\tilde{n})\,\frac{e^{-jn(\phi-\frac{\pi}{2})}}{H_n^{(2)\prime}(kr_o)} \tag{29}$$

The harmonic series in Equation (29) represents the AEP and is evaluated numerically. It converges well only for not too large values of kr_o.

(2b) The Unit Cell Method in Infinitely Extended Angular Space.

A. Phase Sequence Excitation

In the alternate representation, the physical domain is imbedded in an infinitely extended angular space $-\infty < \phi < \infty$, in which the 2π angular periodicity of fields is enforced by periodically distributed image elements and image sources (excited image waveguides) the latter spaced 2π apart (11) (19) (Fig. 35b). Since the resultant fields satisfy: Maxwells equations, boundary conditions on the surface of the cylinder as well as 2π periodicity, by uniqueness, they represent the correct AEP. As a result, the AEP becomes a superposition of contributions observed for $-\pi \leqslant \phi \geqslant \pi$ and due to the individual sources (s = 0, ±1, ±2, . . .) radiating in the unbounded angular space.

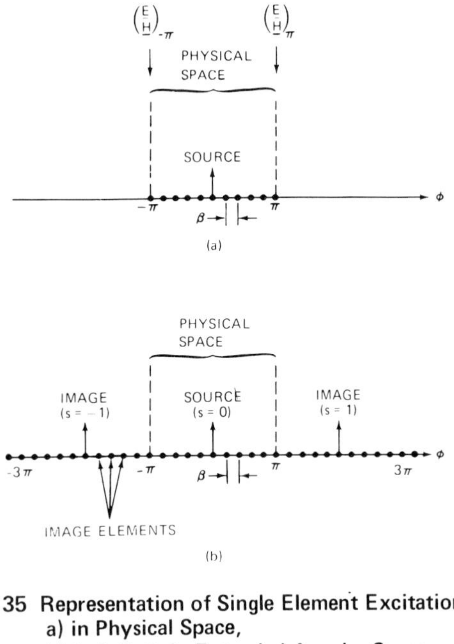

Fig. 35 Representation of Single Element Excitation
a) in Physical Space,
b) in Infinitely Extended Angular Space.

In this space, each source is represented in terms of a band spectrum of angular wave numbers, via the identity

$$\delta_\ell^{sN} = \frac{\beta}{2\pi}\int_{-\pi/\beta}^{\pi/\beta} d\nu\, e^{-j\nu(\beta\ell-2\pi s)}\ (\ell, s = 0, \pm 1, \pm 2 \ldots)\ , \tag{30}$$

where $\beta = 2\pi/N$.

In view of (30), the incidence vector $\underline{a}^s$ for the s-th image source becomes

$$\underline{a}^s = \{\delta_\ell^{sN}\} = \frac{\beta}{2\pi}\int_{-\pi/\beta}^{\pi/\beta}\{e^{-j\nu(\beta\ell-2\pi s)}\}\,d\nu = \frac{\beta}{2\pi}\int_{-\pi/\beta}^{\pi/\beta}\underline{a}^s(\nu)\,d\nu\ . \tag{31}$$

In (31)

$$\underline{a}^s(\nu) = \{a_\ell^s(\nu)\} = \frac{1}{N}\{e^{-j\nu(\beta\ell-2\pi s)}\}\,,\quad (\ell = 0, \pm 1 \ldots)\ (s = 0, \pm 1 \ldots) \tag{32}$$

represents the ν-th phase-sequence incident-wave vector (with ν no longer an integer) of all real and virtual elements in the infinite angular space. It should be noted that the phase sequence fields do not, in general, individually satisfy the 2π periodicity condition in Equation (9). In contrast the entire field due to all periodically spaced sources will satisfy Equation (9). The field in the physical space, due to a singly excited real element is now

$$\underline{E}(r,\phi) = \sum_{s=-\infty}^{\infty}\underline{E}_s(r,\phi)\ , \tag{33}$$

where the summation extends over the source and all its images. The source contribution in Equation (33) is obtained by superposition according to Equation (30)

$$\underline{E}_o(r,\phi) = \frac{1}{N}\int_{-\pi/\beta}^{\pi/\beta}\underline{E}(r,\phi;\nu)\,d\nu \tag{34}$$

where $\underline{E}(r,\phi;\nu)$ is the electric field due to the ν-th phase sequence excitation for s = 0, while the image terms are

$$\underline{E}_s(r,\phi) = \underline{E}_o(r,\phi-2\pi s)\ . \tag{35}$$

It can be shown (11) that the dominant contribution in Equation (33) is due to the s = 0 term, the image terms becoming important only in the back lobe region.

Equations (33)-(35) yield similarly as in (19) without recourse to a Watson transformation (20), an integral representation, which for large values of kr_o, lends itself to a highly convergent asymptotic evaluation. While the harmonic series furnishes no basis for interpretation of the AEP, the asymptotics provides a clear picture of the physical phenomena involved in formation of the AEP, and permits prediction of the AEP's changes with variation of physical parameters.

B. Electromagnetic Problem for a Phase Sequence Excitation.

Due to the angular periodicity of the extended infinite array, the electromagnetic problem reduces to the field solution within a single unit cell (Fig. 36). The Floquet boundary conditions at the radial guide walls become now Equation (9) with n replaced by an arbitrary real ν Validity of these modified relations for the extended infinite array is easily shown following the lines of the Appendix.

The entire text, from eq. (9) to eq. (25) may now be repeated verbatim with n replaced by ν and n_m by $\nu_m = \nu + \frac{2\pi}{\beta}m$. The matching value $\nu=\bar{n}$ and the symbol $\bar{n}$ will be retained.

From Equations (33)-(35), using Equations (20) and (24), with n replaced by ν the integral representation for the far field becomes

$$E_\phi(r,\phi) \sim \frac{V_o^{inc}\,\bar{n}}{2\pi r_o} \sum_{s=-\infty}^{\infty} \int_{-\pi/\beta}^{\pi/\beta} \sum_{m=-\infty}^{\infty} T(\nu_m,\hat{n}) \frac{V_o(r_o;\nu_m)}{V_o^i(r_o;\nu_m)} \frac{H_{\nu_m}^{(2)\prime}(kr)}{H_{\nu_m}^{(2)\prime}(kr_o)} e^{-j\nu_m(\phi-2\pi s)} d\nu =$$

$$= \frac{V_o^{inc}}{2\pi r_o}\bar{n} \sum_{s=-\infty}^{\infty} \int_{-\infty}^{\infty} T(\nu,\tilde{n}) \frac{V_o(r_o;\nu)}{V_o^i(r_o;\nu)} \frac{H_\nu^{(2)\prime}(kr)}{H_\nu^{(2)\prime}(kr_o)} e^{-j\nu(\phi-2\pi s)} d\nu \quad , \tag{36}$$

where the identity

$$\int_{-\pi/\beta}^{\pi/\beta} \sum_{m=-\infty}^{\infty} f(\nu + \frac{2\pi}{\beta} m)\, d\nu = \int_{-\infty}^{\infty} f(\nu)\, d\nu \tag{37}$$

was employed.

From (36) one obtains the formal far field expression

$$E_\phi(r,\phi) \sim -j\frac{V_o^{inc}}{2\pi r_o}\bar{n}\sqrt{\frac{2}{\pi kr}}\, e^{-j(kr-\frac{\pi}{4})} \int_{-\infty}^{\infty} \frac{V_o(r_o;\nu)}{V_o^i(r_o;\nu)}\, T(\nu,\tilde{n})\, \frac{e^{-j\nu(\phi-\frac{\pi}{2})}}{H_\nu^{(2)\prime}(kr_o)}\, d\nu \quad , \tag{38}$$

where use has been made of the fact that only the s = 0 term in the sum yields a non-negligible contribution everywhere, except in the back lobe region.

The steepest descent evaluation of Equation (38), for $kr_o \gg 1$ yields, as the saddle point contribution, the space-wave

$$E_\phi^{spw}(r,\phi) = \frac{kr_o}{2} V_o^{inc}\bar{n} \frac{V_o(r_o, kr_o\sin\phi)}{V_o^i(r_o, kr_o\sin\phi)} T(kr_o\sin\phi, \tilde{n}) \sqrt{\frac{2}{\pi kr}}\, e^{-j(kr-\frac{\pi}{4}-kr_o\cos\phi)} \quad , \tag{39}$$

which, in view of the phase term $k(r-r_o\cos\phi)$ appears to originate from the driven element. The space wave pattern is proportional to $T(kr_o\sin\phi;\tilde{n})$, the factor $\frac{V_o}{V_o^i}$ being a slowly varying function of ϕ.

The creeping-wave contributions arise formally from residues at the poles crossed during the contour deformation from the real axis in the ν plane into the steepest descent path. These pole contributions represent the natural resonances of circular arrays.

This completes the scope of the present survey; for further details, the reader is referred to (11).

IV. APPENDIX

Floquet's Theorem for Uniformly Spaced Circular Arrays

Each electromagnetic field component $\phi_i(r,\theta,\phi;n)$ for the n-th phase-sequence excitation of a uniformly-spaced ring array, satisfies Floquet's Theorem, i.e.

$$\psi_i(r,\theta,\phi+\frac{2\pi}{N};n) = e^{-j\frac{2\pi}{N}n}\,\psi_i(r,\theta,\phi;n) \quad . \tag{A1}$$

Validity of (A1) is demonstrated, with the help of the rotational symmetry of the array, by considering the n-th phase sequence excitation as a superposition of single element excitations.

Let $\psi_i^{(e)}(r,\theta,\phi)$ be the i-th field component generated by the AEP excitation of a uniformly-spaced circular array, due to a unity incident wave at the reference plane in the $\ell=0$ feed waveguide.

By symmetry, a unity AEP excitation of the ℓ-th feed waveguide produces the field

$$\psi_{i\ell}^{(e)}(r,\theta,\phi) = \psi_{io}^{(e)}(r,\theta,\phi - \frac{2\pi}{N}\ell) \qquad \text{(A2)}$$

Consequently, by superposition

$$\psi_i(r,\theta,\phi;n) = \sum_{\ell=0}^{N-1} \psi_{i\ell}^{(e)}(r,\theta,\phi)\, e^{-j\frac{2\pi}{N}n\ell} = \sum_{\ell=0}^{N-1} \psi_{io}^{(e)}(r,\theta,\phi - \frac{2\pi}{N}\ell)\, e^{-j\frac{2\pi}{N}n\ell} \qquad \text{(A3)}$$

$$\begin{aligned} \psi_i(r,\theta,\phi + \frac{2\pi}{N}) &= \sum_{\ell=0}^{N-1} \psi_{io}^{(e)}(r,\theta,\phi - \frac{2\pi}{N}(\ell - 1))\, e^{-j\frac{2\pi}{N}n\ell} = \\ &= e^{-j\frac{2\pi}{N}n} \sum_{\ell=0}^{N-1} \psi_{io}^{(e)}(r,\theta,\phi - \frac{2\pi}{N}(\ell - 1))\, e^{-j\frac{2\pi}{N}(\ell - 1)n} = \\ &= e^{-j\frac{2\pi}{N}n} \sum_{\ell'=-1}^{N-1} \psi_{io}^{(e)}(r,\theta;\phi - \frac{2\pi\ell'}{N})\, e^{-j\frac{2\pi}{N}n\ell'} = \\ &= e^{-j\frac{2\pi}{N}n} \sum_{\ell=0}^{N-1} \psi_{io}^{(e)}(r,\theta,\phi - \frac{2\pi}{N}\ell)\, e^{-j\frac{2\pi}{N}n\ell} = \\ &= e^{-j\frac{2\pi}{N}n}\, \psi_i(r,\theta,\phi;n) \;. \end{aligned} \qquad \text{(A4)}$$

In Equation (A4), use has been made of the 2π periodicity in ϕ of $\psi_{io}^{(e)}(r,\theta,\phi)$. Equation (A1) justifies the phase shift boundary conditions in Equation (15). A proof of Floquet's theorem for the infinitely extended angular space maybe constructed along similar lines.

From,(A1) one now has

$$e^{+j(\phi + \frac{2\pi}{N})n}\, \psi_i(r,\theta,\phi + \frac{2\pi}{N};n) = e^{jn\phi}\, \psi_i(r,\theta,\phi) \qquad \text{(A5)}$$

which indicates that $e^{+jn\phi}\psi_i(r,\theta;\phi)$ is a periodic function $P_i(r,\theta,\phi)$ of ϕ with the period $\frac{2\pi}{N}$.

Expanding $P_i(r,\theta,\phi)$ via Fourier series one obtains

$$\psi_i(r,\theta,\phi) = \sum_{m=-\infty}^{\infty} a_m^{(i)}(r,\theta)\, e^{-j(n+\frac{2\pi}{N}m)\phi} \qquad \text{(A6)}$$

which justifies the modal expression (16) in the radial waveguide.

REFERENCES

1. Patton, W. T. and Tillman, J. D.: An Analysis of the Terminal Properties of a Circular An-atenna Array, Scientific Report No. 2, Contract AF 19 (604)-1557, Electrical Engineering Dept., The University of Tennessee, June 1958.

2. Tillman, J. D.: The Theory and Design of Circular Antenna Arrays, The University of Tennessee, Engineering Experiment Station, 1968.

3. King, R. W., Mack, R. B. and Sandler, S. S.: Arrays of Cylindrical Dipoles, Chapter 4, Cambridge University Press, London 1968.

4. Fenby, R. G.: Limitations on Directional Patterns of Phase Compensated Circular Arrays, Radio Electronic Eng. (GB) v. 30, p. 206, 1965.

5. Longstaff, I. D., et al: Directional Properties of Circular Arrays, Proc. IEE(GB) v. 114, pp. 713-718, 1967.

6. Neff, H. P., Patton, W. T., Pierce, J. F. and Tillman, J. D.: An Amplitude Scanned Circular Array, The University of Tennessee Final Report, Oct. 1958, AD 160881.

7. Sureau, J. C. and Hessel, A.: Element Pattern for Circular Arrays of Axial Slits on Large Conducting Cylinders, IEEE Trans. on Antennas and Propagation, v. AP-17, p. 799, Nov. 1969 (Communication).

8. Borgiotti, G. V. and Balzano, Q.: Mutual Coupling of a Conformal Array of Elements on Cylindrical Surface, IEEE Trans. on Antennas and Propagation AP-18, pp. 55-63, Jan. 1970.

9. Gladman, B R.: Mutual Coupling on Cylindrical Array Antennas, Proceedings of the Conformal Array Antenna Conference held at NELC, San Diego, 13-15 January 1970.

10. Hessel, A. and Sureau, J. C.: Element Patterns of Arrays on Cylindrical Surface, Proceedings of the Conformal Array Antenna Conference held at NELC, San Diego, 13-15 January 1970.

11. Sureau, J. C. and Hessel, A.: Element Pattern for Circular Arrays of Waveguide-Fed Axial Slits on Large Conducting Cylinders. To be published in the Jan. 1971 issue of IEEE Trans. on Antennas and Propagation.

12. Hessel, A. and Sureau, J.C.: On the Realized Gain of Arrays, To be Published as a Communication in the Jan. 1971 issue of IEEE Trans. on Antennas and Propagation.

13. Sureau, J.C. and Hessel, A.: Realized Gain Function for a Cylindrical Array of Open-Ended Waveguides, Program and Digest, 1970, Phased Array Antenna Symposium, June 2-5, 1970, Polytechnic Institute of Brooklyn, L.I. Graduate Center, Farmingdale, N.Y., To be published in the Symposium Proceedings.

14. Borgiotti, G.V. and Q. Balzano, Conformal Arrays on Surfaces with Rotational Symmetry, Program and Digest, 1970 Phased-Array Antenna Symposium, June 2-5 1970, Polytechnic Institute of Brooklyn, Graduate Center, Farmingdale, N.Y. To be published in the Symposium Proceedings.

15. Munger, A.D.: Effects of Mutual Coupling on the Performance of a Cylindrical Array Antenna, NELC Technical Note TN 1697, 6 July 1970. Also Private Communication.

16. Diamond, L.B.: Private Communication.

17. Munger, A.D.: Circular-Array Radar Antenna, Cylindrical-Array Theory, NELC Report 1608, Jan. 1969.

18. Tseng, D.K. and Cheng, F.E.: Pattern Synthesis of Circular Arrays with Many Directive Elements, IEEE Trans. on Antennas and Propagation, AP-16 pp. 758-759, Nov. 1968 (Communication).

19. Felsen, L.B. and Marcuvitz, N.: Modal Analysis and Synthesis of Electromagnetic Fields, Polytechnic Institute of Brooklyn, MRI Report PIB MRI-1083-62, Nov. 1962.

20. Jones, D.S.: The Theory of Electromagnetism, Pergammon Press, N.Y. 1969, Section 8.7.

21. Amitay, N., Galindo, V. and Wu, C.P.: Method of Phased Array Analysis, Program and Digest, 1970, Phased-Array Antenna Symposium, June 2-5 1970, Polytechnic Institute of Brooklyn, Graduate Center, Farmingdale, N.Y. To be published in the Symposium Proceedings.

A Survey of Circular Symmetric Arrays

J. H. PROVENCHER

Naval Electronics Laboratory Center
San Diego, California

Introduction

Considerable interest in circular, cylindrical and conical array antennas in recent years has prompted development of feed systems and components to provide rapid and highly agile beam positioning. Many of the problems associated with planar array design also exist in non-planar array design and a natural outcome is to apply solutions of the planar array design problems to circular arrays with appropriate changes. Beam steering generally requires a method to commute the amplitude and phase distribution around the array and this is usually accomplished by one of the following means:

1. Mechanical or electronic switch
2. Lens-switch combination
3. Hybrid-matrix phase shifter combination

The theory of the circular or ring array has been covered in the literature and some of the earlier arrays, primarily of the mechanically scanned type, will not be discussed here. It is only necessary to say that these arrays employed many of the concepts to be discussed and more recent applications will be presented.

Switched Arrays

A necessary requirement to steer the beam is that the amplitude and phase distribution be varied in such a manner that the excited portion or active sector of the array have the same distributions at each beam position. Several techniques make use of diode switches and phase shifters to effect the permutation. Figure 1 gives a schematic diagram of an array of this type developed by Wheeler Laboratories of Smithtown, New York.

The array has 32 elements with 8 elements activated at a time. In this case, the sector selection is provided by eight single-pole four-throw switches. The proper amplitude-and-phase ordering of the elements is maintained by the inter-connection of transfer switches, permitting cyclic permutation of the outputs from the distribution network. Phase shifters are included for fine steering; their location in the network requires as many phase shifters as there are excited elements. Eight phase tilt steerings are provided in the demonstration model to obtain steering to 0.1 beamwidth. The distribution network is used to obtain desired sum and difference pattern illuminations. The excitations are then carried through the equal path-length switching network without distortion.

The switching network activates a contiguous set (usually $\frac{1}{4}$ the total number) of the elements on the cylinder. The network consists of two parts; one part chooses the desired sector, and the other maintains the desired ordering of the elements. The switching network effects a "coarse" steering, moving the beam in angular increments corresponding to the element spacing (approximately 0.8 beamwidth steps). For any setting of the switches phase tilt can be introduced by the phase shifters to provide "fine" steering. A distribution network is used to derive the aperture illumination as in the case of a linear array; fixed line lengths are used to correct for the curvature of the aperture. Sum and difference excitations are provided. For a 32 element array, 56 PIN diodes are required for the switches and 64 diodes for 4-bit phase shifters. If the modal approach, which is described elsewhere (Reference 1, 2), were to be implemented for the same size array, 256 diodes would be required with

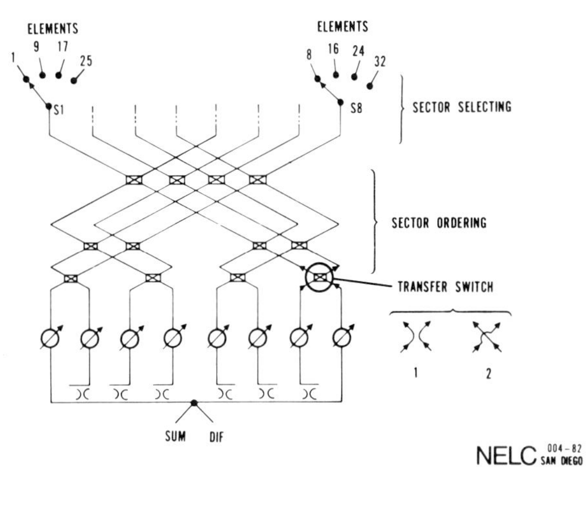

Fig. 1 Schematic Diagram of Diode Switched Array Developed by Wheeler Laboratories.

4-bit phase shifters. The array produced a 16° beam having sidelobes at -20 dB, from a 5' diameter cylinder at L-band. The radiating elements were dipoles mounted for vertical polarization, as shown in Figure 2.

Fig. 2 Circular Array Developed by Wheeler Laboratories.

Another approach for a diode switched array has been developed by Naval Electronics Laboratory Center, San Diego, California. The radiating array has 128 waveguide elements, horizontally polarized, of which 32 are excited at one time. The beam steering method uses phase and amplitude boards which are also capable of inserting amplitude taper. On transmit, the power is equally divided 32 ways and fed to 32 vector (phasor) boards. Each is connected to a single-pole, four throw (SP4T) switch with equal-length cables running to four radiating elements separated by 90° as shown in Figure 3. Any 32 adjacent elements

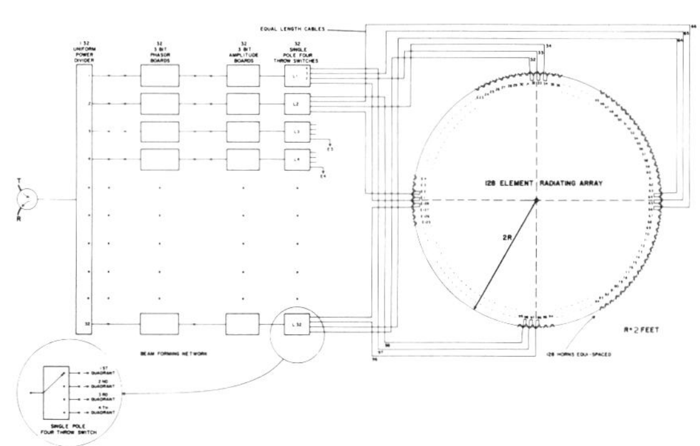

Fig. 3 Schematic Diagram of Diode Switched Array Developed by Naval Electronics Laboratory Center.

in a 90° sector can be excited at the same time since no two elements of a 90° (32-element) sector of the ring array are connected to the same phasor board.

The microwave SP4T switches are built with shunt-mounted PIN diodes on stripline circuits. The circuit has the form of a corporate divider with shunt diodes placed at the junctions. The forward-biased diodes prevent transmission, the reverse-biased diodes permit transmission. Typically, a switch has a loss of 0.8 dB and a VSWR of 1.2:1.

The vector boards consist of a 3-bit phasor and an attenuator, controlled by a 3-bit logic driver. The stripline circuit board has four cascaded hybrid couplers to which are connected four pairs of matched switches - three pairs for the phase switch and one pair for the attenuator. The array yielded a beamwidth of 3.6° and sidelobes on the order of -18 dB. The system operated over a 600 MHz band in S-band. The array is shown in Figure 4 with its associated diode components.

Fig. 4 Circular Array Developed by Naval Electronics Laboratory Center.

Another array of the switched-type shown in Figure 5 was developed by Scanwell Laboratories of Springfield, Virginia. The array elements consist of probe-fed flared horns with chokes. The chokes are necessary in the vertical plane in order to achieve low-back-radiation and, also, to prevent excessive coupling between the upper array and the lower array. In the azimuth plane, the elements radiate as a circular array of vertical probes backed by a ground plane. The elements are fed through a special network of RF cable and distributed RF switch boxes. It required a total of 4900 feed of aluminum jacketed, semi-rigid, coaxial cable to implement the feed network. There are 94 RF switch boxes which are distributed around the in-

Fig. 5 Circular Array Developed by Scanwell Laboratories.

side of the antenna ring. Each box contains two (2) RF switches, SPDT, in which the switching elements consist of magnetically energized, mercury-wetted reed capsules.

The heart of the feed network is a device termed a "Pass-Around", which is contained in the RF scanner unit. It has 31 input ports, 31 output ports, and consists of 155 interconnected SPDT switches. Its function is to switch the input distribution of 30 element signals into the proper sequence at its output ports so that they will be transferred by the feed network to correctly illuminate the particular group of 30 elements required in forming a given radiated beam. The amplitude illumination tapers required for selectable azimuth beamwidths are set up by a dual directional coupler power divider.

Beam steering is accomplished by feeding a seven-bit digital signal into the electronics cabinet from the remote beam controllers. They permit automatic beam command control from external digital or synchro inputs, and they can also be switched into a manual stepping or slewing of the beams via a binary analogue-to-digital encoder. The antenna was designed for the IFF frequencies and has a beam-to-beam switching speed of 3 milliseconds. The azimuth beamwidth is variable; 4.5° or 8.8°, and has sidelobes on the order of -26 dB.

By virtue of its symmetry, a cylindrical array has obvious appeal when an antenna is needed for high-speed scanning of a pencil beam 360 degrees in azimuth and over some range of elevation angles. With such an array, the radiation pattern is formed by establishing the proper currents on the elements (ordinarily on just a sector of the cylinder), and is then scanned in aximuth by moving the excitation around the cylinder, and in elevation by varying the frequency or phases of the currents. Recent work at the Naval Research Laboratory on circular arrays resulted in the development of a matrix and diode switches capable of performing the first function, permitting switching from beam to beam in azimuth in less than a microsecond. A schematic is shown in Figure 6. To demonstrate 3-D scanning with a

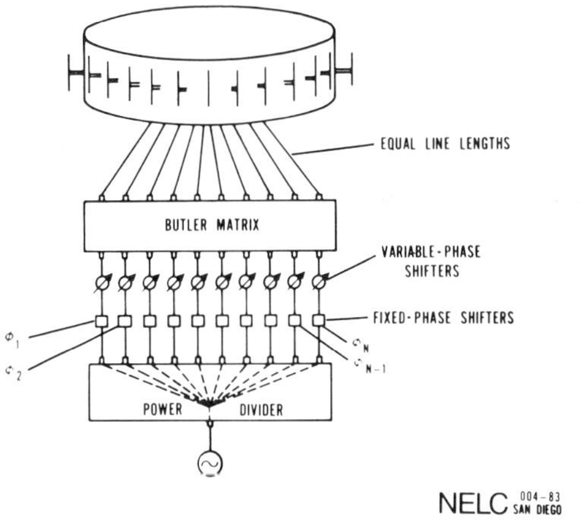

Fig. 6 Schematic Diagram of Cylindrical Array Developed by Naval Research Laboratory.

cylindrical array, a 256-element, S-band array consisting of 32 8-element linear arrays has been built and tested.

The 32 arrays making up the cylinder each consist of 8 vertically polarized dipoles series fed by a serpentine transmission line by means of directional couplers. The radiating elements are sandwich dipoles fed directly from the transmission line and occupying a single surface. Eight of the 32 linear arrays are excited at a time. The currents to be applied are first established by a corporate structure having 8 outputs equal in phase and having a 25 dB Tchebyscheff amplitude distribution. The switch system has a total of 44 transfer switches. An array of 32 elements around the ring would have 5 switches in series in each path. A 64 element array would require 6 switches in series. A pass-around network is made up of eight 4 x 4 matrices with 4 transfer switches per matrix.

The array is scanned in elevation from -20 to +40 degrees by varying the frequency from 3.07 to 3.37 GHz, a larger frequency excursion than the switch matrix was originally designed to cover

(3.1 to 3.3 GHz). Elevation pattern sidelobes were generally 16 to 18 dB down. Azimuth patterns for 0 degrees elevation were 19 to 21 dB down. Figure 7 shows the NRL Array.

Fig. 7 Cylindrical Array Developed by Naval Research Laboratory.

Lens Techniques

The second group of feed techniques encompasses various types of microwave lenses and switching methods to permute the radiated energy about the structure. Early work on the lens approach was done by Cambridge Research Laboratories, Hughes Aircraft Company and Chu Associates and others. Some recent developments will now be described.

A detailed description of the geodesic Luneberg lens antenna and some of its characteristics is given by Johnson (3). A recent investigation at Georgia Tech indicated that a rapid, 360 degree scan antenna could be developed by the design of feeds and switching networks to permit electronic scanning of a geodesic Luneberg lens. The system would employ two lenses, a secondary radiator and a switching network consisting of latching ferrite circulators arranged in a tree-type configuration. Although an inertialess antenna could be realized without employing two lenses, the use of two lenses would result in characteristics superior of those of a system employing only one lens. Specifically, adjacent beam overlaps of less than 3 dB are readily obtainable only through the use of two lenses.

With the aid of Figure 8, operation of the system

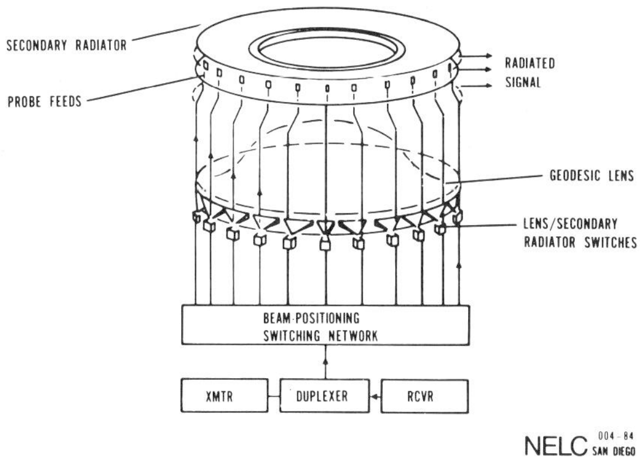

Fig. 8 Schematic Diagram of Lens Antenna Developed by Georgia Institute of Technology.

during transmission can be explained as follows: A signal from the transmitter passes through the duplexer and enters the switching network. The switching network is of the proper state to route the signal to the selected lens/radiator switch which in turn is in the proper state to direct the signal to the desired lens feed. The signal radiated from the selected input feed propagates through the lens and enters the feeds in the opposite side of the lens. The lens/radiator switches associated with the output lens feed then route the signal to corresponding probe feeds on the secondary radiator. Ideally, the aperture distribution of the lens will be transferred to the secondary radiator and its radiation pattern will be the same as that which would be obtained for radiation from the lens itself. Since the radiating system is circularly symmetric, the same principle of operation holds for any direction of propagation. Thus, as each feed around the periphery of the lens is sequentially activated, the radiated beam moves through the full 360°. The system yielded an azimuth beamwidth of 3.8° at a frequency of 10 GHz. Although the lens feed technique has potential application for many systems, the particular system requirements can limit the broad band characteristics. For example, the choice of switching devices is dependent upon frequency, or the pattern coverage (crossover) requirements can complicate the feed system.

The Naval Electronics Laboratory Center, San Diego, California has developed a parallel plate lens and diode switch combination array (4). This method of feeding and scanning the circular array uses the well-known R-2R parallel-plate lens feed system. The lens is a parallel plate region of radius R, with the spacing between parallel plates less than $\frac{1}{2}$ free-space wavelength in order to restrict propagation to the electric field component perpendicular to the plates. Energy is launched and extracted from the lens by means of monopoles mounted $\frac{1}{4}$ wavelength in front of the circumferential ground plane enclosing the parallel plate region. From Figure 9, it can be seen that energy intro-

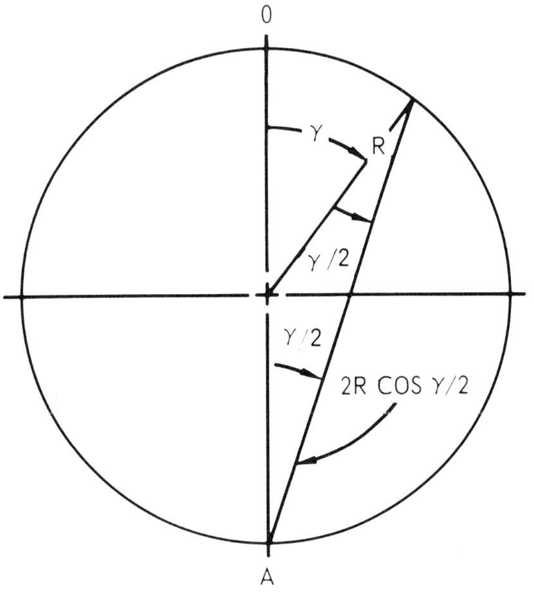

Fig. 9 R-2R Lens Coordinates.

duced at point A travels a distance $2R \cos \gamma/2$ when received by a pickoff probe at γ. For a beam co-phasal distribution on a ring array of radius ρ the energy must be delayed a distance $\rho \cos \alpha$ for an antenna element located at α measured from the beam direction. Thus, the distribution will be provided by the lens if

$$R = \frac{1}{2} \rho$$

and

$$\gamma = \frac{1}{2} \alpha \ .$$

The lens must be one half the radius of the ring array. The lens angle will be twice the array angle if all of the lens ports are used to illuminate one half of the array aperture (180°). The illumination of the ring is accomplished using 64 equispaced probes on the lens, each switchable to one of two diametrically opposite elements on the array. Figure 10 shows typical amplitude and phase distributions obtained from the R-2R lens, where R=4'. Four lens input probes are excited simultaneously to achieve aperture amplitude taper as shown in Figure 11. Three-hundred-sixty-

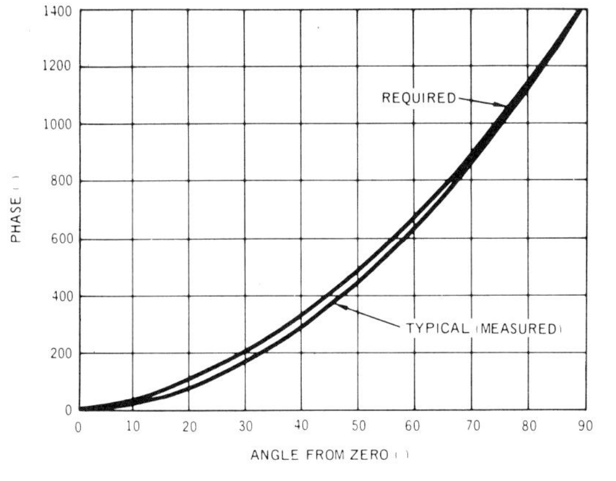

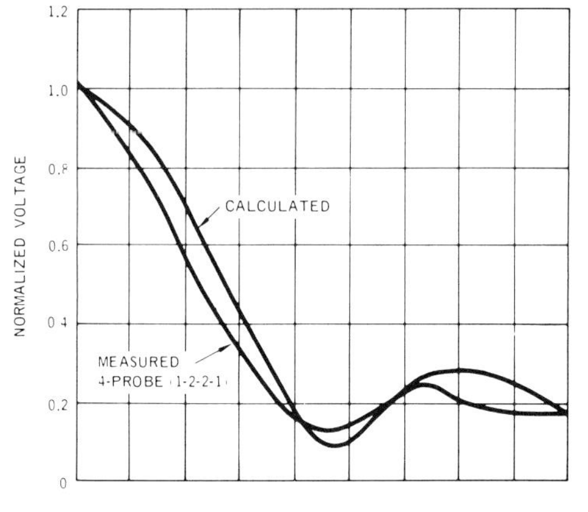

Fig. 10 R-2R Lens Amplitude and Phase Distributions.

degree azimuth beam steering is implemented by means of a SP4T switch at each input-output port of the lens. This switch selects one of the two radiating elements to which it is connected, or a matched load when the terminal is inactive, or the input-output probe.

The radiating array shown in Figure 4 was used with the lens feed and had an azimuth beamwidth of 5° at 3.2 GHz. Figure 12 shows a typical radiation pattern from the array.

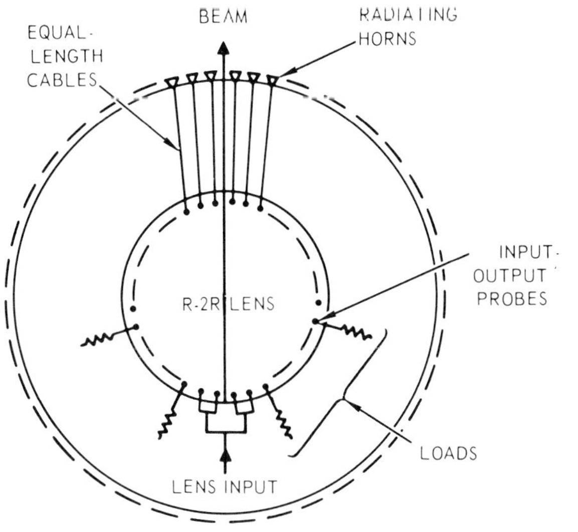

Fig. 11. R-2R Lens Schematic Diagram

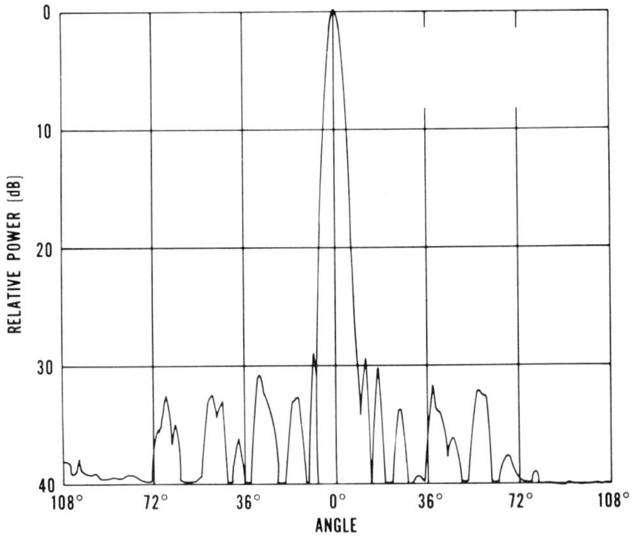

Fig. 12 Radiation Pattern, R-2R Lens Antenna.

Matrix Techniques

The application of the hybrid matrix technique provides another method to feed and steer a ring or cylindrical array. This technique allows multiple or steerable high gain beams to be formed over 360 degrees of azimuth angle. It has been shown that essentially equal degrees of freedom exist with linear and circular arrays insofar as beam shaping, multiple beams, network errors, and bandwidth are concerned. Many people have contributed to hybrid matrix techniques and typical examples will be given.

The matrix circular array concept is based on the formation of N pseudo omni-directional modes with the N element array. These modes are phase orthogonal and available at the outputs of a "mode" forming matrix. These outputs have approximately the same mathematical format as the mathematical contribution of a single element in a linear array. A typical matrix ring array implementation has been described by Sheleg of Naval Research Laboratory (5). Figure 13 gives a typical schematic diagram.

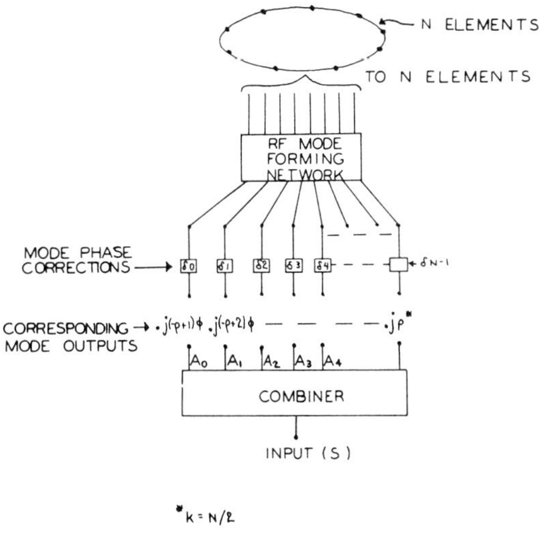

Fig. 13 Typical Matrix Circular Array Schematic.

The matrix techniques previously discussed concern ring arrays, but they are readily extended to produce pencil and other type beams. In general, any beam can be formed using matrix techniques which can be expressed in a Fourier expansion containing mode patterns. Usually, the most directive beam allowed by the highest order Fourier terms is formed. The radiation pattern of the beam in a plane perpendicular to the axis of the cylinder has pattern characteristics which would be expected from a uniformly illuminated linear array of the same extent as the diameter of the cylindrical array. There are instances, such as fan beams and the spatially modulated TACAN antenna pattern, where the modes are not combined for maximum directivity (6).

The development of modern strip-transmission-line technology has allowed array concepts which are presented here to be implemented into operational systems. Prior to this technology, workable high order hybrid matrices were not practical from a size and performance standpoint.

The possibility of forming and electronically rotating the complex radiation pattern generated by the TACAN Transponder antenna has received considerable interest in the past. Radiation Systems, Inc., McLean, Virginia has recently conducted a development program to demonstrate some of the performance advantages which are possible with this type of TACAN antenna. The benefits which accrue through the combination of circular phased array techniques and the unique biconical radiator developed by RSi are:

a. Pattern rotation with a fixed antenna

b. Continuous full band coverage from 960 to 1220 MHz without adjustment

c. Improved elevation plane patterns wherein the modulation remains unchanged for elevation angles up to 60°

d. Improved cross polarization

e. Around the mast installation if required.

It should be recalled that the TACAN pattern consists of a limacoid with nine lobes superimposed upon it which rotates at 900 rpm. This pattern is shown in

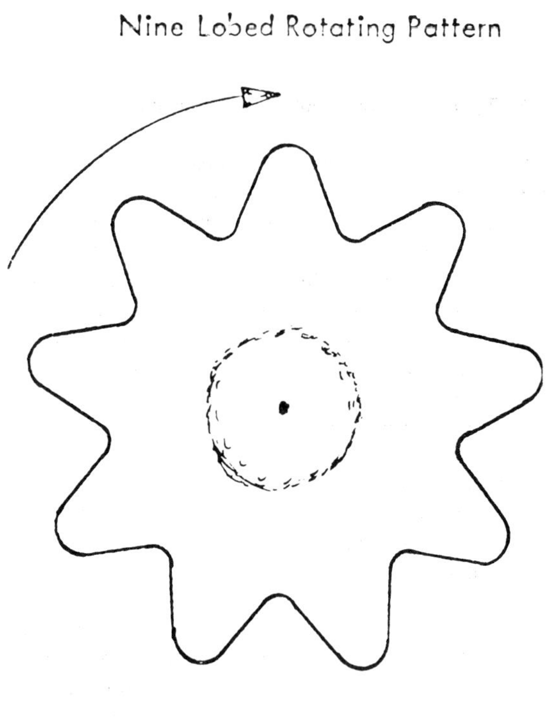

Fig. 14 TACAN Rotating Pattern.

Figure 14 and has the mathematical form:

$$E(\theta) = A_o + B_1 \sin\theta + B_9 \sin 9\theta$$

where θ = azimuth angle
A_o = omni pattern content and
B_1 and B_9 represent the ripple content of the fundamental and 9 lobed patterns respectively.

The trigonometric functions can be formed by combining symmetric modes of the proper mode number as done for fan beam antennas. In this case, the onmi term is provided with the 0 order mode. the θ term is provided by combining the +1 and -1 modes and the 9θ term is supplied by combining the +9 and -9 modes. This pattern is rotated about the fixed array by inserting variable phase shifters in the lines feeding the first and ninth order modes.

A $\frac{1}{2}$ scale model, shown in Figure 15, was con-

Fig. 15 Experimental TACAN Antenna Developed by Radiation Systems, Inc.

structed of both the antenna array and the RF processing networks. The radiator consists of thirty-two probes positioned in a circle at the throat of a biconical horn. The RF processing networks are fabricated in strip-transmission-line. A composite of measured azimuth patterns of this experimental TACAN antenna for frequencies of 1.9 to 2.5 GHz is shown in Figure 16. Developmental work to optimize the array antenna and the feed networks across the full band are incomplete. However, these pattern traces demonstrate the full TACAN bandwidth capability at the scaled frequencies. The amplitude modulation of the pattern is virtually unchanged over a 60° range of elevation angles.

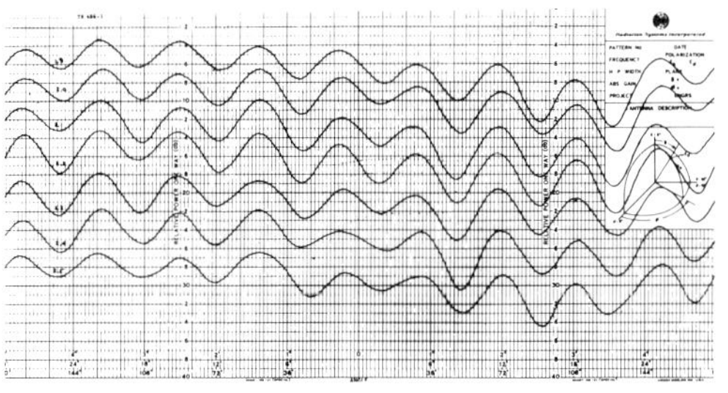

Fig. 16 Radiation Pattern, Measured, TACAN Antenna.

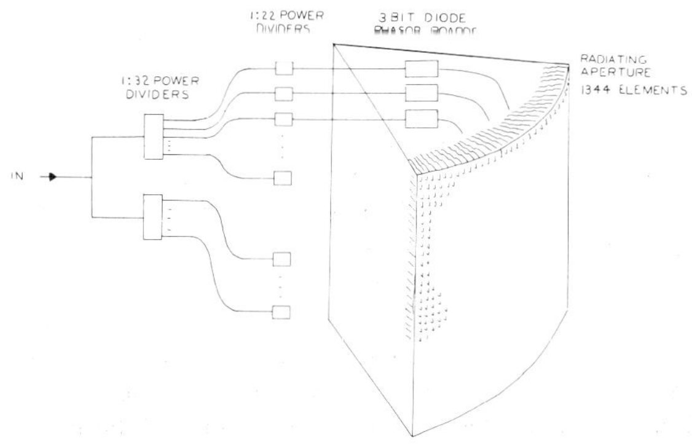

Fig. 17 Schematic Diagram of Partial Cylindrical Array Developed by Naval Electronics Laboratory Center.

For a cylindrical array, the azimuth phases should ideally be changed as the beam is scanned in elevation; unlike the planar array for which azimuth and elevation scans are independent. However, if the array diameter is small enough, then the azimuth phase adjustment with elevation scan can be neglected. For larger arrays, variable phase shifters are required to adjust the azimuth phase for some of the elements as the beam is scanned in elevation.

In order to make necessary phase corrections as the beam is scanned in elevation, a phasor is required at each radiating element. A partial cylindrical array of this type is presently being developed by the Naval Electronics Laboratory Center, San Diego, California. The array consists of 42 linear arrays each with 32 waveguide elements. The linear arrays are mounted on a 16' diameter and make up a 60° sector of the cylinder. The active sector has 1344 elements and each will be implemented with a 3 bit integrated circuit diode phase shifter for elevation steering. The array is designed to operate over a 600 MHz band in S band. A schematic of the array is shown in Figure 17. Amplitude tapering is provided in both planes and the array has a pencil beam of about 3° and side lobes on the order of 22 dB. The present implementation will provide for 400 beam positions in the sector, computer driven, and provides for sum and difference beams for monopulse operation. The array is shown in Figure 18.

Fig. 18 Partial Cylindrical Array Developed by Naval Electronics Laboratory Center.

Array Characteristics

The array antennas which have been described illustrate the variety of techniques which are available to the systems designer. A particular application for an antenna determines to some degree the type of feed system to be used, the switching and phasors to be used and the geometry to be used. It is difficult to compare the various types of circular symmetric arrays described in terms of losses, bandwidth and complexity since the arrays which have been built were tailored to meet specifc applications, and at several different frequency bands.

The use of these types of arrays for wide band systems has been established by experimental models. A matrix fed array has been developed by RSi to cover a 5 to 1 band; Naval Electronics Laboratory Center has tested a circular array for the lens-fed type to cover an octave bandwidth and the arrays discussed of the switched-type have been operated over a 20% band. A major problem area appears to be in the choice of feeding and switching devices to minimize the losses of the switch and phasor components. Considerable losses occur in the element connectors and cabling required to commute the amplitude and phase distributions. Many of the arrays described suffer from this condition and serious consideration should be given to other types of techniques to simplify the design. The utility of the asymmetric arrays has been demonstrated and as in planar phased arrays, the cost is a most important factor. The feeding techniques are, in general, more complicated than those for the planar phased array. However, the inherent advantages of the beam symmetry, wide band capability and physical stability offer attractive trade-offs for some types of systems.

Conformal Arrays on Surfaces with Rotational Symmetry

GIORGIO V. BORGIOTTI
and QUIRINO BALZANO

Raytheon Company
Bedford, Massachusetts

ABSTRACT

The study of a conformal array having rotational symmetry can be drastically simplified by considering the array pattern (for aribtrary excitations at the element input ports) as the linear combination of the patterns relative to a set of fundamental excitations matching the symmetry of the structure. As a numerical example a cylindrical array of circumferential slots on a cylinder is discussed in detail.

I. INTRODUCTION

It is well known that the recognition and judicious utilization of the geometrical symmetry present in a structure may substantially simplify the analysis of a physical problem. This idea has been exploited here to simplify the analysis of arrays of apertures flush mounted on a surface of revolution.

Stated in a general way, the essence of the method consists of decomposing an arbitrary excitation of the structure into component "fundamental" excitations, which as direct consequence of the symmetry, have certain properties of orthogonality and of invariance with respect to the operator describing the physical behavior of the system. Then a number of partial problems, concerning the response of the structure to the fundamental excitations, must be solved. Because of their lesser dimensionality, they are simpler than the original problem. Superposition then allows the solution of the general case.

This idea has already been used for the study of the idealized case of a cylindrical array of uniform axial slits (1) (2) (3). In this paper a more general formulation is presented, leading to effective methods for the analysis of practical antennas.

As an application, a finite array of circumferential slots on a circular cylinder is analyzed in detail.

2. ARRAY CHARACTERIZATION

2.1 General - Mutual Couplings and Realized Gain

Patterns

Consider an array of apertures flush mounted on a surface of revolution. The array consists of M "rings" each consisting of N equispaced elements whose centers lie on a plane orthogonal to the symmetry axis of the structure. The elements in two adjacent rings may be staggered, as shown in Figure 1 depicting an array of radiators with a tri-

Rotationally Symmetric Arrays

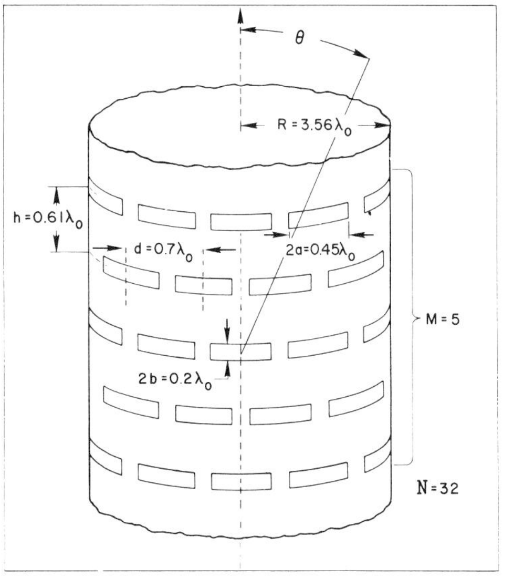

Fig. 1 Array Geometry.

angular grid on a cylinder. Elements belonging to different rings may possibly be different. If the array radiators possess two input ports (as it is the case, for example, for switchable polarization elements) no change in the treatment is necessary, provided the two sets of ports are considered belonging to separate rings (yet geometrically coincident). Thus, in this case, the number of rings to be considered is twice the number of "physical" rings.

Considering the array as a multiport network its "free" excitation (4) (i.e., the set of waves incident to the element input ports) is denoted, as customary, by the column vector $\underline{a}$, having M x N components. The set of reflected waves is similarly represented by the vector $\underline{b}$, related to $\underline{a}$ by:*

$$\underline{b} = \underline{\underline{S}}\,\underline{a} \tag{1}$$

where $\underline{\underline{S}}$ is the scattering matrix. In order to write the linear equations equivalent to the vector relation Equation (1) explicitly, the set of incident and reflected waves at the element ports must be grouped to form the column vector $\underline{a}$ and $\underline{b}$ or, equivalently, the element ports must be ordered. A convenient method is: indicate by a superscript q and a subscript k the ring to which an element belongs and its position in the ring, respectively. Assuming vertical the axis of symmetry of the structure, q will take the values from 1 to M going from the highest to the lowest ring. The subscript k will run from 0 to N-1, starting from a reference element chosen in each ring. The excitation of the ring q is an N dimensional column vector $\underline{a}^q$:

$$\underline{a}^q = \left\{ a_k^q \right\} \quad (k = 0, 1 \ldots N-1 \; ; \; q = 1, 2 \ldots M) \tag{2}$$

where q is kept fixed, and the subscripts k are taken in natural order. The M x N dimensional vector $\underline{a}$ is then given by:

$$\underline{a} = \begin{bmatrix} \underline{a}^1 \\ \underline{a}^2 \\ \cdot \\ \cdot \\ \underline{a}^M \end{bmatrix} \tag{3}$$

As mentioned before, if the elements have two input ports, each ring must be counted twice. Similar element grouping applies also to the vector $\underline{b}$. With this element ordering the scattering matrix takes the form:

$$\underline{\underline{S}} = \begin{bmatrix} \underline{\underline{S}}^{11} & \underline{\underline{S}}^{12} & \ldots & \underline{\underline{S}}^{1M} \\ \cdot & \cdot & \cdot & \cdot \\ \underline{\underline{S}}^{M1} & \underline{\underline{S}}^{M2} & & \underline{\underline{S}}^{MM} \end{bmatrix} \tag{4}$$

where the $\underline{\underline{S}}^{pq}$ are square matrices of Nth order, each of them representing the interaction among the elements of the pth and qth rings. As a mathematical counterpart of the N-fold rotational symmetry of the array, the matrices $\underline{\underline{S}}^{pq}$ are cyclic, that is each row is obtained from the previous one through a cyclic permutation as indicated below:

$$\underline{\underline{S}}^{pq} = \begin{bmatrix} S_o^{pq} & S_1^{pq} & \ldots & S_{N-1}^{pq} \\ S_{N-1}^{pq} & S_o^{pq} & \ldots & S_{N-2}^{pq} \\ \cdot & \cdot & \cdot & \cdot \\ \cdot & \cdot & \cdot & \cdot \\ S_1^{pq} & S_2^{pq} & \ldots & S_o^{pq} \end{bmatrix} \tag{5}$$

The element S_k^{pq} of the matrix $\underline{\underline{S}}^{pq}$ is the mutual coupling (or scattering coefficient) between two elements belonging to the pth and q^{th} ring respectively, for which the numbers representing the azimuthal locations differ by k (mod N).

While Equation (1) describes the circuital properties of the system, its radiative properties can be characterized by the set of element realized gain patterns, i.e., the patterns of the singly excited elements, when the others are match terminated. For a given direction of observation θ, ϕ, the θ component (say) of the realized gain of the various elements (with the phase referred to the same origin) form a column vector $\underline{f}_\theta(\theta, \phi)$. The overall array pattern is thus given by the scalar product:

$$F_\theta(\theta, \phi) = \langle \underline{\tilde{f}}_\theta(\theta, \phi) \cdot \underline{a} \rangle \tag{6}$$

* A column vector, like $\underline{a}$, is indicated in this paper by a bar underlying a letter, while a vector in the "physical space", (like a radiation pattern) will be indicated by an arrow $\vec{f}$ overlining a letter.

where the tilde (~) indicates the transpose. A similar equation of course can be written for the ϕ component of the field:

$$F_{\phi}(\theta, \phi) = < \tilde{\underline{f}}_{\phi}(\theta, \phi) \cdot \underline{a} > \quad (7)$$

In the following when dealing with a set of radiation patterns, forming a column vector, the subscripts θ or ϕ will generally be dropped with the understanding that the developments apply to both the components of the field.

At this point certain simple preliminary considerations are appropriate to point out what is the nature of the mathematical problem occurring in conformal array analysis. In order to utilize the formal relationships in Equations (1), (6) and (7), M different boundary value problems must be solved. Each consists of determining the transverse electric field distribution on all the array apertures when a single element of one of the rings is excited. These very difficult problems can be approached as follows. The transverse electric field distribution of each aperture is approximately represented by using a limited number Q of modes of the waveguides feeding the elements. Then the continuity of the tangential electric field (on the entire surface) and of the tangential magnetic field on the apertures is (approximately) enforced, through an application of moment method, Equations (5) (6). This procedure leads then to a set of MxNxQ linear equations with the same number of unknowns. The formal expression of their coefficients are straight forwardly obtained from the classical solution of the "exterior" electromagnetic problem, known for spherical, conical and cylindrical geometries, but their numerical computations may be very involved in some cases, and may require considable programming skill. The inversion of the matrix of the system is, however, the most difficult task, even using a large computer. For example even for "one mode" elements (Q=1) an array composed of 5 rings each with 32 elements, requires the inversion of a matrix of 160 order with complex elements, an almost impossible task.

Once the transverse electric field at the apertures has been found, the scattering coefficients follow immediately (from the fundamental mode voltages at the various apertures), and, on the other hand, the solutions of classical "exterior" electromagnetic problems provide the set of realized gain patterns.

In the following sections a method is introduced to avoid the direct inversion of the matrix by using the geometrical symmetry of the structure.

2.2 Change of the Vector Basis - Alternate Characterization of the Array Properties

Linearity and superpositon allow the expression of any array excitation $\underline{a}$ as a weighted sum of a set of M x N "fundamental" excitations, constituting a basis for the M x N dimensional vector space to which $\underline{a}$ belongs (the "input space" of the array). A convenient choice for the vector basis is the following:

$$\underline{m}(s, i) = \left| \delta_{ps} N^{-1/2} \exp. (j\, 2\pi ik/N) \right| \quad (8)$$

$$(p, s = 1, 2 \ldots M;\ k, i = 0, 1 \ldots N-1)$$

where the notation in Equation (8) means that only the sth ring is excited, with azimuthal number i, i.e., an incremental phase of $2\pi\, i/N$ between two contiguous elements (k indicates the element position in a ring). The vectors (8) form an orthonormal complete set (in the Hermitian sense):

$$< \underline{m}^{+}(s, i)\, \underline{m}(t, n) > = \sum_{p,k} m^{+}_{pk}(s, i)\, m_{pk}(t, n) = \delta_{st}\, \delta_{in} \quad (9)$$

where the cross denotes the conjugate transpose and δ_{st} is the Kronecker's symbol. The input column vector $\underline{a}$ can be expressed thru a linear combination of the vectors (8):

$$\underline{a} = \sum_{s=1}^{M} \sum_{i=0}^{N-1} a'_{si}\, \underline{m}(s,i) \quad (10)$$

where

$$a'_{si} = < \underline{m}^{+}(s,i)\, \underline{a} > \quad (11)$$

and the set of M x N numbers a'_{si} constitute a column vector $\underline{a}$ representing the array excitation in the vector basis (8). Each component a'_{si} of $\underline{a}$ is recognized to be the excitation of a single ring with a certain azimuthal number.

Each of the N subsets of M vectors $\underline{m}(s, i)$ having the same azimuthal number i span a M dimensional subspace invariant under the operation represented by $\underline{\underline{S}}$. In order to show this let us notice first that the $\underline{m}(s, i)$ are eigenvectors of the submatrices $\underline{\underline{S}}^{st}$ in (4), and that the following orthogonality relationships hold (App. I):

$$< \underline{m}^{+} (s, i) \underline{\underline{S}}^{st} \underline{m} (t, n) > = \Gamma(i)^{st} \delta_{in} \qquad (12)$$

where $\Gamma(i)^{st}$ is the eigenvalue of $\underline{S}^{st}$ corresponding to the eigenvector $\underline{m}$ (s, i). It is expedient to introduce the $(M \times N)^2$ matrix $\underline{\underline{M}}$ whose columns are the vectors $\underline{m}$ (s, i) ordered as follows:

$$\underline{\underline{M}} = |\underline{m}(1,0), \ldots, \underline{m}(M,0), \underline{m}(1,1), \ldots, \underline{m}(M,1), \ldots, \underline{m}(1,N-1), \ldots, \underline{m}(M,N-1)| \qquad (13)$$

The vectors $\underline{m}$ (s, i) have been thus grouped in N sets of M contiguous columns corresponding to the same azimuthal numbers. The matrix $\underline{\underline{M}}$ is unitary (See (9) :

$$\underline{\underline{M}}^{+} \underline{\underline{M}} = \underline{\underline{U}} = \underline{\underline{M}}\,\underline{\underline{M}}^{+} \qquad (14)$$

where $\underline{\underline{U}}$ is the unit matrix of order M x N. The vectors $\underline{a}$ and $\underline{b}$ in the new basis can be written as:

$$\underline{a}' = \underline{\underline{M}}^{+}\underline{a} \qquad (15)$$

equivalent to (11), and

$$\underline{b}' = \underline{\underline{M}}^{+}\underline{b} \qquad (16)$$

In the basis (8) the scattering matrix takes the form [7] :

$$S' = \underline{\underline{M}}^{+}\underline{\underline{S}}\,\underline{\underline{M}} \qquad (17)$$

whose elements are given by (12). The same set of equations (12) shows that $\underline{\underline{S}}'$ has the following structure:

$$\underline{\underline{S}}' = \begin{bmatrix} \underline{\underline{S}}_0 & 0 & \ldots & 0 \\ 0 & \underline{\underline{S}}_1 & \ldots & 0 \\ \ldots & \ldots & \ldots & \ldots \\ 0 & \ldots & \ldots & \underline{\underline{S}}_{N-1} \end{bmatrix} \qquad (18)$$

where the N block matrices $\underline{S}_k$ of order M, according to (12) are given by:

$$\underline{\underline{S}}_k = \begin{bmatrix} \Gamma(k)^{11} & \Gamma(k)^{12} & \ldots & \Gamma(k)^{1M} \\ \ldots & \ldots & \ldots & \ldots \\ \Gamma(k)^{M1} & \Gamma(k)^{M2} & \ldots & \Gamma(k)^{MM} \end{bmatrix} \qquad (19)$$

Equation (18) shows that the input space is the direct sum of N subspaces invariant under the operation represented by $\underline{\underline{S}}$ [7]. In Appendix I it is also shown that the radiation patterns corresponding to different azimuthal numbers are orthogonal. From a physical viewpoint it can be said that the (arbitrary) array excitation has been decomposed into components corresponding to different azimuthal number which are energetically uncoupled.

The use of the basis (8) reduces the original electromagnetic problem, involving the evaluation of mutual couplings and radiation patterns of a M x N element array, to N separate problems of the same nature, each consisting of the analysis of a different M port radiating structure, having as radiators the entire array rings all excited with the same phase sequence. Each element $\Gamma(k)^{st}$ of the matrix (19) represents the mutual coupling between the rings s and t each excited with uniform amplitude and the azimuthal number k. * The reduced problems cannot be solved rigorously but suitable approximation techniques can be used, as discussed in section 3. In either case, the problem is drastically simplified. The solution of the set of reduced problems will provide the various $\Gamma(k)^{st}$, i.e., the matrix $\underline{\underline{S}}$. The scattering coupling matrix $\underline{\underline{S}}$, can be determined from Equations (17) and (14) through only matrix multiplications, an operation simply performed on a digital computer. The antenna pattern for an arbitrary excitation can be found then, by using superposition, from the set of the array patterns $\vec{g}(s, i; \theta, \phi)$ relative to the fundamental excita-

*$\Gamma(k)^{st}$ can be evaluated by computing the complex amplitude coupled in the 0 element of the ring s by the ith phase sequence excitation of the ring t.

tions Equation (8). Thus by recalling Equations (10) and (11):

$$\vec{F}(\theta,\phi) = \sum_{i=0}^{N-1} \sum_{s=1}^{M} a'_{si}\ \vec{g}\ (s, i; \theta, \phi) \tag{20}$$

If $\underline{a}$ represents in the vector basis (8) the excitation of a single radiator, then Equation (20) yields the element realized gain pattern.

It is worth explicitly noticing that the set of patterns $\vec{g}\ (s, i; \theta, \phi)$ (called hereafter "ring patterns") provides a way of characterizing the radiative properties of the array, which is alternative to the customary use of the realized gain patterns. In fact, this characterization can be shown to be much more convenient in beam synthesis problems, because of the properties of orthogonality of the ring patterns corresponding to different azimuthal numbers. However, discussion of this question is outside the scope of this paper.

An interesting relationship following from eq. (14), (15) and (20) can be promptly established:

$$< \tilde{\underline{g}}\ \underline{a}' > = < \tilde{\underline{g}}\ \underline{\underline{M}}^{+}\ \underline{\underline{M}}\ \underline{a}' > = < \tilde{\underline{f}}\ \underline{a} > \tag{21}$$

The left side of Equation (21) is (20) written in a different form. The last equality in Equation (21) shows that the ring patterns and the realized gain patterns are related by the basis transformation

$$\underline{\underline{M}}^{*}\ \underline{g} = \underline{f} \tag{22}$$

where the star denotes the complex conjugate.

3. ONE MODE ELEMENT APPROXIMATION

The elements considered here are apertures fed by waveguides (with the same cross sections) whose axes are orthogonal to the array surface. Because of the size of the waveguide section only a single propagating mode exists, and it will be postulated that the influence of the higher order modes on the radiative and mutual coupling properties of the element is relatively minor. Thus, a "one mode" element model can be made (8), which assumes that the transverse field distribution at the aperture has an invariable functional form, equal to that of the fundamental waveguide mode. The complex amplitudes of the modal voltages of the elements are of course affected by mutual couplings.

For single mode elements the expressions for the self and mutual admittances among the radiators are found in a relatively simple fashion. In fact, application of Lorentz reciprocity theorem leads to the following expression for the mutual admittance between two elements located in the pth and qth positions of the m-th and n-th rings, respectively (9).

$$Y_{pq}^{nm} = \iint_{A_q^n} \hat{n} \times \underline{E}_{ot}^{n}\ (\phi - \frac{2\pi q}{N}) \cdot \underline{H}_{ot}^{m}\ (\phi - \frac{2\pi p}{N})\ dA \tag{23}$$

where the integral is over the aperture of the element in position n, q and *:

$\hat{n}$ is the unit vector normal to the surface

$\underline{E}_{ot}^{n}(\phi)$ is the transverse electric field distribution on the element in position 0 of the n-th ring corresponding to a unit aperture mode voltage.

$\underline{H}_{ot}^{m}(\phi)$ is the tangential component of the magnetic field on the array surface due to a unit voltage at the aperture in the position 0 of the m-th ring..

It is convenient to normalize each admittance Equation (23) to the characteristic admittance of the

*Only the dependence of the field upon the azimuthal angle has been explicitly shown in Equation (23)

propagating mode in the waveguide. The relationship between the normalized admittance matrix Y and the scattering matrix S is then:

$$\underline{\underline{S}} = (\underline{\underline{y}}-\underline{\underline{U}})\ (\underline{\underline{y}}+\underline{\underline{U}})^{-1} \qquad (24)$$

where $\underline{\underline{U}}$ is the identity matrix of order M x N. The change of basis

$$\underline{\underline{y}}' = \underline{\underline{M}}^{+}\ \underline{\underline{y}}\ \underline{\underline{M}} \qquad (25)$$

transforms the admittance matrix into the block diagonal form:

$$\underline{\underline{y}}' = \begin{bmatrix} \underline{\underline{y}}_0 & 0 & \cdots & 0 \\ 0 & \underline{\underline{y}}_1 & \cdots & 0 \\ \cdots & \cdots & \cdots & \cdots \\ 0 & \cdots & \cdots & \underline{\underline{y}}_{N-1} \end{bmatrix} \qquad (26)$$

where the elements of the N matrices $\underline{\underline{y}}_i$ of M-th order are found from (23) and (25) to be given by (App. II):

$$y^{nm}_{(i)} = \iint_{A^n_o} \hat{n} \times \underline{E}^n_{ot}(\phi) \sum_{k=o}^{N-1} H^m_{ot}\left(\phi - \frac{2\pi k}{N}\right) e^{+jki\frac{2\pi}{N}}\ dA \qquad (27)$$

The fundamental set of patterns $\vec{g}\ (s, i; \theta, \phi)$ are found by using the following procedure. The "ring voltage vector" vector $\underline{v}'$ (defined as the set of aperture voltages represented in the basis (8)) are determined by using the relationship (obtained from Equations (1), (24) and (25))*.

$$\underline{v}' = 2\ (\underline{\underline{y}}' + \underline{\underline{U}})^{-1}\ \underline{a}'\ (s, i) \qquad (28)$$

where $\underline{a}'$ (s, i) is the column vector representing the excitation of the sth ring only, with azimuthal number i, i.e., (Section 2):

$$\underline{a}'\ (s, i) = \underline{\underline{M}}^{+}\ \underline{m}\ (s, i) \qquad (29)$$

and recalling the structure of $\underline{\underline{M}}^{+}$, it can be easily seen that $\underline{a}'$ (s, i) has only one component different from zero. Since $\underline{a}'$ (s, i) and $\underline{v}'$ both belong to the subspace corresponding to the azimuthal number i, from Equations (26) and (28), the simpler relation follows:

$$\underline{v}' = 2\left(\underline{\underline{y}}_i + \underline{\underline{U}}_M\right)^{-1}\ \underline{a}'\ (s, i) \qquad (30)$$

where $\underline{\underline{U}}_M$ is the identity matrix of Mth order.

The components of $\underline{v}'$ are thus obtained through an inversion of a matrix of order M, a much simpler task than the inversion of a M x N order matrix, which would have been necessary if the basis transformation Equation (25) had not been done. Then, because of the one mode hypothesis, the transverse electric field is (approximately) known on the array surface. The ring patterns can be evaluated then, because for the geometries considered the solution of the external electromagnetic problem is known. The array response to an arbitrary excitation is found through the development of the previous section.

Explicit expression for a particular structure will be given in the next section.

4. FINITE ARRAY OF CIRCUMFERENTIAL SLOTS ON A CIRCULAR CYLINDER

*$\underline{v}'$ is normalized to have the same dimensions as $\underline{a}$, i.e., the square root of a power.

4.1 Analytical Development

The method developed in the previous section will be used for the study of an array of rectangular circumferential slots arranged on a triangular grid on a conducting cylindrical surface. The geometry is indicated in Figure 1. The spacing among the elements of a ring is $d = 0.7\lambda$ and the distance between the rings is $h = 0.61\lambda$. The slot dimensions are $2a = 0.45\lambda$ and $2b = 0.2\lambda$, and the element waveguides are filled with a material with relative dielectric constant $\epsilon_r = 2.54$. Because of the radiator small size it is postulated that all the waveguide modes other than the fundamental TE_{10} (polarized in the direction of the cylinder axis) have negligible magnitude on the apertures.

To complete the geometrical reference in addition to the cylindrical system ρ, ϕ, z (associated with the rectangular coordinates x, y, z) it is convenient to consider also a curvilinear abscissa on the cylinder section:

$$t = R\phi \tag{31}$$

where R is the cylinder radius. The double Fourier transform of the transverse electric field distribution (for unit modal voltage) of the element of the center ring (p=3) located in the position k = 0 (whose aperture center is at $z = \phi = t = 0$) is given by the scalar

$$\mathcal{E}(u, w) = \sqrt{\frac{ab}{2}}\,\frac{\sin bw}{bw}\,\frac{\cos au}{\left(\frac{\pi}{2}\right)^2 - a^2u^2} \tag{32}$$

where u and w are the variable conjugate to t and z respectively.

In order to utilize the expression (23) for the mutual admittance between the mth and nth rings (excited with the azimuthal number i) the ϕ component of the magnetic field supported by the magnetic current of a single aperture must be determined. The procedure is relatively standard and is based on the expansion of the field in a Fourier series and a Fourier integral for the ϕ and z dependence, respectively (10). A brief account of the method is given in Appendix 3. In this section only the results are given. To simplify, the notations put:

$$Q_s(w,\rho) = \mathcal{E}\left(\frac{s}{R}, w\right)\left[\frac{k}{R\eta\sqrt{k^2-w^2}}\,\frac{H_s^{(2)\prime}(\rho\sqrt{k^2-w^2})}{H_s^{(2)}(R\sqrt{k^2-w^2})} + \frac{s^2w^2H_s^{(2)}(\rho\sqrt{k^2-w^2})}{k\eta\rho R^2(k^2-w^2)^{3/2}H_s^{(2)\prime}(R\sqrt{k^2-w^2})}\right] \tag{33}$$

where η and k are the characteristic impedance and propagation constant of free space $H_s^{(2)}$ is the Hankel function of second kind and sth order and $H_s^{(2)\prime}$ its derivative with respect to the argument. The square root in the argument of the Hankel function is such to satisfy radiation condition for ρ tending to infinity. The ϕ and z components of the magnetic field are then given by:

$$H_\phi(\rho, \phi, z) = -\frac{j}{2\pi}\sum_{s=\infty}^{+\infty} e^{js\phi}\int_{-\infty}^{+\infty} Q_s(w, \rho)\, e^{jwz}\, dw \tag{34}$$

$$H_z(\rho, \phi, z) = -\frac{j}{2\pi}\sum_{s=\infty}^{+\infty} e^{js\phi}\int_{-\infty}^{+\infty} \mathcal{E}\left(\frac{s}{R}, w\right)\frac{swH_s^{(2)}(\rho\sqrt{k^2-w^2})}{k\eta R^2\sqrt{k^2-w^2}}\,\frac{e^{jwz}}{H_s^{(2)\prime}(R\sqrt{k^2-w^2})}\, dw \tag{35}$$

By using (34) the admittance between two rings (for the azimuthal number i) is found to be given by:

$$Y_{(i)}^{mn} = -jN\sum_{s=\infty}^{+\infty}\int_{-\infty}^{+\infty} Q_{i+sN}(w, R)\,\mathcal{E}\left(\frac{s}{R}, w\right) e^{jwh(m-n)}\, e^{j(i+mN)\frac{d\delta}{2R}}\, dw \tag{36}$$

In the second exponential of (36):

$\delta = 1$ if m - n odd

$\delta = 0$ if m - n even

Once the elements of the matrix $\underline{\underline{Y}}_i$, given by Equation (36) are obtained, the inversion of the Mth order matrix in Equation (30) provides the set of ring voltages. In order to simplify the notations denote by $\tau_{(i)}^{sq}$ the elements of $2\,(\underline{\underline{y}}_i + \underline{\underline{U}}_M)^{-1}$. It is convenient also to introduce the radiation pattern of the isolated ring s, excited with the azimuthal number i, which will be called $\vec{g}_o\,(s, i; \theta, \phi)$. Its expression is established from Equations (34) and (35) through standard manipulations, omitted here for brevity:

$$\vec{g}_o(s,i;\theta,\phi) = \frac{j}{k\,R\,\sin\theta} \sum_{n=-\infty}^{+\infty} e^{jn\left(\phi + \frac{\pi}{2} - \frac{\delta d}{2R}\right)} \mathcal{E}\left(\frac{n}{R},\, k\cos\theta\right) \left\{ \hat{\theta}\frac{1}{H_n^{(2)}(k\,R\,\sin\theta)} + \hat{\phi}\frac{n\,\mathrm{cotg}\,\theta}{k\,R\,H_n^{(2)'}(k\,R\,\sin\theta)} \right\} e^{j\,s\,k\,h\,\sin\theta} \qquad (37)$$

where $\hat{\theta}$ and $\hat{\phi}$ are unit vectors in the θ and ϕ directions of a spherical system r, θ, ϕ associated in a standard way with the x, y, z reference. Notice that for all the patterns the phases are referred to the origin of the coordinates.

The pattern of the singly excited sth ring in array environment is the superposition of the isolated rings each weighted with the ring voltages (the components of $\underline{v}'$). Thus from equation (30):

$$\vec{g}(s,i;\theta,\phi) = \sum_{q=1}^{M} T(i)^{sq}\, \vec{g}_o(q,i;\theta,\phi) \qquad (38)$$

and the realized gain pattern of the element in the position ○ of the sth ring is given by:

$$\vec{f}_r(s;\theta,\phi) = \frac{1}{\sqrt{N}} \sum_{i=0}^{N-1} \sum_{q=1}^{M} T(i)^{sq}\, \vec{g}_o(q,i;\theta,\phi) \qquad (39)$$

It is worth noticing that the structures considered by this section are finite arrays on infinite cylinders, while practical antennas are of course finite also in the longitudinal direction. It may be expected that for θ close to zero no reliable information is provided by our model concerning the radiation pattern of a practical antenna. The expression in Equation (37) corresponds to the standard saddle point contribution of $H_n(x)$ and is no more valid near the point $\theta = 0°$ and has to be replaced there by a refinement due to the presence of a branch point of the Hankel function near the saddle point. Through analytical or heuristic reasonings it could be shown that the width of the region of θ where our model approximates a practical antenna on a cylinder, increases with the cylinder radius.

4.2 Numerical Results

For the array of Figure 1 a number of realized gain pattern has been computed, which exhibit features already found in previous studies of simpler arrays of uniform axial slits on a cylinder (1) (2) (3) and also certain new effects due to the finite array size. As expected, elements in different rings have different realized gain patterns, the difference being particularly pronounced for elevation planes.

Figure 2 shows the azimuthal pattern for θ polarization for an element of the center ring. The ripple near broadside, characteristic of cylindrical array is here only a minor perturbation (not visible in the scale of Figure 2), because of the small antenna radius (3) and the axial polarization of the slots.

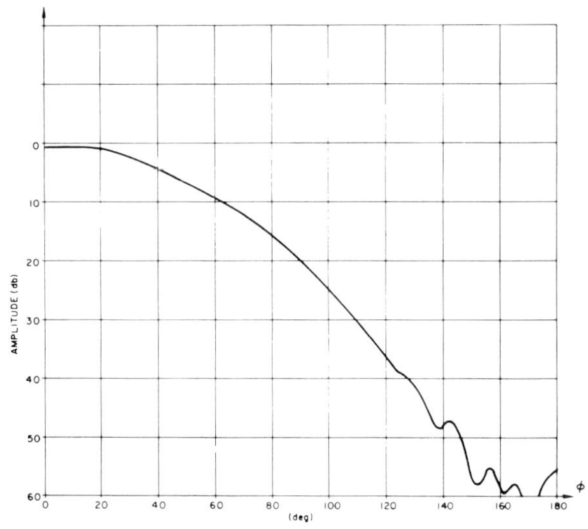

Fig. 2 Realized Gain Pattern, Azimuthal Plane, Center Ring Element θ Polarization.

In Figure 3 the phase of the pattern is shown (referred to the element center). The behavior is regular in the lit region with a characteristic phase oscillation (not visible in the scale of Figure 3), paralleling that of the amplitude pattern. For obvious symmetry reasons the polarization in the $\theta = \pi/2$ plane is always in θ direction. Figure 4 plots the elevation pattern (i.e., in the plane $\phi = 0$) for the θ component of the field (the only one existing).* The differ-

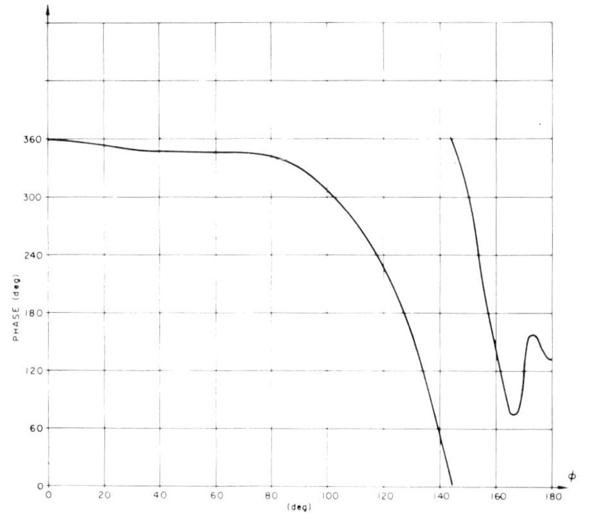

Fig. 3 Phase of Realized Gain Pattern, Azimuthal Plane, Center Ring Element θ Polarization.

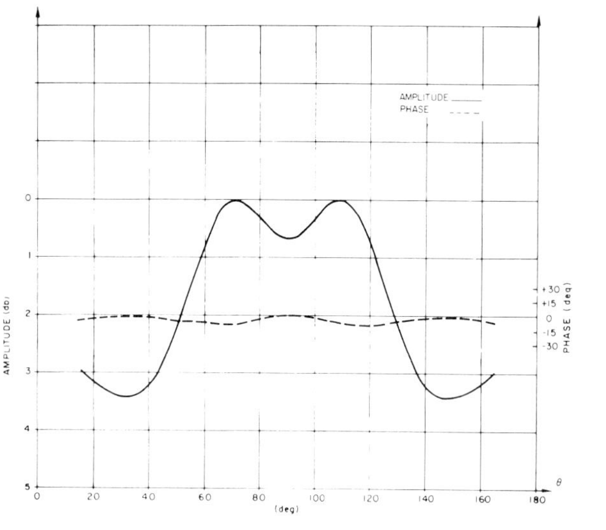

Fig. 4 Realized Gain and Phase Pattern, Elevation Plane, Center Ring Element θ Polarization.

ence with the pattern of an isolated slot on the cylinder, indicated for reference in Figure 12, is substantial. In fact the latter exhibits a very smooth behavior (except in the close neighborhood of $\theta = 0$ and $\theta = 180°$) which when the element is in the array environment the longitudinal periodicity of the structures causes two notches to be present in the pattern. Their physical explanation is qualitatively the same as in a planar array, being related to the presence of fast leaky wave excited on the structure, which interfere destructively with the direct element radiation (11). The realized gain pattern and the isolated slot patterns are much more alike for $\theta = \pi/2$ azimuthal plane (Figure 2 and 11). Figure 5 plots the θ and ϕ components of the realized gain pattern in the conical section $\theta = 45°$ and shows the interesting fact that approaching the shadow region the "cross polarized" component of the field becomes much higher than the θ component.

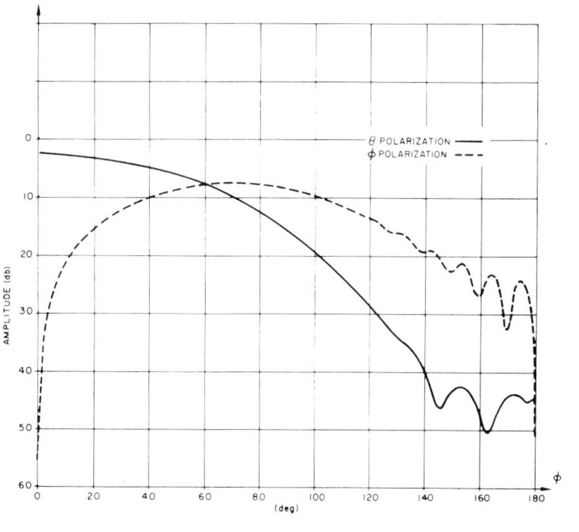

Fig. 5 Realized Gain Pattern, Section $\theta = 45^{o}$, Center Element, θ and ψ Polarizations.

*All the patterns are shown with their maxima normalized at the maximum of Figure 4 (taken as 0 dB level).

Figure 6 and 7 show the patterns in the planes $\theta = \pi/2$ and $\varphi = 0$ respectively for an element located in the lowest ring. In the elevation plane the pattern is substantially different from that of an element of the center ring. Also different is the relative magnitude of the θ and ϕ field component in the shadow region (Figure 8).

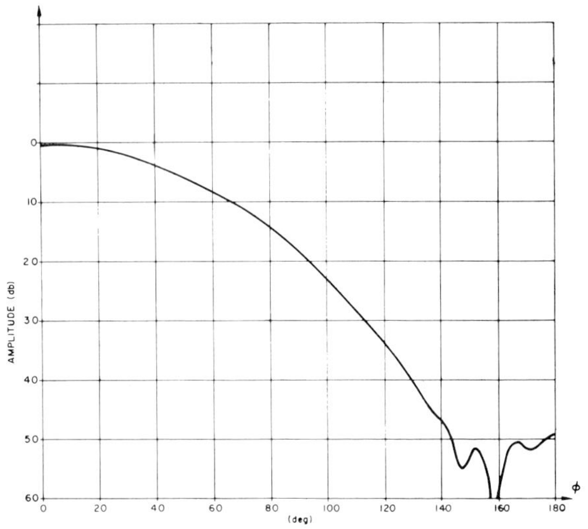

Fig. 6 Realized Gain Pattern, Azimuthal Plane, Lowest Ring Element θ Polarization.

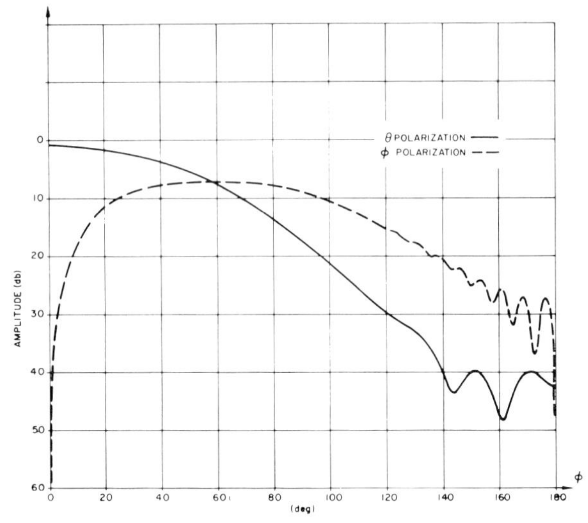

Fig. 8 Realized Gain Pattern, Section $\theta = 45^0$, Lowest Ring Element, θ and ψ Polarization.

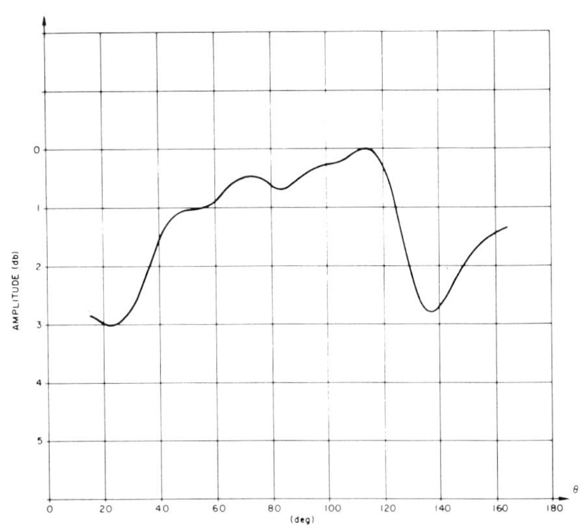

Fig. 7 Realized Gain Pattern, Elevation Plane, Lowest Ring Element, θ Polarization.

Figures 9 and 10 finally plot the azimuthal and elevation patterns for elements located in the fourth (or second) ring.

5. CONCLUSION

This paper has discussed a method of analysis of conformal arrays on surfaces having rotational symmetry. It has been shown that the matrix representing the coupling among the elements can be partially diagonalized by expressing in a suitable vector basis the column vector representing the array excitation. Thus, instead of the original problem a set of N "reduced" problems must be solved each consisting in finding the interaction phenomena among M equivalent elements. The solution is particularly simple for one mode elements as it has been shown by an example of a cylindrical array consisting of 5 rings each of 32 slots.

The technique presented here can be extended into different directions. The study of an infinite cylindrical array, periodic along the cylinder axis can be put in the

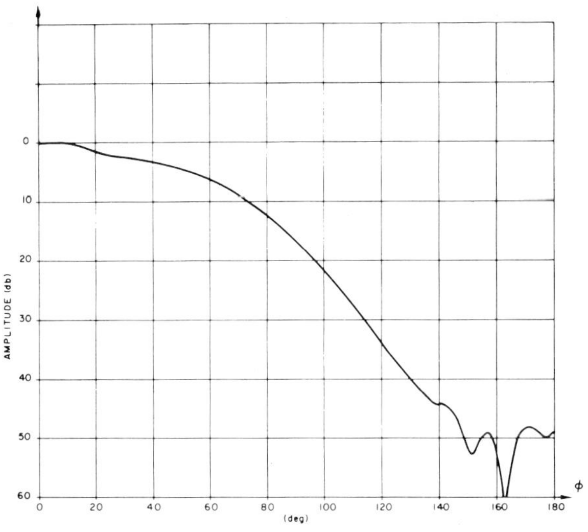

Fig. 9. Realized Gain Pattern, Azimuthal Plane, Fourth Ring (From the Top) Element, θ Polarization.

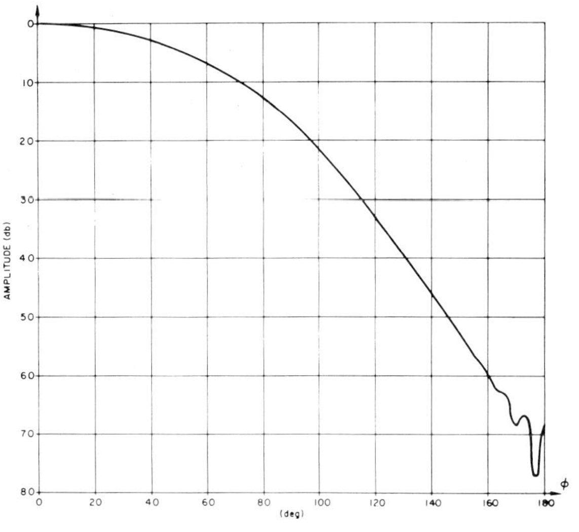

Fig. 11 Isolated Element Pattern, Azimuthal Plane, θ Polarization.

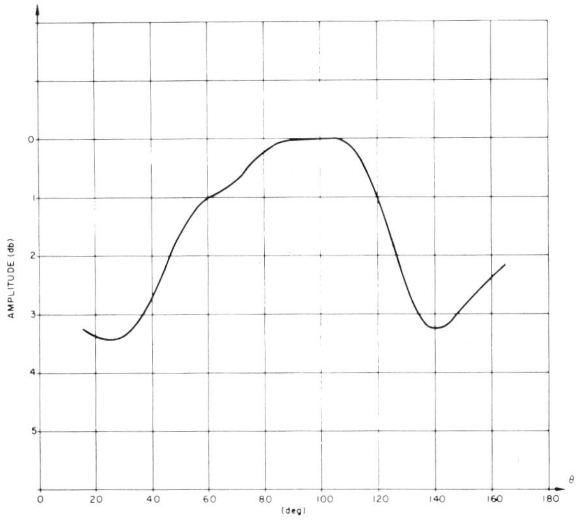

Fig. 10 Realized Gain Pattern, Elevation Plane, Fourth Ring (From the Top) Element, θ Polarization.

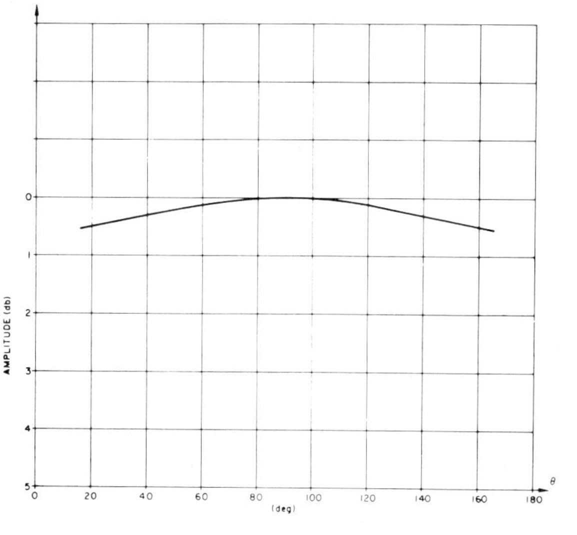

Fig. 12 Isolated Element Pattern, Elevation Plane, θ Polarization.

same theoretical fremework. The additional symmetry simplifies the problem making it easy to immediately diagonalize the submatrices (now of infinite order!) representing the scattering among rings. This model may prove to be useful for the study of large arrays, in the same way as the infinite periodic structure is a useful model for large planar array. On the other hand if the edge effect is wanted but the number of rings M is too large for the inversion of the various submatrices of the reduced problems, the diagonalization of these submatrices can be attempted through a perturbation technique, by using the results for the infinite structure as a zero approximation of the solution (12).

6. ACKNOWLEDGEMENT

The authors gratefully acknowledge the substantial help given by T. B. Dowling for the numerical computations.

APPENDIX 1. ORTHOGONALITY PROPERTIES OF THE FUNDAMENTAL EXCITATIONS

The orthonormalized eigenvectors of a cyclic matrix of Nth order, like $\underline{\underline{S}}^{st}$, are given by*:

$$\underline{e}(i) = \left\{ N^{-1/2} \exp\left(j2\pi ik/N\right) \right\} \quad (i, k = 0, 1 \cdots N-1) \tag{40}$$

whose components coincide with those of $\underline{m}(s, i)$ for p=s. The eigenvalues of $\underline{\underline{S}}^{st}$ are given by

$$\Gamma(i)^{st} = \sum_{k=0}^{N-1} S_{ok}^{st} \exp\left(j\, 2\pi ik/N\right) \quad (i = 0, 1 \ldots N-1) \tag{41}$$

i.e. are the discrete Fourier transform of the first row of $\underline{\underline{S}}^{st}$. Equation (12) is obtained from the very definition of eigenvector and from the orthonormality condition Equation (9).

The radiation patterns corresponding to excitations with different azimuthal numbers are orthogonal. This could be proven directly by writing the Fourier expansions of the field. However, the same result can be obtained in a simpler and more instructive way through conservation of energy considerations.

Denote by $\underline{a}(i)$ and $\underline{a}(q)$ two array excitations belonging respectively to the subspaces corresponding to the azimuthal numbers i and q. If $\vec{F}\left[\underline{a}(i); \theta, \phi\right]$ is the radiation pattern of the array when $\underline{a}(i)$ is applied to the structure we have the energy conservation equation:

$$\left|\underline{a}(i)\right|^2 - \langle \underline{a}^+(i) \underline{\underline{S}}^+ \underline{\underline{S}}\, \underline{a}(i) \rangle = \iint_{4\pi} \left| \vec{F}\left[\underline{a}(i); \theta, \phi\right] \right|^2 d\Omega \tag{42}$$

where the integration on the left side is on the entire solid angle 4π Similar equation is written for $\underline{a}(q)$. When the array is excited by both $\underline{a}(i)$ and $\underline{a}(q)$ simultaneously, conservation of energy leads to an expression containing cross term in $\underline{a}(i)$ and $\underline{a}(q)$. Taking into account Equation (42) and the fact that a(q) and a(i) are orthogonal (because of Equation (9)) the following equality is obtained:

$$\mathrm{Re}\left[\underline{a}^+(i)\, S^+ S\, \underline{a}(q)\right] = \mathrm{Re} \iint_{4\pi} \vec{F}^{\,*}\left[\underline{a}(i); \theta, \phi\right] \vec{F}\left[\underline{a}(q); \theta, \phi\right] d\Omega \tag{43}$$

If $\underline{a}(i)$ and $\underline{a}(q)$ are expressed as a linear combination of $\underline{m}(s, i)$ and $\underline{m}(s, q)$, respectively, by invoking Equations (9) and (12) it is immediately found that the left side of Equation (43) is zero. The orthogonality of the patterns is thus proved.

APPENDIX 2. MUTUAL ADMITTANCES BETWEEN RINGS

From Equations (25) and (27) the following expression is obtained for $\overset{mn}{Y}(i)$:

$$\overset{mn}{Y}(i) = \sum_{p=0}^{N-1} e^{-j\frac{2\pi ip}{N}} \sum_{q=0}^{N-1} e^{j\frac{2\pi iq}{N}} \frac{1}{N} \iint_{A_p^m} \hat{n} \times \vec{E}_{ot}^{m}\left(\phi - \frac{2\pi p}{N}\right) \cdot \vec{H}_{ot}^{n}\left(\phi - \frac{2\pi q}{N}\right) dA \tag{44}$$

*A brief discussion of this property in connection with conformal array theory is in Reference (3).

with the significance of the symbols given in section 3. Equation (44) can be transformed into:

$$Y_{(i)}^{mn} = \frac{1}{N} \iint_{A_o^m} \hat{n} \times \vec{E}_{ot}^m(\phi) \sum_{p=0}^{N-1} \sum_{q=0}^{N-1} \vec{H}_{ot}^n \left[\phi - \frac{2\pi}{N}(\phi - q)\right] e^{j\frac{2\pi i}{N}(\phi - p)} \tag{45}$$

Because of the periodicity:

$$\sum_{p=0}^{N-1} \sum_{q=0}^{N-1} \vec{H}_{ot}^n \left[\phi - \frac{2\pi(q-p)}{N}\right] e^{j\frac{2\pi i (q-p)}{N}} = N \sum_{q=0}^{N-1} H_{ot}^n \left(\phi - \frac{2\pi q}{N}\right) e^{j\frac{2\pi i q}{N}}$$

which, inserted in Equation (45), gives Equation (27).

APPENDIX 3. FIELD OF AN ISOLATED SLOT

The field external to the conducting cylinder can be expressed through two scalar potentials A and F (10)

$$\underline{H} = \nabla A \times \hat{z} + \frac{1}{j\omega\mu} \nabla \frac{\partial F}{\partial z} - j\omega\epsilon F \hat{z} \tag{46}$$

$$\underline{E} = \hat{z} \times \nabla F + \frac{1}{j\omega\epsilon} \nabla \frac{\partial A}{\partial z} - j\omega\mu A \hat{z} \tag{47}$$

where A and F can be represented in the form of a Fourier series and Fourier integrals (with respect to the ϕ and z coordinates respectively) (10):

$$A(\rho, z, \phi) = \frac{1}{2\pi} \sum_{n=-\infty}^{+\infty} e^{jn\phi} \int_{-\infty}^{+\infty} f_n(w) H_n^{(2)}\left(\rho\sqrt{k^2 - w^2}\right) e^{jwz} dw \tag{48}$$

$$F(\rho, z, \phi) = \frac{1}{2\pi} \sum_{n=-\infty}^{+\infty} e^{jn\phi} \int_{-\infty}^{+\infty} g_n(w) H_n^{(2)}\left(\rho\sqrt{k^2 - w^2}\right) e^{jwz} dw \tag{49}$$

The functions $f_n(w)$ and $g_n(w)$ are found in the standard way, by requiring the tangential electric field given by Equation (47) to be coincident with that of the slot (10). This requirement gives (for no $E\phi$ in the slot):

$$f_n(w) = \frac{j\omega\epsilon \frac{1}{R} \mathcal{E}\left(\frac{n}{R}, w\right)}{\left(k^2 - w^2\right) H_n^{(2)}\left(R\sqrt{k^2 - w^2}\right)} \tag{50}$$

$$g_n(w) = \frac{nw \frac{1}{R} \mathcal{E}\left(\frac{n}{R}, w\right)}{R^2 \left(k^2 - w^2\right)^{\frac{3}{2}} H_n^{(2)'}\left(R\sqrt{k^2 - w^2}\right)} \tag{51}$$

From Equation (46) to (51) the expressions (34) and (35) are obtained.

REFERENCES

(1) Borgiotti, G. V.: Mutual coupling effects on arrays of slots on a cylinder 1969 U.R.S.I. Spring Meeting - Washington, D.C. April 1969.

(2) Sureau, J.C. and Hessel, A.: Element Pattern for Circular Arrays of Axial Slits on Large Conducting Cylinder IEEE Transactions on Ant. and Prop. vol. AP. 17 n.6 pp 799-803 November 1969.

(3) Borgiotti, G.V. and Balzano, Q.: Mutual Coupling Analysis of a Conformal Array of Elements on a Cylindrical Surface, IEEE Trans. on Ant. and Prop. Vol. AP-18, No. 1, pp. 55-63, January 1970.

(4) Hansen, R.C.: Microwave Scanning Antennas Academic Press, New York 1966 ch 3, pp. 209 and foll.

(5) Harrington, R. F.: Matrix Method for Field Problems, Proc. IEEE vol. 55, pp. 136-149, February 1967.

(6) Borgiotti, G. V.: Modal Analysis of Periodic Planar Phased Arrays of Apertures, IEEE Proceedings, Vol. 56, No. 11, November 1968 pp. 1881-1892.

(7) Friedman, B.: Principles and Techniques of Applied Mathematics, Wiley, New York 1956, ch. 2.

(8) Wheeler, H. A.: The Grating-lobe Series for the Impedance Variation in a Planar Phased Array Antenna, IEEE Transactions on Ant. and Propagation, Vol. AP-14, pp. 707-714, November 1966.

(9) Borgiotti, G. V.: A Novel Expression for the Mutual Admittance of Planar Radiating Elements, IEEE Transactions on Antennas and Propagation Vol. AP-16, No. 3, May 1968, pp. 329-333.

(10) Harrington, R. F.: Time Harmonic Electromagnetic Field, McGraw Hill, New York, 1961, ch. 5.

(11) Knittel, G. H., Hessel, A. and Oliner, A. A.: Element Pattern Nulls in Phased Arrays and Their Relation to Guided Waves, IEEE Proceedings, Vol. 56, No. 11 pp. 1822-1836, November 1968.

(12) Borgiotti, G. V.: Edge Effects on Large Arrays of Uniform Slits on A Ground Plane, URSI Spring Meeting, Washington, D. C., April, 1970.

Realized Gain Function for a Cylindrical Array of Open-Ended Waveguides

J. C. SUREAU
Grumman Aerospace Corporation
Bethpage, N.Y.

A. HESSEL
Polytechnic Institute of Brooklyn
Farmingdale, L.I., N.Y.

ABSTRACT

An analysis is presented of the realized gain element pattern of rectangular-waveguide-fed axial slots on a large conducting cylinder. The charactertistic features of the patterns are described and their dependence on radius and spacings is discussed.

I. INTRODUCTION

Stacked ring arrays of waveguide-fed slots on a large conducting cylinder have the capability of constant performance through 360° in azimuth as well as pattern control in elevation. Alternatively, a portion of such a cylindrical surface may be used as a conformal phased array. The aperture design of such arrays requires, as in the case of a planar phased array, the knowledge of mutual coupling or the performance of a radiating element in a match-terminated array environment.

Previous work in arrays on conducting cylinders has been limited to the E and H plane patterns of infinite axial slits (1-4) and to the E-plane pattern of an in-phase excited column of a stacked ring array of rectangular-waveguide-fed apertures. (5) To the author's knowledge, there does not appear in the literature any previous work on the pattern of stacked ring array of aperture elements including mutual coupling. In particular, the role which curvature plays in the case has not been explicitly exhibited, although some indication has already been provided for infinite slits. (1, 2)

In this paper, the realized gain element pattern for an array of rectangular waveguide-fed axial slots (See Fig. 1) is presented. Formal expressions for this pattern are obtained via a modal approach. These expressions are approximate only to the extent that the shape of the aperture fields is assumed to be that of the dominant waveguide mode ("single mode" theory), and that the curvature over each aperture may be neglected. Numerical results are interpreted in terms of wave types whose individual dependence on curvature is clearly identified. Also given is a graphical procedure by which the main pattern features in the forward region are predictable.

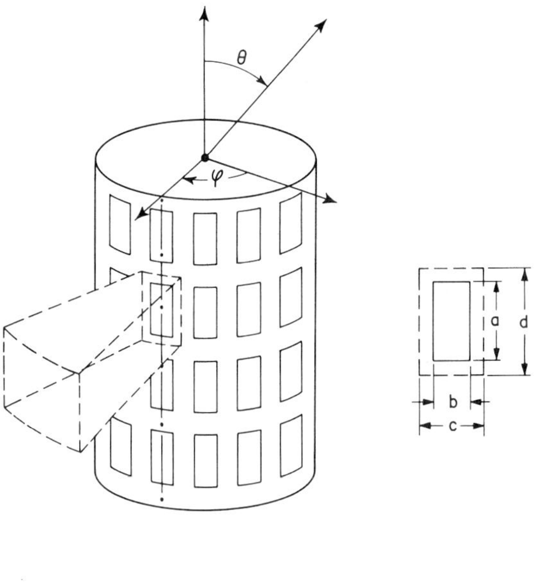

Fig. 1 Cylindrical Array Geometry and Unit Cell.

II. FORMAL SOLUTION AND ITS EVALUATION

It can be shown, by a simple extension of previous work on two-dimensional circular arrays (3), that the element pattern for such a cylindrical array with the geometry defined in Fig. 1, is formally given by:

$$\underline{\mathcal{E}}(\theta,\phi;r_o) = \frac{\beta}{2\pi}\cdot\frac{d}{2\pi}\int_{-\infty}^{+\infty} d\zeta \sum_{n=\infty}^{+\infty} \underline{E}(\theta,\phi;n,\zeta,r_o) \qquad (1)$$

where $\underline{E}(\theta,\phi;n,\zeta,r_o)$ is the dominant mode * radiated far field resulting from a linear phase taper excitation of the entire array, of the form

$$a_{pq} = e^{-j(n\phi_p + \zeta z_q)} \qquad \begin{matrix} p = 0, 1, \ldots, N-1 \\ q = 0, \pm 1, \ldots \end{matrix} \tag{2}$$

and is to be determined.

In (2), ϕ_p and z_q are the angular and axial coordinates, respectively, of the (p, q) lattice point.

It can similarly be shown that, except in the back lobe region, the infinite summation in (1) may be replaced by the integral resulting in

$$\underline{\mathcal{E}}(\theta,\phi;r_o) \sim \frac{a}{2\pi} \cdot \frac{d}{2\pi} \int_{-\infty}^{+\infty} d\zeta \int_{-\infty}^{+\infty} d\nu \, \underline{E}(\theta,\phi;\nu,\zeta,r_o) \tag{3}$$

where ν is now a continuous angular wavenumber.

For a given set of phasing parameters $\{\nu,\zeta\}$, the problem reduces to determining the field radiated by a rectangular waveguide into a sectoral waveguide whose walls are characterized by "phase shift" boundary conditions for all field components, i.e.

$$\Delta \text{ phase in } z \text{ per cell} = \zeta d$$

$$\Delta \text{ phase in } \phi \text{ per cell} = \nu \beta \,.$$

The determination of this radiated field requires the knowledge of the aperture field and in general involves the solution of a boundary value problem. An approximate solution is obtained here by assuming the shape of the aperture field to be that of the feed waveguide dominant mode (only H-type modes are thus excited) and by determining its complex amplitude from matching of complex power on either side of the aperture. This procedure yields the following expression for the active admittance $Y_a(r_o;\nu,\zeta)$ as "seen" by the dominant waveguide mode in the aperture plane:

$$Y_a(r_o;\nu,\zeta) = \frac{8\eta_o}{\pi^2} \frac{ba}{dc} \sum_{m=-\infty}^{+\infty} \sum_{n=-\infty}^{+\infty} \left| \frac{\sin \nu_m \alpha/2}{\nu_m \alpha/2} \cdot \frac{(\pi/2)^2 \cos \zeta_n a/2}{(\pi/2)^2 - (\zeta_n a/2)^2} \right|^2 Y'(r_o;\nu_m,\zeta_n) \tag{4}$$

where N_o is the free space admittance, and where the normalized external modal admittance is

$$Y'(r_o;\nu,\zeta) = G'(r_o;\nu,\zeta) + j B'(r_o;\nu,\zeta) = -j\frac{\kappa}{k} \frac{H_\nu^{(2)}(\kappa r_o)}{H_\nu^{(2)\prime}(\kappa r_o)} \tag{5}$$

with

$$\nu_m = \nu + \frac{2\pi}{\beta} m = \nu + k r_o m \frac{\lambda}{c} \,,$$

$$\zeta_n = \zeta + \frac{2\pi}{d} n \,,$$

$$\kappa = \sqrt{k^2 - \zeta^2} \,.$$

Plots of G' and B' for $kr_o = 100$ are shown in Fig. 2

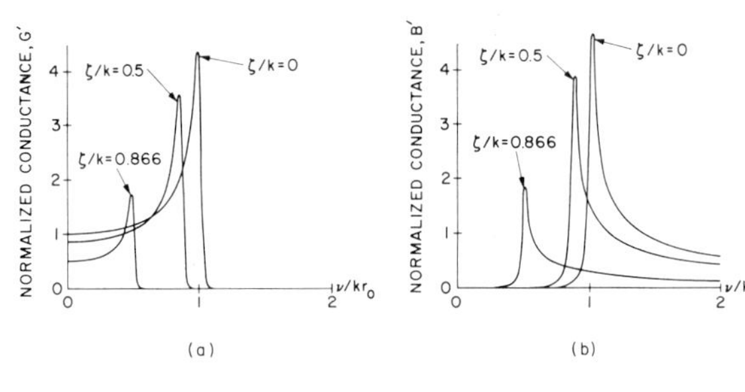

Fig. 2 H-type Mode Admittance Crater, $kr_o = 100$ (a) Conductance, (b) Susceptance

as a function of the normalized phasing parameters. In contrast to similar plots for planar arrays, G' and B' are continuous and finite everywhere. It has already been shown (1, 2) that the peaks in

---

* The more familiar expression

$$\underline{\mathcal{E}} = \int_{-\pi/d}^{\pi/d} d\zeta \sum_{n=0}^{N-1} \left\{ \sum_{p'=-\infty}^{+\infty} \sum_{q'=-\infty}^{+\infty} \underline{E}^{p'q'}(n,\zeta) \right\} ,$$

where $\underline{E}^{p'q'}(n,\zeta)$ denotes the $\{p'q'\}$ harmonic of the n,ζ Floquet mode, has been contracted to the simpler form (1) in which only the fundamental harmonic $\{0,0\}$ of the infinitely extended $(-\infty < n,\zeta < \infty)$ set of Floquet modes appears.

G' or B' are proportional to $\left(\eta r_o)^{1/3} \right)$

The modal series representation for Y_a given by Equation (4) is a weighted sampled summation of the cylindrical admittance crater and may be expected to have features similar to that for planar arrays, i.e. the grating lobe series. In particular, it will exhibit a dip (as opposed to a perfect null) wherever a sample point is such that:

$$\nu + \frac{2\pi}{\beta} m = \varkappa_n r_o = \sqrt{k^2 - (\zeta + \frac{2\pi}{d} n)^2}\, r_o \,. \tag{6}$$

The resulting input equivalent circuit for the aperture including a network selected so as to provide match at some value ν_o, ζ_o, is shown in Fig. 3. Defining the transmission coefficient by:

$$T(r_o;\nu,\zeta) = \frac{V^i(r_o';\nu,\zeta)}{V^{inc}} = \frac{2G_a(r_o;\nu_o,\zeta_o)}{Y_a^*(r_o;\nu_o,\zeta_o) + Y_a(r_o,\nu,\zeta)} \,, \tag{7}$$

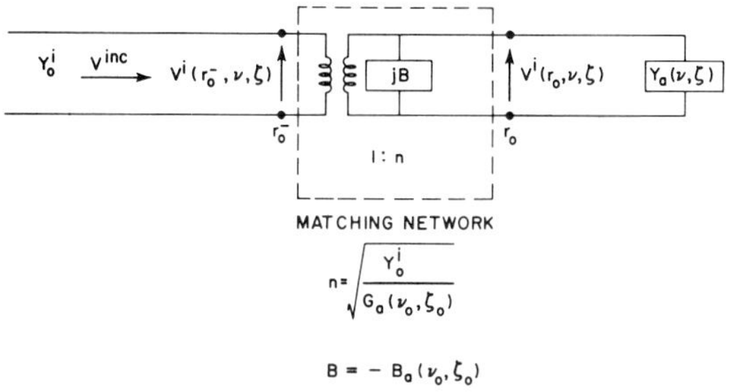

Fig. 3 Equivalent Network of a Unit Cell in a Cylindrical Array.

the far field radiated dominant mode which appears in (1) is:

$$E_\phi(\theta,\phi;\nu,\zeta,r_o) = -j\frac{V^{inc}}{r_o\beta}\sqrt{\frac{2}{\pi\varkappa r}}\, e^{-j(\varkappa r - \frac{\pi}{4})} \frac{V^e(\nu,\zeta)}{V^i(\nu,\zeta)} \sqrt{\frac{Y^i}{G_a(r_o;\nu_o,\zeta_o)}}\, T(r_o;\nu,\zeta) \frac{e^{-j[\zeta z + \nu(\phi - \frac{\pi}{2})]}}{H_\nu^{(2)}(\varkappa r_o)} \,, \tag{8}$$

where $V^e(\nu,\zeta) / V^i(\nu,\zeta)$ is the ratio of the aperture dominant mode voltages of the radial and feed waveguides and is given by

$$\frac{V^e(\nu,\zeta)}{V^i(\nu,\zeta)} = \frac{2\sqrt{2ab}}{d\pi} V(\nu,\zeta) \,, \tag{9}$$

with

$$V(\nu,\zeta) = \frac{\sin(\nu a/2)}{\nu a/2} \frac{(\pi/2)^2 \cos(\zeta a/2)}{(\pi/2)^2 - (\zeta a/2)^2} \,. \tag{10}$$

The element pattern can be recovered by substituting Equation (8) into Equation (1) or Equation (3). In the far field, the integration over ζ can be carried out first. The essential result is that, for a given angle θ only the spectral component $\zeta = k\cos\theta$ contributes. The radiated field expression Equation (3) becomes:

$$\mathcal{E}_\phi(\theta,\phi;r_o) = V^{inc} \frac{\sqrt{2ab}}{r_o \pi^3} \sqrt{\frac{Y^i}{G_a(r_o;\nu_o,\zeta_o)}} \frac{e^{-jkR}}{R} F(\theta,\phi;r_o) \,, \tag{11}$$

with

$$F(\theta,\phi;r_o) = \int_{-\infty}^{+\infty} V(\nu, k\cos\theta)\, T(r_o;\nu, k\cos\theta) \frac{e^{-j\nu(\phi-\pi/2)}}{H_\nu^{(2)}(kr_o\sin\theta)}\, d\nu \,. \tag{12}$$

A similar infinte sum (harmonic series) is also obtained, according to (1) and (3), by replacing the integral over ν in (12) by an infinite summation over n. The element gain function is related to the radiated far field by:

$$G(\theta,\phi;r_o) = 4\pi R^2 \frac{\eta_o |E_\phi(\theta,\phi;r_o)|^2}{Y_o^i |V_o^{inc}|^2} \,. \tag{13}$$

Substituting Equation (11) into Equation (13) after some manipulations yields:

$$G(\theta,\phi;r_o) = \frac{4\pi dc}{\lambda^2} \left| \frac{F(\theta,\phi;r_o)}{\pi k r_o} \right|^2 . \tag{14}$$

The evaluation of a suitable truncated harmonic series expression can be directly performed. Although this method possesses certain advantages* from purely computational viewpoint, it does not provide any basis for interpretations of the results, nor a clear link with similar results for planar arrays.

The virtues of asymptotic evaluation of the integral representation have already been demonstrated in the two dimensional case. (1-3) The integral of interest here is almost identical to that encountered before and the details of its evaluation will not be repeated. The essential result is that, <u>in the forward region,</u> the overall element pattern is the sum of a "space wave" or direct contribution, and "creeping waves" or diffraction contributions.

The space wave is given by:

$$F_{Sp.W}(\theta,\phi;r_o) \sim \pi k r_o V(\nu_s, k\cos\theta) T(r_o;\nu_s, k\cos\theta) \sin\theta\, e^{jkr_o \sin\theta\cos\phi} \tag{15}$$

where

$$\nu_s = kr_o \sin\theta \sin\phi \,.$$

As a consequence of its phase factor, it appears to originate at the element. Its amplitude is proportional to the transmission coefficient evaluated at values of the phasing parameters determined by the observation direction. This particular association is identical to that encountered in planar arrays, although it is not as readily apparent as in the latter case. It is interesting to note that, the matching condition $\nu_o = \zeta_o = 0$, corresponds to "broadside" match in the sense that the space wave is maximized at $\theta = 90°, \phi = 0$, and if it were the only contribution to Equation (14), would result in a maximum gain equal to the unit cell gain $\frac{4\pi dc}{\lambda^2}$.

The effect of curvature on this contribution is relatively mild and is due exclusively to the modification of the admittance crater. The space wave will therefore exhibit characteristic dips in precisely the same direction as in planar arrays and which can be predicted graphically from a "grating lobe" grid superimposed on an admittance crater displayed on a wave number plane, as illustrated in Fig. 4.

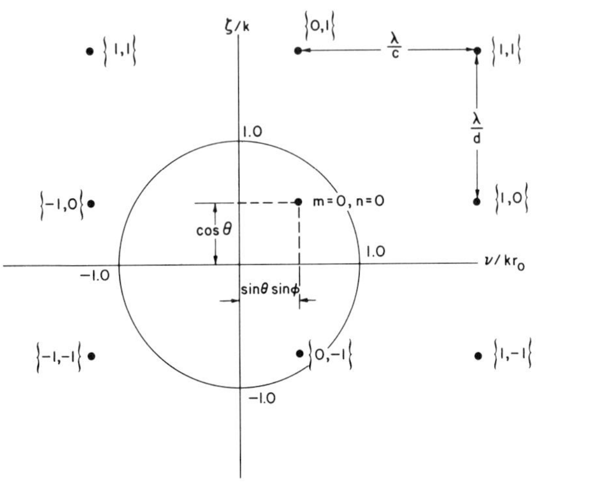

Fig. 4 "Grating Lobe" Grid Location on Wave number Plane.

For a given observation direction (θ, ϕ), the set of sample points, which characterize the associated active admittance and transmission coefficient, are determined by:

$$\frac{\nu_m}{kr_o} = \sin\theta\sin\phi + m\frac{\lambda}{c} \quad (m = 0, \pm 1, \ldots) \tag{16}$$

$$\frac{\zeta_n}{k} \doteq \cos\theta + n\frac{\lambda}{d} \quad (n = 0, \pm 1, \ldots) \tag{17}$$

Combining (16) (17) and (6), the angular location of the dips may be predicted from:

$$\sin\theta\sin\phi + m\frac{\lambda}{c} = \sqrt{1-(\cos\theta + n\frac{\lambda}{d})^2} \,. \tag{18}$$

Condition (18) may be graphically interpreted as corresponding to any one of the sample points (ν_m, ζ_n) lying on the edge of the circle,

$$(\nu/kr_o)^2 + (\zeta/k)^2 = 1$$

which in planar arrays, defines the edge of real space.

*Since the present discussion focuses on the forward (lit) region, the accuracy problems associated with harmonic series evaluation in the shadow region are much less severe.

For a given elevation angle θ, a variation in ϕ (conical cut) is reflected as a horizontal translation of the entire "grating lobe" grid of an amount $\Delta\nu/kr_o = \sin\theta \sin\phi$.

The main effect of curvature on the element pattern is due to the excitation of fast creeping waves which propagate around the cylinder and radiate away from the cylinder in the forward region. Fig. 5 shows the pro-

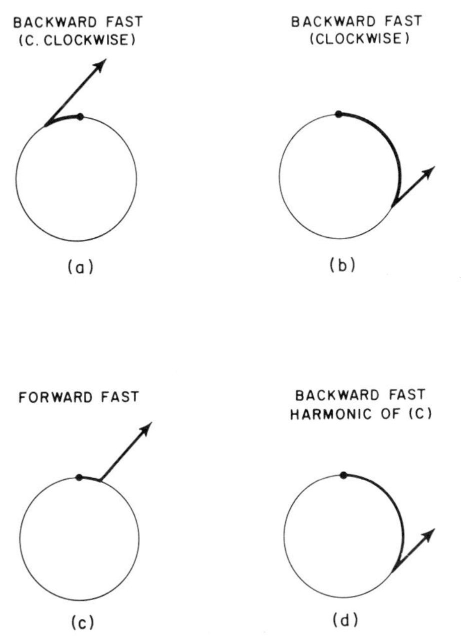

Fig. 5 Typical Contributing Creeping Rays.

jection on the r-ϕ plane of the ray paths followed by various creeping wave types; the actual trajectory includes a z-component (axial) of propagation along the cylinder for $\theta \neq 90°$. These creeping waves are formally, the contributions from poles of $T(r_o, \nu, \zeta)$ ("free" resonances of the structure) which are captured when the integration contour in Equation (12) over the real ν axis is deformed into the steepest descent path. (3) The space wave, on the other hand, corresponds to the saddle point contribution from the integral along the s.d.p.. These poles are a direct consequence of the curvature and double periodicity of the structure. The resulting creeping waves, which depending on their types, contribute in certain restricted areas of the forward region, interfere with the space wave resulting in a rippled beam. A detailed description of these contributions is beyond the scope of this paper. Some information has been previously provided for the two-dimensional case (1-3) and will be extended to the present case in a future publication. The main characteristics of these creeping waves, to the extent that they affect the pattern are as follows:

For a given conical cut, there is associated with each dip in the space wave pattern a creeping wave whose contribution is maximum in the vicinity of this dip resulting in substantial filling and distortion of the latter.

Each creeping wave ceases to contribute on one side of the dip and its amplitude decays exponentially on the other side, at a rate determined by $kr_o^{1/3}$, causing a ripple of progressively decreasing magnitude.

The regions in which they contribute are determined by the element cell size*, their relative excitation amplitude decreases as the radius increases and tends to vanish as $kr_o \to \infty$.

These creeping wave contributions account directly in a predictable fashion for the differences between the actual gain function, as computed from Equation (14), and the space wave contribution. This is clearly demonstrated by the numerical examples given subsequently.

III. NUMERICAL RESULTS

Typical results for the element gain function are shown in Figs. 6 through 9. The patterns shown are conical cuts (θ = constant), and the ordinate is the realized gain normalized to the unit cell gain expressed as a voltage ratio $\sqrt{\frac{G}{G_o}}$. Two sets of curves are generally shown: the solid ones are the actual patterns computed from the harmonic series expression while the dashed ones are the space wave contribution as given by Equation (15). Any differences between these two sets in the forward region are therefore, results of creeping wave effects.

In Fig. 6 the patterns corresponding to $\theta = 90°$ (principal E-plane) exhibit characteristics already encountered in the two-dimensional case. The dip in the space wave occurs near 42°, as predicted from

*and on the phase velocity of the wave (forward fast or backward fast).

the 0.6λ spacing, and the creeping wave effects increase away from broadside and disappear after the dip. The pattern corresponding to $\theta = 60°$ exhibits features which are not found in the two-dimensional case. The space wave experiences two dips, which are

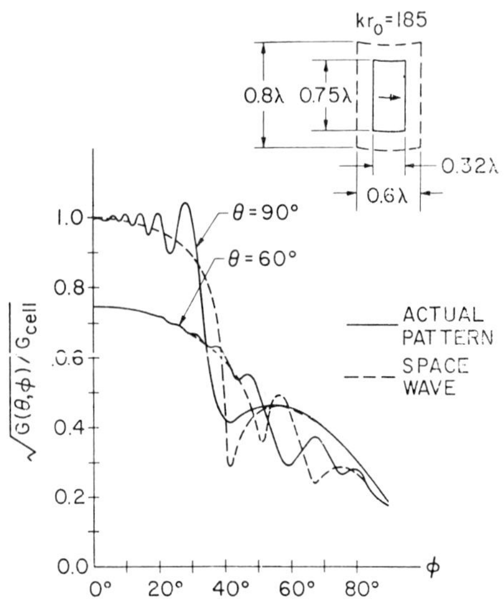

Fig. 6 Element Gain Function; kr_0 = 185, c/λ = 0.6

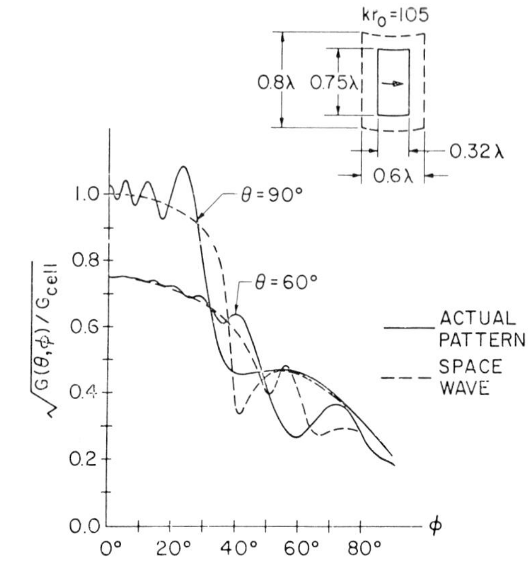

Fig. 7 Element Gain Function; kr_0 = 105, c/λ = 0.6

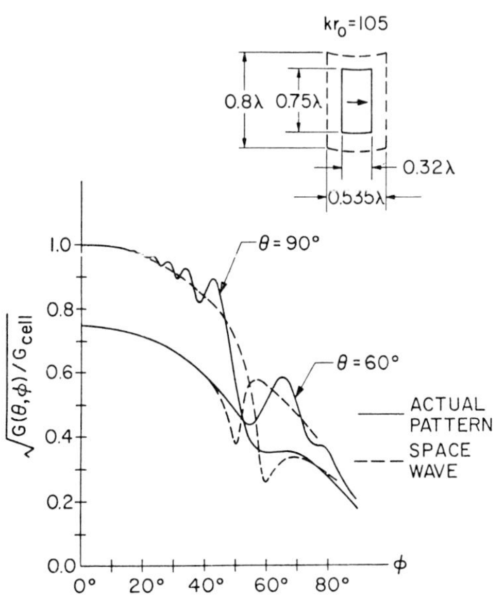

Fig. 8 Element Gain Function; kr_0 = 105, c/λ = 0.535

predictable from considerations of the wave-number plot. The first dip near 50° is attributed to a true elevation grating lobe (m = 0, n = -1). The second one at 68° is due to the cylindrical equivalent of an incipient azimuth grating lobe (m = -1, n = 0). Associated with these two dips is the formation of two different creeping waves. One creeping wave contribution increases away from broadside resulting in the rapid ripple as in the case 90° and disappears rapidly after 68°; the other begins near 50° and decreases away from broadside resulting in the slower ripple. In the region between 50° and 68°, the two interact with each other and with the space wave and cause substantial distortion of the pattern.

Patterns for the same cell and aperture size, but smaller radius are shown in Fig. 7. Generally speaking, the ripples are larger in amplitude and of lower frequency. The patterns shown in Fig. 8 correspond to a smaller E-plane spacing and differ in character from the previous ones in only one significant manner: there are no ripples near broadside for the $\theta = 60°$ conical cut. This is attributed to the fact that no azimuth grating lobe effect is present and the associated creeping wave, which would have caused the ripple, is not excited. (Only the elevation grating lobe dip and the corresponding creeping wave have remained).

The patterns shown in Fig. 9 for the case of

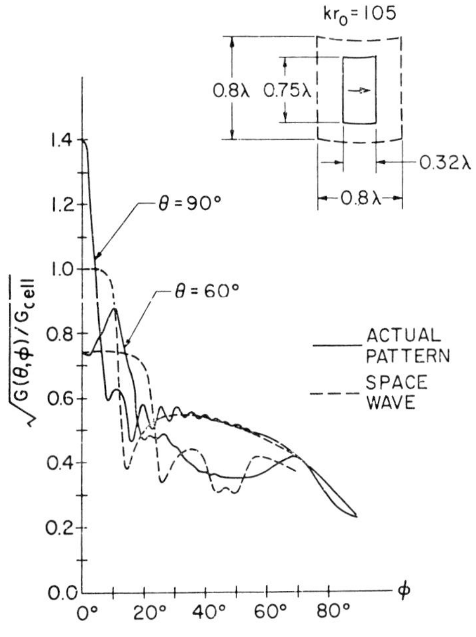

Fig. 9 Element Gain Function; kr_0 = 105, c/λ = 0.8

0.8λ E-plane spacing illustrate better than any of the previous examples the drastic effects which creeping waves can have on the element pattern. In particular they cause the peak gain to be 3dB above the unit cell gain and result in an unusually narrow beamwidth in the principal E-plane. For $\theta = 60°$, the space wave exhibits 3 predictable dips (two of them are almost coincident). In the actual pattern, these are almost totally smoothed out by the 3 associated creeping waves.

As a function of increasing elevation angle, the conical cut patterns tend to be broader and of lower amplitude; the creeping wave effects near broadside decrease and sometimes disappear, while others appear far from broadside.

IV. DISCUSSION

Although cylindrical and planar arrays are formally very similar, the results presented show that there can be significant numerical differences. In general, the perturbations introduced by curvature decrease as the radius increases, as expected, but increase with increasing spacing. Whether or not this effect is considered significant is of course very much dependent on the application. The principal source of these perturbations has been identified to be creeping waves induced by the periodicity of the structure. It is anticipated that, because of their rapidly-varying nature, their contribution to the overall pattern of an array is more difficult to control than that of the space waves; they may therefore ultimately limit the synthesis possibilities.

Although it has not been discussed here, it should be mentioned that the phase variation in the element pattern may be a significant consideration in non-planar arrays since the elements point in different directions and do not have a common phase. The principal source of this effect is the phase variation of the space wave and is most pronounced in the vicinity of the pattern dips.

As in planar arrays, the experimental design of an element, may be carried out with the help of waveguide simulation.* Since this method does not provide directly the dominant effect of curvature, the simulation may just as well be carried out as if the array were planar without significant errors.

Of all the modifications introduced by curvature, perhaps the most striking is the fact that, even for moderately large cylinders and unit cell size less than 1λ, the peak gain can be very much different from the unit cell gain.

From the results presented it is now also possible to extrapolate to element grids other than rectangular. For the forward region, the grating lobe plot associated with a particular element grid, when superimposed on the admittance crater displayed in the wave number plane, provides a direct graphical means of predicting the broad pattern features (space wave) as well as the basic nature of the creeping waves.

V. ACKNOWLEDGMENTS

The research reported in this paper was sponsored in part by the U.S. Army Advance Ballistic Missile Defense Agency under Contract No. DAHC 60-69-C-0110.

REFERENCES

1. Sureau, J.C. and Hessel, A.: Circular Arrays

*In planar arrays, the element gain function in the direction corresponding to the particular simulator may be deduced directly by this technique. However, in the cylindnrical case, the results obtained in a valid simulator are related only to the space wave contri bution.

of Axial Slits on Large Con ducting Cylinders, Trans. Antennas and Propagation, Vol. AP-17, No. 6, November 1969, pp. 800-803.

2. Sureau, J.C. and Hessel, A.: Element Pattern in Circular Arrays of Waveguide-Fed Axial Slits on Large Conducting Cylinders, to be published.

3. Hessel, A. and Sureau, J.C., Element Pattern of Arrays on Cylindrical Surface, Proceedings of the Conformal Array Antenna Conference, Naval Electronics Laboratory Center, San Diego, California, January, 1970.

4. Borgiotti, G. V. and Balzano, Q.: Mutual Coupling Analysis of a Conformal Array of Elements on a Cylindrical Surface, IEEE Trans. Antennas and Propagation, Vol. AP-18, No. 1, January 1970.

5. Gladman, R. B.: Mutual Coupling on Cylindrical Array Antennas, Proceedings of the Conformal Array Antenna Conference, Naval Electronics Laboratory Center, San Diego, California, January 1970.

Basic Theoretical Aspects of Spherical Phased Arrays

H. E. SCHRANK

Westinghouse Electric Corp.
Baltimore, Maryland

ABSTRACT

An array of radiating elements which lie on the surface of a sphere provides a natural antenna configuration for obtaining hemispherical coverage with nearly identical high-directivity beams. This paper presents the results of an analytical study of the beam forming properties of a spherical array consisting of thousands of circularly polarized elements on a sphere over 20 wavelengths in diameter. Elements are distributed by means of an icosahedron geometry, resulting in the best possible uniformity of element spacing, which varies less than ±9 percent from the mean value of 0.76 wavelength. Computed scalar diffraction patterns for various conditions of element excitation are presented in terms of normalized patterns. In particular, the effects of how much of the sphere is active for each beam are presented, along with the effects of element pattern shapes and amplitude distributions. It is shown that an active cap of 140° subtended angle can be used to obtain directive beams with moderately low sidelobes (-20 to -23 dB) over hemispheric coverage regions. Also, spherical arrays allow the use of significantly larger element spacings before grating lobes become objectionable, as compared with spacings required in planar or cylindrical arrays. The methods for generating the element excitation distributions are not within the scope of this paper.

SPHERICAL PHASED ARRAYS

INTRODUCTION

Recent interest in spherical arrays (1-3) for achieving hemispherical antenna coverage is based on the natural ability of spherical arrays to provide good uniformity of pattern and gain over wide angular regions. Other array configurations, as illustrated in Figure 1, suffer from beam degradation as the beam is steered over wide angular regions. For example, when multiple planar arrays are used to obtain hemispheric coverage, Knittel (4) has shown that at least 3 planar apertures or faces are required, and each face must be capable of steering the beam to an angle 63° from broadside. This is shown in the coverage diagram on the left side of Figure 2, and results in beam-

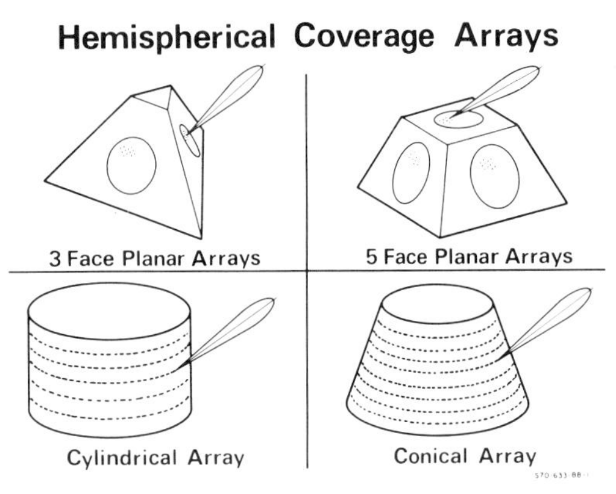

Fig. 1 Hemispherical Coverage Arrays.

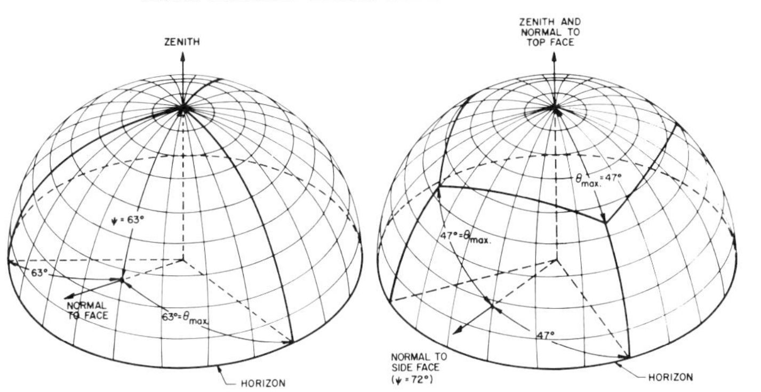

Fig. 2 Scan Requirements for Multi-planar Arrays, (a) for 3-Face Array (left), (b) for 5-Face Array (right); after Knittel, reference (4).

width broading of 2.2 with a gain reduction of 4.1 dB at the maximum scan angle of 63°. This beam degradation can be reduced by using more than 3 planar faces. For instance, a 5-face configuration as shown in the upper right of Figure 1 (with its corresponding coverage diagram on the right side of Figure 2) reduces the maximum steering angle to 47°, but still has a 1.9 dB gain variation at maximum scan.

Cylindrical or conical arrays can reduce beam variations in azimuth, but still suffer beam variations in the elevation scan direction. When these pattern and gain variations are unacceptable, a spherical array can be used to cover the hemisphere with practically identical beams, provided the array is large in terms of wavelength.

This paper discusses an analytical study of the basic beam forming properties of a spherical array, consisting of thousands of circularly polarized elements on a sphere many wavelengths in diameter ($D > 20\lambda$).

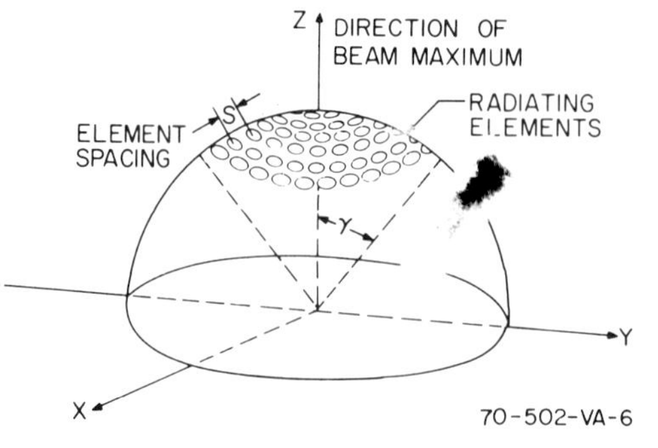

Fig. 3 Spherical Array.

The object of the basic portion of the study reported in this paper was to determine how the far-field patterns of a spherical array are affected by the following factors:

(a) The portion of the sphere active for each beam

(b) Element pattern function

(c) Element spacing

(d) Amplitude distribution

Computed scalar diffraction patterns for various conditions of element excitation are presented in terms of normalized patterns showing beamwidth factors and near sidelobes.

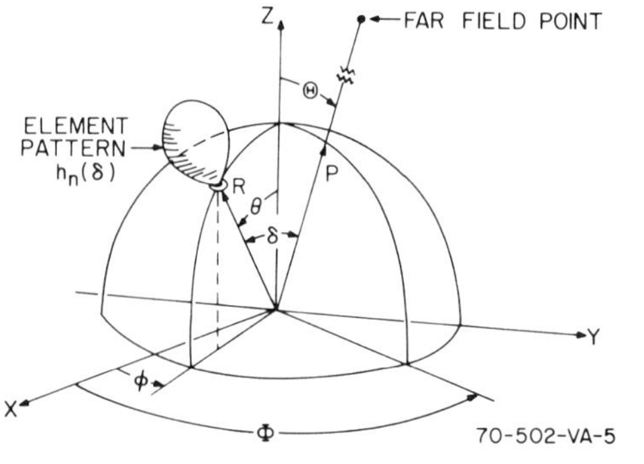

Fig. 4 Coordinate System.

ELEMENT DISTRIBUTION AND PATTERN

In computing scalar diffraction patterns, the coordinate system shown in Figures 3 and 4 was used, and the array elements were located by means of an icosahedron geometery shown in Figure 5. The derived distribution of elements resulted in a near-optimum uniformity of element spacings, which vary less than ± 9 percent from a mean value of $S = 0.73$ wavelengths. This variation can be shown to be the minimum possible for large numbers of elements ($N >> 12$).

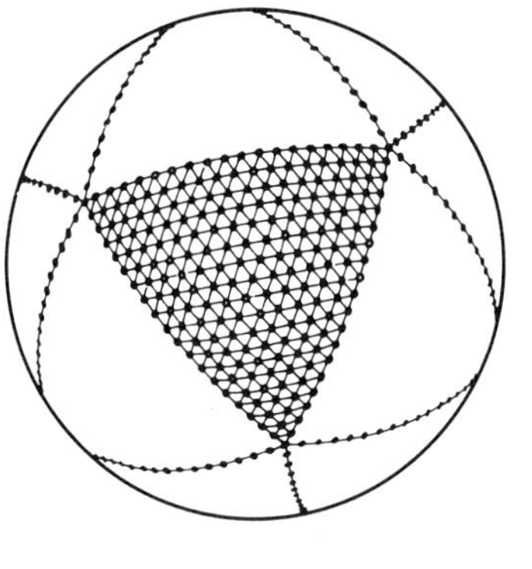

Fig. 5 Icosahedron Geometry of Element Locations.

Each circularly-polarized element was assumed to have a rotationally-symmetrical radiation pattern about its axis as indicated by the pattern function $h(\delta)$ in Figure 4.

BASIC EQUATIONS AND SIMPLIFYING ASSUMPTIONS

The basic equation used for scalar diffraction computations is in terms of the resultant far-field voltage intensity, given by

$$E(\Theta, \Phi) = \sum_{1}^{N} A_N(\theta)\, h_N(\delta) \exp j\left(\varphi(\theta) + \pi D (\cos\delta)/\lambda\right)$$

Where Θ, Φ = field point coordinates
θ, ϕ = element coordinates
N = number of active elements
D = sphere diameter
λ = wavelength
$A_N(\theta)$ = amplitude distribution
$\varphi(\theta)$ = phase distribution
$h_N(\delta)$ = element pattern function

Although Figures 3 and 4 show only half a spherical array, it should be understood that hemispherical angular coverage requires more than half a sphere to radiate uniform beams in all directions. However, for any one beam direction, the active portion of the sphere is less than half (an active cap of half-angle $\gamma = 70°$ was chosen, as discussed below). Also, for the basic computations covered in this paper, only zenith-directed beams were computed (i.e. in the direction of the Z axis of Figure 3), since all beams are assumed to be essentially identical for large values of N (an assumption which was later verified).

For all the basic patterns discussed in this paper, an ideal element phase distribution was used.* That is, all elements were phased to produce an equi-phase front normal to the beam axis. Thus, the phase distribution used was:

$$\varphi(\theta) = -\pi D (\cos\theta)/\lambda$$

where the minus sign denotes lagging phases for all elements above the equator, which was taken as the reference phase, i.e. $\varphi(90°) = 0$.

Some other simplifying assumptions were made in the computations, such as using axially symmetric element patterns with ideal circular polarization, and no mutual coupling effects. Also assumed is that non-uniform amplitude distributions are symmetrical about the beam direction, i.e. the Z-axis.

The computation of scalar diffraction rather than vector diffraction patterns represents a simplifying assumption, in that it assumes all element contributions at the far-field point have their electric fields in the same spatial direction, i.e. are co-polarized. Thus, the resultant field is taken as a phaser summation, disregarding any differences in polarization between contributions. This, together with the assumption that all elements are ideally circularly polarized in all directions, may be considered the most restrictive assumptions in this analysis. Nevertheless, the results are informative and show some interesting basic characteristics of spherical arrays.

COMPUTED PATTERNS

Reference patterns were first computed for assumed isotropic elements with uniform amplitude distribution and ideal phase excitation, as a function of how much of the spherical array is active. Under these conditions, an entire hemisphere ($\gamma = 90°$) of elements results in a pattern shown in Figure 6. It has a half-power beamwidth factor

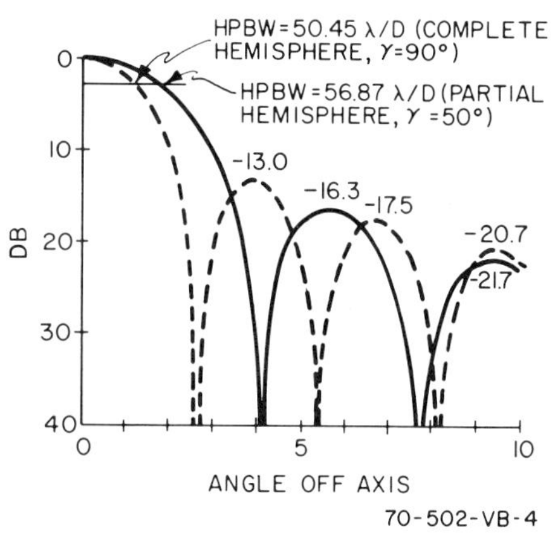

Fig. 6 Normalized Spherical Array Patterns (for isotropic elements and uniform amplitude).

of 50.45 degrees per wavelength of array diameter, and first sidelobes at -13 dB. When less than the hemisphere is active, the beamwidth broadens and the sidelobes reduce, as shown for a typical case, in Figure 6; i.e. an active cap subtending a half-angle of $\gamma = 50°$.

Similar computations for other cap angles are summarized in Figure 7 showing first sidelobe

*in subsequent computations, not covered in this paper, various systematic and random phase errors were introduced to determine their effects.

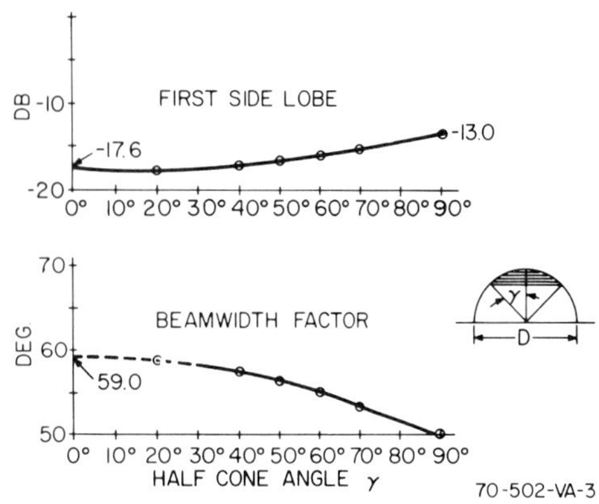

Fig. 7 Variation of Pattern Parameters with Active Cap Size (for isotropic elements and uniform amplitude).

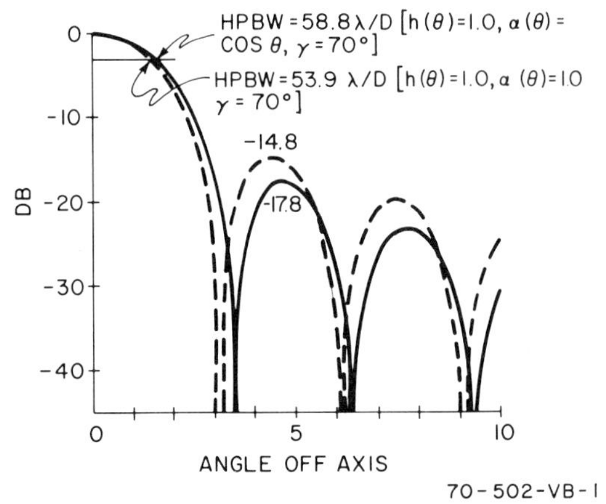

Fig. 8 Effect of Cosine Amplitude Taper (Taper = -9.3 dB at Edge, Isotropic Elements).

levels and beamwidth factors as a function of active cap angle, γ. It is noteworthy that as the active portion of the hemisphere is reduced toward the limiting case of a nearly planar disc of elements, the beamwidth factor and sidelobe level approach those of a first-order Bessel function pattern, namely $59\lambda/d$ and -17.6 dB respectively.

From the effects studied above, an active cap angle of 140° ($\gamma = 70°$) was selected as a reasonable compromise between maximum pattern directivity and minimum number of active elements. For the remainder of the study, this parameter was held constant while other effects were determined.

The effects of a tapered amplitude distribution are shown in Figure 8, where both patterns are for an active cap angle of 140° ($\gamma = 70°$) and isotropic element patterns ($h(\theta) = 1.0$), but the solid pattern represents a cosine amplitude distribution, and the dashed reference pattern represents uniform amplitude. As expected, the tapered amplitude broadens the main beam and lowers the sidelobes.

Of greater interest are far-field patterns computed for more realistic element patterns. From empirical measurements on circular waveguide elements in partial spherical arrays, element patterns were synthesized using a three term cosine series, i.e.

$$h(\delta) = A + B\cos\delta + C\cos^2\delta.$$

Using this form of element pattern with an amplitude taper of -7.5 dB at the edge of the active cap, a typical "realistic" pattern is shown in Figure 9. The

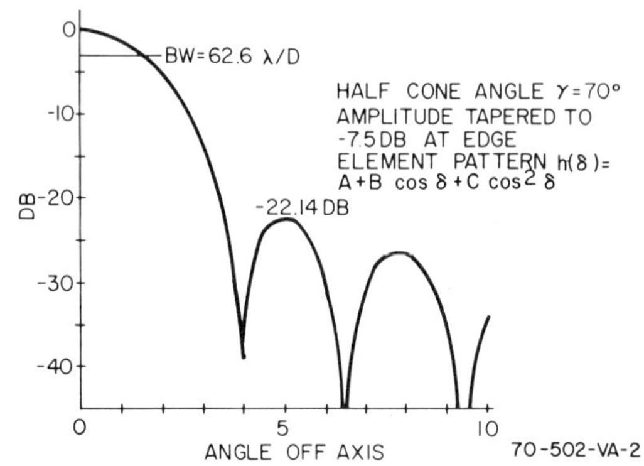

Fig. 9 Spherical Array of Circular Waveguide Elements.

beamwidth factor for this case is 62.6 degrees per wavelength of active diameter, and first sidelobes are at -22.1 dB. This represents about the best sidelobe level to be expected from a spherical array with "simple" (easily obtained) amplitude distributions.

A typical pattern representing the same conditions as in Figure 9, but extending to wider angles, is shown in Figure 10. This pattern again shows a first sidelobe level in the order of -22 dB, and well-behaved sidelobes out to 90°. It was calculated for a spherical array 24 wavelengths in diameter, and represents the most realistic element pattern within the assumptions of this study.

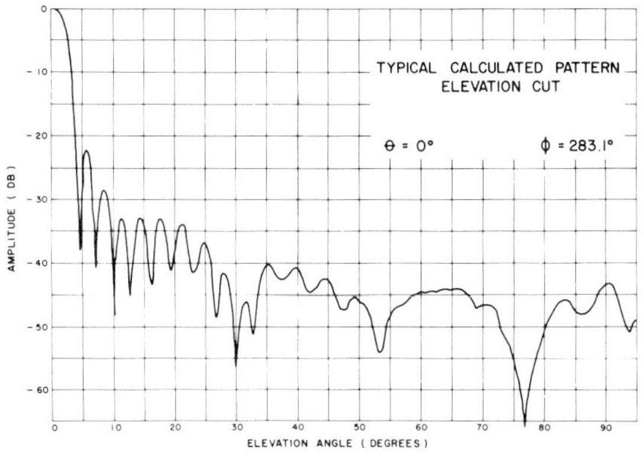

Fig. 10 Typical Computed Pattern Showing Wide-Angle Sidelobes.

GRATING LOBES

Also of interest is the dependence of pattern characteristics on element spacing, which was studied by computing wide-angle far-field patterns for various array diameters, with all other conditions held constant. When the diameter is such that the average element spacing is one wavelength, a broad grating lobe appears at an angle of approximately 75° from the main beam, as predicted from theory. This is illustrated in Figure 11, which shows patterns computed out to 90°

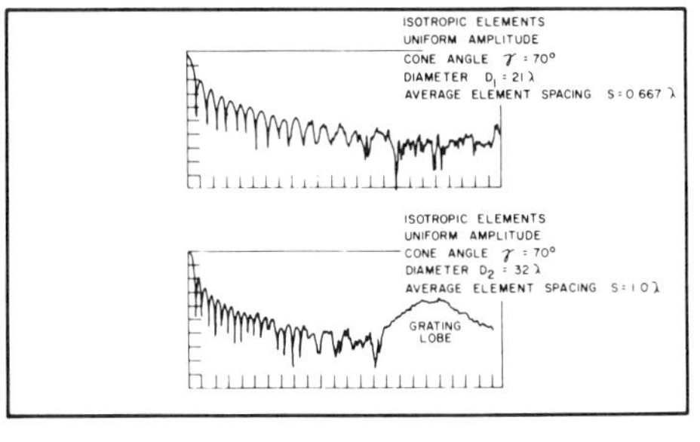

Fig. 11 Calculated Patterns for Grating Lobe Study (First Sidelobes at -15 dB, Grating Lobe Center at 75°, Peak -27 dB).

from the beam center for two diameters, namely $D_1 = 21\lambda$ and $D_2 = 32\lambda$. Although the grating lobe appeared at the predicted angle, its amplitude is 27 dB below the main lobe, which is much lower than grating lobe levels obtained from planar or cylindrical arrays for corresponding element spacings. This represents a distinct advantage in favor of spherical arrays, i.e. wider element spacings can be used before secondary maxima become objectionably high.

CONCLUSIONS

For systems requiring wide-angle or hemispheric coverage with uniform beams, the spherical array has distinct advantages over cylindrical or multi-planar arrays. Large numbers of elements can be located on the sphere with nearly uniform spacings using an icosahedron geometry, and average element spacings can approach one wavelength before grating lobes become objectionable. An active cap portion of the sphere subtending a cone-angle of 140° seems to be a reasonable compromise between maximum directivity and minimum active elements per beam. With moderate amplitude tapering to the edges of the active region (7 to 10 dB), half-power beamwidth factors of about 62 degrees per wavelength of active diameter, and first sidelobe levels of -22 dB are typically obtainable.

REFERENCES

1. Sengupta, D. L. et al: Radiation Characteristics of a Spherical Array of Circularly Polarized Elements, IEEE Trans. Antennas and Prop., Vol. AP-16, Jan. 1968, pp. 2-7.

2. Chan, Ishimaru, and Sigelmann,: Equally Spaced Spherical Arrays, Radio Science, Vol. 3 No. 5, May 1968, pp. 401-404.

3. Chan and Sigelmann: Experimental Investigation on Spherical Arrays, IEEE Trans. Antennas and Prop., Vol. AP-17 May 1969, pp. 348-349.

4. Knittel, G. H.: Choosing the Number of Faces of a Phased-Array Antenna for Hemisphere Scan Coverage, IEEE Trans. Antennas and Prop., Vol. AP-13, Nov. 1965, pp. 878-882.

Evaluation of Large Phased-Array Antennas

CARL BLAKE

Lincoln Laboratory
Massachusetts Institute of Technology
Lexington, Massachusetts

L. SCHWARTZMAN
and F. J. ESPOSITO

Sperry Gyroscope Division
Great Neck, New York

ABSTRACT

Antenna measurements to evaluate the performance characteristics of large phased arrays are discussed as to their merits. Various diagnostic and monitoring techniques which provide information as to the state of the antenna are described.

INTRODUCTION

The far field radiation pattern of an antenna is usually the standard vehicle which characterizes the electrical performance of the antenna. Since a phased array antenna can form thousands of beams, a complete set of radiation pattern data will not only cost an exorbitant amount but will also inundate the user with data. In addition, when the phased array is a large immovable structure, the ordinary type of radiation pattern, a rotational pattern, is costly and difficult to obtain. To minimize the cost, and the amount of data that is necessary for a valid judgment of performance, several techniques have been proposed. These techniques will be described in this paper. Also described are other pattern measurement methods that are or have been employed and various diagnostic and monitoring techniques which provide information as to the state of the antenna.

FAR FIELD PATTERNS

There are two equally valid types of far field radiation patterns. One is the conventional rotational pattern which is defined in this paper as a "beam pattern." (1) The other, which has come into use with large immovable phased arrays is called an "antenna pattern." (2) (3) The "beam pattern" is obtained by forming a single beam from the antenna and moving a probe through its entire field. The "antenna pattern" is obtained by fixing the probe position and electronically scanning the array beams through its field of view. These are illustrated in Fig. 1. A full set of one type defines the other.

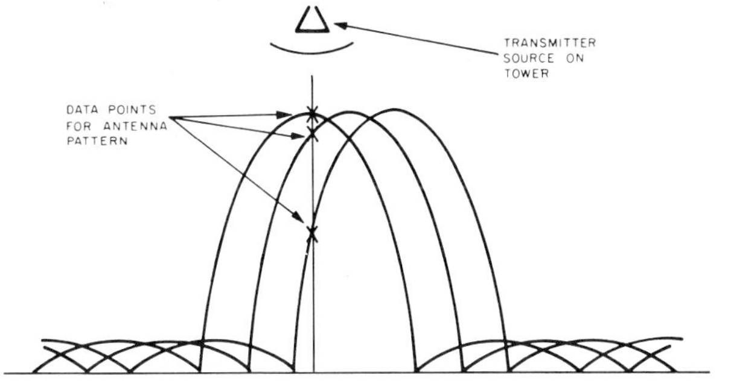

Fig. 1 Antenna Pattern.

The manner by which the data is collected for both or for either is explained in the following paragraphs.

BEAM PATTERN

As stated above, this is the more conventional far-field radiation pattern. With the large immovable phased array an airplane with a transmitter is usually employed to fly through the field of view. The positional data of the airplane is supplied either by an optical tracker or another radar. This data must be time tagged and communicated to the antenna under test to correlate the received signal strength with angular position.

The test is quite expensive since the airplane is usually required for many days in order that a sufficient angular sector is covered. In addition, the data from day to day must be correlated as well as the time tags. This has led, in many cases, to misleading or erroneous data.

ANTENNA PATTERNS

These patterns which are similar in appearance to beam patterns can be collected in a number of different ways. One way is to employ an instrumented test tower which should be an integral part of almost every phased-array antenna system. Data obtained in this manner is the easiest and least expensive to come by. In addition to providing information for the initial evaluation, this "antenna pattern" can be repeated and used for day-to-day monitoring.

Another way this data can be collected is by the use of a reflective balloon. This balloon can either be illuminated by the antenna under test or by another radar. One or two way pattern data can be obtained.

NEAR FIELD MEASUREMENT

Stanford Research Institute has proposed that the far field radiation pattern be measured by focusing the aperture to a point in the near field. Again both "beam patterns" or "antenna patterns" can be obtained. To measure the "beam pattern" will still present large problems. However, no correlation with another radar or utilization of an aircraft would be required. Rather, a large boom can be placed on the array and rotated. "Antenna patterns" for various source angles are possible by adjusting the boom to different angular positions.

Some measurements have been made by SRI on the Electronic Scanning Pencil Beam Antenna (4) (ESPBA) at the Sperry Great Neck plant. The initial results indicate the validity of the technique.

APERTURE PROBING

At the National Bureau of Standard studies and tests have been made of a measurement method for probing the aperture field (5). The far field radiation pattern is then found from the transform of the aperture field. For the method to be valid, complete knowledge of the aperture field is necessary.

The aperture field is measured with the use of a carriage upon which are mounted two orthogonal probes. The amplitude and phase of each is measured as the probe is moved along the aperture. The aperture sample points are separated by one-half wavelength for each beam position. By synchronizing the probe motion with the beam switching a number of aperture distributions corresponding to different beam positions may be sampled. The technique is still in its infancy.

DATA PRESENTATION

For an array of approximately 4000 elements spaced a half wavelength apart in a circular aperture the antenna beamwidth is approximately 30 millisines. To obtain a valid set of data, the far field sample points should be separated no greater than a fifth of the beamwidth. This would result in approximately 105 pieces of data per beam with only a small fraction in the near-inside lobe region.

Depending upon who is going to use the data, a number of different graphical presentations exist. One method is to present a contour map, such as shown in Fig. 2.

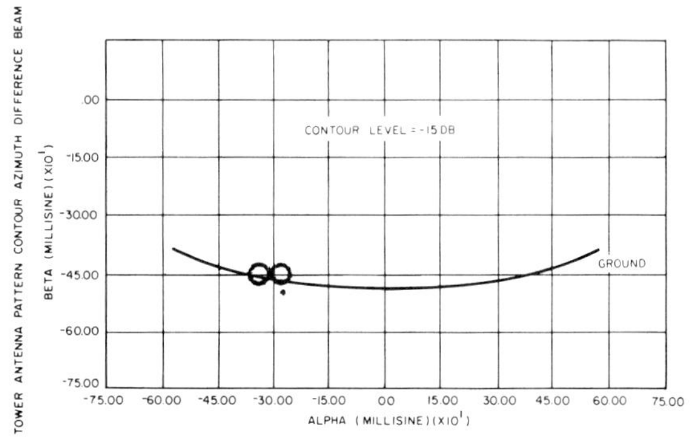

Fig. 2 Contour Plot.

Such maps are of little use to the system user. On the other hand, the statistics of the side lobes could be presented as shown in Fig. 3.

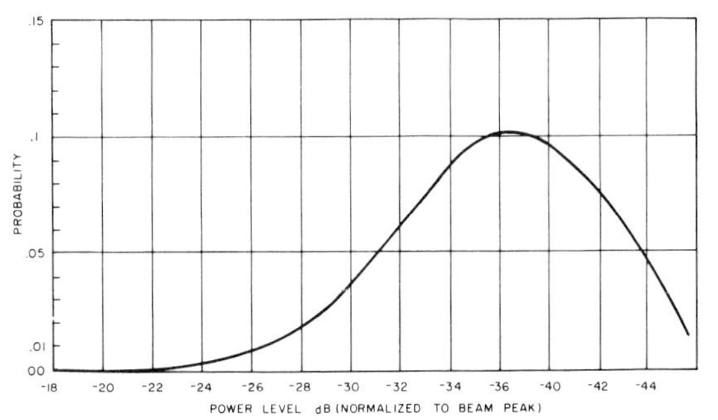

Fig. 3 Sidelobe Probability Density Curve.

Here the data indicates to the system user what the probability is that a certain side lobe will occur. A third method is to plot the main lobe and near in side lobes. This is of interest because it is this presentation which can be most readily correlated to analytical predictions.

AUTOMATED MONITORING AND FAULT FINDING TECHNIQUES

In general, phased arrays contain a large number of elements each of which requires an associated phase shifter and driver circuit. Due to the extremely large number of components employed, economical automated fault finding techniques are mandatory in order to facilitate array maintenance and to establish the operational status of the driver and phase shifter circuitry. Two such techniques which have been implemented on existing phased arrays are:

*A real time status monitoring

**A maintenance system check.

Real Time Status Monitoring (6) - This technique is utilized to continuously test for phase shifter failures during system operation. A particular driver circuit is addressed by a switching matrix which allows the phase shifter diode voltage to be sampled by a special test circuit. The switching matrix is addressed by the computer, which also controls the switching commands (forward or reverse bias) at the input to the driver circuit. The test circuit reads back into the computer, via the Beam Steering Unit, a fault signal, which when related to the switching command indicates the type of malfunction (shorted or open circuit ed condition in a driver circuit or phase shifter circuit). During one system repetition rate period, one element is undergoing a particular test. Therefore, the time to conduct a complete test of all phase shifters and drive circuitry is the product of the number of elements and number of tests per element divided by the system repetition rate.

Maintenance System Check (6) - This technique is employed to provide an RF check of all the phase shifters in the array. The technique is described in the following paragraph.

A transmitter on a test tower directs a CW signal towards the array. With the antenna beam in a fixed direction, the phase shift of one element of the array is varied between values, for example, between 0° and 45°, at a rapid rate. Timing for this switching is provided by a clock in the Beam Steering Unit. This switching of a phase bit on one element generates both an amplitude and phase modulation of the sum signal from all the elements of the array. These amplitude and phase components are digitized by the phase shift test unit and transmitted to the system computer. In the computer, the components are analyzed and used to compute the phase shifts of the tested element. In addition to providing phase information, the test determines shorted or open diode phase shifter conditions and indicates faulty phase shifters when excessive loss occurs. A complete checkout of all 3 bits on a 2165 element array using this technique required 16 minutes.

*Utilized on Hard Point Defense Array Radar (HAPDAR)

**Technique developed at Raytheon by O. Lowenschuss and R. Arsenault, and at Sperry by D. John and D. Stettin.

LIST OF REFERENCES

(1) ARPA Project HAPDAR(U), Final Engineering Report, Sperry Gyroscope Company for Army Missile Command, May 1967, ARPA Work Order NR516, Contract DA-01-021-AMC-11169(Z).

(2) ARPA Project HAPDAR(U), Special Report on Antenna Pattern Measurements, Sperry Gyroscope Company for Army Missile Command, May 1968, Contract DAAH01-68-C-0115.

(3) A. I. Sinsky, "The AN/FPS-85 Test Results" (U), Fifteenth Annual Tri-Service Radar Symposium Record, 17-19 June 1969, Secret.

(4) Electronic Scanning Pencil Beam Antenna (U), Design Plan Technical Report ECOM-CO672-2. Sperry Gyroscope Division, Sperry Rand Corporation for United States Army Electronics Command, December 1967. Contract DA AB07-67-C-0672, Confidential.

(5) ARPA Project Defender, Annual Report Radar Standards Program, National Bureau of Standards, NBS Report 9200, March 1966, ARPA Work Order 515.

(6) ARPA Project HAPDAR(U), Operation and Maintenance Manual for Hard Point Demonstration Array Radar (U), Sperry Gyroscope Company for Advanced Research Project Agency Department of Defense, April 1966. Contract DA-01-021-AMC-11169(Z), Secret.

Limited Scan Arrays

W. T. PATTON

Radio Corporation of America
Missile and Surface Radar Division
Moorestown, New Jersey

Abstract

Limited scan arrays fill the gap between fixed beams antennas and phased arrays of conventional design. Antenna systems capable of electronic beam deflection from a few beamwidths to the order of twenty degrees are discussed.

I. INTRODUCTION

One of the primary missions of instrumentation radars over the past 15 years has been to acquire and track a single target within a small, well-defined sector. The majority of radars developed for this objective were pedestal mounted paraboloid devices, emphasizing probability of detection and acquisition and high precision tracking. More recently, requirements in this area have been expanded to include multiple target clusters contained within sectors much larger than the coverage of a single radar beam. Although the number of targets to be handled and their dispersion have expanded enormously, the requirements for precision tracking have not changed, nor are these requirements restricted to space-object tracking and re-entry missions. Midcourse evaluation air and AICBM defense, fire control systems, and air traffic control systems all demand more in terms of complete hemispheric coverage and multiple target tracking.

The obvious answer is the four-faced fixed phased array system which fulfills all the requirements and includes advanced information processing subsystems to handle as much traffic as the radar can see. The immediate problem that arises is that the cost of one such system is very high; the cost of replacing all existing pedestal mounted paraboloid tracking radars is unthinkable. The compromise, which combines the flexibility of electronic scanning with the basically sound precision pedestal mounted antenna, may prove to be the only truly satisfactory solution to the problem.

Various approaches to the design of an economical electronic scanning antenna system have been conceived and investigated. The different approaches have been determined largely by the amount of scan required of the antenna system. Early investigations involved the paraboloidal antenna with a laterally displaced feed. Ruze (1) has shown that deflection of 10 beamwidths or more, depending upon the f/D ratio of the parabola, can be obtained. The most serious limitation of this approach is the coma sidelobe, which grows rapidly to greater than 10 dB, and the aperture blockage sidelobes occuring when a large cluster of feed horns is installed in the feed region to provide multiple beams.

Beam deflection by lateral feed displacement in the dual reflector cassegrain antenna system was studied numerically by Nihen and Kay. (2) Scan angles much in excess of about 2.5 degrees were not found to be practical in that study.

More recently, an array feed for a single reflector system has been investigated by C. Winter. (3) This technique, illustrated in Figure 1, is limited to scan angles of the order of 10 beamwidths or less. One of the more interesting features of this approach is the apparent absence of comatic aberation which is the bane of the displaced feed techniques. The scanning array feed, like the multiple point feed, is characterized by large blockage ratios and, hence, high sidelobe (of the order of 10 dB) unless an asymmetrical offset feed design is utilized.

The dual reflector configuration, Figure 2, using a parabolic subreflector illuminated by a small phase array has been investigated by W. Fitzgerald. (4) This shares freedom from coma lobe with the single reflector system. It is limited to scan angles much less than 5 degrees and to main beamwidths much less than one degree.

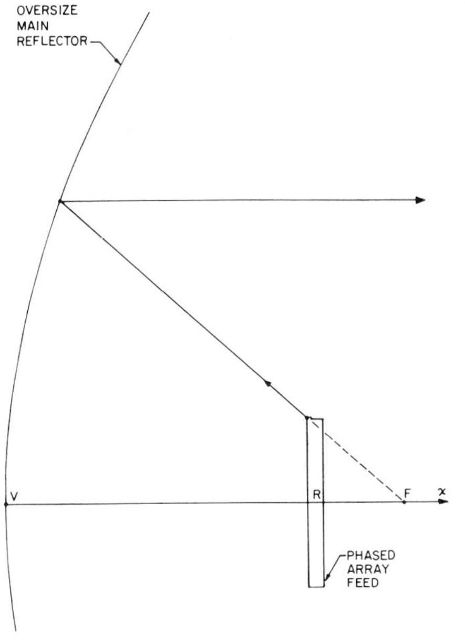

Fig. 1 Electronic Scan single reflector antenna.

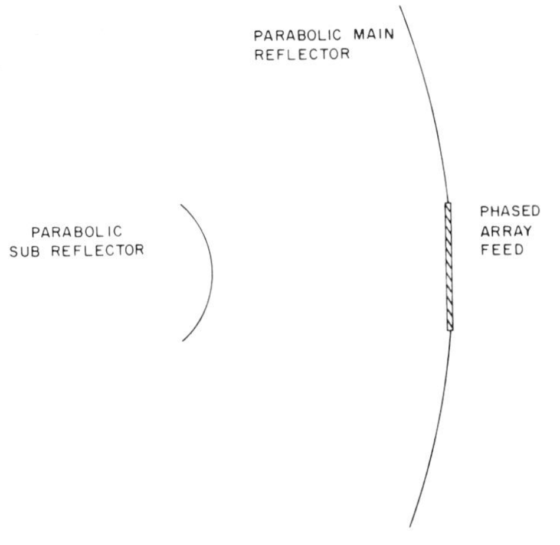

Fig. 2 Electronic Scan double reflector antenna.

When a limited scan range of more than a few degrees is required, an array of the type investigated by W. Patton (5) illustrated in Figure 3 may be used.

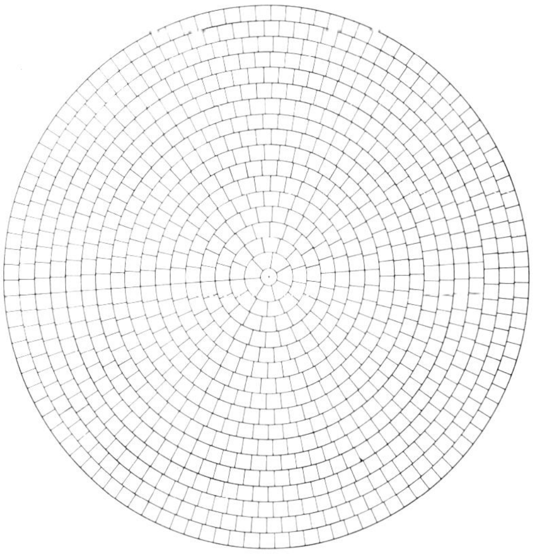

Fig. 3 Circular grid consisting of 18 concentric rings and 1000 antenna elements.

This configuration utilizes electrically large elements to reduce the number of phase shifters required for scanning. It is the size of these elements that fundamentally limits the angle of scan. The scan angle is generally taken to be the half power beamwidth of the component elements.

Grating lobe formation is avoided in limited scan arrays of this type by using an aperiodic arrangement of elements. The array in Figure 3

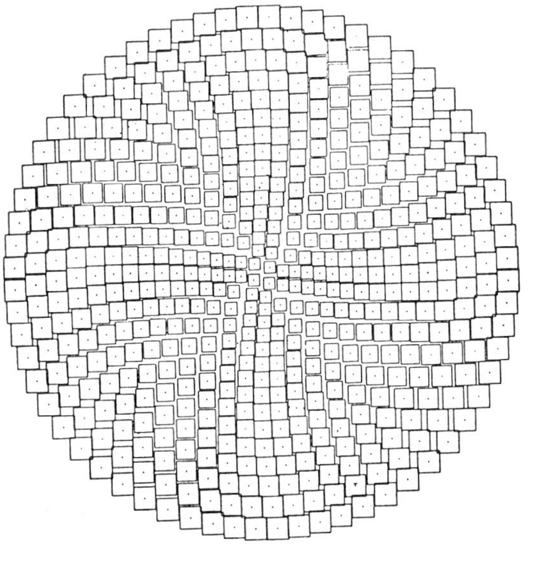

Fig. 4 Aperiodic arrangement of unequal size elements.

accomplishes this with a circular grid of equal area elements. Other techniques for achieving an aperiodic element grid include the use of elements of varying sizes suggested by L. Mansky (6) as shown in Figure 4 and the use of random groupings of smaller elements.

II. SCANNED REFLECTOR SYSTEMS

Single Reflector Displaced Feed

The use of lateral feed displacement to obtain beam agility in a reflector system may not fit strictly within the context of limited scan arrays, but it is never-the-less interesting as an example of the early quest for beam agility and as an indication of some of the problems encountered in scanned reflector systems. Problems shared by all limited scan designs are aperture efficiency, gain or gain variation with scan, and sidelobe level. Some definition of limit on these parameters must be established to determine the limit of scan capability. John Ruze (1) working with an optimum feed at each displaced feed location fully illuminating the main reflector determined that a 1 dB gain loss relative to the normal beam together with a 10.5 dB sidelobe due to coma and aperture blockage would define for him the limit of scan. On this basis he gives for the number of beam widths, N_B, maximum scan by lateral feed displacement

$$N_B = 0.44 + 22\,(f/D)^2. \tag{1}$$

Using this criterion with an f/D ratio of 0.5 reflector could be scanned about 6 beamwidths while a relatively flat reflector with an f/D ratio of 1.0 can be scanned about 22 beamwidths away from the normal beam position. This shows that, for this case at least, scanning capability is enhanced by long focal distances. Substantially larger scan angles can be obtained using 3 dB gain loss as the limit criteria. Such a system would be particularly attractive for satellite applications where earth coverage is required but where sidelobes particularly those falling off the earth, are not of great concern.

The feed displacement technique can be placed more nearly in the context of limited scan arrays by considering a cluster of feeds with electronic switching between them to achieve beam agility. This is accomplished at the considerable pain of increased aperture blockage. Unlike most other limited scan techniques there is little sharing of power among the several elements of the feed cluster and these, together with the switching matrix, must operate at substantially full system power.

Double Reflector Displaced Feed

An investigation of lateral feed displacement in a folded (cassegrain) reflector system was reported by Nihen and Kay (2). This investigation covered a large range of reflector sizes, f/D ratios, and main to subreflector diameter ratios. Unlike the single-reflector system the scan limit is not expressible in beam widths, but rather a practical scan limit of about 2.5 degrees independent of beam width exists for this system. It is much more difficult to maintain a fully effective illumination under scan conditions with the dual-reflector system since there is a tendency for the illumination to "walk off" the edge of the main reflector as the feed is displaced.

Single Reflector Array Feed

An interesting example of the use of an array to feed a reflector was reported by C. Winter (3) in 1968. This technique shown in Figure 1 differs from most of the others considered here in that he does not make full use of the main reflector for any beam position. The main content of the illumination function moves about on the main reflector as it is scanned. Thus the intrinsic aperture efficiency depends upon the number of beamwidths maximum scan for the design as well as the f/D ratio and other design parameters but the gain of the system remains substantially constant within the scan limits. Some of the trade-off considerations associated with this design are illustrated in Table I extracted from reference 7. This table shows that for a given maximum scan angle the effective aperture decreases significantly as the f/D ratio increases but the angle subtended by the main reflector also decreases and fewer radiating elements are needed to effect the required scan angle.

TABLE I

Trade off for 8° Scan capability with 9% blockage ratio

f/D	Effective Aperture	Angle Over Useful Aperture	Angle Over Full Aperture
.4	.28	46.9	74.8
.5	.25	37.3	65.0
.67	.18	----	----

The number of elements required for pattern control in this design depends on the ratio of the array diameter to the useful aperture on the main reflector (blockage ratio) and the element spacing. The element spacing is chosen to provide control of the illumination over the main reflector. Early experiments (8) provided control only over the useful aperture. This was shown to be inadequate since large sidelobes appear at angles well away from the main beam. This has been attributed to radiation outside the controlled sector being partially collimated by the over-sized reflector. This side lobe level varied from about -17.5 dB for the normal beam to approximately -8 dB for a beam direction near maximum scan. It is clear from this experiment that it is necessary to control the illumination of the entire reflector, not just the useful aperture, to minimize sidelobe degradation at the scan limits. More elements are required in the feed with the exact number depending upon sidelobe level allowable. The angular sector of half angle θc over which an array with element spacing d can provide illumination control is given by

$$\sin \theta_c = \frac{\lambda}{2d,} \tag{2}$$

The experiment described above (8) used a 14 foot diameter reflector with a focal length of 69.7 inches operating at 10.25 GHz. The phased array feed was 30 inches in diameter and consisted of 988 elements arranged on a square grid with 0.73λ spacing. The system was designed to scan a 0.7° beam 7° off the normal axis. The array feed caused 9 percent blockage on the normal beam and provided an intrinsic aperture efficiency (aperture utilization factor) of .36. About 10 dB gain loss relative to a uniformly illuminated aperture was observed in this experiment with about 6 dB attributed to aperture efficiency, 1 dB to blockage, and the remaining 3 dB to spillover and other factors.

Double Reflector Array Feed

A phased array feed in a double reflector system using paraboloids for both main and sub-reflector, Figure 2, has been investigated by W. Fitzgerald. (4) This configuration employs a fully illuminated main reflector for the normal beam. Progressively larger spillover is permitted as the beam is scanned. This configuration avoids some of the problems of element spacing found by Winter in the single reflector system through the magnification inherent in the double reflector system. Fitzgerald found that the double reflector is practical only for very large apertures where maximum scan angles up to 5° could be achieved. An obvious advantage of this configuration is the location of the weight of the phased array feed near the vertex of the main reflector. This will provide for a much smaller turning moment than can be achieved with an array feeding a single reflector.

III. LARGE ELEMENTS ARRAYS

Aperiodic Arrays

An aperiodic element grid is attractive for limited scan requirements because it permits the use of larger elements and hence fewer elements than a regular periodic grid designed for the same scan requirements. Figure 5 shows a comparison between required element areas in an aperiodic grid

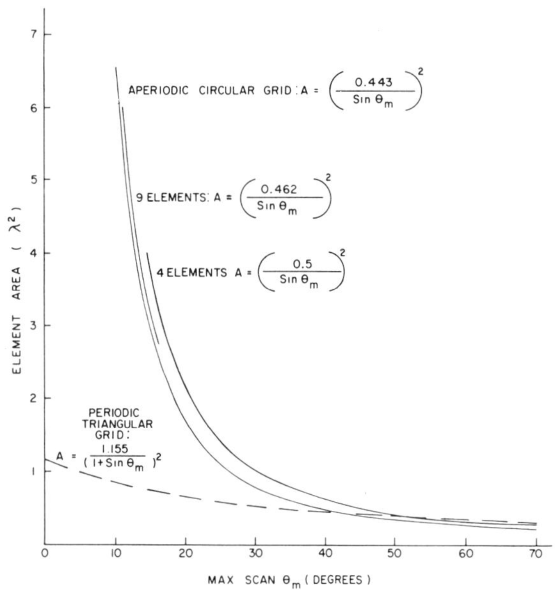

Fig. 5 Comparison of array element areas for a given maximum scan requirement using a regular periodic grid or an equal area aperiodic circular grid.

and those in the most efficient triangular periodic grid. It is clear that the aperiodic grid has a significant advantage for maximum scan angles less than 20° while no advantage exists for maximum scan angles beyond 40° . For maximum scan angles between 20° and 40° the choice is not clear and other considerations may well favor the regular grid.

In phased arrays of conventional design, the radiating elements are arranged neatly in rows and columns on a regular rectangular grating. When elements on such a grating are phased to produce a beam in a direction, θ_s a second beam occurs at the angle, θ_L, given by

$$\sin\theta_L = (\lambda/d) - \sin\theta_s \quad (3)$$

where d/λ is the period of the grating i.e., the fractional wavelength spacing between elements of the array. When the array forms this second beam or grating lobes its gain may be reduced by as much as 3 dB. Perhaps more serious than the loss of power, however, is the fact that the radar can receive signals from two different directions with no way to distinguish between them. If the period of the grating is less than one-half wavelength, there will be no grating lobe for any scan angle, but when the period is greater than one wavelength, at least one grating lobe will be present for any scan angle. The spacing between elements on a regular periodic grating should be between one-half and one wavelength depending upon the angle of maximum scan or the field of view of the array.

Within the constraint of a regular periodic grating, the number of elements required in the array can vary by at most four to one as the field of view varies from one array beamwidth to a maximum of 120 degrees. By employing an aperiodic circular grid, the problem of ambiguous grating lobes is eliminated and elements much larger than one square wavelength in area can be used. The constraint on element size in an aperiodic grid is that the element beamwidth equal the electronic scan field of view of the array. Under this constraint, the width of a uniformly illuminated element is given by

$$W/\lambda = 0.44/\sin\theta_m \quad (4)$$

where θ_m is the angle of maximum scan. With this design criterion, the gain at maximum scan is 3 dB below the gain at broadside. When smaller elements are used, the gain at maximum scan will be higher; however, the design criterion given above will result in the maximum gain at maximum scan for a fixed number of elements.

Although the array is globally aperiodic, Figure 3, it is approximately periodic locally, with a period equal to the element size. The local periodicity produces a vestigial grating lobe. The power level of the vestigial grating lobes varies with the total number of elements in the array approximately as 10/N. The element pattern will appreciably reduce the level of the vestigial grating lobe except when the array is scanned to the limits of the field of view of the element.

The maximum element area in the circular grid array is determined by the scan coverage requirements and the allowable gain degradation at maximum scan. The total array area is determined by the broadside gain. The ratio of these areas determines the total number of elements, and from this the vestigial grating lobe level can be estimated. If the level of the vestigial lobes is excessively high, more elements of smaller area may be used. This reduction in elements of smaller area may be used. This reduction in element area will tend to increase the gain at the scan limits and to decrease the effective level of the vestigial lobe at the scan limit.

In the circular grid array, it is possible to use either elements of equal area or elements whose area increases with radius (Figure 4). As long as the same number of elements is used to cover the same total aperture area, the performance of the two approaches will be equivalent. However, when elements of nearly equal area are used in the circular grid, only three different sizes of square elements are required to cover the array with a minimum of uncovered area.

An indication of the effectiveness of an aperiodic

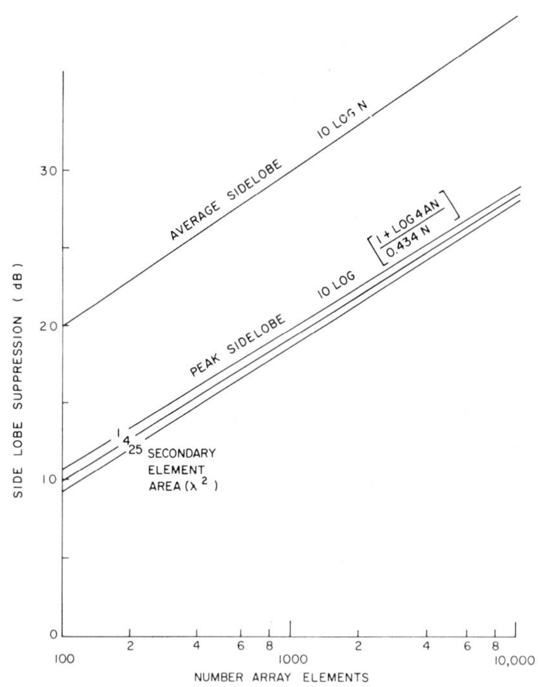

Fig. 6 Expected vestigial grating lobe level produced by an aperiodic array of radiating elements.

location of elements in suppressing the vestigial grating lobe is given in Figure 6. This figure uses the theory of YT Lo (9) derived for randomly located array elements.

Element Efficiency

The design of the radiating element is especially critical to good performance by the array. This is so because the total aperture efficiency of the antenna is the product of the aperture efficiency of the array and the aperture efficiency of the large radiating elements. Thus, it is important that the radiating element have as high a total efficiency as possible.

The aperture efficiency of an antenna can be defined as the ratio of the maximum gain of the antenna to the maximum gain of a uniformly illuminated antenna having the same aperture. It can be shown that the aperture efficiency is given by the ratio of the square of the average illumination intensity to the average power density across the antenna aperture.

$$n_a = \frac{\left[\frac{1}{A}\int_A\int f(x,y)dxdy\right]^2}{\frac{1}{A}\int_A\int f^2(x,y)dxdy} \tag{5}$$

where n_a is the aperture efficiency of an antenna with aperture area A and an illumination distribution function f (x, y). If the aperture is divided into a number of sub-apertures -- all with the same distribution function but differing in area E_n and maximum illumination intensity a_n - - the aperture efficiency becomes:

$$n_a = \frac{\left[\frac{1}{A}\sum_N a_nE_n \frac{\int_{E_n}\int f(x,y)dxdy}{E_n}\right]}{\frac{1}{A}\sum_N a_n^2E_n \frac{\int_{E_n}\int f^2(x,y)dxdy}{E_n}} \tag{6}$$

Since it is assumed that all sub-apertures have the same illumination distribution, the same average illumination intensity, and the same average power density, this expression can be factored into:

$$n_a = \frac{\left[\frac{1}{A}\sum_N a_nE_n\right]^2}{\frac{1}{A}\sum_N a_n^2E_n} \cdot \frac{\left[\frac{1}{E_n}\int_{E_n}\int f(x,y)dxdy\right]^2}{\frac{1}{E_n}\int_{E_n}\int f^2(x,y)dxdy} \tag{7}$$

$$= n_N n_e$$

where n_N is the efficiency of the array and n_e is the efficiency of the element; their product is the total efficiency of the aperture.

IV. LIMITED SCAN ARRAY

Feed System

A limited scan array experiment using an aperiodic array of equal area elements connected to a spherical planar transmission lens feed system, Figure 7, was described by W. Patton. (5) The salient features of that design and the results of the experiment are summarized below.

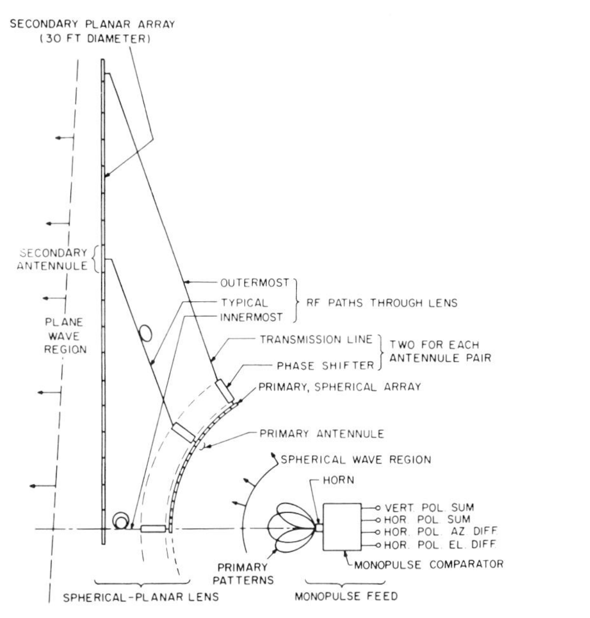

Fig. 7 The spherical primary face of the spherical-planar transmission lens array is illuminated by a multimode monopulse feed horn located at its center of curvature. Equal length coaxial transmission lines carry the energy to the secondary (radiating) face through electronically controlled phase shifters used to steer the radar beam.

The feed system is a transmission lens because it receives energy on one face and retransmits it from the other face. It is spherical-planar because the input face on transmission has a spherical curvature, while the output face on transmission is planar. The spherical input face is called the primary face; the planar output face, the face which launches the radar beam into the surrounding space, is called the secondary face. The spherical curvature of the primary face gives the lens an extremely short mechanical focal length and makes it possible to mount this antenna on a two-axis pedestal system.

The focus of the lens system is at the center of curvature of the spherical primary face. Energy emanating from a feed horn located at the focal point reaches every point on this spherical primary face in phase. Equal length coaxial cables conduct the RF energy from the primary face to the planar secondary face. The energy arrives there everywhere in phase and is launched into a single beam. The shape of the primary feed-horn pattern controls the distribution of illumination across the primary array, and this in turn controls the shape of the beam launched from the secondary face.

A lens geometry such as this gives the designer a great deal of freedom in designing optimum illumination functions, and it gives a degree of independence in the design of sum and difference illumination functions. The feed-horn arrangement with this antenna is very similar to that used for conventional reflector antenna systems, and the techniques developed for feeding the conventional antenna, are directly applicable to the spherical-planar transmission lens antenna.

Several advantages accrue from the spherical-planar transmission lens geometry. Since the antenna is a transmission lens, the feedhorn is mounted on the side of the antenna away from the radiating face. This eliminates aperture blockage by the feed -- a factor that contributes to loss of gain and increased sidelobe levels in reflector type antennas. Since every point on the spherical primary face is equidistant from the feedhorn, there is less spillover loss for a given illumination taper. Every surface element on the spherical primary face is normal to the ray incident from the feedhorn, eliminating loss due to oblique incidence, as would be the case for a planar primary array. This also reduces spillover loss.

A Typical Design

The design of a 30-ft diameter spherical planar transmission lens array providing a 10-degree electronic scan field of view at C-band discussed below will illustrate the design principle. The 30-ft. aperture is filled with 1000 elements arranged on a concentric circular grid. The elements which form the elements of the array are capable of radiating orthogonal polarizations from two isolated channels. Each channel of a secondary element is connected to the appropriate channel of the corresponding primary element by a semi-rigid coaxial transmission line through a digital latching ferrite phase shifter.

In an array of conventional design, 27,000 elements and phase shifters would be required to cover an area 30 ft in diameter. The aperiodic circular grid element arrangement, used here however, provides a filled aperture with only 1000 phase shifter pairs. The saving in weight resulting from the 25 to 1 reduction in the number of phase shifters, coupled with the short moment of inertia, provides mechanical and structural compatibility with an electromechanical scanning pedestal.

The basic array structure (shown in Figure 8) consists of

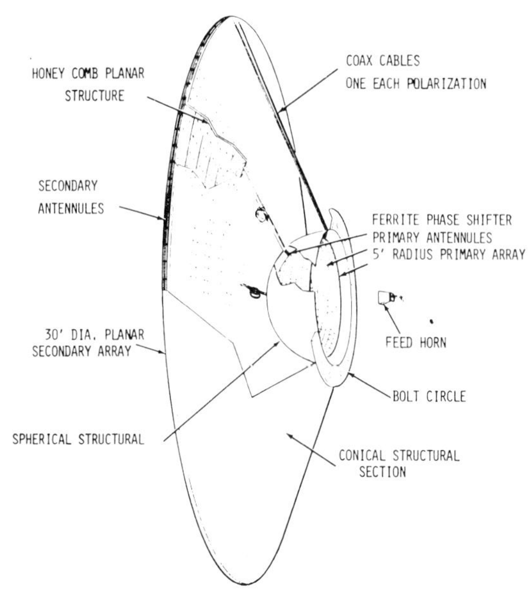

Fig. 8 Cutaway view of spherical - planar lens.

1) A planar face to which the secondary elements are attached;
2) A spherical structure to which the primary elements are attached; and
3) A truncated conical section joining 1) and 2).

The conical section encloses all electronic components except the secondary elements. This array structure can be installed in a radar system now operating in the field by unbolting the existing reflector structure and mounting the transmission lens antenna in its place with no major modification to the pedestal. The structural configuration of this antenna will make the wind torque and inertial loads smaller than those imposed on the servo drive system by the existing parabolic reflector antenna.

The primary array is a sector of a 10-ft. diameter sphere intercepting an included angle of 60 degrees at the center of the sphere. The primary face is illuminated by a multimode feedhorn located at the center of the sphere. The beamwidth of the feedhorn is 60 degrees between 10-dB points on the sum mode pattern. This feed provides sum mode, traverse difference mode, and elevation difference mode illumination for both horizontal and vertical polarizations.

Polarization diversity is generally rather difficult to obtain in array type antenna systems. It ordinarily involves at least doubling the number of antenna components. The antenna system being described here, however, lends itself to a dual orthogonal polarization capability and this capability is used to simplify the design of the antennas. In the hybrid radar being described, each cell of the antenna can handle dual orthogonal polarizations and thus can handle any polarization applied to it independent of its orientation and independent of the polarizations of the surrounding cells. This gives the entire transmission lens antenna the capability of transmitting any polarization that is applied to it by the feedhorn.

The dual polarization capability is provided in each lens cell in the following way. An element of the secondary array is made up of two arrays of dipoles, one orthogonal to the other as shown in Figure 9. These two arrays are each connected to two orthogonal dipoles in the primary array through separate coaxial feed line and its separate phase shifter. The dipoles on the primary array have polarizations corresponding to the dipoles on the secondary array. The total phase length from a primary dipole to a secondary dipole array is the same for each polarization. A field of any polarization incident upon the primary array is resolved there into two orthogonal components. These components are transmitted through phase-coherent channels (in the lens structure) to the secondary array where they are reradiated to reconstitute the incident polarization. Since each lens cell can independently transmit any polarization, it is not necessary to align the polarizations of one antenna cell with respect to other cells. This greatly simplifies the arrangement of elements in the antenna face.

A secondary element providing a 10-degree field of view in the frequency band from 5.4 to 5.9 GHz was built for the model program. The element was 9.5 x 9.5 inches square and contained 32 dipoles in each of two orthogonal polarizations. Each set of 32 dipoles was fed by a separate rectangular-coax corporate feed system. Split-tube balun dipoles spaced approximately 0.84 wavelengths at 5.9 GHz were used, and the dipoles were carefully matched to the 50-ohm impedance of the rectangular-coax feed system across the 5.4 to 5.9 GHz frequency band in a waveguide array simulator. This unit achieved a total loss of less than 1 dB and an isolation between polarization channels of greater than 40 dB.

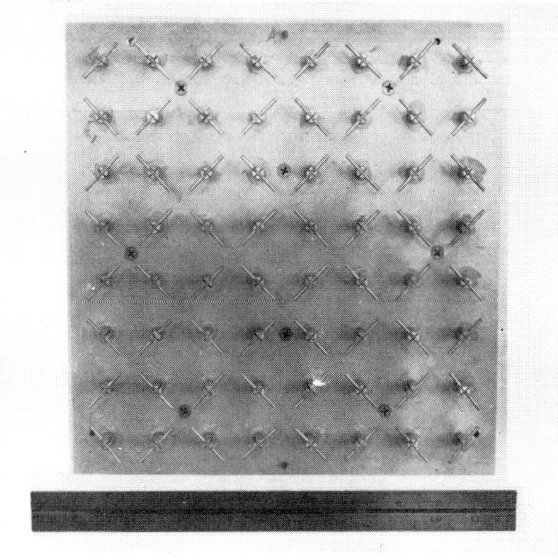

Fig. 9 A dually-polarized subarray is used as the secondary element. This element has a measured efficiency loss of less than 1 dB and has greater than 40 dB isolation between polarization channels.

Experimental Model

A 10-ft diameter engineering breadboard model consisting of the first six rings of the spherical-planar transmission lens array was built and tested. The antenna performance measurements made with the aid of this model are in substantial agreement with the theoretical prediction. The model, Figure 10, is a full scale realization of the center portion of the 30-ft-diameter C-band antenna system. The effect of the circular geometry, equivalent to aperture blockage, can be seen by the portion of the array face not covered with dipole radiators. This is most pronounced in the first three rings of the model and is negligible beyond ring five.

Fig. 10 The 10-ft-diameter engineering breadboard model is a full-scale realization of the central portion of the 30-ft-diameter C-band antenna system.

Since the performance of this antenna depends upon the total number of elements it contains, calculated performance figures will be given for both the 30-ft diameter antenna design and for the 10-ft diameter breadboard model design. The latter calculated performance data can be compared with the measurements actually made on the 10-ft-diameter engineering breadboard model. Antenna patterns have been calculated for two configurations of circular grid arrays designed to scan a 10-degree cone. The 30-ft-diameter antenna design has a pattern at 5° scan shown in Figure 11. The

Fig. 11 Calculated circular grid 1000 element array antenna patterns: $\theta_s = 5.0°$; $\phi_s = 0°$; $\phi = 0°, 180°$ cut.

The calculated pattern for the 10-foot model array at 5° scan is shown in Figure 12. Both arrays are illuminated with a 10-dB Gaussian taper distribution function. The antenna-pattern characteristics com-

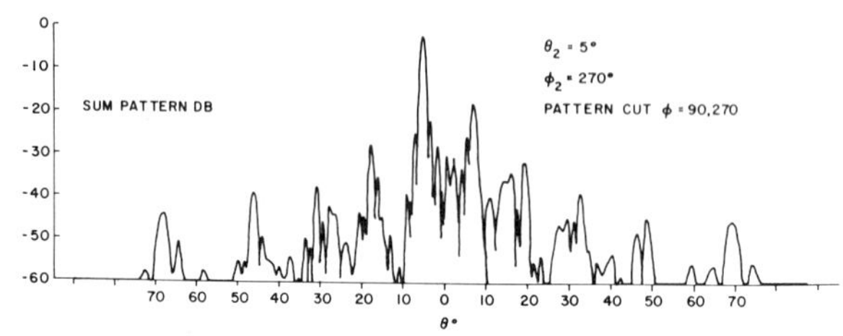

Fig. 12 Calculated circular grid 100 element array antenna patterns: $\theta_s = 5.0°$; $\phi_s = 270°$; ϕ 90°, 270° cut.

puted for these arrays are summarized in Table II.

TABLE II

Antenna Pattern Characteristics

Array Size	Scan Angles s	ϕ_s	Worst Level(dB)	Sidelobe Position(deg)	Remarks
1000 Elements	2.5	160	-24	1.9	First sidelobe
	5.0	133	20.9	-6.3°	Vestigial Grating Lobe
100 Elements	2.5	0	-21	4.5°	First sidelobe
	5.0	90	-15	-7°	Vestigial Grating Lobe

Typical patterns measured on the model antenna are shown in Figures 13 and 14. It may be noted

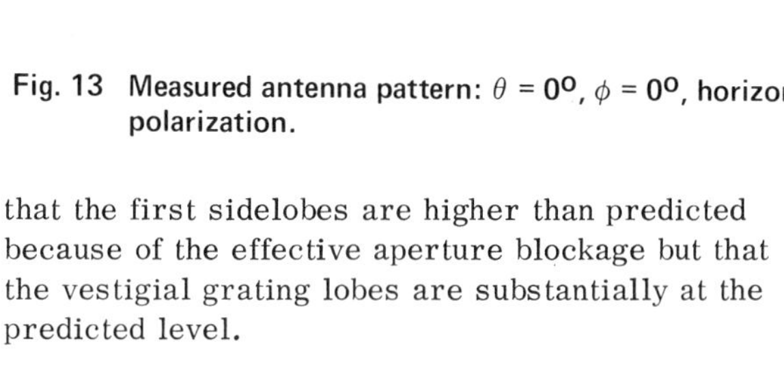

Fig. 13 Measured antenna pattern: $\theta = 0°$, $\phi = 0°$, horizontal polarization.

that the first sidelobes are higher than predicted because of the effective aperture blockage but that the vestigial grating lobes are substantially at the predicted level.

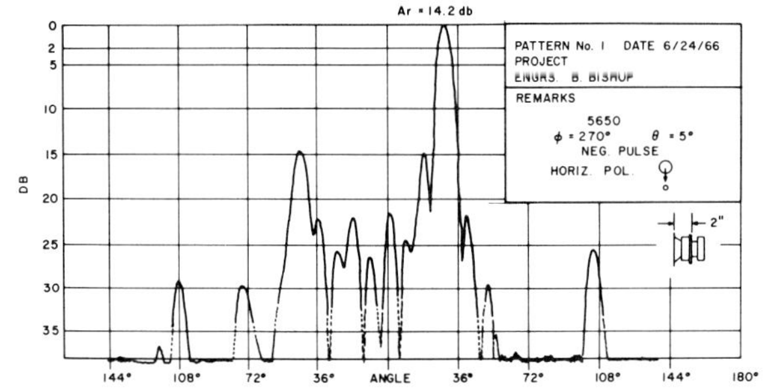

Fig. 14 Measured antenna pattern: $\theta = 5^{o}$, $\phi = 270^{o}$, horizontal polarization.

The efficiencies of these two antenna designs have been calculated in terms of loss relative to an ideal uniformly illuminated aperture. These calculations are shown in Table III.

In Table III, the feedhorn spillover loss represents the amount of energy radiated by the feedhorn which is not intercepted by the primary array. This loss is larger for the model antenna than for the full scale antenna because the feed horn used in the experiment had an actual measured beamwidth greater than 60° at the 10-dB points. The gaps between radiating elements represent a larger proportion of the total array aperture for the model than for the 30-ft diameter antenna and represent a larger loss on both the primary face and the secondary face of the model. The feedhorn mismatch loss for the model represents measured results on the feedhorn actually used in the experiment and does not represent a final feedhorn design. Since the cables for the model for the 10-ft diameter model are shorter than the cables in the 30-ft diameter model the loss due to this item is somewhat smaller for the model. Other factors contributing to the inefficiency of the array are essentially the same for both designs.

A total loss of 5.94 dB has been calculated for the model resulting in an estimated gain of 39.16 dB. The gains measured for the breadboard antenna model at various angles of scan are shown in Table IV.

The measured gain at mid-band of 39.54 dB is in good agreement with the estimated gain for the model. It will be noted also in the Table that the gain decreases with scan in accordance with the design prediction.

TABLE III

Antenna System Losses

Optical Power Divider	Model	Full Scale
Feedhorn Spillover	0.60 dB	0.45 dB
Element Losses	0.28	0.20
Gaps Between Elements	0.50	0.10
Feedhorn Mismatch	0.30	0.05
Monopulse Comparator	0.10	0.10
	1.78	0.90
Transmission System		
Cables	0.60 dB	0.70 dB
Four-Bit Ferrite Phase Shifters	0.75	0.75
	1.35	1.45
Planar Face		
Secondary Elements	1.00 dB	1.00 dB
Phase and Amp. Errors	0.50	0.25
Gaps between Elements	0.80	0.10
Phase Shifter Quantization	0.06	0.06
Illumination Taper	0.45	0.45
	2.81 dB	1.86 dB
Array Losses	5.94 dB	4.21 dB

TABLE IV

Measured Gain of Scanned Array

Beam Position				
θ°	φ°	5.4 GHz	5.56 GHz	5.9 GHz
0	0	38.98	39.54	38.19
2.5	0	38.08	38.64	37.49
2.5	45	38.08	38.84	37.69
2.5	90	37.98	38.69	37.79
5.0	0	36.78	37.04	36.00
5.0	45	36.38	37.09	35.80
5.0	90	36.48	37.09	35.70

IV. CONCLUSIONS

The basic motivation behind the investigations described in this paper has been the reduction of array cost through the accomplishment of a limited amount of agile beam motion with few control components. The objective is a low cost antenna system that is light enough and of sufficiently low moment of inertia that it can be mounted on an electro-mechanical pedestal to obtain the combined advantages of electro-mechanical scanning and beam agility. One measure of success toward this objective is the number of control elements needed to achieve a given degree of scan.

To establish a reasonable criterion for comparison, consider the fact that the number of independent beams that can be formed by a regular array is equal to the number of elements in the array. From this one can argue that the number of beamwidths in the scan sector is a reasonable measure of the minimum number of control points required. The exact form that the required relationship assumes depends upon the shape of the scan coverage as well as the shape of the aperture producing the beam. For simplicity here we will take for N min, the minimum number of control elements

$$N_{min} = \left(\frac{\sin \theta_{max}}{\sin \frac{\theta_3}{2}}\right)^2 , \qquad (8)$$

where θmax is the maximum scan from the normal beam position, and define an element use factor N/N_{min} as a figure of merit for comparing designs.

Using this relation we see that the limited scan aperiodic array using about 1000 elements to scan a 0.36 degree beam to 5 degrees has an element use factor of about 1.3. The single reflector array feed using about 980 elements to scan a 1.0 degree beam 7 degrees has an element use factor of about 2.4. Some approximate numbers for the double reflector array feed system (1500 elements to scan a 0.3 degree beam about 3°) suggests an element use factor of 3.75.

Aperiodic limited scan arrays approach the minimum number of elements for a given beam width and maximum scan angle requirement. They also tend to concentrate weight near the pedestal axis for minimum turning moment. These features make them attractive for pedestal mounted array applications. Although array feed reflector systems seem to require more control elements from data currently available, it is likely that with further development these element requirements can be made to approach that of the aperiodic limited scan array. They will then become attractive from cost considerations for those cases where the main reflector already exists and the added off-axis weight of the feed can be accommodated by an existing pedestal.

A wide variety of means for obtaining a limited amount of scan, have been described in this paper. There are undoubtedly others that have not been mentioned here. However, the feasibility of such antennas has been adequately demonstrated and there is sufficient flexibility in their variety to satisfy most needs for agile beam scanning where the maximum scan requirement does not exceed 20 degrees.

References

1. Ruze, John: Lateral-Feed Displacement in a Paraboloid, IEEE Trans. on Antennas and Propagation, Vol. AP-13, No. 5, p. 660-665, Sept. 1965.

2. Nihen, J. and Kay, A.: Optimum Dual-Reflector Antenna Design Investigation, Tech. Rept. RADC-TOR-63-318, June 1963.

3. Winter, C.: Phase Scanning Experiments with Two-Reflector Antenna Systems, Proc. IEEE, Vol. 56, No. 11, p. 1984-1999, Nov. 1968.

4. Fitzgerald, W.: Private Communication.

5. Patton, W. T.: Analysis of Radar Electronic Scan Technique, Tech. Rept. ESD-TR-66-355, July 1966.

6. Mansky, L., Simson, L., Simson, M., et al: A Development Study of Techniques to an Electronic Scanning and Multiple Target Tracking to a Fixed Beam Radar, Tech. Rept., ESD-TR-66-360, July 1966.

7. Limited Scan Phased Array Antenna, Phase I Final Report Covering August 30, 1967 to January 30, 1968 Prepared by Raytheon Company, Wayland Laboratories under Contract No. N00123-68-C-0389.

8. Winter, C. F.: Limited Scan Phased Array Antenna, Phase II Final Report Covering March 18, 1968 to November 15, 1968 under Contract No. N00123-68-C-0389.

9. Lo, Y. T.: Theory of Arrays with Randomly Spaced Elements, IEEE Transactions on Antennas and Propagations, Vol. AP-12, No. 3, p. 261, May 1964.

Pattern Measurements of Phased- Arrayed Antennas by Focusing into the Near Zone

W. E. SCHARFMAN
G. AUGUST
Stanford Research Institute
Menlo Park, California

ABSTRACT

Radiation patterns for a phased-array antenna focused to within a few aperture diameters from the array are compared with far-field radiation patterns for both theoretical models and an actual antenna. The focused patterns are nearly identical to the far-field patterns over a wide angular range. The exact degree of similarity depends on the ratio of focal length to aperture length, the amplitude taper, the scan angle, and the phase error distribution. The focusing technique permits short-range monitoring of far-field radiation patterns, gain, and beamwidth.

I. INTRODUCTION

In the operation of large phased-array antennas, monitoring of performance is an essential function. These arrays commonly have thousands of separate radiating elements, each fed by computer-controlled phase shifters. By assuming that failures occur randomly throughout the array, the degradation in antenna gain and the increase in sidelobe levels may be estimated. However, for a given array, it is not known a priori whether the errors are indeed random. Systematic failures may produce high sidelobes in limited regions of space. The most direct check on antenna performance is to monitor the antenna characteristics of interest--gain, beamwidth, side-lobe levels, etc. When any of these parameters is degraded to a specified level, the array may be shut down for repairs. It would be desirable to perform the monitoring function at regular intervals without the expense, time, and complexity involved in pattern measurements made by flying an airplane in a prescribed course. Ground-based towers are of only limited use for monitoring, however. At the distance from the array at which a tower must be located to be in the far field, only very low elevation angle measurements can be made, and ground reflections may introduce errors. Another means of monitoring is needed.

In this study we have investigated a technique for pattern measurement in which the phased array is adjusted so that it focuses to a distance of the order of several aperture diameters. By focusing to such close ranges, a test antenna can be moved over a wide range of azimuth and elevation angles, and ground-reflection effects are minimized. The test antenna can be on a telescoping tower to give variable elevation angles and on a track to give azimuth variation.

For monitoring antenna performance the test antenna arm is set at prescribed angles, the beam-steering computer sets the phases of the array elements to focus in the desired direction, and the patterns are recorded in whatever form is desired. This focused-array technique is much simpler and cheaper than either of the techniques mentioned above. This technique also has the advantage of the ranges being so short that attenuation due to rain or snow would not prevent monitoring in almost any weather.

II. THEORETICAL ANALYSIS

The fact that an aperture may be phased so that it focuses to a region much closer than the usual far-field range ($2D^2/\lambda$) is well known and has been discussed by several authors. Sherman computed patterns for several aperture sizes and showed that the angular patterns for the main lobe and first few sidelobes in the focused case were identical to the far-field patterns. (1) Ricardi showed that the gain from linear focused arrays was identical to the far-field gain for ranges as close as about one aperture length. (2) However, little work had been done on the sidelobe levels beyond the first few away from the main lobe, nor on performance at ranges of only several aperture dimensions, even for the first few sidelobes, until recent work at SRI. (3)

We have calculated the antenna patterns of arrays focused at ranges of several aperture diameters and compared them with the far-field patterns of the same arrays (i.e., the far-field patterns of the arrays phased to focus at infinity).

For a fixed array size, we have investigated theoretically the effect of focal length, scan angle, amplitude distribution, and random phase errors.

We have now analyzed the following cases and compared the far-field patterns with the near-field patterns (all the element patterns being isotropic):

Array Size	Amplitude Distrubution	Scan Angle	Focal Distance in Aperture Sizes (F)
30 λ	Uniform	0,20,40,60	1, 2, 4
30 λ	Cosine	0,20,40,60	1, 2, 4
30 λ	Cosine with 0.285-rms quasi-random phase errors	0,20,40,60	1, 2, 4
30 x 15λ	Cosine on both planes	0	2
15 λ, 30 λ, 60 λ	Cosine	0	1, 2, 4

Figure 1 is a plot showing the difference between

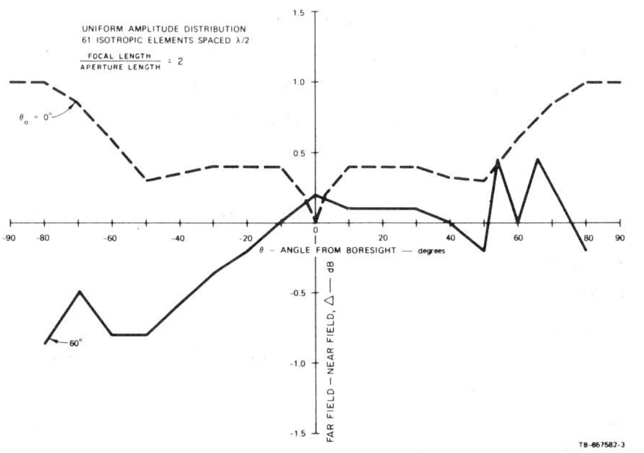

Fig. 1 Difference Between Far-Field and Near-Field Patterns— Uniform Amplitude Distribution, F/L = 2.

the far-field pattern and the focused near-field pattern of a 30-wavelength-long, uniformly illuminated, linear array of 61 isotropic radiators spaced λ/2 apart as a function of angle from boresight for a focal length equal to two aperture lengths. This figure was obtained by drawing an envelope over the peaks of the sidelobes for each case (i.e., far-field and focused near-field). At each angle the difference between this envelope and the main lobe was found for both the far-field and near-field cases. The quantity plotted, Δ is the difference between these two cases. For example, at $\theta = 10°$. the envelope of sidelobes for the far-field pattern may be 30 dB below the maximum gain on the main lobe. At the same angle, the pattern for the focused near-field case may have an envelope 29 dB below the main-lobe gain. Therefore, Δ is 1 dB.

The difference is plotted for two values of scan angle, θ_0, equal to 0 and 60 degrees. It should be noted that the focused near-field pattern was calculated at a constant range from the array center equal to the focal length and not at a constant distance perpendicular to the aperture. It can be seen from Figure 1 that the two patterns agree to less than 0.5 dB over a wide range of angles.

Figure 2 illustrates the effect of changing the

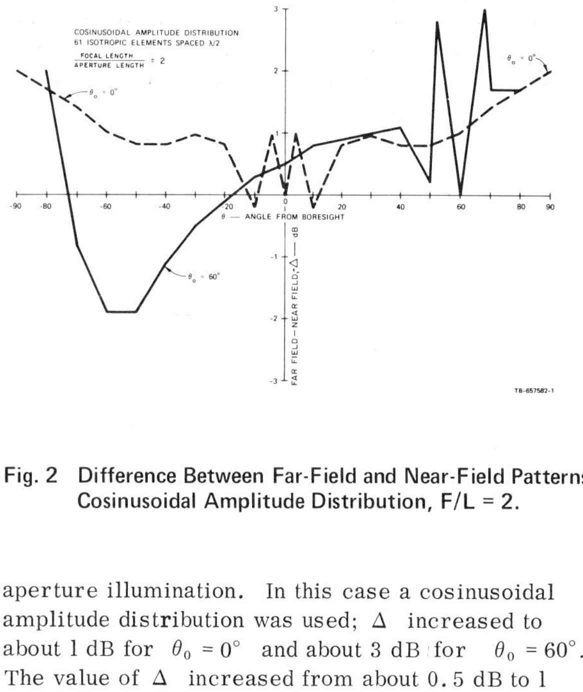

Fig. 2 Difference Between Far-Field and Near-Field Patterns— Cosinusoidal Amplitude Distribution, F/L = 2.

aperture illumination. In this case a cosinusoidal amplitude distribution was used; Δ increased to about 1 dB for $\theta_0 = 0°$ and about 3 dB for $\theta_0 = 60°$. The value of Δ increased from about 0.5 dB to 1 dB in changing the amplitude distribution; while the first sidelobe level decreased from 13 dB to 23 dB.

Since most phased arrays use digital phase shifters, the phase distribution will only approximate the desired phase. This effect was studied. Figure 3 shows the results when a cosinusoidal amplitude distribution and quasi-random phase errors with an rms value of 0.285 radian are assumed. The phase distribution was taken from a Sperry report that calculated the phase distribution on a 51-element

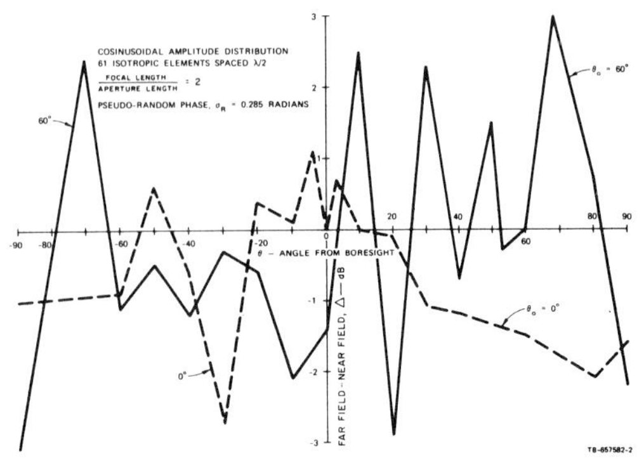

Fig. 3 Difference Between Far-Field and Near-Field Patterns—Cosinusoidal Amplitude Distribution, D/L = 2 Pseudo-Random Phase, σ_R = 0.285 Radians.

linear array, including mutual coupling effects and 45-degree phase quantization. (4) Since our array had 61 elements, the Sperry results were augmented by the phase distribution for ten elements for another scan angle. In this case Δ is between 1 and 2 dB for $\theta_0 = 0°$ and between 2 and 3 dB for $\theta_0 = 60°$.

Figure 4 shows the effect of changing the focal

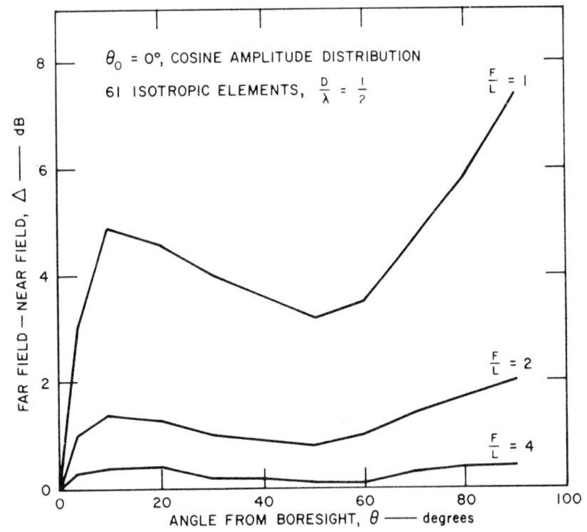

Fig. 4 Difference Between Far-Field and Near-Field Patterns as a Function of F/L—No Random Phase Error.

length for a given aperture size and amplitude distribution without phase quantization. A significant improvement can be seen from increasing the focal length from one to two aperture lengths. At a focal length of four aperture lengths, the errors are less than 0.4 dB for all angles.

Figure 5 shows the effect of changing focal length for a given aperture size and amplitude distribution when the phase quantization errors used in Figure 3

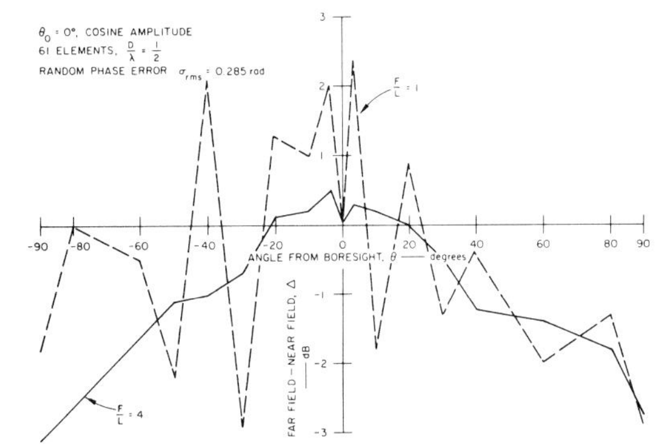

Fig. 5 Difference Between Far-Field and Near-Field Patterns, as a Function of F/L—Random Phase Error.

are applied to the aperture. For F/L = 1, the difference between the far-field pattern and the near-field pattern is less with the phase quantization than with no phase quantization. While for F/L = 4 there is little change for the first 20 degrees near the main lobe, at wide angles the results with phase quantization are considerably degraded from those without phase quantization.

The phase distribution applied to the array for scanning the beam in the far-field case and for focusing it into the near field at various scan angles is shown in Figure 6. At $\theta_0 = 90°$, which is the end-fire direction, the phase distributions for the two cases are identical, and this is true for all focal lengths. That is, at $\theta_0 = 90°$, the array is focused at all ranges. As the scan angle approaches broadside, the phase distributions differ by an increasing amount. At broadside the far-field phase is uniform, while the focused phase distribution is parabolic.

The results of comparing the calculated far-field patterns with focused near-field patterns may be summarized as follows:

(1) The difference between the far-field pattern and near-field pattern, Δ, for the uniformly illuminated apertures was only about 0.4 dB for a focal distance

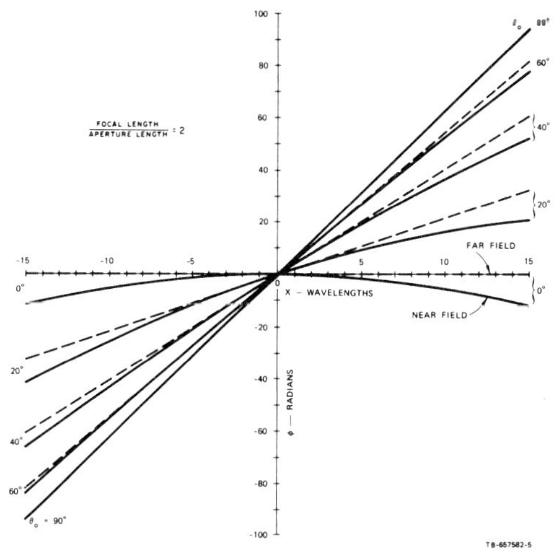

Fig. 6 Distributions along Aperture for Radiation Far Field and Focused Near Field.

of two aperture lengths, $\theta_0 = 0$, and angles off boresight of less than 50 degrees. This difference is sufficiently small for this technique to be useful for monitoring the performance of phased-array antennas. The following items show the variation of Δ found as parameters were varied.

(2) The value of Δ increased when the amplitude distribution was tapered from a uniform to a cosine form. For example, at $\theta_0 = 0°$, F/L = 2 (focal distance/array length), Δ for the close-in lobes was about 0.4 dB with a uniform distribution (first sidelobe, 13.2 dB) and about 0.8 dB with a cosine distribution (first sidelobe, 23 dB). Although this shows that Δ increases with amplitude taper, the increase is at a relatively modest rate. In the example cited, a 10-dB decrease in sidelobe level changed Δ by only 0.4 dB.

(3) Increasing the focal distance, even from one to two aperture dimensions, decreased Δ significantly. For a cosine amplitude distribution $\theta_0 = 0°$, the value of Δ at the first sidelobe decreased from 3.1 to 1.0 to 0.3 dB as the focal distance was increased from one to two to four aperture lengths.

(4) The value of Δ generally increased with increasing scan angle, but the increase was moderate. For a cosine amplitude distribution focused at a distance of two aperture lengths, the value of Δ at the first sidelobe increased from 1.0 to 1.9 dB as θ_0 increased from 0 to 40 degrees. It should be noted that at $\theta_0 = 40°$ the pattern was somewhat asymmetrical and that the value of Δ for one sidelobe was 1.9 dB while that for the other was only 0.8 dB.

(5) Introducing quasi-random phase errors did not significantly change the values of Δ for close-in sidelobes. For the two smaller focal lengths, it generally decreased the value of Δ for wide-angle sidelobes. The reason for the last result is not clear.

(6) The few calculations with a two-dimensional array show a value of Δ a little smaller than for the one-dimensional case.

The theoretical work described above showed that the focused measurement would be useful. To demonstrate the effectiveness of the technique and to confirm that the theoretical model was sufficiently accurate, measurements were made on a phased array.

III. EXPERIMENTAL RESULTS

The array used was the Electronically Scanned Pencil Beam Antenna (ESPBA) developed by Sperry Gyroscope for ECOM. This is a C-band space-fed array consisting of 1721 radiating elements arranged in an ellipse approximately 5 by 8 ft along the minor and major axes. It utilizes 3-bit, ferrite phase shifters for both collimation and scanning.

Measurements were made at ranges of 12 and 36 ft with the array focused at these ranges. Immediately after these measurements, far-field patterns were taken using a tower approximately 1000 ft. away and the array focused at infinity. The array (lens) was mounted over the center of rotation of the turntable so that patterns could be taken at a constant range simply by rotating the turntable. The experimental results are given below.

Beamwidth: The far-field azimuth beamwidth was approximately 2.5 degrees. Measurements made in the near field agree with this value to within 0.1 degree. Beam patterns obtained by scanning the beam past the test antenna gave similar results.

Sidelobe Levels: Because the array performance was temperature sensitive, especially in the severe winter environment, patterns taken with the array set up to nominally the same condition might exhibit differences up to several decibels. This limited the accuracy with which near-zone and far-zone patterns could be compared. Figure 7 shows the repeatability of far-zone boresight patterns taken on consecutive days. Notice that the first sidelobe varied 2 dB while wide-angle lobes at the 30-to-35 dB level varied as much as 6 dB.

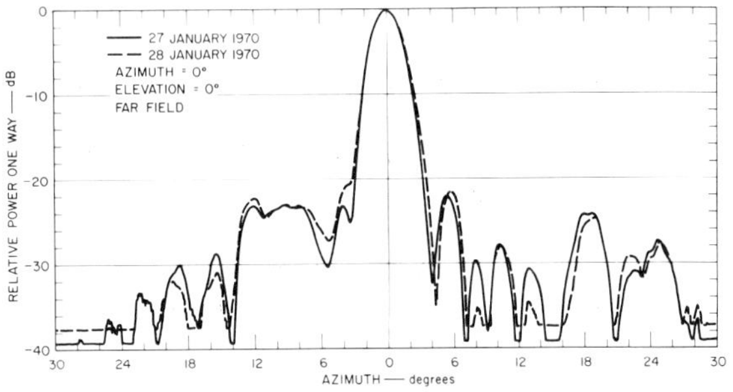

Fig. 7 Repeatability of Far-Field Patterns.

A comparison of far-zone and near-field patterns shows that first sidelobes (~ 20 dB) and wide-angle sidelobes (25 to 30 dB) generally agree to within the variations due to temperature and other effects that limit far-field repeatability (~2 to 3 dB). Agreement between far-field and near-field patterns for a given array condition may have been better than this, but the data are not sufficiently repeatable to confirm this. This comparison is for the sidelobe level over a region of several degrees of angular space. At a given angle the far-field and near-field sidelobe level may differ by a considerably greater number of decibels.

Figure 8 shows a comparison between a far-field

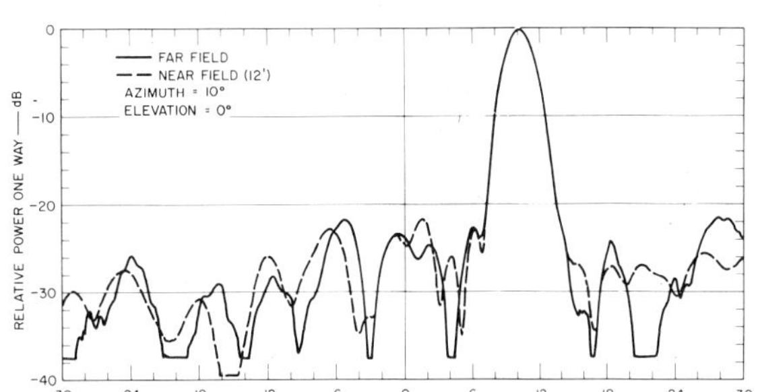

Fig. 8 Comparison of 12-ft. and Far-Field Patterns.

pattern and a near-field pattern for the beam scanned 10 degrees from boresight; the focal length was approximately 12 ft. The agreement here is better than 2 dB for all lobes down to the 30-dB level, except for the far-field lobe at 28 degrees, which is about 4 dB high. These patterns were taken about two weeks apart.

Figure 9 shows a similar comparison for the beam scanned to 10 degrees at a focal length of 36 ft.

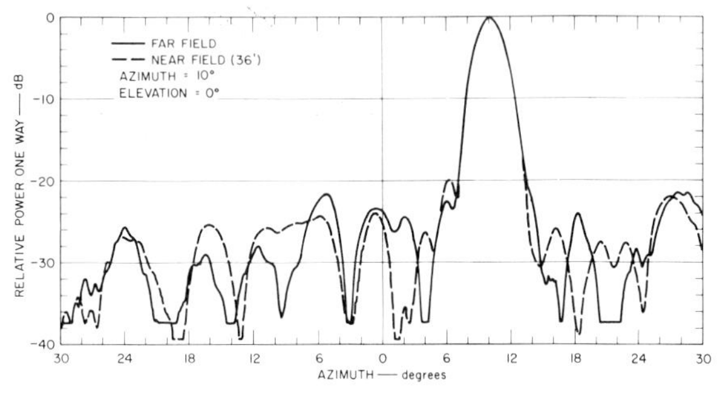

Fig. 9 Comparison of 36-ft. and Far-Field Patterns.

This near-field pattern is not appreciably closer to the far-field pattern than is the 12-ft. pattern. These patterns were taken about one month apart and under quite different temperature conditions.

Gain: The gain measured at both near-field ranges (12 and 36 ft.) agrees with the far-field gain within a few tenths of a decibel. The near-field measurement was especially interesting, because it indicated a gain approximately 2 dB less than a far-field measurement made several months earlier. Far-field gain measurements made immediately after the near-zone measurements agreed with the near-zone value and indeed showed a 2-dB lower gain. Recent checks showed that 1 dB of loss was due to water in the waveguide feed system. This measurement demonstrates the ability of the focused measurement technique to detect antenna degradation.

Row and Column Disabling: Measurements were made in both the focused near field and the far field under conditions where from one to eight adjacent rows or columns were disabled. This had the effect of putting a phase error on a column or an entire row. Near-field and far-field measurements agreed quite well (1 to 2 dB). Disablement of even one column (~ 50 elements) could be seen by an increase of several decibels in the sidelobe levels. This is shown in Figure 10 for a focal length of 12 ft. with Column 23 (near the array center) disabled. The same array condition was measured with a focal length of 36 ft., with the results shown in Figure 11. The agreement was still good but not as good as for the 12-ft. focal length, contrary to what would be expected. Since the 12-ft patterns were measured closer in time to the far-field patterns than were the 36-ft. patterns, we assume that the array condition was not sufficiently stable to allow comparisons between the patterns of different focal length. Disablement of four columns produced a sidelobe level of about 13 dB in regions where the level had been 25 to 30 dB when the columns were phased correctly.

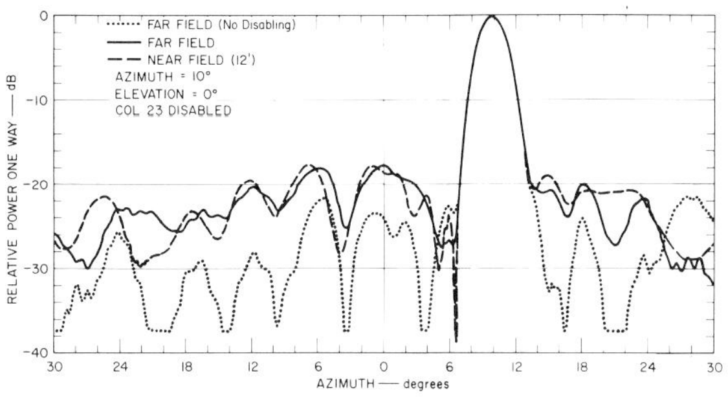

Fig. 10 Comparison of 12-ft. and Far-Field Patterns—Column Disabled.

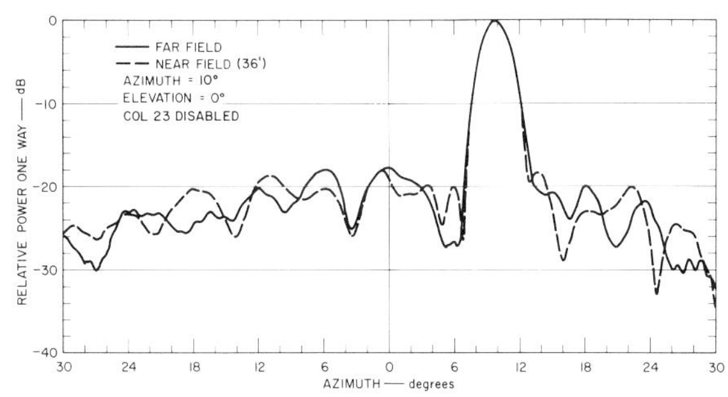

Fig. 11 Comparison of 36-ft. and Far-Field Patterns—Column Disabled.

Antenna degradation due to disabled columns was clearly discernable in the near-field measurement.

Random Amplitude and Phase Disabling: Individual phase shifters were disabled by randomly selecting units and disconnecting the drive. In both near-field and far-field patterns the effect was small, amounting to about 2 dB for 150 phase shifters disabled. By contrast, the column disabling case, with only 50 elements coherently disabled, produced significant pattern deterioration. See Figure 12.

Metallic tape was placed over individual elements on the lens to effectively short out those elements. This was done for up to 450 elements. Measurable increases of about 2 dB in sidelobes were detected in both the near-field and far-field patterns. See Figure 13.

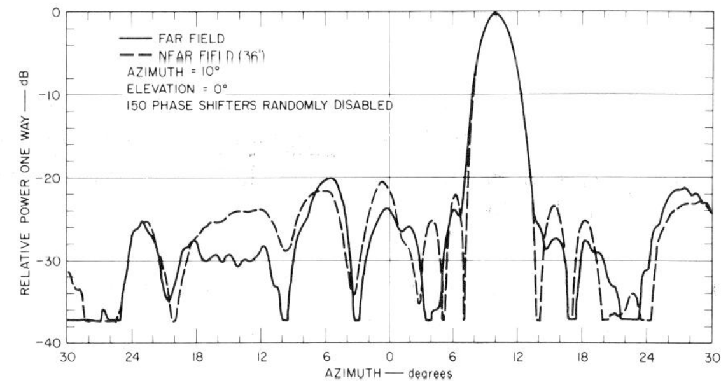

Fig. 12 Comparison of 36-ft. and Far-Field Patterns—Random Disabling.

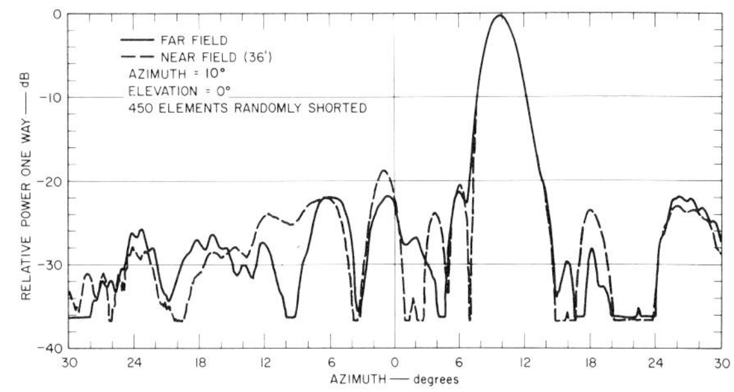

Fig. 13 Comparison of 36-ft. and Far-Field Patterns—Random Shorting.

IV. CONCLUSIONS

Monitoring of antenna performance by focusing a phased array to a distance of several aperture diameters can be done with sufficient accuracy to make the technique worthy of consideration for incorporation into phased-array radar systems. Since the gain and sidelobe levels are directly measured, the output of the monitoring systems can be used as a basis for deciding when antenna performance has degraded to a point where corrective action must be taken.

The focusing of a phased array is relatively easy to accomplish, but the capability must be designed into the array.

REFERENCES

(1) Sherman, J.E.: Properties of Focused Apertures in the Fresnel Region, IRE Trans. Antennas and Propagation, p. 139 (July 1962)

(2) Ricardi, L. J.: Near-Field Characteristics of a Linear Array, in Electromagnetic Theory and Antennas, ed. E. C. Jorden, P. 1301 (Pergamon Press, New York, 1963).

(3) Hansen, R. C.: Microwave Scanning Antennas, Vol. I, Academic Press, 1964, Chap. I, pp 21-46.

(4) Galegro, D. and Esposito, F.: Final Report on Impedance and Pattern Characteristics of Lens Array Antennas, Contract DAAH-01-68-C-0115, Sperry Gyroscope Division, Great Neck, New York (1968).

A Method of Locating Defective Elements in Large Phased Arrays

P. L. RANSOM and R. MITTRA

University of Illinois
Department of Electrical Engineering
Urbana, Illinois

Abstract

The operation of large phased arrays, often consisting of thousands of elements, leads to the practical problem of locating the defective elements. Presently this is accomplished by individual element inspection. A method is suggested here whereby the near radiated field is sampled over a plane parallel to array, numerically processed according to an inverse diffraction transformation to reconstruct the aperture field. Comparison with the known correct aperture field locates the defective elements.

I. INTRODUCTION

The operation of large phased arrays will eventually lead to deteriorated performance due to the presence of defective elements. Presently these elements are found by individual element inspection, either random or systematic. Since large arrays may contain thousands of elements the former method is unreliable and the latter inefficient.

The solution proposed here recognizes that the radiated field of the array is related to the aperture field by the well-known Rayleigh diffraction formula. When the far-field patterns become unacceptable, it is proposed to measure the radiated field (both amplitude and phase) over some plane parallel to the array aperture and from that, to reconstruct the aperture field by applying an inverse diffraction transformation. A comparison of this reconstructed field with the known correct aperture field should reveal the location of the defective elements. Actually, it is proposed to process the error fields (i.e. the difference between the deteriorated and correct fields) thereby increasing the resolution of the method. This is possible since Rayleigh's integral formula is a linear transformation. The proposed method of locating the defective elements, the theory of the inverse diffraction transformation, and some results of computer simulated cases are discussed below.

2. THE METHOD

When the array far-field patterns become unacceptable it is proposed that the deteriorated field g be recorded in both amplitude and phase over a plane parallel to the array as shown in Figure 1, where z may be of the order of the array size. It is only necessary that the plane of recording be sufficiently separated from the array so that the aperture field is not disturbed by the probe mechanism. The correct field g_o can be computed from the desired aperture distribution f_o by the known transformation given in Equation (1) below. The error field $g_\epsilon = g_0 - g$ is numerically processed according to the inverse transformation (2) to obtain the aperture error field $f_\epsilon = f_0 - f$. The maxima of $|f_\epsilon|$ locate the defective elements. However since the inverse transformation gives f_ϵ only approximately, a threshold level criterion must be used to predict the presence of a defective element.

The transformation from the aperture distribution f to the field g over the "recording" plane parallel to the array (see Figure 1) is given by Rayleigh's

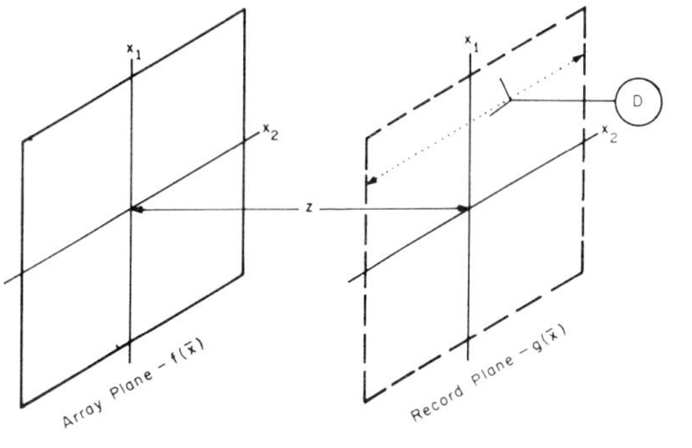

Fig. 1 Orientation of the Array and Recording Planes.

diffraction formula

$$g(\underline{x}) = d_z(\underline{x}) * f(\underline{x}) \qquad (1)$$

where $\underline{x}$ is the transverse position vector with coordinates (x_1, x_2), z is the separation between the array and record planes, and the asterisk (*) implies the convolution. The diffraction kernel d_z is given by

$$d_z(\underline{x}) = -(1/2\pi)\ \partial[\exp(-jkr)/r]\partial z$$

where $r = \sqrt{z^2 + |\underline{x}|^2}$ and k is the free space wave number $2\pi/\lambda$. This transformation has been studied by several authors (1-4) and various methods of inversion (at least approximately) have been reported. The simplest of these is the approximate inverse kernel $d_z{}^*$, where the super-asterisk implies the conjugate. This kernel will be shown later to recover the "visible range" part of f from g. That is, the aperture field can be expressed as the sum of two partial fields; a "visible" part associated with propagation and an "invisible" part associated with evanescence. So the approximate inverse transformation used here is

$$f(\underline{x}) \simeq d_z{}^*(\underline{x}) * g(\underline{x}). \qquad (2)$$

While the diffraction transforms Equations (1) and (2) are shown as convolutions they can be numerically evaluated faster by computing the "fast Fourier transform" (5) (FFT) of the convolved functions, forming the product of the transforms, and evaluating the inverse FFT. If f_ϵ and g_ϵ are two dimensional square arrays with N^2 complex values the convolution integral requires N^4 operations (i.e. a multiplication and an addition). Alternatively each two-dimensional FFT requires $2N^2 \ell n_2 N$ operations, and since the Fourier transform of the kernel can be pre-computed, only two of these FFT's would be required. The saving in time is proportional to $N^2/4\ell n_2 N$ where N^2 is the number of elements in the array (e.g. for N = 60, the FFT evaluation of Equation (1) or (2) is 150 times faster than the convolution method).

3. THE INVERSE DIFFRACTION TRANSFORMATION

Clearly the problem of recovering f from g is equivalent to inverting the diffraction transformation (1). If the inverse kernel d_{-z} is defined by the equation

$$d_z(\underline{x}) * d_{-z}(\underline{x}) = \delta(\underline{x}) \qquad (3)$$

comparison with Equation (1) shows that d_{-z} may be interpreted as that aperture field which produces a δ field a distance z away. Obviously $d_{-z}(\underline{x})$ is no ordinary function. While no expression exists for $d_{-z}(\underline{x})$, the fourier transformation $D_{-z}(\underline{\xi})$ is well defined.

Under the Fourier transform† (two-dimensional)

$$F(\underline{\xi}) = \int d^2x\ \exp(j\underline{\xi}\cdot\underline{x})\ f(\underline{x}) \qquad (4)$$

equations (1) and (3) become respectively

$$G(\underline{\xi}) = D_z(\underline{\xi})\ F(\underline{\xi}) \qquad (5)$$

$$D_{-z}(\underline{\xi})\ D_z(\underline{\xi}) = 1 \qquad (6)$$

The Fourier transform of the kernel is

$$D_z(\underline{\xi}) = \exp[-j\ z\ \zeta\ (\underline{\xi})] \qquad (7)$$

where

$$\zeta(\underline{\xi}) = \begin{cases} \sqrt{k^2 - |\underline{\xi}|^2}\ , & |\underline{\xi}|^2 \le k^2 \\ -j\sqrt{|\underline{\xi}|^2 - k^2}\ , & |\underline{\xi}|^2 > k^2 \end{cases} \qquad (8)$$

† The transverse wave vector $\underline{\xi}$ has cartesian components (ξ_1, ξ_2).

So the longitudinal wave number ζ is real in the region of the spectrum $|\underline{\xi}|^2 \leqslant k^2$ and complex in the region $|\underline{\xi}|^2 > k^2$. Reference to Equation (5) and (7) reveals that those spectral components of $f(\underline{x})$ (i.e. $F(\underline{\xi})$) in the region $|\underline{\xi}|^2 \leqslant k^2$ undergo a phase change (which is associated with propagation) in the transformation from f to g. Meanwhile those $F(\underline{\xi})$ in the region $|\underline{\xi}|^2 > k^2$ are exponentially attenuated (associated with evanescence). On this basis the spectral region $|\underline{\xi}|^2 \leqslant k^2$ will be called the "visible region" and $|\xi|^2 > k^2$, the "invisible region". Furthermore any function may be separated into its visible and invisible parts. Specifically the inverse kernel is

$$D_{-z}(\xi) = D_{-zV}(\xi) + D_{-zI}(\xi) \tag{9}$$

where (6) and (7) give

$$D_{-zV}(\xi) = \begin{cases} \exp\left(j z \sqrt{k^2 - |\xi|^2}\right), & |\xi|^2 \le k^2 \\ 0, & |\xi|^2 < k^2 \end{cases}$$

and (10)

$$D_{-zI}(\xi) = \begin{cases} 0, & |\xi|^2 \le k^2 \\ \exp\left(z \sqrt{|\xi|^2 - k^2}\right), & |\xi|^2 < k^2 \end{cases}$$

Now it can be seen that $d_{-z}(\underline{x})$ is not defined since the inverse Fourier integral of $D_{-z}(\xi)$ is divergent due to the exponential growth of $D_{-zI}(\xi)$. Furthermore, while the expression

$$F(\xi) = D_{-z}(\xi)\, G(\xi) \tag{11}$$

is well defined, any attempt to restore the invisible range components of $F(\xi)$ will result in numerical instabilities because of the behavior of D_{-zI}. However, the visible range part of $F(\xi)$ can be restored from $G(\xi)$ since D_{-zV} given above is numerical stable in computation. So instead of (11) we settle for

$$F_V(\xi) = D_{-zV}(\xi)\, G(\xi) \tag{12}$$

and finally $f_V(\underline{x})$ is obtained by taking the inverse Fourier transform of F_V.

A quantitative feel for the effectiveness of visible range reconstruction can be obtained by considering the aperture field $f(\underline{x}) = \delta(\underline{x})$. Then by (1) $g(\underline{x}) = d_z(\underline{x})$ and so (12) becomes

$$F_V(\xi) = D_{-zV}(\xi)\, D_z(\xi) = \begin{cases} 1, & |\xi|^2 \le k^2 \\ 0, & |\xi|^2 > k^2 \end{cases}$$

now

$$f_V(\underline{x}) = 1/(2\pi)^2 \int d^2\xi \, \exp(-j\xi\cdot\underline{x}), \quad |\xi|^2 \le k^2.$$

Changing to polar coordinates (k, θ), of ξ and letting $r = |\underline{x}|$, the integral becomes

$$f_V(r) = 1/(2\pi)^2 \int_0^k dK\, K \int_0^{2\pi} d\psi\, \exp(-jKr\cos\psi)$$

$$= 1/(2\pi)^2 \int_0^k dK\, K\, J_0(Kr)$$

$$f_V(r) = k^2 J_1(kr)/(kr)$$

Since $k = 2\pi/\lambda$, the first zero occurs at a circle of radius

$$r/\lambda = 3.88/2\pi = 0.61$$

It can be assumed then that if one element in an array was defective and $g(\underline{x})$ was measured exactly, and $G(\xi)$ computed exactly, then the visible range reconstruction would locate the element to within the circle above. Assuming an element separation of at least half a wavelength, there would be at most four elements in the circle.

The approximate inverse kernel d_{-zV} can be shown to be proportional to $J_1(kr)/(kr) * d_z$, which is a complicated expression involving a convolution. It will now be demonstrated that the simple kernel $d_z{}^*$ effectively recovers f_V from g. The expression $d_z{}^* * g$ is reflected in the transform domain by

$$D_z^* \cdot G = D_z^* \cdot D_z \cdot F$$

$$= [D_{zV}^* + D_{zI}] \cdot [D_{zV} + D_{zI}] \cdot F$$

$$= [1_V + D_{2zI}]\, F$$

$$= F_V + D_{2zI} F$$

Since D_{2zI} is given by $\exp\left[-2z\sqrt{|\xi|^2 - k^2}\right]$ where $|\xi|^2 > k^2$, the second term is negligible for modest values of z and $|\xi|$. For example if $\xi_1 = 2k$ and $\xi_2 = k$ then $D_{2zI} = \exp(-4z/\lambda)$. For $z/\lambda = 1/2$, this factor is 3.4×10^{-6} and for $z/\lambda = 5$, about 7.6×10^{-53}.

So the approximation

$$D_z^* \cdot G \simeq F_V$$

is justified and the result is

$$f_V \simeq d_z^* * g$$

Finally it is assumed that f_V approximates f and the approximate inverse diffraction transformation (2) is obtained, subject to this assumption.

4. NUMERICAL RESULTS

Several computer simulated cases were run to evaluate the feasibility of the method. The sequence of steps in the computation are listed below:

1. Simulate the measured field g by (1) where f is the desired aperture distribution f_o deteriorated by random phase error in a specified number of random locations.

2. Simulate the effect of measurement noise by introducing a bounded random error to g.

3. Compute the correct field g_o from the desired aperture distribution f_o by (1).

4. Form the error field $g_\epsilon = g - g_o$.

5. Apply the approximate inverse transformation (2) to obtain f_ϵ.

6. Predict locations of defective elements by the inequality $|f_\epsilon| > a$ threshold level.

For computational purposes a 61 x 61 element planar square array with uniform $\lambda/2$ spacing was considered. The aperture amplitude distribution chosen was Gaussian with a -20 dB edge illumination. A far-field pattern of this array is shown in Figure 2 where

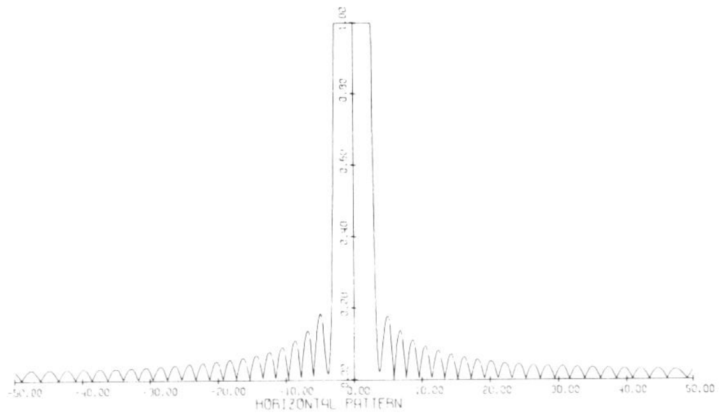

Fig. 2 Sidelobes of the Far-field Pattern of the Array. Maximum Level -35 dB Down from Peak.

the beam width is about 2.3° and the sidelobe levels are at least 35 dB down from peak. Since the beamwidths did not change, only the sidelobes of the far-field patterns will be shown. Also the peaks of all patterns were normalized with respect to the peak of this error free aperture pattern. So the maximum sidelobe level in dB down from peak can not be read from the figures but will be listed in all cases.

To simulate element deterioration, a random phase error $(-\pi \leqslant \phi \leqslant \pi)$ was introduced into 5% (186) of the array elements at random locations. A resulting far-field pattern is shown in Figure 3, where the maximum sidelobe level has risen to -31.5 dB. This rise of 3.5 dB may be sufficient to render the array unacceptable in some applications.

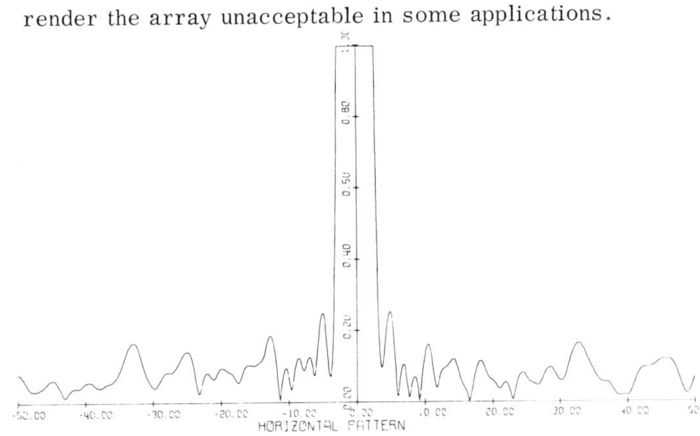

Fig. 3 Sidelobes of the Far-field Pattern of the Array with 5% of the Elements Defective. Maximum Level -31.5 dB Down from Peak.

In applying the method proposed above for locating the defective elements it should be noted that if the inverse kernel were known exactly, the magnitude of the aperture error field at any array location would be

$$|f_\varepsilon| = a\sqrt{2(1 - \cos \phi)}$$

where a is the element amplitude and ϕ is the element phase error, if any. So if a constant threshold level is to be used as the criterion for predicting the location of a defective elements (it is the simplest kind), $|f_\epsilon|$ must be untapered between steps 5 and 6 listed above. Also the probability of finding a defective element will be directly proportional to the amount of phase error. Consequently the elements found in the following cases will be listed by four ranges of the phase error, 0-45°, 45-90°, 90-135°, 135-180°. The threshold level (T. L.) used was

$$\text{T.L.} = |f_\varepsilon|_{ave} + C(|f_\varepsilon|_{max} - |f_\varepsilon|_{ave}) \tag{13}$$

where C is a constant to be chosen.

As a first example the method was applied where it was assumed that the near field g was recorded without noise. Using a value of 0.2 for C in Equation (13), 384 locations were predicted as defective, of which 104 were actually defective. These results are shown in Figure 4 where the locations of the 186 defective elements (denoted by the symbol X) and the 384 predicted locations (□) are plotted to scale on the array aperture. While it may appear that an excessive number of locations were predicted erroneously (and hence would have to be experimentally checked for phase), note that the predicted locations are only about 10% of the entire array and of those predicted 28% were actually defective. It should also be noted that the predicted locations are not scattered aimlessly over the aperture but are clustered around actual defective locations. Table 1 gives the breakdown of the 104 correct predictions as a function of the four phase error ranges. It is apparent that most of the more defective elements were found. To evaluate the effect of the remaining defective elements the far-field pattern was computed after the 104 elements found were "corrected". The pattern is shown in Figure 5 where the sidelobes are -34.4 dB.

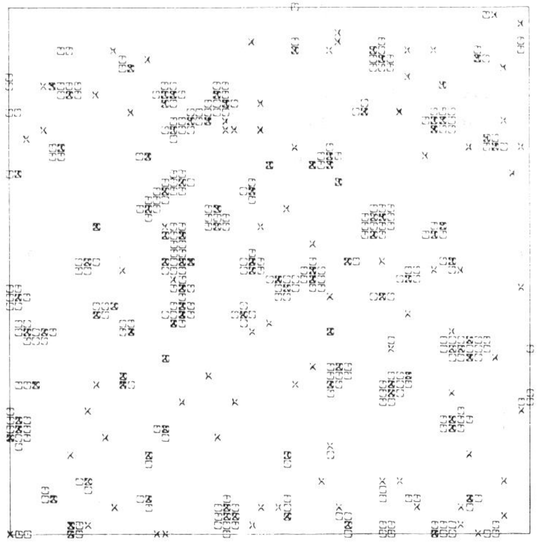

Fig. 4 Actual (x) and Predicted (□) Locations of the Defective Elements with no Measurement Noise.

TABLE 1

DEFECTIVE ELEMENTS FOUND FOR VARIOUS COMPUTER SIMULATED CASES

Phase Error Ranges	0-45°	45-90°	90-135°	135-180°	Total
No. of Defective Elements	34	58	44	50	186
No. Found with 0% Noise (c = 0.2)	6	21	35	42	104
No. Found with 5% Noise (c = 6.2)	4	24	35	42	105
No. Found with 0% Noise and 2 Successive Applications					
1st Application (c = 0.3)	2	10	26	35	73
2nd Application (c = 0.4)	4	29	16	9	58
Total Found	6	39	42	44	131

A second example was computed where a 5% measurement noise was simulated for g by the expression g(1 + .05 r), where r is a random number in the range (-1, 1). Using the same value of C in Equation (13), 420 locations were predicted of which 105 were actually defective giving a prediction accuracy rate of 25%. The actual and predicted locations are plotted in Figure 6, and the phase error ranges of the 105 correct predictions are given in Table 1. After these 105 elements were corrected the far-field pattern shown in Figure 7 had sidelobe levels of -34.2 dB. So a 5% measurement noise simulation only reduced the prediction accuracy slightly.

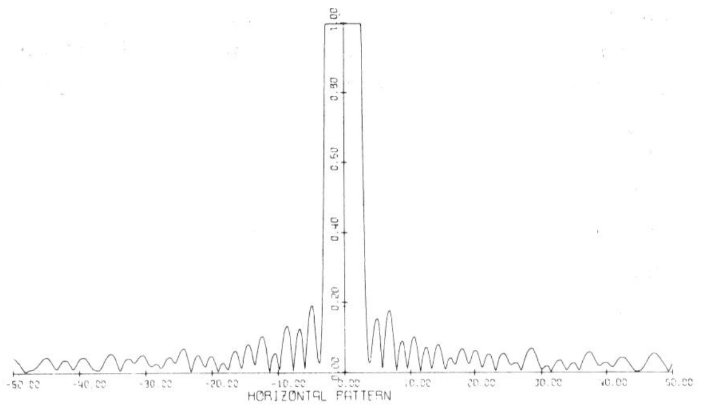

Fig. 5 Sidelobes of the Far-field Pattern of the Array when the Elements Predicted Defective in the Noiseless Case are Corrected. Maximum Level -34.4 dB Down from Peak.

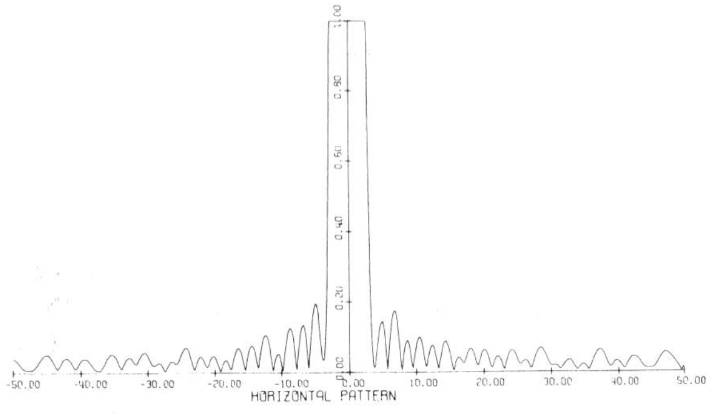

Fig. 7 Sidelobes of the Far-field Pattern of the Array when the Elements Predicted Defective in the 5% Noise Case are Corrected. Maximum Level -34.2 dB Down from Peak.

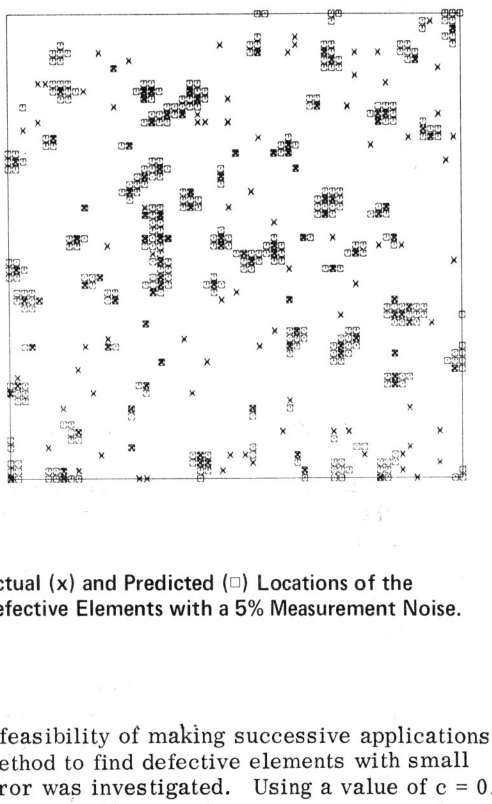

Fig. 6 Actual (x) and Predicted (□) Locations of the Defective Elements with a 5% Measurement Noise.

The feasibility of making successive applications of this method to find defective elements with small phase error was investigated. Using a value of c = 0.3 for the first application, the 0% noise case was repeated resulting in 188 predictions, of which 73 were correct. After these 73 elements were corrected the method was applied again this time with c = 0.4. The higher threshold levels were used in an attempt to increase the prediction accuracy. This time 240 locations were predicted of which 58 were correct. So the prediction accuracy for the two applications was 30%. The breakdown according to the phase error range of the 131 elements found is given in Table 1. The far-field pattern was computed after these elements were corrected. The sidelobes near the main beam were regular as in Figure 2 and at least 34.9 dB down from peak.

5. SUMMARY

Additional work is being done to determine how extensive the measurement of g should be. That is, how many scan paths (or probe points) are required for different results; over what area; how far from the array?

However on the basis of the results reported, the method described above appears to offer a feasible means of locating defective elements in large phased arrays.

REFERENCES

(1) Ransom, P. L.: The Diffraction Transformation of Electromagnetic Fields Between Two Parallel Planes, University of Illinois, Antenna Laboratory Report No. 69-8, June 1969.

(2) Mittra, R. and Ransom, P. L.: Imaging with Coherent Fields, Proc. Sym. on Modern Optics, Polytechnic Institute of Brooklyn Press, 1967, Vol. 17, p. 633.

(3) Sherman, G.: Integral-Transform Formulation of Diffraction Theory, Jour. Opt. Soc. Am., Vol. 57, p. 1040, December 1967.

(4) Shewell, J. and Wolf, E.: Inverse Diffraction and a New Reciprocity Theorem, Jour. Opt. Soc. Am., Vol. 58, p. 1596, December 1968.

(5) Brigham, E. and Morrow, R.: The Fast Fourier Transform, IEEE Spectrum, p. 63, December 1967.

Dual Beam Antenna A Unique Waveguide Phased Array with Independently Steered Beams

D. K. ALEXANDER
B. A. SICHELSTIEL

Westinghouse Electric Corporation
Aerospace and Electronic Systems Division
Baltimore, Maryland

ABSTRACT

A new class of array antennas that is characterized by the formation of two simultaneous, independently-steered beams from a common aperture area is presented. These arrays have become an important addition to the systems that operate with simultaneous multiple modes. The arrays combine single-axis electronic beam steering with mechanical gimballing to provide two beams with completely independent, two-dimensional beam agility. Each beam has a gain that is equivalent to the gain of a single antenna having the same effective aperture. In addition, the beam shape, gain and sidelobe level of each radiation pattern can be tailored to match the requirements of the particular function which it serves. By using a combination of mechanical and electronic steering, the beams can be steered over a total sector of 240 degrees.

A representative form of the concept was built recently to verify the predicted performance. In this case, the two arrays were both edge-slotted waveguide arrays. Mechanizing the concept presented numerous unique design considerations that are described in the paper.

DUAL BEAM ANTENNA

INTRODUCTION

The achievement of peak performance in modern electronic systems is sometimes restricted by compromises in the antenna design. The requirements of multiple function and multiple channel systems are the most difficult to fulfill. Often these systems can attain their ultimate capacity only if more than a single radiated beam can be provided simultaneously. However, when the coexistence of beams is measured in real time and the space for additional aperture area is not available, a gain reduction corresponding to the number of beams usually can be expected. A dual-beam antenna design which offers full aperture gain for two independent beams from a common area has been demonstrated. The features of this design and its unique problems are presented in this paper together with the measured performance.

The basic design uses two edge-slotted waveguide arrays; one superimposed on the second at right angles in such a way that the slots of the rear array are located at the spaces between the waveguides of the front array. (See Figure 1). The radiation patterns of the two arrays are independently formed as a result of the "decoupling" provided by the orthogonal polarization characteristic of this design. In the version of the concept described in this paper, an electronic beam steering feature has been added to increase the versatility of the technique for a number of systems. Independent pointing control of the two beams is accomplished by electronically steering the beams along one axis and mounting the complete dual array assembly on a two-axis gimbal.

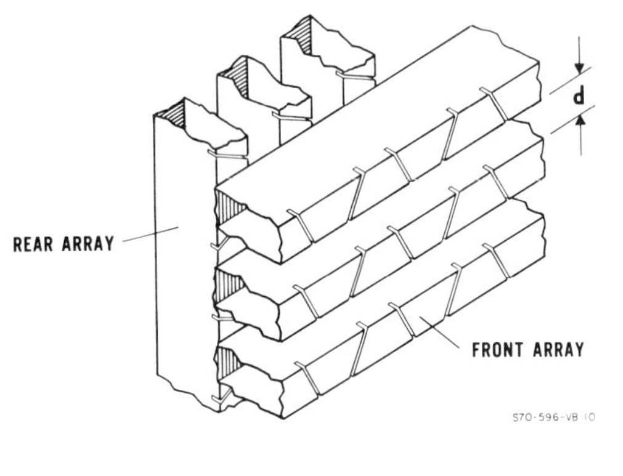

Fig. 1 Array Face Detail – Conceptual Design.

FUNCTIONAL DESCRIPTION

The dual array represented by the pictorial diagram shown in Figure 2, is capable of placing the two

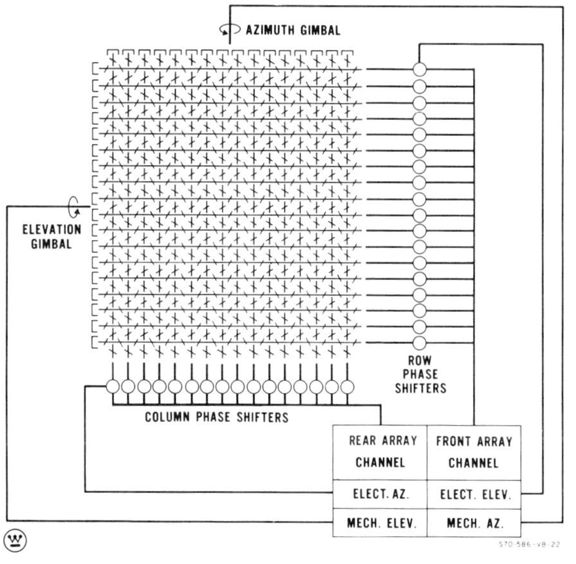

Fig. 2 Dual Beam Antenna – Schematic Diagram.

beams independently within the scanning limits of each array. The row phase shifters of the front array, as depicted, steer the beam electronically in elevation together with an aximuth gimbal for two plane positioning. Conversely, the column phase shifters steer the rear array beam electronically in azimuth and the elevation gimbal handles the second axis.

The effect achieved by combining mechanical and electronic beam steering in the two cardinal planes will permit, for example, one beam to be held on a point in space while the second beam is performing some other scanning function. Mechanical gimballing to move one beam in one plane may be counteracted by electronically steering the second beam to hold its position fixed.

A scan limit of ±100 degrees has been demonstrated; ±45 degrees of electronic steering with ±60 degrees of mechanical gimballing. The electronic limit is based on the amount of radiation pattern degradation that can be tolerated and is largely affected by mutual coupling and grating lobe problems. The latter is discussed in a subsequent paragraph.

APPLICATIONS AND ADVANTAGES

The two beam feature clearly is useful for multi-channel communications systems with dispersed non-static terminal locations. But more significant, are the radar applications where functions operate simultaneously. In comparison with two axes electronic scanning arrays were high cost and weight have hampered their progress toward widespread acceptance, there are several points that make the dual beam array concept distinctively superior. In the weight category, a two-axes, 1-meter, X-band array may have several thousand phase shifters compared to less than one hundred for the dual array. Also, by virtue of the required two-axes gimbal, the dual beam array offers an angular coverage of greater than 200 degrees. Finally, the individual patterns of the two arrays can be designed to custom-fit the requirements of the particular function which they serve, and further, can be mechanically or electronically steered depending on the relative need for minimum beam degradation or rapid beam agility.

There is also the possibility of producing circularly polarized radiation with this array. By connecting the two arrays to a common point through the proper time-quadrature phasing circuit, the combination would act in a way similar to two crossed dipoles displaced by 90 degrees. The basic approach has been expanded to also include systems which operate in two separate frequency bands. For some systems, versions without the electronic steering feature have been found to be useful. Tailored radiation patterns for co-incident beams is all that is needed for functional enhancement in those cases.

DEMONSTRATION MODEL

The validity of the dual array principle was tested using a rectangular model, a low sidelobe scanning feature in the front array. The model, shown in Figure 3, initially used mechanical phase shifters for steering the two beams in the H-planes of each array. An analog ferrite phase shifter* was developed concurrent with demonstration model program to ultimately test the practical constraints of electronically steering a low sidelobe pattern. The tolerances on phase setting accuracy were analyzed and set at ±3 degrees to maintain a maximum peak sidelobe degradation of 5 dB to the steering angle limits.

*R. Kasevich, et. al., "High-Performance High-Power Analog Phase Shifter; Proc. IEEE Vol. 57, No. 8, pp. 1427-1429; August 1969.

The rear array illumination amplitude is tapered slightly in order to produce -18 dB peak sidelobes. The center frequency of the rear array is approximately 5 percent higher than the front array. A nonstandard waveguide size is employed in the construction of the rear array so that the waveguide wavelength is compatible with the spacing between rows of the front array.

Fig. 3 Feasibility Demonstration Model.

DESIGN PROBLEMS

In developing the dual beam antenna, a number of design problems were encountered. Significant among them:

(1) Obtaining sufficiently high conductance values from the rear array slots.

(2) Matching the aperture of the rear array to free space.

(3) Suppressing cross-polarized energy to acceptable levels.

The first two of these have been essentially solved; the third is the subject of continuing development.

As originally conceived, the dual beam antenna would have the configuration shown in Figure 1. Notice that the see-through space, d, between the waveguides of the front array is determined by the element spacing of the front array and the height or "a"-dimension of the front array waveguide. For the case of the demonstration model, d was found to be 0.23 inches. Considering the near field environment of a typical slot in the rear array, we find two conducting walls, separated by the space d, immediately external to the slot. As expected, the impedance of the slot is significantly affected by the proximity of the walls. In particular, achievable values of slot conductance are severely reduced. For maximum slot angles (approximately 20°) normalized slot conductance values of about 0.02 were measured. Thus, the utility of this slot configuration would seem to be severely limited as, even for the resonant array, the summation of the slot conductances must equal one. Clearly, the original concept had to be modified.

In addition, the radiated energy from the rear array experiences a step discontinuity as it passes through the plane of the front array face and enters free space. Left uncompensated, this discontinuity would cause intolerable reflections.

The design used to solve both the above problems is shown in Figure 4. The spacing between the front

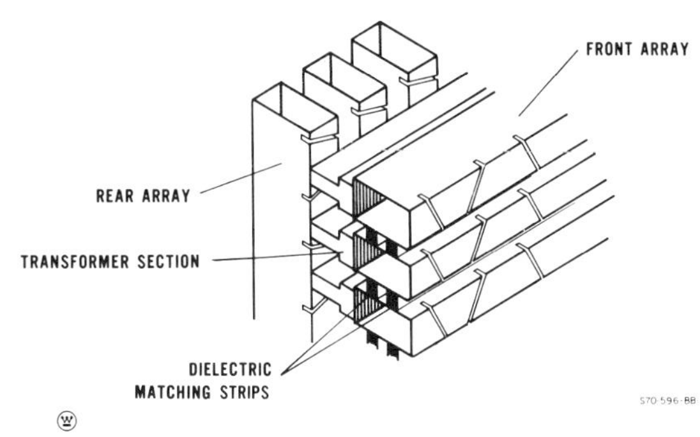

Fig. 4 Array Face Detail – Demonstration Model.

array waveguides is the same as before. However, the spacing of the metallic walls immediately external to the rear array slots has been increased allowing conductance values to be more than doubled. Treating this region and that between the front array waveguides as parallel plate waveguide, a quarter wavelength transformer is provided joining the two regions. In addition, dielectric strips are positioned as shown. Again, using parallel plate transmission line theory, the strips are dimensioned and positioned to exactly cancel the energy reflected from the interface with free space. Both of these aperture matching techniques (the quarter-wave and dielectric transformers) were designed to provide optimum performance for a broadside beam on the demonstration model. As beam steering takes place, their effectiveness decreases somewhat. It is felt that performance over wide steering angles could be improved by using more sophisticated versions of the same basic techniques.

The angling or tilting of the slots on edge-slotted arrays gives rise to cross-polarized radiation lobes. The magnitude and position of these lobes is dependent upon the particular slot angles, element spacing, and the programmed steering angle. On other edge slot arrays built at Westinghouse, a shorting bar, occupying the space between the slotted waveguides, has been used effectively to cancel cross-polarized energy. These arrays were either of the fixed beam type or steered over limited angles. For the dual beam antenna, however, this space must remain clear in order for the rear array to radiate. It is conceivable, though, that the face and supporting structure of the rear array, being generally unreceptive (in incidence angle and frequency) to energy radiated by the front array, might be configured in a manner to effectively replace the shorting bars. The mechanization of this concept is quite involved and has yet to be implemented. In anticipation of its eventual use, however, the front array of the dual beam antenna was fitted with adjustable shorting bars between the waveguides and the level of cross-polarized energy was measured. The results of these measurements for a short spacing of 0.225 inches is summarized in Figure 5. The 0.225 dimension was found to be optimum with the beam steered to 25 degrees and yielded the indicated performance at other steering angles.

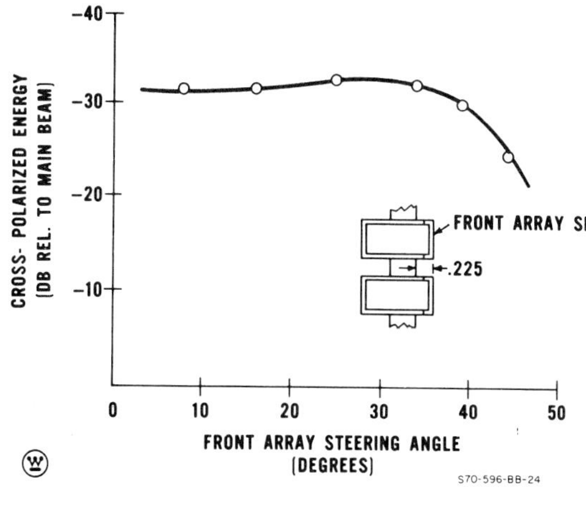

Fig. 5 Maximum Cross-Polarized Energy Level as a Function of Main Beam Steering Angle.

While this level of cross-polarization suppression is acceptable for many antenna applications, it is incompatible with a low sidelobe array where peak sidelobe levels of the intended polarization are held in the range of -35 to -40 dB. In this regard, one additional factor should be considered. On the demonstration model, standard WR-90 waveguide was used for the front array and the operating band was considerably removed from cutoff. This necessitated relatively high slot angles (maximum value of approximately 20 degrees) in order to obtain the required slot conductances. The higher slot angles naturally contributed an excessively high cross-polarized component. In future designs, therefore, a waveguide size closer to cutoff will be chosen (possibly of special manufacture) in order to alleviate the cross-polarization problem. If, for example, maximum slot angles could be held to 10 degrees, one would expect a 6 dB improvement in the curve of Figure 5.

FEATURES OF THE FEASIBILITY MODEL

The test model was built to demonstrate the feasibility of the basic concept and to identify and solve some of the unique problems associated with it. Emphasis was placed on the design of the actual radiating structure. Peripheral considerations such as manifold configuration, phase shifter placement,, supporting structure, etc. were designed for engineering convenience rather than end-use, weight, volume, and form factor.

The radiating aperture of the model is approximately 21 inches square. Each array is made up of 27 edge-slotted waveguides. Each waveguide in the front array has 27 slots and in the rear array, 26. A waveguide manifold is used to end-feed each array. The rear array manifold is a conventional tee-junction type while the front array employs a unique variation of the directional coupler manifold, which is discussed in greater detail below. The model is structured such that both arrays may be fitted with either mechanical phase shifters or developmental analog ferrite phase shifters.

Due to the limited scope of the development program, it was necessary to use standard size waveguide wherever possible to minimize test equipment and material costs and design complexity. Waveguide of special manufacture was used only for the slotted waveguides in the rear array. As a result the waveguide spacing of the front array is such that a grating lobe appears when the main beam is steered to approximately 40 degrees. The position and amplitude of this lobe conforms closely to theoretical predictions. For future arrays, waveguide sizes and array configuration will be chosen to allow larger steering angles.

The two arrays were designed to have quite different radiation characteristics. The front array is designed to have very low sidelobes, on the order of 40 dB down, with no more than 5 dB degradation during beam steering. The rear array has near maximum gain with first sidelobes only 16 to 18 dB down from the main beam. The measured performance confirmed the unique capability of individually tailoring the characteristics of the two arrays to meet different requirements.

The front array low sidelobe performance was accomplished through the use of a specially derived computer program which takes into account all mutual terms in the impedance matrix for the slot elements. Accurately compensated antenna parameters were computed for the aperture distribution to achieve the sidelobe suppression level of greater than 40 dB. The antenna was built directly from the computer generated dimensions and no adjustments were required to correct errors in design or construction.

The design technique is based on slot data acquired by precisely measuring representative slot characteristics. Three fundamental constants are determined. These constants are related to three parameters: (1) the depth to which the slots must be cut into the broadwall for a resonant zero coupling slot, (2) the rate of change of slot susceptance with the slot depth near resonance, and (3) the coupling variation with slot angle.

The computer program combines the raw slot data with sampling data for the desired far-field radiation pattern to compute the required driving point impedances at the elements. The actual slot angles and depths are then calculated for the complete array. The aperture illumination function used in the design can be expressed by the relationship

$$F(x) = (0.08 + 0.92 \cos^2 \frac{\pi x}{a})$$

The theoretical maximum peak sidelobe level for this function is 42 dB.

Beam-steering a low sidelobe array antenna demands that element mismatches be dealt with. For broadside arrays, these mismatches primarily affect only the input impedance of the array and normally have a minimal influence on the radiation pattern. Beam steering, however, introduces a phase bias in the energy reflected from the elements which, with a manifold made up of simple three-port junctions, gives rise to uncontrolled re-radiated energy. In general, this re-radiation has a degrading effect on pattern characteristics but particularly so when sidelobes are being held to very low amplitudes. If four port junctions are used in the manifold, out-of-phase reflections are absorbed in the terminated port. The desired performance of the test model's front array necessitated the use of such a manifold. The manifold configuration chosen consisted of two identical tee-junction manifolds coupled to the array waveguides through short-slot waveguide hybrids. This configuration is particularly attractive from a mechanical packaging standpoint since the array port spacing is the same as for the individual tee-junction manifold, thus eliminating a problem often encountered with conventional directional coupler manifolds. Coupling is also readily controlled. Each tee-junction uses an edge slot in the manifold waveguide and the angle of the slot (which determines the coupling) is a straight-forward, repeatable machining operation. The scheme is depicted in the block diagram of Figure 6. The measured patterns over a range of steering angles confirm its effectiveness.

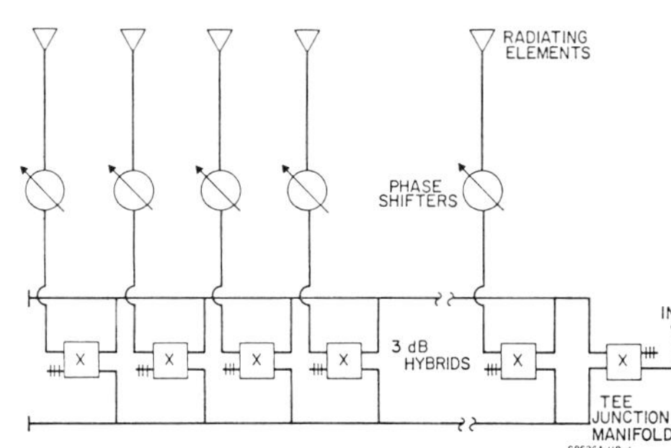

Fig. 6 Reflectionless Manifold System.

MEASURED PERFORMANCE

The measurements program on the test model was extensive but of particular interest were the following general goals:

(1) Obtaining very low sidelobe performance from the front array.

(2) Scanning the front array while maintaining low sidelobes.

(3) Obtaining a predictable radiation pattern from the rear array.

These goals were essentially achieved as evidenced by the measured data. Looking first at the front array, Figure 7 shows the E-plane pattern with all sidelobes more than 40 dB down from the main beam peak. Following that, Figures 8, 9 and 10 show H-plane or "scanning plane" patterns at broadside and five steering angles. Peak sidelobe levels of -38dB at broadside degrade slowly to about -34 dB at 34 degrees steering angle. At 39 degrees steering angle, the expected grating lobe has emerged.

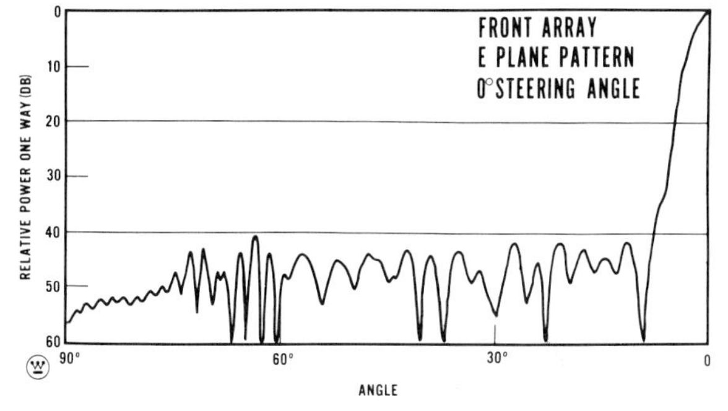

Fig. 7 Front Array, E-Plane Pattern, 0° Steering Angle.

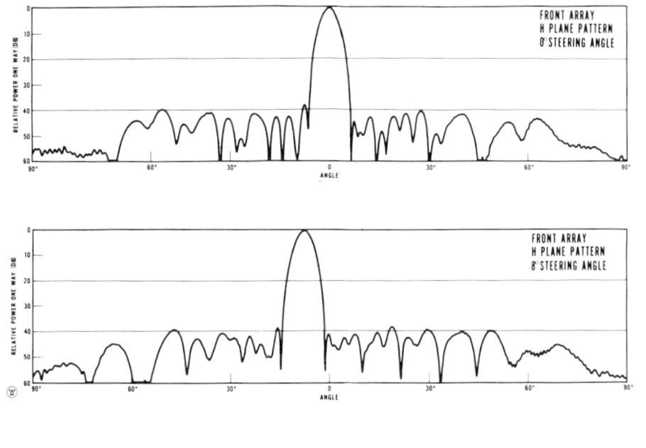

Fig. 8 Front Array, H-Plane Pattern, 0° Steering Angle and 8° Steering Angle.

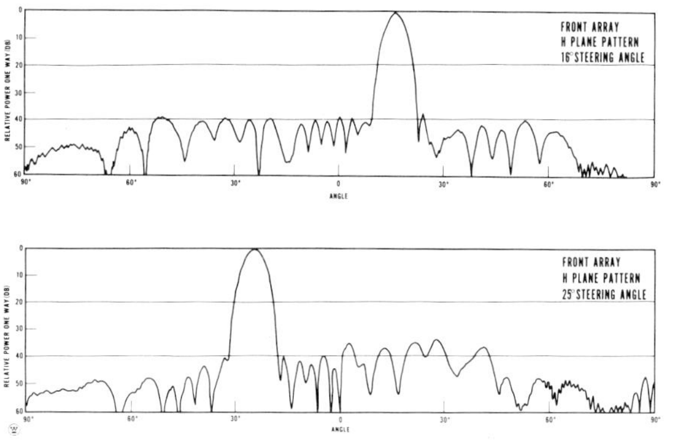

Fig. 9 Front Array, H-Plane Pattern, 16° Steering Angle and 25° Steering Angle.

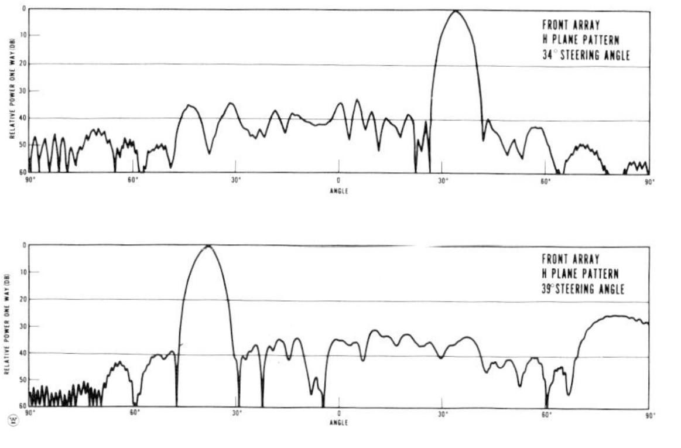

Fig. 10 Front Array, H-Plane Pattern, 34° Steering Angle and 39° Steering Angle.

The rear array produced broadside patterns of very regular and near-theoretical shape shown in Figures 11 and 12. As the main beam is steered, sidelobes rise slowly and gain falls evenly until a loss of about 3 dB is noted at 36 degrees steering angle. This seemingly large gain reduction is due, primarily, to the "first cut" aperture matching of the rear array discussed above. The measured patterns for the steered beam are shown in Figures 13, 14, 15, 16, and 17.

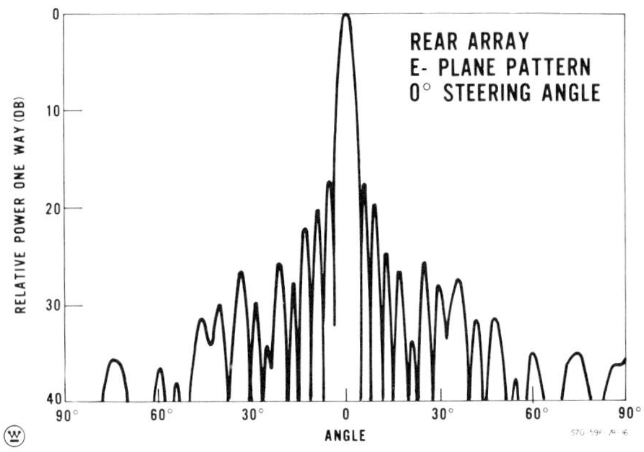

Fig. 11 Rear Array, E-Plane Pattern, 0° Steering Angle.

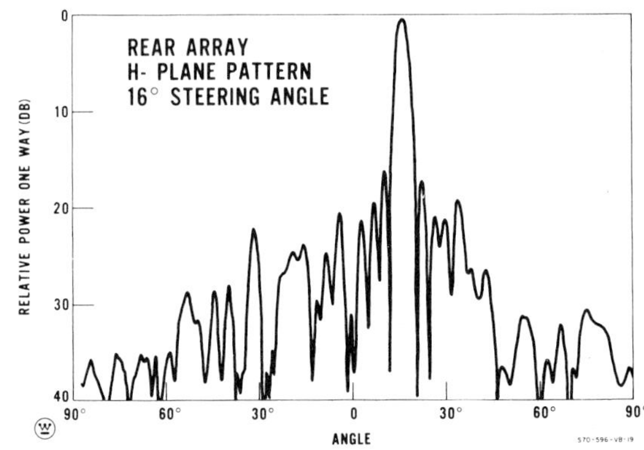

Fig. 14 Rear Array, H-Plane Pattern, 16° Steering Angle.

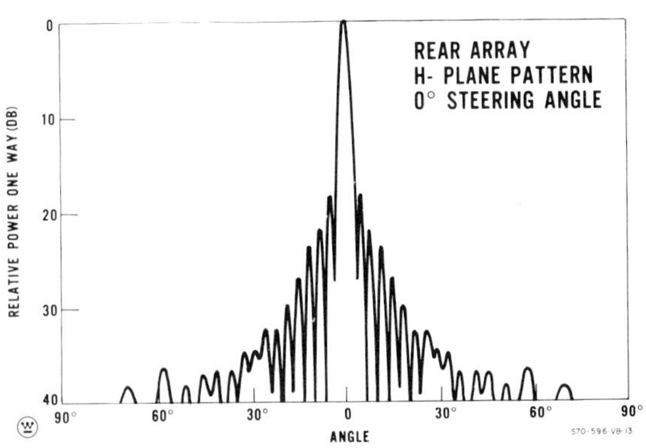

Fig. 12 Rear Array, H-Plane Pattern, 0° Steering Angle.

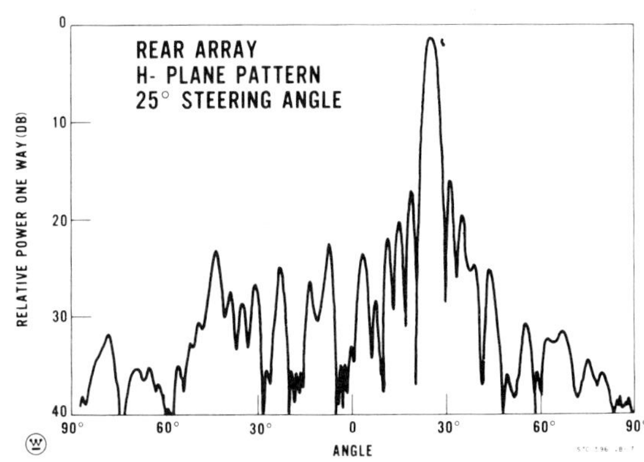

Fig. 15 Rear Array, H-Plane Pattern, 25° Steering Angle.

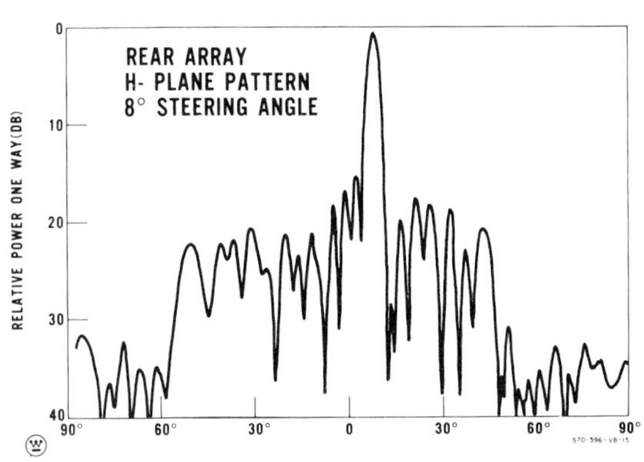

Fig. 13 Rear Array, H-Plane Pattern, 8° Steering Angle.

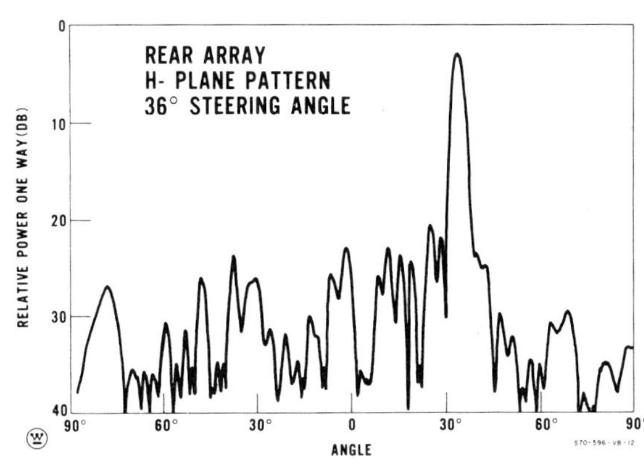

Fig. 16 Rear Array, H-Plane Pattern, 36° Steering Angle.

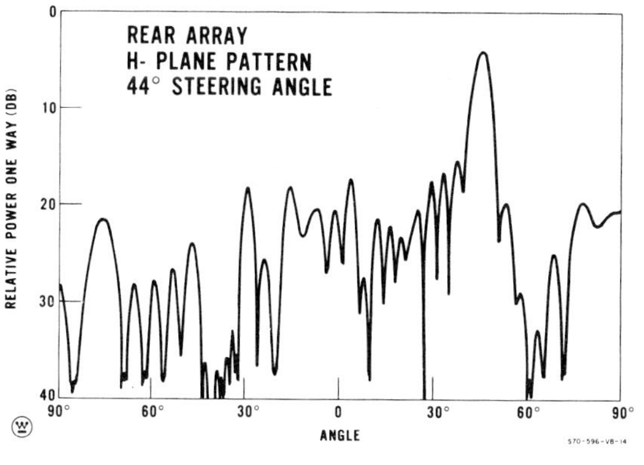

Fig. 17 Rear Array, H-Plane Pattern, 44° Steering Angle.

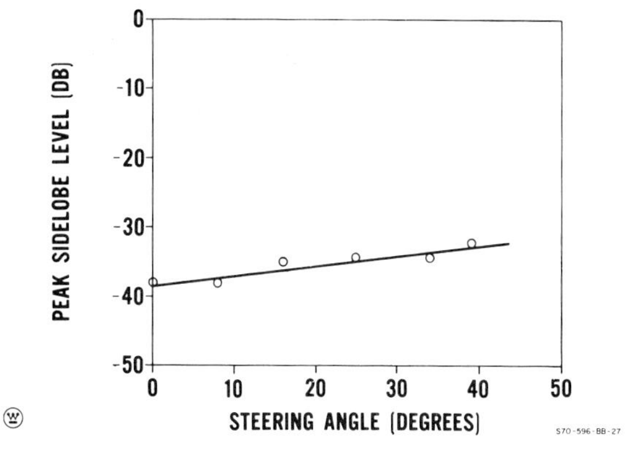

Fig. 19 Peak Sidelobe Level vs. Steering Angle.

Two performance parameters seem particularly worthy of summarization in graphic form. Figure 18 shows gain degradation of the front array with steering angle. A less than 2 dB gain reduction is incurred at steering angles out to 35 degrees. The dashed line shows expected performance if the design were modified to avoid the early emergence of a grating lobe. Figure 19 notes the regular trend of peak sidelobe levels as the front array undergoes beam steering. (The grating is neglected here.)

The dual independent beam array investigation may be readily and briefly summarized by comparing parameter goals with the values actually achieved. This is tabulated in Figure 20.

PARAMETER	FRONT ARRAY		REAR ARRAY	
	GOAL	ACTUAL	GOAL	ACTUAL
MAXIMUM STEERING ANGLE	45°	36°	45°	45°
PEAK SIDELOBE LEVEL				
BROADSIDE (-dB)	40	38	18	16
STEERED (-dB)	35	34	–	12
GAIN				
BROADSIDE (dB)	32.9	32.5	34.9	34.8
STEERED (dB)	31.4	31.0	–	30.8

Fig. 20 Parameter Summary Table.

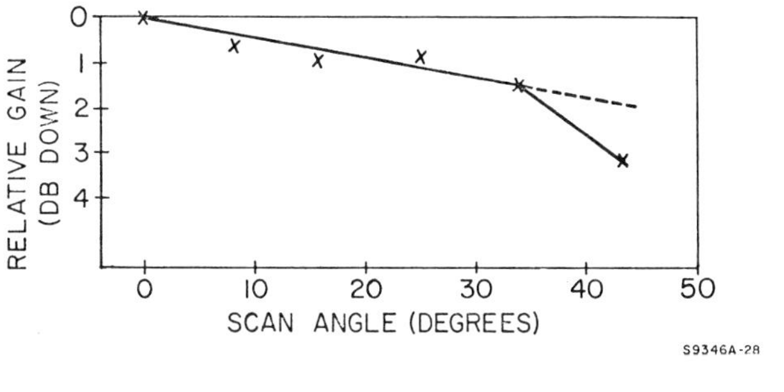

Fig. 18 Gain vs. Scan Angle, X-Band.

ACKNOWLEDGEMENT

The work reported in this paper was supported, in part, by Contract N62269-68-C-0776 between the Naval Air Development Center and Westinghouse Defense and Space Center.

A Single-Plane Electronically Scanned Antenna for Airborne Radar Applications

D. J. LEWIS,
J. R. LEE,
D. K. McCARTHY
Aerospace Group
Hughes Aircraft Company
Culver City, California

INTRODUCTION

The antenna system described in this paper was designed, fabricated and tested to demonstrate the feasibility of a gimbal-mounted phase scanned antenna for use in airborne radar systems. It was felt that an antenna providing single-plane electronic scanning capability and light enough to mount on a conventional two-axis radar gimbal system, would offer a useful alternative to a full two-plane electronic scanning system. By coupling the electronic scanning capability in one plane, with mechanical scanning in the other, sufficiently high scan rates can be provided to obtain many of the desirable performance characteristics of two-plane phase-scanned arrays, but without their complexity and cost. In addition, mechanical and electronic scanning can be combined in a single plane to realize a scanning range that exceeds that obtainable with either scanning system alone.

DETAILED DESIGN DESCRIPTION

The design configuration selected for the antenna utilizes digital, latching, ferrite phase shifters to provide electronic scanning in the H-plane. The array structure was fabricated from lightweight aluminum waveguide and assembled with structural epoxy. The array is designed to operate over a 400 MHz frequency range in the X-band region with a phase-scanning range of ±45 degrees from the normal to the array.

Figure 1 is a picture of the planar array mounted in the nose of a B-26; a photograph of the array mounted on the pattern range is shown in Figure 2. The antenna is very nearly 26 inches square and weighs approximately 21 pounds (including the phase shifters, wiring, connectors and mounting hardware). To provide full two-plane monopulse capability, the array is divided into four quadrants. Each quadrant is constructed from a set of resonant waveguide linear arrays of narrow wall, radiating slots. Baffles approximately one-half wavelength high extend across the face of the array (horizontally in Figure 2a) between each row of radiating slots. The function of these baffles is to eliminate the cross-polarized field component associated with the radiating slot configuration used.

Fig. 1 Array Mounted in B-26 Aircraft for Flight Test.

Each quadrant of the array contains eighteen linear-array elements, the associated phase shifters, and a traveling-wave feed structure that serves to distribute the input power among the elements. The linear-array elements are bonded together with a structural epoxy. U-shaped aluminum spacers are used between each linear array to maintain the correct inter-element spacing. The baffles, spacers and waveguide shorts are made from 0.006-inch aluminum foil, and the waveguides are 0.020 wall aluminum extrusions. The entire quadrant is bonded to a lightweight honeycomb panel that serves as a mounting plate for assembly of the full array.

*The research reported in this paper was sponsored by Naval Air Systems Command under Contract NASC 66-0736-di.

Fig. 2 Array Mounted on Range for Pattern Tests.

The phase shifters are attached to the individual radiating elements with special lightweight flanges; the traveling-wave feed is, in turn, attached to the phase shifters with a similar set of flanges. A view of a quadrant with a complete set of phase shifters and the traveling-wave feed installed is shown in Figure 3.

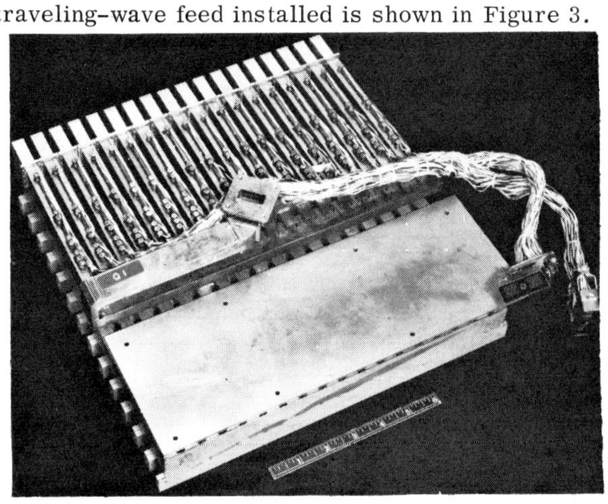

Fig. 3 Rear View of Array Quadrant.

RADIATING ELEMENTS

The linear radiating element consists of a resonant linear array of 13 narrow-wall inclined waveguide slots. The choice of narrow wall slots was dictated by the close element spacing required to avoid the formation of grating lobes. A 25 dB peak sidelobe level Taylor weighting function was used for the aperture distribution.

TRAVELING WAVE FEED

The power is distributed to the phase shifters in each quadrant by means of a traveling-wave feed line that runs across the back of the quadrant. The traveling-wave design was selected because its simplicity leads to a substantial savings in weight, volume, and complexity. The principal disadvantage to the use of the traveling-wave feed is that it produces a phase taper across the face of the array that must be taken into account when computing the proper phase-shifter settings.

PHASE SHIFTERS

Digital latching ferrite phase shifters were used in the array. A photograph of several phase shifters manufactured for the array is shown in Figure 4.

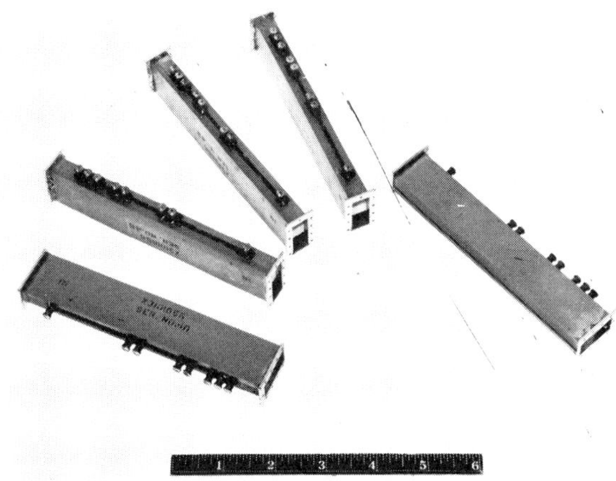

Fig. 4 Array Phase Shifters.

Basically, the latching ferrite phase shifter consists of a rectangular ferrite tube mounted coaxially in the center of a section of X-band waveguide. A single turn control loop threaded through the center of the tube is used to switch the magnetic state of the ferrite and thus change the phase shift through the section. If four ferrite cores of appropriate lengths are provided, it is possible to obtain phase steps of 180°, 90°, 45°, and 22.5°, or any combination of these states. The insertion loss of each phase shifter is approximately 1 dB.

The above type of phase shifter is a latching device; that is, a single pulse of current through the control loop suffices to establish a given phase setting; no bias currents are required. However, the device is inherently nonreciprocal and it is, therefore, necessary to reset the array phase shifters between the transmission and reception of each radar pulse.

A semi-conductor control rectifier (SCR) circuit was used to control the current pulses used to drive each phase shifter. Two SCRs were required for each ferrite section in each phase shifter. A switching time of approximately two microseconds was required to reset the phase shifter.

EXPERIMENTAL RESULTS

A set of pencil-beam patterns recorded at the design frequency in both the E- and H-plane is given in Figure 5. Scan angles in the H-plane of 0, 20, and 45

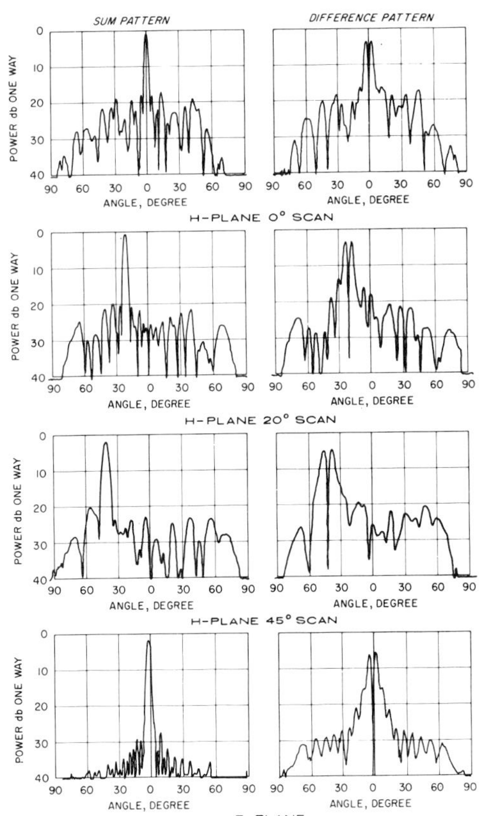

Fig. 5 Typical Radiation Patterns.

degrees are shown. These patterns were recorded with the antenna acting as a receiver. Similar sets of patterns recorded with the array acting as a transmitter revealed no significant changes. During the transmitting tests, a radiation pattern was recorded with the array operating with an input power of 80 kilowatts peak and 40 watts average. Changes in the pattern recorded under these high-power conditions were negligible.

The antenna pattern characteristics as a function of both frequency and scan angle are summarized in Figures 6 through 12 which show pointing error, 3 dB beamwidth, gain, depth of null, and peak sidelobe levels as functions of scan angle.

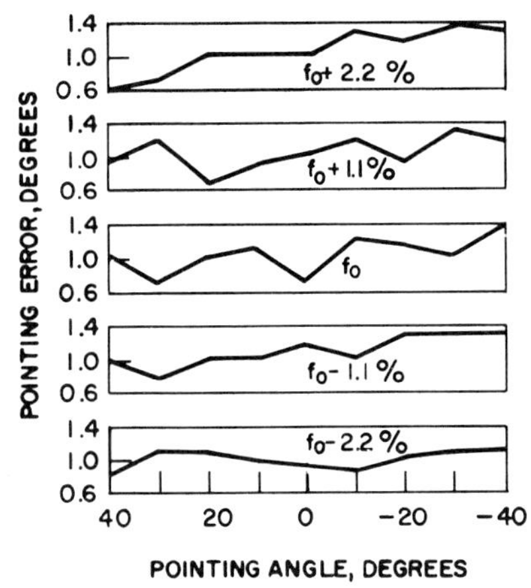

Fig. 6 Pointing Error as a Function of Scan Angle (H-Plane).

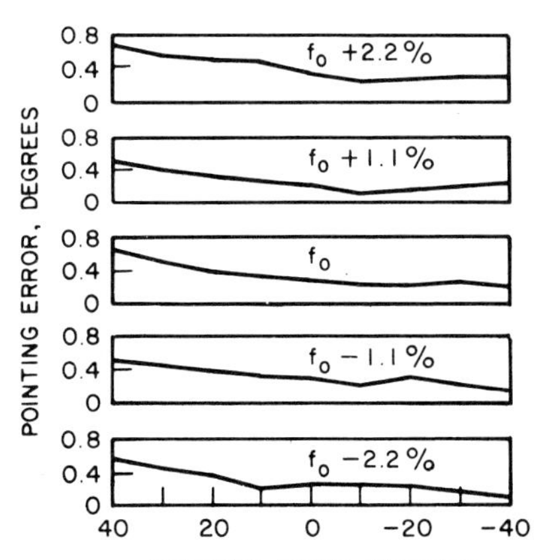

Fig. 7 Pointing Error as a Function of Scan Angle (E-Plane).

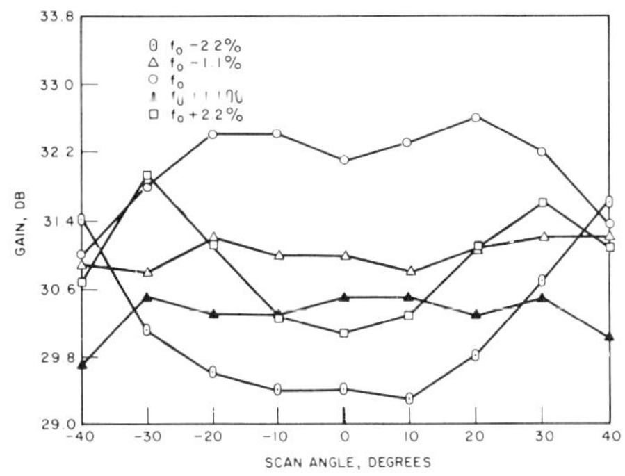

Fig. 8 Gain as a Function of Scan Angle.

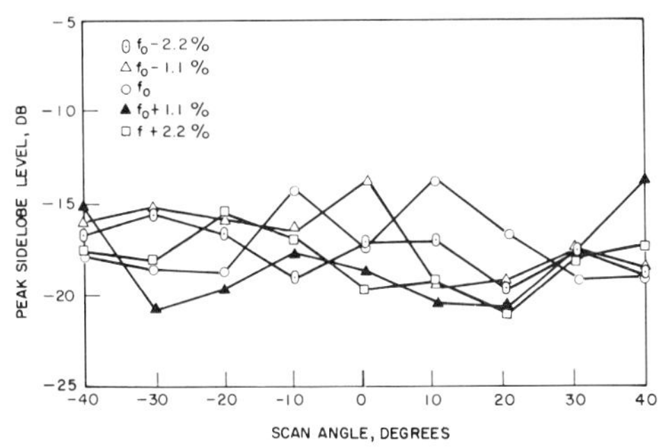

Fig. 9 Peak Sidelobe Level as a Function of Scan Angle.

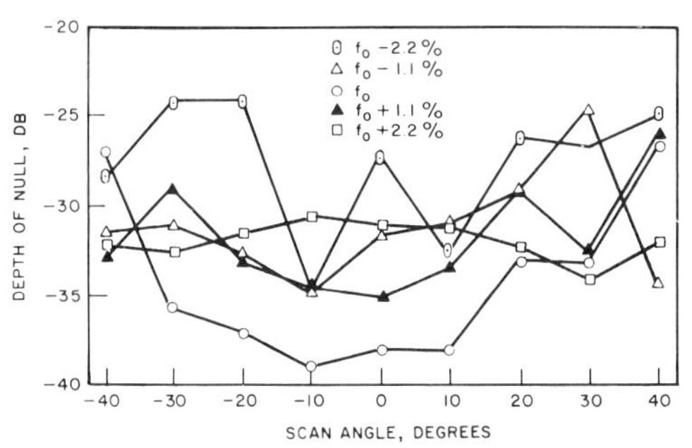

Fig. 10 Depth of Monopulse Null as a Function of Scan Angle (H-Plane).

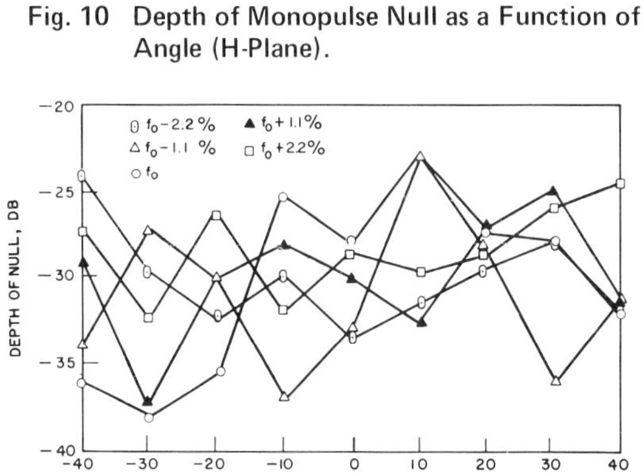

Fig. 11 Depth of Monopulse Null as a Function of Scan Angle (E-Plane).

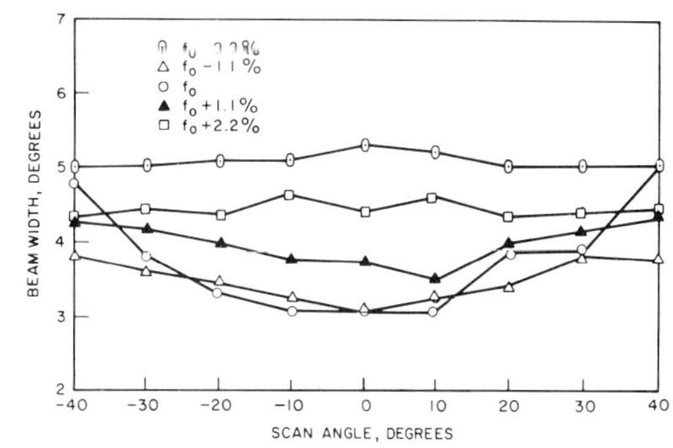

Fig. 12 Three DB Beamwidth as a Function of Scan Angle.

The input VSWR to the antenna was measured as a function of frequency and scan angle during the pattern measurements. The results are shown in Figure 13. Because of the inherently good frequency characteristics of the traveling-wave feed, an excellent match was maintained everywhere with the prescribed operating limits.

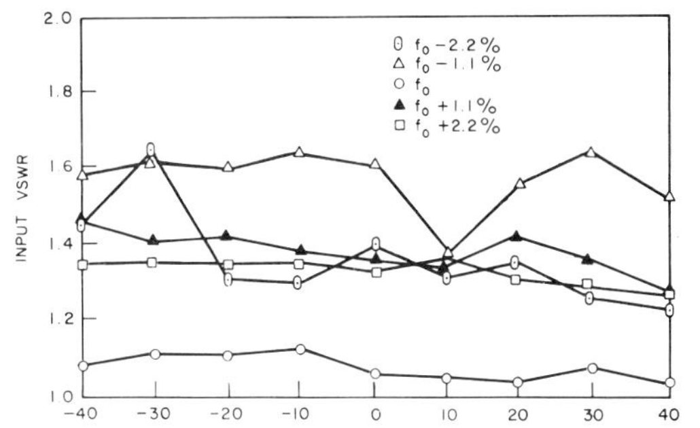

Fig. 13 VSWR as a Function of Scan Angle.

Because the ferrite phase shifters are temperature sensitive devices, the array was heated and roughly stabilized at a temperature of 90° F during the pattern measurements. However, a set of patterns recorded as the array temperature was varied between 79° F and 160° F revealed no serious performance degradation. If larger temperature variations than indicated above were anticipated, it would be desirable to compensate for the corresponding phase shifter errors. This can be accomplished in the computer used to program the phase shifters.

SHAPED BEAM PATTERN

In addition to the pencil-beam patterns, a mapping beam mode of operation was tested with the array. Using phase control only, a csc^2 beam was synthesized. The resultant pattern is shown in

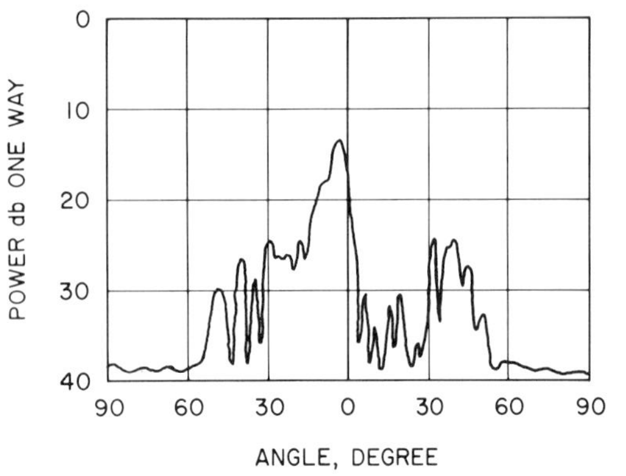

Fig. 14 H-Plane Mapping Beam.

DYNAMIC BEAM STEERING PATTERN TESTS

The patterns depicted in Figures 5 and 14 were recorded in a conventional or "static" fashion; that is, the beam was electronically steered to a given angle with respect to the aperture, and the pattern then recorded by mechanically rotating the aperture.

A second set of patterns that were recorded with a dynamic technique is shown in Figure 15a and 15b. To obtain the patterns shown in Figure 15, the antenna was positioned at a fixed angle relative to the range transmitter, and then electronically scanned to create the radiation pattern.

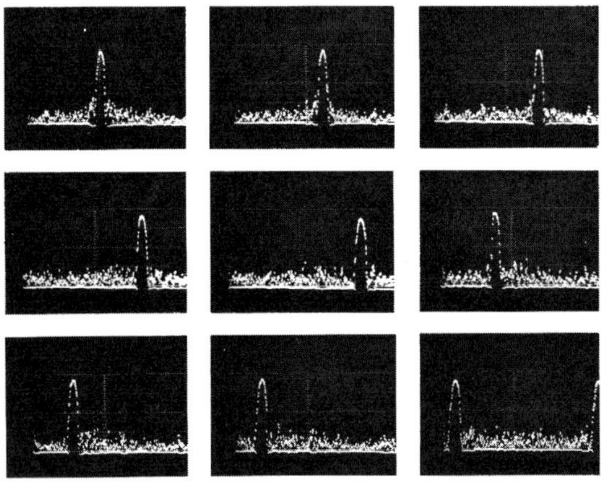

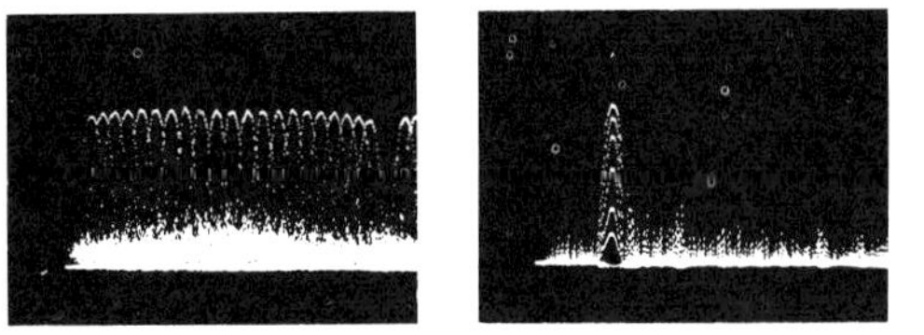

Fig. 15 Dynamic Patterns.

To record the dynamic patterns, a microwave receiver was connected to the output of the antenna sum port. An oscilloscope was connected to the output of the receiver and a camera was used to photograph the radiation pattern on the face of the oscilloscope. The receiver was over driven so that pattern structure 25 to 30 dB below the peak of the beam would be perceptible on the oscilloscope. The array beam steering computer was programmed to scan the beam from -45° to $+45^\circ$ in 0.2° steps at the rate of 1100 steps per second with several steps of turnaround time at the end of each scan. As the computer scanned the array, the phase shifters were set first in the transmitting state for 17 microseconds and then switched to the receiving state for the remaining 900 microseconds of each cycle. The scanning computer and driver circuits were synchronized to the range transmitter so that the transmitter was on only when the antenna was in the receiving state.

A composite pattern was recorded as the antenna was rotated while simultaneously electronically scanning the beam (Figure 15b). The composite pattern permits a rapid evaluation of the array behavior at many scan angles and verifies the proper behavior of the scanning control circuits as well. Note the envelope of the peaks of the beams in the composite pattern traces the gain versus scan angle curve for the array.

CONCLUSIONS

The basic feasibility of the antenna system has been demonstrated. Modification of the aperture shape to meet any particular installation requirements should present no significant difficulty. Further development work would be required, however, to provide a greater frequency bandwidth because of the limitations imposed in the present model by the linear standing wave elements.

Although the microwave characteristic of the array has been the primary subject of this paper, it may be observed that a model of the array has been successfully subjected to mechanical testing and that the array described is presently being used in flight tests.

Physical Design Considerations for Airborne Electronic-Scanning Antennas

M. H. Rosengard

Aerospace Group
Hughes Aircraft Company
Culver City, California

I. INTRODUCTION

The total physical design problem involved in an airborne electronic-scanning antenna system is described in this paper. Differences between airborne and the well-established ground design techniques are considered. The critical elements such as the effect of aircraft installation on antenna surface deflection, thermal densities, the structural integrity of the total installed system, interconnections, materials, maintenance, reliability, manufacturability, and the use of advanced microelectronic techniques are discussed in light of problems peculiar to airborne military environments and requirements. The very number of elements involved in such antennas poses a new set of problems to the physical designer faced with the production of a cost effective system that meets specifications. "Hard" design decisions have to be made early and, because of the cost involved, these decisions become nearly irreversible.

II. COMPARISON OF GROUND, SHIP AND AIRBORNE REQUIREMENTS

It is not the purpose in this discussion to imply that designing an electronically-scanned antenna for airborne use is more difficult than designing one for ground or ship use. More than that, the designer of airborne electronically-scanned antennas would be derelict in his duty if he did not take advantage wherever possible of all the ground and ship design effort that has gone on in the last decade. The Hughes Ground Systems Group has been historically at the forefront of this design effort. Their experience has been invaluable to us in the Hughes Aerospace Group designing airborne arrays.

Probably the most significant difference in design is the factor of weight. Weight, combined with the geometry limitations of an aircraft, requires that the designer be particularly aware of material selection and structural integrity from the onset. Many materials available for ground or ship systems are prohibited in airborne systems because of their high density.

Compactness of design is required, as every square inch of available antenna surface in an airframe is at a premium. This means that services to the array elements must, by definition, come into the rear of the antenna as opposed to its periphery. Such services include interconnections, coolants, and structural support.

Since aircraft in general have limited access for maintenance one must make early decisions regarding maintenance philosophy and arrangement of equipment. In particular, these must be tied to the expected maintenance actions required during the lifetime and use of the system. Failure rates must be allocated such that rapid and effective maintenance action, when required, can be taken without inordinant assembly and disassembly of parts. In fact, one must decide whether field repairs to the element level such as phase shifters are even necessary. The removal of a radome from a large side-looking antenna could very well be considered an unacceptable field maintenance action; hence, acceptable limits of degradation of antenna performance must be very carefully evaluated.

The result is that the designer of airborne electronic-scanning antennas must consider the system reliability and maintenance requirements much more carefully than the designer of ground or shipboard systems.

III. AIRCRAFT INSTALLATION

It is doubtful that an interface specification can be generated for a usable physical-design concept for an airborne antenna without working directly with the airframe designer. The reasons are numerous. Many adjacent thermal zones exist in an airframe whose interface represent a high Δt. Electronic-scanning antennas are so large that when mounted rigidly to an airframe they can become load bearing members. Radome attachment methods play a large role in the aforementioned maintenance considerations. Available aircraft services, particularly in terms of cooling, are all part and parcel of the antenna physical design.

Aircraft "hard points," or mounting surfaces must be carefully selected so that an antenna that is factory boresighted will remain boresighted when installed. Normally, the locations selected for antenna system use within the airframe are not structurally adequate to be used as "hard points." Therefore, the total antenna/airframe configuration must be looked at from the standpoints of the antenna pattern desired, the airframe structure, and last but not least, the antenna mechanical design.

The class of the aircraft also plays a role in antenna design decisions. A fighter obviously has quite different loads during its mission than a bomber. It is not impossible to envision side-looking antennas so large that their hard points would be so far apart they would not stay in the same plane during maneuver loads. At this point, the physical designer must work closely with the systems designer to determine whether or not system operation is necessary during violent or evasive maneuvers. Smaller forward-looking or rear-looking antennas are in most cases bulkhead mounted and, indeed, do have good structurally sound "hard points" eliminating this problem. However, large side-looking arrays require careful design and close coordination among the three parties involved (airframe, system, and antenna designer) or desired system performance will not be obtained.

The requirements for airframe cooling are discussed in the next section; however, it might be well to point out here that these are of no little consequence. Electronically-scanned antennas present high heat density thermal loads, which must be considered during the airframe design so that adequate environmental control is supplied.

The vibration inputs to the antenna from the airframe during the mission must also be known and their impact on the antenna design minimized. Of particular interest would be the gun fire environment that exists in some classes of aircraft. Even though the equipment might not have to operate during gun fire, its effect on the antenna phase shifter life would have to be determined.

IV. THERMAL CONSIDERATIONS

At Hughes it is customary to base our equipment design on thermal/reliability relationships. The RADC Reliability Handbook, containing excellent curves, to which Hughes has contributed, is used as a basis for our thermal and physical design. Our approach is to select the most desirable operating semiconductor junction temperature for our microelectronic elements, determine a Δt between the junction and our coolant, and maintain the thermal interface between the coolant and junction such that the selected junction temperature is not exceeded. It should be noted that we use junction, not case, temperatures in our designs. While this is a bit of a hardship on the equipment designer because he must then design for selected components, taking into account the effect of all of the possible enclosures, it ultimately provides greater reliability.

The radar antenna may be cooled by means of a cold plate through which cooling fluid flows. All heat dissipated at the antenna will be conducted to the cold plate. From a thermal design standpoint, the critical components are the phase shifters. Because of the mounting requirements of the phase shifters and the radiating elements, the thermal conduction path to the cold plate is necessarily complex, and has a large number of thermal interfaces. This complex thermal path requires a high cold plate performance and suitable low cooling fluid temperatures to maintain the junction temperatures of the diodes on the phase shifters at or below the permissible limit. A liquid coolant, such as Coolanol 25, is required to obtain the required cold plate performance.

The liquid flow rate and supply temperature requirements depend upon the total heat dissipation of the antenna. They also depend upon the location of the antenna in the aircraft. The various zones on a typical military aircraft and their respective temperatures ranges are given in Figure 1. Zones I and V may have surrounding temperatures as high as 305°F because they are generally closed compartments surrounded by hot stagnant air (caused by aerodynamic heating). Zones II, IIIB, and IV have lower maximum surrounding temperatures of 180°F since cooling air is generally exhausted from the electronics into these compartments and is allowed to circulate. Zone IIIA, the cockpit, is maintained at air temperatures and pressures suitable to the comfort of the crew.

The thermal load density of each zone given in Figure 2 is in kilowatts per cubic foot. The thermal

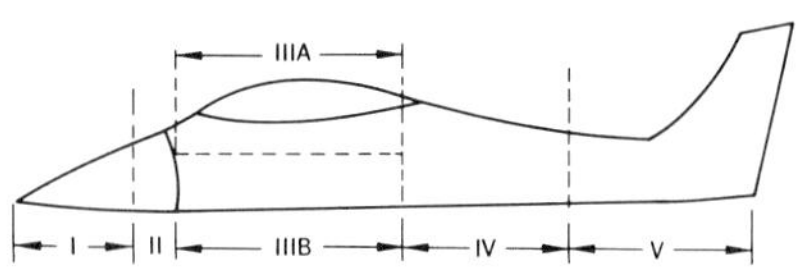

ZONE	DESCRIPTION	CONDITION	COOLING METHOD	TEMP RANGE, °F
I	RADOME	NO AIR COND UNPRESSURIZED	LIQUID COOLED	-65 TO +305
II	FORWARD EQUIP BAY	NO AIR COND UNPRESSURIZED	FORCED AIR AND/OR LIQUID COOLED	-65 TO +180
IIIA	COCKPIT	PRESSURIZED AND AIR COND	TEMPERATURE CONTROLLED AIR	+65 TO +85
IIIB	UNDER COCKPIT BAY	AIR COND UNPRESSURIZED	FORCED AIR AND/OR LIQUID COOLING	+50 TO +180
IV	AFT BAY	AIR COND UNPRESSURIZED	FORCED AIR AND/OR LIQUID COOLING	+50 TO +180
V	AFT SECTION	AIR COND UNPRESSURIZED	FORCED AIR AND/OR LIQUID COOLING	-65 TO +305

Fig. 1 Compartment Temperature Characteristics of Various Sections of a Typical Aircraft.

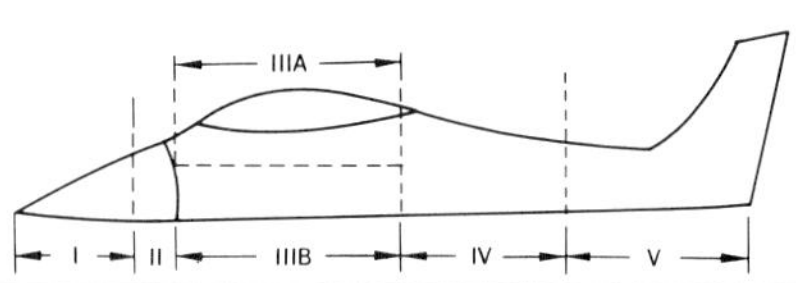

ZONE	DESCRIPTION	TYPICAL THERMAL LOAD DENSITY, KW/CU FT
I	RADOME	2.5
II	FORWARD EQUIPMENT BAY	3.0
IIIA	COCKPIT	1.0
IIIB	UNDER COCKPIT BAY	1.5
IV	AFT EQUIPMENT BAY	1.5
V	AFT SECTION	2.5

Fig. 2 Typical Thermal Load Density for Various Sections of a Military Aircraft.

load densities vary from 1.0 kw/ft^3 to 3.0 kw/ft^3. The lower value (1.0 kw/ft^3) is typical of pressurized cockpit equipment, which is cooled by conditioned air discharged into the cockpit. Zones IIIB and IV, which have equipment with a load density of 1.5 kw/ft^3, usually consist of low dissipation rack-mounted equipment and are generally cooled by forced air. Zones II and V which usually contain high power dissipation equipment, such as the radar transmitter and power supplies, are liquid or forced air cooled or both.

As a rule, the radar antenna is located in Zone I. Additional antennas may be located in Zones IV and V depending on the aircraft. Typical radar antenna equipment produces a thermal load array density of 3.0 watts/in^2. An electronic-scanning antenna located in Zones I and V (where the surrounding temperature can be as high 305°F) will require approximately 20 percent more cooling than an antenna located in Zone II (where the maximum surrounding temperature is only 180°F). The increase in cooling required comes from the additional heat absorbed by radiation and natural convection from the surroundings. This additional cooling required will impose certain penalties on aircraft performance because of the increase of the demand on the aircraft's prime environmental control system.

For some transient conditions, high ambient temperatures, such as may occur in Zone I, may be tolerated without requiring additional cooling. This will depend, however, on the duration of this high temperature. In general, only short high temperature transients will be acceptable.

Airborne, ground, and shipboard applications of radar antenna have different cooling requirements. For the same method of cooling, such as by forced liquid convection, the cooling requirements for the airborne radar antenna will be more severe than those for ground and shipboard radars due to the higher airborne radar surrounding temperature. Table I gives the surrounding temperature ranges for various radar applications. Because of the complexity of the electronic scanning radar antenna, cooling of the ground and shipboard radar by natural convection to surrounding air alone may not be sufficient. As in the airborne radar, liquid cooling may be required also. For lower power dissipation radar antennas, forced air cooling may be adequate. The liquid or forced air cooling may be provided by ground cooling carts or shipboard water. (The latter may require an intermediate transport coolant loop.)

APPLICATION	AMBIENT TEMPERATURE RANGE, °F	APPLICABLE MILITARY SPECIFICATION
AIRBORNE	-65 TO 275*	MIL-E-5400 MIL-E-18927D
GROUND	-65 TO 120	MIL-STD-210
SHIPBOARD	-20 TO 104	MIL-STD-210

* FOR CLASS 4 EQUIPMENT DESIGNED FOR 100,000 FT. ALTITUDE AND CONTINUOUS SEA LEVEL OPERATION OVER THE TEMPERATURE RANGE OF -65° TO 275 °F. 302 °F INTERMITTENT OPERATION FOR 30 MINUTES.

Table I Ambient Temperature Range for Various Radar Applications (during operation).

V. STRUCTURAL CONSIDERATIONS

As mentioned earlier, it is possible that unless careful consideration is given to the mounting location of the antenna it will inadvertantly become an airframe member. In fact one might conceivably design an antenna as a load-bearing member.

Electronic scanning arrays must be designed

and supported in such a manner that degradation of microwave performance, resulting from structural deformation of the array during the service environment, is minimized. This will require high rigidity of the array to withstand inertial loads, and for aircraft installation, isolation of the array from support deformation. Typical inertial loads are shown in Table II, for airborne, ground, and shipboard installations.

AIRBORNE (MIL-E-5400)
- SHOCK
  15-G SHOCK OF 10 MILLISECOND DURATION
- VIBRATION
  ±5-G SINE VIBRATION IN 5 TO 500 Hz RANGE

GROUND
- SHOCK (MIL-STD-810) TRANSPORTATION
  15-G SHOCK OF 10 MILLISECOND DURATION
- VIBRATION (MIL-STD-810) TRANSPORTATION
  ±2.5-G SINE VIBRATION IN 5 TO 500 Hz RANGE

  NUCLEAR BLAST
  12 PSI EXPONENTIALLY DECAYING OVERPRESSURE OF 1.3 SECOND DURATION

SHIPBOARD
- SHOCK (MIL-S-901)
  LESS THAN 20G'S OF 5 TO 20 SECOND DURATION
- VIBRATION (MIL-STD-167)
  ±0.75 G SINE VIBRATION IN 5 TO 20 Hz RANGE
- NUCLEAR BLAST (MIL-E-16400)
  12 PSI EXPONENTIALLY DECAYING OVERPRESSURE OF 1.3 SECOND DURATION

Table II Summary of Critical Shock and Vibration Environments for Airborne, Ground, and Shipboard Equipment.

A particularly promising electronic-scanning array consists of many radiating elements sandwiched between two thin plates (Figure 3). It may therefore be expected to behave as a thick orthotropic sandwich plate. The thin plates serve as the face sheets and the radiating elements serve as the shear core of the equivalent sandwich plate. The term orthotropic is used to describe a plate whose flexural rigidity, i.e., resistance to bending, differs in two mutually perpendicular directions. Such a plate may be rigid about one transverse axis and relatively flexible about the other transverse axis.

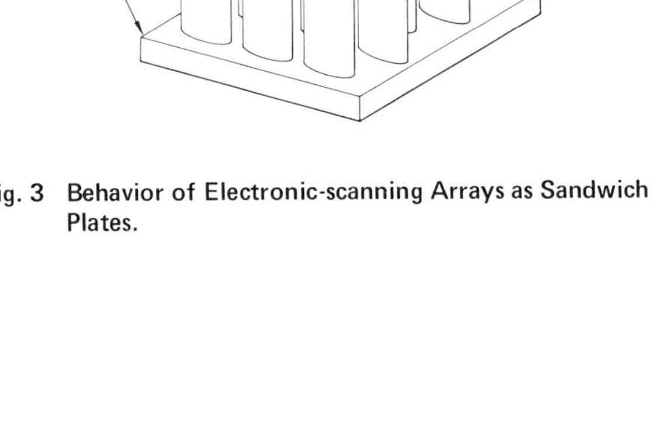

Fig. 3 Behavior of Electronic-scanning Arrays as Sandwich Plates.

To minimize the deformation of this sandwich plate resulting from inertial loads, the plate's rigidity must be maximized. This requires that the rigidity of both the core in shear and the face sheets in tension/compression be maximized and that a satisfactory edge closure be provided. The relatively low core shear rigidity may be significantly increased by mechanically joining the radiating elements together at their ends, thereby providing additional end restraint for each radiating element. A satisfactory edge closure, usually a rectangular frame having a channel-shaped cross-section, must be provided to tie the face sheets and the core together. The edge closure allows the array to resist efficiently twisting deformation as a thin-walled rectangular torque tube. Without it, twisting deformation would be resisted by the relatively flexible core in shear.

In the aircraft, the manner and location of structural supports require careful consideration if we seek simultaneously to minimize deformation of the array and maximize its isolation from support deformations. Aircraft, unlike ground or naval structures, are relatively flexible. Furthermore, the vibration level, which is the critical load environment, is more severe in an aircraft (see Table II). Consequently, all points on the aircraft can be expected to undergo significant deformations and no true hard-points really exist.

To isolate the array from these deformations, it must be supported in a determinate manner. Three point simple support offers a most attractive solution to this problem because it allows the array to move as a rigid body as the support deforms. The location of these three supports may now be considered. First, we note that the supports must be confined to the perimeter because room is not only needed for attachment provisions but also for a backup structure - the edge closure. Second, we note that some degree of symmetry of the supports with respect to the antenna geometry is desirable. A reasonable starting point is, therefore, to place one support at the midpoint of the longest side and the other two supports equidistant from the midpoint of the opposite side (Figure 4). It can be shown that the internal deformation of the array will, in general, be minimized when these supports are located at these quarter points.

This type of design, because of its high rigidity and because of isolation from support-induced loads, will be more than structurally adequate. In addition, the structural resonant frequencies of the array will be quite high, which will result in low dynamic deflection responses. This is an important consideration for airborne installations where vibration environments are critical.

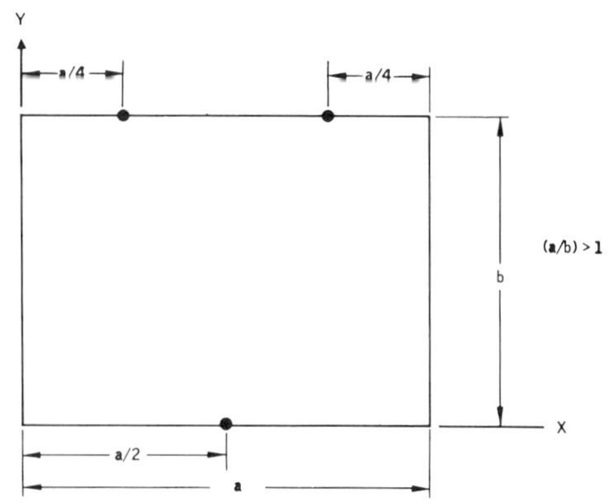

Fig. 4 Antenna Attach Points.

VI. INTERCONNECTION CONSIDERATIONS

Considering that one small, two by three-foot, X-band array containing 2,500 phase-shifters requires about 40,000 interconnections, the connections per square foot of antenna surface amount to 7,000! Thus, by sheer numbers alone interconnections present new problems to the designer of electronic scanning arrays. There are two classes of interconnections to be considered: first, the actual microwave interconnections and second, the signal and power interconnections.

The microwave interconnections are important because every one of them is a possible point of RF power loss. There are several methods of making microwave interconnections. Unfortunately the least lossy is the heaviest. In the evaluation of the method to be selected, one must decide which factor is most important: weight, RF loss, manufacturability, or maintenance (if required). The resulting decision indeed, may be the heart of a good design.

The installation of the phase shifter into the RF structure and its RF interconnection are factors that we at Hughes have considered as important as any other single item of antenna design. There are obviously several selections that are available to us: stripline, RF coupling, and, of course, waveguide.

The signal and power distribution lines, on the other hand, are a different matter. The methods used to make these connections include multilayer boards in a busbar arrangement, flex cable (really a form of multilayer boards), and conventional wire. The important considerations are volume, geometry, weight, and current carrying requirements. Some of the current requirements are high enough to require that cooling be supplied to the bus. Signal and power circuits must be isolated, and clock pulse circuits must also be isolated from the other signal circuits.

The choice of insulation material depends in some degree on the shielding requirements, if any. The two choices, other than EPG, are FEP teflon and "H" film KAPTON. Both are exceptionally good heat resistant materials with a 200°C continuous temperature rating for the FEP teflon and a 150°C temperature rating for the KAPTON. KAPTON cable can be made as thin as FEP (0.10 inch thick), but is quite "notch" sensitive. There is a real probability that KAPTON slots between groups of conductors will propagate. FEP teflon on the other hand is somewhat tougher and more resistant to abrasion and notching. Cables can also be shielded, if necessary, with a thin conductive coating to provide EMI protection. The interconnection problem is not so much materials and methods but the sheer number of interconnections that must be made in an antenna. They can easily become the most severe reliability problem associated with the antenna.

VII. MAINTAINABILITY/RELIABILITY RELATIONSHIP

As mentioned earlier, the relationship between designing for maintenance and for access is a direct function of the reliability. There is no need to design ready access to a part at great cost and complexity if the part itself has a high reliability factor. Therefore, the emphasis in design should be on placing the working electronics, i.e., phase shifters, in as benign an environment as possible. This, in great measure, explains the emphasis at Hughes in designing our equipment for low semiconductor junction operating temperatures. The need for this type of environment in the design of electronically-scanned antennas is because the phase-shifting elements generally are removable only from the front surface of the antenna. In an aircraft this means removal of a radome, which, as stated earlier, could be exceedingly awkward for a field situation and a large side-looking antenna.

Furthermore, the assembly procedure must be such that when the part is removed, the thermal interface is not in any way deteriorated and replacement of the part will result in as good a thermal interface as for the original assembly. Thermal lubricants, silicon greases, indium washers, etc., are undesirable maintenance items because there is no assurance of their replacement when the part is reassembled. Interconnections must be such that, not only are they one-way assemblies but the interconnection points themselves must be completely protected from handling damage.

Another maintenance feature of a good airborne design is one that allows the basic array to remain in the aircraft and the parts requiring maintenance to be removed individually from the basic structure. The design requires built-in test features that isolate faults to the piece part.

VIII. MATERIALS

Material selection, as in any other design, is critical. Not only must the material meet the structural requirements, be economical to fabricate, lightweight, electrically compatible (e.g., microwave skin thickness, grounding), but it also must meet the humidity, sand and dust, salt atmosphere, altitude, etc., requirements of MIL-E-5400. Candidates for a basic structure are shown in Table III.

MATERIAL	COST, $/LBS	TENSILE STRENGTH (1000 PSI)		YIELD STRENGTH (1000 PSI)		EL'g't'n %		M OF E INT, 100,000 PSI		MELT OR SEV, °F		THERM EXP, 10^{-6} IN./IN./°F		THERM COND, BTU/HR/SQ FT/°F		WT
		HIGH	LOW	HIGH	LOW	HIGH	LOW	HIGH	LOW	HIGH	LOW	HIGH	LOW	HIGH	LOW	S G
ALUMINUM 6061	$ 0.52	60	22	50	22	15	1.5	106	100	1215	935	13.7	11.7	92.5	51.0	2.7
BERYLIUM COPPER	2.75	180	160	35	25	50	35	440	-	2341	--	9.3	--	110	100	8.1
CUPRONICKEL	1.25	205	162	73	-	45	42	240	210	2260	2020	9.3	9.0	15	9.5	8.9
PHOSPHOR BRONZE	1.15	130	65	75	50	70	48	170	150	1970	1550	10.2	9.6	120	29	8.8
MAGNESIUM	0.35	54	-	44	19	15	7	65	64	1200	830	16	14	80	24	1.7
NICKEL (MONEL)	1.60	190	130	170	92	24	14	260	-	2606	2600	9.8	7.7	36	8.7	8.8
STAINLESS STEEL (400)	0.35	220	110	165	67	55	15	290	280	2790	2600	10.4	6.4	14.5	8.2	8.0
TITANIUM	5.70	170	60	220	135	25	8	190	145	3040	2730	7.1	4.9	9.8	4.3	4.7
ZIRCONIUM	14.00	108	82	98	58	25	5	140	138	3355	3300	3.6	3.1	9.6	8.1	3.9
TIN AND ALUM (BRASSES)	1.60	84	75	60	53	40	15	160	150	1780	1590	11.8	10.3	28	-	8.5
ALUMINA (CERAMIC)	8.00*	39	20	23	18	12	3	500	320	3362	--	3.7	3.1	10.7	6.2	3.8
										SERVICE						
P V C	26.00	4	1.5	4	1.5	500	250	0.03	0.005	220	140	33	28	0.1	0.07	1.5
POLYSTYRENE (L D)	22.00	8	5	8	5	725	80	5	4	180	140	24	12	0.08	0.06	1.07
POLYSTYRENE (H D)	17.00	4.4	2.9	4.4	2.9	400	12	13	11	190	120	24	22	0.08	0.06	0.96
DIALLYL PHTHALATE	12.00	7	4	7	4	40	20	12	6	500	300	35	15	2.1	1.1	1.6
NYLON (610)	120.00	12.6	7.1	12.6	7.1	2.3	1.5	4.1	1.6	300	225	55	--	0.14	0.10	1.1
PHENOLIC (GL FILLED)	25.00	10	3.5	10.0	3.5	0.8	0.3	50	30	400	300	25	8.3	0.4	0.1	1.7
KOVAR (NICKEL)	4.00	77	65	60	4.0	35	20	250	230	2650	--	5.4	4.5	8.8	7.6	9.1

*CUBIC INCH   REF: MATERIAL ENGINEERING (OCTOBER 1969 ISSUE)   NOTE: SPECIFIC GRAVITY x 0.036 = LBS/CU IN.

Table III Comparison of Materials.

It is interesting to note how well aluminum stands up in this tradeoff despite the appearance of many new materials in the last decade. However, the successful use of electroless-coated plastics throughout the automotive industry makes plastics hard to ignore as a candidate material for any mass production design. Several designs using electroless-plated pastic for other devices have been generated over the years at Hughes. However, their application has been strictly limited by the problem of plate and plastic adherence. As a result, we have not used this technique for any airborne military application.

Some of the materials used by the automotive industry seem prime candidates for phased-array antennas; investigation of them should be started. While strength-to-weight ratio is poor compared to aluminum, the manufacturability and dimensional stability of some of these automotive materials are excellent. Stamped or extruded plastics become automatic candidates because of the quantities of piece parts required to manufacture just one array. On the other hand, the thermal conductivity of plastics leaves a lot to be desired, and any design implementation must take this factor into consideration.

IX. MASS PRODUCTION

Most of the problems previously discussed affect manufacturability in one way or another. This particularly applies to the problem of material selection. The most significant part of the mechanical design for manufacturability of airborne electronically-scanned antennas is that the great number of elements involved makes reversal of design decisions extremely expensive. For example, a 36-inch diameter antenna operating at Ku-band may require 5800 phase shifters in associated radiating elements and may have over 91,000 interconnections. Hard design in tooling is required even for developmental quantities. Thus, if the total design is to be economically feasible, it is necessary that manufacturing engineers who are familiar with mass production techniques, materials, and processes aid in the initial design. No other item of military airborne equipment approaches the quantities of fabricated parts, assemblies, components, and interconnections that are required for electronically-scanned antennas. These figures are of such a magnitude that, even for a very few antennas, mass production design techniques are required. Historically, the physical designer of airborne weapon systems has not had to face this problem. For electronically-scanned arrays, however, even in small quantities, production techniques, facilities, equipment, and cost control must be applied from the initiation of the design.

X. MICROELECTRONICS

Microelectronics must be used for the drive and logic circuits if the volume and weight constraints are to be met. These techniques are well known and will not be discussed in detail as part of this paper. The rationale for their employment is rather straightforward -- low power, weight, volume, and hopefully, cost.

Designs have been generated at Hughes wherein the entire logic for control of many phase-shifter driver circuits of an antenna is contained on four wafers. These wafers are low power LSIs. The enclosure of these wafers permits positive reliable interconnections from wafer to phase shifters. These interconnections present critical reliability problems if chip and wire techniques are employed; however, it is neither necessary nor desirable to use chip and wire techniques. Beam lead, spider bonding, and other positive chip-to-substrate interconnecting techniques are being developed at Hughes and in the industry with sufficiently gratifying results that warrant their employment even at a slight increase in cost.

The effect of the advances made in LSI has been to allow the design of built-in test so that fault isolation is positive. One area of concern to the author has been the thermal and structural interface between the LSI wafer and its mounting substrate. The method of attachment is important because of the large area-to-thickness ratio of an LSI wafer. Any material used to attach the wafer to the substrate must allow for both mechanical and thermal shock at little or no sacrifice in thermal conductivity. This is an engineering problem that has not received adequate attention and must be promptly and carefully solved.

XI. CONCLUSIONS

It is apparent that the field of physical design for airborne electronic scanning antennas requires not only a different approach from the ground or shipboard designs but in addition that the weapon systems designer must utilize high quantity manufacturing techniques and truly perform as a systems equipment designer as opposed to a "black box" designer. Uncoordinated attempts to design an airborne electronic-scanning antenna of any resonable size unilaterally will most likely result in unsatisfactory performance, if not in disaster. It is particularly important to emphasize the close relationship that must be established early between the antenna and the airframe designer.

ACKNOWLEDGEMENTS

I want to thank Robert E. Holman and Landon A. Strattan of Hughes Aircraft Co. for their aid in preparing the structural and thermal sections of this manuscript.

Mechanical Design Considerations of a Phased-Array Antenna

L. T. EMBREKTS

Sperry Gyroscope Division
Great Neck, New York

The problems encountered in the mechanical design phase of phased array antennas are many and they are radically different from conventional antenna problems. The conventional antenna can be considered as a passive component in a microwave system with relatively few design parameters. On the other hand, an electronic scanning phased array is a dynamic electronic system with the characteristic power, wiring, cooling and structural requirements. The unique precision and structural integrity requirements in terms of the many phase shifters and wiring interconnections can cause havoc with schedule and budget if a design error manifests itself into hardware.

For ease of analysis the mechanical design philosophy is shown to be delineated into two classes of requirements, primary or functional performance requirements and secondary or service requirements. Any deliverable item necessarily satisfies both classes of requirements, however, the flexibility for tradeoffs varies greatly in the two.

A baseline for this paper is a recently completed portable phased array which contains 1721 ferrite phase shifter modules. The array is 10 feet high, 6 feet wide and one foot deep and weighs 1300 pounds. The phased array antenna is a space fed, two dimensional electronically scanned microwave lens. From inception through design, build and debugging the project was completed in 18 months. The product necessarily reflects many design and manufacturing innovations because of its first of a kind application.

Fig. 1 Antenna Element Insertion.

Fig. 1 shows the completed array without the collecting ground plane to exhibit array wiring and an insertion of an array element. Each element is serviceable from the collecting side by removal of a matched waveguide radiating element.

The design of the antenna is governed by a dimensional tolerance versus antenna performance study. This serves to establish design objectives and restraints. The fact that ferrite phase shifters are used in the array introduces a unique thermal control criterion which is different from the conventional, heat removal problem. A counterflow forced air cooling technique was developed where each phase shifter is cooled in two distinct areas by air stream at different temperatures. The resultant of this technique exerts temperature control to reduce a thermal gradient within the array to acceptable limits.

In the build or construction phase a modular technique of assembly was employed where each individual piece was controlled to tight tolerances. The total assembly was reduced to five significant building blocks to minimize tolerance buildups. All riveted assembly methods were utilized to eliminate possible distortions compromising the precision achieved on individual pieces.

Fig. 2 Completed Array Structure.

Fig. 2 shows the completed array structure without any electrical components.

The hardware demonstrated that the method of approach to the overall mechanical design objective is valid since all design goals were achieved or surpassed by a significant margin. This was achieved by a realistic antenna performance budget in terms of hardware physical tolerances. This resulted in improved antenna performance and a reduction of production schedule and cost as well.

AUTHOR INDEX

Alexander, D. K., 356
Amitay, N., 68
Andersen, J. B., 112
Arredondo, G. A., 120, 173
August, G., 342
Balzano, Q., 300
Beachler, L. W., 232
Blake, C., 43, 327
Borgiotti, G. V., 300
Buck, G. J., 219
Butzien, P. E., 173
Byron, E. V., 187
Cheston, T. C., 215
Diamond, B. L., 127, 148
Embrekts, L. T., 377
Esposito, F. J., 327
Frank, J., 243
Galindo, V., 68
Giarrizzo, V. A., 239
Hanfling, J. D., 239
Hass, F., 112
Hering, K. H., 237
Hessel, A., 273, 314
Hill, R. T., 193
Horvath, A. L., 261
Humes, R. W., 173
Jones, W. R., 178
Kahn, W. K., 84
Kampinsky, A., 30
Kent, R. C., 227
Kinsey, R. R., 261
Knittel, G. H., 148, 157
Knuckey, R. L., 120
Kummer, W. H., 30
Lee, J. R., 364
Lee, S. W., 92, 178
Lewis, D. J., 364
Lyons, B. J., 232
McCarter, R. S., 6
McCarthy, D. K., 364
McCarthy, J. A., 1
Mittra, R., 349
Oliner, A. A., 107
Parsons, K., 232
Patton, W. T., 330
Provencher, J. H., 291
Ransom, P. L., 349
Rippin, J. F., Jr., 21
Rosengard, M. H., 370
Scharfman, W. E., 342
Schell, A. C., 9
Shrank, H. E., 322
Schwartzman, L., 327
Sichelstiel, B. A., 356
Skolnik, M. I., 15
Stark, L., 51
Sureau, J. C., 314
Tang, R., 92, 178, 254
Temme, D. H., 208
Vogel, G. J., 232
Wasylkiwskyj, W., 84
Weir, J. P., Jr., 227
Wheeler, H. A., 131
Wong, N. S., 92, 178
Wu, C. P., 68